NANOCOSMETICS

NANOCOSMETICS

Fundamentals, Applications and Toxicity

Edited by

Arun Nanda

Department of Pharmaceutical Sciences, Maharshi Dayanand University, Rohtak, India

Sanju Nanda

Department of Pharmaceutical Sciences, Maharshi Dayanand University, Rohtak, India

Tuan Anh Nguyen

Institute for Tropical Technology, Vietnam Academy of Science and Technology, Hanoi, Vietnam

Susai Rajendran

Department of Chemistry, St Antony's College of Arts and Sciences for Women, Dindigul, India

Yassine Slimani

Institute for Research and Medical Consultations, Imam Abdulrahman Bin Faisal University, Saudi Arabia

ELSEVIER

Elsevier
Radarweg 29, PO Box 211, 1000 AE Amsterdam, Netherlands
The Boulevard, Langford Lane, Kidlington, Oxford OX5 1GB, United Kingdom
50 Hampshire Street, 5th Floor, Cambridge, MA 02139, United States

British Library Cataloguing-in-Publication Data
A catalogue record for this book is available from the British Library

Library of Congress Cataloging-in-Publication Data
A catalog record for this book is available from the Library of Congress

ISBN: 978-0-12-822286-7

For Information on all Elsevier publications
visit our website at https://www.elsevier.com/books-and-journals

Publisher: Matthew Deans
Acquisitions Editor: Simon Holt
Editorial Project Manager: Charlotte Rowley
Production Project Manager: Debasish Ghosh
Cover Designer: Greg Harris

Typeset by MPS Limited, Chennai, India

Contents

List of contributors xiii

1

Basic principles

1. Nanocosmetics: an introduction

Pierfrancesco Morganti

1.1 Introduction 3
1.2 Consumer requests and cosmetics 6
1.3 Polymers, nanocomposites, and nonwoven tissues 8
1.4 Conclusive remarks 10
1.4.1 Micro/nanoemulsions 11
1.4.2 Packaging material 12
References 13

2. Transdermal and bioactive nanocarriers

Nikhishaa Sree Raju, Venkateshwaran Krishnaswami, Sivakumar Vijayaraghavalu and Ruckmani Kandasamy

Abbreviations 17
2.1 Introduction 17
2.2 Skin 18
2.2.1 Epidermis 18
2.2.2 Dermis 18
2.2.3 Dermis—epidermis junction 19
2.2.4 Hypodermis 19
2.3 Transdermal delivery 19
2.4 Evolution of transdermal delivery 20
2.5 Development of transdermal delivery in cosmetics 21
2.6 Advantages of transdermal delivery 21
2.7 Components of transdermal patch 21
2.7.1 Polymer matrix 22
2.7.2 Biologically active substances 22
2.7.3 Permeation enhancers 22
2.7.4 Adhesive 23
2.7.5 Backing laminate 23
2.7.6 Release liner 23
2.7.7 Other excipients 23
2.8 Novel technologies toward the development of the transdermal system 23
2.8.1 Iontophoresis 23
2.8.2 Electroporation 24
2.8.3 Microneedles 24
2.8.4 Microdermabrasion 25
2.8.5 Laser radiation 25
2.9 Bioactive nanocarriers 25
2.9.1 Liposomes 25
2.9.2 Niosomes 27
2.9.3 Solid lipid nanoparticles 27
2.9.4 Nanoemulsions 27
2.9.5 Nanostructured lipid carriers 28
2.10 Discussion 28
2.11 Conclusion 29
Acknowledgments 29
References 29

3. Transdermal and bioactive nanocarriers for skin care

Federico Svarc and Laura Hermida

3.1 Introduction 35
3.2 The skin 36
3.3 Nanocarriers and skin penetration 37
3.4 Nanocarrier system 39
3.4.1 Liposomes and related particles 39
3.4.2 Solid lipid nanocarriers and structured lipid nanocarriers 42
3.4.3 Nano- and microemulsions 44
3.4.4 Inorganic nanocarriers 47
3.4.5 Dendrimers and other dendritic structures 50
3.4.6 Other polymeric nanoparticles 52
3.4.7 Polysaccharide nanocarriers 53
3.5 Conclusions and future perspectives 54
References 54

4. Nanoemulsions for cosmetic products

Ana Catarina Faria-Silva, Ana Margarida Costa, Andreia Ascenso, Helena Margarida Ribeiro, Joana Marto, Lídia Maria Gonçalves, Manuela Carvalheiro and Sandra Simões

4.1 Introduction 59
4.2 Emulsion delivery systems in cosmetics 60
4.2.1 Emulsion generalities 60
4.2.2 Microemulsions and nanoemulsions—aren't both nanosystems? 61
4.3 Formulation and production of nanoemulsions 62
4.4 Characterization of nanoemulsions 65
4.5 Why nanoemulsions in cosmetics? 69
4.5.1 Skin care 71
4.5.2 Hair care 71
4.6 Challenges and future perspectives 72
References 72

5. Nanomaterials for cosmeceuticals: nanomaterials-induced advancement in cosmetics, challenges, and opportunities

Bilal Haider Abbasi, Hina Fazal, Nisar Ahmad, Mohammad Ali, Nathalie Giglioli-Guivarch and Christophe Hano

5.1 Introduction to nanotechnology 79
5.2 Nanotechnology is multidisciplinary field 80
5.3 Historical perspective of nanotechnology in cosmetics 83
5.4 Nanotechnology-based cosmetics 87
5.4.1 Liposomes 87
5.4.2 Nanoemulsions 89
5.4.3 Niosomes 90
5.4.4 Solid lipid nanoparticles 91
5.4.5 Nanocapsules 92
5.4.6 Nanocrystals 92
5.4.7 Gold and silver nanoparticles 92
5.4.8 Dendrimers 93
5.4.9 Cubosomes 94
5.4.10 Nanomedicine 95
5.4.11 Hydrogels 95
5.4.12 Fullerenes/Buckyballs 95
5.4.13 Polymersomes 96
5.4.14 Carbon nanotubes 96
5.4.15 Nanostructured lipid carriers 96
5.4.16 Nanospheres 97
5.5 Major classes of nanocosmeceuticals 97
5.5.1 Nail care 97
5.5.2 Lip care 97
5.5.3 Skin care 97
5.5.4 Hair care 98
5.6 Challenging aspects 98
5.6.1 Human health insecurities 98
5.6.2 Ecological hazardous issues 100
5.7 Future prospects and opportunities 101
References 102

6. Polymeric nanocarriers for topical drug delivery in skin cream

M. Malathi, B.N. Vedha Hari and D. Ramyadevi

6.1 Introduction 109
6.2 Materials and methods 110
6.2.1 Materials 110
6.2.2 Formulation design 110
6.2.3 Characterizations 111
6.3 Results and discussion 113
6.3.1 Evaluation of saffron-loaded polymeric nanoparticles 113
6.3.2 Evaluation of cream formulation 119
6.4 Conclusion 121
References 125

7. Organic UV filter loaded nanocarriers with broad spectrum photoprotection

Lucía López-Hortas, María D. Torres, Elena Falqué and Herminia Domínguez

7.1 Introduction 127
7.2 Protection filters 128
7.3 Conventional products for sun protection 130
7.4 Nanocarriers for sun protection 132
References 136

8. Cosmetic nanoformulations and their intended use

Surbhi Dhawan, Pragya Sharma and Sanju Nanda

8.1 Introduction 141
8.2 Nanocarriers/nanomaterials in cosmetics 142
8.2.1 Liposomes 142
8.2.2 Nanoemulsions 150
8.2.3 Solid lipid nanoparticles 151
8.2.4 Nanostructured lipid carriers 151

8.2.5 Cubosomes 151
8.2.6 Nanosponges 152
8.2.7 Dendrimers 152
8.2.8 Nanosilver 152
8.2.9 Nanogold 153
8.2.10 Nanospheres 153
8.2.11 Carbon nanotubes 153
8.2.12 Nanopigments/nanoparticles 154
8.3 Categories of nanotechnology-based cosmetics 154
8.3.1 Skin care 154
8.3.2 Lip care 160
8.3.3 Oral care 160
8.4 Consumer concerns and regulatory guidances 163
8.4.1 Routes of nanocosmetics exposure 163
8.4.2 European Union Guidelines 165
8.4.3 Guidance document issued by FDA for industries 166
8.5 Conclusion 167
References 168
Further reading 169

2

Emerging applications

9. Water-based nanoperfumes

Małgorzata Miastkowska and E. Lasoń

9.1 Introduction 173
9.2 Water-based or/and alcohol-free perfumes 174
9.3 Nanodispersions as a carrier for fragrances 179
9.4 The advantage of nanoemulsions over microemulsions 180
9.5 Conclusions 181
References 181

10. Nanocosmetics for broadband light protection sun care products

Paulo Newton Tonolli, Thiago Teixeira Tasso and Maurício S. Baptista

10.1 Introduction 185
10.2 Visible light: should we protect ourselves? 186
10.2.1 Interaction of visible light with the skin 186
10.2.2 Effects of visible light on the skin 187
10.3 Functional analysis methods to detect solar damage 191
10.4 Strategies for protection from visible light 191
10.4.1 Filters 191
10.4.2 Membrane protection 195
10.4.3 Antioxidants 196
10.5 Final remarks 198
References 198

11. Nanomaterials for hair care applications

Megumi Nishitani Yukuyama, Gabriel Lima Barros De Araújo and Nádia Araci Bou-Chacra

11.1 Introduction 205
11.2 Hair structure 206
11.2.1 Hair shaft 206
11.2.2 Hair follicles 207
11.3 Nanostructured systems for hair treatment 208
11.3.1 Types of nanostructured systems 209
11.3.2 Hair treatment 215
11.4 Future perspectives 221
Acknowledgments 221
References 221

12. Nanoparticles in hair dyes

S. Senthil Kumaran, R. Joseph Rathish, S. Johnmary, M. Krishnaveni, Susai Rajendran and Gurmeet Singh

12.1 Human hair 227
12.1.1 Function of human hair 227
12.1.2 Chemical composition of hair 227
12.1.3 Living being found in human hair 228
12.1.4 Three major components of the hair shaft 228
12.1.5 Anatomy of human hair 228
12.2 Hair colors 229
12.3 Melanins 229
12.4 Building blocks of eumelanins and pheomelanins 229
12.5 Parkinson's and Alzheimer's diseases 229
12.6 Gray hair 229
12.7 Plucking gray hair is bad 229

12.8 Natural turning of gray hair into black hair 230
12.9 Composition of hair dyes in olden days 230
12.10 Common chemicals used in hair dyes 230
12.11 Harmfulness of hair dyes 230
12.12 Side effects of the chemicals used in common hair dyes 231
12.13 Toxic chemicals in hair dye 231
12.14 Graphene hair dye 231
12.15 Gold nanoparticles as hair dye 231
12.16 p-Phenylenediamine-incorporated nanoparticles as hair dye 232
12.17 Tips for faster growth of hair 232
12.18 Best foods to promote hair growth 232
12.19 Fruits for hair growth 233
12.20 Recent research on the use of nanoparticles in hair dyes 233
12.20.1 Estimation of nanoparticles' human skin penetration in vitro by confocal laser scanning microscopy 233
12.20.2 Phototherapy and multimodal imaging 233
12.20.3 Assembly of polymer-grafted proteins 234
12.20.4 Detection of zinc in human hair by self-assembled NPs 234
12.20.5 Penetration of nanocarriers into the hair follicles 235
12.20.6 Lipid nanoparticles for follicular targeting 235
12.20.7 Nanoparticles-based formulation for hair follicle targeting 235
12.20.8 Nanoparticles as a gene (or drug) carrier to the inner ear 236
12.20.9 Hair-dye assay using graphene/ionic liquid electrochemical sensor 237
12.20.10 A targeted approach for the treatment of psoriasis using polymeric micelle nanocarriers 237
12.20.11 Quantification of nanoparticle uptake into hair follicles in pig ear and human forearm 238
12.20.12 Presentations in conferences 238
12.20.13 Polymeric nanoparticles-embedded organogel for roxithromycin delivery to hair follicles 239
12.20.14 Permanent hair dye-incorporated hyaluronic acid nanoparticles 239
12.20.15 Hair fiber as a nanoreactor in controlled synthesis of fluorescent gold nanoparticles 239
12.20.16 Hair dye-incorporated poly-γ-glutamic acid/glycol chitosan nanoparticles based on ion-complex formation 240
12.20.17 Translocation of cell penetrating peptide engrafted nanoparticles across skin layers 240
12.20.18 Safety assessment of personal care products/cosmetics and their ingredients 241
12.20.19 Potential novel drug carriers for inner ear treatment: hyperbranched polylysine and lipid nanocapsules 242
12.20.20 Investigation of polylactic acid nanoparticles as drug delivery systems for local dermatotherapy 242
12.20.21 Nanoparticles—an efficient carrier for drug delivery into the hair follicles 242
12.20.22 In vivo drug screening in human skin by femtosecond laser multiphoton tomography 243
12.21 Concluding remarks 243
References 244

13. Nanomaterials in fragrance products

N. Vijaya, T. Umamathi, A. Grace Baby, R. Dorothy, Susai Rajendran, J. Arockiaselvi and Abdulhameed Al-Hashem

13.1 Introduction 247
13.2 Nano-ingredients 248
13.3 Home fragrance products 248
13.3.1 Astonishing ways to make your home smell amazing 248
13.3.2 Use of orange peels in making your house smell good 249
13.3.3 Tips to keep a kitchen smelling so fresh 249
13.3.4 Toilet freshener 249
13.3.5 Fragrance for body 249

13.4 Natural fragrances 249
13.5 Fragrance affects your skin 250
13.6 Chemicals present in fragrance 250
13.7 Ingredients to be avoided in skin care 250
13.8 Parabens 251
13.9 Nanotechnology in perfumes 251
13.10 Applications of nanotechnology in perfumes 251
13.10.1 Production of perfume (aroma) compounds 252
13.10.2 Time-controlled and prolonged release of scents 252
13.10.3 Use of nanoencapsulation procedures in development of nanotechnology 252
13.10.4 Probable risks of nano-based perfumes 253
13.10.5 High tech of small serving giant thrills with giant threats 254
13.11 Sunscreen 254
13.11.1 Uncoated zinc oxide 254
13.11.2 Use of titanium dioxide as sunscreen 255
13.12 Recent trends in research on electronic noses, nanoparticles, and fragrant products 256
13.12.1 Titanium dioxide nanoparticle-based indoor antiodor product 256
13.12.2 Electronic noses in meat quality assessment 256
13.12.3 Flavor of traditional soup of Chinese Yellow-Feather Chickens 257
13.12.4 Odor intensity assessment 257
13.12.5 Computer-controlled odor generator 258
13.12.6 Nanoparticle-enzyme sensors for detection of bacteria with olfactory output 258
13.12.7 Recognition of Chinese Herbal Medicines with electronic nose technology 258
13.12.8 Area identification of Zhongning Goji berries by an electronic nose 259
13.12.9 Assessment of the indoor odor impact in a naturally ventilated room 259
13.12.10 Odor standardization by bioelectronic noses 259
13.12.11 Comparison of volatile fraction of vodkas made from different botanical materials by electronic nose-based technology 260
13.12.12 Olfaction as a soldier 260
13.12.13 Aromatic wallpaper 260
13.12.14 Bioelectronic noses 261
13.12.15 A bioelectronic sensor using human olfactory and taste receptors 261
13.12.16 Fragrance profiling of *Jasminum Sambac* Ait. flowers using electronic noses 262
13.12.17 Evaluation of an electronic nose for odorant and process monitoring of alkaline-stabilized biosolids production 262
13.12.18 Electronic nose to detect freshness of eggs 263
13.12.19 Verification of odorants in rose 263
13.12.20 Predicting the growth situation of *Pseudomonas aeruginosa* on agar plates and meat stuffs using gas sensors 263
13.13 Future perspectives 264
References 264

14. Natural polymers for natural hair: the smart use of an innovative nanocarrier

Pierfrancesco Morganti and G. Morganti

14.1 Introduction: the hair system 267
14.2 Hair structure and damage 269
14.3 Cosmetic hair care treatments 270
14.4 Natural fibers for natural hair 271
14.4.1 Polysaccharides 272
14.4.2 Chitin 274
14.5 Chitin and hair: final considerations 277
References 283

15. Skin and pollution: the smart nano-based cosmeceutical-tissues to save the planet' ecosystem

P. Morganti, G. Morganti and M.B. Coltelli

15.1 Introduction 287
15.2 Skin aging 288
15.3 Inflammaging and oxidative stress 290
15.4 Life expectance and cosmetic treatments 292
15.5 Air pollution, waste, and human health 293
15.6 Innovative smart cosmeceutical-tissues 295
15.7 Extracellular matrix and active ingredients 296
15.8 Conclusive remarks 298
Aknowledgments 300
Conflicts of interest 300
References 300

16. Nanocarrier-mediated follicular targeting

B. Betül Gökçe and Sevgi Güngör

16.1 Introduction 305
16.2 The barrier characteristics of human skin 306
16.3 Penetration pathways across the skin 308
16.4 The rationale behind nanocarrier-mediated follicular targeting 309
16.5 The role of hair follicles in nanocarrier-mediated skin delivery 310
16.5.1 Follicular ducts as reservoirs of nanocarriers 310
16.5.2 Follicular deposition ability of nanocarriers 310
16.5.3 The extent of nanocarrier particle size for follicular delivery 311
16.6 Commonly used methods to examine follicular targeting of nanocarriers 312
16.6.1 Imaging techniques 312
16.6.2 Quantitative techniques 314
16.7 Nanosized delivery systems for follicular targeting 315
16.8 Iontophoretic technique combined with nanosized delivery systems for follicular targeting 316
16.9 Conclusion 319
References 320

17. Nanoemulsion in cosmetic: from laboratory to market

Vikas Pandey, Rajesh Shukla, Ashish Garg, Mohan Lal Kori and Gopal Rai

17.1 Introduction 327
17.1.1 Cosmetics: the paradigm shift 327
17.1.2 Overview of nanoemulsions 328
17.1.3 Nanoemulsions: satisfying cosmetic demands 330
17.1.4 Nanoemulsions as a potential carrier in cosmeceuticals 330
17.2 Challenges associated with cosmetic nanoemulsions 332
17.3 Methods of production 333
17.3.1 High-energy emulsification methods 334
17.3.2 Low-energy emulsification method 336
17.4 Nanoemulsion in cosmetic: from laboratory to market 339
17.4.1 Dermal application 339
17.4.2 NEs as a novel vehicle in skin care 340
17.4.3 NEs in hair system care 341
17.4.4 As potential carrier 342
17.5 Conclusions 343
References 343

18. Nanomaterials in sun-care products

Jennifer Gubitosa, Vito Rizzi, Paola Fini and Pinalysa Cosma

18.1 Introduction 349
18.2 Solar radiations: characteristics and skin effects 351
18.3 Sunscreen: functional ingredients for sun protection products 353
18.3.1 Organic and inorganic sunscreens 355
18.3.2 The sun protection factor 357
18.3.3 Sunscreen regulatory status: safety and effectiveness 359
18.3.4 Sun-care products labeling 360
18.3.5 Controversy between sunscreen application and decrease in vitamin D synthesis 361
18.4 Nanomaterials in cosmetics 362

18.4.1 Regulatory status of nanomaterials in cosmetics 363
18.4.2 Nanomaterials under REACH and CLP 364
18.4.3 ECHA activity 365
18.4.4 Nanoingredients labeling 365
18.4.5 Nanomaterials safety concerns 365
18.4.6 Nanotoxicology 366
18.5 Mineral-based nanosunscreens: nano-TiO_2 and nano-ZnO 367
18.5.1 Titanium dioxide 367
18.5.2 Zinc oxide 369
18.5.3 Titanium dioxide and zinc oxide ecotoxicity studies 370
18.6 Conclusions 371
References 371

19. Nanomaterials for lip and nail cares applications

Carla Aparecida Pedriali Moraes and Aline Rocha Vieira

19.1 Introduction 375
19.2 Nail care: from conventional nail polish to nowadays innovations 376
19.3 Lip care: traditional formulations and innovations 378
19.4 Other nanotechnological innovations used in nail and lip care formulations 380
References 387

3

Toxic risks of nanocosmetics

20. Current legal frameworks and consumer protection in nanocosmetics

Sonia Selletti

20.1 Legal definition of nanomaterial 393
20.2 Nanomaterials in EC Regulation No. 1223/2009 on cosmetic products 395
20.2.1 Cosmetic definition of “nanomaterials” 395
20.2.2 The use of nanomaterial in cosmetic products 397
20.2.3 Declaration of nanomaterial use and nanonotification 398
20.2.4 Labeling 400
20.2.5 Safety assessment (report) of nanomaterials 400
20.3 Catalog of nanomaterials 401
20.4 Conclusions 402

21. Nanoparticle toxicological risks on intact-skin dermal exposures

Tasleem Arif and Konchok Dorjay

21.1 Introduction 403
21.2 Factors mediating toxicity 404
Particle factors 404
Host factors 404
21.3 Routes of entry and kinetics 405
21.4 Nanoparticles and their hazards 405
Conclusion 407
References 408

22. In vitro standard methods for cellular toxicity of nanocosmetic

Swati Gajbhiye

22.1 Introduction 411
22.2 In vitro skin irritation test 414
22.2.1 Initial considerations and limitations 414
22.2.2 Principle 414
22.2.3 Procedure 415
22.3 In vitro skin corrosion test 416
22.3.1 General description 416
22.3.2 Initial considerations 416
22.3.3 Principle of the test 417
22.3.4 Procedure 417
22.4 In vitro phototoxicity test 419
22.4.1 Initial consideration 419
22.4.2 Principle of the test method 419
22.4.3 Description of the test method 420
22.5 In vitro dermal absorption measurements of cosmetic ingredients 422
22.5.1 General principle 422
22.5.2 Principle of the test 422
22.5.3 Factors affecting dermal absorption and methodology 422
22.5.4 Result 424
22.6 Genotoxicity/mutagenicity testing 424
22.6.1 Ames test (bacterial reverse mutation test) 424

22.6.2 In vitro mammalian cell gene mutation test 427
22.6.3 In vitro mammalian cell micronucleus test 429
22.7 Embryotoxicity testing via three test: embryonic stem cell line, micromass, whole embryo culture 433
22.7.1 Embryonic stem cell test 433
22.7.2 Micromass test 435
22.7.3 Whole embryo culture 436
References 439

23. Current commercial nanocosmetic products

Tuan Anh Nguyen and Susai Rajendran

23.1 Introduction 445
23.2 Current commercial nanocosmetic products 446
References 452

24. Nanocosmetics: future perspective

P. Morganti, Hong-Duo Chen and G. Morganti

24.1 Introduction 455
24.2 Polymeric natural compounds and biopolyesters 457
24.2.1 Chito-oligosaccharides and polysaccharides 458
24.2.2 Biopolyesters 469
24.3 Progress and future perspective of the cosmetic products 471
24.3.1 Natural green cosmetics and biodiversity 471
24.3.2 Advanced nanomaterials and nanotechnologies 473
24.3.3 Smart nanocosmetics and future perspectives 474
References 476

Index 483

List of contributors

Bilal Haider Abbasi Department of Biotechnology, Quaid-i-Azam University, Islamabad, Pakistan

Nisar Ahmad Center for Biotechnology and Microbiology, University of Swat, Swat, Pakistan

Abdulhameed Al-Hashem Petroleum Research Centre, Kuwait Institute for Scientific Research, Al Ahmadi, Kuwait

Mohammad Ali Center for Biotechnology and Microbiology, University of Swat, Swat, Pakistan

Gabriel Lima Barros De Araújo Faculty of Pharmaceutical Sciences, University of Sao Paulo, Sao Paulo, Brazil

Tasleem Arif Ellahi Medicare Clinic, Kashmir, India

J. Arockiaselvi PG and Research Department of Chemistry, SRM University, Chennai, India

Andreia Ascenso Nanostructured Systems for Overcoming Biological Barriers (Nano2B) Group, Research Institute for Medicines (iMed. ULisboa), Faculty of Pharmacy, Universidade de Lisboa, Lisboa, Portugal

A. Grace Baby Corrosion Research Centre, St Antony's College of Arts and Sciences for Women, Dindigul, India

Maurício S. Baptista Biochemistry Department, Institute of Chemistry, University of São Paulo, São Paulo, Brazil

Nádia Araci Bou-Chacra Faculty of Pharmaceutical Sciences, University of Sao Paulo, Sao Paulo, Brazil

Manuela Carvalheiro Nanostructured Systems for Overcoming Biological Barriers (Nano2B) Group, Research Institute for Medicines (iMed. ULisboa), Faculty of Pharmacy, Universidade de Lisboa, Lisboa, Portugal

Hong-Duo Chen Dermatology Department, Key Laboratory of Immunodermatology, Hospital No. 1, China Medical University, Shenyang, P.R. China

M.B. Coltelli Department of Civil and Industrial Engineering, University of Pisa, Pisa, Italy

Pinalysa Cosma Consiglio Nazionale delle Ricerche CNR-IPCF, UOS Bari, Bari, Italy; Dipartimento di Chimica, Università degli Studi "Aldo Moro" di Bari, Bari, Italy

Ana Margarida Costa Nanostructured Systems for Overcoming Biological Barriers (Nano2B) Group, Research Institute for Medicines (iMed. ULisboa), Faculty of Pharmacy, Universidade de Lisboa, Lisboa, Portugal

Surbhi Dhawan Department of Pharmaceutical Sciences, Maharshi Dayanand University, Rohtak, India

Herminia Domínguez Department of Chemical Engineering, Faculty of Sciences, University of Vigo, Ourense, Spain

Konchok Dorjay Department of Dermatology, Dr. Ram Manohar Lohia Hospital and Post Graduate Institute of Medical Education and Research, New Delhi, India

R. Dorothy Department of EEE, AMET University, Chennai, India

Elena Falqué Department of Analytical Chemistry, Faculty of Sciences, University of Vigo, Ourense, Spain

Ana Catarina Faria-Silva Nanostructured Systems for Overcoming Biological Barriers (Nano2B) Group, Research Institute for Medicines (iMed. ULisboa), Faculty of Pharmacy, Universidade de Lisboa, Lisboa, Portugal

Hina Fazal Pakistan Council of Scientific and Industrial Research (PCSIR) Laboratories Complex, Peshawar, Pakistan

Paola Fini Consiglio Nazionale delle Ricerche CNR-IPCF, UOS Bari, Bari, Italy

Swati Gajbhiye Department of Cosmetic Technology, LAD & SRP College for Women, Nagpur, India

Ashish Garg Department of Pharmacy, Rani Durgavati Vishwavidyalaya, Jabalpur, India

Nathalie Giglioli-Guivarch Biomolecules et Biotechnologies Vegetales (BBV) EA2106, Universite Francois-Rabelais de Tours, Tours, France

B. Betül Gökçe Department of Pharmaceutical Technology, Faculty of Pharmacy, Istanbul University, Istanbul, Turkey

Lídia Maria Gonçalves Nanostructured Systems for Overcoming Biological Barriers (Nano2B) Group, Research Institute for Medicines (iMed. ULisboa), Faculty of Pharmacy, Universidade de Lisboa, Lisboa, Portugal

Jennifer Gubitosa Consiglio Nazionale delle Ricerche CNR-IPCF, UOS Bari, Bari, Italy

Sevgi Güngör Department of Pharmaceutical Technology, Faculty of Pharmacy, Istanbul University, Istanbul, Turkey

Christophe Hano Laboratoire de Biologie des Ligneux et des Grandes Cultures (LBLGC), INRA USC1328, Université d'Orléans, Chartres, France

Laura Hermida Chemistry Department, National Institute of Industrial Technology—INTI, Buenos Aires, Argentina

B.N. Vedha Hari Department of Pharmaceutical Technology, School of Chemical and Biotechnology, SASTRA Deemed University, Thanjavur, Tamil Nadu, India

S. Johnmary PG and Research Department of Chemistry, Loyola College, Chennai, India

Ruckmani Kandasamy Centre for Excellence in Nanobio Translational REsearch (CENTRE), Department of Pharmaceutical Technology, University College of Engineering, Anna University, BIT Campus, Tiruchirappalli, India

Mohan Lal Kori Department of Pharmacy, Vedica College of B. Pharmacy, Bhopal, India

Venkateshwaran Krishnaswami Centre for Excellence in Nanobio Translational REsearch (CENTRE), Department of Pharmaceutical Technology, University College of Engineering, Anna University, BIT Campus, Tiruchirappalli, India

M. Krishnaveni PG and Research Department of Chemistry, MVM Government College for Women, Dindigul, India

S. Senthil Kumaran School of Mechanical Engineering, VIT University, Vellore, India

E. Lasoń Faculty of Chemical Engineering and Technology, Institute of Organic Chemistry and Technology, Cracow University of Technology, Cracow, Poland

Lucía López-Hortas Department of Chemical Engineering, Faculty of Sciences, University of Vigo, Ourense, Spain

M. Malathi Department of Pharmaceutical Technology, School of Chemical and Biotechnology, SASTRA Deemed University, Thanjavur, Tamil Nadu, India

Joana Marto Nanostructured Systems for Overcoming Biological Barriers (Nano2B) Group, Research Institute for Medicines (iMed. ULisboa), Faculty of Pharmacy, Universidade de Lisboa, Lisboa, Portugal

Małgorzata Miastkowska Faculty of Chemical Engineering and Technology, Institute of Organic Chemistry and Technology, Cracow University of Technology, Cracow, Poland

Carla Aparecida Pedriali Moraes Laboratório de Pesquisa, Tecnologia em Cosméticos, FATEC Diadema—Luigi Papaiz, São Paulo, Brazil

G. Morganti ISCD Nanoscience Center, Rome, Italy; Dermatology Unit, Campania University, "L. Vanvitelli", Naples, Italy

P. Morganti Accademy of History of Health Care Art, Rome, Italy; China Medical University, Shenyang, P.R. China; Dermatology Unit, Campania University, "L. Vanvitelli", Naples, Italy; ISCD NanoScience Center, Rome, Italy

Pierfrancesco Morganti Academy of History of Health Care Art, Rome, Italy; China Medical University, Shenyang, P.R. China; ISCD NanoScienceCenter, Rome, Italy

Sanju Nanda Department of Pharmaceutical Sciences, Maharshi Dayanand University, Rohtak, India

Tuan Anh Nguyen Institute for Tropical Technology, Vietnam Academy of Science and Technology, Hanoi, Vietnam

Vikas Pandey Guru Ramdas Khalsa Institute of Science and Technology (Pharmacy), Jabalpur, India

Gopal Rai Guru Ramdas Khalsa Institute of Science and Technology (Pharmacy), Jabalpur, India

Susai Rajendran Department of Chemistry, St Antony's College of Arts and Sciences for Women, Dindigul, India; Corrosion Research Centre, St Antony's College of Arts and Sciences for Women, Dindigul, India

Nikhishaa Sree Raju Centre for Excellence in Nanobio Translational REsearch (CENTRE), Department of Pharmaceutical Technology, University College of Engineering, Anna University, BIT Campus, Tiruchirappalli, India

D. Ramyadevi Department of Pharmaceutical Technology, School of Chemical and Biotechnology, SASTRA Deemed University, Thanjavur, Tamil Nadu, India

R. Joseph Rathish PSNA College of Engineering and Technology, Dindigul, India

Helena Margarida Ribeiro Nanostructured Systems for Overcoming Biological Barriers (Nano2B) Group, Research Institute for Medicines (iMed. ULisboa), Faculty of Pharmacy, Universidade de Lisboa, Lisboa, Portugal

Vito Rizzi Dipartimento di Chimica, Università degli Studi "Aldo Moro" di Bari, Bari, Italy

Sonia Selletti Studio Legale Astolfi e Associati, Milan, Italy

Pragya Sharma Department of Pharmaceutical Sciences, Maharshi Dayanand University, Rohtak, India

Rajesh Shukla Guru Ramdas Khalsa Institute of Science and Technology (Pharmacy), Jabalpur, India

Sandra Simões Nanostructured Systems for Overcoming Biological Barriers (Nano2B) Group, Research Institute for Medicines (iMed. ULisboa), Faculty of Pharmacy, Universidade de Lisboa, Lisboa, Portugal

Gurmeet Singh Pondicherry University, Puducherry, India

Federico Svarc School of Exact and Natural Sciences. DQIAQF-INQUIMAE, Buenos Aires University - UBA, Buenos Aires, Argentina

Thiago Teixeira Tasso Chemistry Department, Institute of Exact Sciences, Federal University of Minas Gerais, Minas Gerais, Brazil

Paulo Newton Tonolli Biochemistry Department, Institute of Chemistry, University of São Paulo, São Paulo, Brazil

María D. Torres Department of Chemical Engineering, Faculty of Sciences, University of Vigo, Ourense, Spain

T. Umamathi Department of Chemistry, Yadava College, Madurai, India

Aline Rocha Vieira Laboratório de Pesquisa, Tecnologia em Cosméticos, FATEC Diadema—Luigi Papaiz, São Paulo, Brazil

N. Vijaya Department of Chemistry, Vellalar College for Women, Erode, India

Sivakumar Vijayaraghavalu Central Research Facility, Sri Ramachandra Institute of Higher Education and Research (Deemed to be University), Chennai, India

Megumi Nishitani Yukuyama Faculty of Pharmaceutical Sciences, University of Sao Paulo, Sao Paulo, Brazil

PART 1

Basic principles

CHAPTER

1

Nanocosmetics: an introduction

Pierfrancesco Morganti[1,2,3]

[1]Academy of History of Health Care Art, Rome, Italy [2]China Medical University, Shenyang, P. R. China [3]ISCD NanoScienceCenter, Rome, Italy

1.1 Introduction

Consumers are showing interest in holistic and technological approaches in buying cosmetics for their beauty and health [1,2]. The major interest is toward natural products made by nanotechnology and bionanotechnology, selected for their effectiveness and safeness regarding wellbeing and the environment also [2]. Nanotechnology is a science of manipulating atoms and molecules at the nanoscale, considering that a nanometer (nm) is a billionth of meter. Just to better understand, dust mite has a dimension of 200,000 nm, hair has a wide between 10,000 and 50,000 nm, red blood cells have a wide between 2000 and 5000 nm, a bacterium is around 1000 nm long, and DNA has an average length between 46 and 150 nm and a diameter of 2 nm (Fig. 1.1) [3]. On the other hand, nanobiotechnology, as interfacing with the biotechnology, represents a new field of science that introduces the special physicochemical and biological properties in different sectors (Fig. 1.2), including biomedical nanomedicine, environment, consumer goods, electronics, and agriculture [4,5]. This the reason why the world market for products containing nanomaterials was evaluated at US$ 14,741.6 million in 2015 and is expected to reach US$ 55,016 million by 2022, supported by a Compound Annual Growth Rate (CARG) of 20.7% [6]. However, the emerging trends of nanotechnology include the design, characterization, production, and application on structures, such as the skin, by the control of shape and size at the nanometers scale, carefully verifying their effectiveness and safeness. Nanoparticles (NPs), in fact, possess physicochemical properties which, altered when compared with their larger counterparts, create an increased uptake and interaction with the biological tissues. NPs and nanocarriers (NCs) offer, therefore, particular and unique advantages, enhancing the delivery of active ingredients at the different skin strata (Fig. 1.3) [7–9]. They, in fact, protect the active ingredients from degradation prior to their delivery, controlling also the rate, time, and dose absorbed. Thus NCs may elicit the desired and designed cosmetic response to keep the functional molecules in the specific

Nanocosmetics
DOI: https://doi.org/10.1016/B978-0-12-822286-7.00001-2

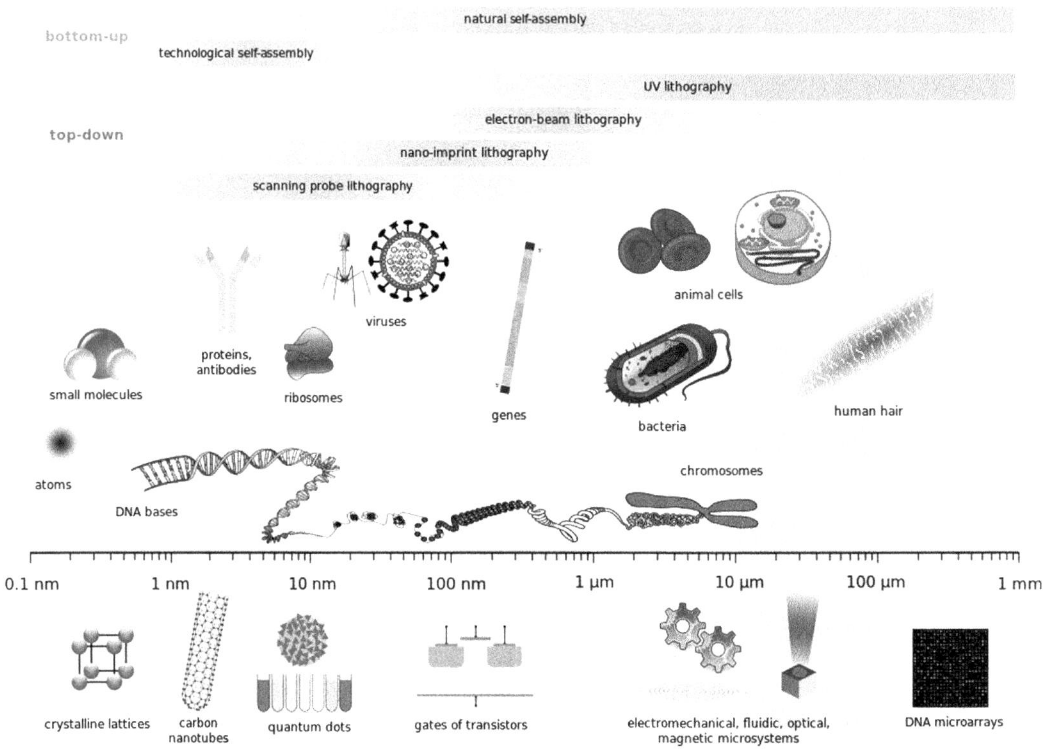

FIGURE 1.1 Nanodimension and its use in nature and technology. Source: *Reproduced with permission from Nanoscopic scale, <https://en.wikipedia.org/wiki/Nanoscopic_scale>. Licensed under the Creative Commons Attribution-Share Alike 2.5 Generic license.*

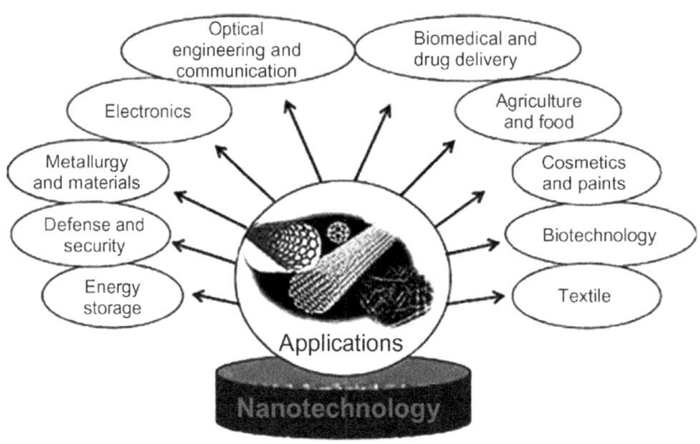

FIGURE 1.2 Sectors involving nanotechnology. Source: *Reproduced with permission from Nasrollahzadeh M, Sajadi SM, Sajjadi M, Issaabadi Z. Chapter 4 - applications of nanotechnology in daily life. Interface Sci Technol 2019;28:113–43. https://doi.org/10.1016/B978-0-12-813586-0.00004-3. Copyright 2019 Elsevier.*

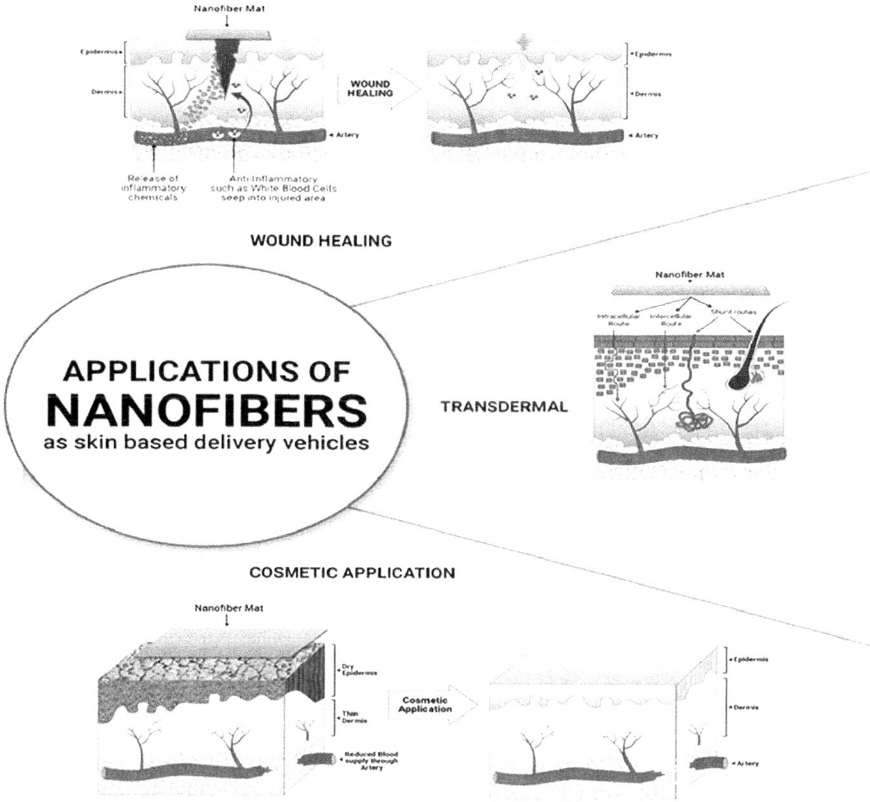

FIGURE 1.3 Innovative nanocarriers may facilitate the penetration of active ingredients through the skin. Source: *Reproduced with permission from Kamble P, Sadarani B, Majumdar A, Bhullar S. Nanofiber based drug delivery systems for skin: a promising therapeutic approach. J Drug Deliv Sci Technol 2017;41:124–33. https://doi.org/10.1016/j.jddst.2017.07.003. Copyright Elsevier 2017.*

skin layer, according to the 4R Approach (right chemical, right site, right concentration, correct period of time) [9]. Naturally, the polymer biocompatibility, the active nanoingredients' polarity, and their possible toxic, immunologic, and genotoxic side effects together with the physicochemical characteristics of micro/nanoemulsions selected as carriers are of supreme importance for the safeness and effectiveness of the nanocosmetics [10]. So far, the continuous increase and in-progress studies and experimental models to evaluate their interactions with tissues and individual cells are justified and necessary. Due to their small size, NPs may easily enter the human body, crossing the various biological barriers, facilitated also by the use of innovative nanocarriers. For all these reasons and before to be commercialized, the cosmetic products containing nanoingredients have to be controlled by five steps: worldwide regulations, easy reproducibility, multidisciplinary researches, sufficient safety data, and confidence building of consumers, industry, and

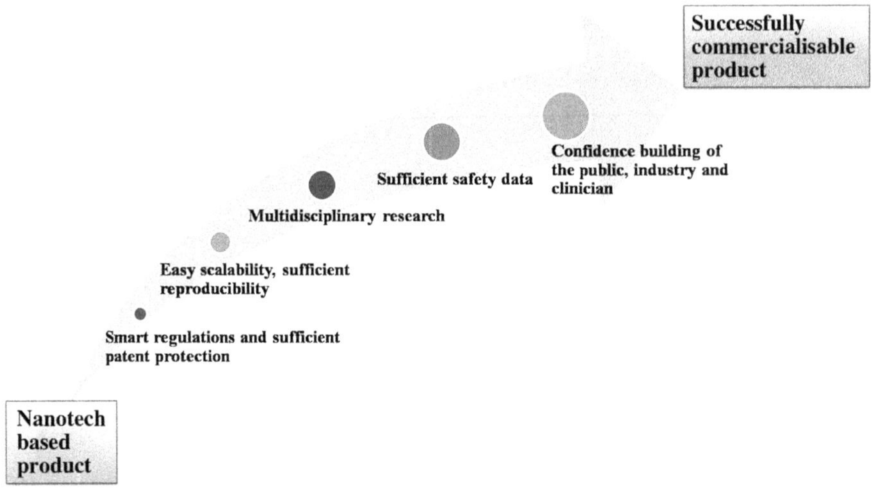

FIGURE 1.4 Necessary controls for cosmetic products based on the use of nanoingredients. Source: *Reproduced with permission from Kaur IP, Kakkar V, Deol PK, Yadav M, Singh M, and Sharma I. Issues and concerns in nanotech product development and its commercialization. J Control Release 2014;193:51–62. Available from: https://doi.org/10.1016/j.jcontel.2014.06005. Copyright Elsevier 2014.*

clinicians (Fig. 1.4) [11]. However, nanotechnology has increased its presence in many new cosmetic products launched worldwide, as recently reported by Jane Henderson in São Paulo [12].

Accordingly, consumers are looking for holistic and the cosmetic use of natural ingredients, such as lignin and chitin, that could be utilized in high-tech nanotechnology-oriented industries which for example, specialized for the production of colors, trying to mimic the same methodology adopted from butterfly and birds [13]. The bright iridescent colors of some butterfly wings, in fact, are produced by light scattering from photonic nanostructured crystals of chitin which, in combination with pigments, give rise to different colors such as blue, turquoise, and green as seen in peacock's tail or in butterfly wings [14–16]. Thus a similar approach could be useful to satisfy the natural request of consumer who prefers to buy natural cosmetics, formulated and produced by innovative technologies. Uncovering the precise mechanisms of these natural color tuning could have many interesting implications for the industrial biomimetic applications to produce, for example, innovative and smart make-up nanocosmetics.

1.2 Consumer requests and cosmetics

According to different sources, the Global Beauty and Personal Care market, valued at US$ 532.43 in 2017, is projected to grow at a CARG of 7.2% during the forecast period of 2019–24 for reaching US$ 805.61 billion in 2023 [2,17–19]. Thus the global niche of natural/technological cosmetic care products reached around US$ 200 million (notably increasing year by year), whereas the global nanotechnology market is expected to exceed US$ 125 billion by 2024 [1,20,21]. This global growth is a reply to the request of natural cosmetics from consumers, 71% of whom consider their use important in the daily life, 72% think that they have a positive effect on how they feel, while 87% recognize as a priority their quality, efficacy, and naturalness [22]. Improvement in the lifestyle of the consumers, in fact, is affecting even more the cosmetic market because people are looking to step up the style quotient and their overall personality [22]. Additionally, the cosmetic growth has been further increased by the request of male grooming products for the major consciousness of these consumers who take in great consideration the smart and natural bionanotechnologies used to produce the actual skin care products. These considerations are held in great esteem especially from population, two-third of which will live in urban areas by 2050 [23,24]. In any way, the natural cosmetic ingredients are ongoing privileged from worldwide consumers, because considered more effective to control their life and wellbeing. Moreover, the innovative nanocosmetics incorporating active and functional natural agents could be based on a new technical and *wholistic* approach, which is able to provide protection from UV rays, pollution, blue light, allergies, and other sensitivities [22–24]. Owing to all these factors, manufacturers are focusing on launching new and specialized products for hair, face, and body based on natural new ingredients obtained when possible from waste materials and made especially by smart bionanotechnologies. Among all the industries, the category of cosmetics and beauty products represents one sector that globally remains impervious to the economic ups and downs because of the continuing worldwide growing usage of these products by women of any age and increasingly by men also [23,24]. Thus manufacturers are focusing on developing new and innovative products and exploring different strategies for sustaining and maintaining their market position. The global cosmetic market, in fact, is segmented and based on various categories of cosmetics, different modes of sale, differences in gender and geography, as clearly shown in some reports [22–24].

The aging population is, actually, the major driver of this market for the strong and general desire to retain a youthful appearance, possibly . . . forever! The number of people aged 65 or older, in fact, is projected to grow from an estimated 524 million in 2010 to nearly 1.5 billion in 2050, with most of the increase in developing countries [25]. For these reasons, it is in prevision a robust demand of antiaging cosmetics in order to prevent wrinkles, fine lines, age spots, dry skin, and hair damages [19]. Moreover, the cosmetic industry is also benefiting and continues to benefit from the increasing popularity of social media channels such as Instagram, YouTube, and Google. These platforms have created a demand for beauty products, helping to fill the gap between cosmetics brands and consumers. Naturally, the nanotechnology market, with a provisional annual growth for the innovative ingredients of around 17% during the forecast period of 2017–24, is strongly influencing the cosmetic market, as previously reported [26]. For all these reasons and to achieve the requested aims, nanotechnology and bionanotechnology are trying to develop and combine natural new materials by precisely engineering atoms and molecules. Thus new biodegradable polymeric materials and nanocomposites are becoming to be used as smart carriers, molecularly assembled on the scale of individual skin cells, with the scope of providing personalized cosmetic dermatology [27].

1.3 Polymers, nanocomposites, and nonwoven tissues

Polymeric materials that are organized as biodegradable nonwoven tissues have been shown to be ideal and innovative carrier systems for biomedical and cosmetic applications [28–30]. They may be used to make tissue scaffolds which, able to restore the skin microenvironment, improve both delivery and penetration of the active ingredients into the right skin layer [30]. Moreover, they may serve as a reservoir system to gradually release active ingredients with their degraded components, which sometimes may also activate specific biological signals [30,31]. Thus the nonwoven tissues, realized by our group as reported later, seem to accumulate and release the selected active ingredients at the site of action, when applied topically on healthy, aged, or diseased skin [32,33]. These smart tissues, made by biodegradable polymers and NPs, can be produced without the use of chemicals, such as emulsifiers, preservatives, colors, and fragrances offering advantages and further safeness over the normal emulsions [28–30]. In addition, their physicochemical properties may be modified in order to obtain scaffolds able to achieve a controlled release of the active ingredients bound to their micro/nanofibers. However, carriers and ingredients used in cosmetic dermatology can be selected differently in the case of healthy, aged, or diseased skin. As also reported and well designed from other authors [31], the micro/nanocarriers may load and release the active ingredients, crossing the stratum corneum (SC) barrier and entering into epidermis and dermis or, alternatively, penetrating across the hair follicle canals. Naturally, the grade of penetrability is totally different for healthy or diseased skin, as clearly reported in Ref. [31]. But which kind of polymers and NPs are used to realize these nonwoven tissues? Among the polymers, industrially utilized, the bio-based biodegradable polysaccharides and biopolyesters are more requested from the consumers because of their natural origin and the possibility to be obtained from waste materials. The more used biopolymers are cellulose, starch, chitin, lignins, and their derived compounds, extracted as polysaccharides from waste feedstock polylactic acid and poly-hydroxy-alkanoates (PHA and PHB), man-made compounds, obtained as biopolyesters from renewable resources. Naturally, all these natural products are used especially in their micro/nanodimension for increasing their effectiveness, according to the first results obtained from our group and the in-progress European polybioskin (www.polybioskin.eu) research project [32–38]. Thus the ability to handle selectively the right materials of nanosize has led and is leading the industry to develop new and smart ingredients with innovative properties and significant advantages, as compared to the microscopic world [39]. Thanks to the bionanotechnology, in fact, it is possible to modify chemistry and topography of natural molecules, NPs, nanocomposites, and biopolymers, such as chitin nanofibrils (CNs), for mimicking the natural extracellular matrix (ECM), for example, by the electrospinning technology (Fig. 1.5). On the other hand, by the spray drier technique, it is possible to make in powder the NPs of CNs and nanolignin (NG), as well as their complexes. In addition, the CN–NG NPs may encapsulate many different active ingredients both hydro-soluble and oil-soluble. These NPs may be sprayed on the natural fiber's surface or bound to its structure for producing and characterizing biodegradable nonwoven tissues, according to the designed and requested activity [30,37]. These innovative tissues, applied on the skin, are able to act as templates to guide cell growth of nutrients, oxygen, and waste, interconnecting also the released active ingredients and cell signals

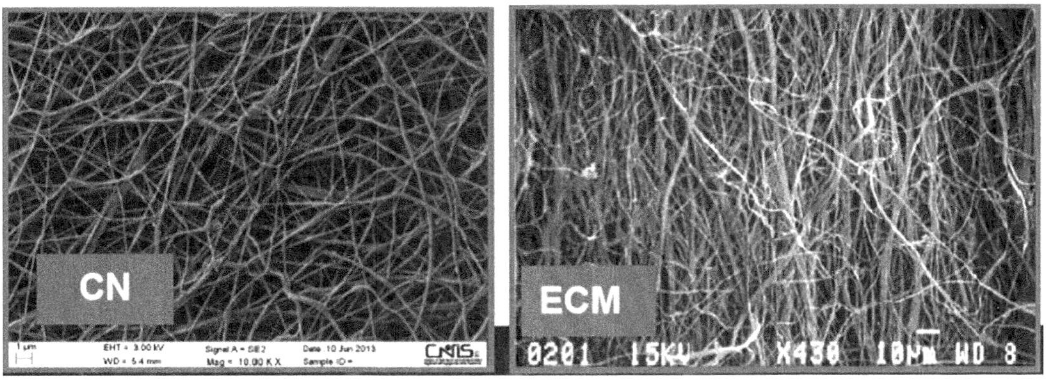

FIGURE 1.5 Chitin nanofibrils tissue (left) and ECM (right). Source: *Reproduced with permission from Morganti P, Morganti G, Colao C. Biofunctional textiles for aging skin. Biomedicines 2019;7(3):E51. https://doi.org/10.3390/Biomedici-nes7030051. Licensed under Creative Commons Attribution (CC BY 4.0) license.*

with the native ECM in vivo [37–39]. As evidenced and summarized from many authors [40,41], the natural ECM molecular structure and composition has a fundamental function and role at the skin level and in tissue engineering. This complex connective network, in fact, provides a physical scaffold for cells and tissues, determining not only the cell shape and mechanic but also its motility during the natural tissue development [41]. According to our recent research studies, the basic carriers, made by innovative biodegradable nonwoven tissues and produced by electrospinning biopolymeric nanomaterials, seem to act as a dynamic bioactive environment, releasing the entrapped active ingredients in the designed dose and time [28–39]. Thus these biopolymers seem to positively influence and enhance the penetration capacity of the selected active ingredients, increasing their antibacterial, anti-inflammatory, immunomodulatory, and tissue repairing activity, slowing down fine lines, wrinkles, and aged spots, thus ameliorating appearance and health of face and body [29,30,37,38]. As previously reported, the smart tissues, realized from our group, are made by natural sugar-like fibers bound to CN, NG, and/or to nanochitin–nanolignin (CN–NG) complexes, obtained by the simple gelation methodology (Fig. 1.6) [30,33,42]. The particular composition and organized structure of both fibers and tissues, because of the activity of the encapsulated ingredients bound at the polymer surface, provide the right structural and mechanical support for cells attachment and subsequent skin development [29,43–45]. Moreover, the easy biodegradability of these smart tissues induces the production of active ingredients which, providing cues and growth factors, regulate the cell functions, supporting both epidermis/dermis formation and repair the skin of people healthy, aged, or affected by mild diseases [37,46–48]. It is important to remember that while the interaction between cells and functional properties of NPs and biodegradable tissues has to be considered fundamental, it is also necessary to demonstrate their safety when applied on the skin [37]. Nanosized ingredients, in fact, can migrate into the tissues, leading to unknown reactions [49]. Thus it has to be clearly shown the biocompatibility and safeness of the CN–NG tissues, evidencing also their effectiveness and absence of side effects such as allergic and/or sensitizing reactions [28–30,37,38]. Both safeness and effectiveness have been shown

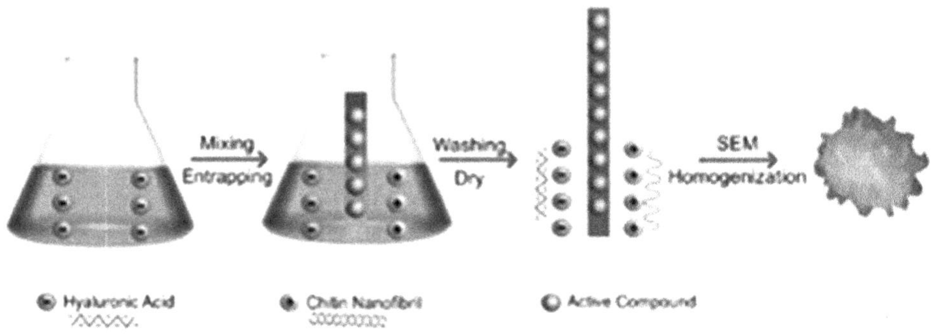

FIGURE 1.6 The chitin nanofibril-hyaluronic acid block polymer obtained by the gelation method. Source: *Reproduced with permission from Morganti P, Coltelli MB. A new carrier for advanced cosmeceuticals. Cosmetics 2019;6:10. https://doi.org/10.3390/cosmetics6010010. Licensed under the Creative Commons Attribution Attribution 4.0 International License (CC BY 4.0).*

in vitro by human keratinocytes and fibroblasts and in vivo by clinic tests [28,37,45]. Moreover, as reported, it has been evidenced the ability of these tissues to modulate the inflammatory response by decreasing the expression of different proinflammatory cytokines, playing also an important positive role in the innate immune response [37,44].

1.4 Conclusive remarks

Beyond question natural nanocosmetics will represent the future of an innovative Cosmetic Dermatology being able to give clear answers to the consumer demand of major naturalness and embodied-advanced technology applied on cosmetic products, considered more effective.

The formulation of nanocosmetics based on natural nanosized ingredients, such as CNs and NG, used as carrier and active ingredients go in this direction. These and other selected natural polymers, as previously shown, may be easily complexed by the simple and sustainable gelation technique, to form innovative NPs or nanosheets without the consumption of high energy and water. Moreover, the obtained complexes act as a carrier, binding and loading different active ingredients for releasing them at different skin layers [30,50,51]. Additionally, the natural CN, NG, and CN−NG, being catabolized from the human enzymes to release biological compounds, such as acetyl glucosamine, glucosamine, and glucose, are free of toxic effects and safe because used from the cells as active ingredients and energy for their survival and signaling functions [52−54]. These innovative nanocarrier tissues and nanoingredients, therefore, have shown to be free of side effects, differently from other nanoingredients man-made, formulated and used by nanoemulsion carriers. On the other hand, the majority of the synthetic ingredients require more safety evaluations because of the toxicity or immunotoxicity their catabolites could determine [55]. However, if topical skin exposure to nonnatural NPs could be a potential route of penetration for cosmetic products, it has been shown that nanotitanium dioxide, used extensively as sun screen, evidenced both in vitro and in vivo a very low derma absorption capacity, causing in reality no substantial toxicity [56]. In any way, because the possibility NPs have to pass the biological barriers through different routes, they could produce toxicity effects in different human systems, including the reproductive ones. Thus more studies are necessary to determine their safeness because NPs, especially when man-made, could evidence side effects for the production of secondary toxic compounds [56,57]. Completely different are the problems regarding, for example, the use of natural NPs, such as chitin-nanofibers which, as previously reported, are catabolized to ingredients not only free of toxicity but also useful for the cell growth and reproduction [36,37,44]. On the other hand, about the reported use of NG [58], it is important to select and characterize the source of its recovery, as well as the purity, morphological characteristics, and the mean size of the polymer selected, together with its stability when complexed with CNs [37]. When these parameters have been well defined, NG and its complexes with CNs have shown not only to have none of toxicity but also to downregulate a panel of antiinflammatory cytokines [37].

1.4.1 Micro/nanoemulsions

Apart the smart activity of the described CN–NG nanotissues and NPs to be used for the formulation of nanocosmetics, it seems important to report the fundamental function of nanoemulsions, compared with the normal ones [31,59,60]. Differently from the normal emulsions, these particular emulsions working at nanodimension and referred to as sub-micron emulsions or hydrogels are made generally by natural polymeric polysaccharides and represent specific carriers able to facilitate the skin penetration and release of loaded ingredients. By their use, it is possible to produce multiphase colloidal oil in water nanoemulsions and microemulsions which, stabilized by microfacial films of surfactant molecules, contain oil droplets between a diameter ranging 5–50 and 50–100 nm, respectively [59–61]. It is interesting to underline that nanoemulsions and microemulsions are achieved through low-energy technique, resulting in droplet size which depends on the weight ratio between surfactant and oil, while the emulsifier plays a fundamental role as a stabilizing agent. These specialized biocarriers, characterized for their thermodynamic stability and safety, are preferred in the pharmaceutical and cosmetic sector because they overcome some of the common problems associated with the active molecules, such as low solubility, enzymatic degradation, and short half-life [59–61]. Moreover, the small droplet size of the nanoemulsions ensures a closer contact with the SC, improving the skin layer penetration, just as it seems to happen by the nonwoven tissues, as previously reported [28–30,62]. Thus both nanoemulsions and biodegradable nonwoven tissues may be used as a suitable biomaterial for fabricating scaffolds which, mimicking the native ECM, are fundamental to regenerate diseased/aged's skin functional tissues. For their biocompatibility, biodegradability, and physical properties, they may be used as a promising material for biomedical and cosmetic applications in the form of hydrogels, sponges, films, and tissues as previously shown [28–38]. Moreover, these scaffolds, when correctly designed and structured with the right active ingredients, are able to gradually replace the diseased tissue with a new one tissue, porous enough to allow cell growth, and promote its adhesion and differentiation [38,47,48,63]. Naturally, in my opinion as previously reported, the dry tissues made by natural fibers embedded by active ingredients represent the best carrier to produce innovative and smart nanocosmetics and advanced medications for aged, sensitive, and diseased skin [28–39].

1.4.2 Packaging material

Last but not least, it is necessary to not forget the packaging polymers used for cosmetics and nanocosmetics normally made too frequently by nonbiodegradable materials, necessary for their conservation and distribution in the market.

Packaging and the great use of nonbiodegradable plastic polymers represent a great problem for the dangerous waste that remains in land and the oceans without being degraded so that, according to the Ellen McArthur Foundation, 1 tonne of plastic for every 3 tonnes of fish is actually living in the sea [64]. If this consumption of plastics will continue, it is in prevision that by 2050 there will be 12 billion of tonnes of plastic in landfills, that is, the equivalent to 35,000 Empire State Buildings! [65]. Unfortunately, it has been estimated that today only 5% of the 120 billion units of packaging consumed are reused [64]. Moreover, the global personal care industry every year produces not only nonbiodegradable waste for bottles and plastic particles used in cleansers, toothpastes, and scrubs, but also plastic bags for a booming value of US$ 500 billion. An additional problem is represented from the increasing use of wipes [66] such as face wipes, baby wipes, and wet toilet paper wipes, 46% of which result virtually indestructible! They are made of synthetic nonwoven textiles and do not tear or disintegrate in the sewage system as toilet paper does. As a consequence, Sydney Water removed about 1 million kilograms of wet wipes from its wastewater systems over the past 2 years! [66]. On the other hand, it is not to be forgotten that beauty products are a bit like food and can deteriorate much faster when they are not stored correctly, thus the necessity to use recyclable eco-friendly packagings and bottles, selected properly by research studies. For all these reasons, it seems urgent to

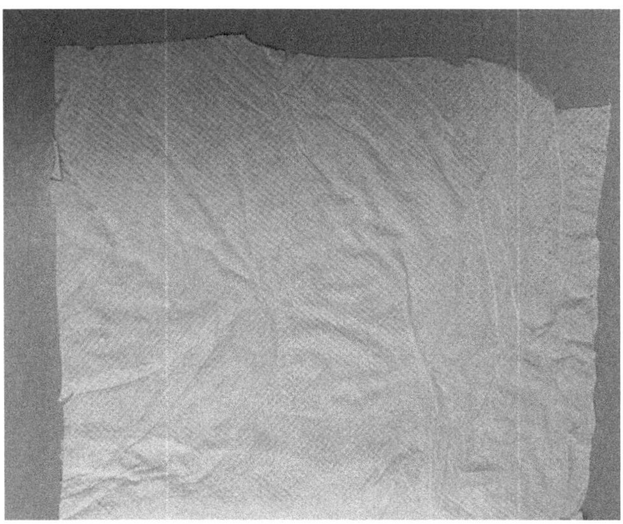

FIGURE 1.7 Skin-friendly and eco-friendly tissues as alternative carriers for cosmetic products. Source: *(Chitin nanofibrils tissue [30], Licensed under the Creative Commons Attribution Attribution 4.0 International License).*

find the way to produce and use bottles, wipes, and bags not only eco-friendly but also safe in their daily use, as the proposed nonwoven tissues (Fig. 1.7) [28–39]. In conclusion, by the right use of the bionanotechnology, it seems possible to produce nanocosmetics made by the nanoemulsion technology and packed by soft and hard recyclable materials. Alternatively or contemporary, a new category of cosmetic tissues could be produced and introduced in the normal skin care market. These smart cosmeceutical tissues, characterized for their effectiveness and safeness for humans and the environment, may be easily packed by simple biodegradable and recycling paper [67]. In conclusion, by the use of natural biodegradable nanocosmetics and a more intensive use of biodegradable materials for their packaging, it will be possible not only to make more effective and safe products, but also to save around 100,000 mammals and 1 million of birds that today die for the over 5 trillion pieces plastic clotting in the oceans [68], thus preserving the biodiversity of our Planet.

References

[1] Silpa R, Shoma J, Sumod US, Sabitha M. Nanotechnology in cosmetics: opportunities and challenges. J Pharm Bioallied Sci 2012;4(3):186–93.

[2] Mintel. Global beauty and personal care trends 2019. London: Mintel Group Ltd; 2019.

[3] Nanoscopic scale, <https://en.wikipedia.org/wiki/Nanoscopic_scale>.

[4] Alam Khan F, editor. Biotechnology fundamentals. Boca Raton, FL: CRC Press; 2012. ISBN: 978-1-4398-2009-4.

[5] Nasrollahzadeh M, Sajadi SM, Sajjadi M, Issaabadi Z. Chapter 4 - applications of nanotechnology in daily life. Interface Sci Technol 2019;28:113–43. Available from: https://doi.org/10.1016/B978-0-12-813586-0.00004-3.

[6] Allied Market Research. Nanomaterial market overview report. Vima, Maharashtra: Allied Group; 2019.

[7] Souto E. Challenging cosmetics-solid lipid nanoparticles (SLN) and nanostructured lipid carriers (NLC). In: Wiechers JW, editor. Science and applications of skin delivery systems. Carol Stream, IL: Allured Publishing Co; 2008. p. 228–50.

[8] Kamble P, Sadarani B, Majumdar A, Bhullar S. Nanofiber based drug delivery systems for skin: A promising therapeutic approach. Journal of Drug Delivery Science and Technology, 41. 2017. p. 124–33. Available from: https://doi.org/10.1016/j.jddst.2017.07.003.

[9] Wiechers JW. Skin delivery: what it is and why we need it. In: Wiechers JW, editor. Science and applications of skin delivery systems. Carol Stream, IL: Allured Publishing Co; 2008. p. 1–21.

[10] Bahadar H, Maqbool F, Niaz K, Abdollahi M. Toxicity of nanoparticles and an overview of current experimental models. Iran Biomed J 2016;20(1):1–11.

[11] Kaur IP, Kakkar V, Deol PK, Yadav M, Singh M, Sharma I. Issues and concerns in nanotech product development and its commercialization. J Control Release 2014;193:51–62. Available from: https://doi.org/10.1016/j.jconrel.2014.06005.

[12] Henderson J. Nanotechnology in cosmetics. Mintel report, São Paulo, Brazil, May 31; 2019. <https://www.mintel.com/blog/beauty-market-news/nanotechnology-cosmetics>.

[13] Ball P. Nature's fantastical palette: color from structure, <http://photobiology.info/bal-l.html>; 2019 [accessed 30.06.19].

[14] Elbaz A, Lu J, Gao B, Zheng F, Mu Z, Zhao Y, et al. Chitin-based anisotropic nanostructures of butterfly wings for regulating cell orientation. Polymers 2017;9:386. Available from: https://doi.org/10.3390/polym9090386.

[15] Piszter G, Kertesz K, Vertesy Z, Balknt Z, Biro LP. Color based discrimination of chitin-air nanocomposites in butterfly scales and their role in conspecific recognition. Anal. Methods 2011;3:78–83. Available from: https://doi.org/10.1039/C0AY00410C.

[16] Parnell AJ, Washington AL, Mykhaylyk OO, Hill CJ, Bianco A, Burg SL, et al. Spatially modulated structural colour in bird feathers. Sci. Rep. 2015;5:18327. Available from: https://doi.org/10.1038/srep18317.

[17] Orbis Research. Global Cosmetics Products Market expected to reach USD 805.61 billion by 2023. Industry size & share analysis, <www.orbisresearch.com/reports/index/global-cos-metics-products-marker-analysis-of-growth-trends-and-forecasts-2018-2023>; 2018 [accessed 30.06.19].

[18] Statista. Cosmetics industry: statistics & facts, <www.statista.com/topics/3137/cosmetics-industry/>; 2018 [accessed 30.06.19].

[19] Mordor Intelligence. Beauty and Personal Care Products Market—growth, trends, forecast (2019–2024). Industry report. Sachibouli, Hyderabad, India,; 2019 [accessed 30.06.19]. <https://www.mordorintelligence.com/industry-reports/global-beauty-and-personal-care-products-market-industry>.

[20] Gran View Research. Organic skin care market growth. <www.grandviewresearch.com/industry-analysis/organic-skin-care-market>; 2017 [accessed 30.06.19].

[21] Market Whatch. Nanotechnology Market 2019. Global industry-key players, size, trends, opportunity, growth analysis and forecast 2024, <www.wiseguyreports.-com/reports/3141909-global-nanotechnology-market-by-component-and-applications-fundings>; 2019 [accessed 30.06.19].

[22] Allied Market Research. Cosmetic industry overview: global cosmetics market. London, UK: Allied Narjet; 2016. <www.alliedmarketresearch.com/consumer-goods/personal-care-and-Cosmetic-market-report>.

[23] Raconteur. The global cosmetic market, <http://www.raconteur.net/inphographics/the-global-cosmetics-market#>; 2019 [accessed 30.06.19].

[24] MarketWatch. Global Cosmetics Market 2018—Industry analysis, size, share, strategies and forecast to 2025. New York, NY: MarketWatch Inc, <www.marketwatch.com//press-release/global-cosmetics-...alysis-size-strategies-and-forecast-to-2025-2018-09-04>; 2019 [accessed 30.06.19].

[25] WHO. Global health and aging. New York:: World Health Organization, National Institute of Aging; 2011.

[26] Costello H. Global Nanotechnology Market. Outlook 2018–2024. Dallas, TX: Orbis Research; 2017 [accessed 30.06.19].

[27] Basavaray KH. Nanotechnology in medicine and relevance to dermatology. Indian J Dermatol 2012;57:169–74.

[28] Morganti P, Palombo M, Carezzi F, Nunziata ML, Morganti G, Cardillo M, et al. A green nanotechnology bioeconomy serving: natural beauty masks to save the environment. Cosmetics 2016;3:41.

[29] Donnarumma G, Fusco A, Morganti P, Palombo M, Anniboletti T, Del Ciotto P, et al. Advanced medications made by green nanocomposites. Int J Res Pharm Nano Sci 2018;5:261–70.

[30] Morganti P, Coltelli MB. A new carrier for advanced cosmeceuticals. Cosmetics 2019;6:10. Available from: https://doi.org/10.3390/cosmetics6010010.

[31] Rancan F, Blume-Peytavi V, Vogt A. Utilization of biodegradable polymeric materials as delivery agents in dermatology. Clin Cosmet Investig Dermatol 2014;27:23–34.

[32] Morganti P, Coltelli MB, Danti S, Bugnicourt E. The skin goal of EU Polybioskin project. Global Res J Pharm Pharmacol 2017;2(1):7–13.

[33] Morganti P, Coltelli MB, Danti S. Biobased tissues for innovative cosmetic products: polybioskin as an EU Research Project. Glob J Nano 2018;3(4):555620. Available from: https://doi.org/10.19080/GJN.2018.03.555620.

[34] Morganti P, Febo P. Innovative tissue engineering for an enlarget market. J Clin Cosmet Dermatol 2018;1(4). Available from: https://doi.org/10.16966/2576-2826.119.

[35] Morganti L. Chitin nanofibrils: Tuning fishery's waste into goods. J Clin Rev Case Rep, 3. 2018. p. 1–3. ISSN: 2573-9565.

[36] Morganti P, Hong-Duo C, Gao X, Morganti G, Febo D. Chitin & lignin: turning foodwaste into cosmeceuticals. J Clin Cosmet Dermatol 2018;3(1). Available from: https://doi.org/10.16966/2576-2826.135.

[37] Danti S, Trombi L, Fusco A, Azimi B, Lazzeri A, Morganti P, et al. Chitin nanofibrils and nanolignin as functional agents in skin regeneration. Int J Mol Sci 2019;20:2669. Available from: https://doi.org/10.3399/ijms20112669.

[38] Morganti P, Morganti G, Colao C. Biofunctional textiles for aging skin. Biomedicines 2019;7(3):E51. Available from: https://doi.org/10.3390/Biomedici-nes7030051.

[39] Morganti P, Yuanhong Lee, Morganti G. Nano-structured products: technology and future. J Appl Cosmetol 2007;25:161–78.

[40] Frantz C, Stewart KM, Weaver VM. The extracellular matrix at a glance. J Cell Sci 2010;123(Pt24):4195–200.

[41] Kim Y, Ko H, Kwon IK, Shin K. Extracellular matrix revised: roles in tissue engineering. Int Neurourol J 2016;20(Suppl. 1):S23–9.

[42] Desai KG. Chitosan nanoparticles prepared by ionotropic gelation: an overview of recent advances. Crit Rev Ther Drug Carrier Syst 2016;33(2):107–58. Available from: https://doi.org/10.1615/CriRevTherDrug-CarrierSyst.2016014850.

[43] Morganti P, Del Ciotto P, Stoller M, Chianese A. Antibacterial and anti-inflammatory green nanocomposites. Chem Eng Trans 2016;47:61–6. Available from: https://doi.org/10.3303/CET1647011.

[44] Morganti P, Fusco A, Paoletti I, Perfetto B, Del Ciotto P, Palombo M, et al. Anti-inflammatory, immunomodulatory and tissue repair activity on human keratinocytes by green innovative nanocomposites. Materials 2017;10:843.

[45] Casadidio C, Peregrina DV, Gigliobianco MR, Deng S, Censi R, Di Martino P. Chitinand chitosans, eco-friendly processes, and applications in cosmetic science. Mar Drugs 2019;17:369. Available from: https://doi.org/10.3390/md17060369.

[46] Morganti P, Fabrizi G, Guarneri F, Palombo M, Palombo P, Cardillo M, et al. Repair activity of skin barrier by chitin-nanofibrils complexes. SOFW J. 2011;137:10–26.

[47] Anniboletti T, Palombo M, Moroni S, Bruno A, Palombo P, Morganti P. Activity of innovative polymeric nanoparticles and non-woven tissue. In: Morganti P, editor. Bionanotechnology to save the environment. Plant and fishery's biomass as alternative to petrol. Basel, Switzerland: MDPI; 2019. p. 340–60. ISBN: 979-3-03842-693-6.

[48] Jui YY, Chin TL, Ing TS, Cardillo M, Morganti L. Cross-sectional study design and data analysis of the effect of chitin nanofibrils-lignin micro/nano particles on Malaysia's subject with disorders. J Clin Cosmet Dermatol 2019;3(2):1–6. Available from: https://doi.org/10.16966/2576.139.

[49] Morganti P, Palombo M, Tishchenko G, Yudin VE, Guarneri F, Cardillo M, et al. Chitin-hyaluronan nanoparticles: a multifunctional carrier to deliver anti-aging active ingredients through the skin. Cosmetics 2014;1:149–58.

[50] Morganti P. Chitin nanofibrils in skin treatment. J Appl Cosmetol 2009;27:251–70.

[51] Morganti P, Palombo M, Palombo P, Fabrizi G, Cardillo A, Carezzi F, et al. Cosmetic science in skin aging: achieving the efficacy by chitin nano-structured crystallites. SOFW J 2019;136(3):14–24.

[52] Shikhman AR, Brinson DC, Valbracht J, Lotz MK. Differential metabolic effects of glucosamine and *N*-acetyl-glucosamine in human articular chondrocytes. Osteoarthritis Cartilage 2009;17(8):1022–8. Available from: https://doi.org/10.1016/j.joca.2009.03.004.

[53] Konopka JB. *N*-acetylglucosamine functions in cell signaling. Scientifica 2012;2012:489208. Available from: https://doi.org/10.6064/2012/489208.

[54] Salazar J, Bello L, Chavez M, Anez R, Rojas J, Bermudez V. Glucosamine for osteoarthritis: biological effects, clinical efficacy, and safety on glucose metabolism. Arthritis 2014;2014:432463. Available from: https://doi.org/10.1155/2014/432463.

[55] Brand W, Noorlander CW, Guannakou C, De Jong WH, Kooi MW, Park MVDZ, et al. Nanomedicinal products: a survey on specific toxicity and side effects. Int J Nanomedicine 2017;12:6107–29. Available from: https://doi.org/10.2147/IJN.SI39687.

[56] Wang R, Song B, Wu J, Zhang Y, Chen A, Shao LQ. Potential adverse effects of nano-particles on the reproductive system. Int J Nanomed 2018;13:8487–506. Available from: https://doi.org/10.2147/IJN.S170723.

[57] Brohi RD, Wang L, Talpur HS, Wu D, Anwar Khan F, Bhattarai D, et al. Toxicity of nanoparticles on the reproductive system in animal models: a review. Front Pharmacol 2017;8:606. Available from: https://doi.org/10.3389/fphar.2017.00606.

[58] Richter AL, Barthi B, Armstrong HB, Brown JS, Plemmons D, Paunov VN, et al. Synthesis and characterization of biodegradable lignin and particles with tunable surface properties. Langmuir 2016;32(25):6468–77. Available from: https://doi.org/10.1021/acs.langmuir.6b01088.

[59] Gupta A, Eral HB, Hatton TA, Doyle PS. Nanoemulsions: formation, properties and applications. Soft Matter 2016;12:2826–41. Available from: https://doi.org/10.1039/c5sm02958a.

[60] Hassan K. Formulation development of micro and nano emulsions of Fenugreek oil. Indo Am J Pharm Sci 2017;4(05):1333–7 ISSN: 2249-7750.

[61] Chaudary H, Gautam B, Kumar V. Nanoemulsions versus lyotropic liquid crystals. Asian J Pharm 2014;8(1). Available from: https://doi.org/10.22377/ajp.v8i1.321.

[62] Morganti P. Use and potential of nanotechnology in cosmetic dermatology. Clin Cosmet Investig Dermatol 2010;3:5–13.

[63] Parisi L, Toffoli A, Ghiacei G, Macaluso GM. Tailoring the interface of biomaterials to design effective scaffolds. J Funct Biomater 2018;9(3). pii:E50. Available from: https://doi.org/10.3390/jfb9030050.

[64] Leone G, Tosti I, Borras JM, de Villamore Martin E. The new plastic economy: rethinking the future of plastics & catalysing action. Mac Arthur Foundation; 2016. Available from: https://www.ellenmacarthurfoundation.org/assets/downloads/publications/NPEC-Hybrid_English_22-11-17_Digital.pdf.

[65] Nouril P The truth about beauty packaging and the environment. Stylist, March 2019, <www.stylist.co.uk>; 2019.

[66] Mitchell RL and Thamsen PU. Investigations into wastewater composition focusing on nonwoven wet wipes. Technical Transactions 1/2017. Environment Engineering, <https://doi.org/10.4467/2353737XCT.17.010.6107>; 2017.

[67] Sonker AS, Richa, Pathak J, Rajneesh, Pandey A, Chattejee A, et al. Bionanotechnology: past, present and future. In: Sinha BP, Richa, editors. New approach in biological research. New York: NOVA Science Publishers; 2017. p. 99–141.

[68] Rossler M Earth Day 2018: end plastic pollution. UNESCO report. UNESCO World Heritage Centre, April 19, <http://whc.unesco.org>; 2018.

CHAPTER

2

Transdermal and bioactive nanocarriers

Nikhishaa Sree Raju[1], *Venkateshwaran Krishnaswami*[1], *Sivakumar Vijayaraghavalu*[2] *and Ruckmani Kandasamy*[1]

[1]Centre for Excellence in Nanobio Translational REsearch (CENTRE), Department of Pharmaceutical Technology, University College of Engineering, Anna University, BIT Campus, Tiruchirappalli, India [2]Central Research Facility, Sri Ramachandra Institute of Higher Education and Research (Deemed to be University), Chennai, India

Abbreviations

NLCs	Nanostructured lipid carriers
nm	Nanometer
PEG	Polyethylene glycol
PIB	Polyisobutylene
PVC	Polyvinylchloride
SC	Stratum corneum
SLNPs	Solid lipid nanoparticles
UV	Ultraviolet
V	Volt

2.1 Introduction

The term cosmeceuticals was coined by Dr. Albert Kligman (1984) by combining cosmetic and pharmaceutical [1]. It represents the wide range of products that are aimed to enhance the health and beauty of the skin [2]. With an increase in average life expectancy of the world population, the consumer's desire to tradewealth for youth also increased; this demand surged the pharma/skincare industries to respond with full vigor to develop multitude of ingredients and technologies toward this goal [3]. Earlier creams and gels were considered as cosmeceuticals; with rapid innovation, the iontophoresis, microneedles, microdermabrasion, electroporation, and laser radiation were enlisted [4].

DOI: https://doi.org/10.1016/B978-0-12-822286-7.00002-4

Further advancements in nanomedicine lead to the invention of several transdermal systems that could deliver the desired molecules at the desired concentration and site of the skin in a passive/active manner [5,6]. These nanocargos include and not limited to liposomes, niosomes, nanoemulsions, solid lipid nanoparticles (SLNPs), and nanostructured lipid carriers [7]. This book chapter overviews the physiology of the skin and describes the transdermal and nanocarriers that could improve the overall health of the skin.

2.2 Skin

Skin is the largest organ of the body and integumentary system; it is composed of three major layers (epidermis, dermis, and hypodermis) which are described in the forthcoming sections [8,9]. It performs many vital functions, including protection (physical, chemical, and biologic) and thermo- and hydro-regulation [10]. It also acts as a major route for drug delivery by permitting the topically applied drugs to pass through stratum corneum (SC) to reach other sites deep within the skin; thereby, the drug could reach systemic circulation [11]. Its large surface area enhances the greater absorption of the drug and hence higher payloads can be delivered effectively using transdermal delivery systems [12]. However, SC of the skin also inhibits the passive absorption of certain classes of poorly permeable and/or drugs with high molecular weight [13].

2.2.1 Epidermis

The epidermis is subdivided into basal-, squamous-, granular-, and horny-cell layers [14]. The epidermis is a renewing layer and the cells in the epidermis are constantly in nonsynchronous motion, making it a dynamic tissue. The most predominant function of the epidermis is that it serves as a physiologic barrier to microbial invasion, chemical penetration from the environment, and fluid and also solute loss within the body [15].

The stratified outermost layer contains various types of cells including keratinocytes (keratin synthesizers), melanocytes, and Langerhans cells. The keratinocytes possess intercellular bridges and plenty of stainable cytoplasm, making it different from the clear dendritic cells [16]. Melanocytes can elaborate melanin, the light-absorbing pigment that plays a pivotal role in the protection of skin from UV radiation. The metabolism of complex antigenic materials is handled by the third major resident of the epidermis called the Langerhans cells [17]. These three types of cells, after activation, leave the epidermis and move toward the lymph nodes where they are involved in the antigen presentation during the induction of immunity [18]. Merkel cells, being present in lesser populations, contain neuroendocrine peptides present within the intracytoplasmic granules [19].

2.2.2 Dermis

The structural and nutritional support is provided by a vascularized dermis present beneath the epidermis. The presence of collagen and elastin containing fibrous matrix contributes to the mucopolysaccharide gel, which is a major component of the dermis.

Nutrition, cutaneous sensation, and recirculating cells are provided to the dermis by the nerves and mast cells along with vascular structures [20]. The permanent residents of the dermis are completed by fibroblasts, dermal dendritic cells, and macrophages. Hyaluronic acid is present in a lesser amount, but it is a major mucopolysaccharide that accumulates in diseased conditions.

The collagen present in a huge amount within the dermis acts as a stress-resistant material of the skin, whereas elastin plays a vital role in maintaining elasticity. The collagen fibers exist in a constant state and can be degraded by proteolytic enzymes (collagenases) [21]. A specific helical polypeptide chain of fibroblasts that integrates the procollagen molecule further assembles into collagen fibrils. Amino acids such as hydroxyproline, glycine, and hydroxyl-lysine enrich the collagen. The most abundant proteins in the body, fibrillar collagens, are present in the dermis part of the skin. The elastic fibers are composed of protein filaments and an amorphous protein. The elastic fiber is fused to the extracellular matrix by the fibroblasts. These fibers occur as fine particles in the papillary dermis, while in the reticular dermis they occur as coarse particles [14].

2.2.3 Dermis–epidermis junction

A porous basement membrane zone forms the interface between the epidermis and dermis that holds the two layers together by permitting the exchange of fluids and cells. The basal keratinocytes are the most vital constituents of the dermal–epidermal junction through dermal fibroblasts that are involved to a lesser extent. This junction acts as a semipermeable barrier between the layers, provides support for the epidermis, and establishes the direction of growth; and cell polarity provides developmental signals and directs the organization of the cytoskeleton in basal cells [22].

2.2.4 Hypodermis

Hypodermis is the deepest part of the skin comprising fatty tissues that plays a pivotal role in thermoregulation, protection from mechanical injuries, and insulation and acts as an energy store. The hypodermis is majorly composed of adipocytes constituting the lipid-laden cytoplasm of rounded cells, including triglycerides and fatty acids. These adipocytes appear optically empty in stained sections due to their dissolving nature during fixation and dehydration [10]. The adipocytes are systematically arranged as primary and secondary lobules separated by connective tissue septa containing cells that include dendrocytes, fibroblasts and mast cells, vessels, and nerves and the deepest part of the sweat glands that contributes to the formation of dermal plexuses [23] (Fig. 2.1).

2.3 Transdermal delivery

"Transdermal delivery" refers to the site of delivery of a biologically active substance (cosmetic agent or drug). According to this delivery system, the cosmetic agent can either be water-soluble or -insoluble, permeable, or impermeable [24]. Transdermal delivery

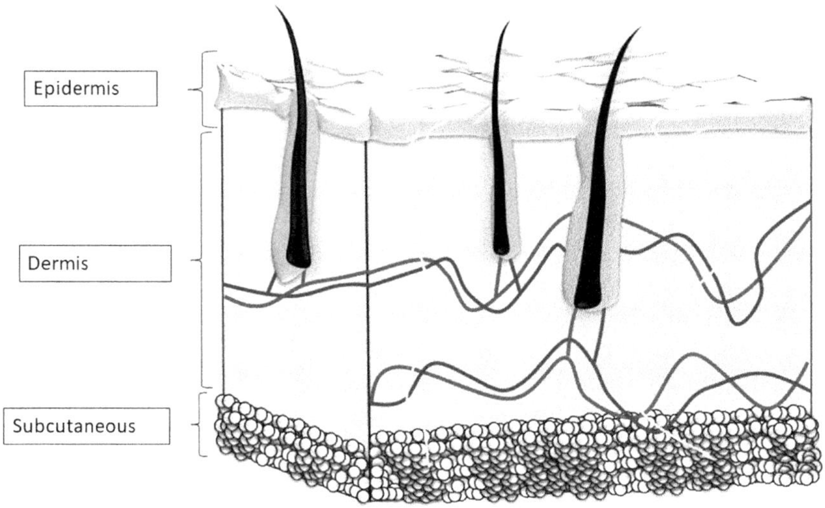

FIGURE 2.1 Anatomical position of the skin and its associated layers where the transdermal delivery system will deliver the active medicament.

includes patches that exhibit the delivery of the active substance through the intact skin. Cosmeceuticals include active ingredients that provide a therapeutic response for dermatological conditions and maintain the health of the skin. In this field, transdermal delivery includes novel technologies for better permeation of active substances into the skin. The evolution of transdermal delivery systems contributes to several theories than the ancient techniques followed.

2.4 Evolution of transdermal delivery

Since the evolution of man, topical remedies administered over the skin have been practiced. About 100,000 years ago, an ocher-rich liquefied mixture was found for skin protection in the Blombos Cave of South Africa [25]. The ancient products of lead were used for eye diseases and have been proved to achieve beneficial roles for eye infection in recent times [26]. A Greek physician named Galen (Father of Pharmacy) introduced herbal ingredients into the formulation of drugs for different dosage forms. The cold creams that are being used today are based on his formulations [27]. The forerunner for the transdermal patches of today is medicated plasters of ancient China containing herbal active ingredients with an adhesive to stick onto the skin that was used for dermatological applications [28].

A Persian physician (IbnSina), introduced the concept of categorizing topical drugs into hard (impermeable) and soft (permeable). He also conceptualized that the drugs applied topically could affect the tissues present beneath the skin [29].

The major drawback of the research focus in the late 19th century was to prove the skin permeation of lipophilic substances without assessing their systemic toxicity. Beginning of the 20th century marked the development of pharmacokinetic and pharmacodynamic studies, thereby the systemic effect of many topical applications and their peak levels in the body compartments were analyzed [30]. Dale Wurster, a renowned pharmacist, defined the transdermal delivery system in pharmacokinetic parameters. His studies on the design of the diffusion cell act as a predecessor for the recent studies on the biological activity of the active ingredient for transdermal delivery [31].

In the recent decades, the transdermal delivery system has achieved its heights by addressing the unsolved problems of ancient research that includes effective drug property, its effective permeation, and attaining the desired concentration at the target site with lesser toxicity. To attain these objectives, novel technologies were developed and treated several dermatological conditions including psoriasis, eczema, vitiligo, and dermatitis.

2.5 Development of transdermal delivery in cosmetics

Introduction of biologically active substances in the cosmetic products led to an improvement in the appearance and health of the skin [32]. The major setback observed was the delivery of creams and pastes effectively across the dermal layers, which in turn led to the development of devices such as microneedles, microdermabrasion, laser radiation, and iontophoresis that could enhance the permeation of payloads across the skin and reach the target site. The biologically active ingredients are loaded in the device along with the excipients and tested for their efficacy and stability. The ease of self-use and lack of major health hazard made these devices a commercial success among population and found its place in the retailers' shelves. However, solving one problem led to another problem, the devices used impaired the skin's natural healing process [33]. To combat this problem, the pharma and cosmetic industries surged to develop more advanced technological systems that could effectively aid the skin restoration in dermatological conditions.

2.6 Advantages of transdermal delivery

The transdermal delivery has vast advantages when compared with other routes. This delivery system avoids the painful injections, reduces the dose, and weakens the risk of diseases caused by needle reuse [34]. Additionally, this system is noninvasive, possesses self-administration, provides a release for sustained action, and it is inexpensive. At the therapeutic level, it includes painless treatment with no injections involved and no dangerous needles to dispose of. The transdermal system has opted for the drugs that have increased first-pass effects through the liver, poor oral intake, frequent administration, and interaction with stomach acid [24].

2.7 Components of transdermal patch

The components involved in the transdermal delivery system are described in Fig. 2.2.

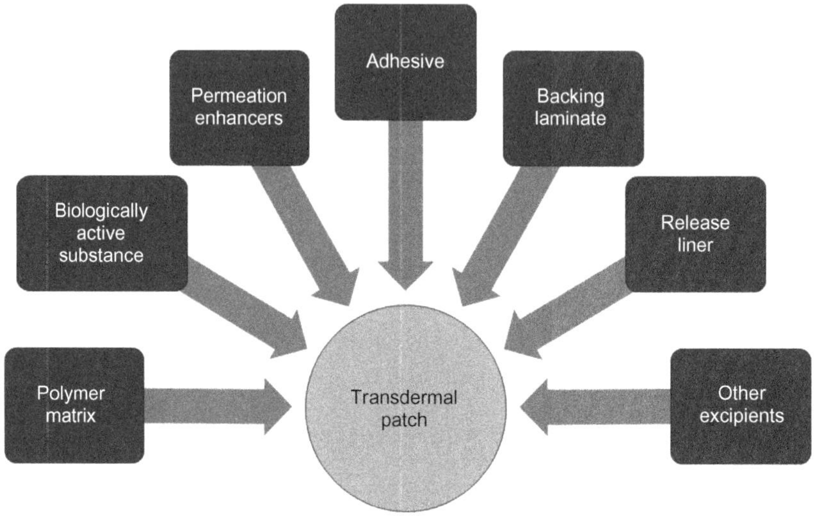

FIGURE 2.2 Components required for the development of transdermal patches.

2.7.1 Polymer matrix

Polymer matrix is the major component of the transdermal patch fabricated with multiple layers of polymer in which a reservoir is sandwiched between two polymeric layers. Polymer is composed of an outer impregnable layer preventing the loss of active substances and an inner polymeric layer functioning as a controlling membrane. The polymer matrix is designed based on the active substance used for the patch formulation, adhesion-cohesion balance, compatibility, and stability with other components of the patch [35]. The polymer that is commonly used in cosmetics is natural polymers (rosin, collagen, hyaluronic acid, etc.), synthetic polymers [polyvinylchloride (PVC), polyethylene glycol (PEG), polypropylene, etc.], and synthetic elastomers [polyisobutylene (PIB), silicon rubber, acrylonitrile, etc.] [36].

2.7.2 Biologically active substances

The biologically active substances exhibit appropriate physiochemical and pharmacokinetic properties since these properties have a direct effect on skin permeation [37]. These substances emerged from natural sources or synthetic compounds that possess a therapeutic effect upon reaching the site of action. The patches are designed for those that fail to perform their action through other routes of administration and causing noncompliance due to frequent dosing [38]. Application of novel technologies like microneedles permits the usage of macromolecules such as proteins and peptides as active ingredients.

2.7.3 Permeation enhancers

The main role of permeation enhancers is to increase the permeability of the SC to attain an increased therapeutic level of the biological substance [39]. These enhancers perform three main functions including lipid disruption of the SC, protein modification, and partitioning promotion. These enhancers are further classified into natural enhancers (aloe vera, eucalyptus, olive oil, etc.) and synthetic enhancers (poloxamer, pyrrolidone, isopropyl myristate, etc.) [40,41].

2.7.4 Adhesive

The adhesive maintains the contact between the patch and the skin. It possesses the ability to adhere to the skin with not more than applied finger pressure, to be tacky, and to exert a strong holding force. The selection of adhesive is based on the patch design and the formulation. Adhesives should be physiochemically and biologically compatible with other components of the patch [42].

2.7.5 Backing laminate

A backing laminate is present to provide support and possess chemical resistant and is excipient compatible. This layer has a low moisture vapor transmission rate, optimal elasticity, flexibility, and good tensile strength. The commonly used backing layers are plastic films, PIB, heat seal layer, and aluminum vapor coated layer [43].

2.7.6 Release liner

Upon storage, the release liner performs its function of prevention of loss of active substance that has migrated into the adhesive layer and contamination. It is considered as a part of packing material. The liner is composed of a base layer and a release coating layer of silicon or Teflon. Fluoropolymers are the commonly used release liners [44].

2.7.7 Other excipients

The excipients of the transdermal patch include solvents such as chloroform, methanol, acetone, and dichloromethane to prepare a reservoir. Plasticizers are used to provide plasticity to the patch, for example, triethyl citrate, polyethylene glycol, and dibutyl phthalate. The concentration of these plasticizers ranges from 5% to 20% [6].

2.8 Novel technologies toward the development of the transdermal system

The novel technologies that are employed currently for the transdermal delivery system are described in Fig. 2.3.

2.8.1 Iontophoresis

Iontophoresis includes the application of low electric current applied directly or indirectly to the skin to enhance the permeability of the biologically active substance [45]. The iontophoresis system involves electrorepulsion for charged solutes, electroosmosis for uncharged solutes, and

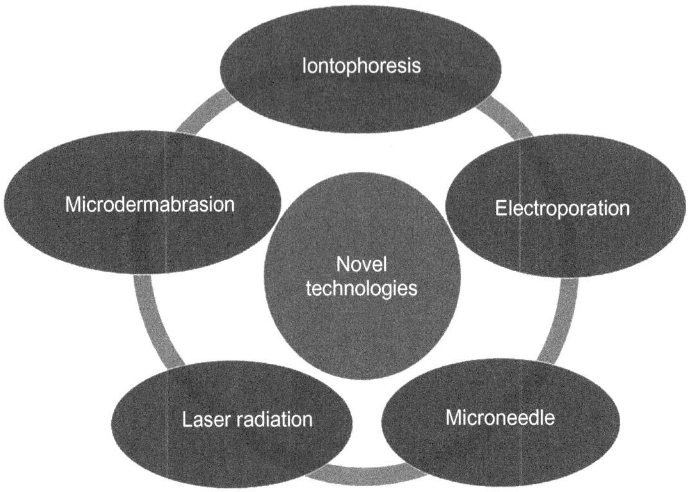

FIGURE 2.3 Current novel technologies utilized for enhanced efficacy in transdermal delivery of cosmetics.

electroperturbation for both charged and uncharged solutes [33] Recently, spironolactone and its metabolite canrenone were formulated in an iontophoretic setup for the treatment of acne. The method was proved to be valuable due to its accuracy and precision from the permeation studies and aims to control the quality of skincare products [46]. L'Oreal patented an iontophoresis technology of delivering vitamin C through the skin. This method involves applying vitamin C and its derivatives to the skin along with a polymer upon direct current [47].

2.8.2 Electroporation

Electroporation is attributed to the application of high-voltage pulse to the skin to induce the formation of the transient pore. Herein, high voltage of about 100 V and short time durations of about milliseconds are employed [48]. The electrical parameters that are considered to enhance the permeability are waveform, rate, and number of the electrical pulse. This technology has been applied and is commercially successful for molecules with different lipophilicity and size [49,50]. Recently, transdermal delivery of neostigmine was carried out by the electroporation technology permitting delivery of lower doses of hydrophilic compounds making it promising for clinical practice [51]. Another work suggests the novel in-vivo monitoring system of electroporation based on fluorescently labeled molecules aiming for the quantification of transdermal delivery. This technology provides rapid determination of active substances delivered at the site of action [52].

2.8.3 Microneedles

The initial microneedle system consisted of a reservoir and multiple projections to deliver the substance through the skin. The development in technology involves either a reservoir or a microprojection array [53]. The microneedle is commonly used to deliver macromolecules through the skin to increase the permeability of the substance. Normally, the needle size ranges from 50 to 100 mm length for the active substance delivery [54]. A recent report suggests that microneedles deal with wrinkle improving adenosine that showed better efficacy than adenosine cream and has the scope of developing into novel cosmetics products [55]. On the other hand, skin microneedle technique has been proved to be effective for the treatment of atrophic acne scars [56].

2.8.4 Microdermabrasion

This method includes the disruption of the topmost layers of the skin to achieve the permeation of the biologically active substance. These devices are similar to the one employed by the dermatologists for the treatment of acne, scars, etc. Skin resurfacing can be achieved by abrasion technology and has been proved to play a pivotal role in skin rejuvenation [57]. This modality is used in treating acne vulgaris and avoids complications when compared to the traditional acne creams [58].

2.8.5 Laser radiation

Laser radiation includes controlled exposure of laser to the skin resulting in the ablation of SC causing meager damage to the epidermis [59]. The main mechanism of this technology transdermally is the selective removal of the SC that enhances the permeability of highly lipophilic and hydrophilic substances. Radiations are administered for dermally for acne leading to disruption of the affected cells over a less time frame of about 300 ns [60]. The dermatological conditions including acne scars, burns, wrinkles, and hypertrophic scars can be treated by low-level laser therapy with heat through transdermal delivery [61]. A patent was granted to the International business machine corporation, New York in 2015 on laser-assisted transdermal delivery of cosmetic agents for skin resurfacing and tightening [62].

2.9 Bioactive nanocarriers

For the delivery of the biologically active substance, nanocarriers are employed to achieve increased permeation through the skin and transport to the affected area. Nanocosmeceuticals are fast emerging for conditions such as acne, wrinkles, and pigmentation. The bioactive nanocarriers are advantageous over other ancient carriers in exhibiting enhanced skin permeation, sustained release of the active substance, improved stability, specificity, and better entrapment efficiency. The bioactive nanocarriers that are currently used for cosmetics are described in Fig. 2.4 and Table 2.1 explains the carriers and the cosmetic agents with their therapeutic uses.

2.9.1 Liposomes

Liposomes are vesicular structures composed of the lipid bilayer of phospholipids suitable for both hydrophilic and lipophilic substances with size ranging from 20 nm to micrometers [73]. Phosphatidylcholine, a major component of the liposomes, is widely

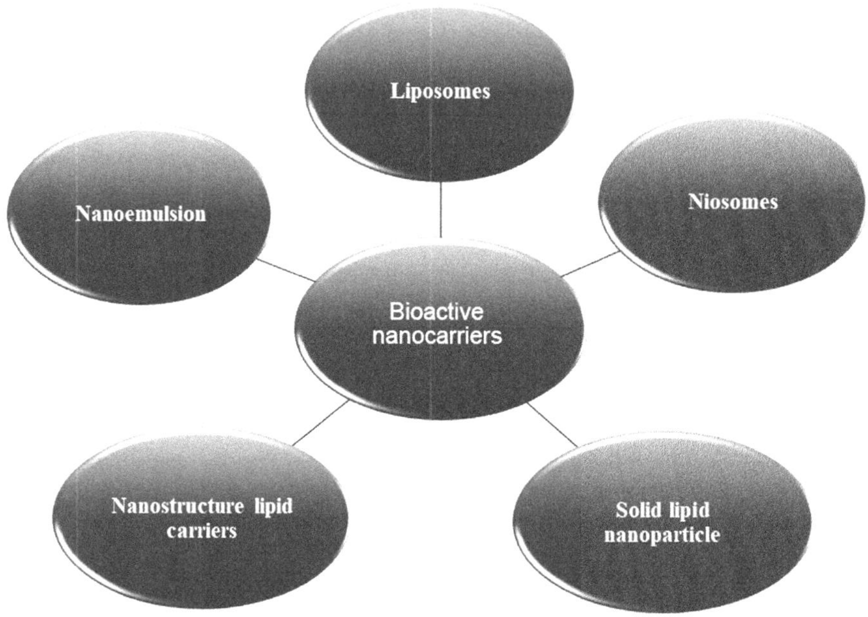

FIGURE 2.4 Bioactive nanocarriers for cosmetic products utilized in the transdermal delivery system.

TABLE 2.1 Cosmetic agents loaded in nanocarriers along with their therapeutic uses.

S. no.	Nanocarriers	Size range (nm)	Biologically active substance	Therapeutic uses	Reference
1	Liposomes	20–10,000	Folic acid	Skin regenerative and nutritional agent	[63]
			Vitamin D3	UV protective and antiaging agent	[64]
2	Niosomes	100–2000	Quercetin	Antioxidant and skin whitening agent	[65]
			Asiaticoside	Antipsoriasis, antiaging, and burn and wound healing agent	[66]
3	Solid lipid nanoparticles	50–100	Idebenone	Antioxidant and hydrating agent	[67]
			Tazarotene	Antipsoriasis, acne scare healing, and photoaging agent	[68]
4	Nanoemulsions	50–200	Hyaluronan	Hydrating and antiaging agent	[69]
			Eicosapentaenoic acid and docosahexaenoic acid	Antiinflammatory and antiproliferative agent	[70]
5	Nanostructured lipid carriers	10–1000	Isoliquiritigenin	Antioxidant and skin whitening agent	[71]
			Hydroquinone	Antioxidant and skin bleaching agent	[72]

used in skincare formulations due to its softening properties. Liposomes can encapsulate the active substance and hence used widely for dermatological applications [74]. In the cosmetic industry, liposomes are commonly employed and a stream of researchers are working toward the development of cosmetic products. Recently, Kapoor et al. [63] demonstrated the transdermal delivery of folate fortification along with liposomes in cosmetic products. This delivery was performed without the use of surfactant and external energy for permeation [63].

Bi et al. [64] used liposomes as a carrier to transdermally deliver vitamin D3 and found to improve stability. This product can be used to restore the damage in the photoaging condition [64].

2.9.2 Niosomes

Vesicles possessing a bilayer structure composed of self-assembly of hydrated nonionic surfactants are termed as niosomes. These are enclosed by a membrane that is formed when the macromolecules appear as a bilayer [75]. The size of niosomes ranges from about 100 nm to 2 μm in diameter. The components of niosomes include nonionic surfactants and cholesterol. These niosomes are suitable for hydrophilic and lipophilic substances [76]. This acts as a vehicle and provides encapsulation to the active substance. Niosomes show better stability than liposomes and possess decreased resistance of the horny layer, allowing the active substance to reach the living tissues [77]. Lu et al. [65] worked on quercetin loaded niosomes for whitening and antioxidant ability and was found to possess better skin permeation and sustained release of quercetin over time by transdermal delivery [65]. *Centella asiatica* extract-loaded niosomes and hyaluronic acid were investigated by Wichayapreechar et al. [66] and found to penetrate through the transdermal system. This technique can be employed for hydrophilic biologically active compounds [66].

2.9.3 Solid lipid nanoparticles

SLNPs are composed of a single layer of shells with a lipoic core of size ranging from 50 to 1000 nm. SLNPs consist of biodegradable lipids with low toxicity [78]. The smaller size permits contact with SC increasing the penetration of active substances. SLNPs have better stability to liposomes when joined with the active substance [79]. The solid nature and decreased mobility inhibit the leakage of active substances from the carrier [80]. In a recent study, idebenone, an antioxidant was loaded into SLNPs. The study suggested that the approach could act as a developing strategy to obtain transdermal formulations for skincare products [67]. Another work demonstrated the formulation of transdermal delivery of tazarotene loaded into SLNPs for the treatment of psoriasis. The loaded formulation showed better tolerability than the marketed formulation showing the enhanced potential of transdermal delivery of tazarotene [68].

2.9.4 Nanoemulsions

Nanoemulsions are kinetically stable dispersion of liquid where the oily and aqueous phase are in coordination with a surfactant with size ranging from 50 to 200 nm [81,82].

These emulsions are in dispersed phase with smaller particles having decreased oil/water interfacial tension. Nanoemulsions are transparent and exhibit the properties of low viscosity, high interfacial area, high kinetic stability, and high solubilization capacity [83]. Nanoemulsion was reported to act as a carrier for lipophilic hyaluronic acid and can be successfully used as a transdermal delivery carrier with cosmetic applications [69]. Nanoemulsion loaded with miglyol, salmon, and rapeseed oil was reported as a matrix. The study results proved that the transdermal formulation is one of the best cosmetic products in consideration of its stability, turbidity, and size [70].

2.9.5 Nanostructured lipid carriers

Nanostructured lipid carriers (NLCs) are developed to overcome the disadvantages associated with SLNPs. The size of these carriers ranges from 10 to 1000 nm and possesses better entrapment efficacy to SLNPs [84]. NLCs follow a biphasic release pattern and ensure close contact to the SC causing increased permeability of the active substance through the skin. NLCs are stable upon storage and possess increased UV protection with minimum side effects [85]. An NLC-based transdermal formulation was developed and reported for isoliquiritigenin with ceremide as a solid lipid that improved the efficacy of the cosmetic agent [71]. NLCs were investigated with emphasis on hydroquinone transdermal delivery to enhance the skin permeability [72].

2.10 Discussion

Over the years of development, the transdermal delivery system has achieved tremendous success. Exploration of the system has brought about novel technologies to forego by overcoming the drawbacks. Recent research in transdermal delivery includes macromolecular delivery by microneedle technology. Hu et al. [86] developed a glucose-responsive insulin delivery platform. They initially integrated H_2O_2-responsive vesicles with a microneedle array patch. This smart insulin patch demonstrated glucose-mediated disassembly with rapid responsiveness [86]. They were focusing on the development of a transdermal patch of insulin with chemical permeation enhancers.

Stimuli-responsive patches are emerging in cosmeceuticals where cosmetic agents are loaded with responsive polymers based on temperature, pH, and hypoxia-sensitive mechanisms. Dissolving microneedles also are being investigated by many researchers with several advantages and seem to be nonirritant. Hong et al. [87] determined the safety of the combination of wrinkle cream dissolving microneedles. Their results proved that this novel technology is a favorable tool to be used for skin rejuvenation [87].

Predominant development has been observed over the past few decades on iontophoresis with emphasis on a safety level. Patents have been granted to the inventors for novel formulations. Recently, a patent was awarded to In Cube Labs, United States, for the iontophoretic system, kits, and method for transdermal delivery of cosmetic agents [88]. These two technologies are most commonly employed due to their advantages over other novel modalities.

Nanocarriers play a pivotal role in the development of the transdermal delivery system irrespective of the novel technologies. Nanocarriers possess a wide range of applications making it a key component in the delivery system. In cosmeceuticals, bioactive nanocarriers are boon to the researchers as it provides improved therapeutic response of the active ingredient, increased specificity and faster results can be observed. A major drawback in the nanocarriers is the instability in water due to the aggregation of nanoparticles making it a challenging task to the researchers [89].

Among several nanocarriers, niosomes and nanostructured lipid nanoparticles gain the spotlight. Recently, the properties of quercetin were enhanced by developing niosomal patch encapsulated with quercetin with improved antioxidant and skin whitening properties [65]. Nanostructured lipid nanoparticles are employed in the development of the product along with several excipients to achieve the controlled release of vitamin E, thereby improving their antioxidant and photoprotecting properties [72].

2.11 Conclusion

The transdermal delivery system has made progress and crossed several barriers imposed by the human system. In cosmeceuticals, transdermal delivery marked exceptional growth in formulating products with cosmetic agents by employing novel technologies such as microneedles and iontophoresis. The introduction of nanocarriers into the system has made it a step ahead of the remaining delivery systems. This chapter emphasized by providing information on the ongoing research work in cosmetology and pharmaceuticals. Focus on the updates on the bioactive nanocarriers in transdermal has been emphasized. We have covered the novel technologies employed in cosmetic products and listed out a few examples. A glimpse of the recent trends in transdermal delivery and the importance of the bioactive nanocarriers has also been discussed.

Acknowledgments

The authors gratefully acknowledge the Department of Science and Technology (GoI), New Delhi supported National Facility for Drug Development for Academia, Pharmaceutical and Allied Industries (NFDD) (Ref. No. VI-D&P/349/10-11/TDT/1 Dt: 21.10.2010) and National Facility for Bioactive Peptides from Milk (NFBP) Project (Ref. No. VI-D&P/545/2016-17/TDT; Dt: 28.02.2017).

References

[1] Brandt FS, Cazzaniga A, Hann M. Cosmeceuticals: current trends and market analysis. Semin Cutan Med Surg 2011;30:141–3 3.

[2] Choi CM, Berson DS. Cosmeceuticals. Semin Cutan Med Surg 2006;25(3):163–8.

[3] Datta HS, Paramesh R. Trends in aging and skin care: Ayurvedic concepts. J Ayurveda Integr Med 2010;1 (2):110.

[4] Swain S, Beg S, Singh A, Patro N, Bhanoji Rao ME. Advanced techniques for penetration enhancement in transdermal drug delivery system. Curr Drug Deliv 2011;8(4):456–73.

[5] Vitorino C, Almeida A, Sousa J, Lamarche I, Gobin P, Marchand S, et al. Passive and active strategies for transdermal delivery using co-encapsulating nanostructured lipid carriers: in vitro vs. in vivo studies. Eur J Pharm Biopharm 2014;86(2):133–44.

[6] Alexander A, Dwivedi S, Giri TK, Saraf S, Saraf S, Tripathi DK. Approaches for breaking the barriers of drug permeation through transdermal drug delivery. J Control Release 2012;164(1):26–40.
[7] Rigano L, Lionetti N. Nanobiomaterials in galenic formulations and cosmetics. Nanobiomaterials in Galenic Formulations and Cosmetics. William Andrew Publishing; 2016. p. 121–48.
[8] Gantwerker EA, Hom DB. Skin: histology and physiology of wound healing. Clin Plastic Surg 2012;39(1):85–97.
[9] Goebel AS, Schmaus G, Neubert RH, Wohlrab J. Dermal peptide delivery using enhancer molecules and colloidal carrier systems–part I: carnosine. Skin Pharmacol Physiol 2012;25(6):281–7.
[10] Kanitakis J. Anatomy, histology and immunohistochemistry of normal human skin. Eur J Dermatol 2002;12 (4):390–401.
[11] Mueller J, Oliveira JS, Barker R, Trapp M, Schroeter A, Brezesinski G, et al. The effect of urea and taurine as hydrophilic penetration enhancers on SC lipid models. Biochim Biophys Acta 2016;1858(9):2006–18.
[12] Schoellhammer CM, Blankschtein D, Langer R. Skin permeabilization for transdermal drug delivery: recent advances and future prospects. Expert Opin Drug Deliv 2014;11(3):393–407.
[13] Ruela AL, Perissinato AG, Lino ME, Mudrik PS, Pereira GR. Evaluation of skin absorption of drugs from topical and transdermal formulations. Braz J Pharm Sci 2016;52(3):527–44.
[14] Berger TG, Elston DM, James W. Andrews' diseases of the skin: clinical dermatology. Saunders Elsevier; 2006.
[15] Chu DH. Overview of biology, development, and structure of skin. Fitzpatrick's dermatology in general medicine. 7th ed. New York: McGraw-Hill; 2008. p. 57–73.
[16] Yousef H, Sharma S. Anatomy, skin (integument), epidermis. StatPearls. Treasure Island, FL: StatPearls Publishing LLC; 2018.
[17] Jaitley S, Saraswathi TR. Pathophysiology of Langerhans cells. J Oral Maxillofac Pathol 2012;16(2):239.
[18] Lowes MA, Suárez-Fariñas M, Krueger JG. Immunology of psoriasis. Annu Rev Immunol 2014;32:227–55.
[19] Fenner J, Clark RA. Anatomy, physiology, histology, and immunohistochemistry of human skin. Skin Tissue Eng Regen Med 2016;1:1–7.
[20] Vandergriff TW, Bergstresser PR. Anatomy and physiology. Dermatology. 3rd ed. Philadelphia, PA: Saunders; 2012. p. 43–54.
[21] Kolarsick PA, Kolarsick MA, Goodwin C. Anatomy and physiology of the skin. J Dermatol Nurses Assoc 2011;3(4):203–13.
[22] Burgeson RE, Christiano AM. The dermal–epidermal junction. Curr Opin Cell Biol 1997;9(5):651–8.
[23] Kanitakis J, Vassileva S, Woodley D. Immunohistochemistry of normal human skin. Diagnostic immunohistochemistry of the skin: an illustrated text. London: Chapman & Hall Medical; 1998. p. 38–51.
[24] Levin G, Sacks H, Rudaev S, inventors; TransPharma Medical Ltd, assignee. Transdermal delivery system for cosmetic agents. United States patent US 8,016,810, 2011 September 13.
[25] Henshilwood CS, d'Errico F, Van Niekerk KL, Coquinot Y, Jacobs Z, Lauritzen SE, et al. A 100,000-year-old ochre-processing workshop at Blombos Cave, South Africa. Science. 2011;334(6053):219–22.
[26] Tapsoba I, Arbault S, Walter P, Amatore C. Finding out Egyptian Gods' secret using analytical chemistry: biomedical properties of Egyptian black makeup revealed by amperometry at single cells. Anal Chem 2009;82(2):457–60.
[27] Gaskell E Great moments in medicine and pharmacy: a history of medicine and pharmacy in pictures, by GA Bender and RA Thom, Detroit, Northwood Institute Press, for Parke, Davis & Co., 1966, 2 vols, p. 421, 238, $12.00, $10.00. Medical history. 1967 July;11(3):320.
[28] Chien YW. Development of transdermal drug delivery systems. Drug Dev Ind Pharm 1987;13(4-5):589–651.
[29] Moghimi HR, Shafizade A, Kamlinejad M. Drug delivery systems in Iranian traditional pharmacy. Tehran: Traditional Medicine and MateriaMedica Research Center, SBMU; 2011.
[30] Pastore MN, Kalia YN, Horstmann M, Roberts MS. Transdermal patches: history, development and pharmacology. Br J Pharmacol 2015;172(9):2179–209.
[31] Wurster DE, Bhattacharjya S, Flanagan DR. Effect of curing on water diffusivities in acrylate free films as measured via a sorption technique. AAPS PharmSciTech 2007;8(3):E152–7.
[32] Lupo MP. Antioxidants and vitamins in cosmetics. Clin Dermatol 2001;19:467–73.
[33] Scott JA, Banga AK. Cosmetic devices based on active transdermal technologies. Ther Deliv 2015;6 (9):1089–99.
[34] Draelos ZD. Cosmeceuticals: undefined, unclassified, and unregulated. Clin Dermatol 2009;27(5):431–4.

[35] Wolff HM. Optimal process design for the manufacture of transdermal drug delivery systems. Pharm Sci Technol Today 2000;3(5):173–81.
[36] Kandavilli S, Nair V, Panchagnula R. Polymers in transdermal drug delivery systems. Pharm Technol 2002;26(5):62–81.
[37] Alkilani AZ, McCrudden MT, Donnelly RF. Transdermal drug delivery: innovative pharmaceutical developments based on disruption of the barrier properties of the SC. Pharmaceutics 2015;7(4):438–70.
[38] Trommer H, Neubert RH. Overcoming the SC: the modulation of skin penetration. Skin Pharmacol Physiol 2006;19(2):106–1021.
[39] Chen Y, Quan P, Liu X, Wang M, Fang L. Novel chemical permeation enhancers for transdermal drug delivery. Asian J Pharm Sci 2014;9(2):51–64.
[40] Das AS, Ahmed AB. Natural permeation enhancer for transdermal drug delivery system and permeation evaluation: a review. Asian J Pharm Clin Res 2017;10:5–7.
[41] Pham QD, Björklund S, Engblom J, Topgaard D, Sparr E. Chemical penetration enhancers in SC—relation between molecular effects and barrier function. J Control Release 2016;232:175–87.
[42] Ameen D, Michniak-Kohn B. Development and in vitro evaluation of pressure sensitive adhesive patch for the transdermal delivery of galantamine: effect of penetration enhancers and crystallization inhibition. Eur J Pharm Biopharm 2019;139:262–71.
[43] Ebert CD, Heiber W, Andriola R, Williams P. Development of a novel transdermal system design. J Control Release 1987;6(1):107–11.
[44] Kasperek R, Poleszak E. Modern polymers in matrix tablets technology. Polim Med 2014;44(3):189–96.
[45] Banga AK. Transdermal and intradermal delivery of therapeutic agents: application of physical technologies. CRC Press; 2011.
[46] Ferreira-Nunes R, Ferreira LA, Gratieri T, Cunha-Filho M, Gelfuso GM. Stability-indicating analytical method of quantifying spironolactone and canrenone in dermatological formulations and iontophoretic skin permeation experiments. Biomed Chromatogr 2019;33(11):e4656.
[47] Delgadillo JC, Planard-Luong TH, inventors; L'Oreal SA, assignee. Iontophoresis method of delivering vitamin c through the skin and iontophoresis device comprising: an electrode assembly including at least one electrode and an aqueous active agent. United States patent application US 16/062,167, 2018 December 6.
[48] Rastogi R, Anand S, Koul V. Electroporation of polymeric nanoparticles: an alternative technique for transdermal delivery of insulin. Drug Dev Ind Pharm 2010;36(11):1303–11.
[49] Sharma M. Transdermal and intravenous nano drug delivery systems: present and future. Applications of Targeted Nano Drugs and Delivery Systems. Elsevier; 2019. p. 499–550.
[50] Brown MB, Martin GP, Jones SA, Akomeah FK. Dermal and transdermal drug delivery systems: current and future prospects. Drug Deliv 2006;13(3):175–87.
[51] Berkó S, Szűcs KF, Balázs B, Csányi E, Varju G, Sztojkov-Ivanov A, et al. Electroporation-delivered transdermal neostigmine in rats: equivalent action to intravenous administration. Drug Des Devel Ther 2016;10:1695.
[52] Blagus T, Markelc B, Cemazar M, Kosjek T, Preat V, Miklavcic D, et al. In vivo real-time monitoring system of electroporation mediated control transdermal and topical drug delivery. J Control Release 2013;172 (3):862–71.
[53] Donnelly RF, Singh TR, Woolfson AD. Microneedle-based drug delivery systems: microfabrication, drug delivery, and safety. Drug Deliv 2010;17(4):187–207.
[54] El-Khordagui LK. Microneedles: an emerging approach for active transdermal delivery of insulin. J Bioequiv Availab 2012;4(7).
[55] Kang G, Tu TN, Kim S, Yang H, Jang M, Jo D, et al. Adenosine-loaded dissolving microneedle patches to improve skin wrinkles, dermal density, elasticity and hydration. Int J Cosmet Sci 2018;40(2):199–206.
[56] El-Domyati M, Barakat M, Awad S, Medhat W, El-Fakahany H, Farag H. Microneedling therapy for atrophic acne scars: an objective evaluation. J Clin Aesthet Dermatol 2015;8(7):36.
[57] Fernandes M, Pinheiro NM, Crema VO, Mendonça AC. Effects of microdermabrasion on skin rejuvenation. J Cosmet Laser Ther 2014;16(1):26–31.
[58] Kempiak SJ, Uebelhoer N. Superficial chemical peels and microdermabrasion for acne vulgaris. Semin Cutan Med Surg 2008;27(3):212–20.
[59] Barry BW. Novel mechanisms and devices to enable successful transdermal drug delivery. Eur J Pharm Sci 2001;14(2):101–14.

[60] Baron ED, Harris L, Redpath WS, Shapiro H, Hetzel F, Morley G, et al. Laser-assisted penetration of topical anesthetic in adults. Arch Dermatol 2003;139(10):1288–90.

[61] Avci P, Gupta A, Sadasivam M, Vecchio D, Pam Z, Pam N, et al. Low-level laser (light) therapy (LLLT) in skin: stimulating, healing, restoring. Semin Cutan Med Surg 2013;32(1):41.

[62] Felsenstein JM, Hedrick JL, Wynne JJ, Zupanski-Nielsen DS, inventors; International Business Machines Corp, assignee. Laser-assisted transdermal delivery of nanoparticulates and hydrogels. United States patent application US 13/966,974, 2015 January 22.

[63] Kapoor MS, D'Souza A, Aibani N, Nair SS, Sandbhor P, Banerjee R. Stable liposome in cosmetic platforms for transdermal folic acid delivery for fortification and treatment of micronutrient deficiencies. Sci Rep 2018;8 (1):16122.

[64] Bi Y, Xia H, Li L, Lee RJ, Xie J, Liu Z, et al. Liposomal vitamin D3 as an anti-aging agent for the skin. Pharmaceutics. 2019;11(7):311.

[65] Lu B, Huang Y, Chen Z, Ye J, Xu H, Chen W, et al. Niosomal nanocarriers for enhanced skin delivery of quercetin with functions of anti-tyrosinase and antioxidant. Molecules. 2019;24(12):2322.

[66] Wichayapreechar P, Anuchapreeda S, Phongpradist R, Rungseevijitprapa W, Ampasavate C. Dermal targeting of *Centella asiatica* extract using hyaluronic acid surface modified niosomes. J Liposome Res 2019; 1–11.

[67] Montenegro L, Panico A, Santagati L, Siciliano E, Intagliata S, Modica M. Solid lipid nanoparticles loading idebenone ester with pyroglutamic acid: In vitro antioxidant activity and in vivo topical efficacy. Nanomaterials. 2019;9(1):43.

[68] Aland R, Ganesan M, Rao PR. In vivo evaluation of tazarotene solid lipid nanoparticles gel for topical delivery. Int J Pharm Sci Drug Res 2019;11(1):45–50.

[69] Kong M, Chen XG, Kweon DK, Park HJ. Investigations on skin permeation of hyaluronic acid based nanoemulsion as transdermal carrier. Carbohydr Polym 2011;86(2):837–43.

[70] Kabri TH, Arab-Tehrany E, Belhaj N, Linder M. Physico-chemical characterization of nano-emulsions in cosmetic matrix enriched on omega-3. J Nanobiotechnol 2011;9(1):41.

[71] Noh GY, Suh JY, Park SN. Ceramide-based nanostructured lipid carriers for transdermal delivery of isoliquiritigenin: development, physicochemical characterization, and in vitro skin permeation studies. Korean J Chem Eng 2017;34(2):400–6.

[72] Wu PS, Lin CH, Kuo YC, Lin CC. Formulation and characterization of hydroquinone nanostructured lipid carriers by homogenization emulsification method. J Nanomater 2017;2017.

[73] Arora N, Agarwal S, Murthy RS. Latest technology advances in cosmaceuticals. Int J Pharm Sci Drug Res 2012;4(3):168–82.

[74] Sundari PT, Anushree H. Novel delivery systems: current trend in cosmetic industry. Eur J Pharm Med Res 2017;4(8):617–27.

[75] Duarah SA, Pujari KU, Durai RD, Narayanan VH. Nanotechnology-based cosmeceuticals: a review. Int J Appl Pharm 2016;8(1):8–12.

[76] Kazi KM, Mandal AS, Biswas N, Guha A, Chatterjee S, Behera M, et al. Niosome: a future of targeted drug delivery systems. J Adv Pharm Technol Res 2010;1(4):374.

[77] Chandu VP, Arunachalam A, Jeganath S, Yamini K, Tharangini K, Chaitanya G. Niosomes: a novel drug delivery system. Int J Nov Trends Pharm Sci 2012;2(1):25–31.

[78] Bangale MS, Mitkare SS, Gattani SG, Sakarkar DM. Recent nanotechnological aspects in cosmetics and dermatological preparations. Int J Pharm Pharm Sci 2012;4(2):88–97.

[79] ZurMühlen A, Schwarz C, Mehnert W. Solid lipid nanoparticles (SLN) for controlled drug delivery–drug release and release mechanism. Eur J Pharm Biopharm 1998;45(2):149–55.

[80] Pardeike J, Hommoss A, Müller RH. Lipid nanoparticles (SLN, NLC) in cosmetic and pharmaceutical dermal products. Int J Pharm 2009;366(1–2):170–84.

[81] Shah P, Bhalodia D, Shelat P. Nanoemulsion: a pharmaceutical review. Syst Rev Pharm 2010;1(1).

[82] Patel RP, Joshi JR. An overview on nanoemulsion: a novel approach. Int J Pharm Sci Res 2012;3(12):4640.

[83] Özgün S. Nanoemulsions in cosmetics. Anadolu Univ 2013;1(6):3–11.

[84] Patel DK, Tripathy S, Nair SK, Kesharwani R. Nanostructured lipid carrier (NLC) a modern approach for topical delivery: a review. World J Pharm Pharm Sci 2013;2(3):921–38.

[85] Shailesh L, Snehal RP, Ashwini P, Manoj S, Arvind B. Nanostructured lipid carriers in stability improvement for cosmetic nanoparticles. Int J Pharm Pharm Res 2016;6(1):168–80.

[86] Hu X, Yu J, Qian C, Lu Y, Kahkoska AR, Xie Z, et al. H_2O_2-responsive vesicles integrated with transcutaneous patches for glucose-mediated insulin delivery. ACS Nano 2017;11(1):613–20.

[87] Hong JY, Ko EJ, Choi SY, Li K, Kim AR, Park JO, et al. Efficacy and safety of a novel, soluble microneedle patch for the improvement of facial wrinkle. J Cosmet Dermatol 2018;17(2):235–41.

[88] Imran MA, inventor; InCube Labs LLC, assignee. Iontophoretic systems, kits and methods for transdermal delivery of cosmetic agents. United States patent application US 16/134,445, 2019 March 21.

[89] Van Tran V, Nguyen TL, Moon JY, Lee YC. Core-shell materials, lipid particles and nanoemulsions, for delivery of active anti-oxidants in cosmetics applications: challenges and development strategies. Chem Eng J 2019.

CHAPTER

3

Transdermal and bioactive nanocarriers for skin care

Federico Svarc[1] *and Laura Hermida*[2]

[1]School of Exact and Natural Sciences. DQIAQF-INQUIMAE, Buenos Aires University - UBA, Buenos Aires, Argentina [2]Chemistry Department, National Institute of Industrial Technology—INTI, Buenos Aires, Argentina

3.1 Introduction

Many papers and reviews have been published during the last 10 years about transdermal bioactive nanocarriers [1–3]. Nevertheless, most of those publications have been directed toward medical and therapeutic applications, where it is necessary to introduce drugs into the systemic circulation of humans (or mammals) to reach specific sites such as cancerous cells or for imaging agents. Instead, very few reports have been written about their usage in cosmetics. The reason has its roots in the definition of cosmetics accepted by the majority of countries and regions. The Federal Food, Drug & Cosmetic Act (FD&C Act) defines cosmetics as *articles intended to be rubbed, poured, sprinkled, or sprayed on, introduced into, or otherwise applied to the human body ... for cleansing, beautifying, promoting attractiveness, or altering the appearance,* while drugs are those articles *intended for use in the diagnosis, cure, mitigation, treatment, or prevention of disease ... and articles (other than food) intended to affect the structure or any function of the body of man or other animals.* This fundamental concept, with small variants, is also accepted by the European Community [4], Japan, Mercosur, etc. Maybe the exception concerns sunscreens that are considered OTC drugs in the United States. Thus, instead of bolstering penetration, nanocosmetics try to keep the penetration of active molecules not farther than the stratum basale or at the most the dermis. Still, actives such as antioxidants (polyphenols, vitamins C and E, coenzyme Q10), anticellulite compounds (caffeine, iodine substituted organic molecules, some vegetable extracts), whiteners (Kojic acid, hydroquinone), hydrating compounds, as the natural moisturizing factor (NMF) or hyaluronic acid, "antiaging" molecules such as Retinol, molecular oxygen, and also small polypeptides that are able to trigger a signaling

Nanocosmetics
DOI: https://doi.org/10.1016/B978-0-12-822286-7.00003-6

cascade increasing the proliferation of collagen, elastin, proteoglycan, glycosaminoglycan, and fibronectin [5] must penetrate enough, trespassing the stratum corneum (SC), to produce visible effects on the condition of the skin. Also, a controlled release is suited both to increase bioavailability and, in some cases, to reduce irritation effects by limiting their local concentration. This effect can be obtained by transporting the active compounds with nanocarriers across the epidermis, but also by favoring the passive diffusion of small molecules into the inner layers of the skin. In other cases, for example, inorganic and organic UV blockers such as ZnO, TiO_2, and PLGA entrapped UVA and UVB filters, the size of the nanoparticles is directly tuned to keep them outside the inner layers.

Due to the aforementioned reasons, nanocompounds smaller than 20 nm are seldom found in the art. Moreover, the sizes usually found are in the range of 50–300 nm.

A recent review [6] using a theoretical approach demonstrated that intact nanoparticles within the size range mentioned above and topically applied could not penetrate the full healthy human skin. Nevertheless, they were able to increase skin permeation of the entrapped active ingredients. This may reassure some concerns on safety issues that regulations' authorities and cosmetic manufacturers could have.

3.2 The skin

The skin is a stratified organ composed of three layers, namely, the epidermis, the dermis, and subcutaneous tissue. The outermost level (nonviable epidermis), known as SC, is the primary barrier of the skin. Under normal conditions, it protects the body and underlying tissue from environmental threats, such as infections by pathogens, dehydration, chemicals, and different kinds of particles. In fact, SC creates a barrier that is impermeable to most hydrophilic molecules above 500 kDa and particles bigger than 100 nm [7].

SC is formed by a conglomerate of dead cells called corneocytes and is approximately 10–15 μm thick. Corneocytes are flat stacked cells filled with a network of keratin filaments, water, and NMF. They are surrounded by a densely crosslinked protein cornified envelope and a complex mixture of intercellular lipids, such as ceramides and cholesterol. The lipid layers serve as an interface between the hydrophilic corneocytes and the lipophilic intercellular lipid matrix, which directs the penetration of most substances along a tortuous pathway between them. Indeed, SC has a very high density and low hydration around 15%–20%.

The viable epidermis is located underneath the SC. Being around 50–150 μm thick, it is responsible for the regeneration of the SC. The loss of cells from the SC is compensated by cell growth in the innermost layer of the epidermis, the stratum basale, where lamellar bodies are initiated. Therefore, the epidermis is formed by cells at different degrees of differentiation, and a complete turnover of the SC occurs once every 2 or 3 weeks. There are also other structures that cross the epidermis, namely, the skin appendages. These are hair follicles and sweat glands, which form shunt pathways through the intact epidermis, occupying only less than 1% of its total surface. This complex architecture is closely related to the interaction of nanomaterials with skin and its barrier properties [8]. Even though skin, and specifically SC, is prepared to avoid the penetration of nanoparticles, there are some nanocarriers specially designed to deliver their contents, that is, drugs or bioactives, beyond the epidermis, which could mean their access to the systemic circulation [3].

3.3 Nanocarriers and skin penetration

Nanocarriers are defined as nanomaterials used to transport and deliver drugs or bioactives to a target tissue. These structures can be made of a huge variety of materials and may differ in structure and chemical nature. They have been designed and developed as promising delivery systems for topical administration, offering advantages over conventional passive systems, such as improved stability, protection from degradation, reduced skin irritancy, controlled release of active ingredients, improved permeation of actives into the skin, and versatile properties related to their increased surface area [9].

Several reviews have been published concerning different aspects of nanocarriers intended for transdermal delivery [10–14]. The development and characterization of submicron particles and other nanostructures have emerged in the last decades for both pharmaceutical and cosmetic applications.

The historical development of permeation research is well described by Hadgraft and Lane [15]. Over time, the skin has become an important route for drug delivery in which topical, regional, or systemic effects are desired. Nevertheless, skin constitutes an excellent barrier and presents difficulties for the transdermal delivery of active agents, since few drugs possess the characteristics required to permeate across the SC in sufficient quantities to reach a therapeutic concentration in the blood. To enhance drug transdermal absorption, different methodologies have been investigated, developed, and patented [10,11].

Several investigations have been done to determine the penetration pathways of topically applied substances into the skin. These include the diffusion through the intact epidermis and through the skin appendages, such as hair follicles. Two pathways through the intact barrier have been identified: the intercellular and the transcellular route (Fig. 3.1).

The follicular pathway has been proposed for the topical administration of nanoparticles and microparticles. It has been investigated mainly in porcine skin due to its similar behavior with respect to humans. For instance, after the topical application of a fluorescent dye onto porcine skin mounted in Franz diffusion cells, the fluorescence was detected on the surface, within the horny layer, and in most of the follicles, confirming a similar penetration profile between porcine and human skin [16]. Nanoparticles have also been studied in porcine skin revealing polystyrene nanoparticles accumulated preferentially in the follicular openings. This preferential distribution increased in a time-dependent manner, and the follicular localization was favored by the smaller particle size [17]. In other investigations, it has been shown by differential stripping the influence of microparticle size in skin penetration. They can act as efficient drug carriers or can be utilized as follicle blockers to stop the penetration of topically applied substances [18].

It has to be considered that there are currently a limited number of methods available for quantifying drugs localized within the skin or various layers of the skin. To date, direct, noninvasive quantification of the amount of substance penetrated through the follicles after a topical application has not been possible. Stripping techniques, such as tape

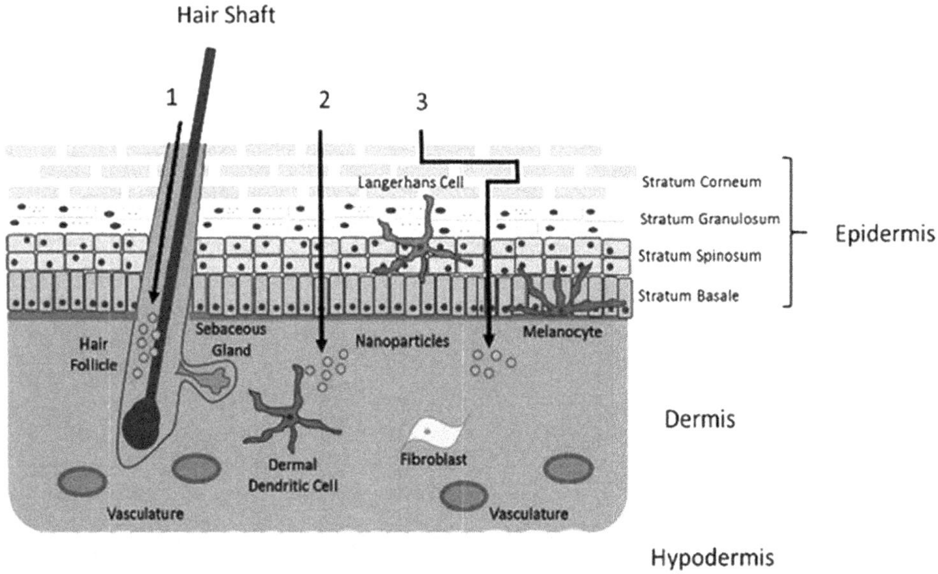

FIGURE 3.1 Nanoparticle skin penetration pathways: (1) appendageal route, (2) intracellular route, or (3) intercellular route. Source: *Adapted from Palmer BC, De Louise LA. Molecules 2016;21:1719. doi:10.3390/molecules21121719.*

stripping (TS), and cyanoacrylate skin surface biopsy have been used to remove the part of the SC containing dye topically applied [19]. The combination of TS and cyanoacrylate skin surface stripping, known as differential skin stripping, has been proposed as a new noninvasive and selective method to study the penetration of topically applied substances into the follicles [20,21].

The intracellular route contemplates the crossing through the corneocytes and the intervening lipids [22]. The intracellular macromolecular matrix within the SC abounds in keratin, which does not contribute directly to the skin diffusive barrier but supports its mechanical stability, and thus intactness of the SC. The narrow aqueous transepidermal pathways have been observed by confocal laser scanning microscopy. It has been demonstrated that transcellular diffusion is not important for transdermal drug transport [23].

The intercellular lipid route is the pathway between the corneocytes, as shown in Fig. 3.1 (path 3). Interlamellar regions in the SC, including linker regions, contain less ordered lipids and more flexible hydrophobic chains. This is the reason for the nonplanar spaces between crystalline lipid lamellae and their adjacent cells' outer membranes. Fluid lipids in skin barrier are crucially important for transepidermal diffusion of both the lipidic and amphiphilic molecules, occupying those spaces for the insertion and migration through intercellular lipid layers of such molecules [24,25]. Hydrophilic molecules predominantly diffuse "laterally" along surfaces of the less abundant, water-filled interlamellar spaces or through such volumes. Polar molecules can also use the free space between a lamella and a corneocyte outer membrane to the same end [26].

Regions of poor cellular and intercellular lipid packing coincide with wrinkles on the skin surface and are simultaneously the sites of the lowest skin resistance to the transport of hydrophilic entities.

The intracluster/intercorneocyte is the best sealed and more transport resistant pathway [27]. However, it is crossed by hydrophilic channels with small width openings between $\geq 5\,\mu m$ (skin appendages) and ≤ 10 nm (narrow intercorneocyte pores). Sweat ducts ($\geq 50\,\mu m$), pilosebaceous units (5–70 μm), and sebaceous glands (5–15 μm) represent the lowest resistance to diffusion. Junctions of corneocytes–clusters and cluster boundaries fall within the range [28].

Transdermal drug transport can be increased by widening or increasing the number of pathways. This effect can be achieved by exposing the SC to a strong electrical field (electroporation/iontophoresis), by the use of sound to modify cell permeability (sonoporation/sonophoresis), by a thermal stimulus, or through the incorporation of suitable skin penetrants [24]. A study revealed that the width of the hydrophilic transepidermal pores expanded by transdermal iontophoresis was around 22–48 nm [29].

The lipophilic cutaneous barrier is governed by molecular weight and distribution coefficient rather than molecular size [30]. The difficulty to cross the cutaneous lipophilic barrier consequently decreases with lipophilicity of permeant, but molecules heavier than 400–500 Da cannot find sufficiently wide defects in the intercellular lipidic matrix to start diffusing through the lipidic parts of the cutaneous barrier [28–30].

3.4 Nanocarrier system

Several different kinds of nanoparticles have been used to transport active substances through the epidermis in a controlled manner to obtain different penetration and effects. Nanoparticles are composed of diverse materials and/or their mixtures. These materials can be classified into four distinct families: [31–33] lipids or surfactant self-assembly structures, dendrimers, polymer-based, and polysaccharide-based nanostructures. A fifth category should be mentioned, namely, porous inorganic nanocompounds, which is capable to transport active molecules. The first group is represented by liposomes, niosomes, nanoemulsions, solid lipid nanoparticles (SLNs), and nanostructured lipid carriers (NLCs). Within the liposomal group, we can also mention ultradeformable liposomes (UDLs) and ethosomes [34–36].

3.4.1 Liposomes and related particles

The existence of lipidic dispersions with the lamellar structure was described several decades ago [37]. Originally, the existence of these vesicles was associated with the presence of phosphatidylcholine associated with cholesterol, and eventually dicetyl phosphate, conducting to bilamellar phases (Fig. 3.2).

To be able to obtain such type of structures, the amphiphilic molecule must be relatively insoluble and the polar head cross section must be similar to the hydrocarbon tail one. First, hydrocarbon chains align spontaneously to form bilayers in aqueous environments.

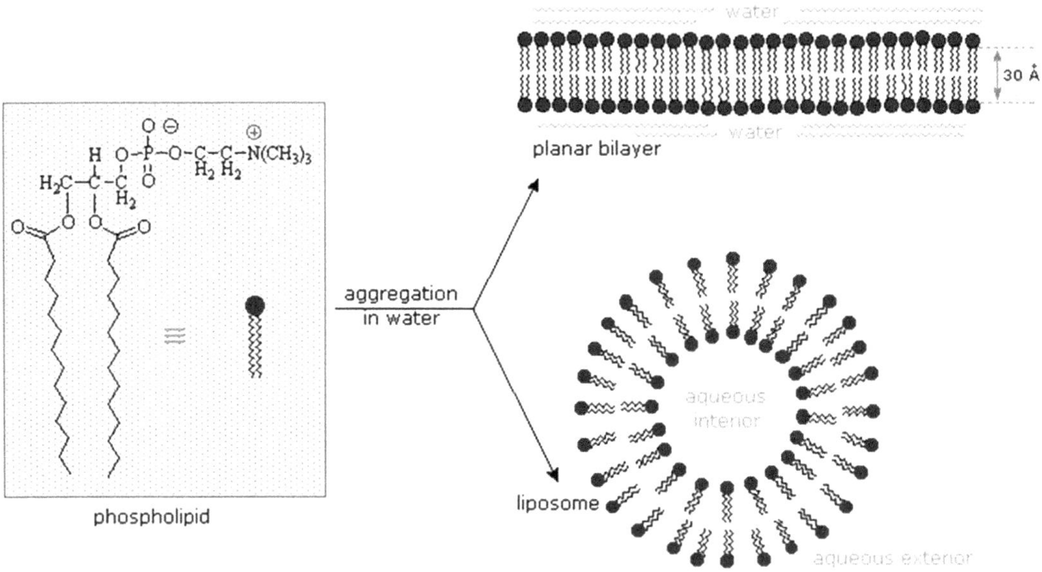

FIGURE 3.2 Aggregation of a phospholipid to generate bilamellar phases.

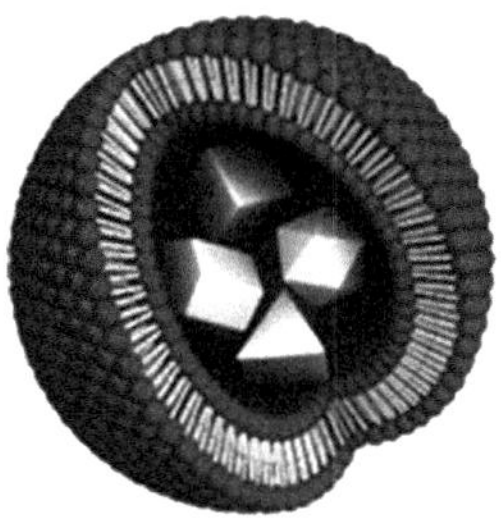

FIGURE 3.3 Schematic cross section of a loaded liposome.

In this energetically most-favorable arrangement, the hydrophilic heads face the water at each surface of the bilayer, and the hydrophobic tails are shielded from the water in the interior.

When submitted to high shear (by ultrasound, extrusion, etc.) curvature of the bilayers is obtained, adopting spherical or ellipsoidal shapes. Multilamellar bilayers from 100 to 5000 nm are formed, which need to be further down-sized if necessary.

Usually, cholesterol is added to increase physical stability. It locates itself between the lipid tails of the phospholipids, increasing the temperature transition range between liquid crystal and solid gel, which may cause the release of loaded actives. Phospholipids such as dicetyl phosphate are added as ionic surface-active agents to avoid the flocculation of particles by changing their surface charge. Hydrophilic active molecules transportation by liposomes occurs when those are dissolved into the water nucleus (Fig. 3.3), while lipophilic molecules insert themselves into the peripheric phospholipid crown.

The replacement of phospholipids with nonionic synthetic lipids designed by molecular engineering made possible to obtain more stable vesicles, called *niosomes*. X-ray diffraction patterns showed a more compact external shell due to external paraffinic chains in a liquid state [38,39]. Dicetyl phosphate was replaced by nonionic emulsifiers specially designed to favor curved surfaces. These R&D developments resulted in the famous commercial cream *Niosome*.

At L'Oreal factories, c.1995, to produce niosomal vesicles we used high-pressure Gaulin homogenizers. Typically, sizes obtained were between 25 and 250 nm.

Once applied to the skin, both types of vesicles release their active contents, while the external lipidic bilayer is either easily absorbed onto SC or integrated into the epidermis because of its structural affinity.

Several liposomal commercial cosmetic formulations containing active molecules such as retinol, tocopherol, ascorbyl palmitate, coenzyme Q10, vitamins A, E, and C, panthenol, grape seed extract, and green tea extract are sold under different trade names for wrinkle reduction, antiaging, antioxidative, cells regeneration, UVA protection, revitalizing, antiaging, and other marketing claims [40]. In some cases [41,42], an enhancement of skin penetration was observed compared to unencapsulated bioactives such as ascorbyl palmitate [43].

As the penetration into the skin of traditional liposomes and niosomes is limited, two distinct solutions have been devised to increase their permeability through it.

Ethosomes are soft vesicles obtained by replacing a significant (20%–50%) part of the water in the formulation by ethanol or isopropyl alcohol. Their size range varies from tens of nanometers to a few microns. They permeate through skin layers more rapidly and possess a higher transdermal flux [34,35]. The high concentration of ethanol affects the skin lipid organization, giving the vesicles the ability to penetrate the SC. Their lipid bilayer is packed less tightly than those of more conventional liposomes (CLs) allowing a stable and more malleable structure and improving drug distribution throughout the SC. For that reason, they have been used not only for the transportation of cosmetic components such as peptides and proteins but also for active ingredients such as Minoxidil, antiviral drugs, hormones, among others.

UDLs (also called Transferosomes) is another approach designed to penetrate across the SC, instead of aggregate or coalesce on the skin surface, thus becoming a powerful tool for the controlled/targeted delivery of cosmetic actives [36].

In this case, there is no need to include any solvent in the formulation. On the other hand, UDLs are composed of a mixture of lipids with low phase transition temperatures and an appropriate amount of an edge activator, typically a single chain surfactant. The surface-active molecule acts as a membrane destabilizer, producing an increase in membrane deformability. This effect has been proposed to be the cause of their ability to penetrate the skin, reaching SC deeper layers and the viable epidermis. They have been proposed to deliver from small molecules to proteins [44,45].

The lipid matrix of ultra deformable liposomes UDL has a decreased elastic modulus as compared to conventional liposome (CL), specially designed to increase the transcutaneous delivery of actives. As CL, they are capable to carry both hydrophobic and hydrophilic load. Unlike CL, that have a rigid structure, they deform and pass through narrow constrictions 10 times smaller than their own diameter. Specific types and concentrations of edge activators are required for providing the maximum deformability to the vesicle membranes [44]. Typically Sodium cholate, Span 80, Tween 80 and dipotassium glycyrrhizinate are amongst the most widely used compounds to obtain UDL.

To evaluate their capability of penetration, as well as the risk of systemic access of actives, Romero, and collaborators [44] studied the in vitro skin penetration of UDL of both hydrophilic and hydrophobic fluorescent probes with a Saarbrücken Penetration Model. The charged UDL, of circa 100 nm diameter, were applied on human skin explants excised from abdominal reductive surgery. It was verified that the penetration of UDL during the first hour of the test was almost seven times higher than those of CL. The hydrophobic probe was found into the stratum corneum up to 14 μm depth, while the hydrophilic one at a mean depth of 24 μm near the viable epidermis.

The same research group encapsulated extracts from three blueberry varieties in UDL, which showed good antioxidant activity and low cytotoxicity on a human keratinocyte cell line (HaCaT) [45]. Very recently (7th Argentine Congress of Cosmetic Chemistry. Carilo. November 2019), Montanari compared the pigskin model penetration of hydrophilic molecules carried by UDL incorporated into several different vehicles (gels, fluid or consistent creams), with surgery explants of human skin. He found that the penetration to the viable epidermis and dermis was much higher with the pigskin model, introducing a question mark about its extensive usage.

Romero and col. further proposed a novel, more stable nanocarrier, with similar penetrability properties as UDL [46]. Ultradeformable archaeosomes (UDA) are UD vesicles composed of totally polar lipids from microorganisms of the archaea domain [46]. The polar lipids consist of regularly branched, and usually fully saturated, phytanyl chains of 20, 25, or 40 carbon length. The phytanyl chains are attached via ether bonds to the glycerol backbone. These ether bones improve lipid stability and therefore structural stability of the nanocarrier, to oxidative stress, high temperature, pH, the action of phospholipases, bile salts, and serum proteins. Moreover, Caimi et al [47] demonstrated that UDA behaved as superstable nanovesicles because of its high endurance during heat sterilization, storage under cold-free conditions and lyophilization in presence of crio-protectors, which could lead to simpler manufacture processing for topical vaccination solid, dispersible formulations.

3.4.2 Solid lipid nanocarriers and structured lipid nanocarriers

Another important family of lipidic carriers is constituted by solid lipid nanocarriers (SLNs) and NLCs. The first generation of lipidic carriers, SLN, is composed of solid mixtures of fats and waxes dispersed in water, while the second generation, NLC, is based on a mixture of solid and liquid lipids (Fig. 3.4). SLNs were first developed by replacing the liquid oil of emulsions by highly purified solid lipids, which remained solid at a physiological temperature [48], while NLCs were proposed as an alternative to SLN, to overcome their low loading capacity and content release during storage [49]. Although some advantages have been described in comparison to other colloidal systems, lipid crystallization during cooling diminished encapsulation percentages and could promote the active release during storage. The use of a mixture of solid and liquid lipids avoids crystal formation,

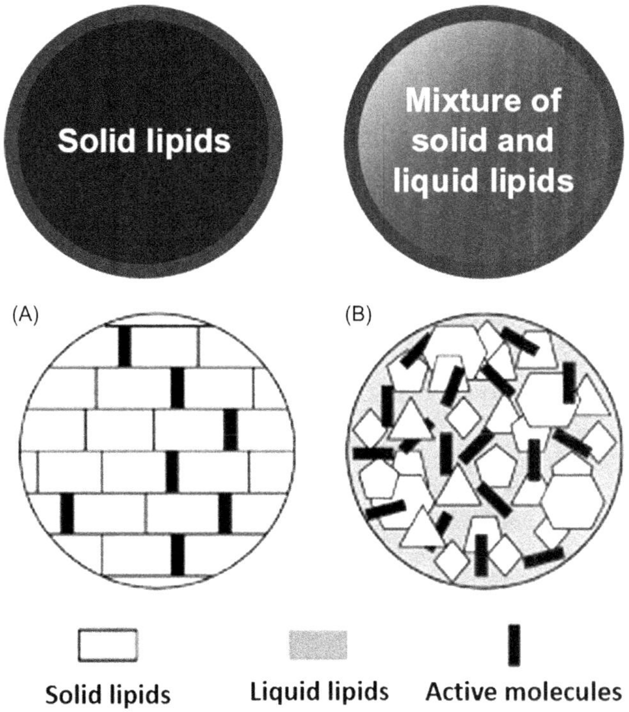

FIGURE 3.4 Evolution of lipid carriers from SLN (solid lipids) to NLC (a mixture of solid and liquid lipids). Below, a schematic representation of active molecules inserted in an ordered structure of solid lipids (A) and a nanostructured, disordered structure (B).

resulting in a nanostructured matrix which can easily accommodate the active and provide long-term stability [49].

Cosmetic benefits of using NLC have also been described [50–52]. Due to their small size, lipid nanoparticles adhere to the skin, forming a film that leads to occlusion effects [50]. This occlusion is generally followed by increased hydration due to a reduced water loss, which was observed after in vivo application of a cream enriched with SLN [51]. Moreover, the use of lipid nanocarriers has demonstrated to enhance the skin bioavailability of active compounds. For instance, increased penetration of coenzyme Q10 was observed after the application of Q10-loaded SLN suspension, in comparison to its solutions in isopropanol or liquid paraffin [51]. Another advantage of both SLN and NLC is that they are generally formulated with biocompatible low-toxicity materials.

SLNs have also been used by several authors [50] to deliver retinol and Vitamin A palmitate to the skin to improve pharmacokinetics and stability. Also, tocopheryl acetate has been encapsulated with high encapsulation efficiency in these nanocarriers. No major benefit over control formulations in terms of delivery was detected, but improved stability was demonstrated. Penetration studies were carried out with porcine and human skin using TS to extract material from the skin surface. These experiments showed that porcine skin retained less recovery of the total dose applied through a silicon oil/SLN formulation than human skin. The difference could be attributed to the specific orientation and thickness of lipids in their respective SC, supporting the hypothesis that human skin is less

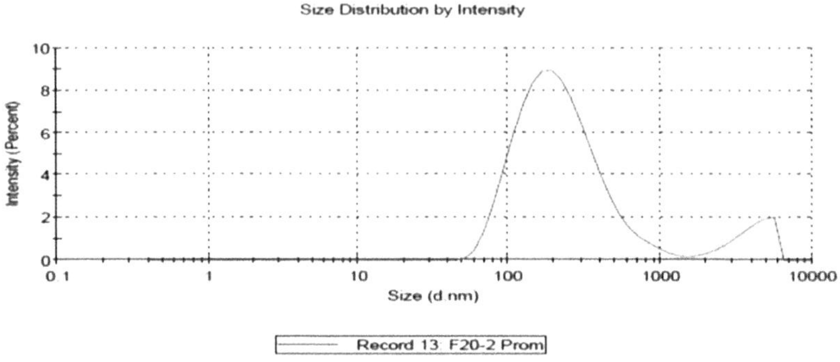

FIGURE 3.5 Typical size distribution of PFD-NLC obtained by dynamic light scattering.

permeable than pigskin. Lademann et al. [53] described the smaller size of human hair follicles as its possible origin.

We formulated NLC to transport perfluorodecaline (PFD), a molecule used as an oxygen carrier to the skin [54]. PFD is hard to stabilize in emulsion-based formulations due to its both hydrophobic and lipophobic properties. Moreover, it possesses high density (1.93 g/cm^3), which tends to destabilize colloidal systems. PFD-NLC nanoparticles were developed using conventional cosmetic ingredients. The optimal lipidic matrix was composed of cetostearyl alcohol as the solid support, glyceryl-oleate as the liquid phase, and Ceteareth 20 as the emulsifier. PFD was included in the lipidic phase. First, a predispersion of the fused lipids in water was produced with a high-speed homogenizer and then processed by sonication or through a high-pressure homogenizer (1000 bar) to size-down the particles to a mean size of 200–250 nm with a polydispersion index of around 0.5 (Fig. 3.5).

The nanoparticles were characterized by both atomic force microscopy (AFM) (Fig. 3.6) and scanning electron microscopy (SEM) (Fig. 3.7), provided with an EDAX probe to quantify F (PFD) retained inside.

Even though particles appear elongated due to the contact with the probe, the size observed by AFM in the z-axis was coherent with that obtained by DSL. Images obtained by SEM under vacuum with a gold surface deposit evidenced the heterogeneity of the sample.

DSC studies performed on the samples hinted that PFD was included in the defects of the NLC lipidic matrix. FTIR studies revealed the characteristic peak of C-F stretching corresponding to PFD.

A preliminary in vivo test was performed with 4% PFD containing NLC, applied daily on the nasobuccal area of Caucasian volunteers, ranging between 50 and 60 years old, which was compared to a similar placebo NLC without PFD. Statistical treatment of images after 10 weeks of treatment showed an improvement in the condition of the skin (unpublished results).

3.4.3 Nano- and microemulsions

Nano- and microemulsions are isotropic dispersions of two immiscible liquid phases with droplet size in the nanometric scale, typically between 20 and 200 nm. They may

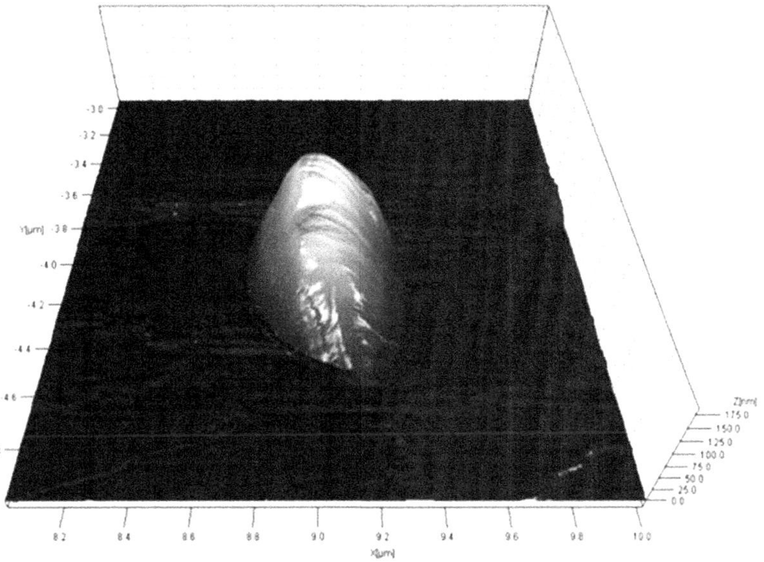

FIGURE 3.6 AFM image of a single NLC particle.

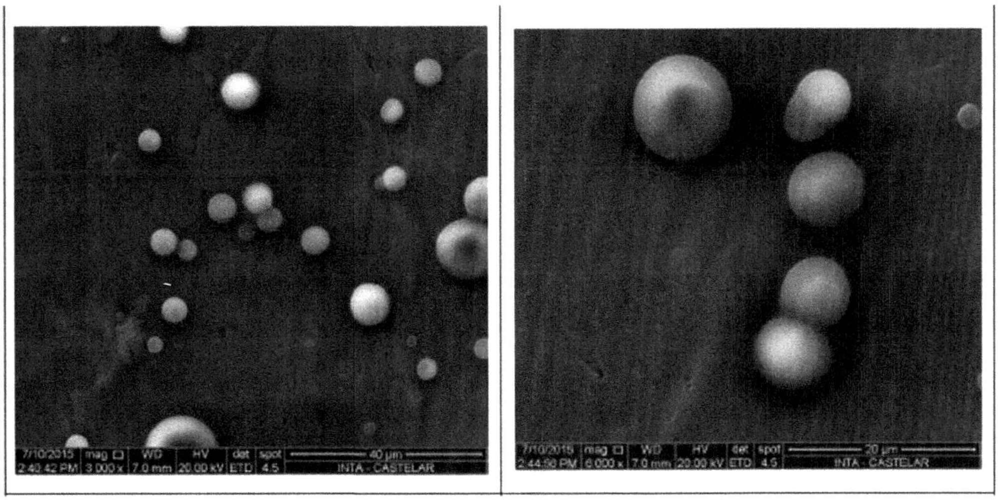

FIGURE 3.7 SEM images of NLC at 3000× (left) and 6000× (right).

present themselves both as O/W or W/O types, depending on the nature and the concentration of emulsifiers and coemulsifiers. Originally, they were conceived as distinctly different types of colloidal dispersions: a microemulsion was a thermodynamically stable dispersion, whereas a nanoemulsion was not. Following McClements [55], nanoemulsions are obtained by high mechanical energy methods to down-size droplets to <200 nm. They usually contain a mixture of surfactants (<10%) and charged compounds to stabilize the dispersed phase. On the other hand, microemulsions are transparent or translucent, with <100 nm droplets, stabilized by high concentrations of surfactant (>10%). Generally speaking, to be produced they need much more emulsifier present in their interphase, being the chemical and not the mechanical work the major driving force that contributes to disperse and stabilize them. Under certain conditions and even at room temperature, they may be produced spontaneously (low-energy methods), thus being thermodynamically stable. Their small droplet size gives extra stability against sedimentation and/or Ostwald ripening. Nowadays, however, researchers tend to avoid that distinction and unify the definition of nanoemulsion as a colloidal system, with a droplet size lower than 100–200 nm, preferably kinetically and thermodynamically stable, which require the use of high mechanical energy, high surfactant levels, or both in order to be produced [2,35].

Depending on the type of nanoemulsion, namely, O/W or W/O, hydrophobic or hydrophilic actives can be formulated, respectively. Its ability to penetrate the skin due to the small size of droplets has been demonstrated. In fact, the presence of the emulsifier at the interface helps to cross the skin barrier by reversible alteration of the cellular arrangement, both by disruption of the lipidic mortar and by fusion of the lipids bilayer interface with the cell walls.

Based on the surface charge, nanoemulsions can be categorized as anionic, cationic, and nonionic. The oily phase has a vital role when lipophilic active compounds are encapsulated. The emulsifiers and coemulsifiers are selected on the basis of their solubility and hydrophile–liphophile balance (HLB) value. In many instances nonionic surfactants are preferred because of their low toxicity and irritation profile. Usually, the coemulsifiers are C3-C8 alcohols such as glycerine, propylene glycol, ethylene glycol, or propanol.

It has to be mentioned that nanoemulsions are generally considered nontoxic, though some irritation can be present on the skin due to the high concentration of emulsifiers needed.

As mentioned before, nanoemulsions can be produced by high energy methods, such as high-pressure homogenization or microfluidization, or by low-energy methods. In the first case, severe turbulence leads to uniform droplets smaller than 100 nm, resulting in a better texture and prolonged shelf-life products. Ultrasonication or high shear rotor/stator equipment is frequently used too [56,57]. When low-energy methods are used, the emulsification is obtained by changing temperature and/or the composition of the formulation, and thus its HLB. Through energy-saving and softer process conditions, labile encapsulated bioactives are better preserved.

As for the other nanocarriers, characterization of nanoemulsions includes size determination, storage stability, polydispersity index, and zeta-potential, as an indication of scaling-up feasibility.

It should be mentioned that the phase inversion temperature (PIT) method is based on the specific ability of certain emulsifiers to change from hydrophilic to hydrophobic or vice versa with temperature conditions [56]. For example, ethoxylated emulsifiers are capable of this kind of behavior due to the hydration of the polar groups at low temperatures. Below transition temperature, o/w nanoemulsions are formed, while above, its surface area is occupied by the nonpolar hydrocarbon chains and the emulsion is inverted.

Sunscreens have been incorporated in nanoemulsions produced by PIT to facilitate an even application on the skin by mechanical dispensing valves. A typical composition, with a particle size of 59.4 nm, was patented by L'Oreal containing water, ethylhexyl palmitate, PG5 laurate, and vitamin B3 [58].

Nanoemulsions have also been used in cosmetics to deliver fragrances and some other natural and synthetic molecules.

The mode of transport through dermal and transdermal routes (Fig. 3.1) varies with size, polydispersity, and zeta-potential of the specific nanoemulsion. Furthermore, the penetration also depends on the hydration level of the skin. The elastic nature of nanoemulsion droplets can change their spherical shapes and, under stressed conditions, can penetrate the narrow intercellular gap junctions. However, if the size of the droplets exceeds 50 nm, they are less likely to penetrate through passive diffusion, requiring external assistance to enter the deeper layers of the skin. The fusion of the lipidic layers of the nanoemulsion interface with those of the cellular matrix leads to the release of the encapsulated active compounds. Transdermal delivery is possible when the size of the dispersed phase is below 50 nm.

Droplet surface charge drives their transport through different skin layers favored by ionic interactions: anionic and cationic surfactants have been used to modify the surface charge and design dermal or transdermal delivery depending on the available charge on the skin surface. Hydrated SC layer favors a better interaction of hydrophilic molecules with the epidermal and dermal cells. Hydrophobic drugs are less likely to interact and permeate through the skin.

3.4.4 Inorganic nanocarriers

Three nanosized inorganic oxides are most frequently used in cosmetic formulations: TiO_2, ZnO, and SiO_2. The first one is extensively used, coated or uncoated, in sunscreen formulations, with dimensions in the range of 20 nm. It acts as Physical UV-filter and is not used as a carrier of other molecules [59].

Instead, silica and hydrated silica have been extensively used due to their chemical inertness and immense capacity to adsorb/desorb hydrophilic molecules. Silica can be obtained as nanospheres of controlled size by two different pathways: the pyrolytic route and the sol-gel route. The products of the first one are commercially available under Aerosil, Cab-o-sil, and other registered trade names [60]. Depending on their specific surface (which is enormous), they are further classified in quality ranges. For example, Aerosil 200 from Degussa has a specific surface of 200 m^2/g and a mean size of 12–14 nm. When observed with an SEM, a cluster structure can be noticed due to intermolecular adhesion forces between individual particles (Fig. 3.8) whose hydrodynamic radius is about 100 nm.

At the *26th IFSCC Congress 2010* [61], we presented permeation data of bioactives from an anticellulite complex adsorbed on pyrolytic silica and formulated into an emulsion. The anticellulite complex contained caffeine (CAF), triethanolamine hydroiodide (TEAH), algae extract, Hedera helix extract, and *Ruscus aculeatus* root extract. CAF and TEAH release profiles were obtained at 32°C, with Franz diffusion cells. Three different skin model membranes were tested: (1) a polycarbonate (PC) membrane of 50 nm pore size (Millipore VMTP04700), (2) the same PC membrane but treated with a lipidic equimolar mixture of

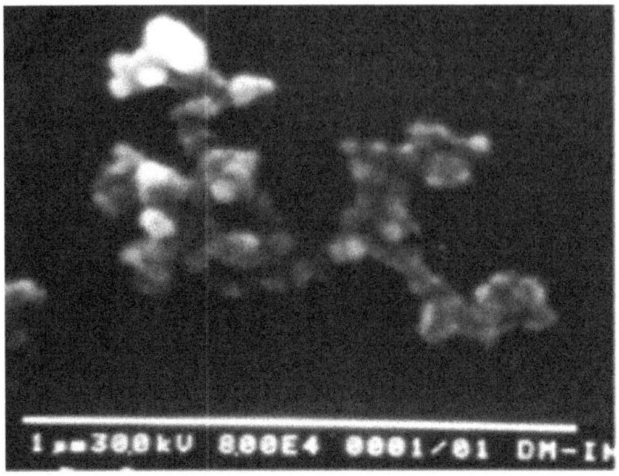

FIGURE 3.8 SEM image of commercial pyrolytic silica.

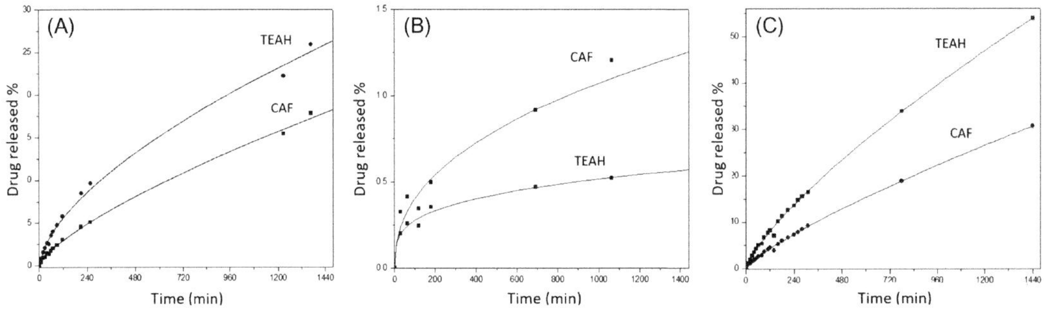

FIGURE 3.9 CAF and TEAH diffusion through Franz diffusion cells using different skin models: PC membranes (A), PC treated with a lipidic layer with a similar composition to SC (B), and a full-thickness porcine skin (C). Source: *Extracted from Svarc F, Daraio ME, Arnejo N, Carballo O, Debandi M, François N, et al. Anticellulitic microspheres: in vitro assessment of release behavior. In: 26th IFSCC Congress, Buenos Aires. AAQC; 2010. p. 263–4.*

cholesterol, C18/C22 fatty acids, and ceramides III, IIIB, and VI following the Gooris and Bouwstra protocol [8], and (3) a full-thickness porcine ear skin (Fig. 3.9). Treated PC membrane was specially designed to mimic the barrier function and the compact orthorhombic lateral packing of the SC lipids.

It may be noted that the passage through the PC membrane covered by a lipidic layer severely hinders the diffusion of the two hydrophilic lipolytic molecules (CAF and TEAH), specially TEAH, while the full-thickness porcine skin allowed after 1 day the passage of 30% and 54%, compared to 1%, 5%, and 0.5%, respectively.

CAF and TEAH permeated slowly on the treated PC, after 3 days we found between 5% and 10% of the originally loaded quantity. Instead, in this case, the full-thickness porcine skin let pass-through between 40% and 50% of the total quantity charged to the donor compartment.

Comparing with our own results of previous studies [62], we concluded that our results are in agreement with the proposal that the SC is a very efficient barrier to the penetration of drugs, particularly ionic or ionizable substances. Thus their permeation may occur through the pilous appendages of the skin. Moreover, we detected that when we employed an O/W emulsion to vehiculate the charged silica nanocarriers, the diffusion was significantly hindered compared to those formulated in an aqueous media (data not shown). It must be noted that this kind of carrier keeps on the surface of the membranes (50 nm controlled pores) or, at worst, in the inner borders of the pilosebaceous pores. This observation is in total accord with the conclusions we cited above from the Nanoderm project [59] obtained employing much more sophisticated techniques.

A further conclusion in this particular case is that drug delivery is obtained by passive diffusion and not by nanoparticle penetration into the inner layers of the skin.

Another possible way to obtain silica nanocarriers is by the Stober sol-gel process [63]. Many active compounds have been trapped by variations of TEOS (tetraethyl orthosilicate) or TMOS (tetramethyl orthosilicate) hydrolysis on a nanoemulsion template. The sol-gel routes allow to design and produce nanoparticles with highly ordered structures and controlled pore-size distribution, pore geometry, and pore network tortuosity [1].

One of the challenges of our group was to entrap PFD into silica nanoparticles. Recently, silica PFD nanocapsules have been synthesized as oxygen carriers used for ultrasound imaging [64]. However, in the field of cosmetics there are only a few scientific reports, even if it has been shown that after introducing PDF saturated with $O_2(g)$ into an emulsion, a reduction of wrinkles and an increase of skin moisture content were observed [65]. We recently described a new route to obtain silica nanocapsules by one-pot controlled hydrolysis of sodium silicate and PFD-loaded silica nanocapsules, by performing the controlled hydrolysis on a PFD template nanoemulsion [66]. As observed in Figs. 3.10 and 3.11, monodisperse spherical nanoparticles can be obtained, in a controlled range of 20–100 nm.

PFD-loaded nanocapsules obtained by our group are shown in Fig. 3.11. As may be seen, particles cluster in a similar way as those obtained by the pyrolytic process. In this case, the image was obtained with a transmission electron microscope (TEM) provided with an EDAX probe, capable to quantify fluorine associated with particles. The optimized PFD encapsulation (20% w/w) gave rise to nanoparticles with a diameter of (23 ± 3) nm and a BET (Brunauer, Emmett & Teller) surface area of (38.9 ± 1.1) m^2/g, as determined from

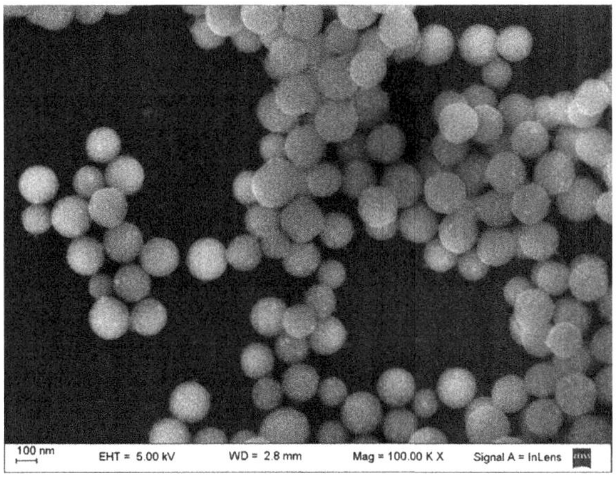

FIGURE 3.10 SEM image of empty silica nanoparticles produced by hydrolysis of Na_2SiO_3.

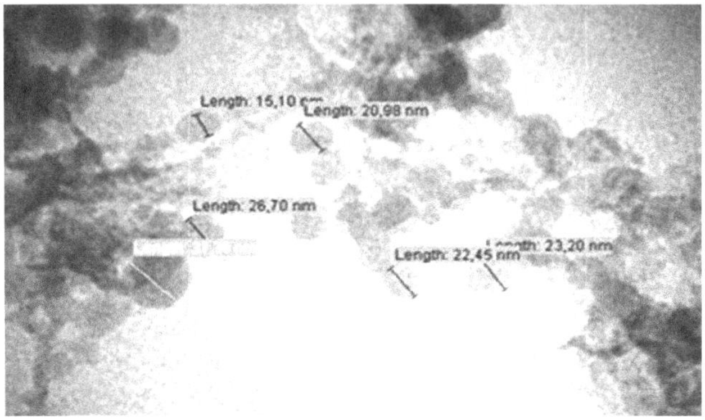

FIGURE 3.11 TEM image of silica nanoparticles loaded with PFD.

N_2(g) adsorption isotherm. The measured adsorption average pore width was 11.6 nm, showing both the adsorption and desorption plots very low hysteresis, which suggests weak intermolecular forces between the gas and the silica nanoparticles, allowing gas exchange with the skin.

Due to the low potential penetration via topical application into the systemic circulation, no toxicity effects are expected. Commercial silica nanoparticles as described above have been used in cosmetic formulations for long. CIR Expert Panel concluded that silica and hydrated silica are safe to be used in cosmetics and personal care products [67]. They are also considered safe by the FDA and accomplish general provisions of the Cosmetic Directives of the European Union [4].

3.4.5 Dendrimers and other dendritic structures

Dendrimers are symmetrical, highly branched synthetic nanoscale polymers with a layered architecture. Their biocompatibility and pharmacokinetics may be tuned through a smart route of synthesis, modifying both the molecular weight and the composition [68]. These particularities confer them unique properties and behavior, different from the linear polymers, such as monodispersity, polyvalency, self-assembly, stability, and solubility.

They consist of three parts (Fig. 3.12): a central core; generations of tree-like branching chains attached to the central core, called dendrons (from the Greek: *dendra* which means tree); and terminal functional groups attached to the terminal group of the branches. Therefore, the arrangements of the branches are in a layer-by-layer arrangement, surrounding the central core, and each layer defines its "generation" as G-1, G-2, G-3, etc. For example, outer dendrons shown in Fig. 3.12 should be classified as G-3.

Because dendrimers are prepared stepwise, it is expected to obtain monodisperse populations, as opposed to traditional polymers where polydisperse products are normally synthesized. This fact is important to reduce variability when used as carriers for active molecules.

Monodisperse products can be easily obtained up to G-3, but for higher generations, minor defects result in slight deviations from absolute monodispersity.

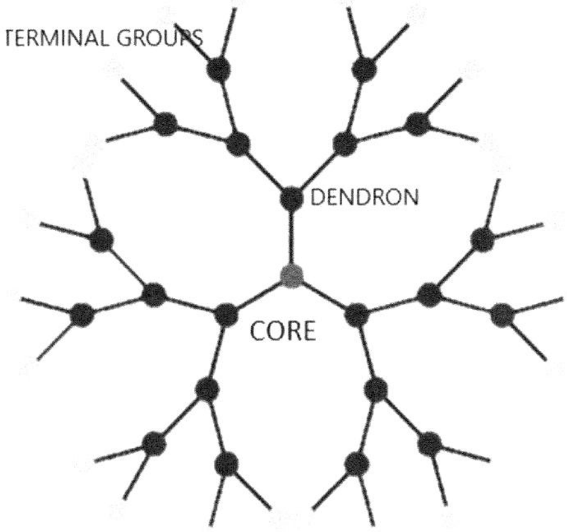

FIGURE 3.12 Basic structure of a dendrimer.

Dendrimers can be based on many types of chemistry, whose nature determines their solubility, degradability, and biological activity. They are most commonly based on polyamines, polyamides, polyamidoamines, carbohydrates, polyesters, and poly (aryl ethers). The most common scaffold is polyamidoamines, called PAMAM, which are available commercially in a wide range of generations and functionalities from Polymer Factory [69], Sigma Aldrich, and other companies.

PAMAM dendrimers are water-soluble molecules and their ability to entrap hydrophobic molecules makes them good solubility enhancers. However, there is some concern about in vivo applications due to their cytotoxicity. In fact, the toxicity of cationic dendrimers such as PAMAM can be explained by the interaction between negatively charged cell membranes and the positively charged dendrimer surface, which leads to the formation of nanopores in the cell membrane. Surface modification with negatively charged moieties or polyethylene glycol can significantly reduce cytotoxicity [70].

The dendrimer architecture has three main sites for drug entrapment: void spaces (by molecular entrapment), branching points (by hydrogen bonding), and outside surface groups (by electrostatic interactions) [71].

In spite of the great number of reports describing drug encapsulation in dendrimers, there are still no tools or rules to select the adequate dendrimer for a specific molecule. Hence, screening with different dendrimers is suggested to find the right combinations.

Drug entrapment is size-dependent and increases with the increased generations, that is, molecular weight. Considering one of the main mechanisms for drug entrapment is electrostatic interaction, the pH plays a vital role: it should be in a range where both dendrimer and drug are fully ionized with the opposite charge. On the other hand, preferentially small hydrophobic molecules can be trapped in the interior voids, in such a way to enhance drug dissolution, stability, and bioavailability.

Kraeling et al.[72] recently examined the skin penetration of generation 3 (G3) to generation 6 (G6) PAMAM amine-terminated dendrimer nanoparticles conjugated with a fluorophore with Franz diffusion cells. After 24 hours, most fluorescence appeared in the SC or in hair follicles of both pig and human skin. Upon binding of glycolic acid to dendrimers, they found that dendrimers, especially G4, increased the penetration of the active and demonstrated that dendrimer terminal group functionality and charge may alter skin absorption of associated chemicals.

Folic acid (vitamin B9) has been used in cosmetics to improve skin, nails, and hair conditions. Used topically for skin treatment, it provides increased hydration by bolstering skin-barrier function. This can improve moisture-retention and alleviate skin dryness [73]. Several in vitro and in vivo studies (imaging and quantification of collagen density performed by multiphoton laser scanning microscopy) were carried out by Beiersdorf R&D to elucidate the effects of topical folic acid to counteract the age-dependent reduction in the amount of collagen [74,75]. It was stated that *topical application of folic acid- and creatine-containing formulation significantly improved firmness of mature skin in vivo*. Other companies, like L'Oreal, make extensive use of folic acid in their cosmetic formulations and marketing communication.

A self-assembled nanostructure was very recently devised to anchor folic acid to amine groups by gelation of anionic albumin with 4.0 G PAMAM dendrimers [71]. Most probably this kind of nanocarriers will be more usual for cosmetics applications once the issues about toxicity, pharmacokinetics, and biocompatibility are completely clarified.

Hyperbranched polymers are another class of dendritic polymers that merit attention because they possess some dendrimer-like properties and can be prepared in a one-pot synthesis. Typically, they are polydisperse, but they have the advantage of low cost and are commercially available in large quantities as Hybrane from DSM, Lupasol from BASF, and other companies [68].

3.4.6 Other polymeric nanoparticles

Polymeric nanoparticles can be obtained from natural or artificial biodegradable polymers.

The polymer most commonly used is poly-L-lactic acid and copolymers with glycolic acid (PLGA) [41,76,77]. PLGA has been approved by the FDA for the last 30 years for a huge variety of applications. The amount and rate of drug release can be controlled through mechanisms that are well known and depend basically on the composition, molecular weight, and nanoparticle characteristics. A wide range of large and small as well as hydrophobic and hydrophilic molecules have been encapsulated into PLGA nanoparticles. For instance, PLGA nanospheres encapsulating ascorbyl tetraisopalmitate, tocopheryl acetate, and retinyl palmitate were tested for antiaging effects. When applied topically before UVA irradiation, a marked reduction of wrinkle formation was observed [78].

Sunscreen creams based on morin-loaded (a natural polyphenol with photoprotection and antioxidant properties) PLGA nanoparticles were developed, which combined with titanium and zinc dioxide nanoparticles, exhibited a very good SPF (sun protection factor) and an effective antioxidant effect [79].

Vitamin K1, an active antioxidant with recognized antiaging effects, was encapsulated PLC (polycaprolactone) nanocapsules, showing a significant skin permeation into dermis [80].

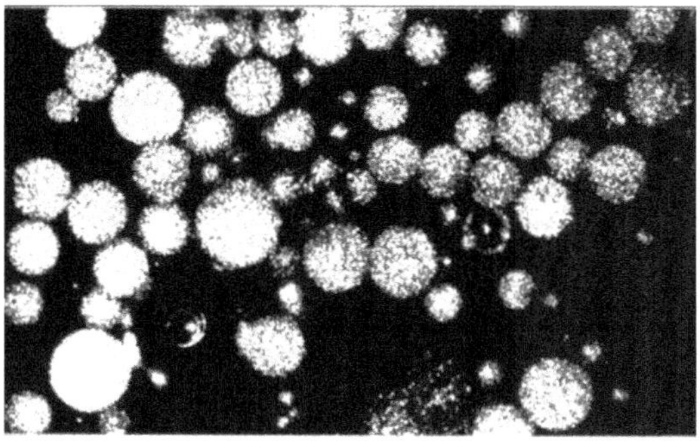

FIGURE 3.13 Image of Polytrap microparticles. Unit particles have less than 1 μm, but are associated as agglomerates of 20–80 μm.

Lauryl methacrylate/glycol dimethacrylate crosspolymer particles, sold as a free-flowing powder under the brand Polytrap (Fig. 3.13), can adsorb high levels of lipophilic materials while providing an elegant skin feel. They are employed as sebum control in facial cleansers, color cosmetics, fragrances, and toiletries, but are also capable of a sustained release of active molecules while reducing irritation, as, for example, salicylic acid and other $\alpha-$hydroxy acids used for mild chemical peeling and acne care [81]. When rubbed on the skin surface, the weakly associated micrometric agglomerates dissociate and release their contents. Lipophilic drugs which are difficult to transport, such as ubiquinone or vitamin C, may also be protected and incorporated easily into cosmetic formulations using this strategy.

3.4.7 Polysaccharide nanocarriers

Natural products are searched by consumers due to their good efficacy with minimal toxicity.

In recent years, much attention has been paid to marine resources as a new source of inexpensive and safe substances. Among them, alginate and chitosan (CHI) nanoparticles have become attractive to design novel nanocarriers.

For example, an alginate nanoemulsion was prepared by ultrasound techniques and then stabilized by the addition of divalent cations such as Cu^{+2} or Ca^{+2} to nanocarriers [82]. The obtained nanocapsules average size could be tuned between 100 and 200 nm, which was suitable for European cosmetic regulations, with a low polydispersity index (around 0.15) and negative z-potential (−28 to −22 mV). The toxicity on human HaCaT cells (human keratinocytes) was tested in vitro, while ex vivo corrosion studies were assessed on pig ear skin with a Franz cell, showing a nontoxic and noncorrosive profile at 10% concentration. Curcumin-loaded nanocarriers were incorporated into three different gels. After 2 weeks, curcumin remained encapsulated, as shown by its unchanged fluorescence profile.

The development of nanocarriers prepared from CHI and its derivatives is also promising for cosmetic applications [83]. Particulate CHI structures are 3D crosslinked networks. The linear CHI heteropolymer, formed by *N*-acetyl-D-glucosamine and D-glucosamine, must be crosslinked. The crosslinking density determines the main properties of the obtained nanoparticles, such as drug release and mechanical integrity. Several different crosslinkers can be employed: ionic, covalent, polyelectrolytes, or hydrophobic self-assembly, affecting the design and properties of delivery systems, such as loading capacity, shape, size, and biocompatibility. However, their applications were limited due to their low solubility at neutral pH and their moderate toxicity associated with terminal ionizable amino groups. Water CHI derivatives have been synthesized in a variety of structures and are commercially available to overcome these limitations.

CHI also has a variety of reactive groups that can be chemically modified to obtain amphiphilic properties, leading to systems as nanomicelles, microspheres, or liposomes with excellent biocompatibility and biodegradability, which can improve drug solubility and stability of hydrophobic molecules, as well as help to overcome toxicity problems.

CHI-based nanocarriers have been widely used for its well-known antibacterial activity. They have been proposed for dental products against oral pathogens, with fewer side effects than chlorhexidine gluconate [84] and for the controlled release of fluor [85].

There are also many reports of CHI nanoparticles for hair care, for example, as minoxidil carriers [86] and as carriers of bioactives for skin care and protection, such as photoprotective and antioxidant compounds [87].

3.5 Conclusions and future perspectives

Due to the fact that human life has been prolonged, that we expect this tendency to continue in the future and the social need to maintain a youthful appearance, there is a need for better and more effective cosmetics. New concepts such as *cosmeceuticals* and integration of cosmetics with healthy food and lifestyle upsurged. Thus the old concepts assigning cosmetics only a superficial action are being gradually replaced by the acceptance that a more in-depth penetration of actives is necessary, while keeping at a minimum the toxicological and irritation risks.

Development and proliferation of tuned nanocarriers of different compositions and the possibility to design them to reach specific targets and to deliver different kinds of substances in a controlled manner are changing the state of the art, opening new roads to the formulators and to the cosmetic industry.

Maybe, with time, the very definition of cosmetics should be rediscussed under the light of scientific advancement.

References

[1] Mihranyan A, Ferraz N, Strømme M. Current status and future prospects of nanotechnology in cosmetics. Prog Mater Sci [Internet] 2012;57(5):875–910.

[2] Kaul S, Gulati N, Verma D, Mukherjee S, Nagaich U. Role of nanotechnology in cosmeceuticals: a review of recent advances. J Pharm 2018;1–19 2018.

[3] Escobar-Chávez JJ, Rodríguez-Cruz IM, Domínguez-Delgado CL, Díaz-Torres R, Alma Luisa Revilla-Vázquez AL, Casas Aléncaster N. Nanocarrier systems for transdermal drug delivery. Recent advances in novel drug carrier systems [Internet]. InTech; 2012. p. 41. Available from: <http://www.intechopen.com/books>.

[4] Regulation (CE) No. 1223/2009 of the European Parliament and of the Council of 30 November 2009 on cosmetic products. Off J Eur Union 2009; 342:59–209.

[5] Schagen SK. Topical peptide treatments with effective anti-aging results. Cosmetics 2017;4(2).

[6] Watkinson AC, Bunge AL, Hadgraft J, Lane ME. Nanoparticles do not penetrate human skin—a theoretical perspective. Pharm Res 2013;30(8):1943–6.
[7] Bos JD, Meinardi MMHM. The 500 Dalton rule for the skin penetration of chemical compounds and drugs. Exp Dermatol 2000;9(3):165–9.
[8] Bouwstra JA, Gooris GS, Ponec M. Skin lipid organization, composition and barrier function. IFSCC Mag. 2007;10(4):297–307.
[9] Abramovits W, Granowski P, Arrazola P. Applications of nanomedicine in dermatology: use of nanoparticles in various therapies and imaging. J Cosmet Dermatol 2010;9(2):154–9.
[10] Barry BW. Novel mechanisms and devices to enable successful transdermal drug delivery. Eur J Pharm Sci 2001;14(2):101–14.
[11] Rizwan M, Aqil M, Talegaonkar S, Azeem A, Sultana Y, Ali A. Enhanced transdermal drug delivery techniques: an extensive review of patents. Recent Pat Drug Deliv Formul 2009;3(2):105–24.
[12] Papakostas D, Rancan F, Sterry W, Blume-Peytavi U, Vogt A. Nanoparticles in dermatology. Arch Dermatol Res 2011;303(8):533–50.
[13] Schroeter A, Engelbrecht T, Neubert RH, Goebel AS. New nanosized technologies for dermal and transdermal drug delivery: a review. J Biomed Nanotechnol 2010;6(5):511–28.
[14] Neubert RH. Potentials of new nanocarriers for dermal and transdermal drug delivery. Eur J Pharm Biopharm 2011;77(1):1–2.
[15] Hadgraft J, Lane ME. Skin permeation: the years of enlightenment. Int J Pharm. 2005;305(1–2):2–12.
[16] Jacobi U, Kaiser M, Toll R, Mangelsdorf S, Audring H, Otberg N, et al. Porcine ear skin: an in vitro model for human skin. Skin Res Technol 2007;13(1):19–24.
[17] Alvarez-Román R, Naik A, Kalia YN, Guy RH, Fessi H. Skin penetration and distribution of polymeric nanoparticles. J. Control Release 2004;99:53–62.
[18] Teichmann A, Ossadnik M, Richter H, Sterry W, Lademann J. Semiquantitative determination of the penetration of a fluorescent hydrogel formulation into the hair follicle with and without follicular closure by microparticles by means of differential stripping. Skin Pharmacol Physiol. 2006;19:101–5.
[19] Escobar-Chávez JJ, López-Cervantes M, GanemRondero A. Conventional methods for cutaneous drug sampling. In: Narasimha Murthy, editor. Dermatokinetics of therapeutic agents. Boca Raton, FL: CRC Press, Taylor and Francis Group; 2011. p. p. 81–130.
[20] Escobar-Chávez JJ, Melgoza-Contreras LM, López-Cervantes M, et al. The tape stripping technique as a valuable tool for evaluating topical applied compounds. In: Caldwell Gary W, Atta-ur-Rahman, Yan Z, Iqbal Choudhary M, editors. Frontiers in Drug Design & Discovery, Vol. 4. Utrech: Bentham Science Publishers; 2009. p. p. 189–227.
[21] Teichmann A, Jacobi U, Ossadnik M, et al. Differential stripping: determination of the amount of topically applied substances penetrated into the hair follicles. J Investig Dermatol 2005;125:264–9.
[22] Potts RO, Guy RH. Predicting skin permeability. Pharm Res 1992;9(5):663–9.
[23] Cevc G, Vierl U. Nanotechnology and the transdermal route. A state of the art review and critical appraisal. J Control Release 2010;141(3):277–99.
[24] Xiang TX, Anderson BD. Influence of chain ordering on the selectivity of dipalmitoyl phosphatidylcholine bilayer membranes for permeant size and shape. Biophys J 1998;75:2658–71.
[25] Geinoz S, Guy RH, Testa B, Carrupt PA. Quantitative structure–permeation relationships (QSPeRs) to predict skin permeation: a critical evaluation. Pharm Res 2004;21:83–92.
[26] Cevc G. Drug delivery across the skin. Expert Opin Investig Drugs 1997;6:1887–937.
[27] Schätzlein A, Cevc G. Non-uniform cellular packing of the stratum corneum and permeability barrier function of intact skin: a high-resolution confocal laser scanning microscopy study using highly deformable vesicles (Transfersomes). Br J Dermatol 1998;138:583–92.
[28] Mitragotri S. Modeling skin permeability to hydrophilic and hydrophobic solutes based on four permeation pathways. J Control Release 2003;86:69–92.
[29] Aguilella V, Kontturi K, Murtomiiki L, Ramírez P. Estimation of the pore size and charge density in human cadaver skin. J Control Release 1994;32:249–57.
[30] Johnson ME, Blankschtein D, Langer R. Evaluation of solute permeation through the stratum corneum: lateral bilayer diffusion as the primary transport mechanism. J Pharm Sci 1997;86:1162–72.
[31] Strauss G. Liposomes: from theoretical model to cosmetic tool. J Soc Cosmet Chem 1989;40(1):51–60.

[32] Magdassi S. Delivery systems in cosmetics. Colloids Surf A Physicochem Eng Asp 1997;123–124:671–9.
[33] Angajala G, Subashini R. Transdermal nanocarriers: new challenges and prospectives in the treatment of *Diabetes mellitus*. J Chem Appl Biochem 2014;1(2):1–10.
[34] Kumar GP, Rao PR. Ultra deformable niosomes for improved transdermal drug delivery: the future scenario. Asian J Pharm Sci 2012;7(2):96–109.
[35] Eswar Kumar A. A review on novel trends in vesicular carrier for enhanced transdermal delivery: ethosomes. IJSR Int J Sci Res 2017;4:268–70.
[36] Aggarwal D, Nautiyal U. Ethosomes: a review. Int J Pharm Med Res 2016;4(4):354–63.
[37] Cevc G, Gebauer D, Stieber J, Schätzlein A, Blume G. Ultraflexible vesicles, transfersomes, have an extremely low pore penetration resistance and transport therapeutic amounts of insulin across the intact mammalian skin. Biochim Biophys Acta 1998;1368(2):201–15.
[38] Rieger MM. Current aspects of cosmetic science. III. Liposomes and their uses. Cosmet Toilet 1981;96 (8):35–8.
[39] Handjani-Vila RM, Ribier A, Rondot B, Vanlerberghie G. Dispersions of lamellar phases of non-ionic lipids in cosmetic products. Int J Cosmet Sci 1979;1:303–14.
[40] Sebag H, Vanlerberghe G Bio-mimetic surface-active agents. In: XIII Congress of the International Federation of Cosmetic Chemists (IFSCC), Buenos Aires; 1984. p. 111–27.
[41] Montenegro L. Nanocarriers for skin delivery of cosmetic antioxidants. J Pharm Pharmacogn Res 2014;73–92.
[42] Fočo A, Gašperlin M, Kristl J. Investigation of liposomes as carriers of sodium ascorbyl phosphate for cutaneous photoprotection. Int J Pharm 2005;21–9.
[43] Gopinath D, Ravi D, Rao BR, Apte SS, Renuka D, Rambhau D. Ascorbyl palmitate vesicles (Aspasomes): formation, characterization and applications. Int J Pharm 2004;271(1–2):95–113.
[44] Montanari J, Morilla MJ, Romero L. Ultradeformable liposomes for topic application of cosmetic actives can modulate their destination across the stratum corneum. In: AAQC, editor. 26th IFSCC Congress, Buenos Aires. 2010. p. 45.
[45] Montanari J, Vera M, Mensi E, Morilla M, Romero E. Nanoberries for topical delivery of antioxidants. J Cosmet Sci 2013;64(6):469–81.
[46] Higa LH, Schilrreff P, Perez AP, Iriarte MA, Roncaglia DI, Morilla MJ, et al. Ultradeformable archaeosomes as new topical adjuvants. Nanomed Nanotechnol Biol Med 2012;8(8):1319–28.
[47] Caimi AT, Parra F, de Farias MA, Portugal RV, Perez AP, Romero EL, et al. Topical vaccination with super-stable ready to use nanovesicles. Colloids Surf B Biointerfaces 2017;152:114–1123.
[48] Müller RH, Radtke M, Wissing SA. Solid lipid nanoparticles (SLN) and nanostructured lipid carriers (NLC) in cosmetic and dermatological preparations. Adv Drug Deliv Rev 2002;S131–55.
[49] Müller RH, Petersen RD, Hommoss A, Pardeike J. Nanostructured lipid carriers (NLC) in cosmetic dermal products. Adv Drug Deliv Rev 2007;522–30.
[50] Souto EB, Müller RH. Cosmetic features and applications of lipid nanoparticles (SLN®, NLC®). Int J Cosmet Sci 2008;30(3):157–65.
[51] Pardeike J, Hommoss A, Müller RH. Lipid nanoparticles (SLN, NLC) in cosmetic and pharmaceutical dermal products. Int J Pharm 2009;170–84.
[52] Purohit DK, Nandgude TD, Poddar SS. Nano-lipid carriers for topical application: current scenario. Asian J Pharm 2016;10(1):S1–9.
[53] Lademann J, Richter H, Schanzer S, Knorr F, Meinke M, Sterry W, et al. Penetration and storage of particles in human skin: perspectives and safety aspects. Eur J Pharm Biopharm 2011;77(3):465–8.
[54] Zannoni V, Quintana Lazópulos SS, Defain Tesoriero MV, Peter Gauna R, Ferreira A, Hermida L, et al. Carriers lipídicos para la elaboración de productos cosméticos anti-envejecimiento con perfluorodecalina. In: Proceedings of XXII COLAMIQC, Punta del Este, Uruguay; 2015.
[55] McClements DJ. Nanoemulsions versus microemulsions: terminology, differences, and similarities. Soft Matter 2012;8(6):1719–29.
[56] Rai VK, Mishra N, Yadav KS, Yadav NP. Nanoemulsion as pharmaceutical carrier for dermal and transdermal drug delivery: formulation development, stability issues, basic considerations and applications. J Control Release 2018;270:203–25.

[57] Akram S, Alamilla-Beltrán L, Álvarez-Bermúdez, Anton N, Bourbon AI, Chung C, Dons F. Nanoemulsions. In: Jafari SM, McClements DJ, editors. Academic Press; 2018. 642 pp.
[58] Bernard AL, Ikeda Y, El Akkari R, Simmonet JT Cosmetic composition. International application published under the Patent Cooperation Treaty (PCT): Loreal [FR/FR]; 14, Rue Royale, F-75008 Paris, France. WO 2014/098264 Al; 2014. p. 1–49.
[59] Butz T, Reinert T, Pinheiro T, Moretto P, Pallon J NANODERM quality of skin as a barrier to ultra-fine particles—final report; 2007 (November):55.
[60] CTFA Monograph ID: 2793, Personal Care Products Council, Washington, D.C.
[61] Svarc F, Daraio ME, Arnejo N, Carballo O, Debandi M, François N, et al. Anticellulitic microspheres: in vitro assessment of release behavior. 26th IFSCC Congress, Buenos Aires. AAQC; 2010. p. 263–4.
[62] Svarc F, Daraio M, Arnejo N, Carballo O, Blanco A, François N, et al. Microesferas con activos anticelulíticos: cinética de liberación a través de membranas que simulan el estrato córneo humano. XIX COLAMIQC. Guayaquil: Sociedad Ecuatoriana de Químicos Cosméticos; 2009.
[63] Ibrahim IAM, Zikry AAF, Sharaf MA. Preparation of spherical silica nanoparticles: stober silica. J Am Sci 2010;6(11):985–9.
[64] Chin LS, Lim M, Hung TT, Marquis CP, Amal R. Perfluorodecalin nanocapsule as an oxygen carrier and contrast agent for ultrasound imaging. RSC Adv. 2014;4(25):13052–60.
[65] Stanzl K, Zastrow L, Röding J, Artmann C. The effectiveness of molecular oxygen in cosmetic formulations. Int J Cosmet Sci 1996;18(3):137–50.
[66] Svarc FE, Ranocchia RP, Perullini M, Jobbágy M, Bilmes SA. A new route to obtain perfluorodecalin nanocapsules as an oxygen carrier in cosmetic formulations. J Dermatol Study Treat 2018;(1):1–9 2018.
[67] <https://www.cir-safety.org/ingredients>.
[68] Lee CC, MacKay JA, Fréchet JMJ, Szoka FC. Designing dendrimers for biological applications. Nat Biotechnol 2005;23(12):1517–26.
[69] <https://www.polymerfactory.com/dendritic-materials>.
[70] Janaszewska A, Lazniewska J, Trzepiński P, Marcinkowska M, Klajnert-Maculewicz B. Cytotoxicity of dendrimers. Biomolecules. 2019;9(8):330.
[71] Chauhan AS. Dendrimers for drug delivery. Molecules. 2018;23(4).
[72] Kraeling MEK, Topping VD, Belgrave KR, Schlick K, Simanek E, Man S, et al. In vitro skin penetration of dendrimer nanoparticles. Appl Vitr Toxicol 2019;5(3):134–49.
[73] Debowska R, Rogiewicz K, Iwanenko T, Kruszewski M, Eris I. Folic acid (folacin)—New application of a cosmetic ingredient. Kosmet Medizin 2005;26(3):123–9.
[74] Knott A, Koop U, Mielke H, Reuschlein K, Peters N, Muhr GM, et al. A novel treatment option for photoaged skin. J Cosmet Dermatol 2008;7(1):15–22.
[75] Fischer F, Achterberg V, März A, Puschmann S, Rahn CD, Lutz V, et al. Folic acid and creatine improve the firmness of human skin in vivo. J Cosmet Dermatol 2011;10(1):15–23.
[76] Essa S Development and characterization of polymeric nanoparticles (NPs) made from functionalized poly (D, L-lactide) (PLA) polymers [Thesis]. Montreal: Universite de Montreal; 2011.
[77] Diz VE, Leyva G, Zysler RD, Awruch J, Dicelio LE. Photophysics of an octasubstitutedzinc(II) phthalocyanine incorporated into solid polymeric magnetic and non-magnetic PLGA-PVA nanoparticles. J Photochem Photobiol A Chem 2016;316:44–51.
[78] Tsukada Y, Sasai A, Tsujimoto H, Yamamoto H, Kawashima Y. PLGA nanosphere technology for novel nanomedicine and functional cosmetics. Nanoparticle technology handbook. Elsevier; 2018. p. 461–7.
[79] Shetty PK, Venuvanka V, Jagani HV, Chethan GH, Ligade VS, Musmade PB, et al. Development and evaluation of sunscreen creams containing morin-encapsulated nanoparticles for enhanced UV radiation protection and antioxidant activity. Int J Nanomed 2015;10:6477–91.
[80] da Silva ALM, Contri RV, Jornada DS, Pohlmann AR, Guterres SS. Vitamin K1-loaded lipid-core nanocapsules: physicochemical characterization and in vitro skin permeation. Skin Res Technol 2013;19(1):10–11.
[81] Urgen MK. Patent application publication (10) Pub. No.: US 2010/0224498A1.; 2010; Vol. 1.
[82] Munnier E, Eddaoudi I, Perse X, Chourpa I. Novel fully characterized polysaccharide nanocarriers: a tool for skincare innovation. In: IFSCC, editor. 28th IFSCC Congress. Paris: Société Française de Cosmetologie (SFC); 2014. p. 3353–62.

[83] Farah F, Doloff C, Anderson DG LR Crosslinked chitosan nanoparticles. In: Polymeric drug delivery techniques translating polymer science for drug delivery; 2013. p. 18–21.
[84] Aliasghari A, Rabbani Khorasgani M, Vaezifar S, Rahimi F, et al. Evaluation of antibacterial efficiency of chitosan and chitosan nanoparticles on cariogenic streptococci: an in vitro study. Iran J Microbiol 2016;8:93–100.
[85] Sanko N, Escudero C, Sediqi N, Smistad G, Hiorth M. Fluoride loaded polymeric nanoparticles for dental delivery. Eur J Pharm Sci 2017;104:326–34.
[86] Matos BN, Reis TA, Gratieri T, Gelfuso GM. Chitosan nanoparticles for targeting and sustaining minoxidil sulfate delivery to hair follicles. Int J Biol Macromol 2015;75:225–9.
[87] Cerqueira-Coutinho C, Santos-Oliveira R, dos Santos E, Mansur CR. Development of a photoprotective and antioxidant nanoemulsion containing chitosan as an agent for improving skin retention. Eng Life Sci 2015;15:593–604.

CHAPTER

4

Nanoemulsions for cosmetic products

Ana Catarina Faria-Silva, Ana Margarida Costa, Andreia Ascenso, Helena Margarida Ribeiro, Joana Marto, Lídia Maria Gonçalves, Manuela Carvalheiro and Sandra Simões

Nanostructured Systems for Overcoming Biological Barriers (Nano2B) Group, Research Institute for Medicines (iMed. ULisboa), Faculty of Pharmacy, Universidade de Lisboa, Lisboa, Portugal

4.1 Introduction

Nanoemulsions containing products are a growing market in cosmetics as they allow an enhanced delivery of active compounds [1]. As carriers they can move through the stratum corneum (SC), promoting the delivery of actives into the dermis [2]. Although nanoemulsions present several advantages, they are not widely used due to the expensive equipment needed for their preparation and the difficulty in understanding the production mechanism and their stabilization [3].

Furthermore, a nanoemulsion is sometimes called "submicron emulsion" by cosmetic companies due to the fact that the term "nano" is frequently associated with harmful effects. However, this misconception is not based on scientific evidences. The ingredients used in nanoemulsions are safe and they cannot penetrate the skin as intact particles or achieve deep skin penetration. Therefore the concern should be focused on the chemical composition, regardless of the size.

The skin acts as an effective barrier against external aggressions due to its anatomy and physiology and it is also very important to maintain body homeostasis [4,5]. The skin is constituted by different layers: the SC, the viable epidermis, the dermis, and the subcutaneous tissue (Fig. 4.1). The SC controls drug penetration, being the main barrier to topical delivery. Only compounds with less than 500 Da, and with adequate hydrophobicity, can cross the SC successfully [4]. There are three possible penetration routes through the skin barrier for topically applied products (Fig. 4.1): (1) the transcellular pathway, by direct transportation of the active compound to the cells through the lipid bilayers and corneocytes; (2) the intercellular pathway, when the active compound diffuses through the SC by

Nanocosmetics
DOI: https://doi.org/10.1016/B978-0-12-822286-7.00004-8

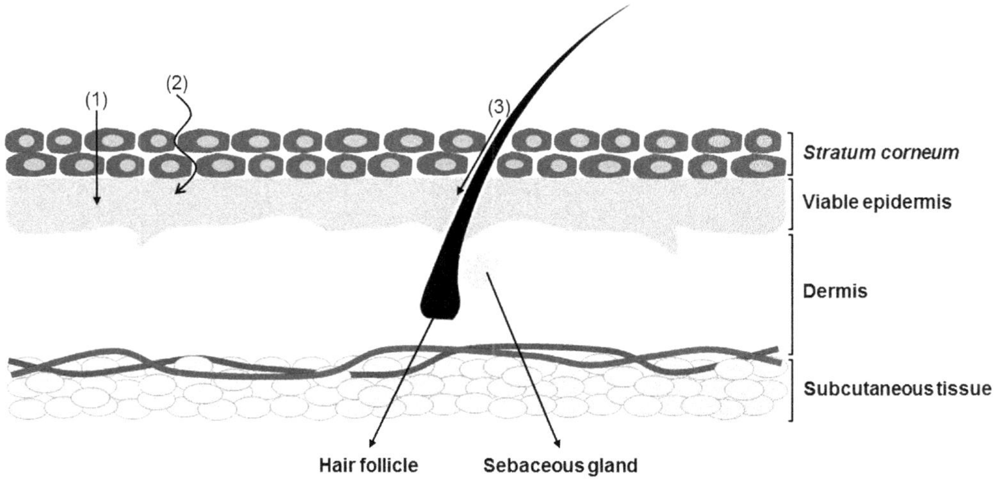

FIGURE 4.1 Schematic representation of the different skin layers with the possible penetration routes for topically applied products: (1) transcellular pathway; (2) intercellular pathway; and (3) follicular pathway.

the lipid layers around the corneocytes; (3) the follicular pathway via hair follicles and the associated sebaceous glands and sweat ducts, as there are increased blood capillaries around them, helping in the active compound penetration and accumulation in the skin [5,6].

Considering all these properties of the skin barrier, a lipid-based formulation with a low size would be the most suitable product for topical applications, namely, nanoemulsions that will be described in detail in this chapter.

4.2 Emulsion delivery systems in cosmetics

4.2.1 Emulsion generalities

Emulsions are colloidal systems, in which one liquid is dispersed in another liquid. These two-phase systems are based on the dispersion of fine droplets of small sizes as internal phase (also designed as disperse, dispersed, or discontinuous phase) in an external phase (or continuous phase) [7,8]. Emulsions are constituted by an oil and water phase, whose miscibility between both is only possible by the addition of a suitable emulsifying agent. When an oil is dispersed in water, it forms an oil-in-water (O/W) emulsion, but when water is dispersed in oil, it forms a water-in-oil (W/O) emulsion. The size of these systems presents a high heterogenicity, and the diameter of the droplets may range between 0.1 and 50 μm [7,8].

Emulsions are currently of great interest in the pharmaceutical industry because of their versatility in delivering both lipophilic and hydrophilic drugs, either in semisolid (creams) or in fluidic pharmaceutical preparations (lotions and liniments) [9]. According to the type of emulsion (O/W or W/O), different preparations for skin care can be achieved. Usually,

W/O emulsions present a better occlusive effect due to their ability in reducing the water evaporation in the upper layers of SC [10,11]. These types of emulsions also present good emollient properties, and therefore they can improve the barrier function of the skin, in cases of eczema, psoriasis, or atopic dermatitis [8,12]. Nevertheless, W/O emulsions possess a high content of oil and a greasy texture, being not well accepted by the consumers. On the other hand, O/W emulsions have a better acceptability due to low grease and less sticky feeling when applied to the skin. Due to low oil content, O/W emulsions are absorbed faster and the formulations can be easily washed from the skin when compared with W/O emulsions [9,13]. Despite the skin hydration effect provided by the O/W emulsions, they present limited lipid-replenishing [10].

The conventional emulsions used for cosmetic applications are usually called as macroemulsions; however, there are more complex dispersed systems that also form emulsions, namely, multiple emulsions, Pickering emulsion, microemulsions, and nanoemulsions.

4.2.2 Microemulsions and nanoemulsions—aren't both nanosystems?

Microemulsions and nanoemulsions are both colloidal systems that result from the dispersion of two immiscible liquids, and the formed dispersed system presents fine droplets with sizes below 100 and 200 nm, respectively [14,15].

Hoar and Schulman [16] hypothesized that a mixture of oil, water, alcohol, and cationic surfactant formed a transparent oil-continuous system. Later, in 1959, Schulman et al. [17] suggested the formation of microemulsions, by dispersing uniform and spherical droplets, either of water or oil, into an appropriate continuous phase.

Microemulsions have been considered as a good approach for the delivery of either hydrophilic or lipophilic drugs through the skin barrier. These isotropic systems are characterized as being clear or translucent and present a fluid consistency. They are constituted by a blend of oil, water, surfactant, and cosurfactant, whose internal phase droplets present a uniform size below 100 nm, contrasting with the conventional emulsions that form milky and coarse dispersions with a broad droplet size [8,15]. Another important characteristic of microemulsions is their thermodynamic stability, which is due to the high amount of surfactant (usually higher than 20%), allowing its spontaneous formation [15,18]. Also, the interfacial region is flexible, allowing the curvature that is needed to surround small particles or to allow the transitions from oil-continuous to water-continuous structures, which is characteristic of microemulsions [19].

The definition of microemulsions has become confused with the introduction of nanoemulsions, a colloidal system characterized by a similar droplet size of the dispersed phase (below 200 nm) and the same macroscopic aspect (clear or translucent) as the microemulsions. The main difference between both systems is the amount of surfactant: nanoemulsions are constituted by a lower amount of surfactants compared to the microemulsions (between 3% and 10% [18]), being considered thermodynamically unstable. However, for its formation, energy input methods are required to achieve a kinetically stable formulation [15]; therefore the destabilization of nanoemulsions is considerably slow (months), the Ostwald ripening being the most probable process responsible for their destabilization [20,21]. Another difference between both systems is the behavior when exposed to temperature and water: microemulsions are affected by exposure to water and temperature variation, while nanoemulsions remain stable [4,20].

4.3 Formulation and production of nanoemulsions

In cosmetics, O/W nanoemulsions have been studied more than W/O. Nevertheless, in parallel with the preparation parameters to consider, the composition of the system must be carefully selected. The choices of the oil, or the surfactant-to-oil ratio, among others, are critical parameters in nanoformulation production. The most appropriate method depends mostly on system composition or scale-up requirements. Table 4.1 presents the main nanoemulsion components. Examples of each type of component are given.

Oils are mainly responsible for the emolliency of the nanoemulsions while solubilizing the hydrophobic active molecules.

Surfactants play a critical role in droplet size reduction and in the stability of the nanoemulsion. Nanoemulsions for cosmetic use can be prepared using a large range of surfactants and/or surfactant combinations, especially in the case of nanoemulsions obtained by high-energy homogenization methods. Table 6.1 presents the more traditional nonionic surfactant examples as well as more recent amphiphilic molecules. The selection of the surfactant should consider the effective reduction of the interfacial tension and simultaneously the absence of skin irritancy. For this reason, nonionic surfactants are preferred in cosmetics manufacturing.

Cosurfactants are ingredients commonly used in classical emulsions and in a variety of cosmetic products. Usually polyols, they are able to refine nanoemulsion as they facilitate droplet size reduction and simultaneously act as humectants, solvents for active ingredients, or even penetration enhancers for such active molecules [4]. These cosurfactants do not impact negatively the rheology of the final preparations.

The aqueous phase is also important in nanoemulsion stability as it may include buffer salts, preservatives, viscosity enhancing agents such as carbomers and, in the presence of active ingredients, it may contain penetration enhancers to enhance skin delivery [44].

Nanoemulsions are kinetically stable and thermodynamically unstable and highly dependent on the process of nanoscale droplet formation. Due to the low amount of surfactant, nanoemulsions require high-energy input methodologies to achieve a stable formulation [15]; nevertheless, low-energy techniques can still be used for their preparation [45]. For cosmetic purposes, nanoemulsions can be thus manufactured either by high-energy mechanical dispersion, in which a mechanical device is used, or by low-energy methods, a physicochemical process also designated as spontaneous emulsification. Table 4.2 summarizes the different techniques used to manufacture nanoemulsions identifying the advantages and the limits of each approach.

High-energy preparation methods are commonly used at the laboratory and especially at the industry scale and are less dependent on the composition of the nanoemulsion system [46]. It means that to achieve submicron droplet sizes the range of the composition variables when using mechanical means [47] is bigger than when using spontaneous emulsification, a phenomenon that requires no high-energy input but involves selecting

TABLE 4.1 Chemical categories used in nanoemulsions for cosmetics: type and examples of main nanoemulsion components.

Component		Type	Examples	References
Oil		Triglyceride	Caprylic/capric triglycerides, avocado oil, apricot kernel oil, coconut oil	[22–25]
		Terpenes	Limonene	[26]
		Fatty acids	Oleic acid	[27]
		Esters	Isopropyl myristate, isopropyl palmitate, isosteary neopentanoate	[28,29]
		Alkanes	Isododecane, isohexadecane, vaseline, parleam	[30]
Surfactant	Nonionic surfactants	Etoxilated alkyl alcohols/alkyl acids	PEO-8 isostearate, PEO-20 stearate, Steareth-10	[31]
		Sucrose alkyl ester	Cetearyl glucoside, sucrose distearate	[23]
		Sorbitan alkyl ester	Polysorbate 20, Polysorbate 61	[32]
		Polyglycerol alkyl ester	Decaglycerol monostearate	[33]
	Mixtures of fatty alcohols and surfactants	Phosphoric alkyl	K cetyl phosphate, Trilaureth-9 citrate	[34]
		Citric alkyl ether		
	Phospholipids and peptide-based gemini surfactant	Lecithin	Soybean lecithin	[35,36]
		Gemini	Dilauramidoglutamide lysine	
	Ionic surfactants	Alkyl trimethyl ammonium	Behenyl trimethylAmmonium chloride	[37]
		Alkyl amido propyl trimethyl ammonium		
	Amphiphilic oligomers	Amphiphilic oligomer	Poloxamer 231	[38]
	Silicones		Inulin lauryl carbamate	[39]
Cosurfactant		Polyols	Glycerin, dipropylene glycol, PEG300, PEG400, poloxamer	[35,40]
Antioxidants			Ascorbic acid, alpha-tocopherol	[41]
pH adjusting agent			Sodium hydroxide or hydrogen chloride	[42]
Preservatives			Methyl paraben, propyl paraben	[43]

TABLE 4.2 Nanoemulsion preparation methods, advantages and limitations.

Energy	Emulsification method	Principle of the method	Advantages	Limitations
High	High-pressure homogenization (HPH)	Shear, collision, cavitation	Flexibility on the composition; low process time	High cost; not recommended for sensitive products
	Microfluidization	High-pressure injection	Controlled size droplets	High cost; not recommended for large scale
	Ultrasonication	Cavitation	Flexibility on the composition; Less expensive than HPH	Small batches
Low	Phase inversion temperature (PIT)	Cooling	Low cost; easy to scale-up	Limited to nonionic surfactants
	Phase inversion composition (PIC)	Dilution	Low cost; easy to scale-up	Requires titration
	Spontaneous emulsification	Dispersions and condensation	Low cost; easy to scale-up	Limited amount of oil
	Solvent displacement	Diffusion of organic solvent	Low cost; easy to scale-up	Organic solvents

ingredients that favor spontaneous emulsification. However, high-energy homogenization methods require specific and costly equipment often not available at the laboratory level.

High-pressure homogenization is a technique that combines hydraulic sheer, cavitation, and turbulence to produce nanoemulsions with submicron size [48,49]. To reduce the particle size, intense shear forces generated within the homogenizer and pressure are generally applied in two stages, starting with high-pressure cycles followed by low-pressure ones (Niro Soavi Inc., Niro Soavi: VHP Homogenizer Technology, available at www.niro-soavi.it). Various companies manufacture high-pressure homogenizers and the final product is determined by the model used [50]. The size is controlled mainly by the applied pressure and the number of homogenization cycles in each stage. Moreover, the temperature of homogenization influences the droplet size with higher temperatures being responsible for smaller sizes [50,51].

Microfluidization is another high-energy technique that uses a microfluidizer device. In this method, the coarse emulsion is forced to pass between small channels with a specific configuration, under the influence of high pressure produced by a displacement pump, until an impingement area, where fine particles are produced. Similarly, with high-pressure homogenization, the desirable physicochemical properties [size and broadness of the size distribution—polydispersity index (PdI)] can be achieved by manipulating the operator pressure and by passing several times the coarse emulsion through the microfluidizer [48,52].

Nanoemulsions can also be produced by the ultrasonic emulsification technique. This technique uses a probe that emits ultrasonic waves to disintegrate the coarse emulsions. The droplet size can be controlled by varying the time and the energy input [48], as well as by controlling the oil and surfactant concentrations and the oil–surfactant ratio [53]. High-pressure homogenization and microfluidization can be used to prepare nanoemulsions at a laboratory and industry scale, while the ultrasonic emulsification is mainly used in a laboratory setting [48].

Low-energy techniques present advantages for the industry since they do not require specialized and costly equipment and the energy input for manufacturing microemulsions and nanoemulsions is considered low. To achieve a stable formulation using low-energy methodologies, higher amounts of surfactant may be needed, and the selection of oils and surfactants may also be limited [54].

Spontaneous emulsification for the production of nanoemulsions was first reported by Taylor and Ottewill [55]. It was proposed that through the dilution of an O/W microemulsion with water, the cosurfactant can diffuse throughout the water-continuous phase, achieving a nanoemulsion, which is no longer thermodynamically stable [55]. Upon the dilution of a self-emulsification system, a rapid diffusion of water-miscible compounds (surfactant, cosurfactant, or solvents) from the dispersed phase toward the continuous phase occurs, without modification of the curvature of surfactant [18,45]. Spontaneous emulsification can be used for the preparation of nanoemulsions by the solvent displacement method, which involves the preparation of an organic solution containing oil, a lipophilic surfactant, and a water-miscible solvent (usually ethanol or acetone) and an aqueous solution containing surfactants. The organic phase is then added to the aqueous phase forming an O/W nanoemulsion under magnetic stirring. In this second step, the diffusion of solvent from the organic phase toward the aqueous phase originates the formation of a nanodroplet. Lastly, the water-soluble solvent is removed by evaporation and the nanodroplets of oil are dispersed in an aqueous solution [18].

Nanoemulsions can also be produced by spontaneous emulsification, but through the phase inversion composition (PIC) method. This technique involves the continuous addition of water to a blend of oil–surfactant under continuous stirring. Initially, water is dispersed into the oily phase forming W/O emulsion, but with an increase of water fraction, the spontaneous curvature of surfactant changes and a transition to an O/W emulsion occur. In this case, water-miscible solvents are not used [56–58].

Similarly to microemulsions, nanoemulsions can be prepared through the phase inversion temperature (PIT) technique. According to McClements [59], spontaneous emulsification and PIT techniques can produce either microemulsions or nanoemulsions, depending on the surfactant–oil ratio: high ratios originate microemulsions, while low surfactant–oil ratios form nanoemulsions. In the PIT method, oil, water, and nonionic surfactants are mixed together at room temperature and typically generate O/W emulsions. Upon applying heat gradually, the nonionic surfactant becomes lipophilic and finally is solubilized in the oily phase: the initial O/W emulsion undergoes phase inversion to W/O emulsion.

4.4 Characterization of nanoemulsions

One of the main concerns related to the use of emulsions is its stability. Stability of emulsions is related to the particle size of the internal phase and its distribution, the density of aqueous and oily phases, and the integrity of the dispersed phase [14]. Emulsions can lose their stability by different processes: phase inversion, flocculation, coalescence, creaming, Ostwald ripening, and cracking [7].

Tests and methodologies for quality control of dispersed systems, as other pharmaceutical forms, are regulated and fully described in the pharmacopoeias and in the literature (Table 4.3). In addition, toxicity and efficacy tests should be performed before the more specific skin delivery studies for topical nanoemulsions. The latter includes the release, permeation, and penetration of bioactive ingredients through synthetic membranes or skin models, using Franz diffusion cell apparatus and confocal scanning laser microscopy, as recommended by OECD Guideline 428 [60]. Stability tests on the final packaging should also be carried out in accordance with the ICH (International Conference on Harmonization of Technical Requirements) Q1A standards. The natural skin surface pH is on average around 5 and should be considered when preparing topical formulations [61].

Quality control tests of nanoemulsion-derived products are carried out at predefined times for different conditions of temperature and humidity correspondent to the climatic zone. Accelerated stability studies can be conducted during a shorter period for shelf-life determination. Thermodynamic stability studies are usually carried out as follows: (1) heating-cooling cycles; (2) centrifugation; and (3) freeze-thaw cycles. The formulations that remain stable under these conditions are then subjected to dispersibility studies for evaluating the efficiency of self-emulsification [62]. The incorporation of labile compounds in nanoemulsions has increased their shelf-life at room temperature as reported for silymarin [63].

Regardless of composition and method of preparation, nanoemulsions as other emulsions are characterized by physicochemical, pharmacotechnical, and biological methods. Accordingly, organoleptic properties; entrapment efficiency (EE = bioactive content incorporated in formulation/total amount of bioactive added × 100) and loading efficiency (LE = bioactive content incorporated in formulation/total formulation weight × 100); density; viscosity and texture; droplet size and morphology; ZP; surface tension; friccohesity; refractive index; percent transmittance; dispersibility; pH; osmolarity, among others, are the main parameters evaluated in those quality control assays [62].

In most cases, nanoemulsions are liquid translucent dispersions in which the refractive index is similar to that of water and transmittance is >99% [49,62]. Due to the complex structure, these systems may present alterations that are usually evaluated first by their organoleptic characteristics and later by the analysis of their structure using more sensitive methods. In fact, the thermodynamic instability phenomena are quite dependent on nanoemulsion type and composition. For example, coalescence and Ostwald ripening (coarsening of emulsion droplets over time) are commonly favored by short distances between globules and a considerable water solubilization in the oily phase (with high polarity) in W/O nanoemulsions. A molar fraction of the "insoluble" oil in droplets above the critical value as well as very small droplet sizes and low volume fractions are favorable conditions to prevent destabilization mechanisms, such as flocculation and coalescence with consequent sedimentation/creaming and separation phases, respectively [64]. Sonneville-Aubrun et al. [51] have theoretically and experimentally determined the critical molar fraction of the "insoluble" oil. Experimental data revealed that 50% isocetyl stearate in weight relative to the oily phase was required to stabilize a 40 nm nanoemulsion

TABLE 4.3 Quality control assays of dispersed systems.

Assay	Method	Experimental conditions
Organoleptic properties	Macroscopic and sensorial analysis by the operator	NA
Chemical assays		
Bioactive agent identification and quantification (entrapment efficiency, EE%)	UV–visible spectrophotometry or high performance liquid chromatography (HPLC)	Sample diluted against suitable blank and measured at wavelength of peak absorption of the bioactive agent (after separation of nonencapsulated fraction by centrifugation for EE%)
pH	Potenciometry	NA
Physical assays		
Rheology	Viscosimetry	Dynamic viscosity (correspondent to flow curves of shear rate vs shear stress) and oscillatory frequency sweep tests can be performed using a rheometer at different shear rates and shear stress/frequencies, respectively
Texture	Texture profile analysis (TPA)	TPA (hardness, elasticity, compressibility, adhesiveness, and cohesiveness) can be obtained by a texture analyzer with a probe, which is depressed into the sample at a defined rate to a desired depth
Droplet size distribution	Dynamic light scattering (DLS)	Sample diluted (usually in double-distilled water) and measured at predefined conditions correspondent to formulation properties at a constant temperature and angle of detection
	Atomic force microscopy (AFM)	Sample diluted with water followed by drop coating on a glass slide and air dried. Samples are scanned using peak force tapping mode and scan assisted
Droplet morphology	Scanning electron microscopy (SEM)	Sample observed usually at 20 kV
	Transmission electron microscopy (TEM)	Sample negatively stained with 1% aqueous solution of phosphotungstic acid or by dropping 2% uranyl acetate solution onto a 200 μm mesh size Pioloform-coated copper grid or a microscopic carbon-coated grid
		Sample examined at 80 kV
Zeta potential (ZP)	Laser-Doppler anemometry (LDA)	Sample diluted and measured at predefined conditions correspondent to formulation properties at a constant temperature. Zeta potential is estimated from the electrophoretic mobility of droplets

(Continued)

TABLE 4.3 (Continued)

Assay	Method	Experimental conditions
Nanoemulsion type (W/O or O/W)	Dye solubilization Dilutability Conductivity Fluorescence	O/W nanoemulsion is dyed (external phase) with an aqueous dye (under microscopic observation); diluted in water remaining stable; with high conductance and UV fluorescence in spots
Microbiological assays		
Microbiological tests	ISO 16212:2008 ISO 21149:2006	Membrane filtration and direct transfer to an appropriate media for further incubation
	ISO 21148:2005	Results should be below the limits proposed

NA, Not applicable.

containing isododecane. Therefore these structural changes should be detected during stability studies, allowing characterizing the organization of the different phases and interactions between the various ingredients. The methodology applied in these studies includes viscoelastic rheological determinations and other physicochemical techniques, such as differential scanning calorimetry in addition to the analysis of the size, distribution, and morphology of nanoemulsion droplets by laser diffraction or DLS, AFM, (cryo)-TEM and SEM associated with a specific software. This evaluation is fundamental since the droplet size of the (nano)emulsions is strongly affected by the nature and amount of the surfactant used, with a tendency to decrease with an increase of its concentration [49,65]. Homogeneous and stable nanoemulsion dispersion should present droplets at around 20−200 nm in size with PdI < 0.2 and ZP (i.e., particle surface charge) ± 30 mV obtained by photon correlation spectroscopy that analyses the variations in the intensity of scattering by particles due to the Brownian motion as a function of time. In turn, SEM technique provides a three-dimensional image of the droplets using the appropriate accelerating voltage at different magnifications. TEM offers higher resolution for disclosing the size and surface morphology of droplets by bright-field imaging at increasing magnification combined with diffraction modes [49,62]. Small-angle x-ray scattering may also be used to probe nanoemulsion structures [66].

Fourier-transform infrared spectroscopy can be used for assessing interaction between excipients and bioactive agents, bioactive loading, polymerization besides identification of the functional groups of molecules [62].

Regarding the rheology, nanoemulsions may present a wide variety of textures from lotion to gel forms, which could be evaluated using a viscometer or a rheometer. Survismeter is one of the most extensively used equipment as it measures viscosity, interfacial tension and surface tension, contact angle, dipole moment, particle size, and hydrodynamic volumes of the nanoemulsions [62]. Regarding the interfacial tension, "low values correspond to phase behavior, mainly the coexistence of surfactant phase or middle-phase nanoemulsions with aqueous and oily phases in equilibrium" [62]. Consistency is quite important not only for stability but also for efficient release. Nanoemulsions usually exhibit lower apparent viscosity, and therefore a faster release of active ingredients should be expected. Viscosity decreases with higher water content, while it increases with lower amount of surfactant and cosurfactant [49,64].

The characteristics of final nanoemulsion formulations can be modulated. The texture is dependent on the presence of polymeric thickener or gelling agents, both phases' volume and surfactant fractions, and the presence of crystalline lamellar phases. Polymers can also contribute to obtain other textures and sensorial feelings. Gel texture with lower oil content for lighter products with pseudoplastic flow can be quite desirable since it is easier to apply. Microgel (e.g., carbomers) can be used to thicken nanoemulsions at low oil volume fraction even though it can be quite sensitive to ionic components of formulation. At medium to high-volume oil fractions, the microgel nanoemulsion turns white, and with micron-size irregular structures which are visible through microscopy [64]. Ma et al. [67] developed a structured microgel nanoemulsion system constituted by carrageenan and alginate polymers, and with attractive applications for both protection and controlled release of bioactive molecules. Associative (e.g., hydrophobically modified ethoxylated urethane polymers) and nonassociative polymers (e.g., cellulose and gum derivatives) can also thicken nanoemulsions [64].

4.5 Why nanoemulsions in cosmetics?

Nanoemulsions are of a great interest for the cosmetic field. A crystalline lamellar phase could be formed in the bulk and at the droplet surface through fatty alcohol/surfactant associations. This molecular assembly (α-gel) allows the reconstruction of the network structure responsible for obtaining a stable transparent and low viscosity nanoemulsion without polymer, besides creating an occlusive and protective layer onto the skin with hydration benefits [37,51,64–67]. Moreover, due to nanoemulsions capacity of dispersing small droplets of oil, nanoemulsions can be a suitable approach for the delivery of lipophilic active ingredients with interest in the skin care, such as vitamin E [68], lycopene [69], carotenoids [70], or coenzyme Q-10 [71] or simply nonalcoholic fragrances.

The small droplet size of the dispersed phase is another important factor that confers important physicochemical and biological features. It prevents the flocculation and the coalescence of nanoemulsions, and the Brownian motions are sufficient to overcome the sedimentation of the droplets by gravity. Moreover, the reduced droplet size also provides a better skin deposition and penetration of nanoemulsions, allowing an efficient delivery of active ingredients when compared to the conventional vehicles [18,72]. Lastly, the aspect of nanoemulsions (transparent or translucent) is related to the submicron size of the dispersed phase. The aspect combined with the fluidity provides a pleasant sensorial skin feeling that makes nanoemulsions suitable for being used in fluid cosmetic products (lotions or transparent milk) [73–75]. Nanoemulsions flow easily over the skin, with no creaming and glossy coating, without separating into their constituents after the topical application [51]. Additionally, nanoemulsions have a stronger occlusive effect that keeps skin moisture and their active ingredients have faster skin absorption and better penetration in narrow gaps such as pilosebaceous follicles and hair scale spacings. Despite the inherent fluidity of nanoemulsions, its texture and rheology can be manipulated by adding thickening agents or gelling agents, such as starches, carboxymethyl cellulose [76], carbomers [77], or alginates [78].

The composition of nanoemulsions is a relevant aspect that also confers good characteristics for being used in cosmetics. Those systems can be prepared using low amounts of surfactants, which will limit skin irritation [48,73]. For nanoemulsions formulation, it is possible to use emollient oils, which present compatibility with the SC lipids, providing at the same time, skin hydration and comfort. Nanoemulsions containing medium chain triglycerides [68] and isopropyl myristate [71] as components of the organic/oily phase have already been developed. These lipids present emollient properties [79] and are considered safe for topical applications [80,81]. Another emollient used for the preparation of nanoemulsions is the coconut oil, which is mainly constituted by two fatty acids (lauric and myristic acids) [79,82,83], and with the ability to improve the skin hydration in cases of atopic dermatitis and xerosis [84–86].

Lipids with other properties that improve the skin conditions have also been used for the development of nanoemulsions: palm oil esters present moisturizing and nonirritant effect [87,88] and lanolin presents occlusive properties and the ability to retain water in the skin [88,89]. Olive oil is another substance commonly used in cosmetics. The emollient properties of this excipient are controversial; nevertheless, due to the oleic acid content, this excipient causes disorganization of the lipids present in the SC, increasing the skin permeation of substances [84]. This excipient has been used as the main component of the oily phase for several nanoemulsion preparations [90–92], and Tou et al. [93] already demonstrated that olive oil-based nanoemulsions present a higher skin permeation and drug flux through ex vivo models when compared with liquid-paraffin control. Vegetable oils have also been used as a component of the oily phase for the preparation of nanoemulsions. These oils are biocompatible, safe, and present a complex composition of fatty acids that can confer some important biological properties to the formulation, namely, skin hydration, antiinflammatory, antimicrobial, and photoprotection [61,94,95]. Despite the inherent biologic properties of the lipid selected, it is important to mention that the type of oil used for the nanoemulsions preparation represents a great impact on the physicochemical stability of the formulation [96].

Skin hydration is another important feature of these submicron systems. Higher content of water can improve the hydration of SC and consequently the movement of the dispersed droplet through the skin [62]. Yang et al. [97] demonstrated that nanoemulsions could hydrate the keratin and change the structure of the SC, which promoted the percutaneous absorption of substances. Besides that, other types of nanoemulsions (lanolin-based nanoemulsions [79] and rice bran oil-based nanoemulsions [61]) demonstrated an increase of skin hydration, without changing the skin pH value.

An example of the degree of evolution of nanoemulsions application in cosmetics is the incorporation of gamma aminobutyric acid, a neurotransmitter with relaxing muscle properties, in nanoemulsions with the aim of reducing wrinkles [98].

Nanoemulsions can be easily prepared and are feasible of being scale-up [74]. They have the potential to be the formulation of liquid and stable aerosols (cosmetic wet wipes and sprayable products). All these technological advances make nanoemulsions more suitable for skin and hair formulations, in comparison with other systems, such as macroemulsions and microemulsions.

4.5.1 Skin care

La Maison Chanel claimed some years ago that nanoemulsions could improve the efficacy of several products. The Fresh Body Moisturizing Mist "Brume Fraîche pour le Corps" Coco Mademoiselle from Chanel is an O/W nanoemulsion spray. This fluid sprayable nanoemulsion remains homogeneous (without creaming phenomena in the container) due to the nanosized droplets. In addition, the nanoemulsion provides a fast and strong moisturizing effect by means of skin occlusion. Another example is presented by Chanel Précision with "Solution Déstressante," a calming nanoemulsion containing *Centella asiatica* extract as the soothing active ingredient that provides a destressing feeling upon application onto the skin. The immediate perception of this effect is due to the fast skin penetration of the active ingredients driven by the nanoemulsion form. Another type of product where nanoemulsions are intensively used is the wet cosmetic wipes. Evonik developed a self-emulsifying system containing polyglyceryl-4 laurate and dilauryl citrate, which allows the preparation of nanoemulsions by the PIC technology [99]. Heunnemann et al. [100] studied the phase behavior of this self-emulsifying system under dilution using diethylhexyl carbonate in the oily phase. Sinerga also developed a combination of emulsifiers called Nanocream [101] (potassium lauroyl wheat amino acids, palm glycerides, capryloyl glycine), which allows the formulation of O/W nanoemulsions. This product is intended to be used in sprayable emulsions, hyperfluid emulsions, and wet wipes.

4.5.2 Hair care

Nowadays, shampoo formulations are expected to have other functions beyond hair cleansing, such as conditioning, smoothing the hair surface, improving compatibility, and lather creaminess. The moisture preservation and lubrication of hair is due to silicone oil, the main component of conditioners. Silicone oil is hydrophobic; thus it is poorly absorbed by hair. To overcome this issue, stable nanoemulsions containing silicone oil have been successfully developed [102]. The nonionic surfactants Span 80 (Sorbitan Oleate) and Tween 80 (Polysorbate 80) were found to be suitable emulsifiers to prepare oil-in-water nanoemulsions, being stable and enhancing the absorption of silicone oil into the hair surface, in a nanoemulsion size-dependent manner, thus being a promising application for hair care [102]. One of the problems of silicone oil nanoemulsions is the instability of the system. To overcome this, Park et al. [103] developed a method to produce highly stable silicone oil nanoemulsions using amphiphilic triblock copolymers. These silicone oil nanoemulsions could be charged modulated and thus used to form a multilayer emulsion thin film by layer-by-layer deposition. The silicone oil nanoemulsions produced by this method demonstrated to be highly stable and had the ability to electrostatically interact with hair, completely coating the hair surface with a layer of silicone oil.

Cosmetic sciences are often at the frontier of new topical delivery systems development, as modern cosmetics aim at intradermal penetration. An example of this is the use of nanoemulsions on another common hair concern: hair loss. Hair loss can be caused by hormones and genetics as in androgenetic alopecia or by a chronic inflammatory process affecting the hair follicles on a small scalp region as in alopecia areata [104]. Nanoemulsions composed of squalene (sebum lipid) were developed by Aljuffali et al. [104] to deliver active compounds against hair loss to the hair follicles. Diphencyprone (DPCP) is a topical drug for alopecia areata treatment that induces hair regrowth through local immune response. Minoxidil is a drug for treating androgenic alopecia by promoting vasodilation and hair regrowth. These molecules may cause side effects such as skin irritation and risk of systemic absorption. Encapsulation of these compounds in nanoemulsions reduced their side effects, by targeting into the hair follicles and controlling the release [104]. The nanoemulsions containing DPCP and minoxidil with a mean droplet diameter of 194 nm were tested in vitro and in vivo and showed a reduced systemic absorption and a higher follicular uptake for squarticles (nanoparticles formed from sebum-derived lipids) versus control with free active.

4.6 Challenges and future perspectives

Despite the intense multidisciplinary investigation of chemical phenomena underlying the submicrometric droplet production mechanisms and the role of surfactants and cosurfactants, data demonstrating the benefits of using nanoemulsions are scarce compared with microemulsions and classical macroemulsions, in studies dissociated from their commercial promotion.

Nanoemulsions have been used and claimed in many skin and hair care products. Notwithstanding, specific information about these commercial cosmetic products is quite difficult to obtain since regulatory affairs only demand industry to present the INCI list of ingredients on product packaging. This list is insufficient to identify the pharmaceutical form since the ingredients used to formulate macroemulsions and nanoemulsions can be identical. Furthermore, the manufacturing process is usually not disclosed.

Cosmetic manufacturers and raw material producers are becoming more open to the use of natural ingredients in cosmetics and to the use of environment-friendly ingredients and preparation methods. This can be seen as an opportunity for new developments in disperse system sciences and to reach other markets in a period where socially conscious consumers are a new driving force behind the cosmetic market.

References

[1] Nafisi S, Maibach HI. Nanotechnology in cosmetics. In: Sakamoto K, Lochhead R, Maibach H, Yamashita Y, editors. Cosmetic science and technology: theoretical principles and applications. Elsevier; 2017. p. 337–69. Available from: https://doi.org/10.1016/B978-0-12-802005-0.00022-7.

[2] Rigano L, Lionetti N. Chapter 6—Nanobiomaterials in galenic formulations and cosmetics. In: Grumezescu AM, editor. Nanobiomaterials in galenic formulations and cosmetics: applications of nanobiomaterials. Elsevier; 2016. p. 121–48. Available from: https://doi.org/10.1016/B978-0-323-42868-2.00006-1.

[3] Burguera JL, Burguera M. Analytical applications of emulsions and microemulsions. Talanta 2012;96:11–20. Available from: https://doi.org/10.1016/j.talanta.2012.01.030.

[4] Nastiti CMRR, Ponto T, Abd E, Grice JE, Benson HAE, Roberts MS. Topical nano and microemulsions for skin delivery. Pharmaceutics 2017;9(4):1–25. Available from: https://doi.org/10.3390/pharmaceutics9040037.

[5] Yukuyama MN, Ghisleni DDM, Pinto TJA, Bou-Chacra NA. Nanoemulsion: process selection and application in cosmetics—a review. Int J Cosmet Sci 2016;38(1):13–24. Available from: https://doi.org/10.1111/ics.12260.

[6] Shaker DS, Ishak RAH, Ghoneim A, Elhuoni MA. Nanoemulsion: a review on mechanisms for the transdermal delivery of hydrophobic and hydrophilic drugs. Sci Pharm 2019;87(3):17. Available from: https://doi.org/10.3390/scipharm87030017.

[7] Eccleston G. Emulsions and microemulsions. In: Swarbrick J, editor. Encyclopedia of pharmaceutical technology. New York: Informa Healthcare; 2006. p. 1548–65.

[8] Im-Emsap W, Siepmann J. Disperse systems. In: Banker GS, Rhodes C, editors. Modern pharmaceutics. New York: Marcel Dekker, Inc; 2002. p. 365–436.

[9] Billany M. Suspension and emulsions. In: Aulton ME, editor. Pharmaceutics: the science of dosage form design. London: Churchill Livingstone; 2001. p. 334–59.

[10] Trapp M. Is there room for improvement in the emollients for adjuvant therapy? J Eur Acad Dermatol Venereol 2007;21(Suppl. 2):14–18. Available from: https://doi.org/10.1111/j.1468-3083.2007.02382.x.

[11] Nairn JG. Solutions, emulsions, suspensions, and extracts. In: Gennaro A, editor. Remington: the science and practice of pharmacy. Philadelphia, PA: Lippincott Williams & Wilkins; 2000. p. 721–52.

[12] Lodén M. Role of topical emollients and moisturizers in the treatment of dry skin barrier disorders. Am J Clin Dermatol 2003;4(11):771–88. Available from: https://doi.org/10.2165/00128071-200304110-00005.

[13] Rekhi GS, Eddington ND, Fossler MJ, Schwartz P, Lesko LJ, Augsburger LL. Evaluation of in vitro release rate and in vivo absorption characteristics of four metoprolol tartrate immediate-release tablet formulations. Pharm Dev Technol 1997;2(1):11–24. Available from: https://doi.org/10.3109/10837459709022605.

[14] Maphosa Y, Jideani VA. Factors affecting the stability of emulsions stabilised by biopolymers. In: Karakus S, editor. Science and technology behind nanoemulsions. London: IntechOpen; 2018. p. 65–81.

[15] Lopes LB. Overcoming the cutaneous barrier with microemulsions. Pharmaceutics 2014;6(1):52–77. Available from: https://doi.org/10.3390/pharmaceutics6010052.

[16] Hoar TP, Schulman JH. Transparent water-in-oil dispersions: the oleopathic hydro-micelle. Nature 1943;152:102–3.

[17] Schulman JH, Stoeckenius W, Prince LM. Mechanism of formation and structure of micro emulsions by electron microscopy. J Phys Chem 1959;63(10):1677–80. Available from: https://doi.org/10.1021/j150580a027.

[18] Bouchemal K, Briançon S, Perrier E, Fessi H. Nano-emulsion formulation using spontaneous emulsification: solvent, oil and surfactant optimisation. Int J Pharm 2004;280(1-2):241–51. Available from: https://doi.org/10.1016/j.ijpharm.2004.05.016.

[19] Rosen M. Emulsification by surfactants. In: Rosen MJ, Kunjappu JT, editors. Surfactants and interfacial phenomena. Hoboken, NJ: John Wiley & Sons, Inc; 2012. p. 336–65.

[20] Anton N, Vandamme TF. Nano-emulsions and micro-emulsions: clarifications of the critical differences. Pharm Res 2011;28(5):978–85. Available from: https://doi.org/10.1007/s11095-010-0309-1.

[21] Izquierdo P, Esquena J, Tadros TF, Dederen C, Garcia MJ, Azemar N, et al. Formation and stability of nano-emulsions prepared using the phase inversion temperature method. Langmuir 2002;18(1):26–30. Available from: https://doi.org/10.1021/la010808c.

[22] Ribeiro RC, Barreto SM, Ostrosky EA, da Rocha-Filho PA, Veríssimo LM, Ferrari M. Production and characterization of cosmetic nanoemulsions containing *Opuntia ficus-indica* (L.) mill extract as moisturizing agent. Molecules 2015;20(2):2492–509. Available from: https://doi.org/10.3390/molecules20022492.

[23] Eid AM, Elmarzugi NA, Enshasy HE. Development of Avocado oil nanoemulsion hydrogel using sucrose ester stearate. J Appl Pharm Sci 2013;3(12):145–7. Available from: https://doi.org/10.7324/JAPS.2013.31226.

[24] Mahdi ES, Noor AM, Sakeena MH, Abdullah GZ, Abdulkarim MF, Sattar MA. Formulation and in vitro release evaluation of newly synthesized palm kernel oil esters-based nanoemulsion delivery system for 30% ethanolic dried extract derived from local Phyllanthus urinaria for skin antiaging. Int J Nanomed 2011;6:2499–512. Available from: https://doi.org/10.2147/IJN.S22337.

[25] Pengon S, Chinatangkul N, Limmatvapirat C, Limmatvapirat S. The effect of surfactant on the physical properties of coconut oil nanoemulsions. Asian J Pharm Sci 2018;13(5):409–14. Available from: https://doi.org/10.1016/j.ajps.2018.02.005.

[26] Bei W, Zhou Y, Xing X, Zahi MR, Li Y, Yuan Q, et al. Organogel-nanoemulsion containing nisin and D-limonene and its antimicrobial activity. Front Microbiol 2015;22(6):1010. Available from: https://doi.org/10.3389/fmicb.2015.01010.

[27] Solè I, Maestro A, González C, Solans C, Gutiérrez JM. Optimization of nano-emulsion preparation by low-energy methods in an ionic surfactant system. Langmuir 2006;22(20):8326–32. Available from: https://doi.org/10.1021/la0613676.

[28] Hanifah M, Jufri M. Formulation and stability testing of nanoemulsion lotion containing *Centella asiatica* extract. J Young Pharm 2018;10(4):404–8. Available from: https://doi.org/10.5530/jyp.2018.10.89.

[29] Rodriguez C, Magniez H, Coletta D, Quillet S, Swoboda B. Cosmetic nanoemulsion. France; WO/2016/177704; 2016.

[30] Quemin E Translucent nanoemulsion, production method, and uses thereof in the cosmetic, dermatological and/or ophthalmological fields. France; WO/2002/056843; 2002.

[31] Roessler BJ, Weiner ND, Baker Jr JR, Chandrasekharan R, Wu, HL. Nanoemulsion formulations. USA; WO/2001/091728; 2001.

[32] Fernandes PA, Yukuyama MN. Cosmetic compositions in the form of oil-in-water nanoemulsions, comprising solid fatty alcohol(s), liquid fatty ester(s), other oil(s) and non-ionic surfactant(s). Brazil; WO/2015/066777; 2015.

[33] Simonnet JT, Sonneville O, Legret S. Nanoemulsion based on glycerol fatty esters, and its uses in the cosmetics, dermatological and/or ophthalmological fields. USA; US6541018; 2003.

[34] Simonnet JT, Sonneville O, Legret S. Nanoemulsion based on alkyl ether citrates and its uses in the cosmetics, dermatological, pharmacological and/or ophthalmological fields. USA; US6413527; 2002.

[35] Zhou H, Yue Y, Liu G, Li Y, Zhang J, Gong Q, et al. Preparation and characterization of a lecithin nanoemulsion as a topical delivery system. Nanoscale Res Lett 2009;5(1):224. Available from: https://doi.org/10.1007/s11671-009-9469-5.

[36] Tian Y, Guo Y, Zhang W. Effect of oil type, aliphatic alcohol, and ionic surfactants on the formation and stability of ceramide-2 enriched nanoemulsions. J Dispers Sci Technol 2016;37(8):1115–22. Available from: https://doi.org/10.1080/01932691.2015.1083440.

[37] Okamoto T, Tomomasa S, Nakajima H. Preparation and thermal properties of fatty alcohol/surfactant/oil/water nanoemulsions and their cosmetic applications. J Oleo Sci 2016;65(1):27–36. Available from: https://doi.org/10.5650/jos.ess15183.

[38] Tamarkin D, Besonov A, Berman T, Schuz D, Gazal E. Poloxamer foamable pharmaceutical compositions with active agents and/or therapeutic cells and uses. USA; US8709385B2; 2010.

[39] Simonnet JT. Nanoémulsion Transparente à Base de Tensioactifs Siliconés et Utilisation en Cosmétique ou en Dermopharmacie. France; EP0780114A1; 1997.

[40] Walstra P. Principles of emulsion formation. Chem Eng Sci 1993;48(2):333–49. Available from: https://doi.org/10.1016/0009-2509(93)80021-H.

[41] Rege S, Momin S, Bhowmick D. Effect of ascorbic acid on the oxidative stability of water-in-oil emulsion in the presence of lipophilic antioxidants. Int J Food Prop 2014;18(2):259–65. Available from: https://doi.org/10.1080/10942912.2012.675609.

[42] Ding S, Anton N, Akram S, Er-Rafik M, Anton H, Klymchenko A, et al. A new method for the formulation of double nanoemulsions. Soft Matter 2017;13(8):1660–9. Available from: https://doi.org/10.1039/c6sm02603f.

[43] Sikora E, Jaworska M, Kennard K, Lasoń E. Nanoemulsions as a form of perfumery products. Cosmetics 2018;5(4):63. Available from: https://doi.org/10.3390/cosmetics5040063.

[44] Simões S, Ribeiro HM, Almeida AJ. Microemulsões e nanoemulsões. In: Souto EB, Lopes CM, editors. Novas Formas Farmacêuticas para a Administração de Fármacos. Porto: Edições Universidade Fernando Pessoa; 2012. p. 271–96.

[45] Solans C, Solé I. Nano-emulsions: formation by low-energy methods. Curr Opin Colloid Interface Sci 2012;17(5):246–54. Available from: https://doi.org/10.1016/j.cocis.2012.07.003.

[46] Sharma S, Sarangdevot K. Nanoemulsions for cosmetics. IJARPB 2012;2(3):408–15.

[47] Yuan Y, Gao Y, Zhao J, Mao L. Characterization and stability evaluation of β-carotene nanoemulsions prepared by high pressure homogenization under various emulsifying conditions. Food Res Int 2008;41(1):61–8. Available from: https://doi.org/10.1016/j.foodres.2007.09.006.

[48] Chime SA, Kenechukwu FC, Attama AA. Nanoemulsions—advances in formulation, characterization and applications in drug delivery. In: Sezer AD, editor. Application of nanotechnology in drug delivery. London: IntechOpen; 2014. p. 77–126.

[49] Lovelyn C, Attama AA. Current state of nanoemulsions in drug delivery. J Biomater Nanobiotechnol 2011;2(5):626–39. Available from: https://doi.org/10.4236/jbnb.2011.225075.

[50] Liedtke S, Wissing RH, Mäder K. Influence of high-pressure homogenization equipment on nanodispersions characteristics. Int J Pharm 2000;196(2):183–5. Available from: https://doi.org/10.1016/s0378-5173(99)00417-2.

[51] Sonneville-Aubrun O, Simonnet JT, L'Alloret F. Nanoemulsions: a new vehicle for skin care products. Adv Colloid Interface Sci 2004;108-109:145–9. Available from: https://doi.org/10.1016/j.cis.2003.10.026.

[52] Singh Y, Meher JG, Raval K, Khan FA, Chaurasia M, Jain NK, et al. Nanoemulsion: concepts, development and applications in drug delivery. J Control Release 2017;252:28–49. Available from: https://doi.org/10.1016/j.jconrel.2017.03.008.

[53] Ghosh V, Mukherjee A, Chandrasekaran N. Ultrasonic emulsification of food-grade nanoemulsion formulation and evaluation of its bactericidal activity. Ultrason Sonochem 2013;20(1):338–44. Available from: https://doi.org/10.1016/j.ultsonch.2012.08.010.

[54] Komaiko JS, McClements DJ. Formation of food-grade nanoemulsions using low-energy preparation methods: a review of available methods. Compr Rev Food Sci Food Saf 2016;15:331–52. Available from: https://doi.org/10.1111/1541-4337.12189.

[55] Taylor P, Ottewill RH. The formation and ageing rates of oil-in-water miniemulsions. Colloids Surf A Physicochem Eng Asp 1994;88(2-3):303–16. Available from: https://doi.org/10.1016/0927-7757(94)02853-2.

[56] Fernandez P, André V, Rieger J, Kühnle A. Nano-emulsion formation by emulsion phase inversion. Colloids Surf A Physicochem Eng Asp 2004;251(1-3):53–8. Available from: https://doi.org/10.1016/j.colsurfa.2004.09.029.

[57] Solè I, Pey CM, Maestro A, González C, Porras M, Solans C, et al. Nano-emulsions prepared by the phase inversion composition method: preparation variables and scale up. J Colloid Interface Sci 2010;344(2):417–23. Available from: https://doi.org/10.1016/j.jcis.2009.11.046.

[58] Mayer S, Weiss J, McClements DJ. Vitamin E-enriched nanoemulsions formed by emulsion phase inversion: factors influencing droplet size and stability. J Colloid Interface Sci 2013;402:122–30. Available from: https://doi.org/10.1016/j.jcis.2013.04.016.

[59] McClements D. Nanoemulsions versus microemulsions: terminology, differences, and similarities. Soft Matter 2012;8(6):1719–29. Available from: https://doi.org/10.1039/C2SM06903B.

[60] Praça FSG, Medina WSG, Eloy JO, Petrilli R, Campos PM, Ascenso A, et al. Evaluation of critical parameters for in vitro skin permeation and penetration studies using animal skin models. Eur J Pharm Med Res 2018;111:121–32. Available from: https://doi.org/10.1016/j.ejps.2017.09.034.

[61] Bernardi DS, Pereira TA, Maciel NR, Bortoloto J, Viera GS, Oliveira GC, et al. Formation and stability of oil-in-water nanoemulsions containing rice bran oil: in vitro and in vivo assessments. J Nanobiotechnol 2011;9(1):44. Available from: https://doi.org/10.1186/1477-3155-9-44.

[62] Gurpreet K, Singh S. Review of nanoemulsion formulation and characterization techniques. Indian J Pharm Sci 2018;80(5):781–9. Available from: https://doi.org/10.4172/pharmaceutical-sciences.1000422.

[63] Parveen R, Baboota S, Ali J, Ahuja A, Ahmad S. Stability studies of silymarin nanoemulsion containing Tween 80 as a surfactant. J Pharm Bioallied Sci 2015;7(4):321–4. Available from: https://doi.org/10.4103/0975-7406.168037.

[64] Sonneville-Aubrun O, Yukuyama MN, Pizzino A. Chapter 14—application of nanoemulsions in cosmetics. In: Jafari SM, McClements DJ, editors. Nanoemulsions. Academic Press; 2018. p. 435–75.

[65] Marto J, Gouveia L, Jorge IM, Duarte A, Gonçalves LM, Silva SMC, et al. Starch-based Pickering emulsions for topical drug delivery: a QbD approach. Colloids Surf B Biointerfaces 2015;135:183–92. Available from: https://doi.org/10.1016/j.colsurfb.2015.07.024.

[66] Mason TG, Wilking JN, Meleson K, Chang CB, Graves SM. Nanoemulsions: formation, structure, and physical properties. J Phys Condens Matter 2006;18(41):R635–66. Available from: https://doi.org/10.1088/0953-8984/18/41/R01.

[67] Ma D, Tu Z-C, Wang H, Zhang Z, McClements DJ. Microgel-in-microgel biopolymer delivery systems: controlled digestion of encapsulated lipid droplets under simulated gastrointestinal conditions. J Agric Food Chem 2018;66(15):3930–8. Available from: https://doi.org/10.1021/acs.jafc.8b00132.

[68] Saberi AH, Fang Y, McClements DJ. Effect of glycerol on formation, stability, and properties of vitamin-E enriched nanoemulsions produced using spontaneous emulsification. J Colloid Interface Sci 2013;411:105–13. Available from: https://doi.org/10.1016/j.jcis.2013.08.041.

[69] Butnariu MV, Giuchici CV. The use of some nanoemulsions based on aqueous propolis and lycopene extract in the skin's protective mechanisms against UVA radiation. J Nanobiotechnol 2011;9:3. Available from: https://doi.org/10.1186/1477-3155-9-3.

[70] Moreira JB, Goularte PG, Morais MG, Costa JAV. Preparation of beta-carotene nanoemulsion and evaluation of stability at a long storage period. Food Sci Technol 2019;39(3):599–604. Available from: https://doi.org/10.1590/fst.31317.
[71] El-Leithy ES, Makky AM, Khattab AM, Hussein DG. Optimization of nutraceutical coenzyme Q10 nanoemulsion with improved skin permeability and anti-wrinkle efficiency. Drug Dev Ind Pharm 2018;44(2):316–28. Available from: https://doi.org/10.1080/03639045.2017.1391836.
[72] Marzuki NHC, Wahab RA, Hamid MA. An overview of nanoemulsion: concepts of development and cosmeceutical applications. Biotechnol Biotechnol Equip 2019;33(1):779–97. Available from: https://doi.org/10.1080/13102818.2019.1620124.
[73] Tadros T, Izquierdo P, Esquena J, Solans C. Formation and stability of nano-emulsions. Adv Colloid Interface Sci 2004;108-109:303–18. Available from: https://doi.org/10.1016/j.cis.2003.10.023.
[74] Das RJ, Ray S, Pal P, Das AK, Mazumder B. Nanoemulsions in non-invasive drug delivery systems. In: Mazumder B, Ray S, Pal P, Pathak Y, editors. Nanotechnology: therapeutic, nutraceutical, and cosmetic advances. Boca Raton, FL: CRC Press; 2019. p. 137–72.
[75] Bugaj AM. Intradermal delivery of active cosmeceutical ingredients. In: Donnelly RF, Singh TRR, editors. Novel delivery systems for transdermal and intradermal drug delivery. Chichester: John Wiley & Sons, Ltd; 2015. p. 209–42.
[76] Arancibia C, Miranda M, Matiacevich S, Troncoso E. Physical properties and lipid bioavailability of nanoemulsion-based matrices with different thickening agents. Food Hydrocoll 2017;73:243–54. Available from: https://doi.org/10.1016/j.foodhyd.2017.07.010.
[77] Mou D, Chen H, Du D, Mao C, Wan J, Xu H, et al. Hydrogel-thickened nanoemulsion system for topical delivery of lipophilic drugs. Int J Pharm 2008;353(1-2):270–6. Available from: https://doi.org/10.1016/j.ijpharm.2007.11.051.
[78] Cuomo F, Cofelice M, Lopez F. Rheological characterization of hydrogels from alginate-based nanodispersion. Polymers 2019;11(2):259. Available from: https://doi.org/10.3390/polym11020259.
[79] Rowe R, Sheskey P, Quinn M, editors. Handbook of pharmaceutical excipients. 6th ed London: Pharmaceutical Press; 2009.
[80] Traul KA, Driedger A, Ingle DL, Nakhasi D. Review of the toxicologic properties of medium-chain triglycerides. Food Chem Toxicol 2000;38(1):79–98. Available from: https://doi.org/10.1016/s0278-6915(99)00106-4.
[81] Final report on the safety assessment of myristyl myristate and isopropyl myristate. J Am Coll Toxicol 1982;1(4):55–80.
[82] Miastkowska M, Sikora E, Ogonowski J, Zielina M, Łudzik A. The kinetic study of isotretinoin release from nanoemulsion. Colloids Surf A Physicochem Eng Asp 2016;510:63–8. Available from: https://doi.org/10.1016/j.colsurfa.2016.07.060.
[83] Ramisetty KA, Pandit AB, Gogate PR. Ultrasound assisted preparation of emulsion of coconut oil in water: understanding the effect of operating parameters and comparison of reactor designs. Chem Eng Process 2015;88:70–7. Available from: https://doi.org/10.1016/j.cep.2014.12.006.
[84] Karagounis TK, Gittler JK, Rotemberg V, Morel KD. Use of "natural" oils for moisturization: review of olive, coconut, and sunflower seed oil. Pediatr Dermatol 2019;36(1):9–15. Available from: https://doi.org/10.1111/pde.13621.
[85] Evangelista MT, Abad-Casintahan F, Lopez-Villafuerte L. The effect of topical virgin coconut oil on SCORAD index, transepidermal water loss, and skin capacitance in mild to moderate pediatric atopic dermatitis: a randomized, double-blind, clinical trial. Int J Dermatol 2014;53(1):100–8. Available from: https://doi.org/10.1111/ijd.12339.
[86] Agero AL, Verallo-Rowell VM. A randomized double-blind controlled trial comparing extra virgin coconut oil with mineral oil as a moisturizer for mild to moderate xerosis. Dermatitis 2004;15(3):109–16.
[87] Han NS, Basri M, Rahman M, Rahman R, Salleh A, Ismail Z. Phase behavior and formulation of palm oil esters o/w nanoemulsions stabilized by hydrocolloid gums for cosmeceuticals application. J Disper Sci Technol 2011;32(10):1428–33. Available from: https://doi.org/10.1080/01932691.2010.513301.
[88] Stone L. Medilan: a hypoallergenic lanolin for emollient therapy. Br J Nurs 2000;9(1):54–7. Available from: https://doi.org/10.12968/bjon.2000.9.1.6415.

[89] Pereira TA, Guerreiro CM, Maruno M, Ferrari M, Rocha-Filho PA. Exotic vegetable oils for cosmetic O/W nanoemulsions: in vivo evaluation. Molecules 2016;21(3):248. Available from: https://doi.org/10.3390/molecules21030248.

[90] Yukuyama MN, Kato ETM, de Araujo GLB, Löbenberg R, Monteiro LM, Lourenço FR, et al. Olive oil nanoemulsion preparation using high-pressure homogenization and d-phase emulsification—a design space approach. J Drug Deliv Sci Technol 2019;49:622–31. Available from: https://doi.org/10.1016/j.jddst.2018.12.029.

[91] Katsouli M, Giannou V, Tzia C. A comparative study of O/W nanoemulsions using extra virgin olive or olive-pomace oil: impacts on formation and stability. J Am Oil Chem Soc 2018;95(10):1341–53. Available from: http://doi:10.1002/aocs.12091.

[92] Katsouli M, Tzia C. Development and stability assessment of coenzyme Q10-loaded oil-in-water nanoemulsions using as carrier oil: extra virgin olive and olive-pomace oil. Food Bioprocess Tech 2019;12(1):54–76. Available from: https://doi.org/10.1007/s11947-018-2193-3.

[93] Tou KAS, Rehman K, Ishak WMW, Zulfakar MH. Influence of omega fatty acids on skin permeation of a coenzyme Q10 nanoemulsion cream formulation: characterization, in silico and ex vivo determination. Drug Dev Ind Pharm 2019;45(9):1451–8. Available from: https://doi.org/10.1080/03639045.2019.1628042.

[94] Bajerski L, Michels LR, Colomé LM, Bender EA, Freddo RJ, Bruxel F, et al. The use of Brazilian vegetable oils in nanoemulsions: an update on preparation and biological applications. Braz J Pharm Sci 2016;52(3):347–63. Available from: https://doi.org/10.1590/s1984-82502016000300001.

[95] Musa SH, Basri M, Masoumi HRF, Shamsudin N, Salim N. Enhancement of physicochemical properties of nanocolloidal carrier loaded with cyclosporine for topical treatment of psoriasis: in vitro diffusion and in vivo hydrating action. Int J Nanomed 2017;12:2427–41. Available from: https://doi.org/10.2147/IJN.S125302.

[96] Jaworska M, Sikora E, Ogonowski J. The influence of glicerides oil phase on O/W nanoemulsion formation by PIC method. Period Polytech Chem Eng 2014;58(Suppl.):43–8. Available from: https://doi.org/10.3311/PPch.7299.

[97] Yang M, Gu Y, Yang D, Tang X, Liu J. Development of triptolide-nanoemulsion gels for percutaneous administration: physicochemical, transport, pharmacokinetic and pharmacodynamic characteristics. J Nanobiotechnol 2017;15(1):88. Available from: https://doi.org/10.1186/s12951-017-0323-0.

[98] Nasir A. Nanotechnology and dermatology: part II-risks of nanotechnology. Clin Dermatol 2010;28(5):581–8. Available from: https://doi.org/10.1016/j.clindermatol.2009.06.006.

[99] Meyer J, Polak G, Scheuermann R. Preparing PIC emulsions with a very fine particle size. Cosmet Toilet 2007;122(1):61–70.

[100] Heunnemann P, Prévost S, Grillo I, Marino CM, Meyer J, Gradzielski M. Formation and structure of slightly anionically charged nanoemulsions obtained by the phase inversion concentration (PIC) method. Soft Matter 2011;7(12):5697–710. Available from: https://doi.org/10.1039/C0SM01556C.

[101] Sinerga. NANOCREAM®—Fine particle-size emulsifier, <https://www.sinerga.it/files/materie-prime/nanocream/nanocream-flyer.pdf>; [accessed 16.10. 19].

[102] Hu Z, Liao M, Chen Y, Cai Y, Meng L, Liu Y, et al. A novel preparation method for silicone oil nanoemulsions and its application for coating hair with silicone. Int J Nanomed 2012;7:5719–24. Available from: https://doi.org/10.2147/IJN.S37277.

[103] Park H, Shin K, Lee JY, Kim JE, Seo HM, Kim JW. Highly stable, electrostatically attractive silicone nanoemulsions produced by interfacial assembly of amphiphilic triblock copolymers. Soft Matter 2018;14(27):5581–7. Available from: https://doi:0.1039/c8sm00187a.

[104] Aljuffali IA, Sung CS, Shen FM, Huang CT, Fang JY. Squarticles as a lipid nanocarrier for delivering diphencyprone and minoxidil to hair follicles and human dermal papilla cells. AAPS J 2014;16(1):140–1450. Available from: https://doi.org/10.1208/s12248-013-9550-y.

CHAPTER

5

Nanomaterials for cosmeceuticals: nanomaterials-induced advancement in cosmetics, challenges, and opportunities

Bilal Haider Abbasi[1], *Hina Fazal*[2], *Nisar Ahmad*[3], *Mohammad Ali*[3], *Nathalie Giglioli-Guivarch*[4] and *Christophe Hano*[5]

[1]Department of Biotechnology, Quaid-i-Azam University, Islamabad, Pakistan [2]Pakistan Council of Scientific and Industrial Research (PCSIR) Laboratories Complex, Peshawar, Pakistan [3]Center for Biotechnology and Microbiology, University of Swat, Swat, Pakistan [4]Biomolecules et Biotechnologies Vegetales (BBV) EA2106, Universite Francois-Rabelais de Tours, Tours, France [5]Laboratoire de Biologie des Ligneux et des Grandes Cultures (LBLGC), INRA USC1328, Université d'Orléans, Chartres, France

5.1 Introduction to nanotechnology

The nanotechnology is the apprehension, handling, and reengineering of matter in the language of nanometer (less than 100 nm) to manufacture materials with novel characteristics and superior functioning. The concept of nanotechnology incorporates two basic approaches, top-down approach which refers to the process in which structures with larger sizes are minimized to a size of nanometers while remaining their fundamental properties and bottom-up approach in which matter are constructed from the atomic and subatomic level through assembling and self-assembling [1]. Nanotechnology or nanoscience is an integral discipline of both science and technology, which holds the use of several sources of technologies for the controlled production of material within the limit of nanosize (1–100 nm) [2]. The nanotechnology is advancing exponentially with the increased use of nanomaterials [3]. The novel characteristics, which are introduced into the

Nanocosmetics
DOI: https://doi.org/10.1016/B978-0-12-822286-7.00005-X

materials, include transparency, hardness and toughness, flexibility, hydrophobic nature, and bioavailability [4]. One salient characteristic of nanoscale objects is that their surface has maximum percent of their ingredients atoms as compared to macroscopic bodies [5]. The involvement of nanotechnology can be observed in various dimensions of science including medical sciences, ecological or environmental sectors, electronics and energy [6]. Nanomaterials are those material objects with structural geometry in units, particles, fibers, or any other structural component of size less than 10 nm minimally in a single dimension [7]. A nanomaterial simply could not be explained by the nanoscale geometry as the structural evaluation of most of the materials is in nanoscale [8]. The widely applicable classes of nanomaterials are discrete nanomaterials (nanoparticles and nanofibers), nanoscale device materials, and bulk nanomaterials. The bulk nanomaterials are synthesized in large quantities and they might be composed of the other two types of nanomaterials. In the majority of cases, the nanomaterials are made of many crystals or grains (polycrystalline) [8].

5.2 Nanotechnology is multidisciplinary field

Nanotechnology is an emerging field that paves the way toward many industrial sectors like medicine department (nanomedicine), plastic industry, electronics, and aeronautics [4]. Carbon nanotubes (CNTs) have been seen to exhibit 100 times more strength than steel still retaining 1/6 the weight as steel, which is a revolutionary scientific achievement [4]. Some apparent examples of, where nanotechnology is branching off are functional polymer fillers, energy transformation, chemical gas sensing, sun blocker manufacturing, catalytic processing, nanotubes, nanowires, textiles production, and ceramic MEMS [4]. The six methods for the synthesis of nanomaterials have been excessively discussed. It includes plasma ionization, electrodeposition, sol-gel synthesis, chemical vapor deposition, ball milling, and the use of natural nanoparticles [9]. The role of nanotechnology in the energy industry is that it provides ecofriendly energy resources. The researches basically are stepping toward a platform where those catalysts can be sorted out which have the potential to enhance energy efficiency of fossils fuels. The use of nanoscale materials is well documented in power systems due to their exclusive characteristics super strength and resistivity toward corrosion that makes them a very special choice in mechanical engineering. Nanotechnology is also effective to provide extraconductivity to materials, accessing super conductance at room temperature rather than by reaching superconductivity at very high temperature. The nanotechnological approaches are also utilized in what so-called green energy (water energy, wind and solar energy) based upon their ecofriendly nature, enabling the society to overcome their dependency on petroleum resources and subvert the potentials hazards resulting from the use of fossil fuels [10]. The advancements in nanotechnology have introduced a new era in medical sciences with the development of novel diagnostic tools for the detection of the diseases. Different diagnostic tools for clinical diagnosis of disease include nanobiosensor, nanobarcode, microarray, biochip, and nanorobots. The detection of disease like cancer, cardiac problems, diabetes, muscular, and skeletal diseases has already been possible due to the establishment of nanosize particles such as CNTs, gold nanoparticles, dendrimers, and quantum dots [11]. The revolution in nanotechnology has made it possible to increase food production, improved food preservation techniques as well as quality enhancement of food. The nanomaterials are potent in shelf lives extensions, pathogens detection, flavor, and nutritional value valuation [12]. Pharmaceuticals companies with the help of nanoparticles are looking into controlled drug release by protecting against catalytic and chemical degradation to secure therapeutic efficacy. Certain new nanoparticles have been designed for the target delivery of drugs and therapeutic genes to the central nervous system. This is a significant breakthrough in the treatment of some important neurodegenerative disorder such as Alzheimer's disease, Parkinson's disease, and tumors in brain [13]. In the development of biotechnology and agriculture industries, the nanotechnology provides new ways in a large scope such as it provides us nanofood to provide safe food having high quality of nutritional values and also to improve the ways of digestion and absorption of food [14]. Similarly, the use of nanofertilizers is the best choice to improve the nutrients and to solve the continuously recurring problem of eutrophication. In plants, gene therapy is used to improve the crops, that is, 3-nm (MSN) mesoporous silica nanoparticle is used for the insertion of foreign DNA into the cell. In order to control the condition of soil and growth of crop, nanotechnology provided sensors which are very sensitive to environmental changes and give earlier response. The sensors are connected to the global positioning system. For the removal of heavy metals, ligand-based nanocoating is used. In the management of weeds, an attempt has been made to develop specific herbicide molecules, which are encapsulated with nanoparticle against the target. These molecules bind to receptor of the roots. It moves to all the part and inhibit the breakdown of glucose due to which the specific weed plant gets killed [14]. The applications of nanotechnology are very useful in textile industries. Different product produce by nanotechnology such as nanocomposites and nanofibers has useful applications of high performance. These products are also used in other traditional textiles to improve performance and to provide new applicable ways successfully. The main properties of nanotechnology used by textiles include resistance to soil, resistance to wrinkles, antibacteria, antistatic, water repellence, protection from UV light, flame retardation, and also to enhance the ability of dyes, etc. Nanofinishes and coating have an incredible role in the industry of textiles [15] The use of nanoparticles or coating-based nanotechnology in chemical finishing is more restrain, can exist for a long time period, and can greatly improve the functionality [16]. Nano-Tex enhances fabrics by creating nanowhiskers for water repellence, in which derivatives of carbon and hydrogen and cotton fiber were added to the fabrics in order to peach fuzz-like effect without decreasing the strength of cotton fibers. On the surface of the fabric, the intermolecular space between the whiskers is smaller than the drop of water but is greater than the molecule of water. So, water do not cross the surface of fabric and remain on top of the whiskers [17,18]. From protection from UV light, nanosized titanium dioxide and zinc oxide are the most useful and important ones as compared to other organic and inorganic UV-blockers, because TIO and ZNO have high properties of absorbing and scattering of UV lights [17,19]. Nanosilver particles have antimicrobial properties. The applications of these particles are acceptable in medical textiles, which are developed and commercialized for health care and hygienic safety [20]. It has been proved that nanotechnology is also very useful treatment of various diseases and in diagnostic testing. In the treatment of cancer, nanoparticles are used for chemotherapy in which nanoparticles are designed in such a way that it delivers the drug to the target site (tumor) without affecting normal tissues or organs [21,22]. CNTs are used in the treatment of cancer. These nanoparticles are the allotropic form of carbon having cylindrical framework. These are further divided into two parts such as (SWCNTs) single-walled carbon nanotubes and (MWCNTs) multiple walled-carbon nanotubes [23,24]. Florescent and theranostic nanoparticles are very useful in diagnostic testing at the academic level [25,26]. Nanotechnology also plays an important role in the treatment of HIV and AIDS. Polymeric nanoparticles are used for the delivery of (ARV) antiretroviral drugs inside the cell and to the brain [27]. This technique is also used in vaccination to treat the infection of HIV [28]. Nanoparticles are used in prevention of dentistry. Nanoapatites are used to manage the biofilms at the surface of tooth and also used to reduce the content of mineral in the initial stages of enamel lesion [29]. Casein phosphopeptide and amorphous calcium phosphate are those nanoparticles, which are used to prevention the remineralization. Glass ionomers cement (GIC) is used in the process of dental restoration. Nanotechnology has introduced nanoionomers, which are used in dental application. These nanoionomers are the modified form of GIC such as resin-modified glass ionomers by using nanoclusters and nanoparticles. Other applications of nanotechnology in dental care are nanohydroxyl apatite toothpaste, nanoparticles are also used for periodontics, dental implants, endodontic applications, endodontic regeneration, tissue engineering, and also used as antibacterial [30].

Nanotechnology can be defined as the modification of an atom and molecules in the size range from 1 to 100 nm, producing a novel organizational structure exhibiting various properties and behaviors [31,32]. Nano is a Greek word, which means "dwarf," so it means very miniature or small dimension. Nanotechnology is an advanced science that comprises synthesis, characterization, design, and application of matters by restricting shape and dimension at the nanometer scale. At the nanoscale, the properties and behaviors of matter differ from the larger particles. Thus the melting point, fluorescence properties, conduction of electricity, magnetic permeability, chemical reactivity, and alteration regarding to the particle size utilized in the preparation of products [33]. The exclusive chemical and physical properties of nanoparticles can be exploited for new performance and commercial applications that aids the society. Furthermore, the surface area/volume ratio of matter increased due to the decrease in size, thus the surface area/volume ratio of nanomaterials is greater than the larger materials which boost up their reactivity, that is why nanoparticles are utilized for the production of pharmaceutical products and cosmetics [34]. The discovery of new materials, phenomena, and procedures at the nanoscale and the establishment of novel theoretical and practical procedures for research provide new opportunities for the establishment of innovative nanomaterials and nanosystems at the end of the 20th century [35].

As mentioned above, the particles have new properties such as boosted electrical conductivity, resiliency, and strength [36,37]. The combination of nanotechnology with biological sciences has given shape to a novel field known as Nanobiotechnology. Through Nanobiotechnology one can easily identify and understand disease processes, biomarkers, and drug action mechanism [38]. A well-known example is Abraxane, it is a chemotherapeutic agent established by Abraxis used to cure breast cancer. In the chemotherapy, nanomaterials are directly delivered into tumor cells, as minute particles easily breach cell membrane [39]. Nanotechnology is also utilized to develop various vaccines such as malaria and hepatitis [40]. To develop strong immunity against pathogens, nanoparticle vaccines are directly delivered to particular dendritic cells in the immune system of an individual [41]. Miniature biochips have been developed through which a glucose level of a person is monitored [42]. Through nanomedications early diagnosis and prevention, suitable cure of diseases is possible. Some nanoparticles are used as labels and tags. With the development of nanodevices such as gold nanomaterials, DNA sequencing has become more efficient. It is easy to identify a particular DNA sequence in a sample when these gold particles are tagged with short segments of DNA [43]. The antibacterial potential of Ciprofloxacin can be boost up by using zinc oxide, which interferes with different proteins that are involved in antibiotic resistance and thus reduce the antibiotic resistance [44].

Nanotechnology is a nascent engineering branch that involves the use of nanoscale particles. In this century, nanoparticles have become widespread in workplaces and consumer products [45]. This technology is applicable in many products, such as cosmetics, food, biocidal product, and medical devices. The implementation of this novel technology has revolutionized several skin diseases treatment and diagnostic modalities [46]. It is not known for cosmetic firms that nanotechnology is the direction of the future and is regarded to be the accessible hottest and evolving technology. Cosmetics companies use nanoscale ingredient variants to provide stronger UV protection, more exceptional skin penetrating, long-lasting impacts, higher color, and finish quality, etc. [47]. This extensive use of nanoscale materials in cosmetics is because these nanoparticles acquire stronger features that vary from the particles on a large scale. These altered properties include color, transparency, solubility, and chemical reactivity, making the nanomaterials attractive to the cosmetics and personal care industries [45]. *Nanoparticles are widely exposed to employees who create or use nanoparticles in manufacturing plants is provoking significant concerns regarding their toxicity to the human body by scientists and the public* [48,49]. While the applications of nanomaterials are up-and-coming, their widespread presence in everyday products has raised concerns about potential risks and demand for adequate regulations. Throughout the years, various investigations have analyzed the toxicity profile of nanoparticles on main organs such as respiratory tract, lung, brain, liver, kidney, skin, as well as the immune system. Nanomaterials have been widely adopted by the cosmetics industry, as a way of adding value to the existing products and new products. As such, the inclusion of the nanospecific provisions will apply to many products within the cosmetics industry [49–51].

5.3 Historical perspective of nanotechnology in cosmetics

One of the imminent technologies in the 21st century is Nanotechnology, which is considered as a huge boon in the industry of cosmetics. Nanomaterials are particles that range from 1 to 100 nm and the technology and science that manipulates or develops these nanomaterials is called nanotechnology [52,53]. The Romans, Greek, and Egyptians prepared hair dyes by using nanotechnology, which is recorded during 4000 BC. Nanotechnology is emerged in various fields like biology, chemistry, physics, and engineering since 1959. Nanotechnology has been introduced to the field of dermal preparations, health products, and cosmetics for almost 4 decades [54]. In 1961, Raymond Reed who is the Founding member of Cosmetic Chemists US society coined the "cosmetics" term. The products that promote skin beauty, intensify the cleansing, and amplify the skin appearance are called cosmetics [55]. As stated earlier, in around 4000 BC the cosmetics use was first credited to the Egyptians and later the Americans, Japanese, Chinese, Romans, and Greeks started using it. At the end of the 19th century, the women of western countries secretly used household items as cosmetics and then the cosmetics were used by the 20th century women without concealment. The latest technologies incorporation developed innovative cosmetic formulations in the 21th century that are used enormously nowadays [56–58]. When biologically active therapeutic ingredients are incorporated into cosmetic products to enhance appearance is called Cosmeceuticals [59]. Cosmeceuticals are used to treat different conditions like hyperpigmentation, uneven complexion, and dark spots, skin dryness, photoaging, wrinkles, and hair damage [60]. In the personal care industry, one of the fastest growing sections is Cosmeceuticals [61]. As nanoparticles have immense benefits but little study has been conducted on long- and short-term effects in organisms and environment. Due to some reports of toxicity from nanomaterials, safety concerns are raised regarding its possible dangers. Nanocosmeceuticals have various advantages, providing controlled drug release from its carriers by various factors that include preparation method, ratio, additives and polymers, drug composition, and the chemical or physical interactions between different components. Nanocosmeceuticals are used in the preparation of hair care products such as Nirvel and Origem shampoos. They are also used to make perfume fragrance last longer such as in Allure Parfum. The sunscreens efficacy and other formulations of skin care are also improved by nanoparticles as they increase the UV protection. The active ingredients are actively transported into the skin, as the nanoparticles have very small size, which increase the surface area of cosmetics. Cosmeceuticals are better than conventional cosmetics because of more stability, nice sensorial properties, and high entrapment efficiency. Most nanomaterials have the ability to deliver both hydrophilic and lipophilic drugs. Nanoparticles are extensively used in hair serums, conditioners, and hair repairing shampoos; skin-whitening cream, moisturizing cream, and antiwrinkle creams [62,63]. Various novel nanocarriers used in cosmeceuticals are shown in Fig. 5.1 [64].

As everything has negative and positive aspects, the negative aspects of using nanomaterials in cosmeceuticals are that it may cause damage to membranes, proteins, DNA and produce oxygen species in large quantities, inflammation, and oxidation stress. Some nanoparticles such as silver nanoparticles, copper nanoparticles, TiO_2, carbon-based fullerenes, and CNTs may have toxic effect on human cells and tissues. In sunscreens, titanium dioxide is used which has showed that it causes damage to fats, RNA, and DNA inside cells. The regulatory agencies imposed no strict inspections for regulation and inspection of nanocosmeceutical products. It may also be toxic to the environment. It is also rising concerns that it may cause toxic effects after using as clinical trials are not required to approve cosmeceuticals [65,66].

Cosmetics are substances or products consumed by people for the purpose of polishing body, glorifying exteriors, and improving attractiveness [67]. The cosmetic industries produce a tremendous amount of cosmetics such as creams, hair colors, perfumes, nail polish, body sprays and lotions, various powders, tooth cleaners, conditioners, shaving creams, hair removers, and antiperspirants. The term makeup is applied to the use of any

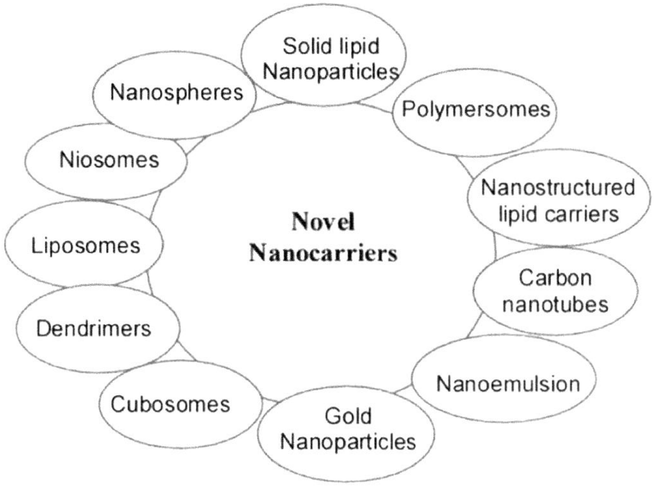

FIGURE 5.1 Novel nanocarriers used in cosmeceuticals. Source: *Reproduced from Kaul S, Gulati N, Verma D, Mukherjee S, Nagaich U. Role of nanotechnology in cosmeceuticals: a review of recent advances. Hindawi J Pharm 2018;2018:3420204 [64].*

substance with the intention of changing physical appearance [68]. It has been reported that probably all the cosmetics industries and firms use nanotechnological ways of manufacturing for their product. The excessive use of nanomaterials in cosmetic products is due to the fact that these nanoparticles achieve properties, which are reproducible than larger size materials [45] (Table 5.1).

Currently, the most searched area for the use of nanomaterials is the production of cosmetics for skin care [69]. Nanotechnology has numerous applications in skin care departments and a large number of cosmetic products contain nanoparticles, which has the ability to transport the active ingredient deeper inside the dermal layers. Sun blockers have been produced which can absorb a wide range of spectrum including UV, by using nanodispersed zinc oxides [70]. The cosmetic industries are claiming that they can with the use of nanomaterials enhance the physiology of skin through the improvement in texture of skin and slowing down the age-related wrinkles. One class of nanoparticles includes nanomaterials made up of gold, silver, titanium dioxide, and oxides of zinc. Gold and silver nanoparticles demonstrate extremely high antimicrobial activity which is clear from the fact that silver nanoparticle is used in deodorants and gold in toothpastes and antiaging creams. TiO_2 nanoparticles and ZnO are used in the manufacturing of sunscreens. They both also exhibit higher efficiency in absorbing UVA and UVB rays as compared to other counterparts. The second class of nanoscale materials is biodegradable and includes nanoemulsions, liposomes, and solid lipid nanoparticles (SLNs). The abundant use of liposomes is due to the reason that it can be loaded with biologically active compounds such as vitamin A and vitamin E and are efficiently acting in antiaging agents and as emollient. SLNs are famous among pharmaceutical firms and cosmetics industries as a best vehicle for delivery systems. SLNs, besides the use as liposomes, also contain UV-resistant activities [71]. It is explored that some nanoparticles like silver can provide protection against certain diseases of skin such as atopic dermatitis and can also act as preservatives in cosmetic products. Due to its antimicrobial potential, it is also applied in the preparation of cosmetic products to impede acne and skin cleansing [72]. Nanoparticles in

TABLE 5.1 List of different nanomaterials used in cosmetics.

EC/List name	Type	CA	Name	E
Barium sulfate	Colorant	7727-43-7	BARIUM SULFATE / CI 7712	231-784-4
Carbon black	Colorant	1333-86-4	CARBON BLACK / CI 7726	215-609-9
Chromium (III) oxide	Colorant	1308-38-9	CI 7728	215-160-9
Diiron trioxide	Colorant	1309-37-1	CI 77491	215-168-2
Triiron tetraoxide	Colorant	1317-61-9	CI 77499	215-277-5
Prussian blue	Colorant	14038-43-8	CI 77510	237-875-5
Titanium dioxide	Colorant	13463-67-7	CI 77891	236-675-5
Copper	Colorant	7440-50-8	COPPER / CI 77400	231-159-6
Gold	Colorant	7440-57-5	GOLD / CI 77480	231-165-9
Zinc oxide (ZnO)	UV filter	8051-03-4	ZINC OXIDE	617-101-6
Zinc oxide (ZnO), calcined	UV filter	93686-58-9	ZINC OXIDE	297-712-9
Zinc oxide dispersion	UV filter	N/A	ZINC OXIDE	910-472-8
Zinc oxide dispersion	UV filter	N/A	ZINC OXIDE	939-819-1
Zinc oxide nano	UV filter	N/A	ZINC OXIDE	933-598-5
Aluminum oxide	Other	1344-28-1	ALUMINA	215-691-6
Silicon dioxide	Other	112926-00-8	HYDRATED SILICA	231-545-4
Hydroxylapatite	Other	1306-06-5	HYDROXYAPATITE	215-145-7
Platinum	Other	7440-06-4	PLATINUM	231-116-1
Platinum	Other	7440-06-4	PLATINUM POWDER	231-116-1
Retinol	Other	68-26-8	RETINOL	200-683-7

Table reproduced from European Chemicals Agency (ECHA). These are also registered with REACH Regulation.

low quantities with minimum toxicity are providing new opportunities for the production of cosmetic products with improved formulations, active contents transport, and higher efficiency with penetrability of skin. Nanomaterials have stabilized different aspects of cosmetics in a wide range of interactions with many targets (hair, skin, teeth, etc.). In addition to these positivity of the nanotechnology in cosmetology, the study on the dark sides of the nanotechnology is the recent subject for researches, that is, nanotoxicology which has made it mandatory to test any kind of formulated product before commercialization for human use [73]. Nanomaterials are used for the purpose of availing their size- and shape-based characteristics [74].

5.4 Nanotechnology-based cosmetics

Cosmetics, according to the Federal Food, Drug, and Cosmetic Act (FD&C Act), are defined as articles intended to be rubbed, poured, sprinkled, or sprayed on, introduced into, or otherwise applied to the human body for cleansing, beautifying, promoting attractiveness, or altering the appearance, such as skin moisturizers, perfumes, lipsticks, fingernail polishes, eye and facial makeup preparations, cleansing shampoos, permanent waves, hair colors, and deodorants, as well as any substance intended for use as a component of the cosmetic product. For a long time, people have been using nanomaterials. The cosmetic implementation of nanomaterials reverted to the Greco Roman period when black hair was dyed with a blend of lime and water. Further tests have shown that the mixture contains 5 nm of lead selenide nanocrystals [75]. Recent regard, however, has been provided to the use of nanomaterials as topical therapy tools. Nanocosmetics achieved prominence in the worldwide cosmetics industry in the 1990s, when Lancôme, a luxury division of L'Oréal, introduced the first cosmetic product based on nanotechnology in 1995, a facial cream made up of vitamin E nanocapsules to fight signs of aging. Several other significant global companies have joined suit since then [76]. The cosmetic sector is interested in nanoparticles used in drug delivery systems. Examples include vesicular delivery systems for nanoencapsulation, including nanoemulsions and nanocrystals, liposomes and niosomes, micelles, polymeric nanocapsules, SLNs and nanostructured lipid carriers (NLCs), CNTs and fullerenes, and dendrimers.

5.4.1 Liposomes

These are coaxial vesicles having two layers in which the aqueous volume is surrounded by lipid (bilayer) comprised of man-made or natural phospholipids which are usually considered nontoxic products. The liposomal bilayer can attach to plasma membrane, which mediates release of its contents, enabling them usable for the systematic delivery of cosmetic contents. The liposomes are well suited for cosmetic preparation as they are easy to synthesize, efficiency in active ingredient absorptions and regular provision of contents/agents into the cells for a specific course of time [77]. Besides liposomes, transferosomes, niosomes, and ethosomes are also nowadays supposed to penetrate active contents through skin [45]. On comparing to inorganic pigments, which contains stable (chemically) solid particles, the constituents in the liposome are loosely bounded. Liposomes are specifically used as a vehicle for the transport of sensitive agents (vitamins) [69]. Certain limitations in the application of liposomes make it less usable for commercialization such as low loading capacity, physically and chemically instability, and minimum reproducibility. Ultrasomes are liposomal bodies with encapsulation of *Micrococcus luteus* endonuclease. The supplementation of this enzyme through away the deleterious effects of ultraviolet radiations on the DNA sequences in cells of skin. It also works as an immune booster as it represses the activation of proinflammatory cytokines. The enzyme endonuclease also minimizes the susceptibility of skin to cancer. This category of liposomes further compensates damages to cells, fights wrinkles formation, and provides elasticity to skin. Another example of liposomes is photosomes, which are widely appreciable in sunscreen products. The photosomes are associated with photolytic enzymes, isolated from a marine plant known as *Anacystic nidulans*. Under the substantial light exposure, the photosomes protect DNA from any damage due to UV radiations, preventing immune network from getting compromised and alleviating the risk of cancer development [78,79].

Liposomes are considered as low risks because the lipid bilayer is composed of phospholipids, which are generally recognized safe (GRAS) ingredients [80]. The liposomes encapsulate the medications, defend them from metabolic degradation, and release active constituents into the cell [77]. Due to the following properties such as higher absorption of active constituents by skin, ease of formulation, and continuous delivery of ingredients into the cells make liposomes potential for cosmetic applications [81]. Liposomes have been established for the delivery of vitamins and fragrance. Liposomes have also been utilized in sunblock creams, antiaging creams, and to cure hair loss. Liposomes release active compounds and encapsulate the drugs in a controlled way to protect it from metabolic degradation [82]. Their sizes range from 20 nm to 2 μm and can deliver both hydrophilic and hydrophobic compounds. They can be both unilamellar and multilamellar in structure [83]. Liposomes were loaded with active components such as vitamins K, E, and A and antioxidants like lycopene, CoQ10, and carotenoids [84]. However, they have some advantages and disadvantages as shown in Fig. 5.2. Due to the conditioning and softening properties of phosphatidylcholine, it is used as a key component in liposomes for a variety of skin care formulations such as creams, moisturizer, and products of hare care like conditioners and shampoos (Table 5.2). Active moiety can be encapsulated inside liposomes, as they are biocompatible, nontoxic, and biodegradable in nature [85]. As vegetable phospholipids have essential fatty acids that are esterified, they are widely used in dermatology and cosmetics for topical applications. These phospholipids also transport linoleic acid inside skin cells [86,87]. Clinical studies have shown that liposomes increase the smoothness of skin, reduce efflorescence in treatment of acne, and help in the reduction of wrinkles [88]. They are also being developed to deliver vitamins, botanicals, and fragrances from anhydrous formulations like lipsticks, deodorants, body sprays, and antiperspirants [87].

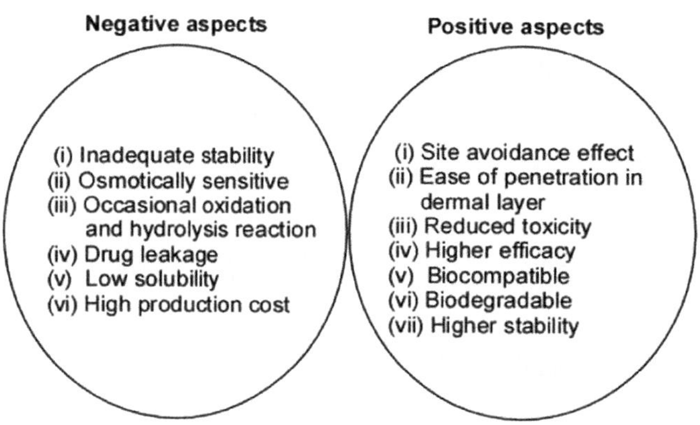

FIGURE 5.2 Negative and positive aspects of liposomes. Source: *Reproduced from Kaul S, Gulati N, Verma D, Mukherjee S, Nagaich U. Role of nanotechnology in cosmeceuticals: a review of recent advances. Hindawi J Pharm 2018;2018:3420204 [64].*

TABLE 5.2 Liposomal marketed formulations.

Product name	Uses	Marketed by
Rehydrating Liposome Day Creme'	Moisturizer	Kerstin Florian
Liposome Face & Neck Lotion	Prevents photoaging and nourishes the skin	Clinicians Complex
Liposome Concentrate	Makes smoother, softer, and firmer skin. Rejuvenating and hydrating	Russell Organics
Lumessence Eye Cream	Firming and antiwrinkle	Aubrey Organics
Fillderma Lips Lip Volumizer	Outlines the lips, skin moisturizer, fills wrinkles contour, and increases lips volume	Sesderma
Advanced Night Repair Protective Recovery Complex	Skin repair	Estee Lauder'
C-Vit Liposomal Serum	Brightens the complexion, enhances skin's firmness and elasticity, boosts collagen synthesis, and hydrates the skin	Sesderma
Liposomal Skin Cream Natural Progesterone	Maintain feminine healthy balance	NOW Solutions
Moisture Liposome Eye Cream	Brightens, firms, and moisturizes the delicate skin near eyes	Decorte
Moisture Liposome Face Cream	Moisturizer	Decorte
Dermosome	Moisturizer	Microfluidics
Capture Totale	Has a glow effect with sunscreen and remove dark spots and wrinkles	Dior

Reproduced from Kaul S, Gulati N, Verma D, Mukherjee S, Nagaich U. Role of nanotechnology in cosmeceuticals: a review of recent advances. Hindawi J Pharm 2018;2018:3420204 [64].

5.4.2 Nanoemulsions

The nanoemulsions also known as microemulsions are nanoscale droplets (liquid), which are dispersed in another liquid. They provide a flexible system, which indicate that their structure and geometry can be manipulated on the basis of the manufacturing method. The components from which nanoemulsions can be developed are generally safe and show less or no toxic effects. The higher stability and being best choice to deliver active agents in cosmetics is purely due its nanosize. One advantage of micromaterials is that it can extend the shelf lives of the products [45]. However, such systems require the help of surfactant and cosurfactant to meet the demand of energy supply, as nanoemulsions show no thermodynamically stabilization [73]. Nanoemulsions as ultrafine emulsions are recently involved in researches to be used in skin care products or as controlled delivery systems. Extensive use of these formulations is in deodorants, sunscreens, and in skin and hair care products. Their remarkable properties like rapid penetration, merging textures, and their biophysical properties especially, and hydrating power make them appropriate candidates for skin delivery of cosmetics [89].

Different nanoparticles are already in use such as nanoemulsions and mineral nanoparticles existing in our natural environment, which includes zinc oxide (ZnO), calcium fluoride, alumina, silver, copper silicon dioxide, and titanium dioxide (TiO_2). Some nanoparticles of minerals, for instance, ZnO and TiO_2 are extremely effective UV filters, which are able to scatter and reflect the visible part of sun radiation while absorbing UV light. Due to these properties, they are mostly utilized in sunblock cream. In the markets, some other nanocosmetic products include exfoliant scrub, styling gel, bronzer, body firming lotion, eyeliner, etc. Recently, nanoemulsions have gained increasingly significance because they are the best carrier for the delivery of cosmetics and enhance spreading of active constituents in a specific skin layer. Emulsions having a dimension of small droplet (20–300 nm) are known as nanoemulsions. The nanoemulsions high stability and best suitability to deliver active constituent are due to their smaller particle dimension (Alloret et al., 2004) [90]. In nanoemulsions, water and oil phase are in combination with surfactant and they are thermodynamically stable liquid dispersion. Nanoemulsions are utilized for the transport of different cosmeceutical products such as lotions, fragrance, sunblock cream, hair serums, shampoos, conditioners, and nail polish [91]. As compared to liposomes, nanoemulsions are more efficient for the delivery of lipophilic compound due to their lipophilic nature. Nanopigments are designed to remain on the skin and are a main constituent of certain sunblock creams [92] [93].

5.4.3 Niosomes

They are vesicles or sacs synthesized from nonionic surfactants, showing high specificity by allocating the active ingredient to the target site. Unlike liposomes, which are unstable, Niosomes exhibit a significant level of stability and are biodegradable [11]. The bilayer of niosomes is formed by nonionic surfactants and systems possess both hydrophilic and hydrophobic terminals. These bodies are inserted in conjugation with cholesterol and polyethylene glycol. Niosomes can be found in different beauty creams and hair related cosmetics [73]. Some of the benefits of skin-related niosomes include improving the strength of trapped active ingredients, improving the bioavailability of poorly absorbed ingredients, and enhancing skin penetration [94]. Main constituents for the preparation of niosomes comprise cholesterol and nonionic surfactants such as alkyl amides, alkyl ether, spans, polyoxyethylene, sorbitan ester, tweens, brijs, steroid-linked surfactants, and crown ester [95]. Niosomes can be used for the transport of less absorbable medications [96]. In niosomes medication transport system, the medication is enclosed in a vesicle and penetration of drugs into the target tissue is improved. Niosomes overwhelmed the issues regarded to liposomes, such as high cost, stability problems, and vulnerability to oxidation [97]. Due to their high penetration ability, niosomes are utilized in skin care applications and cosmetics because it allows the drugs to penetrate into the living tissue at higher rate [98]. Niosomes have been utilized to analyze the immune response nature initiated antigens and also used as an iobitridol carrier which is a diagnostic agent utilized for X-ray imaging [99]. These are large unilamellar, multilamellar, and small unilamellar vesicles that have 0.10 μm, 50 nm, and 0.025–0.05 μm sizes, respectively. These can deliver both

TABLE 5.3 Use of niosomes in marketed formulations.

Product name	Marketed by	Uses
Eusu Niosome Makam Pom Whitening Facial Cream	Eusu	Skin whitening
Shampooing Floral Repair	Identik	Shampoo for hair repair
Masque Floral Repair	Identik	Masque of hair repair
Anti-Age Response Cream	Simply Man Match	Wrinkles treatment
Mayu Niosome Base Cream	Laon Cosmetics	Moisturizing and whitening
Niosome +	Lancome	White and clear skin tone, Foundation cream
Niosome + Perfected Age Treatment	Lancome	Wrinkles remover

Reproduced from Kaul S, Gulati N, Verma D, Mukherjee S, Nagaich U. Role of nanotechnology in cosmeceuticals: a review of recent advances. Hindawi J Pharm 2018;2018:3420204 [64].

hydrophilic hydrophobic compounds and can also be used to deliver drugs that are poorly absorbable [96]. In 1970s, L'Oreal developed first niosomes from synthetic liposomes and were patented by them in 1987. A number of cosmeceuticals niosomes products are available in the market such as conditioner, shampoo, moisturizing cream, skin-whitening creams, and antiwrinkle creams (Table 5.3).

5.4.4 Solid lipid nanoparticles

At the beginning of 1990s, SLN was established and its size ranges from 50 to 1000 nm [100]. SLNs are lipids oily droplets stabilized by surfactant and solid at normothermia. The core of SLN is lipoidal or oily in nature and surrounded by a single layer. SLNs are prepared from purified triglycerides, complex glyceride mixtures, and waxes; solid lipid that is solid at normothermia is used instead of liquid lipid and is stabilized by polymers or surfactants [101]. For preparation of SLN, biocompatible compounds are used to bypass toxicity problems [102]. They can defend the enclosed constituents from destruction and utilized for the meticulous transport of cosmetic agents into the stratum corneum as they have improved penetration ability [103]. As SLNs show low toxicity, they are used extensively in pharmaceuticals and cosmeceutical products. SLNs are also used in sunblock creams because they have UV-resistant properties [104]. As SLNs are solid in nature, so they are more stable than liposomes [105]. SLNs are lipid droplets, which exist as solid at room temperature. The aid of surfactants leads to the stabilization of SLNs. The encapsulation of these particles protects the cosmetic agents from the biodegradation by enzymes, guides the controlled transport of active ingredients for an extended time duration, and enhances the insertion of active agents into stratum corneum [45]. SLNS are well known in the cosmeceuticals due to the presence of intrinsic properties such as nanoscale geometry, magnified skin penetration minimum toxicity, site dependent activity, and higher bioavailability [106]. The efficiency of these nanoparticles is associated with the way of preparation and physicochemical characteristics of the active agents [107]. The occlusive properties of SLNs increase the skin hydration [108].

5.4.5 Nanocapsules

Nanocapsules are composed of solid or liquid core in which the cosmetic agents are contained in a cavity which is entrapped by a polymeric membrane either synthetic or natural in nature [109]. They are submicroscopic materials, which suppress the penetration of UV octyl methoxy cinnamate in skin of pig as compared to traditional emulsions [110]. These systems are usually formed as colloidal suspension which have the ability to penetrate into gels or powders in the pursuit to enhance structural stability. The nanostructures of the nanocapsules are the main reason that why active agents are released under controlled conditions and how the harmful effects are prevented at the site of delivery [73]. The sensory property of nanocapsules, the cosmetic world can now modify the properties in order to amplify the quality and efficiency of the products. The capsulation of nanomaterials in the system safeguards the ingredient from decomposition [11].

5.4.6 Nanocrystals

These are the clusters of several hundred to tens of thousands of atoms, which are in present in aggregation. The size of these nanomaterials usually ranged from 10 to 400 nm and displayed physical and chemical stability between molecules and bulk solids [111]. The nanocrystals have maximum loading capacity and surface area, which advocate larger adhesiveness with higher retention time at thet target site [73]. Nanocrystals are mostly found as individual or polycrystalline form and contain Rutin (flavonoids) as an active compound [112]. Moreover, they also exhibited certain magical properties elevated membrane penetration, superior adhesion, and permeability [113]. Hydroxyapatite is the best example of nanocrystals used in tooth enamel [114]. Cosmetics having, poorly soluble drugs, nanocrystals provide penetration power through dermal application. They allow effective and safe channel through skin [115]. Petersen [116] reported that as compared to water-soluble rutin glycoside, nanocrystals of rutin have 500 times maximum potential of SPF (Sun protection factor). Muller et al. [117] analyzed that both hesperidin and rutin nanocrystals boosted the SPF, which verifies that the penetration into the skin is enhanced by nanocrystals.

5.4.7 Gold and silver nanoparticles

Nanogold and nanosilver have antimicrobial properties. They are effectively used in deodorant, facial creams, antiaging products, and moisturizer creams [11]. There are some reports by the cosmetic industries that silver nanoparticles used in cosmetics will provide protection all day long (underarm deodorant). Similarly, other reports also suggested that nanogold could disinfect bacteria present in mouth, thereby adding it to the toothpaste [45]. The antibacterial activities of silver nanoparticles result in the modification in the permeability of the bacterial cell wall. In the bacterial cell, the silver ions associate with the respiratory chains, which lead to increase production of reactive oxygen species (ROSs),

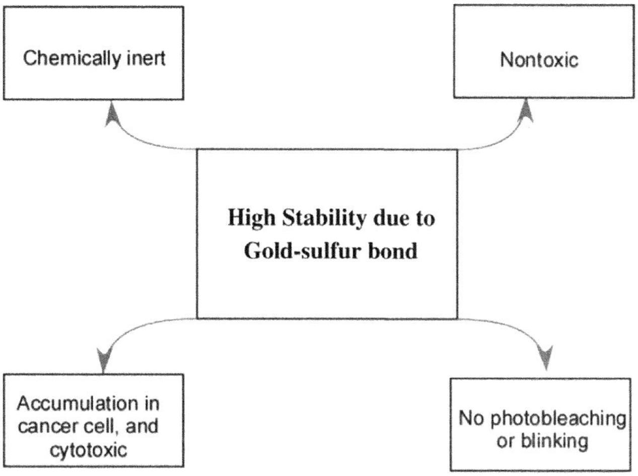

FIGURE 5.3 Gold nanoparticle merits. Source: *Reproduced from Kaul S, Gulati N, Verma D, Mukherjee S, Nagaich U. Role of nanotechnology in cosmeceuticals: a review of recent advances. Hindawi J Pharm 2018;2018:3420204 [64].*

making it difficult for bacteria to survive and grow. The ability of gold nanoparticles to carry active agents across the skin makes it a fascinating strategy in the cosmetic world. The gold NPs have been used for improving skin texture, slowing aging, and lotions due to its antioxidant and antibacterial properties [73]. Gold nanoparticles or nanogold size ranges from 5 to 400 nm and their properties are determined by assembly and interparticle interactions of gold nanoparticles [118]. Nanogold are of various shapes such as nanotriangles, branched, nanocube, nanostar, nanorod, nanocluster, nanoshell, and nanosphere. The resonance frequency is strongly affected by the environmental conditions, dielectric properties, size, and shapes of gold nanoparticles. Gold nanoparticles color may range from purple to red, to blue and due to accumulation of nanogold; the color may appear black [119]. Nanogolds are noncytotoxic, biocompatible, highly stable, and inert in nature and are available in unconjugated form and conjugated form (Fig. 5.3; [120]). Due to their crystallinity, shape, large surface area, and small size, they can be easily delivered into targeted site. Nanogold has a very high drug-loading capacity [121]. As nanogolds have strong antibacterial and antifungal properties, therefore it has been studied in cosmeceutical industries as a very valuable material. Nanogold is being used in different cosmeceutical products such as antiaging creams, deodorant, face pack, lotion, and creams. To manufacture more effective lotions and creams, giant companies in cosmetics like L'Core Paris and L'Oreal are using nanogolds in their products [122]. In beauty care, main nanogold properties are vitalizing skin metabolism, delaying aging process, improvising elasticity and firmness of skin, antiseptic properties, antiinflammatory property, and increasing blood circulation (Table 5.4; [123]).

5.4.8 Dendrimers

The name "dendrimer" is derived from two Greek words: "Dendron" means tree and "Meros" means part. Dendrimers are unimolecular, hyperbranched, monodisperse, and tree-like structures about 20 nm in dimension. Various external groups are present in

TABLE 5.4 Gold nanoparticles used and marketed by cosmetic companies.

Product name	Uses	Marketed by
24 K Gold Gel Cream	Makes shiny and glow skin	O3 +
Nano Gold BB Cream SPF 50 PA+++	Blocks UV rays, decreases wrinkles, and makes the skin fairer	Tony Moly
24 K Nano Ultra Silk Serum	Maintains healthy skin, improves fine lines and wrinkles, and moisturizes the skin	Orogold
Nano Gold Anti-Aging Lifting Serum	Promotes elastin and collagen production, reduces wrinkles and fine lines and other aging signs.	Nuvoderm
Nano Silk Day & Gold Day Cream	Skin protection from premature skin aging induced by light UV rays	LR Zeitgard
Nano Gold Foil Liquid	Promotes skin whitening, moisturizes and repairs skin damage	Ameizii
Nano Gold Energizing Eye Serum	Repairs cell growth, reduces inflammation, promotes collagen, and prevents aging	Chantecaille
Chantecaille Nano Gold Energizing Cream	Firms the skin, repairs sun damaged skin, stimulates collagen production and cell regeneration	Chantecaille

Reproduced from Kaul S, Gulati N, Verma D, Mukherjee S, Nagaich U. Role of nanotechnology in cosmeceuticals: a review of recent advances. Hindawi J Pharm 2018;2018:3420204 [64].

dendrimers, which are appropriate for multifunctionalities [124] [43]. It comprises three different parts: closely packed surface, branching units, and core moiety. Newkome dendrimer was the first ever dendrimer formulated in 1958. Dendrimers are utilized for targeted transport of active constituent to liver and macrophages [125]. The end groups of dendrimers are engineered to attach active constituents for the purpose of targeting. Both hydrophobic and hydrophilic medicines can be fused with dendrimer because of its versatility [126]. In nanotechnology-based cosmeceutical products such as sunblock cream, shampoos, antiacne creams, and hair gels, dendrimers are also being utilized for hair, nails, and skin treatment [127]. They are well-dispersed symmetrical structures that are regularly branched off with a tree-like arrangement with high density of functional groups at terminals [11]. They are versatile in functioning due to the presence of numerous external groups [45]. Upon interacting with skin bilayer, dendrimers facilitate drug skin permeability and such kind of nanosytems is very productive for cosmetics delivery and is used in skin and hair care products [73]. Furthermore, Unilever, Wella The Dow, and L'Oreal companies have filled patents for dendrimer applications in cosmetics.

5.4.9 Cubosomes

They are discrete structures of regular cubic liquid crystalline phase having maximum surface area as compared to parent cubic phase. They exhibit high thermal stability and are capable to maintain moisture in skin [11]. Cubosomes formation is the result of self-assembly of liquid crystalline particles of some surfactants when combined with water and a microstructure at a specific ratio [45]. Cubosomes have unique properties due to the proper water ratio. These are self-assembled advanced nanostructured particles that are submicron and discrete in nature. Microstructure of cubosomes is made at a certain ratio when water is mixed with surfactant system and aqueous lipids [128]. The strong packed structure is formed because of cubic liquid bicontinuous phases that are wrapped around two separate water regions which are divided by controlled surfactant bilayers with a minimal periodic three-dimensional surface [129]. Cubosomes appear like slightly spherical dots, which are cavernous in structure and have a diameter of 10–500 nm. They are capable of encapsulating amphiphilic, hydrophobic, and hydrophilic substances. Their preparation methods are relatively simple; they deliver controlled targeted release of bioactive agents with different drug-loading capacity. Internally, cubosomes have high surface area and possess lipid biodegradability [130]. As cubosomes is an appealing choice for delivering cosmetics, several patents are filed for its applications in cosmetic industries.

5.4.10 Nanomedicine

The beautification of body is performed through cosmetic surgery by the use of nanomedicine. They are suitable to apply on the skin as they turned the body/face into attractive look, provide color uniformity and wrinkle free. One example of the nanomedicines is titanium dioxide contained in antiaging creams, which make the women attractive and juvenile. Due to the culture and societal stress, women are rapidly opting for certain cosmetic surgeries. However, the use of nanomedicine can have deadly effects if it surpasses the boundary line and person can end up with cancers or tumors. Hence, informed choice should be made prior to the use of nanomedicine [131].

5.4.11 Hydrogels

Hydrogels are networks of 3D hydrophilic polymers that become swollen when they come in close contact with water or other biological fluids. They do not get dissolved in water due to the existence of physical cross-links. They are predictable and their property can be changed in order to prevent damage from happening [45]. Hydrogels are polymeric networks that have chemical or physical cross-links with the ability to expand without dissolving in water or other biological fluids. Due to high capacity and other exceptional properties, it can be used as a skin-related delivery system for skin-related cosmetic ingredients. Depending on the need for fresh products, they can pose future changes in their property [132].

5.4.12 Fullerenes/Buckyballs

Fullerenes are the advance type of nanoscale materials manufactured by the application of nanotechnology such as carbon fullerenes. They exhibit antioxidant potential superior to vitamins. Fullerenes demonstrate brightening effect by inhibiting free radicals as a result of UV exposure and also control melanin production. They are used in creams responsible for healthy and fresh look and to clean the dark circles surrounding the eyes [133–135].

5.4.13 Polymersomes

Polymersomes are composed of self-assembled copolymer amphiphiles blocks. They are artificial vesicles having aqueous cavity in the center. Polymersomes are made from lipophilic bilayer and the inner core is hydrophilic, making it capable of loading both hydrophilic and lipophilic drugs [136]. They are highly versatile and biologically stable. Polymersomes drug releasing and encapsulation capabilities are easily modified by adding different stimuli-responsive and biodegradable blocks of copolymers. They have a radius of 50–5000 nm [137]. The use of polymersomes is the proficient way to protect and encapsulate different sensitive molecules such as RNA fragments, DNA, enzymes, peptides, proteins, and drugs. Polymersomes having different properties such as permeability, different membrane thickness, and responsiveness to stimuli can be prepared by using a variety of synthetic block copolymers [138,139]. Polymersomes are more stable than liposomes because of their rigid and thick bilayer. Their flexible membrane makes it capable of controlled and targeted drug release [140]. In cosmeceutical industries, the polymersomes uses are being investigated and several patents are filed for using it to enhance cell activation energy of the skin and improve skin elasticity (US20130171274).

5.4.14 Carbon nanotubes

CNTs are rolled graphene having SP^2 hybridization and are one of the unique inventions in nanotechnology. They are hollow cylindrical fibers, made of graphene walls that are rolled "chiral" angles. CNT lengths are approximately 10s of microns and have a diameter of 0.7–50 nm [141,142]. CNTs are extremely light weighted and have three types that are multiwalled CNTs, double-walled CNTs, and single-walled CNTs. Multiwalled CNTs range from 2 to 50 nm in diameter and are made from multilayer tubes of graphene, double-walled CNTs consist of two carbon concentric nanotubes, while single-walled CNTs range from 1 to 2 nm and consist of self-rolled single graphene sheet [143]. CNTs are produced by the saline solution method, the flame synthesis method, the chemical vapor deposition method, laser ablation, and the arc discharge method [144]. In the cosmeceutical industry, several carbon nanoparticle patents have been filed such as cosmetic compositions and hair colorants (WO2006052276A2 and WO2005117537A3).

5.4.15 Nanostructured lipid carriers

NLCs are also called second-generation lipid nanoparticles [145]. They range from 10 to 1000 nm and three types of NLCs are developed that are multiple, amorphous, and imperfect type [146]. NLCs have increased skin hydration and penetrative properties. They have very low side effects and have enhanced UV protection [147]. NanoRepair Q10 serum and NanoRepair Q10 cream were the first cosmetic products having lipid nanoparticles were induced into market in 2015. In the current market, cosmetic products containing NLCs are more than 30 in numbers [148,149].

5.4.16 Nanospheres

Nanospheres have core-shelled structure with the diameter size of 10–200 nm. Nanospheres can be amorphous or crystalline in nature [150]. Nanospheres can be loaded with drugs, genes, and diverse enzymes [151]. The nanospheres are used to deliver the active ingredients more efficiently and precisely to deep targeted layer of the skin. The nanosphere uses are increasing in cosmetics such as in production of antiacne creams, moisturizing creams, and antiwrinkle creams.

5.5 Major classes of nanocosmeceuticals

Cosmetics are anticipated as the quickest growing part of the personal care industry. Nanocosmeceuticals are abundantly assimilated in lip, skin, nail, and hair creams.

5.5.1 Nail care

Nail care nanocosmetic products have bigger advantage over conventionally used products. Nail polishes constructed by nanotechnology have advantages such as durability, fast dryness, improved toughness, resistance to chip, and due to elastic in nature, are easy in applications [152].

5.5.2 Lip care

The lip care nanocosmeceutical products include lip balm, lipstick, lip volumizer, and lip gloss. Several nanoparticle types can be merged into lipsticks and lip gloss to soften and moisturize the lips by hampering transepidermal water loss [153] that stop the pigment migration from the lips area and keep color for a longer time period. Lip volumizer comprising liposomes increases the volume of lip, fills lip contour wrinkles, outlines and hydrates the lips.

5.5.3 Skin care

The skin care cosmetic products improve the functioning and quality of skin by the collagen growth and fighting harmful consequence of the free radicals. They form the skin better by keeping the keratin composition in good condition. Nanoparticles such as zinc oxides and the titanium dioxides used in sunscreen products are most productive minerals which defend skin through penetrating inside the deep skin layers, making the product less smelly, less greasy, and transparent [154]. Liposomes, niosomes, SLNs, and nanoemulsions are broadly used in nourishing formulations, as they retain moisture for prolonged span and made thin film of humectants. Antiaging nanocosmeceutical marketed products assimilating liposomes, nanocapsules, nanosphere, and nanosomes manifest aids such as rejuvenation of skin, renewal of collagen, and lifting and firming the skin [155].

5.5.4 Hair care

Nanocosmetic products of hair include conditioning agents, shampoos, hair growth stimulants, styling products, and coloring. Shaft targeting, hair follicle, and the increased amount of active ingredients are achieved by unique sizes of the nanoparticles and intrinsic properties. Nanoparticles including in shampoo seal humidity inside the cuticles by improving resident contact time between hair follicles and scalp by creating protective films [156]. Conditioning nanocosmeceutical agents have a purposive role of imparting gloss, silkiness, softness, and shine and improve disentangling hairs. Novel nanocarriers like liposomes, nanoemulsion, nanospheres, niosomes, and microemulsion have a chief role in repairing damaged cuticle, restoring gloss and structure, and forming hairs shiny, less brittlely and nongreasy [157].

5.6 Challenging aspects

5.6.1 Human health insecurities

As far as the use of nanotechnology in cosmetics is concerned, it is highly appreciated for its demanding application. The nanosystems use lipid nanomaterials and polymeric structures generally do not pose any obvious threat due to their biocompatibility and biodegradability [117,158]. There are some negative aspects arising from the use of nanotechnology toward the human health, which need to be kept in mind while formulating a nanoproduct. The study of Borowska and Brzóska [159] explored that the use of metal-based nanoparticles can eventually lead to the increase in the risk factors in cosmetic products. One such example includes the TiO_2 NPs, which promote cytokines secretions leading to inflammation and ultimately to necrosis. Another study on PVP-coated AgNPs confirmed that continuous use of these nanoparticles could lead to deterioration of cellular DNA [160]. The negative aspect of TiO_2 NPs can also be found in keratinocytes and can modify the functioning of mitochondria [161]. Hackenberg and Kleinsasser [162] observed the toxicity of ZnO NPs toward the genome of cells in epidermis in small concentration and led to alteration in mitochondrial processes. The toxicity was also witnessed in the fibroblasts and nasal mucosa. Carbon-driven nanosystems (fullerenes) also exhibited toxicity and many studies have given at alarming signals [163].

Despite the numerous advantages related to the efficacy of nanoparticles, many strong concerns have been raised about their safety [164], including for cosmetic use. The risks that should be considered include those related to consumers, but also, compared to the workers who could be in daily contact with these nanosystems, or the risks associated with the environment and ecosystem contamination. The main concern is related to the diminished size of these particles, which may make them easier to cross the barriers of the human body, entering at different points such as skin, blood, and respiratory pathways [165]. From one point of view, this facility in surpassing these obstacles could be desired, making easier the treatment of the body. However, it may also mean the "opening" of the body to foreign particles. Also, nanomaterials exhibit peculiarities about their reactivities, surface area, physicochemical, and toxicological properties that differ from the more substantial particles [165,166].

It is clear that it is not a simple matter, and that many factors influence the toxicological effects [167] highlights that the species, level of particle aggregation, and surface coatings are essential variables that alter the toxicity of nanoparticles. Discuss that, more than the particle size, the chemistry of the particle is responsible for the toxicity. They based this affirmation, between other indicatives, on the evaluation of TiO_2 nano- and microparticles, showing that in both there was no verified toxic effect for humans [168]. Also, Nohynek and Dufour [169] highlighted the studies carried out by the US FDA (United States Food and Drug Administration) proving the absence of penetration of TiO_2 nanoparticles [169,170].

In spite of their enormous potential benefit, very few studies are conducted on the long- and short-term effects on health in organisms and environment. Due to health threat, environmental concerns, and product functionality, possible hinders may be present [133]. It has been seen that cancer is caused by less soluble nanoparticles and more pronounced toxicity may also be revealed [171]. Due to the large surface area, health threats from nanoparticles may rise when nanoparticles are compared to large particles. The toxicity also relies on the chemical make-up of nanoparticles that are soaked up on the skin surface [172]. Also, there is a connection between toxicity and size of particles: the lesser the size of nanoparticles, the more ratio of volume to surface area, due to this reason there is higher biological and chemical reactivity. Health threat to the humans made by nanoparticles relies on exposure degrees and the way through which they approach the body such as dermal routes, Inhalation, ingestion, etc.

5.6.1.1 *Routes of exposure*

The health risks that nanoparticles pose to humans also depend on the route and extent of exposure to such materials. Nanomaterials enter the body mainly through three routes.

5.6.1.1.1 Inhalation

It is the most common exposure route of airborne nanoparticles, according to the National Institute of Occupational Health and Safety. For example, workers may inhale nanomaterials while producing them if the appropriate safety devices are not used, while consumers may inhale nanomaterials when using products containing nanomaterials, such as spray versions of sunscreens containing nanoscale titanium dioxide. According to officials at the National Institutes of Health, although the vast majority of inhaled particles enter the pulmonary tract, evidence from studies on laboratory animals suggests that some inhaled nanomaterials may travel via the nasal nerves to the brain and gain access to the blood, nervous system, and other organs. A study of inhalation of silicon dioxide put forward that the size of the particle from 1 to 5 nm make more toxicological reaction than 10 nm similar dose. Experiments conducted on CNTs have disclosed that on chronic display of epithelioid granulomatous lesions and interstitial inflammation were caused in the lungs. Few carbon-based fullerenes may oxidize cells or might hazardous when they are inhaled [173]. When 2, 40, and 20 nm gold nanoparticles were exposed to mice by the intratracheal route, where later found in mice's macrophages and liver cells. The 20 nm TiO_2 completely destroy DNA even at very low dose while the ability to break DNA of 500 nm size of TiO_2 very low was also observed [174].

5.6.1.1.2 Ingestion

Ingestion of nanomaterials may occur from the unintentional hand-to-mouth transfer of nanomaterials or the intentional ingestion of nanomaterials. A large fraction of nanoparticles, after ingestion, rapidly pass out of the body; however, according to some of the studies we reviewed, a small amount may be taken up by the body and then migrate into organs. Through the skin, studies have shown that specific nanomaterials have penetrated layers of pigskin within 24 hours of exposure [175]. According to some of the studies reviewed by the US Government Accountability Office (GAO), concerns have been raised that nanomaterials in sunscreens could penetrate and damaged skin. The heart, liver, spleen, pancreas, and bones became the target organs in mice when they were given zinc oxide nanoparticles in two doses. Copper nanoparticles caused internal injuries and many toxicological effects in mice that are used in different cosmetic products [176]. Silver nanoparticles are mostly used in antimicrobial formulations and wound dressing and now used in cosmetics such as face creams, toothpaste, and soaps. The lethal silver concentration for both fibroblasts and keratinocytes is the same concentration, which is lethal for bacteria [177]. When mice ingested 13.5 nm of gold nanoparticles, a notable reduction of body weight, RBCs, and the damaged spleen were observed [178].

5.6.1.1.3 Dermal routes

There are chances of skin barrier changes such as scrapes; dermatitis conditions and wounds can affect the penetration of nanoparticles [179]. When nanoparticles having size less than 10 nm were used in edema, eschar formation and prolonged erythema were reported. Fullerenes are currently being used by cosmeceutical industries to make products such as face creams and moisturizers, but the toxicity linked to them remains less understood.

5.6.2 Ecological hazardous issues

The environment is also at risk due to the exposure of nanomaterials through release into the water, air, and soil, during the manufacture, use, or disposal of these materials. These nanomaterials, if antibacterial in nature and if released in sufficient amounts, could potentially interfere with beneficial bacteria in sewage and wastewater treatment plants and could also contaminate water intended for reuse [180]. The toxic effects of nanotechnology in manufacturing and consumption of cosmetics have been scrutinized. The smaller size of particles, charge and reactivity of the nanomaterials are responsible for activation of toxic pathways with the production of ROSs [181]. The applications of nanotechnology in sunscreens are a milestone in health care department but it also reflects unfriendly mechanisms in aquatic environments [182]. The most abundantly used nanoparticles in the production of sunscreens include oxides of zinc and titanium, both of which are sources of hydrogen peroxide, which induces stress in phytoplankton. These NPs cause potential problem to the species inhabiting in aqueous surroundings [73].

Furthermore, the silver nanoparticles are excessive applicable in the manufacturing of cosmetic products by cosmetic industries. The toxic effects of AgNPS and TiO_2 NPs in the presence and absence of coating were investigated in *Daphania magna*. The coated titanium nanoparticles had nothing to do with survival of the *Daphania* species. However, uncoated TiO_2 NPs resulted in the reduced reproductive rates and stunted growth [183]. The toxicological studies on ZnO NPs have been well documented. These NPs might be an imminent to the ecosystem as small amount of these NPs can damage the environment [184]. Asharani et al. [185] performed toxicity-related studies of Au, Ag, and Pt NPs in developing embryos of zebra fish. The silver nanoparticles fluctuate the circulatory system as well as altering the morphology of heart and lead to death. Similarly, the nanosilver and nanoplatinum delayed the hatching process in zebra fish. The accumulation of gold nanoparticles showed no such detrimental effects.

5.7 Future prospects and opportunities

The possible impact of nanotechnology in the cosmetics sector can be influenced by the fact that great companies like L'Oreal, Dior, and Shiseido are spending massive capital in nanotechnology research. Nanotechnology and innovative cosmetic dermatology can help despite the current international financial crisis. Strengthen a nation's economic system. Although nanotechnology has featured in cosmetic formulations for many years, only a few technologies have been employed so far, mainly liposomes and metal oxide nanoparticles. They have improved characters and several advantages compared with the traditional formulations. The reason for their limited use is the regular bans and halts on nanotechnology-based cosmetic products by many establishments. This has led to reluctance in developing nanotechnology in cosmetics by several companies. For the development of nanotechnology, national as well as international cooperation is required. The Government should require companies to describe their use of manufactured nanomaterials. Potentially unsafe products should be withdrawn from sale. An independent expert group should be constituted to notify the regime on the hazards and benefits of nanosunscreens. Clear information should be supplied to consumers about the utilization of nanomaterials in cosmetic products, as well as nanotechnology more broadly, in this context, proper labeling of the products in order to ensure their safety for commercial applications.

To strengthen the use of nanotechnology in the cosmetic industry, these prospects should be exploited: in-depth toxicity study of TiO_2 and ZnO nanoparticles should be done because of their full application in the cosmetics industry, as surveys so far have demonstrated mixed results. Liposomes and nanoemulsions should be exploited for their stability as they are not removed from the skin even after washing and maintain the existing integrity of the skin lipid bilayers. Acceptability of microemulsions can be increased by the use of safer surfactants, as they do not alter the permeability of membrane even after repeated use. Reliable and low-cost triggers for controlled discharge of cosmetic molecules need to be identified. Drug-carrying ability of lipid-based nanoparticles, like

SLNs, NLCs, and Nanocapsules, needs to be improved. Transport, interaction, and penetration of lipid nanoparticles with the stratum corneum needs to be fully understood. SLN and NLC formulations, as well as the effect of wetting agents used for modifications, should be considered further. In vivo study of the cosmetics containing nanoparticles should be done. Therefore it can be concluded that the use of nanotechnology in cosmetics in of great essence but the negative effects must be taken into consideration prior to the use of the nanomaterials and all the reservations related to this advancing technology should be dealt with great care and caution.

References

[1] Sanchez F, Sobolev K. Nanotechnology in concrete—a review. Const Build Mater 2010;24(11):2060–71.
[2] Zeb S, Ullah I, Karim A, Muhammad W, Ullah N, Khan M, et al. A review on nanotechnology applications in electric components. Nanoscale Rep 2019;2(2):32–8.
[3] Krishna VD, Wu K, Su D, Cheeran MC, Wangand JP, Perez A. Nanotechnology: review of concepts and potential application of sensing platforms in food safety. Food Microb 2018;75:47–54. Available from: https://doi.org/10.1016/j.fm.2018.01.025.
[4] Lines MG. Nanomaterials for practical functional uses. J Alloys Comp 2008;449(1–2):242–5.
[5] Love JC, Estroff LA, Kriebel JK, Nuzzo RG, Whitesides GM. Self-assembled monolayers of thiolates on metals as a form of nanotechnology. Chem Rev 2005;105(4):1103–70.
[6] Adlakha-Hutcheon G, Khaydarov R, Korenstein R, Varma R, Vaseashta A, Stamm H, et al. Nanomaterials, nanotechnology: applications, consumer products, and benefts. In: Linkov I, Steevens J, editors. Nanomaterials: risks and benefits. Dordrecht: Springer; 2009. p. 195–207.
[7] Zhang L, Webster TJ. Nanotechnology and nanomaterials: promises for improved tissue regeneration. Nano Today 2009;4(1):66–80.
[8] Ramesh KT. Nanomaterials. Nanomaterials. Boston, MA: Springer; 2009. p. 1–20.
[9] Guz AN, Rushchitskii YY. Nanomaterials: on the mechanics of nanomaterials. Int Appl Mech 2003;39 (11):1271–93.
[10] Ahmadi MH, Ghazvini M, Alhuyi Nazari M, Ahmadi MA, Pourfayaz F, Lorenzini G, et al. Renewable energy harvesting with the application of nanotechnology: a review. Int J Energy Res 2019;43(4):1387–410.
[11] Hameed A, Fatima GR, Malik K, Muqadas A, Fazal-ur-Rehman M. Scope of nanotechnology in cosmetics: dermatology and skin care products. J Med Chem Sci 2019;2(1):9–16.
[12] Nurfatihah Z, Siddiquee S. Nanotechnology: recent trends in food safety, quality and market analysis. Nanotechnology: applications in energy, drug and food. Cham: Springer; 2019. p. 283–93.
[13] Saeedi M, Eslamifar M, Khezri K, Dizaj SM. Applications of nanotechnology in drug delivery to the central nervous system. Biomed Pharmacother 2019;111:666–75.
[14] Elizabath A, Babychan M, Mathew AM, Syriac GM. Application of nanotechnology in agriculture. Int J Pure Appl Biosci 2019;7(2):131–9.
[15] Asif AKMAH, Hasan MZ. Application of nanotechnology in modern textiles: a review. Int J Curr Eng Technol 2018;8:227–31.
[16] Khan FR, Misra SK, García-Alonso J, Smith BD, Strekopytov S, Rainbow PS, et al. Bioaccumulation dynamics and modeling in an estuarine invertebrate following aqueous exposure to nanosized and dissolved silver. Environ Sci Technol 2012;46(14):7621–8.
[17] Xin JH, Daoud WA, Kong YY. A new approach to UV-blocking treatment for cotton fabrics. Textile Res J 2004;74(2):97–100.
[18] Joshi M, Viswanathan V. High-performance filaments from compatibilized polypropylene/clay nanocomposites. J Appl Polym Sci 2006;102(3):2164–74.
[19] Saito M. Antibacterial, deodorizing, and UV absorbing materials obtained with zinc oxide (ZnO) coated fabrics. J Coat Fab 1993;23(2):150–64. Available from: https://doi.org/10.1177/152808379302300205.
[20] Ali SW, Rajendran S, Joshi M. Synthesis and characterization of chitosan and silver loaded chitosan nanoparticles for bioactive polyester. Carbohydr Polym 2011;83(2):438–46.

[21] Shen B, Ma Y, Yu S, Ji C. Smart multifunctional magnetic nanoparticle-based drug delivery system for cancer thermo-chemotherapy and intracellular imaging. ACS Appl Mater Interfaces 2016;8(37):24502–8.

[22] Huang B, Abraham WD, Zheng Y, López SCB, Luo SS, Irvine DJ. Active targeting of chemotherapy to disseminated tumors using nanoparticle-carrying T cells. Sci Transl Med 2015;7(291):291ra94.

[23] Rastogi V, Yadav P, Bhattacharya SS, Mishra AK, Verma N, Verma A, et al. Carbon nanotubes: an emerging drug carrier for targeting cancer cells. J Drug Deliv 2014;2014.

[24] Sanginario A, Miccoli B, Demarchi D. Carbon nanotubes as an effective opportunity for cancer diagnosis and treatment. Biosensors 2017;7(1):9.

[25] Wolfbeis OS. An overview of nanoparticles commonly used in fluorescent bioimaging. Chem Soc Rev 2015;44(14):4743–68.

[26] Janib SM, Moses AS, MacKay JA. Imaging and drug delivery using theranostic nanoparticles. Adv Drug Deliv Rev 2010;62(11):1052–63.

[27] Mamo T, Moseman EA, Kolishetti N, Salvador-Morales C, Shi J, Kuritzkes DR, et al. Emerging nanotechnology approaches for HIV/AIDS treatment and prevention. Nanomedicine 2010;5(2):269–85.

[28] Khalil NM, Carraro E, Cótica LF, Mainardes RM. Potential of polymeric nanoparticles in AIDS treatment and prevention. Expert Opin Drug Deliv 2011;8(1):95–112.

[29] Hannig M, Hannig C. Nanomaterials in preventive dentistry. Nat Nanotechnol 2010;5(8):565.

[30] Zafar MS, Alnazzawi AA, Alrahabi M, Fareed MA, Najeeb S, Khurshid Z. Nanotechnology and nanomaterials in dentistry. Advanced dental biomaterials. Woodhead Publishing; 2019. p. 477–505.

[31] National Science and Technology Council, Nanotechnology-shaping the world research directions: IWGN, 1999. https://obamawhitehouse.archives.gov/administration/eop/ostp/nstc/docsreports/archives

[32] Salla LD, Cadioli LP. A RELAÇÃO DOS NANOMATERIAIS COM A NOVA REVOLUÇÃO INDUSTRIAL. Rev Interface Tecnol 2006;3(1):11–20.

[33] Daniel MC, Astruc D. Gold nanoparticles: assembly, supramolecular chemistry, quantum-size-related properties, and applications toward biology, catalysis, and nanotechnology. Chem Rev 2004;104(1):293–346.

[34] Oberdörster E. Manufactured nanomaterials (fullerenes, C60) induce oxidative stress in the brain of juvenile largemouth bass. Environ Health Perspect 2004;112(10):1058–62.

[35] Bhushan B. Governance, policy, and legislation of nanotechnology: a perspective. Microsyst Technol 2015;21(5):1137–55.

[36] Consumer Reports, 2009. http://www.consumerreports.org/cro/health-fitness/nanotechnology-7-07/overview/0707_nano_ov_1.htm

[37] National Nanotechnology Initiative, 2009. https://www.congress.gov/bill/111th-congress/senate-bill/1482?s = 1&r = 71

[38] Jain KK. Nanotechnology in clinical laboratory diagnostics. Clin Chim Acta 2005;358(1–2):37–54.

[39] Abraxis oncology. Rx only. Washington, DC US Food and Drug Administration. (2005).

[40] AzoNano.com. The A to Z of nanotechnology. Nanoparticles treatment for glaucoma may stop millions from going blind. Available from: <http://azonano.com/news.asp?news ID = 4290> [accessed 5.01.09].

[41] Science Daily. New nanoparticles vaccine is both more effective and less expensive. Available from: <http://www.sciencedaily.com/releases/2007/09/070917112546.htm> [accessed 5.01.09].

[42] Nanotechwire.com. Smart insulin nanostructures pass feasibility test, UT study repots. Available from: <http://nanotechwire.com/news.asp?nid = 5058&ntid = 130&pg = 1> [accessed 7.01.09].

[43] Nikalje AP. Nanotechnology and its applications in medicine. Med Chem 2015;5(2):081–9.

[44] Naseri N, Valizadeh H, Zakeri-Milani P. Solid lipid nanoparticles and nanostructured lipid carriers: structure, preparation and application. Adv Pharm Bull 2015;5(3):305.

[45] Raj S, Jose S, Sumod US, Sabitha M. Nanotechnology in cosmetics: opportunities and challenges. J Pharm Bioallied Sci 2012;4(3):186.

[46] Bleeker EA, de Jong WH, Geertsma RE, Groenewold M, Heugens EH, Koers-Jacquemijns M, et al. Considerations on the EU definition of a nanomaterial: science to support policy making. Regul Toxicol Pharmacol 2013;65(1):119–25.

[47] Tsang MP, Kikuchi-Uehara E, Sonnemann GW, Aymonier C, Hirao M. Evaluating nanotechnology opportunities and risks through integration of life-cycle and risk assessment. Nat Nanotechnol 2017;12(8):734.

[48] Auffan M, Rose J, Bottero JY, Lowry GV, Jolivet JP, Wiesner MR. Towards a definition of inorganic nanoparticles from an environmental, health and safety perspective. Nat Nanotechnol 2009;4(10):634.

[49] Bowman DM, Van Calster G, Friedrichs S. Nanomaterials and regulation of cosmetics. Nat Nanotechnol 2010;5(2):92.
[50] Ray PC, Yu H, Fu PP. Toxicity and environmental risks of nanomaterials: challenges and future needs. J Environ Sci Health C 2009;27(1):1–35.
[51] Brohi RD, Wang L, Talpur HS, Wu D, Khan FA, Bhattarai D, et al. Toxicity of nanoparticles on the reproductive system in animal models: a review. Front Pharmacol 2017;8:606.
[52] Logothetidis S, editor. Nanostructured materials and their applications. Springer Science & Business Media; 2012.
[53] Maynard AD. A research strategy for addressing risk. Nanotechnology 2006;444:267–9.
[54] Bangale MS, Mitkare SS, Gattani SG, Sakarkar DM. Recent nanotechnological aspects in cosmetics and dermatological preparations. Int J Pharm Pharm Sci 2012;4(2):88–97.
[55] Gautam A, Singh D, Vijayaraghavan R. Dermal exposure of nanoparticles: an understanding. J Cell Tissue Res 2011;11(1):2703–8.
[56] Sharma B, Sharma A. Future prospect of nanotechnology in development of anti-ageing formulations. Int J Pharm Pharm Sci 2012;4(3):57–66.
[57] Sharma S, Sarangdevot K. Nanoemulsions for cosmetics. IJARPB 2012;1(3):408–15.
[58] Dureja H, Kaushik D, Gupta M, Kumar V, Lather V. Cosmeceuticals: an emerging concept. Indian J Pharmacol 2005;37(3):155.
[59] Mukta S, Adam F. Cosmeceuticals in day-to-day clinical practice. J Drugs Dermatol 2010;9(5 Suppl. ODAC Conf Pt 1):s62–6.
[60] Srinivas K. The current role of nanomaterials in cosmetics. J Chem Pharm Res 2016;8(5):906–14.
[61] Brandt FS, Cazzaniga A, Hann M. Cosmeceuticals: current trends and market analysis. Semin Cutan Med Surg 2011;30(3):141–3.
[62] Nohynek GJ, Lademann J, Ribaud C, Roberts MS. Grey goo on the skin? Nanotechnology, cosmetic and sunscreen safety. Crit Rev Toxicol 2007;37(3):251–77.
[63] Mu L, Sprando RL. Application of nanotechnology in cosmetics. Pharm Res 2010;27(8):1746–9.
[64] Kaul S, Gulati N, Verma D, Mukherjee S, Nagaich U. Role of nanotechnology in cosmeceuticals: a review of recent advances. Hindawi J Pharm 2018;2018:3420204. Available from: https://doi.org/10.1155/2018/3420204.
[65] Starzyk E, Frydrych A, Solyga A. Nanotechnology: Does it have a future cosmetics? SÖFW J 2008;134(6):42–52.
[66] Dahiya A, Romano JF. Cosmeceuticals: a review of their use for aging and photoaged skin. Cosmet Dermatol 2006;19(7):479–84.
[67] Khan AD, Alam MN. Cosmetics and their associated adverse effects: a review. J Appl Pharm Sci Res 2019;1–6.
[68] Draelos ZD. Cosmetics: the medicine of beauty. J Cosmet Dermatol 2015;14(2):91.
[69] Huber B, Burfeindt J. Nanotechnology in cosmetics. Nanocosmetics. Cham: Springer; 2019. p. 17–25.
[70] Singh A, Dubey S, Dubey HK. Nanotechnology: the future engineering. Nanotechnology 2019;6(2).
[71] Bowman DM, May ND, Maynard AD. Nanomaterials in cosmetics: regulatory aspects. Analysis of cosmetic products. Elsevier; 2018. p. 289–302.
[72] Gajbhiye S, Sakharwade S. Silver nanoparticles in cosmetics. J Cosmet Dermatol Sci Appl 2016;6(01):48.
[73] Santos AC, Morais F, Simões A, Pereira I, Sequeira JA, Pereira-Silva M, et al. Nanotechnology for the development of new cosmetic formulations. Expert Opin Drug Deliv 2019;16(4):313–30.
[74] Vinod TP, Jelinek R. Inorganic nanoparticles in cosmetics. Nanocosmetics. Cham: Springer; 2019. p. 29–46.
[75] Landriscina A, Rosen F, Adam J. Nanodermatology: nanoescale therapeutics for topical application. In: Quan T, editor. Molecular Mechanisms of Skin Aging and Age-Related Diseases. Boca Raton: CRC Press; 2016. Available from: https://doi.org/10.1201/b21370.
[76] Melo A, Amadeu MS, Lancellotti M, Hollanda LM, Machado D. The role of nanomaterials in cosmetics: national and international legislative aspects. Quím Nova 2015;38(4):599–603.
[77] Jain S, Sapee R, Jain NK. Proultraflexible lipid vesicles for effective transdermal delivery of norgesterol, proceedings of 25th conference of CRS; 1998.
[78] Ashtiani HRA, Bishe P, Lashgari NA, Nilforoushzadeh MA, Zare S. Liposomes in cosmetics. J Skin Stem Cell 2016;3(3).

[79] Hougeir FG, Kircik L. A review of delivery systems in cosmetics. Dermatol Ther 2012;25(3):234–7.
[80] Arora N, Agarwal S, Murthy RSR. Latest technology advances in cosmaceuticals. Int J Pharm Sci Drug Res 2012;4(3):168–82.
[81] Blume G. Flexible liposomes for topical applications in cosmetics. Science and applications of skin delivery systems. Carol Stream, IL: Allured Publishing Co; 2008. p. 269–82.
[82] Hope MJ, Kitson CN. Liposomes: a perspective for dermatologists. Dermatol Clin 1993;11(1):143–54.
[83] Bhupendra G, Prajapati Nilesh K, Menon M, Rakesh P. Topical liposomes in drug delivery: a review. Int J Pavement Res Technol 2012;4(1):39–44.
[84] Tasleem A, Nuzhatun N, Syed SA, Sheikh S, Raheel M, Muzafar RS. Therapeutic and diagnostic applications of nanotechnology in dermatology and cosmetics. J Nanomed Biother Disc 2015;5(3):1–10.
[85] Egbaria K, Weiner N. Liposomes as a topical drug delivery system. Adv Drug Deliv Rev 1990;5(3):287–300.
[86] Rieger MM. Skin lipids and their importance to cosmetic science. Cosmet Toilet 1987;102(7):45–9.
[87] Blume G, Teichmüller E. New evidence of the penetration of actives by liposomal carrier system. Cosmetics and toiletries manufacture worldwide. London: Wimington Media Ltd; 1997. p. 135–9.
[88] Ghyczy M, Nissen H-P, Biltz H. The treatment of acne vulgaris by phosphatidylcholine from soybeans, with a high content of linoleic acid. J Appl Cosmetol 1996;14(4):137–45.
[89] Brugè F, Damiani E, Marcheggiani F, Offerta A, Puglia C, Tiano L. A comparative study on the possible cytotoxic effects of different nanostructured lipid carrier (NLC) compositions in human dermal fibroblasts. Int J Pharm 2015;495(2):879–85.
[90] Özgün S. Nanoemulsions in cosmetics. Anadolu Univ 2013;1(6):3–11.
[91] Nasir A. Nanotechnology and dermatology: part II—risks of nanotechnology. Clin Dermatol 2010;28 (5):581–8.
[92] L'Oréal, 2001. Patent No. FR2680684B1.
[93] Lautenschlager S, Wulf HC, Pittelkow MR. Photoprotection. The Lancet 2007;370(9586):528–37.
[94] Gökçe EH, AlgınYapar E, TuncayTanrıverdi S, Özer O. Nanocarriers in cosmetology. In: Grumezescu A, editor. Nanobiomaterials in Galenic Formulations and Cosmetics, Applications of Nano Bio Materials. United Kingdom: Elsevier; 2016. p. 363–93.
[95] Chandu VP, Arunachalam A, Jeganath S, Yamini K, Tharangini K, Chaitanya G. Niosomes: a novel drug delivery system. Int J Nov Trend Pharma Sci 2012;2(1):25–31.
[96] Kumar GP, Rajeshwarrao P. Nonionic surfactant vesicular systems for effective drug delivery—an overview. Acta Pharm Sin B 2011;1(4):208–19.
[97] Biswal S, Murthy PN, Sahu J, Sahoo P, Amir F. Vesicles of non-ionic surfactants (niosomes) and drug delivery potential. Int J Pharm Sci Nanotechnol 2008;1(1):1–8.
[98] Verma NK, Roshan A. Niosomes and its application: a review. IJRPLS 2014;2(1):182–4.
[99] Puri D, Bhandari A, Sharma P, Choudhary D. Lipid nanoparticles (SLN, NLC): a novel approach for cosmetic and dermal pharmaceutical. J Glob Pharma Technol 2010;2(5):1–15.
[100] Bangale MS, Mitkare SS, Gattani SG, Sakarkar DM. Recent nanotechnological aspects in cosmetics and dermatological preparations. Int J Pharma Pharm Sci 2012;4(2):88–97.
[101] Mohamed HH, Omaima NG. 2nd edition Rice bran solid lipid nanoparticles: preparation and characterization, vol. 1. Universal Research Publications; 2011.
[102] Müller RH, Radtke M, Wissing SA. Nanostructured lipid matrices for improved microencapsulation of drugs. Int J Pharma 2002;242(1-2):121–8.
[103] Arora N, Agarwal S, Murthy RSR. Latest technology advances in cosmaceuticals. Int J Pharma Sci Drug Res 2012;4(3):168–82.
[104] López-García R, Ganem-Rondero A. Solid lipid nanoparticles (SLN) and nanostructured lipid carriers (NLC): occlusive effect and penetration enhancement ability. J Cosmet Dermatol Sci Appl 2015;5(02):62.
[105] Souto EB, Müller RH. Cosmetic features and applications of lipid nanoparticles (SLN®, NLC®). Int J Cosmet Sci 2008;30(3):157–65.
[106] Patwekar S, Gattani S, Giri R, Bade A, Sangewar B, Raut V. Review on nanoparticles used in cosmetics and dermal products. World J Pharm Pharm Sci 2014;3(8):1407–21.
[107] Jensen LB, Magnussson E, Gunnarsson L, Vermehren C, Nielsen HM, Petersson K. Corticosteroid solubility and lipid polarity control release from solid lipid nanoparticles. Int J Pharm 2010;390(1):53–60.

[108] Song C, Liu S. A new healthy sunscreen system for human: solid lipid nannoparticles as carrier for 3,4,5-trimethoxybenzoylchitin and the improvement by adding Vitamin E. Int J Biol Macromol 2005;36 (1–2):116–19.

[109] Gorai S. Nanotechnology in cosmetics. Beats Nat Sci 2014;1(3):1–7.

[110] Hwang SL, Kim JC. In vivo hair growth promotion effects of cosmetic preparations containing hinokitiol-loaded poly (ε-caprolacton) nanocapsules. J Microencapsul 2008;25(5):351–6.

[111] Petersen, R. (2015). US Patent No. 9,114,077. Washington, DC: US Patent and Trademark Office.

[112] Garg A, Visht S, Sharma PK, Kumar N. Formulation, characterization and application on nanoparticle: a review. Der Pharm Sin 2011;2(2):17–26.

[113] Abla MJ, Singh ND, Banga AK. Role of nanotechnology in skin delivery of drugs. Percutaneous penetration enhancers chemical methods in penetration enhancement. Berlin: Springer; 2016. p. 1–13.

[114] Arif T, Nisa N, Amin SS, Shoib S, Mushtaq R. Therapeutic and diagnostic applications of nanotechnology in dermatology and cosmetics. J Nanomed Biother Discov 2015;5:134. Available from: https://doi.org/10.4172/2155-983X.

[115] Karim KA, Mandal A, Biswas N, et al. Niosome: a future of targeted drug delivery systems. J Adv Pharma Technol Res 2010;1(4):374–80.

[116] Petersen, R. Nanocrystals for use in topical cosmetic formulations and method of production thereof. U.S. Patent 12/514,621, 25 February 2010.

[117] Muller RH, Shegokar R, Keck CM. 20 years of lipid nanoparticles (SLN & NLC): present state of development & industrial applications. Curr Drug Discov Technol 2011;8(3):207–27.

[118] Lata K, Jaiswal AK, Naik L, Sharma R. Gold nanoparticles: preparation, characterization and its stability in buffer. Nano Trends 2014;17(1):1–10.

[119] Khan AK, Rashid R, Murtaza G, Zahra A. Gold nanoparticles: synthesis and applications in drug delivery. Trop J Pharm Res 2014;13(7):1169–77.

[120] Verma HN, Singh P, Chavan RM. Gold nanoparticle: synthesis and characterization. Vet World 2014;7(2):72.

[121] Yeh YC, Creran B, Rotello VM. Gold nanoparticles: preparation, properties, and applications in bionanotechnology. Nanoscale 2012;4(6):1871–80.

[122] Khalid A, Tawfik AS. Gold and silver nanoparticles: synthesis methods, characterization routes and applications towards drugs. J Environ Amp Anal Toxicol 2016;6(4):1–10.

[123] Thakor AS, Jokerst J, Zavaleta C, Massoud TF, Gambhir SS. Gold nanoparticles: a revival in precious metal administration to patients. Nano Lett 2011;11(10):4029–36.

[124] F. Michael, Cosmetic compositions for hair treatment containing dendrimersor dendrimerconjugates, 2010. Patent No. 6068835.

[125] Papakostas D, Rancan F, Sterry W, Blume-Peytavi U, Vogt A. Nanoparticles in dermatology. Arch Derma Res 2011;303(8):533.

[126] Awani RK, Ruchi T, Priyanka M, Pooja Y. Dendrimers: a potential carrier for targeted drug delivery system. PharmaBio Eval 2016;3(3):275–87.

[127] Yapar EA, Inal O. Nanomaterials and cosmetics. J Pharm Istanbul Univ 2012;42(1):43–70.

[128] Tilekar K, Khade P, Kakade S, Kotwal S, Patil R. Cubosomes—a drug delivery system. Int J Pharm Chem Biol Sci 2014;4(4).

[129] Rohit RB, Riyaz AO, Bhargav RH, Prasanna PG. Cubosomes: the inimitable nanoparticulate drug carriers. Scholars Acad J Pharma 2013;2(6):481–6.

[130] Madhurilatha T, Paruchuri SK, SuriaPrabha K. Overview of cubosomes: a nano particle. Int J Res Pharma Chem 2011;1:535–41.

[131] Latham M. Nanomedicine and body modification: critical perspectives. In pursuit of nanoethics. Dordrecht: Springer; 2014. p. 61–72.

[132] Gokce EH, et al. Resveratrol-loaded solid lipid nanoparticles versus nanostructured lipid carriers: evaluation of antioxidant potential for dermal applications. Int J Nanomed 2012;7:1841.

[133] Oberdörster G, Oberdörster E, Oberdörster J. Nanotoxicology: an emerging discipline evolving from studies of ultrafine particles. Environ Health Perspect 2005;113(7):823–39.

[134] Zhou Z. Liposome formulation of fullerene-based molecular diagnostic and therapeutic agents. Pharmaceutics 2013;5(4):525–41.

[135] Montenegro L. Nanocarriers for skin delivery of cosmetic antioxidants. J Pharm Pharmacogn Res 2014;2 (4):73–92.

[136] Ambikanandan M. Challenges in Delivery of Therapeutic Genomics and Proteomics. Elsevier; 2011.

[137] Kim SH, Shum HC, Kim JW, Cho JC, Weitz DA. Multiple polymersomes for programmed release of multiple components. J Am Chem Soc 2011;133(38):15165–71.

[138] Bermudez H, Brannan AK, Hammer DA, Bates FS, Discher DE. Molecular weight dependence of polymersome membrane structure, elasticity, and stability. Macromolecules 2002;35(21):8203–8.

[139] Discher DE, Eisenberg A. Polymer vesicles. Science 2002;297(5583):967–73.

[140] Zhang XY, Zhang PY. Polymersomes in nanomedicine—a review. Curr Nanosci 2017;13(2):124–9.

[141] Ibrahim KS. Carbon nanotubes-properties and applications: a review. Carbon Lett 2013;14(3):131–44.

[142] Kaushik BK, Majumder MK. Carbon nanotube: properties and applications. Carbon nanotube based VLSI interconnects. New Delhi: Springer; 2015. p. 17–37.

[143] Hirlekar R, Yamagar M, Garse H, Vij M, Kadam V. Carbon nanotubes and its applications: a review. Asian J Pharm Clin Res 2009;2(4):17–27.

[144] Zhang M, Li J. Carbon nanotube in different shapes. Mater Today 2009;12(6):12–18.

[145] Dilip K, Surendra T, Suresh KN, Roohi K. Nanostructured lipid carrier (Nlc) a modern approach for topical delivery: a review. World J Pharma Pharm Sci 2013;2(3):921–38.

[146] Shailesh L, Snehal RP, Ashwini P, Manoj S, Arvind B. Nanostructured lipid carriers in stability improvement forcosmetic nanoparticles. Int J Pharm Pharm Res 2016;6(1):168–80.

[147] Westesen K, Bunjes H, Koch MHJ. Physicochemical characterization of lipid nanoparticles and evaluation of their drug loading capacity and sustained release potential. J Control Rel 1997;48(2–3):223–36.

[148] Khan S, Baboota S, Ali J, Khan S, Narang RS, Narang JK. Nanostructured lipid carriers: an emerging platform for improving oral bioavailability of lipophilic drugs. Int J Pharm Invest 2015;5(4):182.

[149] Müller RH, Radtke M, Wissing SA. Solid lipid nanoparticles (SLN) and nanostructured lipid carriers (NLC) in cosmetic and dermatological preparations. Adv Drug Deliv Rev 2002;54:S131–55.

[150] Singh A, Sharma PK, Garg VK, Garg G. Hydrogels: a review. Int J Pharm Sci Rev Res 2010;4(2):016.

[151] Mamo B, Abebe W, Gabriel TD. Literature review on biodegradable nanospheres for oral and targeted drug delivery. Asian J Biomed Pharm Sci 2015;5(51):01.

[152] Bethany H. Zapping nanoparticles into nail polish. Laser Ablation Method Makes Cosmetic and Biomedical Coatings in a Flash 2017;95(12):9.

[153] Tripura P, Anushree H. Anushree novel delivery systems: current trend in cosmetic industry. Eur J Pharma Med Res 2017;4(8):617–27.

[154] Smijs TG, Pavel S. Titanium dioxide and zinc oxide nanoparticles in sunscreens: focus on their safety and effectiveness. Nanotechnol Sci Appl 2011;4:95.

[155] Glaser DA. Anti-aging products and cosmeceuticals. Facial Plast Surg Clin North Am 2004;12(3):363–72.

[156] Rosen J, Landriscina A, Friedman AJ. Nanotechnology-based cosmetics for hair care. Cosmetics 2015;2 (3):211–24.

[157] Hu Z, Liao M, Chen Y, Cai Y, Meng L, Liu Y, et al. A novel preparation method for silicone oil nanoemulsions and its application for coating hair with silicone. Int J Nanomed 2012;7:5719.

[158] Jimtaisong A, Saewan N. Utilization of carboxymethyl chitosan in cosmetics. Int J Cosmet Sci 2014;36 (1):12–21.

[159] Borowska S, Brzóska MM. Metals in cosmetics: implications for human health. J Appl Toxicol 2015;35 (6):551–72.

[160] Szmyd R, Goralczyk AG, Skalniak L, Cierniak A, Lipert B, Filon FL, et al. Effect of silver nanoparticles on human primary keratinocytes. Biol Chem 2013;394(1):113–23.

[161] Tucci P, Porta G, Agostini M, Dinsdale D, Iavicoli I, Cain K, et al. Metabolic effects of TiO_2 nanoparticles, a common component of sunscreens and cosmetics, on human keratinocytes. Cell Death Dis 2013;4(3):e549.

[162] Hackenberg S, Kleinsasser N. Dermal toxicity of ZnO nanoparticles: a worrying feature of sunscreen? Nanomedicine 2012;7(4):461–3.

[163] Kolosnjaj J, Szwarc H, Moussa F. Toxicity studies of fullerenes and derivatives. Bioapplications of nanoparticles. New York, NY: Springer; 2007. p. 168–80.

[164] Moore MN. Do nanoparticles present ecotoxicological risks for the health of the aquatic environment? Environ Int 2006;32(8):967–76.

[165] Simkó M, Mattsson MO. Risks from accidental exposures to engineered nanoparticles and neurological health effects: a critical review. Part Fibre Toxicol 2010;7:42.

[166] Handy RD, Owen R, Valsami-Jones E. The ecotoxicology of nanoparticles and nanomaterials: current status, knowledge gaps, challenges, and future needs. Ecotoxicology 2008;17(5):315–25.

[167] Tsuji JS, Maynard AD, Howard PC, James JT, Lam CW, Warheitand DB, Santamaria AB. Research strategies for safety evaluation of nanomaterials, part IV: risk assessment of nanoparticles. Toxicological Sci 2005;89 (1):42–50.

[168] Nohynek GJ, Lademann J, Ribaud C, Roberts MS. Grey goo on the skin?Nanotechnology, cosmetic and sunscreen safety. Crit Rev Toxicol 2007;37(3):251–77.

[169] Nohynek GJ, Dufour EK. Nano-sized cosmetic formulations or solid nanoparticles in sunscreens: a risk to human health? Arch Toxicol 2012;86(7):1063–75.

[170] Sadrieh N, Wokovich AM, Gopee NV, Zheng J, Haines D, Parmiter D, et al. Lack of significant dermal penetration of titanium dioxide from sunscreen formulations containing nano- and submicron-size TiO_2 particles. Toxicol Sci 2010;115(1):156–66.

[171] Buzea C, Pacheco II, Robbie K. Nanomaterials and nanoparticles: sources and toxicity. Biointerphases 2007;2 (4):MR17–71.

[172] Yah CS, Iyuke SE, Simate GS. A review of nanoparticles toxicity and their routes of exposures. Iran J Pharm Sci 2012;8(1):299–314.

[173] Paul J, Roel PFS. Toxicological characterization of engineered nanoparticles. Nanopart Technol Drug Delivery 2006;161–70.

[174] Wakefield G, Lipscomb S, Holland E, Knowland J. The effects of manganese doping on UVA absorption and free radical generation of micronised titanium dioxide and its consequences for the photostability of UVA absorbing organic sunscreen components. Photochem Photobiol Sci 2004;3(7):648–52.

[175] Ryman-Rasmussen JP, Riviere JE, Monteiro-Riviere NA. Penetration of intact skin by quantum dots with diverse physicochemical properties. Toxicol Sci 2006;91(1):159–65.

[176] Nel A, Xia T, Mädler L, Li N. Toxic potential of materials at the nanolevel. Science 2006;311(5761):622–7.

[177] Poon VK, Burd A. In vitro cytotoxity of silver: implication for clinical wound care. Burns 2004;30(2):140–7.

[178] Zhang XD, Wu HY, Wu D, Wang YY, Chang JH, Zhai ZB, et al. Toxicologic effects of gold nanoparticles in vivo by different administration routes. Int J Nanomed 2010;5:771.

[179] Mavon A, Miquel C, Lejeune O, Payre B, Moretto P. In vitro percutaneous absorption and in vivo stratum corneum distribution of an organic and a mineral sunscreen. Skin Pharmacol Physiol 2007;20(1):10–20.

[180] Chen J, Dong X, Zhao J, Tang G. In vivo acute toxicity of titanium dioxide nanoparticles to mice after intraperitioneal injection. J Appl Toxicol 2009;29(4):330–7.

[181] Sajid M, Ilyas M, Basheer C, Tariq M, Daud M, Baig N, et al. Impact of nanoparticles on human and environment: review of toxicity factors, exposures, control strategies, and future prospects. Environ Sci Pollut Res 2015;22(6):4122–43.

[182] Sánchez-Quiles D, Tovar-Sánchez A. Sunscreens as a source of hydrogen peroxide production in coastal waters. Environ Sci Technol 2014;48(16):9037–42.

[183] Das P, Xenopoulos MA, Metcalfe CD. Toxicity of silver and titanium dioxide nanoparticle suspensions to the aquatic invertebrate, *Daphnia magna*. Bull Environ Contam Toxicol 2013;91(1):76–82.

[184] Ma H, Williams PL, Diamond SA. Ecotoxicity of manufactured ZnO nanoparticles—a review. Environ Pollut 2013;172:76–85.

[185] Asharani PV, Lianwu YI, Gong Z, Valiyaveettil S. Comparison of the toxicity of silver, gold and platinum nanoparticles in developing zebrafish embryos. Nanotoxicology 2011;5(1):43–54.

CHAPTER

6

Polymeric nanocarriers for topical drug delivery in skin cream

M. Malathi, B.N. Vedha Hari and D. Ramyadevi

Department of Pharmaceutical Technology, School of Chemical and Biotechnology, SASTRA Deemed University, Thanjavur, Tamil Nadu, India

6.1 Introduction

Nanotechnology is a broad area that mainly covers the nanoparticle synthesis of different sizes, shapes, chemical composition, and their future use for human convenience. Even though well-defined and pure nanoparticles are produced by chemical and physical methods, those are costly and more toxic to the biological systems and the environment. So, the alternative to those designs is produced from natural sources such as plant extract and biological organisms in an eco-friendly manner. In recent years, the natural synthesis and product approaches are gaining much importance for the treatment of various diseases and disorders and also for targeted drug delivery system [1].

Crocus sativus L. is a plant that belongs to the family *Iridaceae* containing more than 150 volatile and nonvolatile compounds. It has been cultivated in India, Iran, and Greece, and the compound derived from *C. sativus L.* such as safranal and picrocrocin are responsible for its bitter taste. The active ingredient which has been derived from *C. sativus L.* has much pharmacological actions such as anticancer, antitussive, antioxidant, anxiolytic, aphrodisiac, antinociceptive, anticonvulsant, antihypertensive, cytotoxic effects, antidepressant, antigenototoxic, antiinflammatory, and relaxant activity. Saffron-based products are available commercially for culinary as well as cosmetic purposes in the form of spice and creams, respectively [2].

Creams are semisolid emulsions such as oil-in-water emulsion or water-in-oil emulsion, which can be applied topically intended for therapeutic activity or cosmetic use [3].

The aim of the present work is to extract the water soluble active ingredients of *C. sativus L.* and develop its polymeric nanoparticles and incorporate into cream base for topical application.

Nanocosmetics
DOI: https://doi.org/10.1016/B978-0-12-822286-7.00006-1

6.2 Materials and methods

6.2.1 Materials

The saffron stigmas were purchased from the local supplier, Thanjavur district, Tamil Nadu, India. Polycaprolactone (PCL) and polyvinylalcohol (PVA) were purchased from Sigma Aldrich Pvt Ltd., Mumbai, India. Stearic acid, Bees wax, Propylene Glycol, Triethanolamine, and other chemicals and buffer reagents were obtained from SD Fine Chem. Ltd, Chennai, India.

Aqueous extract of saffron stigmas was prepared by adding 10 mg saffron in 10 mL of distilled water and keeping it overnight at room temperature, for maceration of the material [4].

6.2.2 Formulation design

6.2.2.1 Factorial design experiment

The Design Expert Software (Version 8.0) was used to design the formulation compositions and was performed using a minimum number of experiments to give maximum information. Two different parameters of nanoparticles such as size (nm) and % Entrapment efficiency were taken as the dependent variable and evaluated using 3^2 factorial design with Response Surface Methodology (RSM). Various trials were made to evaluate the experimental error and optimize the formulation design. The lower and higher level of independent variables, namely, polymer and surfactant was taken as 5–25 mg and 0.5%–2.5%, respectively.

The regression equation was calculated for response (Y) by using the formula

$$Y = b_0 + b_1X_1 + b_2X_2 + b_3X_1X_2 + b_4X_1^2 + b_5X_2^2$$

where X_1 and X_2 are the independent or input variables used when one factor is modified at a time from least to the highest value and b represents the coefficient of term X.

The term X_1X_2 denotes the changes in response when two factors are changed, and ${X_1}^2$ and ${X_2}^2$ are used to study the nonlinearity [5].

6.2.2.2 Formulation of PCL polymeric nanocarriers

The polymeric nanoparticles (nanocarriers) were formulated by solvent evaporation technique using PCL polymer and PVA as surfactant, for the encapsulation of *C. sativus L.* extract. The aqueous extract of saffron stigmas was prepared and the 1% PVA surfactant was dissolved in the aqueous media and sonicated for 5 minutes in a bath sonicator for its complete solubility. The organic phase was prepared by dissolving the weighed quantity of PCL polymer in acetone and sonicated for 5 minutes for its complete dissolving. The organic solution was added dropwise into the aqueous mixture by keeping it in a probe sonicator for 20 minutes to ensure total evaporation of the organic solvent. During the sonication, the nanoparticles were formed and it was stored at room temperature and checked for the stability [6].

6.2.2.3 *Formulation of cream*

The oil-in-water (O/W) emulsion-based cream was prepared using stearic acid, bees wax, and mineral oil acted as the oil phase, which was heated until the ingredient melts completely. The optimized nanosuspension (of three different quantities for C_1, C_2, and C_3 formulation) and water were mixed with triethanolamine and heated up to 85°C to obtain the aqueous phase. The oil phase was added slowly to the aqueous phase with continuous and fast stirring until the smooth cream was formed. Then, it was cooled and stored at room temperature. Formulation C_4 was prepared with the crude saffron extract incorporated into the cream base. Table 6.1 shows the composition of cream formulation [7].

6.2.3 Characterizations

6.2.3.1 *Characterization of polymeric nanocarriers*

The particle size and surface charge were analyzed for the nanoparticle formulation using a zeta sizer (Malvern Nano Series ZS, UK), based on the principle of dynamic light scattering and charge conductivity, respectively. The size distribution and zeta potential were measured for the nanoparticle formulation to ensure the monodispersity and stability [8].

The optimized nanosuspension was ultracentrifuged to separate the nanoparticles as pellet and resuspended again and then freeze dried. The lyophilized nanoparticle was evaluated for the surface morphology using SEM (Jeol, Japan). For SEM analysis, the lyophilized sample was placed on a double-sided tape kept on a metal stub. This was then platinum coated under reduced pressure and was placed on the sample holder for the SEM analysis at 25 kV [9].

XRF (SPECTRA plus) analysis was performed for the elemental characterization of the herbal formulation and was considerably used for the detection of glass and metals. The sample when subjected to the analysis gets excited by making it bombarded due to the high amount of energy and it undergoes ionization and it emits one or more electrons. When the radiation gets emitted, the substance loses its stability. This emission helps to identify the substance with respect to each element of its characteristic energy. For the bulk sample of nanoparticles, the pellet technique was used with the binder in the ratio of 1:1 to 1:10 [10]. The conditions maintained during the study were nitrogen gas

TABLE 6.1 Formulation of nanoparticles incorporated into cream.

Formulation code	Stearic acid (g)	Bees wax (g)	Propylene glycol (g)	Triethanolamine (mL)	Nanosuspension (mL)	Water (mL)	Total (g)
C1	3.75	2.625	3.75	0.75	5	13	30
C2	3.75	2.625	3.75	0.75	10	8	30
C3	3.75	2.625	3.75	0.75	15	3	30
C4 (crude extract)	3.75	2.625	3.75	0.75	3	15	30

atmosphere, 34 mm manual mask, a flow counter, a scintillation counter, a 60 position cup loader with 10 cups mounted with 34 mm masks, analyzer crystals XS-55, PET and LiF200, two collimators, S8 tools and a remote access software are installed, TOUCH, a hardware component, Cu Kα x-ray tube, Goniometer, and the detector.

For evaluation of entrapment efficiency, the aqueous *C. sativus L.* extract was suitably diluted using distilled water and scanned for its absorbance maxima in UV-visible spectrophotometer (Perkin Elmer, USA). The absorbance was measured for different concentrations of solutions and the calibration curve was plotted using concentration against absorbance, to check the linearity and regression coefficient.

The formulations prepared with various concentrations of PCL polymer and PVA surfactant were centrifuged for 10 minutes at 2000 rpm to separate the microparticles. The supernatant was taken and again centrifuged at 12,000 rpm for 30 minutes at 4°C. After the ultracentrifugation, the supernatant was taken and suitably diluted to measure the absorbance at λ_{max} and the drug concentration in supernatant was calculated using the calibration data [11]. The % entrapment efficiency was calculated using the formula

$$\%\ \text{Entrapment Efficiency} = \frac{\text{Total drug content} - \text{Drug content in supernatant}}{\text{Total drug content}} \times 100$$

To evaluate the in vitro anti-oxidant activity, the reducing power for the nanoparticle formulation and crude extract was measured by the Ferric Reducing Antioxidant Power (FRAP) method using ascorbic acid as the standard. The different concentrations of the polymeric nanoparticle formulation and the crude extract (100, 200, 300, 400, and 500 μg/mL) were mixed with 2.5 mL of potassium ferricyanide and 2.5 mL of phosphate buffer and then heated at 50°C for 20 minutes using thermostatic water bath. After heating, 2.5 mL of trichloro acetic acid was added and then centrifuged at 2000 rpm for 10 minutes. Then, 2.5 mL was taken from the supernatant and was mixed with 2.5 mL of distilled water and 2.5 mL of ferric chloride. These samples were analyzed under a UV-Visible spectrophotometer at 700 nm [12].

To study the in vitro anti-inflammatory activity, the saffron extract and the nanoparticle formulation (1 mg/mL) were prepared freshly and examined for its anti-inflammatory activity. Different concentrations of extract and the nanoparticle formulation were taken in separate test tubes. To that sample, 0.2 mL of fresh egg white and 2.8 mL of phosphate buffer (pH 6.4) were added and made up to the final concentrations as 20, 40, 60, 80, and 100 μg/mL. About 2.8 mL of distilled water was taken as the control. Then, the samples were kept for incubation at 37°C ± 2°C for 15 minutes and then heated at 70°C for 5 minutes. Then, it was cooled and the samples were analyzed under a UV-Visible spectrophotometer for its absorbance at 700 nm with respective blank solution.

The reference drug used was Diclofenac sodium, which has been treated similarly with fresh egg white and measured for the absorbance. Then, the percentage inhibition for denaturation of protein was calculated using the formula [13]

$$\%\ \text{Inhibition} = \frac{Vt}{Vc - 1} \times 100$$

where Vc is absorbance of control and Vt is absorbance of test sample.

6.2.3.2 Evaluation of cream containing the saffron-loaded polymeric nanoparticles

The cream formulations were first observed for its physical characteristics such as color, odor, and gritty particles. All the formulated creams were tested for its homogeneity by visual inspection [14]. Besides, the prepared cream formulation which has been stored at room temperature was tested for its pH measurement using pH paper and pH meter (ELICO LI120, Hyderabad, India) [15]. Brookfield Digital Viscometer (Brookfield Engineering Laboratories, USA) was used to measure the viscosity, using spindle No. 64 at 20 rpm and the readings were measured after six complete revolutions. All the readings were measured in triplicate to calculate average and presented with standard deviation [16].

The drug content in the prepared cream was evaluated by taking 1 g of cream which was dissolved in distilled water and was diluted for the required concentration to measure its absorbance at λ_{max}. The concentration of the saffron extract was calculated using the standard calibration data [17].

For the study of in vitro drug release, the dissolution of all the nanoparticle formulations and crude extract was performed by the modified basket method using a USP XII Type 2 dissolution apparatus to study the percentage drug release. Accurately 1 mL sample from each formulation was taken in a small two side open ended tube, one side of which was tied with dialysis membrane and placed inside the basket. The dissolution study was performed for 8 hours using 100 mL distilled water as media. For each time interval, 10 mL of media was withdrawn and replaced with the same amount of the fresh media to maintain the sink condition and the samples were measured for its absorbance at λ_{max} [18].

In vitro dissolution studies were performed for creams using the dialysis bag method for 8 hours using distilled water as media. This was performed using a magnetic stirrer by taking the cream containing 1 mg equivalent of the extract. For each time interval, 5 mL of media was replaced fully with fresh media and the samples were analyzed using a UV-Vis spectrophotometer at λ_{max} [19].

Regarding the drug release kinetics, the kinetic modeling was carried out to understand the drug release mechanism from the crude sample, nanoformulation, and creams. In this analysis, the results of dissolution data were fitted with diverse kinetic models like first order, zero order, Higuchi's model, Korsemeyer–Peppas model, Hixson–Crowell model, Hopfenberg model, Baker–Lonsdale model, Makoid-Banakar model, Weibull model, and Gompertz model. The R^2, K, SS, and n values were determined from the linear curve plots, from which the mode of drug release and kinetics could be correlated [20].

6.3 Results and discussion

6.3.1 Evaluation of saffron-loaded polymeric nanoparticles

6.3.1.1 Factorial design for the formulation of polymeric nanocarriers

A 3^2 factorial design was performed to study about two factors using three levels as −1, 0, and 1, identified as low, medium, and high, respectively. Here, the polymer (X_1) and surfactant concentration (X_2) were used as independent variables and the particle size and entrapment efficiency were measured as responses Y_1 and Y_2, respectively [21].

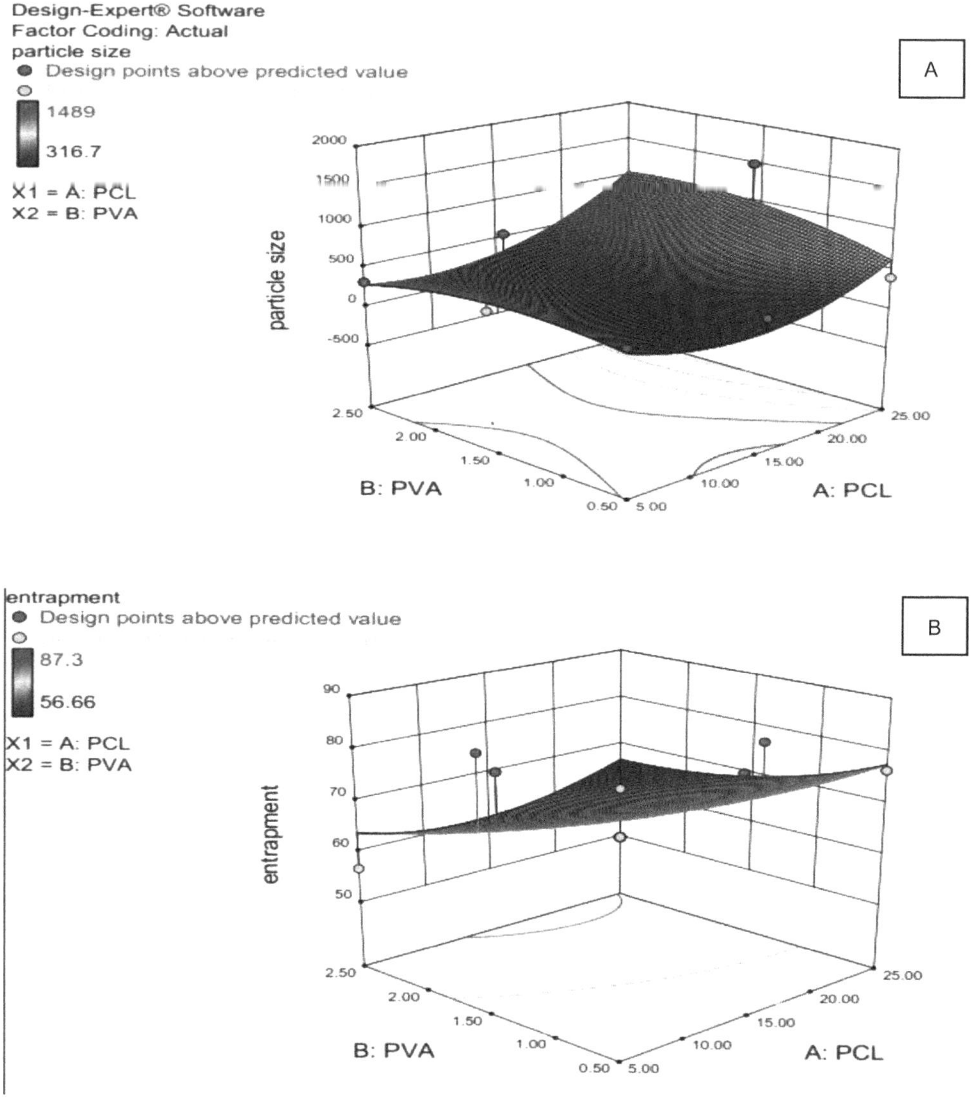

FIGURE 6.1 (A) 3D graph of particle size using RSM; (B) 3D graph of entrapment efficiency using RSM.

Fig. 6.1A and B shows the 3D image depicting the effect of polymer and surfactant on particle size and entrapment efficiency, respectively. The values of the dependent and independent variables were shown in Table 6.2.

6.3.1.2 Measurements of particle size and zeta potential

The average particle size of the formulated nanoparticles of *C. sativus L.* extract was found to be between the range of 300–700 nm, except F8 and F9 formulation, which

TABLE 6.2 3^2 factorial design for the preparation of nanoformulation.

Formulation code	Independent	Variables	Dependent	Variables (response)
	X_1 (PCL)	X_2 (PVA)	Y_1 (particle size) (nm)	Y_2 (Entrapment efficiency) (%)
F1	−1	−1	453.4	85.11
F2	−1	0	403.2	85.115
F3	−1	1	316.7	56.66
F4	0	−1	332.8	87.3
F5	0	0	333.9	63.22
F6	0	1	549.7	69.79
F7	1	−1	404.7	76.36
F8	1	0	1489	69.79
F9	1	1	797.7	56.66
	For X_1		For X_2	
Low	−1 = 5 mg		−1 = 0.5%	
Medium	0 = 15 mg		0 = 1.5%	
High	1 = 25 mg		1 = 2.5%	

contained a higher concentration of surfactant and polymer. The particle size decreased at the lower polymer concentration and was found to be independent of the surfactant level. When both the polymer and surfactant concentration increased, the particle size also increased. The zeta potential was measured as the surface charge of the nanoparticles to check its stability [22]. Polydispersity index (PDI) was checked to confirm the narrow size distribution range. The sizes, zeta potential, and PDI of each trial formulation of nanoparticles were shown in Table 6.3. The optimized formulation F4 exhibited an average size of 300 nm with PDI as 0.032, indicating monodisperse nature of the particles. The zeta potential was higher in F4, compared to all other trials, indicating better stability.

6.3.1.3 Entrapment efficiency

The amount of *C. sativus L.* extract which was entrapped into the polymeric nanoparticles was determined using a UV-Vis spectrophotometer using the calibration plot shown in Fig. 6.2. It was observed that as the surfactant level increased the entrapment efficiency decreased for the constant polymer level. This was due to the higher solubility of extract in the higher surfactant level. Also, the higher concentration of polymer decreased the entrapment efficiency due to the precipitation of excess polymer during the preparation. The highest entrapment efficiency was found to be 87.3, 85.115, and 85.11 for the formulations F4, F2, and F1, respectively, for the lower surfactant and polymer level [23]. The % entrapment efficiency of all the formulations was depicted in Table 6.3.

TABLE 6.3 Particle size, zeta potential, and entrapment efficiency of nanoparticles.

Formulation code	Particle size (nm)	Zeta potential (mV)	Entrapment efficiency (%)	Polydispersity index (PDI)
F1	453.4	−0.49	85.11	0.297
F2	403.2	−0.13	85.115	0.305
F3	316.7	−1.65	56.66	0.303
F4	332.8	−5.61	87.3	0.032
F5	333.9	−2.05	63.22	0.582
F6	549.7	−0.20	69.79	0.620
F7	404.7	−1.05	76.36	0.223
F8	1489	−0.59	69.79	0.578
F9	797.7	−1.92	56.66	2.353

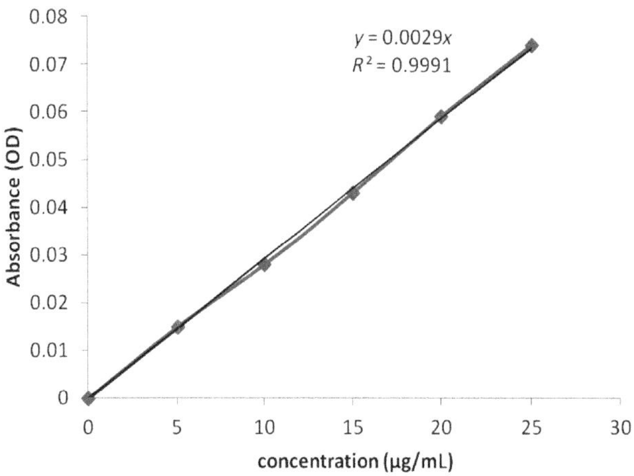

FIGURE 6.2 Calibration curve of saffron in distilled water.

6.3.1.4 *Morphological studies*

SEM images of the optimized nanoparticles was observed at 30,000 × magnification to clearly display the structural and surface characteristics of the extract loaded polymeric nanoparticle formulation. Fig. 6.3 shows the size of the particles in the range of 100–350 nm and the particles were found to be uniformly spherical in shape with smooth surface without any pores [24].

The quantitative analysis by X-ray fluorescence spectroscopy was used to determine the concentration of trace elements present in *C. sativus L.* extract formulation in oxide form. The oxides $SO_3$and CaO were present in the highest percentage concentration as 14.81% and 13.07%, respectively, and the element CuO (0.77%) and ZnO (0.74%) were present in

FIGURE 6.3 SEM for optimized nanoformulation.

the lowest percentage concentration. The elements O and Cl had the higher percentage concentration like 27.22% and 23.22%, respectively, whereas both Cu and Zn had the lowest percentage concentration as 0.61% [25]. The XRF data were shown in Table 6.4.

6.3.1.5 Anti-oxidant activity

The reducing power assay proved that both the nanoparticle formulation and the crude saffron extract possessed anti-oxidant activity at different concentrations and the results are shown in Table 6.5. When comparing the crude extract with the nanoformulation, the crude extract showed more anti-oxidant activity due to the free drug present in it, whereas in nanoformulation the drug was encapsulated into a polymer matrix. Yet the activity was lesser, as compared with the standard Ascorbic acid. As the concentration was increased, the value of absorbance also increased, and hence it can be concluded that the *C. sativus L.* possessed anti-oxidant activity both in crude extract and in nanoformulation, especially at higher concentrations [26].

6.3.1.6 Anti-inflammatory activity

The in vitro anti-inflammatory activity was analyzed for the various concentrations of crude extract and for the nanoparticles and the % inhibition was shown in Table 6.6. The results explicitly illustrate that the crude extract possessed the greater anti-inflammatory activity at a lower concentration due to free drug whereas the nanoformulation showed better activity only at a higher concentration because of the encapsulation of extract in polymer. Its activity was in the range between 66% and 70%. Both crude extracts and the nanoparticle formulation were compared with Diclofenac

TABLE 6.4 Evaluation of elemental content of *Crocus sativus* nanoparticles by XRF.

Element in oxide form Formula	Concentration (%)	Element form Formula	Concentration (%)
SO_3	14.81	O	27.22
CaO	13.07	Cl	23.22
Al_2O_3	9.16	Ca	9.34
PbO	8.27	Pb	7.67
Na_2O	8.04	K	6.67
K_2O	8.03	Na	5.96
SiO_2	6.29	S	5.93
Fe_2O_3	4.18	Al	4.85
MgO	3.43	Si	2.94
CuO	0.77	Fe	2.92
ZnO	0.74	Mg	2.07
–	–	Cu	0.61
–	–	Zn	0.61

TABLE 6.5 In vitro antioxidant activity of nanoparticles.

S. No.	Concentration (μg/mL)	Ascorbic acid	Saffron crude extract	Nanoparticle formulation
1	100	0.462	0.083	0.013
2	200	1.027	0.107	0.049
3	300	1.588	0.134	0.064
4	400	1.798	0.135	0.098
5	500	2.185	0.141	0.119

TABLE 6.6 Anti-inflammatory activity of crude extract and its nanoformulation.

S. No.	Concentration (μg/mL)	% Inhibition saffron extract	% Inhibition nanoformulation	% Inhibition of Diclofenac sodium
1	20	76.91	66.66	79.5
2	40	86.22	67.46	100.99
3	60	93.87	68.39	116.6
4	80	103.06	68.86	142.58
5	100	109.84	69.72	153.30

sodium (standard drug), which showed that they had lesser anti-inflammatory activity than the standard control. Yet, at a lower concentration of 20 μg/mL, the crude extract had a similar effect as the standard whereas at higher concentrations, the activity was found to be lesser than the standard [27].

6.3.2 Evaluation of cream formulation

The formulated creams were observed to be slightly orange in color, odorless, smooth and no gritty particles present in it. All cream formulations were homogeneous and uniform without any bubbles and clumps [28]. The pH of all cream formulations was found to be neutral and it was in the range of 6.8–7. The pH values are shown in Table 6.7 [29].

The viscosity of each cream formulation was shown in Table 6.7 and it was found to be in the range of 4000–12,500 cps. From the data, it was clear that as the polymer and drug concentration increased the viscosity also increased. The cream formulation C4 containing crude extract without any polymer showed lesser viscosity, compared to the other batches [30].

The drug content for each cream formulation was determined using water as a media by measuring the absorbance of the suitably diluted solutions using a UV-Vis spectrophotometer at 439 nm [31]. The results were shown in Table 6.7 and it was found to be in the range of 90%–95%.

6.3.2.1 In vitro dissolution studies

The crude saffron extract showed around 72% drug release up to 8 hours, and the nanoparticles of the extract (F1, F2, F4, and F7) which contains the lowest surfactant level showed the gradual increase in drug release up to the range of 70%–88% at the end of 8 hours. The other formulations (F3, F5, F6, F8, and F9) showed 100% drug release at 30 minutes, 2.5 hours, 3 hours, 6 hours, and 1 hour, respectively. The results suggested that when the surfactant concentration increased the drug release also increased and reached the maximum within a short period of time. The release of *C. sativus L.* extract from the polymeric nanoparticle depends on the surfactant concentration because it enhanced the solubility, so the release was more. The lower concentrations of surfactant and high concentration of polymer in the nanoparticle lead to high entrapment of the extract which resulted in the sustained release of the extract. When the concentration of polymer was further increased, the drug release was found to be increasing because the excess polymer precipitated out during the preparations which lead to lesser

TABLE 6.7 Viscosity, pH, and drug content for creams.

S. No.	Formulation	Viscosity (cps)	pH	% Drug content
1	C1	6375.66	6.8	94
2	C2	10,253	6.8	95
3	C3	12,457	6.9	95
4	C4	4335	7	93

entrapment [32]. The nanoformulation showed maximum release than the crude extract because of the small size particles having a greater surface area, which extends the drug release. The crude extract showed lesser release because there was no uniform size and shaped particles present in it [33].

The cream formulation containing nanosuspension showed maximum drug release, because of the small particle size that enhanced the solubility and permeability. The cream formulation prepared using the crude extract showed less drug release due to its poor solubility. The cumulative percentage of drug release of each cream formulation increased gradually as the time increased and at 7 hours the formulations C1 and C2 reached 100%. The drug release of formulation C3 reached 80% at 8 hours, whereas the formulation C4 which was prepared using the crude saffron extract without any polymer or surfactant showed only 56% at 8 hours. When the amount of nanoparticles added was less, the drug release was more, whereas at higher concentration, the drug release was retarded [34]. Figs. 6.4 and 6.5 show the dissolution profile of the nanosuspensions and creams, respectively.

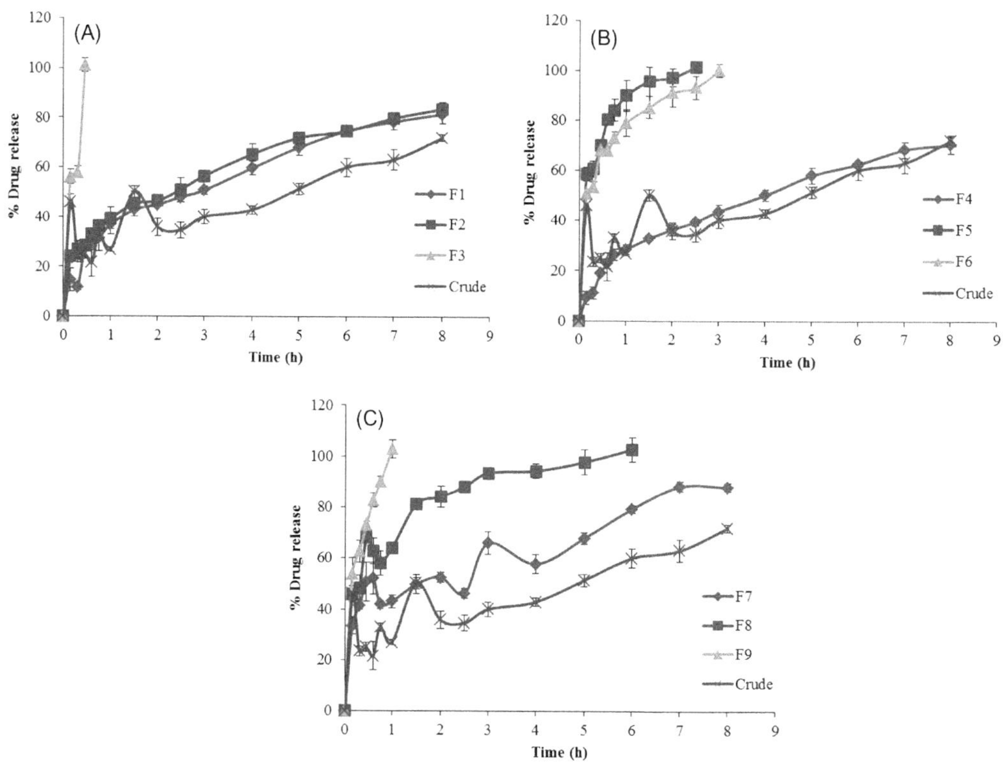

FIGURE 6.4 In vitro Drug release profile of *Crocus sativus L.* extract nanosuspension compared with crude extract.

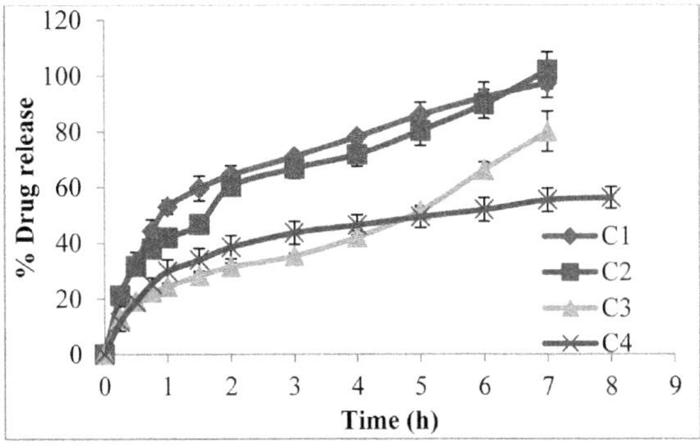

FIGURE 6.5 Drug release profile of creams containing polymeric nanoparticles of *Crocus sativus* extract.

6.3.2.2 *Release kinetics*

Kinetic modeling of the dissolution studies was performed by fitting in vitro release data of the nanoparticles and creams into different kinetic equations. The R^2 value which was obtained from the various models showed that the drug release followed different mechanisms based on Higuchi, Korsemeyer–Peppas, Baker–Lonsdale, Makoid-Banakar, and Weibull kinetic models. The nanoparticles and creams predominantly followed Higuchi and Korsemeyer–Peppas model, which explained the drug release based on diffusion mechanism from the polymeric systems. The "n" value <0.45 was seen in most of the nanoparticles which stipulate the Fickian diffusion transport, whereas "n" value >0.45 in creams indicated the non-Fickian mode of diffusion [35]. The release for the nanoparticles of F1, F2, F4, F5, F6, F8 and C_1, C_2, C_3 also obeyed the Makoid-Banakar and Weibull model which was due to the interaction between drug and polymer that also implied the release from matrix type of system [36]. The formulations F1 and F2 followed the Baker–Lonsdale model which explained the drug release from the spherical matrices [37]. The release kinetic data of the nanoparticles and creams were shown in Tables 6.8 and 6.9, respectively.

6.4 Conclusion

The polymeric nanoparticles provide a significant potential for drug delivery due to their ability of encapsulating the drug. The use of polymeric nanoparticle for topical administration is very challenging and an attractive application area. The polymeric nanoparticles pose the ability to penetrate through the skin barriers providing the expected release and inspiring applications on the field of controlled or sustained drug delivery. In this work, the PCL polymeric herbal nanoparticles from aqueous *C. sativus L.* extract were successfully formulated using the solvent evaporation method and optimized using RSM. The herbal nanoparticles were incorporated into a cream formulation for effective topical application. The physiochemical parameters and characterization study results revealed that the developed system is a suitable tool for improving the efficacy of the herbal extract.

TABLE 6.8 Release kinetics of *Crocus sativus L.* nanoparticle formulation.

Models		F1	F2	F3	F4	F5	F6	F7	F8	F9
Zero	R^2	0.531	0.255	−4.023	0.660	−3.303	−2.115	−0.507	−1.023	−4.013
	K_0	12.87	13.40	18.75	10.93	18.73	18.82	14.08	19.11	17.48
	SS	4203.25	6034.05	54,410.10	2375.17	44,109.57	33,699.91	10,872.11	24,714.55	47,260.64
First	R^2	0.860	0.753	0.925	0.879	0.923	0.905	0.235	0.843	0.866
	K_1	0.273	0.315	4.836	0.192	3.165	2.215	0.389	1.595	3.498
	SS	1249.22	2000.45	812.21	843.43	785.93	1017.43	5510.91	1908.85	1257.08
Higuchi	R^2	0.972	0.927	−1.218	0.992	−0.630	0.013	0.579	0.493	−1.154
	K_H	30.52	32.21	49.97	25.69	49.23	48.35	34.48	47.82	46.73
	SS	244.77	590.04	24,027.05	54.54	16,709.62	10,675.20	3032.49	6186.93	20,311.61
Korsemeyer–Peppas	R^2	0.985	0.992	0.854	0.993	0.899	0.952	0.884	0.977	0.815
	K_{KP}	33.74	39.21	34.52	26.65	82.89	76.21	48.78	69.92	82.56
	SS	130.21	57.47	69.37	44.47	1027.01	512.57	835.90	276.74	1735.78
	N	0.429	0.359	0.392	0.474	0.108	0.161	0.243	0.219	0.070
Hixson–Crowell	R^2	0.792	0.651	−1.166	0.826	−0.484	0.100	0.065	0.495	−0.963
	K_{HC}	0.074	0.084	0.211	0.054	0.206	0.204	0.099	0.207	0.203
	SS	1861.75	2827.97	23,468.81	1209.34	15,222.06	9727.89	6739.25	6166.70	18,506.77
Hopfenberg	R^2	0.860	0.753	0.934	0.879	0.923	0.905	0.235	0.843	0.866
	SS	1249.54	2001.63	713.86	843.62	786.17	1017.81	5511.53	1909.51	1257.48
Baker–Lonsdale	R^2	0.984	0.976	−0.626	0.989	−0.070	0.417	0.718	0.702	−0.337
	K_{BL}	0.023	0.027	0.062	0.015	0.063	0.062	0.034	0.062	0.062
	SS	143.31	194.29	17,615.27	70.89	10,968.67	6301.57	2029.05	3634.99	12,604.05

Makoid-Banakar	R^2	0.985	0.995	0.870	0.993	0.992	0.993	0.942	0.978	0.943
	K_{MB}	33.75	38.21	101.67	26.66	95.93	84.11	42.83	71.27	98.61
	SS	130.21	37.08	1400.29	44.46	73.76	68.99	412.55	259.78	531.20
Weibull	R^2	0.979	0.977	0.926	0.990	0.967	0.984	0.854	0.951	0.887
	A	2.385	1.952	0.173	3.206	0.458	0.612	1.435	0.748	0.493
	β	0.620	0.529	1.178	0.633	0.580	0.532	0.356	0.534	0.206
	SS	182.82	180.82	792.46	68.46	332.58	167.31	1052.20	591.30	1061.35
Gompertz	R^2	0.951	0.933	0.909	0.9659	0.953	0.961	0.817	0.910	0.894
	A	0.985	0.837	0.035	1.239	0.128	0.198	0.646	0.263	0.134
	β	1.252	1.097	3.587	1.130	2.015	1.805	0.806	1.691	1.875
	SS	436.22	540.79	980.85	238.40	478.75	416.10	1316.29	1097.27	994.96

TABLE 6.9 Release kinetics of creams.

Models		C1	C2	C3
Zero	R^2	0.573	0.770	0.889
	K_0	17.277	17.339	12.476
	SS	6307.37	4016.71	1319.17
First	R^2	0.877	0.825	0.824
	K_1	0.582	0.470	0.207
	SS	1810.96	3059.80	2089.05
Higuchi	R^2	0.941	0.923	0.825
	K_H	41.529	40.806	28.409
	SS	869.57	1342.44	2082.08
Korsemeyer–Peppas	R^2	0.951	0.926	0.896
	K_{KP}	46.069	38.030	16.265
	SS	717.30	1290.45	1235.74
	N	0.431	0.546	0.850
Hixson–Crowell	R^2	0.878	0.812	0.844
	K_{HC}	0.200	0.117	0.059
	SS	1799.78	3283.34	1850.90
Hopfenberg	R^2	0.877	0.825	0.894
	SS	1811.42	3059.65	1257.54
Baker–Lonsdale	R^2	0.913	0.841	0.767
	K_{BL}	0.054	0.047	0.018
	SS	1284.40	2781.99	2773.96
Makoid-Banakar	R^2	0.960	0.963	0.721
	K_{MB}	43.974	35.009	34.651
	SS	591.77	642.99	532.64
Weibull	R^2	0.898	0.835	0.902
	A	1.512	3.192	37,278,489.4
	β	0.782	1.157	6.041
	SS	1505.92	2880.67	1164.65
Gompertz	R^2	0.857	0.766	0.735
	A	0.606	0.747	1.599
	β	2.118	2.180	1.817
	SS	2104.19	4088.48	3152.48

References

[1] Mohammad AW, Lau CH, Zaharim A, Omar MZ. Elements of nanotechnology education in engineering curriculum worldwide. Procedia Soc Behav Sci 2012;60:405–12.

[2] Hadi M, Modaghegh MS, Esmaeili H-A, Rajbai O, Hosseinzadeh H. Safety evaluation of saffron (*Crocus sativus*) tablets in healthy volunteers. Phytomedicine 2008;15:1032–7.

[3] Eccleston GM. Functions of mixed emulsifiers and emulsifying waxes in dermatological lotions and creams. Colloids Surf A:Physicochem Eng Asp 1997;123-124:169–82.

[4] Sánchez AM, Carmona M, del Campo CP, Alonso GL. Solid-phase extraction for picrocrocin determination in the quality control of saffron spice (*Crocus sativus* L.). Food Chem 2009;16:792–8.

[5] Derakhshandeh K, Erfan M, Dasashzadeh S. Encapsulation of 9-Nitrocamptothecin, a novel anticancer drug, in biodegradable nanoparticles: factorial design, characterization and release kinetics. Eur J Pharm Biopharm 2007;66:34–41.

[6] Peter Christoper GV, VijayaRaghavan C, Siddharth K, Siva SelvaKumar M, Hari Prasad R. Formulation and optimization of coated PLGA–Zidovudine nanoparticles using factorial design and in vitro in vivo evaluations to determine brain targeting efficiency. Saudi Pharm J 2014;22:133–40.

[7] Kadam PV, Bhingare CL, Nikam RY, Pawar SA. Development and validation of UV spectrophotometric method for the estimation of Curcumin in cream formulation. Pharm Methods 2013;4:43–5.

[8] de pinho Neves AL, Milioli CC, Muller L, Riella HG, Kuhnen NC, Stulzer HK. Factorial design as tool in chitosan nanoparticles development by ionic gelation technique. Colloids Surf A Eng Asp 2014;445:34–9.

[9] Azzam EMS, Sayyah SM, Taha AS. Fabrication and characterization of nanoclay composites using synthesized polymeric thiol surfactants assembled on gold nanoparticles. Egypt J Pet 2013;22:493–9.

[10] Aldwayyan AS. Synthesis and characterisation of CdO nanoparticles starting from organometallic Dmphen-CdI complex. Int J Electrochem Sci 2013;8:10506–14.

[11] Shah U, Joshi G, Sawant K. Improvement in antihypertensive and antianginal effects of felodipine by enhanced absorption from PLGA nanoparticles optimized by factorial design. Mater Sci Eng C 2014;35:153–63.

[12] Dar RA, Brahman PK, Khurana N, Wagay JA, Lone ZA, Ganaie MA, et al. Evaluation of antioxidant activity of crocin, podophyllotoxin and kaempferol by chemical, biochemical and electrochemical assays. Arab J Chem 2013;84.

[13] Chandra S, Dey P, Bhattacharya S. Preliminary in vitro assessment of anti-inflammatory property of *Mikania scandens* flower extract. J Adv Pharm Educ Res 2012;2(1):25–31 ISSN 2249-3379.

[14] Smaoui S, Hlima HB, Chobba IB, Kadri A. Development and stability studies of sunscreen cream formulations containing three photo-protective filters. Asian J Chem 2013;10:S1216–22.

[15] PurushothamRao K, Khaliq K, Kharat SS, Sagare P, Patil SK. Preparation and evaluation o/w cream for skin psoriasis. Int J Pharm Biosci 2010;1(3).

[16] Bono A, Anisuzzaman SM, Ding OW. Effect of process conditions on the gel viscosity and gel strength of semi-refined carrageenan (SRC) produced from seaweed (*Kappaphycus alvarezii*). J King Saud Univ Eng Sci 2014;26:3–9.

[17] Jian H, Zhu L, Zhang W, Sun D, Jiang J. Galactomannan (*from Gleditsiasinensis Lam.*) and xanthan gum matrix tablets for controlled delivery of theophylline: invitro drug release and swelling behaviour. Carbohydr Polym 2012;87:2176–82.

[18] Du Y, Zhai Y, Zhang J, Wu C, Luo C, He Z. Development and evaluation of taste-masked dry suspension of cefuroxime axetil for enhancement of oral bioavailability. Asian J Pharm Sci 2013;8:287–94.

[19] Araújo FA, Kelmann RG, Araújo BV, Finatto RB, Teixeira HF, Koester LS. Development and characterization of parenteral nanoemulsions containing thalidomide. Eur J Pharm Sci 2011;42:238–45.

[20] Bose A, Wong TW, Singh N. Formulation development and optimization of sustained release matrix tablet of ItoprideHCl by response surface methodology and its evaluation of kinetic release. Saudi Pharm J 2013;21:201–13.

[21] Hermans K, den Plas DV, Everaert A, Weyenberg W, Ludwig A. Full factorial design, physicochemical characterisation and biological assessment of cyclosporine A loaded cationic nanoparticles. Eur J Pharm Biopharm 2012;82:27–35.

[22] Song X, Zhao Y, Hou S, Xu F, Zhao R, He J, et al. Dual agents loaded PLGA nanoparticles: Systematic study of particle size and drug entrapment efficiency. Eur J Pharm Biopharm 2008;69:445–53.

[23] Vijayan V, Aafreen S, Sakthivel S, Ravindra Reddy K. Formulation and characterization of solid lipid nanoparticles loaded Neem oil for topical treatment of acne. J Acute Dis 2013;282–6.
[24] Sokolova V, Ludwig A-K, Hornung S, Rotan O, Horn PA, Epple M, et al. Characterization of exosomes derived from human cells by nanoparticle tracking analysis and scanning electron microscopy. Colloids Surf B Biointerfaces 2011;87:146–50.
[25] Omote J, Kohno H, Toda K. X-ray fluorescence analysis utilizing the fundamental parameter method for the determination of the elemental composition in plant samples. Anal Chim Acta 1995;307:117–26.
[26] Dewan SMR, Amin MN, Adnan T, Uddin SN, Shahid-Ud-Daula AFM, Sarwar G, et al. Investigation of analgesic potential and invitro antioxidant activity of two plants of Asteraceae family growing in Bangladesh. J Pharm Res 2013;6:599–603.
[27] Gondkar AS, Deshmukh VK, Chaudhari SR. Synthesis, characterization and invitro anti-inflammatory activity of some substituted 1,2,3,4 tertahydropyrimidine derivatives. Drug Invent Today 2013;5:175–81.
[28] Joshi SS, Barhate SD. Physical characteristic of three component creams containing span (60,80) as surfactants. Der Pharm Sin 2011;2(5):81–7.
[29] Patil MVK, Kandhare AD, Bhisi SD. Pharmacological evaluation of ethanolic extract of *Daucus carota* Linn root formulated cream on wound healing using excision and incision wound model. Asian Pac J Trop Biomed 2012;646–55.
[30] Butani D, Yewale C, Misra A. Amphotericin B topical microemulsion: formulation, characterization and evaluation. Colloids Surf B Biointerfaces 2014;116:351–8.
[31] Kumar KK, Sasikanth K, Sabareesh M, Dorababu N. Formulation and evaluation of diacerein cream. Asian J Pharm Clin Res 2011;4:0974–2441.
[32] Kumar N, Goindi S, Bansal G. Physiochemical evaluation and in vitro release studies on itraconazolium sulphate salt. Asian J Pharm Sci 2014;9:8–16.
[33] VedhaHari BN, Begum Y, Ramya Devi D. Solid state modification for the enhancement of solubility of poorly soluble drug: carrageenan as carrier. Int J Appl Pharmaceutics 2012;4.
[34] Sun L, Zhang W, Liu X, Sun J. Preparation and evaluation of sustained release Azithromycin tablets invitro and invivo. Asian J Pharm Sci 2014;1–7.
[35] Lokhande AB, Mishra S, Kulkarni RD, Naik JB. Influence of different viscosity grade ethylcellulose polymers on encapsulation and invitro release study of drug loaded nanoparticles. J Pharm Res 2013;7(5):414–20.
[36] Dash S, Murthy PN, Nath L, Chowdhury P. Kinetic modelling on drug release from controlled drug delivery system. Acta Pol Pharm – Drug Res 2010;67:217–23.
[37] Costa P, Lobo JMS. Modeling and comparison of dissolution profiles. Eur J Pharm Sci 2001;13:123–33.

CHAPTER

7

Organic UV filter loaded nanocarriers with broad spectrum photoprotection

Lucía López-Hortas[1], María D. Torres[1], Elena Falqué[2] and Herminia Domínguez[1]

[1]Department of Chemical Engineering, Faculty of Sciences, University of Vigo, Ourense, Spain
[2]Department of Analytical Chemistry, Faculty of Sciences, University of Vigo, Ourense, Spain

7.1 Introduction

Ultraviolet (UV) radiation can be responsible for harmful effects on the skin, including photoaging and photocarcinogenesis. To provide broad-spectrum UV protection, topical sunscreens are designed, and modern products have been formulated to protect in the UVA (320–400 nm) and in the UVB (280–320 nm) regions [1]. The efficacy of the sun filters, characterized by the sun protection factor (SPF), is also determined by the carrier components [2], which enhance photoprotection by synergistically combining the advantages [3]. Their active sunscreen ingredients include organic and inorganic compounds that reflect, scatter, or absorb the UV light. Some chemical UV filters could permeate through the skin leading to undesirable effects, such as skin irritation and sensitization, or can permeate into the bloodstream causing systemic toxicity or acting as endocrine disruptors [4]. The UV filter should have a high affinity for the stratum corneum and remain on the outermost skin layers to protect the skin from harmful UV radiation [1]. Traditional sunscreens are not useful if skin damage already occurred after sun exposure, therefore active photoprotection could be provided by topical sunscreens, such as antioxidants and liposome-containing DNA repair enzymes [5].

Suncreens based on emulsions, oils, and gels have some limitations such as water washability, low stability, percutaneous absorption of UV filters and to overcome these limitations innovative micro and nanocarriers have been proposed [1,4].

This chapter presents a survey of the need of sunscreens to prevent some effects associated with UV radiation and describes the interest of incorporating UV filters in cosmetics, both based on conventional carriers and in different nanosystems.

Nanocosmetics
DOI: https://doi.org/10.1016/B978-0-12-822286-7.00009-7

7.2 Protection filters

The extrinsic component that affects the skin appearance represents approximately 80% of the cutaneous aging; this oxidative stress and DNA alteration of the skin cells are due to external factors, mainly the exposure to sunlight [6]. Photoaging is directly associated with the exposure of skin pigments to UV rays of sunlight [7]. The physiological alterations induced by UV light on human skin tissue can produce acute and chronic clinical manifestations such as erythema, hyperpigmentation, or wrinkling, among others, or more important health problems, such as melanoma and nonmelanoma skin cancers [8,9]. For these reasons, the sun protection has an essential relevance in the development of personal care products.

The photoprotective potential of sunscreen actives is based on their capacity to prevent UVB and UVA radiation (280–320 and 320–400 nm, respectively). In consequence, the formulation of the sunscreen products combines different spectrum UV filters in order to provide a broad spectrum skin protection [10,11]. The efficacy of sunscreen products against UV radiation is measured by the SPF and the use of 2 mg of product per cm^2 of skin is recommended; so that an SPF 15 represents an approximate 93% of UV protection, an SPF 30 represents a 97% protection and an SPF 50, recommended for hyperpigmentation or skin cancer, blocks 98% radiation. The SPF level, durability, and water resistance (if the labeled SPF value is retained after 40 minutes of water immersion) and type of vehicle formulation employed are aspects to consider in the design of cosmetic products with photoprotection [12]. The sunscreen products are considered as cosmetic products in the European Union and as Over-the-Counter drug products in the United States [13]. After their topical application, the cosmetic with sunscreen agents should behave as a photostable preserving film during the sunlight exposure. Furthermore, these products also should be kept on the stratum corneum at the skin surface without transference to the epithelial tissue to prevent endocrine disruptive effects [8,14].

Nowadays, several cosmetics for daily use incorporate UV filters in their composition such as moisturizing creams, lipsticks, and makeup. Shampoos and conditioners also incorporate these compounds in order to protect the hair structure from the action of sunlight [15]. These different matrices combine different UV filters to ensure their physical stability and chemical compatibility with the remaining ingredients used [16]. Sunscreen agents can be classified for their physicochemical properties in two main groups (Fig. 7.1). First, inorganic UV filters, also referred to as physical filters or mineral sunscreens or sunblocks, are compounds able to reflect, scatter, and absorb the UV radiation and are available either as dry powders or predispersions. This group is integrated by pigments such as metallic oxides that provide a prolonged and stable photoskin coverage after the solar radiation exposure [17]. The topical application of the micronized forms of these UV filters leave a white layer on the skin surface; this effect is not present when these compounds are used in nanoparticle sized [8,12]. Inorganic UV filters group are integrated by pigments such as calcium carbonate, kaolin, magnesium oxide, talc, and mainly titanium dioxide and zinc oxide. These chemical compounds block principally UVA radiations [18,19]. Second, organic UV filters, also denoted as chemical sunscreens, are compounds that simulate the natural human body melanin mechanism due to their molar absorptivity in the UV range. These compounds absorb UV radiation with excitation to a higher energy state occurs and dissipate the excess of energy as heat, light, or fluorescence through their conjugated chemical bonds and

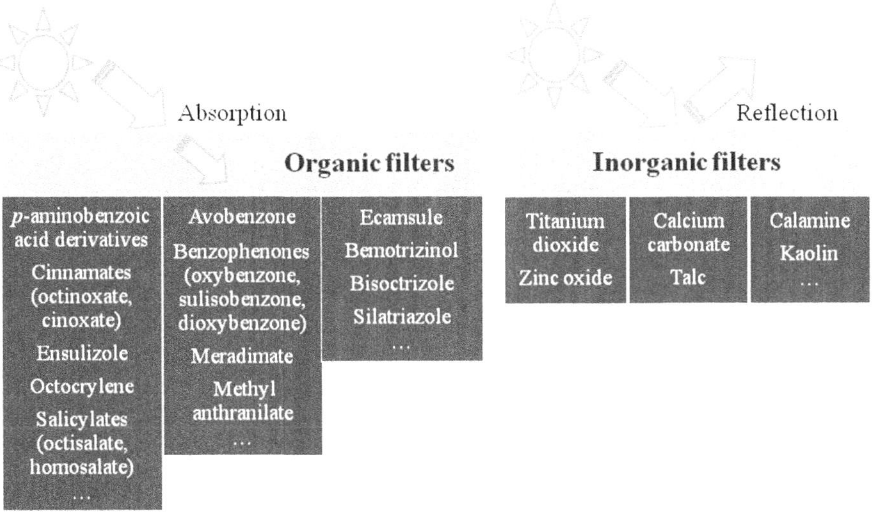

FIGURE 7.1 Action mechanism and types of organic and inorganic UV filters.

carbonyl groups. In comparison to physical UV filters, the chemical sunscreens present a protection against UVA and UVB rays more complete without any undelivered traces of the product on the skin after their application [6,8,12], but they should be applied at least 20–30 minutes before sun exposure and reapplied every 2 hours.

Organic UV filters are characterized by the presence of aromatic structures (one or more) conjugated with carbonyl moieties and/or carbon–carbon double bonds [16]. This type of UV filters can be classified into three groups according to the range of protection: the main chemical compounds that offer UVA prevention are benzophenones (e.g., oxybenzone, sulisobenzone, and dioxybenzone), avobenzone or Parsol 1789, meradimate, methyl anthranilate, etc. The principal organic UVB filters are *p*-aminobenzoic acid (PABA) (e.g., padimate O), cinnamates (such as octinoxate, cinoxate), salicylate (e.g., octisalate, homosalate, and trolamine salicylate) derivatives, octocrylene, ensulizole, ethylhexyl triazone, etc., while ecamsule, silatriazole, bemotrizinol, and bisoctrizole act in a broad effective protection spectrum [18,20]. Note here that these compounds besides their use in cosmetic formulations also protect from light alteration of different matrices as plastic food packaging materials or wood to prevent their degradation by UV light, among others, since they are included in their composition during their manufacturing process [21,22].

The ideal UV filter should safeguard the epidermis and dermis tissues against a wide range of wavelength UV light. This sunscreen agent should have low viscosity and an adequate solubility in order to favor the spreadability of the formulated personal care product. The waterproof capability contribute to the reduction of toxicity to aquatic environmental and inability to cause systemic absorption without endocrine activity are also desirable attributes for sunscreen agents [12,18]. Consumers are in contact with a notable concentration of UV filters since some mixtures of sunscreen agents are used in the formulation of several personal care products to guarantee the protection against sunlight damage [23].

In this context, the widespread use of UV filters causes that these compounds are found in the environment since most of the chemical compounds are not biodegradable and constitute an emerging group of environmental pollutants. Sunscreen products are present in wastewater as a result of anthropogenic activity, as, for example, domestic washing process, showering, bathing, and swimming or wastes not treated from industrial plants. Consequently, aqueous and marine biota ecosystems become polluted [16,24,25]. The bioaccumulation of these hazardous substances, as well as their resistance to degradation, is encouraged by their lipophilic properties (mainly defined by organic UV filters with low molecular weight). For this reason, these compounds are accumulated in nonpolar lipid tissues belonging to food chain so that the process of biomagnification takes place affecting humans [26–29]. The widespread occurrence of sunscreen compounds and derived massive human exposition causes that these substances can be present in human's fluids and tissues [30–33]. The main health risks associated with UV filters are photoallergic by contact, endocrine disruptive effects, and neurotoxic diseases [34–36]. Further studies headed to minimize absorption of UV filters should be carried out to claim the skin nonpenetration of these substances in the human body. The encapsulation of these sunscreen agents is a solution that is being valued with different personal care products [8,37].

7.3 Conventional products for sun protection

Sunscreen cosmetics are a very interesting and complex application, which over the last decade have undergone a number of modifications to increase their efficacy and stability [38,39]. Conventional screens for sun protection can use one or more chemicals including avobenzone, oxybenzone, octocrylene, octisalate, octinoxate, or homosalate [40]. Most common are oxybenzone, methoxycinnamate, and PABA derivatives, which are estrogenic chemicals and are all linked to cancer and/or environmental concerns. They are in about 97% of all conventional sunscreens [41,42]. Owing to increased demand for higher SPFs, and limitations on the use of some UV filters in cosmetics, manufacturing technique plays a key role in maximizing the performance of acceptable substances where the rheological understanding has an essential relevance [43,44].

Effective topical conventional sun protection products are sunscreens that absorb strongly in the adequate UV region, show notable photostability, and present minor spectral modifications upon UV radiation exposure [45]. It should be highlighted that organic UV absorbers feature the potential to form relatively long-lived triplet states, which can stimulate singlet oxygen production, and cause transformations in biological substrates. Above authors concluded that many organic sunscreens function by photoisomerization, where isomeric mixtures serve as the major sun protection components.

Millions worldwide use conventional sunscreen products, and the diversity of available products is expanding [46,47]. Table 7.1 summarizes some of the most representative products and matrices for sun protection containing organic UV filter.

Osterwalder and Herzog [44] studied the SPF for body creams made with conventional organic UV filters. These authors stated that uniform protection profiles lead to sun protection independent of the action spectrum of the endpoint and the UV-radiation source. Ou-Yang et al. [50] investigated about the possibility of preparing sun protection

TABLE 7.1 Conventional products for sun protection.

Product	Matrix	Protection ranges	References
Body creams	Oil-in-water emulsion	UV	[44]
Body butter Body lotion	Moisturizer: oil-in-water emulsion, water-in-oil emulsion	UV	[48]
High altitude face lotion/ body spray	Oil-in-water emulsion	UV	[49]
Face lotion Body sprays	Water-resistant cream	UVA, UVB	[50]
Face capsules	Sol-gel	UV	[37]
Baby lotion	Water-in-oil emulsion	UVA, UVB	[51]
Nail polish	Gel	UVA, UVB	[52]
Nail cosmetics	Gel	UVB	[46]

formulations based on organic filters, which do not interfere with sweat cooling during exercise. These authors developed sunscreen lotions for the face and sprays for the body with high SPF values (around 70) with promising results. Bora et al. [49] proposed the development of a sunscreen cream formulation (i.e., in a standard oil-in-water emulsion) using conventional organic compounds to provide UV protection in high altitude areas. The obtained results pointed out that the formulation possessed 50 + in vitro SPF value and remained stable for 6 and 12 months under storage at 40°C and 75% relative humidity; and −20°C, respectively. Wirunchit et al. [51] studied baby lotion creams focusing on how to make UV materials to protect UV radiation. Organic ones in a wide range of concentration from 5% to 15% were tested with suitable results. Couteau et al. [53] analyzed the photoprotective effects of different clear or colorless nail polishes. They found that colorless nail polish applied in two layers provides UV radiation screen (SPF values about 150) for the nails in the UVB and UVA ranges. This is especially important in patients undergoing chemotherapy since the application of nail polish and/or a colorless base coat is recommended to prevent the adverse effects that can occur on the nails throughout treatment.

Other works [37] made a comprehensive study between sunscreens based on free or encapsulated organic UV filters in terms of skin retention, penetration, and photostability. These authors found that organic UV filters, while proposing relevant advantages, also presented the risk to penetrate the stratum corneum and diffuse into underlying structures with unknown consequences; likewise, their photostability are noted thorny outcomes in sunscreen development and subsequent performance. Latter authors found that encapsulation technology is a promising strategy to improve the efficacy of the sunscreen product using organic UV filters and to reduce the safety problem.

Other authors reported a comprehensive comparative between organic and mineral sunscreen products [43,48]. Early authors found that for sunscreen cosmetics containing mixtures of organic filters and minerals, the compliance rate was around 75%. Latter authors formulated two different types of sunscreen moisturizers with organic and mineral filters indicating that a minimum of 10% zinc oxide should be used in the production of sunscreen moisturizers to provide moderate sun protection.

In recent studies, some authors (see as, e.g., [54,55]) explored conventional sunscreen formulations incorporated with natural extracts. Sopyan et al. [54] proposed the preparation of sun protection lotions enriched with black tea (*Camellia sinensis Linnaeus*) leaves extract to prove their ability against UV rays. The presence of these extracts (0.04%, w/v) jointly with the conventional organic ones led to a relevant enhancement of the SPF value (around 25), achieving lotions with good physical quality and safe for topical use. Banerjee et al. [55] investigated the addition of herbal oils from *Helianthus annuus* to conventional sunscreen creams with favorable viscosity profiles for topical applicants, bestowing beneficial dermatological effects. The presence of used herbal oil promoted the protection against UV radiations due to the antioxidant and antiinflammatory nature of the natural oil constituents. The inherent SPF of the oil represented a value around 6. Authors concluded that this preparation could hence work as alternative for full synthetic chemical-based formulations where low SPF values are involved.

In recent years, there has been some debate over the effectiveness of commercial sunscreen, as well as concerns about the potential harm of some of chemicals used in these products. So far, sun protection meant UV protection. However, visible light (VIS) and near infrared A (IRA) radiation can have harmful effects on our skin, resulting in photoaging and even cancer induction. In this context, more studies elucidating the benefit of photoprotective agents against UV, VIS, and IRA with the aim to be environmentally safe are required [42]. Consequently, there is a growing demand for herbal alternatives for a number of natural sources [56], smart self-assembled microgel films as encapsulating carriers for UV-absorbing molecules [57], presenting the loaded films absorption larger than 50% of the UVA–UVB wavelengths which make them fit to be used for skin protection against sun. As well as other nanocarriers for sun protection cosmetics need further studies to ensure their functionality as safety sun protectors [58].

7.4 Nanocarriers for sun protection

New generation of cosmetics are based on innovative delivery systems for active ingredients [59]. Among different strategies, nanotechnology has been used to improve the stability and solubility of active principles [60]. Nanoencapsulation of sunscreen is also useful to prevent photodegradation and skin accumulation [61] and to increase safety of sun protection products through reduction in the use of chemical UV filters. The contradictory results have raised concerns about the possible penetration through the skin barrier, but no penetration of inorganic nanoparticles, such as zinc oxide and titanium dioxide, beyond the stratum corneum into intact and healthy human skin was observed using two-photon imaging [62]. Also, Mohammed et al. [63] assessed the skin penetration of agglomerated zinc oxide by multiphoton tomography with fluorescence lifetime imaging microscopy of applied to human volunteers but did not observe permeation or cellular toxicity in the viable epidermis. Prow et al. [64] reported that clinical data suggest that nanoparticle penetration is limited for nanoparticles >10 nm in diameter through the stratum corneum into viable human skin. However, accumulation in the hair follicle openings can occur and enhanced uptake can also occur in damaged areas and in certain diseased skin. Different systems have been proposed, such as solid lipid nanoparticles (SLNs), nanoemulsions, nanostructured lipid carriers (NLCs), polymeric nanoparticles, cyclodextrins, nanoemulsions, and vesicular nanosystems. This section presents examples of the utilization of these systems for the formulation of cosmetics with UV radiation protection properties.

Nanocarriers, such as microemulsions, liposomes, and micro- and nanoparticles, are promising systems for dermal and transdermal delivery of bioactives. Microemulsions have a high solubilization capacity even for poorly soluble drugs and combined with their permeation enhancing effect high flux rates can be obtained. Vesicular nanosystems are versatile structures able to encapsulate active photoprotective compounds with different nature, both lipophilic and hydrophilic. Vesicular nanosystems such as niosomes could be applied topically, acting as controlled release systems as they fuse with the lipids of the stratum corneum [60]. Incorporation of antioxidants can be of interest. Muhtadi et al. [65] formulated gel nanoemulsions of extracts from rambutan fruit peel, which provided sunscreen protection and antioxidant properties. Topical administration of antioxidants can enhance the endogenous protection system and lower the UV radiation-induced oxidative damage in the skin. The cosmetic industry is incorporating antioxidants to UV filters, because almost all the post radiation reactions involve directly or indirectly reactive oxygen species [2].

Liposomal carrier systems exhibit a high flexibility and mobility, whereas the micro- and nanoparticular systems show an increase in the follicular penetration depth [66]. Patsinakidis et al. [67] confirmed that a broad-spectrum liposomal sunscreen with high SPF can prevent UV irradiation–induced damage in patients with cutaneous lupus erythematosus. Monteiro et al. [68] proposed the inclusion of sunscreen molecules in liposomes and cyclodextrins. The formulation containing only the lipo/octyl *p*-methoxycinnamate system showed high in vivo protection and water resistance. Severino et al. [69] tried the extrusion of elastic liposomes composed of egg phosphatidylcholine and cholesterol loaded with benzophenone-3 to produce reduced-size (100 nm) elastic liposomes with improved in vitro sun protector factor. Abbas and Kamel [70] prepared surfactant-based elastic vesicles (spanlastics) (201 nm) to encapsulate resveratrol with high efficiency (79%), with a fast release pattern. Abbas et al. [71] prepared colloidal carriers containing resveratrol and a binary mixture of surfactants to improve the physicochemical properties, enhancing drug loading, and showing a uniform particle size distribution. In both studies, they observed amelioration of antioxidant (catalase, reduced glutathione and superoxide dismutase), antiinflammatory, and antiwrinkling markers as well as skin protective effect, after UVB irradiation, compared to those of a positive control. These systems also showed occlusive properties, of interest to keep the skin layers moistened can have some beneficial effects to combat aging and loss of skin elasticity [2].

Cerqueira et al. [72] developed photoprotective niosomes, spherical, unilamellar, and with an entrapment efficiency of more than 45% for different sunscreens (octyl methoxycinnamate, diethylamino hydroxybenzoyl hexyl benzoate, and phenylbenzimidazole sulfonic acid). The formulations were nontoxic in macrophages and the pH was compatible with skin.

Mesoporous silica nanoparticles have widely been used as vehicles for therapeutic agents because of their high storage capacity and the possibility of modulating drug release. Brezoiu et al. [73] encapsulated grape pomace ethanolic extracts containing phenolic compounds into mesoporous MCM-41-type silica matrices to enhance the extract stability and to preserve the radical scavenging activity, with good cytocompatibility on HaCaT keratinocytes due to minimum exposure of ZnO to the cells. Frizzo et al. [61] developed a new sunscreen with the simultaneous encapsulation of zinc oxide nanoparticles and octocrylene in poly-styrene-co-methyl methacrylate nanoparticles with high encapsulation efficiency, good solar protection factor, and low cytotoxicity on human dermal fibroblasts. The incorporation of zinc oxide in nanoparticles to a **mesoporous matrix** can improve the textural, structural, and morphological properties, maintaining the safety of sunscreens. The nanocomposites had an increased in vitro SPF, reduced cytotoxic activity and favorable antimicrobial properties compared to the physical filter alone [74].

SLNs consist of a two-phase drug delivery system comprising a nanoscale solid phase composed of lipids dispersed in an aqueous phase [2]. The lipids used are physiologically compatible, and the solid lipid matrix favors the controlled release [3,75]. The use of natural lipids can result in a carrier with biological activities. NLCs are the second generation of lipid nanoparticles, with enhanced drug encapsulation ability and reduced actives expulsion [76]. Xia et al. [77] incorporated sunscreens into lipid carriers with an increased SPF. They compared SLN and NLC produced by hot high-pressure homogenization. The loading capacities of NLC reached up to 70%, which was appropriate to obtain high protection factors. Nanostructured polymeric lipid carriers (NPLCs) and nanocapsules (NCs) are characterized by a wall of hydrophobic polymer that surrounds their lipid core [4].

NLCs have been proposed as an effective delivery system for both organic and inorganic sunscreens, with advantageous UV-blocking properties. Both SLNs and NLCs allow the entrapment of the organic sunscreen inside the lipid matrices, thus reducing their potential skin irritation decreasing the required amount of the UV filter and maintaining long-term stability. In some formulations, improvements on the SPF due to the synergistic effect of the mixtures have been reported [2]. Gilbert et al. [4] compared percutaneous absorption and cutaneous bioavailability of benzophenone-3 loaded into SLN, NLC, NPLC, and NC and confirmed that NPLC and NC significantly reduced benzophenone-3 skin permeation maintaining the highest SPF. The SPF increased from 3 for aqueous solutions to 5.5–12.3 for nanomaterials, and the partition coefficient in the epidermis/stratum corneum also increased from 0.25 in aqueous solutions to 0.5 in NLC, 0.3 in NPLC, and 0.45 in NC.

Niculae et al. [78] prepared lipid nanocarriers formulated with vegetable oils and low concentrations of synthetic UVA and UVB filters (butyl-methoxydibenzoylmethane and octocrylene). These systems showed spherical shape and physical stability, as well as improved photoprotection (UVA-PF = 40.2 and SPF UVB = 17.3) and antioxidant activity, particularly for those nanocarriers prepared from rice bran oil. Niculae et al. [79] developed sunscreens with a UVA filter and with α-tocopherol to enhance the photostability of SLN and NLC, but the photoprotection of the cosmetics based on lipid nanoparticles was not improved. Natural antioxidants from vegetable oils have been reported to reduce UV filters concentrations in sunscreens, maintaining or elevating the efficacy, resulting in photostable systems.

Lacatusu et al. [80] reported that the NLC based on vegetable pumpkin and amaranth oils to coencapsulate and codeliver avobenzone and octocrylene showed improved antioxidant and photoprotective properties and the release was prolonged compared to the use without the oils. Andréo-Filho et al. [81] developed stable SLN containing the chemical UV filter octyl methoxycinnamate and partially replaced it (20%) by urucum oil without observing toxic responses in the skin, skin penetration, or change in the protection factor. Badea et al. [82] designed new NLC containing vegetable oils (pomegranate seed, wheat germ, blackcurrant seed, sesame seed, carrot root, raspberry seed, and rice bran) to obtain efficient formulations with UV protection and antioxidant properties. These NLCs were used as carriers for the UVA absorbent agent diethylamino hydroxybenzoyl hexyl benzoate and the cream based on pomegranate seed oil presented higher SPF. The combination with wheat germ oil enhanced the entrapment efficiency and photoprotection. Dario et al. [83] proposed the bocaiúva almond oils, with a high content of fatty acids, polyphenols, oligoelements, and β-carotene for effective protection against UV radiation with potential to replace organic filters. They prepared NLCs by high-pressure homogenization of oils and cetyl palmitate, as liquid and solid lipids, respectively, to entrap avobenzone with an efficiency of 75.2% and 33.3% (w/w), respectively. The sunscreen formulation using NLC containing bocaiúva almond oil improved the photoprotective activity of a sunscreen base showing a synergistic effect. Chu et al. [84] developed spherical amorphous NLCs loaded with the organic filters Uvinul T150 and Uvinul A Plus Granular and pumpkin and kenaf seed oil with carnauba wax and beeswax. The formulations showed high entrapment efficiency (>95%) and drug loading, antioxidant activities, as well as UV-absorbing properties.

Pivetta et al. [76] produced homogeneous and stable NLC to efficiently (97.4%) retain quercetin, a flavonoid with poor cutaneous permeation and low stability. These NLCs also showed antioxidant activity, antiallergic potential, and in the reconstructed human skin model was nonphototoxic. Badea et al. [85] confirmed the photostability of NLC formulated with naringenin and UVA filters. The presence of naringenin formulated into hydrogels with UVA filter enhanced the SPF and the antioxidant activity against short- and long-life radicals. At concentrations under 50 μg/ml, the nanocarriers showed no cytotoxicity. Salunkhe et al. [75] developed NLC for topical delivery of idebenone, an antioxidant and has lower molecular weight analog to Coenzyme Q10, to be used as sunscreens with antioxidant properties. The high skin deposition and SPF (23) lead to the more antioxidant effect of formulations, nanoemulsions (NEs), SLN, and NLC.

Asfour et al. [86] have developed NLC formulations containing 6% titanium dioxide to enhance the photostability of all-*trans* retinoic acid and to alleviate its skin photosensitivity, which was confirmed in an in vivo study conducted on mice. Kamel and Mostafa [2] developed photoprotective NLCs containing Apifil with 2% rutin, which showed high drug encapsulation, occlusive effect, release efficiencies, and SPF, which could be enhanced with TiO_2 at 5% reaching more than twofold the value of the standard antioxidant. Suter et al. [87] developed an effective formulation of NLC or SLN improving the skin delivery of the amorphous heptapeptide DEETGEF. The SLN produced by hot high-pressure homogenization offers physical stability, protection of the incorporated labile actives and their controlled release, with an entrapment efficiency of 91%, and in a clinical study decreased DNA damage in UVA-irradiated skin.

Nikolić et al. [3] developed a photoprotective formulation with chemical sunscreens (namely, ethylhexyl triazone, bis-ethylhexyloxyphenol methoxyphenyl triazine, and ethylhexyl methoxycinnamate) incorporated into NLCs produced by hot high-pressure homogenization. The NLC suspension with carnauba wax had superior UV absorbance and increased the SPF by more than 45% compared with the organic UV filters. This synergistic effect of the lipid nanoparticles and organic UV filters may allow reducing the concentrations of the actives.

Puglia et al. [88] compared SLN and NLC as potential carriers for octyl-methoxycinnamate and NLC dispersions were more efficient against UV -mediated photodegradation of the filter. Puglia et al. [1] used NLC and NE for the topical application of UVA or UVB sun filters (ethyl hexyltriazone, diethylamino hydroxybenzoyl hexyl benzoate, bemotrizinol, octylmethoxycinnamate, and avobenzone). The incorporation in NLC lowered the skin permeation of the sun filter, the photostability was maintained except for octylmethoxycinnamate and avobenzone, as expected, and no differences in photoprotective efficacy were observed.

In conclusion, nanotechnology offers several alternative possibilities for the efficient encapsulation of sunscreens, with additional advantages in relation to the high loading capacity, accumulation in the outer layers of skin minimizing skin absorption retarded as well as retarded release and the possibility of achieving synergistic actions, which could reduce the use of UV screens.

References

[1] Puglia C, Damiani E, Offerta A, Rizza L, Tirendi GG, Tarico MS, et al. Evaluation of nanostructured lipid carriers (NLC) and nanoemulsions as carriers for UV-filters: characterization, *in vitro* penetration and photostability studies. Eur J Pharm Sci 2014;51(1):211–17.

[2] Kamel R, Mostafa DM. Rutin nanostructured lipid cosmeceutical preparation with sun protective potential. J Photochem Photobiol B: Biol 2015;153:59–66.

[3] Nikolić S, Keck CM, Anselmi C, Müller RH. Skin photoprotection improvement: synergistic interaction between lipid nanoparticles and organic UV filters. Int J Pharm 2011;414(1-2):276–84.

[4] Gilbert E, Roussel L, Serre C, Sandouk R, Salmon D, Kirilov P, et al. Percutaneous absorption of benzophenone-3 loaded lipid nanoparticles and polymeric nanocapsules: a comparative study. Int J Pharm 2016;504(1-2):48–58.

[5] Megna M, Lembo S, Balato N, Monfrecola G. "Active" photoprotection: sunscreens with DNA repair enzymes. G Ital Dermatol Venereol 2017;152(3):302–7.

[6] Cavinato M. Cosmetics and Cosmeceuticals. In: McQueen CA, editor. Reference module in biomedical sciences. Amsterdam: Elsevier; 2018.

[7] Krutmann J, Bouloc A, Sore G, Bernard BA, Passeron T. The skin aging exposome. J Dermatol Sci 2016;85(3):152–61.

[8] Damiani E, Puglia C. Nano- and microcarriers for enhancing the UV protection of sunscreens: an overview. J Pharm Sci 2019;108:3769–80.

[9] Zhang S, Duan E. Fighting against skin aging: the way from bench to bedside. Cell Transplant 2018;27(5):729–38.

[10] Couteau C, Diarra H, Coiffard L. Effect of the product type, of the amount of applied sunscreen product and the level of protection in the UVB range on the level of protection achieved in the UVA range. Int J Pharm 2016;500(1-2):210–16.

[11] Hung C-F, Fang C-L, Al-Suwayeh SA, Yang S-Y, Fang J-Y. Evaluation of drug and sunscreen permeation via skin irradiated with UVA and UVB: comparisons of normal skin and chronologically aged skin. J Dermatol Sci 2012;68(3):135–48.

[12] Bom S, Jorge J, Ribeiro HM, Marto J. A step forward on sustainability in the cosmetics industry: a review. J Clean Prod 2019;225:270–90.

[13] Klimová Z, Hojerová J, Pažoureková S. Current problems in the use of organic UV filters to protect skin from excessive sun exposure. Acta Chim Slov 2013;6(1):82–8.

[14] Lorigo M, Mariana M, Cairrao E. Photoprotection of ultraviolet-B filters: updated review of endocrine disrupting properties. Steroids 2018;131:46–58.

[15] Dario MF, Baby AR, Velasco MVR. Effects of solar radiation on hair and photoprotection. J Photochem Photobiol B Biol 2015;153:240–6.

[16] Chisvert A, Salvador A. UV filters in cosmetics: regulatory aspects and analytical methods. In: Salvador A, Chisvert A, editors. Analysis of cosmetic products. 2nd ed Amsterdam: Elsevier; 2018.

[17] Romanhole RC, Ataide JA, Cefali LC, Moriel P, Mazzola PG. Photostability study of commercial sunscreens submitted to artificial UV irradiation and/or fluorescent radiation. J Photochem Photobiol B Biol 2016;162:45–9.

[18] Donglikar MM, Deore SL. Sunscreens: a review. Pharmacognosy J 2016;8(3):171–9.

[19] Nair S, Nagarajappa GB, Pandey KK. UV stabilization of wood by nano metal oxides dispersed in propylene glycol. J Photochem Photobiol B Biol 2018;183:1–10.

[20] Geoffrey K, Mwangi AN, Maru SM. Sunscreen products: rationale for use, formulation development and regulatory considerations. Saudi Pharm J 2019;27:1009–18.

[21] Hu L, Tian M, Feng W, He H, Wang Y, Yang L. Sensitive detection of benzophenone-type ultraviolet filters in plastic food packaging materials by sheathless capillary electrophoresis-electrospray ionization-tandem mass spectrometry. J Chromatogr A 2019;1604:460469.

[22] Liu XY, Timar MC, Varodi AM. A comparative study on the artificial UV and natural ageing of beeswax and Chinese wax and influence of wax finishing on the ageing of Chinese Ash (*Fraxinus mandshurica*) wood surfaces. J Photochem Photobiol B Biol 2019;201:111607.

[23] Ramos S, Homem V, Alves A, Santos L. Advances in analytical methods and occurrence of organic UV-filters in the environment—a review. Sci Total Environ 2015;526:278–311.

[24] Abdelraheem WHM, He X, Duan X, Dionysiou DD. Degradation and mineralization of organic UV absorber compound 2-phenylbenzimidazole-5-sulfonic acid (PBSA) using UV-254 nm/H_2O_2. J Hazard Mater 2015;282:233–40.

[25] Tsui MMP, Chen L, He T, Wang Q, Hu C, Lam JCW, et al. Organic ultraviolet (UV) filters in the South China sea coastal region: environmental occurrence, toxicological effects and risk assessment. Ecotoxicol Environ Saf 2019;181:26–33.

[26] Gago-Ferrero P, Díaz-Cruz MS, Barceló D. UV filters bioaccumulation in fish from Iberian river basins. Sci Total Environ 2015;518-519:518–25.

[27] Rainieri S, Barranco A, Primec M, Langerholc T. Occurrence and toxicity of musks and UV filters in the marine environment. Food Chem Toxicol 2017;104:57–68.

[28] Ramos S, Homem V, Santos L. Analytical methodology to screen UV-filters and synthetic musk compounds in market tomatoes. Chemosphere 2020;238:124605.

[29] Vidal-Liñán L, Villaverde-de-Sáa E, Rodil R, Quintana JB, Beiras R. Bioaccumulation of UV filters in *Mytilus galloprovincialis* mussel. Chemosphere 2018;190:267–71.

[30] Broniowska Ż, Pomierny B, Smaga I, Filip M, Budziszewska B. The effect of UV-filters on the viability of neuroblastoma (SH-SY5Y) cell line. NeuroToxicology 2016;54:44–52.

[31] Klotz K, Hof K, Hiller J, Göen T, Drexler H. Quantification of prominent organic UV filters and their metabolites in human urine and plasma samples. J Chromatogr B 2019;1125:121706.

[32] Molins-Delgado D, Olmo-Campos MDM, Valeta-Juan G, Pleguezuelos-Hernández V, Barceló D, Díaz-Cruz MS. Determination of UV filters in human breast milk using turbulent flow chromatography and babies' daily intake estimation. Environ Res 2018;161:532–9.

[33] Valle-Sistac J, Molins-Delgado D, Díaz M, Ibáñez L, Barceló D, Díaz-Cruz MS. Determination of parabens and benzophenone-type UV filters in human placenta. First description of the existence of benzyl paraben and benzophenone-4. Environ Int 2016;88:243–9.

[34] Bechambi O, Chalbi M, Najjar W, Sayadi S. Photocatalytic activity of ZnO doped with Ag on the degradation of endocrine disrupting under UV irradiation and the investigation of its antibacterial activity. Appl Surf Sci 2015;347:414–20.

[35] Hiller J, Klotz K, Meyer S, Uter W, Hof K, Greiner A, et al. Systemic availability of lipophilic organic UV filters through dermal sunscreen exposure. Environ Int 2019;132:105068.

[36] Silveira RS, Rocha BA, Rodrigues JL, Barbosa F. Rapid, sensitive and simultaneous determination of 16 endocrine-disrupting chemicals (parabens, benzophenones, bisphenols, and triclocarban) in human urine based on microextraction by packed sorbent combined with liquid chromatography tandem mass spectrometry (MEPS-LC-MS/MS). Chemosphere 2020;240:124951.

[37] Cozzi AC, Perugini P, Gourion-Arsiquaud S. Comparative behavior between sunscreens based on free or encapsulated UV filters in term of skin penetration, retention and photo stability. Eur J Pharm Sci 2018;121:309–18.

[38] Dukhin AS, Goetz PJ. Chapter 1. Introduction. In: Dukhin AS, Goetz PJ, editors. Characterization of liquids, dispersions, emulsions, and porous materials using ultrasound. Amsterdam: Elsevier; 2017.

[39] Nanda S, Nanda A, Lohan S, Kaur R, Singh B. Chapter 3. Nanocosmetics: performance enhancement and safety assurance. In: Grumezescu A, editor. Nanobiomaterials in galenic formulations and cosmetics: applications of nanobiomaterials. Amsterdam: Elsevier; 2016.

[40] Kockler J, Robertson S, Oelgemöller M, Davies M, Bowden B, Brittain HG, et al. Butyl methoxy dibenzoylmethane. Profiles Drug Subs Excip Relat Methodol 2013;38:87–111.

[41] Nafisi S, Maibach HI. Chapter 22. Nanotechnology in cosmetics. In: Sakamoto K, Lochhead RY, Maibach HI, Yamashita Y, editors. Cosmetic science and technology: theoretical principles and applications. Amsterdam: Elsevier; 2017.

[42] Singer S, Karrer S, Berneburg M. Modern sun protection. Curr Opin Pharmacol 2019;46:24–8.

[43] Coiffard LJM, Couteau C. Sunscreen products: Issues in terms of efficiency. Actual Pharm 2013;52:35–40.

[44] Osterwalder U, Herzog B. Sun protection factors: world wide confusion. Br J Dermatol 2009;161(3):13–24.

[45] Cantrell A, McGarvey DJ, George Truscott T. In: Giacomoni PU, editor. Comprehensive series in photosciences. Amsterdam: Elsevier; 2001.

[46] Dinani N, George S. Nail cosmetics: a dermatological perspective. Clin Exp Dermatol 2019;44:599–605.

[47] Jovanovic Z, Schornstein T, Sutor A, Neufang G, Hagens R. Conventional sunscreen application does not lead to sufficient body coverage. Int J Cosmet Sci 2017;39:550–5.

[48] Obanla OR, Ojewumi ME, Ayoola AA, Omodara OJ, Falope FY, Gbadamosi OH. Comparative and experimental study on the properties and potency of synthesized organic and mineral sunscreen moisturizer. Environ Eng 2019;10:612–25.

[49] Bora SN, Mazumder B, Patowary P, Kishor S, Bhutia DY, Chattopadhyay P, et al. Formulation development and accelerated stability testing of a novel sunscreen cream for ultraviolet radiation protection in high altitude areas. Drug Dev Ind Pharm 2019;45:1332–41.

[50] Ou-Yang H, Meyer K, Houser T, Grove G. Sunscreen formulations do not interfere with sweat cooling during exercise. Int J Cosmet Sci 2018;40:87–92.

[51] Wirunchit S, Apivitcholchat C, Chodjarusawad T, Koetniyom W. The study of UV protection materials. AIP Conf Proc 2018;1:020024.

[52] Cotteau C, Paparis E, Coiffard L. Evaluation of different colorless nail polishes used as supportive care in patients with cancer in terms of photoprotective efficacy and water resistance. J Clin Aesthet Dermatol 2018;11:20–4.

[53] Cotteau C, Paparis E. Coiffard L. Evaluation of different colorless nail polishes used as supportive care in patients with cancer in terms of photoprotective efficacy and water resistance. The J Clin Aesthe Dermatol 11:20–4.

[54] Sopyan I, Permata RD, Gozali D, Syah ISK. Formulation of lotion from black tea extract (*Camellia sinensis linnaeus*) as sunscreen. Int J Appl Pharm 2009;11:205–9.

[55] Banerjee K, Thiagarajan N, Thiagarajan P. Formulation and characterization of a *Helianthus annuus*-alkyl polyglucoside emulsion cream for topical applications. J Cosmet Dermatol 2019;18:628–37.

[56] Rasheed A, Shama SN, Mohanalakshmi S, Ravichandran V. Formulation, characterization and in vitro evaluation of herbal sunscreen lotion. Orient Pharm Exp Med 2012;12:241–6.

[57] Aguirre G, Khoukh A, Taboada P, Chougrani K, Alard V, Billon L. Smart self-assembled microgel films as encapsulating carriers for UV-absorbing molecules. Polym Chem 2018;9:1155–9.

[58] Severino P, Fangueiro JF, Chaud MV, Cordeiro J, Silva AM, Souto EB. Chapter 1 - Advances in nanobiomaterials for topical administrations: new galenic and cosmetic formulations. In: Grumezescu A, editor. Nanobiomaterials in galenic formulations and cosmetics: applications of nanobiomaterials. Amsterdam: Elsevier; 2016.

[59] Golubovic-Liakopoulos N, Simon SR, Shah B. Nanotechnology use with cosmeceuticals. Semin Cutan Med Surg 2011;30(3):176–80.

[60] Cerqueira-Coutinho C, Dos Santos EP, Mansur CRE. Nanosystems in photoprotection. J Nanosci Nanotechnol 2015;15(12):9679–88.

[61] Frizzo MS, Feuser PE, Berres PH, Ricci-Júnior E, Campos CEM, Costa C, et al. Simultaneous encapsulation of zinc oxide and octocrylene in poly (methyl methacrylate-co-styrene) nanoparticles obtained by miniemulsion polymerization for use in sunscreen formulations. Colloids Surf A Physicochem Eng Asp 2019;61:39–46.

[62] Breunig HG, Weinigel M, König K. *In vivo* imaging of ZnO nanoparticles from sunscreen on human skin with a mobile multiphoton tomograph. BioNanoScience 2014;5(1):42–7.

[63] Mohammed YH, Holmes A, Haridass IN, Sanchez WY, Studier H, Grice JE, et al. Support for the safe use of zinc oxide nanoparticle sunscreens: lack of skin penetration or cellular toxicity after repeated application in volunteers. J Investig Dermatol 2019;139(2):308–15.

[64] Prow TW, Grice JE, Lin LL, Faye R, Butler M, Becker W, et al. Nanoparticles and microparticles for skin drug delivery. Adv Drug Deliv Rev 2011;63(6):470–91.

[65] Muhtadi M, Suhendi A, Wikantyasning ER. Gel nanoemulsion of rambutan (*Nephelium lappaceum* L.) fruit peel extracts: formulation, physical properties, sunscreen protecting, and antioxidant activity. Asian J Pharm Clin Res 2017;10(11):220–4.

[66] Schroeter A, Engelbrecht T, Neubert RHH, Goebel ASB. New nanosized technologies for dermal and transdermal drug delivery. A review. J Biomed Nanotechnol 2010;6(5):511–28.

[67] Patsinakidis N, Wenzel J, Landmann A, Koch R, Gerß J, Luger TA, et al. Suppression of UV-induced damage by a liposomal sunscreen: a prospective, open-label study in patients with cutaneous lupus erythematosus and healthy controls. Exp Dermatol 2012;21(12):958–61.

[68] Monteiro MSSB, Ozzetti RA, Vergnanini AL, de Brito-Gitirana L, Volpato NM, de Freitas ZMF, et al. Evaluation of octyl *p*-methoxycinnamate included in liposomes and cyclodextrins in anti-solar preparations: preparations, characterizations and *in vitro* penetration studies. Int J Nanomed 2012;7:3045–58.

[69] Severino P, Moraes LF, Zanchetta B, Souto EB, Santana MHA. Elastic liposomes containing benzophenone-3 for sun protection factor enhancement. Pharm Dev Technol 2012;17(6):661–5.

[70] Abbas H, Kamel R. Potential role of resveratrol-loaded elastic sorbitan monostearate nanovesicles for the prevention of UV-induced skin damage. J Liposome Res 2019;10:1–9.

[71] Abbas H, Kamel R, El-Sayed N. Dermal anti-oxidant, anti-inflammatory and anti-aging effects of compritol ATO-based resveratrol colloidal carriers prepared using mixed surfactants. Int J Pharm 2018;541(1-2):37–47.

[72] Cerqueira C, Nigro F, Campos VEB, Rossi A, Santos-Oliveira R, Cardoso V, et al. Nanovesicle-based formulations for photoprotection: a safety and efficacy approach. Nanotechnology 2019;30(34):345102.

[73] Brezoiu A-M, Matei C, Deaconu M, Stanciuc A-M, Trifan A, Gaspar-Pintiliescu A, et al. Polyphenols extract from grape pomace. Characterization and valorisation through encapsulation into mesoporous silica-type matrices. Food Chem Toxicol 2019;133:110787.

[74] Vieira CO, Grice JE, Roberts MS, Haridass IN, Duque MD, Lopes PS, et al. ZnO:SBA-15 nanocomposites for potential use in sunscreen: preparation, properties, human skin penetration and toxicity. Skin Pharmacol Physiol 2018;32(1):32–42.

[75] Salunkhe SS, Bhatia NM, Pokharkar VB, Thorat JD, Bhatia MS. Topical delivery of Idebenone using nanostructured lipid carriers: evaluations of sun-protection and anti-oxidant effects. J Pharm Investig 2013;43 (4):287–303.

[76] Pivetta TP, Silva LB, Kawakami CM, Araújo MM, Del Lama MPFM, Naal RMZG, et al. Topical formulation of quercetin encapsulated in natural lipid nanocarriers: evaluation of biological properties and phototoxic effect. J Drug Deliv Sci Technol 2019;53:101148.

[77] Xia Q, Saupe A, Müller RH, Souto EB. Nanostructured lipid carriers as novel carrier for sunscreen formulations. Int J Cosmet Sci 2007;29(6):473–82.

[78] Niculae G, Lacatusu I, Badea N, Stan R, Vasile BS, Meghea A. Rice bran and raspberry seed oil-based nanocarriers with self-antioxidative properties as safe photoprotective formulations. Photochem Photobiol Sci 2014;13(4):703–16.
[79] Niculae G, Lacatusu I, Bors A, Stan R. Photostability enhancement by encapsulation of α-tocopherol into lipid-based nanoparticles loaded with a UV filter. C R Chim 2014;17(10):1028–33.
[80] Lacatusu I, Niculae G, Badea N, Stan R, Popa O, Oprea O, et al. Design of soft lipid nanocarriers based on bioactive vegetable oils with multiple health benefits. Chem Eng J 2014;246:311–21.
[81] Andréo-Filho N, Bim AVK, Kaneko TM, Kitice NA, Haridass IN, Abd E, et al. Development and evaluation of lipid nanoparticles containing natural botanical oil for sun protection: characterization and *in vitro* and *in vivo* human skin permeation and toxicity. Skin Pharmacol Physiol 2018;31(1):1–9.
[82] Badea G, Lăcătuşu I, Badea N, Ott C, Meghea A. Use of various vegetable oils in designing photoprotective nanostructured formulations for UV protection and antioxidant activity. Ind Crop Products 2015;67:18–24.
[83] Dario MF, Oliveira FF, Marins DSS, Baby AR, Velasco MVR, Löbenberg R, et al. Synergistic photoprotective activity of nanocarrier containing oil of *Acrocomia aculeata* (Jacq.) Lodd. Ex. Martius-Arecaceae. Ind Crop Products 2018;112:305–12.
[84] Chu CC, Tan CP, Nyam KL. Development of nanostructured lipid carriers (NLCs) using pumpkin and kenaf seed oils with potential photoprotective and antioxidative properties. Eur J Lipid Sci Technol 2019;121 (10):1900082.
[85] Badea G, Badea N, Brasoveanu LI, Mihaila M, Stan R, Istrati D, et al. Naringenin improves the sunscreen performance of vegetable nanocarriers. New J Chem 2017;41(2):480–92.
[86] Asfour MH, Kassem AA, Salama A. Topical nanostructured lipid carriers/inorganic sunscreen combination for alleviation of all-trans retinoic acid-induced photosensitivity: Box-Behnken design optimization, *in vitro* and *in vivo* evaluation. Eur J Pharm Sci 2019;134:219–32.
[87] Suter F, Schmid D, Wandrey F, Zülli F. Heptapeptide-loaded solid lipid nanoparticles for cosmetic anti-aging applications. Eur J Pharm Biopharm 2016;108:304–9.
[88] Puglia C, Bonina F, Rizza L, Blasi P, Schoubben A, Perrotta R, et al. Lipid nanoparticles as carrier for octylmethoxycinnamate: *in vitro* percutaneous absorption and photostability studies. J Pharm Sci 2012;101 (1):301–11.

CHAPTER

8

Cosmetic nanoformulations and their intended use

Surbhi Dhawan, Pragya Sharma and Sanju Nanda

Department of Pharmaceutical Sciences, Maharshi Dayanand University, Rohtak, India

8.1 Introduction

The concept of nanotechnology was first brought into notice by Richard Fenyman in the year 1959. But in 1974, it was Norio Taniguchi who actually coined the term "nanotechnology" [1]. The word "nanotechnology" can be broken into two individual words, namely, technology and a Greek number "nano" meaning dwarf [2]. It is a science of manipulating the structure of atoms to a nanometer range. This leads to the development of a new structure possessing properties and nature different from the parent structure. According to European Union, the term "nanomaterial" means "A natural, incidental or manufactured material containing particles, in an unbound state or as an aggregate or as an agglomerate and where, for 50% or more of the particles in the number size distribution, one or more external dimensions is in the size range 1–100 nm." They also include fullerenes, graphene flakes, and single-wall carbon nanotubes having either one or more external dimensions below 1 nm (European Union, 2011). Ever since 1959 nanotechnology has made its space in multiple fields like engineering, biology, physics, chemistry including medicines, cosmetics, and dermal preparations [3].

In the year 1961, the word "cosmetics" was coined by Raymond Reed, a founding member of US Society of Cosmetic Chemists. According to FDA, cosmetics can be defined as "articles intended to be applied to the human body or any part thereof for cleansing, beautifying, promoting attractiveness, or altering the appearance" (US FDA). Cosmetics may include products for lips (lipsticks, balms, lip gloss), skin (moisturizers, sunscreens, lotions), and nail (nail polish, nail lacquers). Another segment of cosmetics is "cosmeceuticals" that include products having both cosmetic and health related benefits [4]. These cosmeceutical products are known to treat various problems like wrinkles, photoaging, hair damage, hyperpigmentation, acne, dryness, and hair fall [5]. The field of cosmetics is developing ever since it has been introduced. The technological development has brought

Nanocosmetics
DOI: https://doi.org/10.1016/B978-0-12-822286-7.00017-6

innovations in cosmetic formulations. Nowadays, nanomaterials/nanoparticles are extensively used to develop cosmetic products known as nanocosmetic formulations. Many cosmetic companies are investing in developing nanotechnology-based cosmetic products. Firstly, these tiny ingredients are believed to provide better entrapment, improved stability of cosmetic ingredients, better dispersibility, enhanced performance, better protection from UV rays, improved texture and esthetic appeal, better penetration, target specificity, and prolonged effects [6–8]. Secondly, these products look exclusive, influential, and trendy at the same time. The quantity of sale of products containing nanomaterials is projected to reach over US$55.3 billion by the year 2022 [9].

Along with the benefits, these minute materials are associated with some risks. The setbacks experienced while using nanotechnology-based cosmetic products are as follows [10]:

- No stringent rules and regulations are set up by the regulatory bodies for manufacturing, marketing, use, and disposal of nanomaterials used in cosmetics.
- No clinical trials required for their approval, hence raises toxicity concerns.
- Nanomaterial exposure may be harmful to human tissues and organs.
- These minuscule particles may be reactive and can cause damage to DNA, proteins, and membranes.
- They may harm the environment.

8.2 Nanocarriers/nanomaterials in cosmetics

With the technological development, several novel nanocarrier systems and nanomaterials have been developed. Based on these technological developments, many patents have been filed claiming the benefits of nanocosmetics, which are summarized in Table 8.1. Fig. 8.1 illustrates various nanocarriers/nanomaterials used in cosmetics.

8.2.1 Liposomes

The concept of exploiting liposomes for topical drug delivery was given by Mezei and Gulasekharam. Liposomes are vesicular structures of size range from 15 nm to several micrometers in an aqueous environment. These vesicles are composed of an aqueous core that is surrounded by one or more bilayer membranes (unilamellar or multilamellar) particularly formed using natural or synthetic phospholipids. Cholesterol is also added to the preparation in order to increase the stability and to keep the entrapped substance inside the vesicles for a longer time period. The applications of liposomes have expanded to varied fields especially to the cosmetics, where it has been recognized as one of the widely used cosmetic delivery system. Their structural makeup allows the entrapment of both the hydrophilic and lipophilic drugs in the aqueous core and nonpolar portion of the bilayer membrane, respectively. In 1986, Christian Dior brought the first liposomal cosmetic product, that is, Capture antiaging cream into the market [14]. Cosmetic liposomes can be divided into different forms depending upon the composition and purpose of use.

TABLE 8.1 Patents based on nanocosmetics [11–13].

Publication year	Publication number	Title	Assignee	Claim
2019	BR102018004126A2	Cosmetic preparations with hydrating, antioxidating, and photoprotective action containing extracts of the species *Spondia purpurea*, whether or not vehicled on lipid and polymeric nanoparticles	Universidade Federal Do Ceara	The hydroalcoholic, glycolic, or dried extracts of the species used for the said action using lipid or polymeric nanoparticles shows threefold enhancement of their activity
2019	KR20190079276A	Antiaging cosmetics with nanoparticles-ingredient that solubilized insoluble oligoanionic acid and manufacturing method for antiaging cosmetics	–	Nanoparticle useful for solubilizing a poorly soluble oleanolic acid that enhances its solubility and skin permeability for better antiaging effect
2018	CN108401417A	Including improving the cosmetics of nanoparticle and preparation method thereof of active principle containing whitening	–	Nanoparticles enhance cutaneous permeation of the active agents and long-term stability for skin whitening
2018	US20180008521A1	Low viscous cosmetic composition using a natural emulsifying agent	Amore Pacific Corp	The o/w nanoemulsion using natural emulsifier for good feel, stability, fast absorption
2018	WO2018105755A1	Cosmetics containing nanoparticles having encapsulated therein whitening-improving active ingredient, and method for producing said cosmetics	–	Nanoparticles having encapsulated therein whitening-improving active ingredient for long-term stability and skin penetration
2018	BR102015012999A2	Composition, process of preparation, and use of nanocosmetic based on carnauba wax and quercetin with moisturizing, antioxidant and photoprotective action	Univ Federal Do Ceara	The encapsulation of quercetin in carnauba wax lipid nanoparticles gives the product threefold action
2017	US20170281735A1	Compositions and methods for the treatment of photoaging and other conditions	Pro Transit Nanotherapy LLC	Nanoparticle comprises at least one biodegradable polymer and at least one antioxidant enzyme for preventing UV radiation induced skin disorder or skin disease
2017	ES2602107T3	Nanoemulsion		Said nanoemulsion used in the treatment of

(*Continued*)

TABLE 8.1 (Continued)

Publication year	Publication number	Title	Assignee	Claim
			Biofrontera Pharmaceuticals GmbH	dermatological diseases, diseases associated with viruses or cell proliferation, tumor diseases, and/or psoriasis
2017	BR102015032464A2	Topical formulations for photoprotection containing passiflora cincinnata	Universidade Federal Da Bahia	Uses micro, nanoemulsions, micelles preparation of the extract for photoprotection
2016	JP2016108285A	Cationized vesicle and composition thereof	–	To promote the penetration of substances into cells of skin or hair for the purpose of use in cosmetics
2015	US20150223451A1	Nanoformulation of musk-derived bioactive ingredients for nanocosmetic applications	Shaker A. Mousa, Amna Saddiq	Nanoencapsulated musk bioactive compounds using nanopolymers of hyaluronic acid and fatty acids that cross-linked with chitosan for cosmetic use as antiaging, antimicrobial, and fragrance
2013	RU2499406C1	Composition for producing cosmetic products and method for production thereof	–	Protects skin from inflammation and aging, controls and maintains the skin water-oil balance
2013	US20130022655A1	Metal oxide nanocomposites for UV protection	BASF SE	Sunscreen having improved properties with o/w type emulsion
2013	US20130034638A1	biodegradable, biocompatible, and nontoxic material sheets consisting of said material and the use thereof in food, pharmaceutical, cosmetic, and cleaning products	Inis Biotech LLC	The replacement of synthetic polymers by biopolymers
2013	US20130059769A1	Topically administered, skin-penetrating glycosaminoglycan formulations suitable for use in cosmetic and pharmaceutical applications	Eva Turley	For dermal rejuvenation, enhancement, hyaluronan replenishment, and protection therapy
2013	US20130068242A1	Semipermanent mascara and method of applying	Cry Baby Culture	100% waterproof mascara that lasts 2–3 weeks

(Continued)

TABLE 8.1 (Continued)

Publication year	Publication number	Title	Assignee	Claim
2013	EP2583665A2	Cosmetic composition containing retinol stabilized by porous polymer beads and nanoemulsion	Act Co., Ltd	For stabilizing retinol in composition that have antiinflammatory and antiwrinkle properties
2012	KR101326376B1	Cosmetic composition using nanocapsule containing azelaic acid and skim milk for treating acne skin and its manufacturing method thereof	–	Easy to penetrate into the skin to obtain the desired pharmacological effect at low concentration and has no side effect even when used daily, and it does not precipitate on a long-time storage
2010	US20100266649A1	Gel technology suitable for use in cosmetic compositions	Avon Products, Inc.	Preparation of optical gel system by nanoparticles reduces wrinkles
2010	EP2254545A2	Preparation of cationic nanoparticles and personal care compositions comprising said nanoparticles	BASF SE	Cationic nanoparticles impart antimicrobial property for skin and hair care products
2009	US20090175915A1	Nanoparticle compositions providing enhanced color for cosmetic formulations	Avon Products, Inc.	Improving the appearance by altering the optical properties of the biological surface
2009	US20090220556A1	Nanodiamond UV protectant formulations	International Technology Center	Sunscreen that reduces UV exposure by transmitting the rays
2008	EP1986594A2	Coatings for mammalian nails that include nanosized particles	Coty Sas Coty S SA	Method to increase strength and flexibility of nails enamel by adding one or a mixture of nanoclay, nanosilica, nanopigment, etc.
2008	EP1909745A1	Cosmetic pigment composition containing gold or silver nanoparticles	Korea Research Institute of Bioscience and Biotechnology	Preparation of harmless long-lasting colored pigments for various usages without precipitation
2008	WO2008079758A1	Nanocomposite pigments in a topical cosmetic application	Avon Products, Inc.	These pigments alter the optical properties of the biological surface, which results in improving the esthetic appearance
2008	CA2669392A1	Nanocrystals for use in topical cosmetic formulations	AbbVie Deutschland	The nanocrystal suspensions of improved physical stability increase the

(Continued)

TABLE 8.1 (Continued)

Publication year	Publication number	Title	Assignee	Claim
		and thereof method of production	GmbH and Co KG	bioactivity of the molecules in the skin
2006	KR20060029538A	Functional cosmetics using nanosilver	–	Effective against bacteria, fungi, and viruses as cosmetics suitable for acne skin or sensitive skin comprising the nanosilver
2006	KR100623013B1	Nanoemulsion, the use thereof, and preparing method thereof	–	Skin flexible, skin-friendly formulation that improves skin elasticity, effect in atopic dry skin, etc.
2005	US20050220730A1	Nail polish compositions comprised of nanoscale particles free of reactive groups	Alain Malnou, Martinez Francisco	Novel long-lasting nail polishes that have better physical resistance, improved gloss, transparency, and viscosity for application
2003	US20030064086A1	Cosmetic compositions comprising nanoparticles and processes for using the same	LOreal SA	Composition having long wear property by nanoparticles for at least one keratinous substance
2003	US20030003070A1	Use of nanoscale active antidandruff ingredients	Cognis Deutschland GmbH and Co KG	Long-lasting antidandruff action by enhancing its distribution on scalp
2002	WO2002060399A1	A controlled delivery system for hair care products	Salvona Llc	Controlled release of active agents to hairs for protection or conditioning purpose
2002	EP1239823B1	Use of nanoscale deodorants	Cognis Deutschland GmbH & Co.KG	Long-lasting dermatologically compatible and stable
2000	WO2000078281A1	Antimicrobial body care product	The Procter & Gamble Company	Silver nanoparticles for antimicrobial activity

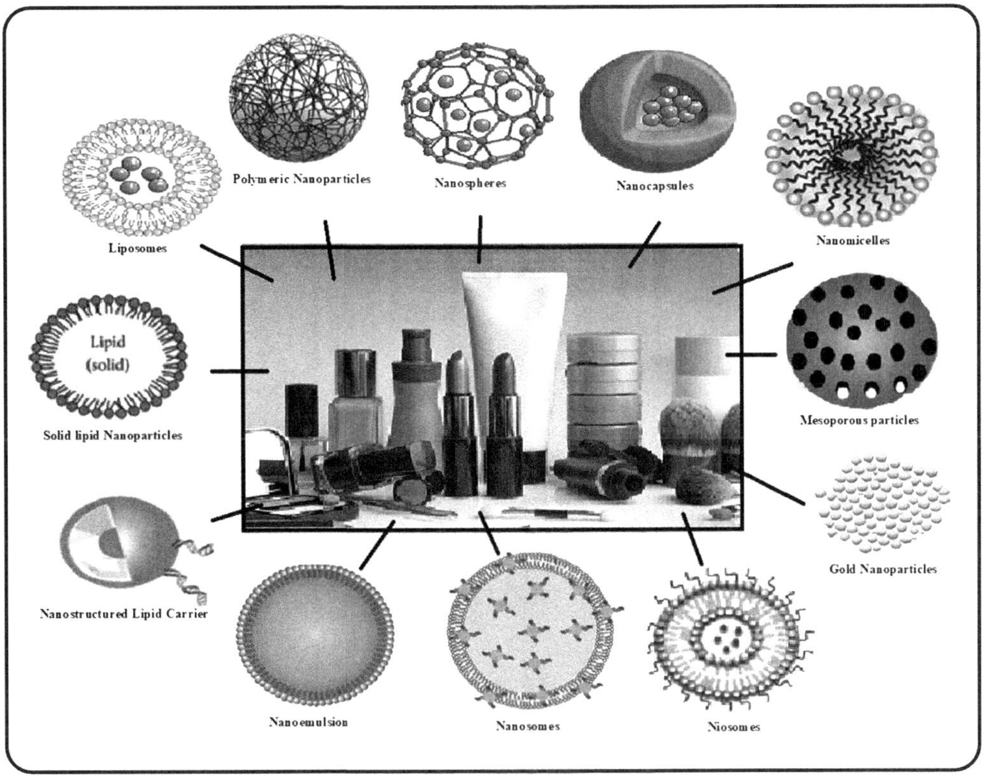

FIGURE 8.1 Various nanocarriers/nanomaterials used in cosmetics.

Transferosomes [15]

- Ultra deformable, reactive, biocompatible, biodegradable, and highly efficient liposomes.
- Their size ranges from 200 to 300 nm.
- They penetrate through the skin using transcellular and intracellular routes.
- Composed of phospholipid, cholesterol, and an edge activator, that is, surfactants like sodium cholate.

Niosomes [16]

- Also known as nonionic surfactant vesicles.
- They are composed of cholesterol and nonionic surfactants like Spans, Tweens, and Brij generally belonging to the alkyl or dialkyl polyglycerol ether class.
- Based on the structure, they can be further classified as small unilamellar vesicles (25–500 nm in diameter), multilamellar vesicles (0.5–10 μm in diameter), or large unilamellar vesicles (0.1–1 μm in diameter).

Novasomes [17]

- It is a patented and innovative encapsulation technique with several advantages like high loading capacity, improved aesthetics, stability, performance, cost-effective, reduced irritation, and easy scale up.
- These are modified liposomes or a variation of niosomes.
- They are nonphospholipid paucilamellar vesicles.
- Their size may range from 0.1 to 1 μm.
- Prepared using a mixture of monoester of polyoxyethylene fatty acids, cholesterol, and free fatty acids at 74:22:4 ratios.
- It consists of high capacity amorphous core surrounded by two to seven bilayer membranes.

Marinosomes [15]

- Prepared using extracts from marine lipid.
- These extracts are rich in omega-3 polyunsaturated fatty acids like eicosapentaenoic acid and docosahexaenoic acid.
- The enzymes present in the epidermis metabolize these lipids and convert them into the metabolites that have antiinflammatory and antiproliferative properties.
- Hence, marinosomes can be used in treating inflammatory problems of the skin.

Ultrasomes [14,15]

- Ultrasomes are multilamellar liposomes prepared using natural egg phospholipids. They entrap UV-endonuclease enzymes.
- Endonuclease enzyme is obtained from *Micrococcus luteus*.
- Ultrasomes possess multiple activities like recognition of damage caused by UV rays and fourfold acceleration of the repair program, protection of the immune system by repairing UV damaged DNA, and reduction in expression of tumor necrosis factor α, interleukins 1, 6, and 8. They are also known to stimulate the melanin production following UV exposure.

Photosomes [14,15]

- These are multilamellar liposomes that release photolyase obtained from a marine plant called *Anacystis nidulans*.
- Highly used in sun-protecting products, as they stop light from damaging the DNA. Hence, they can be beneficial in preventing the immune suppression and risk of skin cancer.

Ethosomes [18]

- Soft and flexible multilamellar vesicles.
- Composed of phospholipid, that is, phosphatidylcholine, water, and 20%–50% ethanol.
- The addition of ethanol to the system helps in deeper and easier penetration of the vesicles into the skin as ethanol disrupts the bilayer structure of the skin.

Asymmetric oxygen carrier system liposomes [19]

- These liposomes are used for skin oxygenation.
- These vesicles consist of perfluorocarbon nucleus in which a variety of gases like oxygen can be dissolved.
- These perfluorocarbons are hydrophobic in nature and the nucleus is enclosed by a phospholipid bilayer membrane.

Yeast-based liposomes [20,21]

- The derivatives from yeast cell possess skin repairing, soothing, and oxygenating abilities.
- Their liposomes stimulate dermal fibroblasts and help in feeling well.

Phytosomes [22]

- To improve the bioavailability of the herbal extracts or isolated herbal active components, a new concept of phytosomes was introduced.
- It is a complex mixture of phospholipid generally lecithin and natural active extracts such as flavonoids, glycosides, and terpenoids.

Nanosomes [23]

- Small liposomes with nanosize range vesicles.
- Prepared using pure phosphatidylcholine.

Glycerosomes [24]

- These are modified form of liposomes that contain glycerol in addition to phospholipids.
- They have high potential to be used in cosmetics to deliver active compounds due to their high performance, healing, and beautification properties.

Oleosomes [25]

- These are sphere-shaped organelles that are 1–3 μm in diameter. Hence, they also termed as natural liposomes.
- These oleosomes are rich reservoirs of oils, pigments, and vitamins.
- These are found in a variety of oil bearing plant seeds or fruits.
- Oleosomes find great applications in the cosmetic field as a drug encapsulating system.

Invasomes [26]

- These are soft liposomal vesicles with high membrane fluidity.
- They contain ethanol and terpenes or a mixture of terpenes. Their presence help in promoting skin permeation and cutaneous drug delivery by causing alteration in packing structure of stratum corneum.

Sphingosomes [27]

- Lipid vesicular systems consisting of an aqueous core enclosed by lipid bilayer membrane.
- Synthetic or natural sphingolipids constitute the lipid bilayer.

- Sphingosomes are composed of cholesterol and sphingolipids.
- They are classified based upon the size and number of lipid bilayers formed.
- Ceramides are one of the most commonly used sphingolipids.
- These ceramides repair the damaged and dehydrated skin.

Polymerosomes [3]

- These are artificial vesicular system with radius ranging from 5 to 50nm or more.
- They are prepared using block copolymers.
- These are biologically stable and highly versatile.
- Structurally they contain hydrophilic inner core and lipophilic bilayer, thus allowing encapsulation of both lipophilic and hydrophilic drugs.
- They can encapsulate and protect sensitive molecules, proteins, enzymes, drugs, peptides, RNA fragments, and DNA.
- Varied properties of different block copolymers prepare polymersomes with different characteristics like membrane thickness, permeability, and stimuli sensitiveness.
- Provide targeted and controlled drug delivery.
- Show more stability than liposomes.
- Patents have been filed that are using polymersomes to improve skin elasticity and for skin cell activation energy enhancement.

8.2.2 Nanoemulsions

Nanoemulsions are isotropic thermodynamically stable dispersions of oil and water with a narrow globular size range from 10 to 100 nm. Nanoemulsions are composed of oily phase, aqueous phase, surfactant, and cosurfactant mixed at an appropriate ratio which is decided using a pseudoternary phase diagram. These are transparent or translucent in appearance. These dispersions are stable due to the presence of proper ratio of surfactant and cosurfactant molecules that form an interfacial film between the oil and water. Based upon the composition, nanoemulsions can be classified as:

- **Oil in water (o/w):** oil forms the dispersed phase and water forms the dispersion medium.
- **Water in oil (w/o):** water forms the dispersed phase and oil forms the dispersion medium.
- **Bicontinuous**: tiny water and oil droplets are dispersed within the system.

Properties like low viscosity, high kinetic stability, ability to carry both hydrophilic and hydrophobic drugs, efficient drug penetration, high interfacial area, controlled release, and high solubilization capacity allow nanoemulsions to be widely used as a medium to deliver cosmeceuticals like deodorants, sunscreens, shampoos, lotions, nail enamels, conditioners, and hair serums [28].

8.2.3 Solid lipid nanoparticles

Solid lipid nanoparticles (SLNs) are first-generation lipoidal drug carrier systems that came up in the beginning of 1990s. This system is composed of physiological and biocompatible lipid that is dispersed in water or in an aqueous surfactant solution. This system contains an oily or lipidic core that is covered by a single layer of shell. This system can entrap a wide range of active compounds with various properties like lipophilic, hydrophilic, or poorly water-soluble. The use of biocompatible ingredients helps in avoiding toxicity issues and this has made SLNs popular in the pharmaceuticals and cosmeceuticals field. Better penetration through the skin due to their small size and skin hydrating ability due to their occlusiveness are some of the reasons that allowed SLNs to be used in various cosmetic products. SLNs act as physical sunscreens by resisting UV rays and hence can be incorporated into sun-protecting products with the advantage of improved photoprotection and reduced side effects. Due to their modified release pattern, SLNs are used in perfumes and day creams as well [29].

8.2.4 Nanostructured lipid carriers

These are second-generation lipid nanoparticles that were introduced to surmount the drawbacks encountered by SLNs. The particle size may range from 10 to 1000 nm. These are prepared using a blend of solid lipid and liquid lipid in the acceptable ratio of 70:30 up to 99.9:0.1. The resulted blend forms an amorphous solid that is solid at body temperature. Physiological and biocompatible lipids are used to formulate the nanostructured lipid carriers (NLCs), hence toxicity issues are reduced. They are proved to be better than SLNs in terms of stability, drug loading, skin hydrating ability, penetration, safety, and sun protection feature. Characteristics like enhanced physical and chemical stability of the drug, improved skin bioavailability, film formation, and controlled occlusion have introduced NLCs into the cosmetics area. In 2005, Dr. Rimpler GmbH, Germany first time launched cosmetic products containing NLCs into the market, namely, NanoRepair Q10 cream and NanoRepair Q10 serum. These products claimed increased skin penetration. At present, more than 30 cosmetic products loaded with NLC are available in the market [30].

8.2.5 Cubosomes

Cubosomes are biocompatible, thermodynamically stable, and bioadhesive nanostructured particles having a size range of 10–500 nm. These are prepared by mixing amphiphilic lipids like glycerol monooleate and surfactants like pluronic 127 in an appropriate ratio. These are liquid crystalline particles that contain three-dimensionally arranged non-intersecting bicontinuous lipid bilayers. Properties like multicompartmental structure, high drug payload, easy method of preparation, use of biodegradable lipids, encapsulation of a variety of molecules like hydrophilic, hydrophobic or amphiphilic moieties, targeted and controlled release make them a highly valuable drug carrier system for topical, transdermal, parenteral, and oral routes [31]. Cubosomes have attracted many cosmetic companies. They have been continuously investing on cubosomes and so far many patents have been filled claiming their cosmetic applications.

8.2.6 Nanosponges

Nanosponges are freely flowing particles having narrow cavities in the nanometer range. These cavities can fill both hydrophilic and lipophilic moieties. These nanosponges possess 3D network of degradable polyester. They are prepared upon mixing these polyesters with a crosslinker in a solution using techniques like liquid-liquid suspension polymerization and quasi emulsion solvent diffusion method. They have high entrapping potential and release the actives in a diffusion controlled manner. Their characteristics have attracted them to be used in dermatological and cosmetic products. Nanosponges can be used to load antifungal, local anaesthetics, antibiotics, etc., for topical use. These nanosponges can thereby be added to a base like gel, lotion, cream, powder, and ointment to be used topically [32].

8.2.7 Dendrimers

Dendrimers are micellar, nanosized, hyperbranched, radially symmetric, globular, monodisperse, three-dimensional synthetic polymers, having very well-defined size, shape, and definite molecular weight. They have a size range of 2–20 nm. The structure of the dendrimers can be divided into three different units, namely, central core, generations also known as branches which are attached to the central unit radically, and functional group terminally attached to the outermost branch. Their end groups can be functionalized in order to achieve desired physicochemical or biological properties. The drug can either be added to the central core or can be attached on the surface. If the drug is added to the central core, then it is released in a controlled manner [33,34]. Dendrimers have made great contribution in nanotechnology-based cosmeceuticals. They have found applications in hair care, nail care, and skincare products. Many renowned companies are using dendrimers in their products like mascara, shampoos, gels, and antiacne products. Several patents have been filed on the dendrimer technology by companies like L'Oreal, The Dow Company, Wella, and Unilever [3].

8.2.8 Nanosilver

Silver nanoparticles (AgNPs) are the nanoparticles of silver having a size range of 1–100 nm. AgNPs have positioned themselves in the field of nanoscience and nanotechnology, most importantly in nanomedicine. Nanosilver is known to possess several therapeutic benefits like antibacterial, antifungal, antiviral, antiinflammatory, antitumor, and antiangiogenic potential. Dermal toxicity studies were performed on rats that established their safety for topical applications. Hence, nanosilver can replace the conventional antimicrobial agents by using them in topical antimicrobial formulations. Nanosilver has been used in deodorants claiming 24-hour antibacterial protection. Other dermal benefits due to which it has been widely accepted in cosmetics include protection against atopic dermatitis, antiacne effects, acts as a preservative, inhibit the growth of dermatophytes, wound healing property, effect against skin cancer, and improvement of skin texture [35]. Nowadays, AgNPs have been incorporated into several dermatological products like soaps, toothpastes, wet wipes, deodorants, lip products, as well as face and body foams.

8.2.9 Nanogold

Gold nanoparticles are generally in the size range of 5–400 nm. The properties of nanogold greatly get affected by interparticle interactions and their assembly. They exist in multiple colors like red to purple, to blue, and almost black due to aggregation. High drug payload capacity, targeted drug delivery, inertness, high stablility, biocompatibility, and noncytotoxicity are some of the characteristics of these nanoparticles [36].

These nanoparticles possess antifungal and antibacterial properties. They have been incorporated into multiple cosmetic products like cream, face pack, deodorant, and antiaging creams. These nanogold themselves have multiple beauty care benefits like antiinflammatory, antiseptic and antiaging effects, maintenance and improvisation of skin elasticity and firmness, blood circulation acceleration, and reviving skin metabolism. Many cosmetic giants like L'Oreal and L'Core are making use of these nanogold into various cosmetic products like creams and lotions [3].

8.2.10 Nanospheres

These are crystalline or amorphous spherical nanoparticles having a core and a shell structure with an average particle size of 10–200 nm. The drug is protected from chemical and enzymatic degradation since it is dissolved, attached, entrapped, or encapsulated to the polymer matrix. They are classified into two categories based on the type of polymer used, namely, biodegradable nanospheres like gelatin, modified starch or albumin nanospheres and nonbiodegradable nanospheres like polylactic. Varied genes, enzymes, and drugs can be enclosed into the core of the spheres. These nanospheres convert poorly soluble, poorly absorbed and unstable bioactive compound into a promising drug [37]. These are being used in cosmetic products especially for skin care products like antiwrinkle, antiacne, and moisturizing creams in order to achieve deep penetration and targeted drug delivery [3].

8.2.11 Carbon nanotubes

Carbon nanotubes (CNTs) are cylindrical, hollow, and seamless shells made by rolling graphene sheets with SP^2 hybridization. Their diameter may vary from 0.7 to 50 nm with lengths in the range of 10s of microns. CNTs usually have a length-to-diameter ratio larger than 1,000,000. Individually, these nanotubes bind themselves through pi-bonding and give a rope-like appearance. CNTs can either have single wall (single-wall nanotubes with diameter ranging from 1 to 2 nm), double wall (double-walled nanotubes), or multiple walls (multiwalled nanotubes with diameter ranging from 2 to 50 nm) of graphene. Their main attraction points include light weight, minuscule size with a high aspect ratio, and good tensile strength and conducting features that have attracted their usage in the field of Pharmacy. They are believed to be promising candidates that can carry active moieties, vaccines, and nucleic acids and deliver them to the unreachable sites by crossing the membranes [38]. This technology has also been explored in cosmeceuticals which has resulted in filing of patents on CNTs and peptide-based CNTs containing hair coloring and cosmetic products [3].

8.2.12 Nanopigments/nanoparticles

Gold and silver in nanocolloidal form exhibit a wide range of applicability. Nanogold and nanosilver possess red and yellow color, respectively. Gold and silver have established a great future in the cosmetics and personal care industry. This is mainly due to their non-toxic nature, high stability, and disinfecting property. Seeing the toxicity caused by the pigments used in lipsticks and nail paints, nanogold and nanosilver pigments have been mixed in appropriate ratios to give different color shades. Titanium dioxide and zinc oxide used in sunscreens give white color to the skin. If the particle size of these agents is reduced to the nanorange, the product becomes transparent with improved stability. There are many nanopigments and nanoparticles that are used in the cosmetic industry [32].

8.3 Categories of nanotechnology-based cosmetics

Nanotechnology has made its position in both medical and cosmetic fields. Nanotechnology-driven cosmetic products are known to provide promising beauty solutions as they can penetrate deep into the skin's surface to provide better results. The major categories of nanotechnology-based cosmetics have been described below:

8.3.1 Skin care

Skin care products involve a variety of categories depending upon their application purpose like moisturizers, sunscreens, cleansers, and antiaging creams. A number of nanocarriers like nanoparticles, nanoemulsions, neosomes, and SLNs have been incorporated into cosmetic bases to improve their results. Table 8.2 lists some marketed skin care and under eye care nanocosmetics. Figs. 8.2 and 8.3 illustrate various benefits of nanocosmetics in skin care and eye care.

8.3.1.1 Sunscreens

Sunscreens form protective shield over the skin and prevent skin damage caused by UV rays. Minerals like zinc oxide (ZnO) and titanium dioxide (TiO_2) act as a natural reflector of both UVA and UVB rays and prevent penetration of these rays into the deeper skin layers. Conventional sunscreens leave white chalky layer onto the skin surface. Nanotechnology was introduced to combat this problem. When the size of these sun-blocking agents was reduced to nanoscale, an efficient sunscreen product was obtained that provided clear and natural appeal [41].

8.3.1.2 Antiaging creams

Ageing affects the dermal layer of the skin that leads to breakdown of collagen and elastin. Nanoparticles possess the potential to penetrate to the deeper layers of the skin; hence antiaging agents like retinol, vitamin E, vitamin C, and coenzyme Q10 are expected to work better when delivered in the nanorange. Many cosmetic companies are using nanotechnology to design antiaging products. For example, L'Oreal's Revitalift antiwrinkle cream is loaded with nanosomes of Pro-Retinol A, Lancome's Hydra Zen Cream is loaded

TABLE 8.2 List of marketed skin care and under eye care products [3,11–13,39,40].

S. no.	Type of nanoformulation	Product name	Company	Key ingredients	Benefits
1.	Liposomes	SesdermaC-Vit Liposomal Serum	Sesderma	Vitamin C, Ascorbyl glucoside, Mulberry extract, Hyaluronic acid, etc.	Antiwrinkle, skin brightening moisturizer
		Russell Organics Liposome Concentrate	Russell Organics	Vegetable oil and floral water loaded with superoxide dismutase, beta glucans, etc.	Hydrating and rejuvenating lotion
		Capture Totale Le serum	Dior	Longoza, Limonene, Rye seed extract, rice protein, lecithin, etc.	Antiwrinkle, brightening sunscreen
		emerginC Hyper-Vitalizer Face cream	emerginC	Alpha lipoic acid, Coenzyme Q10, Lutein, etc.	Antiaging, corrects uneven skin tone
2.	Niosomes	Anti-Age Response Cream	Simply Man Match	Pomegranate seed oil, ribonucleotide monophosphate, Ginseng extract avocado oil. mineral salts, etc.	Antiaging cream
		MayuNiosome Base Cream	Laon Cosmetics	Wild Ginseng, Saponin, etc.	Brightening moisturizer
3.	Solid lipid nanoparticles	Allure Body Cream	Chanel	Squalene, linalool, tocopheryl acetate, hexyl cinnamal, limonene, ascorbic acid, citric acid, etc.	Body Moisturizer
		Soosion Facial Lifting Cream SLN Technology	Soosion	Jojoba oil, herbal extracts, peptides, etc.	Antiwrinkle cream
		Youthful face contour creme	Sany Skincare	Epidermal growth factor, antioxidants, plant extracts, and peptides	Antiwrinkle, skin firming cream
4.	Nanostructured Lipid Carriers	Phyto NLC Active Cell Repair	SirehEmas	Olive oil, *Cucumis sectivus* extract, curcuma xanthorrhiza extract, etc.	Hyperpigmentation reducing, antiwrinkle, antiinflammatory rejuvenating moisturizer
		Cutanova Cream Nanovital Q10	Dr. Rimpler	Coenzyme Q10, *Cannabis sativa* seed oil, hydrolyzed wheat protein, soy protein, *Zea mays* oil, panthenol, ursolic acid, etc.	Antiaging, UV protectant cream

(Continued)

TABLE 8.2 (Continued)

S. no.	Type of nanoformulation	Product name	Company	Key ingredients	Benefits
		IopeSupervital Extra Moist Softner	Amore Pacific	Complex of Selaginella, waxberry extracts, etc.	Antiaging toner for rough and dry skin
5.	Nanoemulsion	Bepanthol Ultra FacialProtect Cream	Bayer HealthCare	Ceramides, lecithin, vit. B3, dexpantenol, lactic acid, glycine, glycerine, etc.	Antiaging, antipollutant, moisturizing cream
		Phyto-Endorphin Hand Cream	Rhonda Allison	Monk's Pepper-Casticin extract, Daisy flower extract, lecithin cholesterol, corn oil, soyabean oil, sweet orange peel oil, sweet almond oil	Revitalizing, age spot moisturizer
6.	Gold Nanoparticles	Chantecaille NanoGold Energizing Cream	Chantecaille	Blend of 24-k gold and silk, vit. C, vit. E, algae extract, plantago extract, etc.	Rejuvenating, antiaging cream, neutralize sun damage
		Tony Moly NanoGold BB Cream SPF 50 PA+++	Tony Moly	*Centella asiatica* extract, madecasoides, tea tree extract, Pantheol, allantoin, silk amino acid	Skin whitening, antiwrinkle sunscreen
		NanoGold 24-Hour Cream	Joyona International Marketing Ltd.	Micro silk proteins, nanogold particles, nanosilver complex, etc.	Regenerating antibacterial, cleansing, and balancing cream
		NanoGold Firming Treatment	Chantecaille	Swiss Apple Stem Cell extract, Microalgae Dunaliella Salina extract, peptides, etc.	Healing, antiinflammatory, antiaging
7.	Nanospheres	Hydralane Ultra Moisturizing Day Cream	Hydralane Paris	Marine collagen, peptides, etc.	Hydrating deep moisturizer
		Clearly It! Complexion Mist	Kara Vita	Complex Origanum, salicylic acid, etc.	Antiacne, pore refining toner
		Enlighten Me!	Kara Vita	Hydroquinone, kojic acid, glycolic acid, etc.	Lightening cream
8.	Nanosomes	Viterol. A (viatrozene gel) 16%	DS Laboratories, Inc.	Viatrozene, 16%, Almond Oil, Vit A, Vit E, Vit C, D-Panthenol, Gingko Biloba, Chestnut, Witch Haze, etc.	Antiwrinkle cream for wrinkles and expression lines
		Hydroviton.CR Liquid Normalizing Soap 80 g	DS Laboratories, Inc.	Chamomile Extract, Rosemary Extract, Hidroviton, etc.	Antiinflammatory, exfoliator, cleanser

(Continued)

TABLE 8.2 (Continued)

S. no.	Type of nanoformulation	Product name	Company	Key ingredients	Benefits
		Revitalift Double Lifting	L'Oreal	Pro-Tensium E, Pro-retinal A, etc.	Antiwrinkle gel
		Oligo DX Cellulite Treatment	DS Laboratories, Inc.	*Nelumbo nucifera* extract, Acacia Gum Extract, Hydrolyzed Casein, Ivy Extract, Carrageenan, etc.	Body firming lotion
9.	Nanocapsules	Lancome Soleil Soft-Touch Anti-Wrinkle Sun Cream SPF 15	L'Oreal	Squaline, zinc oxide, Xanthum gum, Vit A, Vit D3, etc.	Antiaging cream
		Hydra Flash Bronzer Daily Face moisturizer	Lancome	Vitamin E, Hyaluronic Acid, Aloe Water, Vit C. Vit E, etc.	Antioxidant, hydrating Moisturizer
		Soleil Express ProtectionInstant Cooling SunSpritz SPF 15	Lancome	Vitamins, antioxidants,	Sun protection spray
10.	Fullerenes	Sircuit SkinSircuit addict + Firming AntiOxidant Serum	Sircuit Skin Cosmeceuticals Inc.	Kombuchka, Buddleja Stem Cells, Glycine Soja Protein, Vitamin E, Spin trap, Resveratrol, Grape Seed, Wine extract, etc.	Antiaging
		Zelens Fullerene C-60 Night Cream	Zelens	Fullerene C-60	Antiaging cream
11.	Nanoencapsulation	Neova Therapy Dual Action Lotion	ProCyte Corporation	Retinol, GHK copper peptide complex, etc.	Antiaging, brightening lotion
		Hydra Zen Anti-stress Moisturizing Face Cream	Lancome	Triceramides, Rose, Suffruticosa, Moringa extracts, etc.	Rejuvenating, hydrating moisturizer
		Dr. Brandt Time Arrest Laser Tight	Cosmetic Dermatology, Inc.	Platinum Heptapeptides, Sweet Pea, Hyaluronic Acid, etc.	Antiaging cream
12.	Nanoparticles	Nouriva Repair Moisturizing Cream	Ferndale Laboratories, Inc.	Zinc oxide, white petrolatum, lanolin, mineral oil, glycerin, lecithin, glycolic acid, allantoin, etc.	Moisturizing cream
		Cosil Nano Beauty Soap	Natural Korea	NanoColloidal Silver, Arbutin, Aloe Vera, collagen, etc.	Cleanser, pore detoxifier, exfoliator soap

(Continued)

TABLE 8.2 (Continued)

S. no.	Type of nanoformulation	Product name	Company	Key ingredients	Benefits
		LancomeRenergie Microlift	Lancome	Colloidal silica, soy protein, etc.	Antiwrinkle
13.	Lyphazome nanospheres	Celazome O-Plex Target Acne Spot Treatment	Celazome New Zealand Limited	Origanum Complex, Sulfur, Tea Tree Oil, etc.	Antiacne
15.	Nanovitamins	Colorescience Genie Sparkle Bottles	ColoreScience	Bismuth Oxychloride, Iron Oxide, Carmine, Mica. Chromium Oxide Green, Ultramarine Blue, etc.	Bronzer/highlighter
17.	Nanomicelles	Nano-in Deep Cleaning for makeup removal and exfoliation	Nano-Infinity Nanotech Co., Ltd	Zinc oxide, glycerine, etc.	lotion
23.	Nanosilver	Nano Cyclic Cleanser Silver	Nano Cyclic	Nanosilver, sericin other bioactives	Exfoliates dead skin, diminishes age spots, pore cleaner
Under eye care					
24.	Liposomes	Lumessence Eye Cream	Aubrey Organics	Rosa rubiginosa seed oil, *Camellia sinensis* leaf oil, oat extract, rye seed extract, shea butter, carrageenan, grape fruit oil, etc.	Antiwrinkle eye cream
27.	Nanostructured lipid carriers	IopeSupervital Extra Moist Eye Cream	Amore Pacifc	Mineral cations, organic anions, Recoverine, Omega-3, Vit K, etc.	Removes eye wrinkles, dullness, and poor elasticity
29	Nanospheres	Eye Tender	Kara Vita	13 bioactives including peptides	Under eye antiwrinkle treatment
30.	Nanocapsules	Eye Contour Nanolift	Euoko	Rhodiola, sugar beet extract, Ceramide 2, hyaluronic acid, Eyeliss, etc.	Antiwrinkle eye cream
31.	Lyphazome nanosphere	CelazomeEye Treat	Celazome New Zealand Limited	Shea Butter, Olive oil, Squalane, Argireline, Green Tea extract, Hyaluronic acid, Vit E, etc.	Under eye cream
32.	Fullerenes	Sircuit Skin White out + Daily Under Eye Care	Sircuit Skin Cosmeceuticals Inc.	L-Arbutin, L-Malic Acid, L-Tartaric Acid, L-Lactic Acid, Vitamin E, Witch Hazel extract, Calendula oil, Fullersomes, etc.	Puffy eye, dark circle treatment

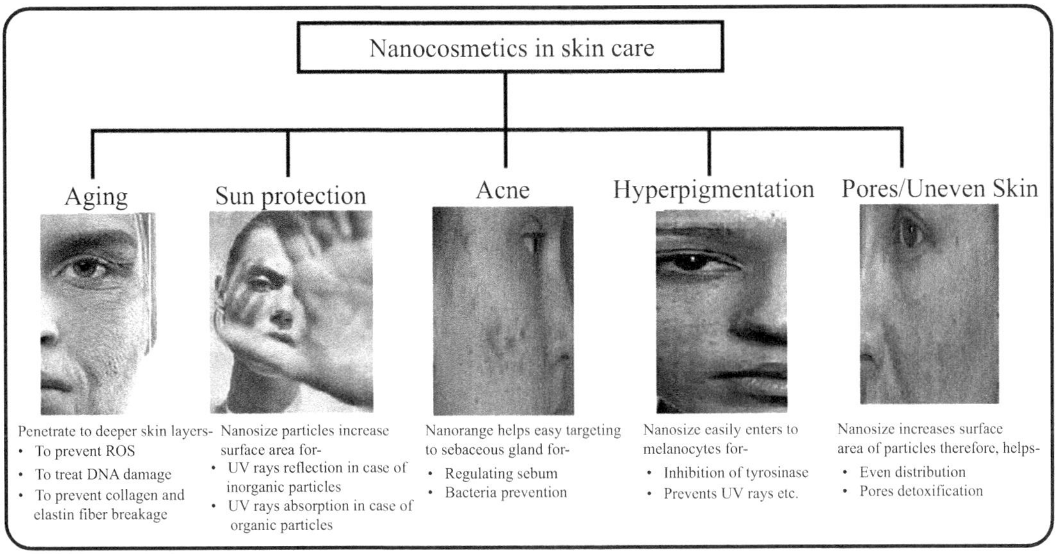

FIGURE 8.2 Various benefits of nanocosmetics in skin care.

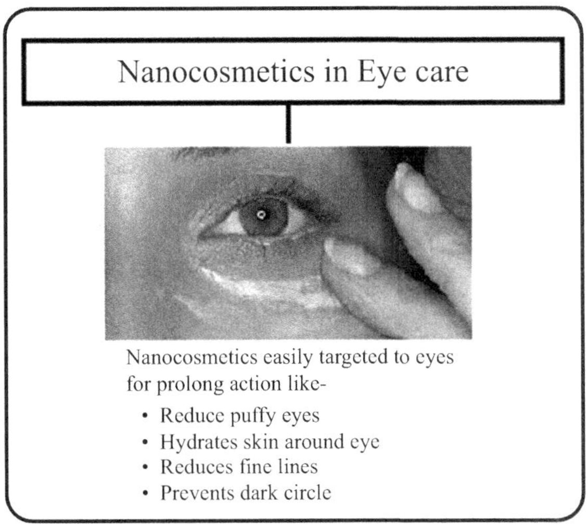

FIGURE 8.3 Various benefits of nanocosmetics in eye care.

with nanoencapsulated Triceramide [12], and Chantecaille's nanogold face energizing cream is loaded with 24 carat gold nanoparticles [32].

8.3.1.3 Moisturizers

The dehydration condition of the skin can be avoided and treated using moisturizers that form a thin film of humectant. Therefore moisturizers can be applied to prevent water loss or to restore the normal functioning of stratum corneum [12]. Nanocarriers like liposomes, nanoemulsions, SLNs, and niosomes are extensively added to the moisturizers as they retain the moisture for a prolonged time [3].

8.3.1.4 *Skin cleansers*

A cleanser act as a personal care product that is used for skin, generally face to remove makeup, oil, dirt, dead skin cells, and other pollutants. Metal nanoparticles act as skin disinfectant and decontamination agents. Nano Cyclic Inc. produces cyclic nanosilver cleanser and cleansing pink bar containing silver nanoparticles and natural ingredients. These products kill harmful bacteria and fungi, fight acne, exfoliate dead skin, remove makeup gently, and diminish age spots and sun damaged skin [12].

8.3.2 Lip care

Nanotechnology-based lip care products generally include lipsticks, lip balm, lip gloss, and lip volumizer. Nanoparticles incorporated into lip care products soften the lips by preventing transepidermal water loss, prevent pigment migration from the lips, and show longer stay. Lip volumizer loaded with liposomes improve volume and hydration of the lips. Pigments used in lipsticks cause lead toxicity. However, it was noticed that nanogold possesses red color and nanosilver possesses yellow color. Therefore gold and silver nanopigments have been incorporated into the lipsticks to get new colored pigments [32]. Silica nanoparticles are also used in lipsticks. These nanoparticles bring homogenous distribution of pigments and prevent migration or bleeding of pigments into the fine lines of lips [42]. Table 8.3 lists some of the marketed lip care nanocosmetics. Fig. 8.4 illustrates the benefits of nanocosmetics in lip care.

8.3.3 Oral care

Products used to maintain oral hygiene involve toothpastes and mouth washes. The trend of preparing these products using nanotechnology in order to improve their efficacy and performance has taken a fast track. Nanocarriers like quantum dots, dendrimers, gold nanoparticles, silver nanoparticles, and hydrogels are generally employed in the oral care products. Table 8.4 lists some of the marketed oral care nanoproducts. Fig. 8.5 illustrates the benefits of nanocosmetics in oral care.

8.3.3.1 *Toothpastes*

Toothpastes loaded with nanotechnology are prepared with the aim to achieve self-healing of the teeth, enamel rebuilding, and protection against bacterial infections. Toothpastes containing nanosized hydroxyapatite and titanium dioxide are prepared to achieve better protection and whitening, respectively. Silver nanoparticles are incorporated into toothpastes that act as potent microbicide. Toothpastes loaded with $CaCO_3$ nanoparticles and 3% nanosized sodium trimetaphosphate promoted remineralization of early carious lesions when compared with conventional toothpaste with no nanoadditives. A research team of SusTech Darmstadt and Henkel prepared Nanitactive-loaded oral products. Nanitactive comprises components present in the teeth, that is,

TABLE 8.3 List of some marketed lip care nanocosmetics.

S. no.	Type of nanoformulation	Product name	Company	Active agent	Used as
1.	Liposomes	Fillderma Lips Lip Volumizer	Sesderma	Hyaluronic acid, fermented black sweet tea, lactic acid, niacinamide, etc.	Hydrates, volumize the lips
2.	Nanospheres	Lip Tender	Kara Vita	Argireline, Matrixyl 3000, Pepha-Tight, Tyrostat-09, Velvet Veil, Olive oil, CanadianWillowherb extract, Green Tea extract, Squalane, Hyaluronic acid, etc.	Lip moisturizer
3.	Nanocapsule	Primordiale Optimum Lip	Lancome	Vitamin E, Gatuline, etc.	Antiwrinkle lip treatment
4.	Nano-zinc oxide	Luscious Lips Rejuvenation Duo	Vortex Health & Beauty Ltd.	Vit C, E, Aloe Vera extract, Algae extract, Stevia extract, Grape Seed extract, Castor oil, Hyaluronic acid, etc.	Reduces lip lines, lip care
5.	Micronized topaz powder	Revlon SkinLightsGlosslights for Lips Translucent Gloss	Revlon	Rose Quartz, Topaz, mother of pearl, etc.	Lipstick, Lip gloss

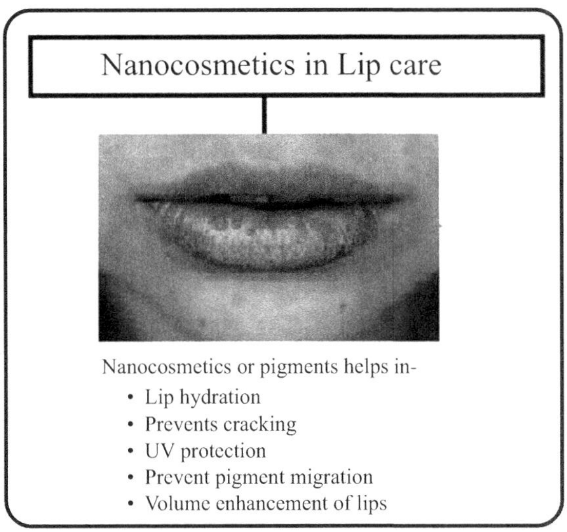

FIGURE 8.4 Various benefits of nanocosmetics in lip care.

calcium phosphate nanoparticles and protein. "Aclaim" from Group Pharmaceuticals is nanotechnology inspired toothpaste used for hypersensitivity and remineralization of teeth.

TABLE 8.4 List of some marketed oral care nanoproducts.

S. no.	Type of nanoformulation	Product name	Company	Active agent	Used as
1.	Nanoparticles	NanoCare gold\	Dental NanoTechnology, Poland	Nanogold or nanosilver, chlorhexidin, isopropyl alcohol, etc.	Cavity disinfectant
2.	Nanohybrid composite	SRPhonaresNHC	Ivoclar Vivadent	Silicon oxide	Denture teeth
3.	Nanohybrid composite	Ketac Nano Light-Curing Glass Ionomer Restorative	3M Science	Zirconia/silica nanofillers nanoclusters	Teeth restoration

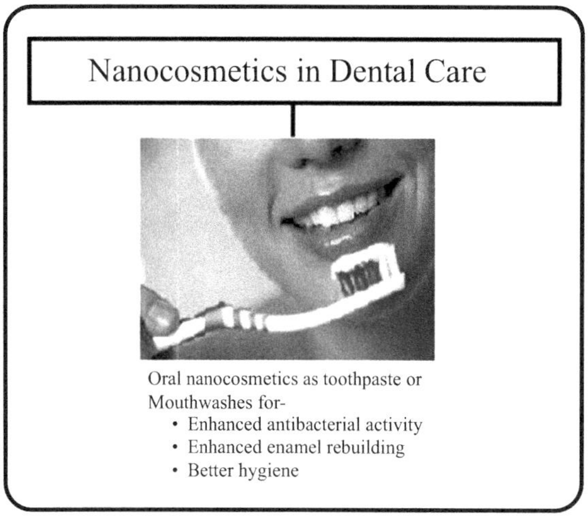

FIGURE 8.5 Various benefits of nanocosmetics in oral care.

8.3.3.2 Mouth wash

Mouth wash helps to maintain oral hygiene by preventing or reducing tartar, plaque, and gingivitis. Sometimes, whitening agents are also added to the mouth wash to whiten the teeth. Many products are present in the market containing nanotechnology inspired mouth washes. For example, UK dentists prepared "Nano Whitening Mouth wash", Enamel Care Technology for whitening and remineralization of the teeth.

8.3.3.3 Hair care

Hair care products like shampoos, conditioners, dyes/colorants, hair growth promoting and styling products have been enthused with nanotechnology. Cosmetic companies are nowadays making an attempt to discover the role of nanoparticles to prevent hair loss, promote hair growth, treat hair-related disorders, and maintain shine, silkiness, and health of hairs. Nanocarriers like nanoemulsion, nanospheres, and liposomes are known to improvise the effects of hair care products [43]. For example, Proxiphen N containing shampoo is used to treat alopecia/baldness, silver nanoparticles are used to treat bacteria caused hair depilation and sericin nanoparticles in hair products treat damaged hair cuticles. Table 8.5 lists some of the marketed hair care nanoproducts. Fig. 8.6 illustrates the benefits of nanocosmetics in hair care.

8.3.4 Nail care

Adding nanotechnology to the nail care products has shown better promising results when compared to the conventional nail paints. Nanotechnology-based nail polishes have improved toughness, mar resistance, chip resistance, durability, ease of application, and fast drying. A provisional patent was granted to Nano Labs Corp. in 2012 for making nanonail polish and nail lacquer with properties like drying to a very hard state, resistant to shock, cracking, scratching and chipping, and high elasticity. One strategy to treat fungal infection of toe nails involves the addition of silver and metal oxide nanoparticles in nail polish, since these nanoparticles possess antifungal properties [12]. Table 8.6 lists some of the marketed nail care nanoproducts. Fig. 8.7 illustrates the benefits of nanocosmetics in nail care.

8.4 Consumer concerns and regulatory guidances

8.4.1 Routes of nanocosmetics exposure

Nanocosmetic formulations involve the use of nanoparticles/nanomaterials for their preparation. Nanomaterials have been present in the cosmetic products since hundreds of years ago. But, the trend of using nanoranged materials in cosmetics has fastened up in the modern age. Though nanotechnology comes with several benefits, but some risks and uncertainties are generally associated with their application in cosmetics. As the rate of production of nanocosmetics has increased, the number of consumers and workers has increased many folds. The risk of exposure of nanomaterials has fallen both on the workers and on the consumers. Therefore a complete awareness about the routes of nanomaterial exposure is required. Various routes through which humans get exposed to the nanomaterials involve inhalation, ingestion, and penetration through the skin. The degree of exposure and the route of exposure greatly measure the seriousness of the health hazard caused by nanomaterials. Workers generally get exposed during production of products containing nanomaterials, during use, disposal, or recycling of these products. These nanomaterials when released to the water, air, and soil during their manufacturing process, use or disposal may pose some environmental risks. Therefore it is required that all these risks must be addressed during manufacturing, utilization, and disposal of nanocosmetics.

According to the National Institute of Occupational Health and Safety, organisms get exposed to the airborne nanoparticles through inhalation. Workers may inhale the nanoparticles during production or disposal. Whereas consumers may inhale nanoparticles while using products like deodorant and perfumes. Some nanoparticles may reach the brain via nasal nerves while other may enter the blood stream and get transported to other organs of the body [44].

TABLE 8.5 List of some marketed hair care nanoproducts.

S. no.	Type of nanoformulation	Product name	Company	Active agent	Used as
1.	Niosomes	Identik Masque Floral Repair	Identik	Adenosine *Punica granatum* seed extract, hydrolyzed yeast extract, etc.	Hair repair masque
2.	Nanoemulsions	Korres Red Vine Hair Sun Protection Serum	Korres	Aloe vera, Citronellol, Coumarin, Sunflower Seed Extract, Citral, Lecithin, Limonene, etc.	Hair color protectant, sun protectant spray
3.	Nanosomes	Spectral DNC-N Hair Loss Treatment	DS Laboratories, Inc.	Nanoxidil 5%	Hair loss treatment
4.	Nanoparticles	Pureology Nano Works ShineLuxe	Pureology	Mica mirrors, Jasmine, vanilla, rose, etc.	Hair polish, color protection
		PureologyNanowax	Pureology	Orange peel Wax, ThermalAntiFadeComplex, silicone technology, bergamot, jasmine, Mushroom blend, etc.	Thermal and color protectant
5.	Micronized titanium dioxide	California Baby Swimmer's Defense Hair Conditioner	California Baby	Vegan, Cabbage-based waxes, lemongrass, Aloe vera, etc.	Hair conditioner to protect hairs from chlorine water or sea water in the presence of sun rays

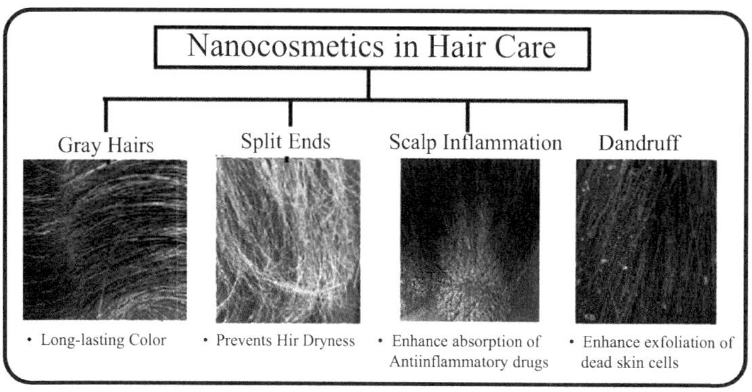

FIGURE 8.6 Various benefits of nanocosmetics in hair care.

TABLE 8.6 List of some marketed nail care nanoproducts.

S. no.	Type of nanoformulation	Product name	Company	Active agent	Used as
1.	Lyphazome	Celazome Tip Treat Cuticle Exfoliator	Celazome New Zealand Limited	Petrolatum, Alcohol, Lecithin, Mineral Oil, Paraffin, Pineapple Extract, Sodium Hyaluronate, etc.	Nail treatment, nail hydration
2.	Micronized polycarbonated resin	Sally Hansen Thicken Up!	Del Laboratories, Inc.	Protein, trehalose, retinol, etc.	Protects weak thin nails
3.	Nanoparticles	Nano-In Hand and Nail Moisturizing Serum and Foot Moisturizing Serum	Nano-Infinity Nanotech	Nano-zinc oxide, natural essence, etc.	Nail moisturizer

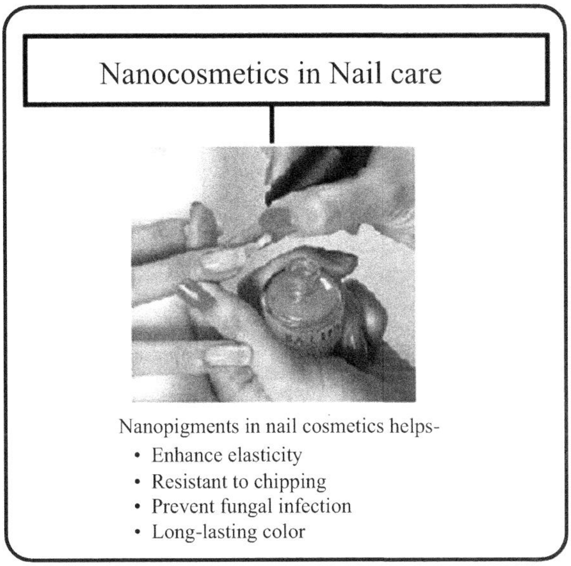

FIGURE 8.7 Various benefits of nanocosmetics in nail care.

There are chances that nanoparticles can get ingested into the body from hand to mouth transfer unintentionally. They can also be ingested into the body from the cosmetic products like lip color, lip balm, and lip gloss that are meant to be used near mouth area or on lips. Studies reveal that the majority of the nanoparticles pass out of the body rapidly while some of them are taken up by the body and migrated to different organs [45].

Skin is another route through which nanoparticles can reach blood circulation. Nanoparticles can penetrate across the skin following any of the three pathways, namely, intercellular, transfollicular, and transcellular. A huge variety of nanocosmetics are available in the market that is meant to be applied on the skin. Concerns have been raised regarding the danger that may arise due to unwanted penetration of these particles into the deeper layers of the skin and systemic circulation. This movement of particles across the skin depends upon the physicochemical characteristics of the nanoparticles, type of vehicle used, nature of the substance, and the skin condition. Skin issues like eczema, acne, and wound enhance the chances of entry of nanoparticles into skin and blood [46].

8.4.2 European Union Guidelines

The European Commission's (EC) Scientific Committee on Consumer Safety (SCCS) published tenth revised version of "The SCCS Notes of Guidance for the Testing of Cosmetic Ingredients and Their Safety Evaluation" on November 7, 2018. It included guidelines to be followed in Europe regarding testing and safety evaluation of cosmetic ingredients. The guidance is specifically designed for the interest of the public authorities and cosmetic industry to improve harmonized compliance with the current cosmetic European Union (EU) legislation. The guidance also addressed nanomaterials used in cosmetics which included (1) definition of nanomaterial, (2) potential safety issues of nanomaterials, and (3) information required by SCCS for evaluation of nanomaterials as cosmetic ingredients (https://ec.europa.eu/health/sites/health/files/scientific_committees/consumer_safety/docs/sccs_o_224.pdf).

8.4.3 Guidance document issued by FDA for industries

The "Task Force" of FDA issued a guidance document for the industry under the title "Guidance for Industry: Safety of Nanomaterials in Cosmetic Products" in the year 2007 (https://www.fda.gov/regulatory-information/search-fda-guidance-documents/guidance-industry-safety-nanomaterials-cosmetic-products). These guidelines were issued in order to ensure that cosmetic products loaded with nanomaterials are safe and free from adulteration. Task force requires the reports and related data from the manufacturers showing the effect of nanomaterial-loaded products that are not subjected to premarket consent like cosmetic products.

In 2014, FDA issued a guidance document entitled "Considering whether an FDA-regulated product involves the application of nanotechnology." These considerations will be followed broadly to all FDA-regulated products, including cosmetic products. According to this, when considering whether an FDA-regulated product involves the application of nanotechnology, the following points need to be asked:

- Whether a material or an end product is tailored in such a way that at least one of its external dimension or an internal or surface structure falls in the nanorange (approximately 1–100 nm).
- Whether a material or an end product is tailored in such a way that it shows properties or phenomena, including physical or chemical properties or biological effects, featuring its dimension(s), even if these dimensions fall outside the nanoscale range, up to 1 μm (1000 nm).

8.4.3.1 Safety assessment of nanomaterials in cosmetic products

There is prohibition of selling adulterated or misbranded cosmetic products during interstate commerce as stated in section 301(a) of the Federal Food, Drug, and Cosmetic Act (FD&C Act) [21 U.S.C. 331(a)]. No premarket approval for marketing of cosmetic products and ingredients used in them is required in US according to FD&C Act (except colorants).

US manufacturers are allowed to formulate cosmetic products using any ingredient except for colorants and ingredients banned by regulatory bodies to be used in cosmetics. It is also noted that only those ingredients will be used that does not make the cosmetic product adulterated [section 601 of the FD&C Act (21 U.S.C. 361)] or misbranded [section 602 of the FD&C Act (21 U.S.C. 362)]. The manufacturer is not required to submit any safety data in order to release the cosmetic product in the market as stated in FD&C Act. But the manufacturers need to realize their responsibility and should gather the reports and data that assure the product safety before launching it to the market.

According to the Federal Register of March 3, 1975 (40 FR 8912 at 8916), "the safety of a product can be adequately substantiated through (1) reliance on already available toxicological test data on individual ingredients and on product formulations that are similar in composition to the particular cosmetic, and (2) performance of any additional toxicological and other tests that are appropriate in light of such existing data and information. Although satisfactory toxicological data may exist for each ingredient of a cosmetic product, it will still be necessary to conduct some toxicological testing with the complete formulation to assure adequately the safety of the finished cosmetic."

All these general principles are applicable to both types of cosmetic products (conventional and nanocosmetics).

8.4.3.2 Primary considerations to assess the safety of nanomaterials in cosmetic products

The physicochemical properties of the materials change upon size reduction to the nanoscale range. This may lead to some biological interactions as stated in section III A. This arises safety concerns regarding nanomaterial-loaded products. Therefore it is required to report changes in the properties as well as in biological effects during safety assessment of the nanomaterials.

- It is required to provide complete information of the nanomaterials. These nanomaterials should be characterized to determine their physicochemical properties. They should also be assessed for the presence of impurities, if any.
- The pharmacodynamic (absorption, distribution, metabolism, and excretion) and toxicity profile of the nanomaterials used in cosmetics should be generated and made available.
- If any characteristic property or biological activity of the nanomaterial-loaded cosmetic product is witnessed, it is required to assess the toxicity through the traditional testing method.

However, it is required to improvise the conventional toxicity testing methods or to develop new strategies to determine (1) the principal physicochemical properties affecting toxicity report of nanomaterials and (2) how these properties affect the cosmetic products. A complete set of information and data should be prepared and made available proving the safety of the product under the intended conditions of use.

8.5 Conclusion

Nanotechnology has made its inevitable place in almost every field. It has brought revolution in several industries including cosmetics. With the development in technology, the growth of cosmetic industry has increased many folds. The invention of novel delivery systems like liposomes, niosomes, SLNs, NLCs, cubosomes and nanomaterials like fullerenes, nanogold, nanosilver, nanopigments has risen up the use of nanotechnology in cosmeceuticals. Therefore, the conventional cosmetic products have been replaced with novel nanotechnology-based products. These nanocarriers/nanomaterials promise controlled and targeted drug delivery, better stability and efficiency, biocompatibility, prolonged action, and higher drug-loading capacity. Nanotechnology-based cosmetic products are highly diverse. These include products majorly for skin care, lip care, hair care, eye care, and nail care. The commercialization of nanotechnology in cosmetics has resulted in availability of nanotechnology-based cosmetic products. With the packet full of benefits, nanotechnology also comes up with some risks for the humans and environment. The use of nanotechnology has raised concern over the health and safety of consumers. It is required that the manufacturers should design and sell nanoproducts in a way that fully respects the health and safety of consumers and the environment. Also, the regulatory bodies are required to set up stringent laws regarding the use of nanomaterial in cosmetics.

References

[1] Nahar L, Sarker SD. Importance of nanotechnology in drug delivery. Open Access J Pharm Res 2017;1(5):000130.
[2] Logothetidis S. Nanostructured materials and their applications. Nanosci Technol 2012;12:220.
[3] Kaul S, Gulati N, Verma D, Mukherjee S, Nagaich U. Role of nanotechnology in cosmeceuticals: a review of recent advances. J Pharm 2018;2018:19.
[4] Dureja H, Kaushik D, Gupta M, Kumar V, Lather V. Cosmeceuticals: an emerging concept. Indian J Pharmacol 2005;37:155–9.
[5] Srinivas K. The current role of nanomaterials in cosmetics. J Chem Pharm Res 2016;8(5):906–14.
[6] Mu L, Sprando RL. Application of nanotechnology in cosmetics. Pharm Res 2010;27(8):1746–9.
[7] Nanda S, Khurana RK, Dhiman S, Sandhu P, Singh B. Nanocosmetics: fundamentals, safety and regulatory aspects. In: Singh B, Singh KK, Rekhi GS, editors. Nanopharmaceuticals. M/s Studium Press, LLC; 2015. p. 453–83.
[8] Nanda S, Nanda A, Singh B. Federal perspectives of nanostructured systems: an update. In: Singh B, Vyas SP, Kaur IP, editors. Nanostructured drug delivery systems. M/s Studium Press, LLC.; 2015. p. 491–525.
[9] Shokri J. Nanocosmetics: benefits and risks. Bioimpacts 2017;7(4):207–8.
[10] Mukta S, Adam F. Cosmeceuticals in day-to-day clinical practice. J Drugs Dermatol 2010;9:62–6.
[11] Kumari K, Sharma PK, Gupta R. Nano-cosmeceuticals: an emerging novel trend towards dermal care. Adv Cosmet Dermatol 2017;3(1):1–12.
[12] Lohani A, Verma A, Joshi H, Yadav N, Karki N. Nanotechnology-based cosmeceuticals. Int Sch Res Not Dermatol 2014;2014:14.
[13] Duarah S, Pujari K, Durai RD, Narayanan VH. Nanotechnology-based cosmeceuticals: a review. Int J Appl Pharm 2016;8(1):8–12.
[14] Ahmadi Ashtiani HR, Bishe P, Lashgari NA, Nilforoushzadeh MA, Zare S. Liposomes in cosmetics. J Skin Stem Cell 2016;3(3):e6581.
[15] Reva T, Vaseem AA, Satyaprakash S, Md.Khalid JA. Liposomes: the novel approach in cosmaceuticals. World J Pharm Pharm Sci 2015;4(6):1616–40.
[16] Kurapati S. The current role of nanomaterials in cosmetics. J Chem Pharm Res 2016;8(5):906–14.
[17] Singh A, Malviya R, Sharma PK. Novasome—a breakthrough in pharmaceutical technology a review article. Adv Biomed Res 2011;5:184–9.
[18] Verma P, Pathak K. Therapeutic and cosmeceutical potential of ethosomes: an overview. J Adv Pharm Technol Res 2010;1(3):274–82.
[19] Patravale VB, Mandawgade SD. Novel cosmetic delivery systems: an application update. Int J Cosmet Sci 2008;30(1):19–33.
[20] Lautenschlager H. Liposomes in dermatological preparations: Part II. Cosmet Toilet 1990;105:63–72.
[21] Suzuki K, Sakon K. The applications of liposomes to cosmetics. Cosmet Toilet 1990;105:65–78.
[22] Karimi N, Ghanbarzadeh B, Hamishehkar H, Keyvani F, Pezeshki A, Gholian MM. Phytosome and liposome: the beneficial encapsulation systems in drug delivery and food application. J Appl Food Biotechnol 2015;2(3):17–26.
[23] Nounou MI, El-Khordagui LK, Khalafallah NA, Khalil SA. Liposomal formulation for dermal and transdermal drug delivery: past, present and future. Recent Pat Drug Deliv Formul 2008;2(1):9–18.

[24] Ganesan P, Choi DK. Current application of phytocompound-based nanocosmeceuticals for beauty and skin therapy. Int J Nanomed 2016;11:1987–2007.
[25] Socaciu C. New techonolgies to synthesize. Extract and encapsulate natural food colorants. Bull Univ Agric Sci Vet Med Cluj-Napoca Animal Sci Biotechnol 2007;64:1–2.
[26] Lakshmi PK, Kalpana B, Prasanthi D. Invasomes—novel vesicular carriers for enhanced skin permeation. Syst Rev Pharm 2013;4(1):26.
[27] Fathima KM, Antony N, Paul A, Nair SC. Sphingosome vescicular system. Int J Pharm Sci Rev Res 2016;41 (1):208–13.
[28] Savardekar P, Bajaj A. Nanoemulsions—a review. Int J Res Pharm Chem 2016;6(2):312–22.
[29] Mukherjee S, Ray S, Thakur RS. Solid lipid nanoparticles: a modern formulation approach in drug delivery system. Indian J Pharm Sci 2009;71(4):349–58.
[30] Muller RH, Petersen RD, Hommoss A, Pardeike J. Nanostructured lipid carriers (NLC) in cosmetic dermal products. Adv Drug Deliv Rev 2007;59(6):522–30.
[31] Bhosale RR, Osmani RA, Harkare BR, Ghodake PP. Cubosomes: the inimitable nanoparticulate drug carriers. Sch Acad J Pharm 2013;2(6):481–6.
[32] Nanda S, Nanda A, Lohan S, Kaur R, Singh B. Chapter 3: Nanocosmetics: performance enhancement and safety assurance. In: Alexandru Mihai Grumezescu (ed.), Nanobiomaterials in galenic formulations and cosmetics. 2016. p. 47–67, Elseiver.
[33] Abbasi E, Aval SF, Akbarzadeh A, Milani M, Nasrabadi HT, Joo SW, et al. Dendrimers: synthesis, applications, and properties. Nanoscale Res Lett 2014;9:247.
[34] Malik A, Chaudhary S, Garg G, Tomar A. Dendrimers: a tool for drug delivery. Adv Biol Res 2012;6 (4):165–9.
[35] Gajbhiye S, Sakharwade S. Silver nanoparticles in cosmetics. J Cosmet Dermatol Sci Appl 2016;6:48–53.
[36] Verma HN, Singh P, Chavan RM. Gold nanoparticle: synthesis and characterization. Vet World 2014;7 (2):72–7.
[37] Guterres SS, Alves MP, Pohlmann AR. Polymeric nanoparticles, nanospheres and nanocapsules, for cutaneous applications. Drug Target Insights 2017;2:147–57.
[38] Hirlekar R, Yamagar M, Garse H, Vij M. Carbon nanotubes and its applications: a review. Asian J Pharm Clin Res Issue 2009;2(4):17–27.
[39] Ajazzuddin M, Jeswani G, Jha AK. Nanocosmetics: past, present and future trends. Recent Pat Nanomed 2015;5:3–11.
[40] Chaudhri N, Soni CG, Prajapati SK. Nanotechnology: an advance tool for nano-cosmetics preparation. Int J Pharm Res Rev 2015;4(4):28–40.
[41] Faunce T. Exploring the safety of nanoparticles in Australian Sunscreens. Int J Biomed Nanosci Nanotechnol 2010;1:87–94.
[42] Viladot PJL, Delgado GR, Fernandez BA. Lipid nanoparticle capsules. European Patent 2549977A2, 2013.
[43] Hu Z, Liao M, Chen Y, Cai Y, Lele M, Liu Y, et al. A novel preparation method for silicone oil nanoemulsions and its application for coating hair with silicone. Int J Nanomed 2012;2012(7):5719–24.
[44] Tsuji JS, Maynard AD, Howard PC, James JT, Lam CW, Warheit DB, et al. Research strategies for safety evaluation of nanomaterials, part IV: risk assessment of nanoparticles. Toxicol Sci 2006;89(1):42–50.
[45] Hoet PHM, Brüske-Hohlfeld I, Salata OV. Nanoparticles—known and unknown health risks. J Nanobiotechnol 2004;2(12). Available from: https://ec.europa.eu/environment/chemicals/nanotech/faq/definition_en.htm >.
[46] Cevc G, Vierl U. Nanotechnology and the transdermal route. A state of the art review and critical appraisal. J Control Release 2010;141(3):277–99.

Further reading

U.S. Food and Drug Administration, "Is it a cosmetic, a drug, or both? (Or is it soap?)," <https://www.fda.gov/cosmetics/cosmetics-laws-regulations/it-cosmetic-drug-or-both-or-it-soap>.

PART 2

Emerging applications

CHAPTER

9

Water-based nanoperfumes

Małgorzata Miastkowska and E. Lasoń

Faculty of Chemical Engineering and Technology, Institute of Organic Chemistry and Technology, Cracow University of Technology, Cracow, Poland

9.1 Introduction

Fragrance is the combination of various odorous substances that evaporate at different periods. Each fragrance has the so-called "top note," which is the odor diffusing first during the application of the perfume or during opening of the container, the "heart note or body," which corresponds to the complete fragrance (emission for several hours after the "top note"), and the "base note" one, which is the most persistent odor (emission for several hours after the "heart note"). For many years, lower aliphatic alcohols, especially ethanol, were considered to be essentially the only chemical compound that allows to dissolve the fragrance compositions that are lipophilic in nature, in one homogeneous and transparent phase. Depending on the concentration of the fragrance compounds in the ethanol solution, we distinguish *extrait de parfum* (up to 40% fragrance composition in 90%–96% ethanol solution), *eau de parfum* (10%–15% fragrance composition in 80%–90% ethanol solution), *eau de toilette* (5%–10% fragrance composition in 60%–85% ethanol solution), *eau de cologne* (3%–8% aromatic compounds in 70%–80% ethanol solutions), or *au fraiche* (up to 3% fragrance composition diluted mainly with water) [1,2]. Ethanol is not only a solvent for hydrophobic aromatic substances but also acts as an antimicrobial agent and dries fast after application on the skin. On the other hand, it can cause skin irritation, drying by removal of the hydrolipidic film covering and protecting the surface of the skin, especially among people with allergies or sensitive skin. Moreover, ethanol is relatively expensive, volatile, and flammable, and thus presents a fire hazard in production and use. It is also worth emphasizing that ethanol is not allowed for religious reasons in some Islamic countries [3–5].

The studies concerning water-based or/and alcohol-free perfumes have been conducted by researchers and cosmetic companies. As an alternative to ethanol-based perfumes, there are perfumes based on oils or solid and semisolid state perfumes (pomades, solid

Nanocosmetics
DOI: https://doi.org/10.1016/B978-0-12-822286-7.00007-3

emulsions, and gels). Oil as a solvent and carrier of aromatic compounds cannot overshadow or change the scent of the fragrance composition, that is why only neutral oil can be used like jojoba oil or fractionated coconut oil. This type of product, commonly known as "skin scents," improves the fragrance longevity and allows the scent to be released gradually. Additionally, the fragrance concentration in oil products is quite high, usually around 20%. Although oil-based perfumes do not dry the skin and even moisturize it, their main disadvantage is that they leave greasy spots on the perfumed surfaces and may cause stains on the clothes [6].

Other examples of alcohol-free perfumery products existing in the literature are based on a safe solvent like water and are represented by emulsions [7,8], liposomes [9], or micelles [10]. Nevertheless, the introduction of lipophilic fragrance systems to water without a cosurfactant (e.g., ethanol) is very difficult, with respect to thermodynamic stability. It requires the use of solubilizers, which can be polyols (glycols and glycerin) or surfactants. Those are the reason why replacing organic solvents with water is a challenging task for researchers. Therefore oil-in-water (O/W) type nanosystems are the best solution to dispersed lipophilic fragrances into the water phase, encapsulate them in micelles, hence protect against external factors and extend their longevity by releasing fragrances in a controlled way and allow to maintain the clear and transparent appearance of the final product.

9.2 Water-based or/and alcohol-free perfumes

To avoid the problem that ethanol is a solvent with a defined irritating potential which can cause skin vexation and inflammation, especially among people with allergies or sensitive skin, the studies concerning water-based perfumes have been conducted. Patents concerning water-based or/and alcohol-free perfumes are presented in Table 9.1 [11–25].

As we can see patents concerning alcohol-free perfumes are present in the cosmetic market since almost 30 years. World cosmetic companies like Yves Saint Laurent, Coty, Chanel, or L'Oreal own patents concerning perfumes in the form of O/W type micro- and nanoemulsions. If we look at the composition of perfumes described in the patent literature (Table 9.1) some of them are composed of cationic and anionic surfactants which can be an irritant to the skin. They contain additional solubilizers (e.g., glycereth-7-triacetate and dimethyl sulfoxide) and solvents like isoprene glycol, 1,2-hexanediol, or isohexadecane. That is the reason why the studies concerning water-based perfumes are still carry on. Preferably, the primary surfactants should be predominantly or even exclusively nonionic and in the lowest possible amount, if the nanodispersions are intended for skin or hair application. Suitable nonionic surfactants in the preparation of water-based nanoformulations as a fragrance carrier are ethoxylated alkylphenol ethers, ethoxylated aliphatic alcohols, and ethoxylated castor oil or hydrogenated castor oil derivatives (Table 9.1). Other improvements of the formulation composition also concern the elimination of additional solvents and solubilizers.

There are also some literature reports concerning micro- and nanoemulsions as a carrier for fragrances and essential oils.

TABLE 9.1 Patents concerning water-based or/and alcohol-free perfumes.

Patent claims	Perfume composition	Assignee	Patent number/ earliest application date
Nonalcoholic perfuming product consisting of a lipophilic phase based on a concentrate of odoriferous substance or perfume and a water-soluble phase combined in a microemulsion	Odorous materials 5% and 50% w/w, surfactant (C8–C10 polyglycolyzed glyceride, glyceryl stearate polyglycolyzed) 10% and 50% w/w, first cosurfactant-active agent (polyglycerol isostearate, polyglycerol dioleate) 10%–20% w/w, second cosurfactant-active agent (polyethoxylated oleyl alcohol ether phosphate DEA, polyethoxylated ether phosphate sodium dioleate) 10%–15% w/w, and an aqueous phase up to 100% w/w	YVES SAINT LAURENT PARFUMS	EP0516508 A1/ 1991
O/W microemulsions comprising perfume oil, aqueous phase, and one or more surfactants with HLB between 9 and 18, and cosurfactants of which at least 0.5% of ionic cosurfactant	Perfume oil 0.01% and 40% w/w, of water at least 40% w/w, 1% or less lower aliphatic alcohol, one or more primary surfactants with HLB between 9 and 18 are present and the weight ratio of perfume oil to total surfactant is between 0.75 and 2.5	GIVAUDAN	EP0571677 A1/ 1992
An alcohol-free transparent perfume consisting essentially of an alcohol-free perfume base, oil-in-water microemulsion	Water, at least one hydrophobic perfume oil, at least one cationic surfactant, and at least one nonionic surfactant	INTERNATIONAL FLAVORS & FRAGRANCES	US5468725 A/1993
A nonalcoholic fragrance carrier with good fragrance solubility, product stability and clarity, and silky, nontacky rub out when applied to the skin	A hydrophobic perfume base 0.05%–50% w/w, water 40%–95% w/w, a nonionic surfactant less than 6% w/w, glycereth-7-triacetate 0.1%–15% w/w, water 50%–80% w/w	DRAGOCO GERBERDING	US5736505 A/1996
Aqueous perfume composition, without alcohol, nongreasy and nontacky, containing aqueous medium, one or more nonionic surfactants and a solvent	Solvent (isoprene glycol) 5%–30% w/w, oleyl alcohol ethoxylated with 10 moles of ethylene oxide 0.5%–5.0% w/w, hydrogenated castor oil ethoxylated with 40 moles of ethylene oxide 0.5%–5.0% w/w, aromatics 1%–10% w/w, water up to 100% w/w	LABORATOIRE B F INTERNATIONAL	EP1294350 A1/ 2000

(Continued)

TABLE 9.1 (Continued)

Patent claims	Perfume composition	Assignee	Patent number/ earliest application date
Ethanol-free, nonfatty, nonsticky perfumed aqueous cosmetic microemulsion composition, including one or more fragrance materials, a solvent, an aqueous medium, and optionally, one or more surfactants	Solvent (1,2-hexanediol) 5%–30% w/w, essential oil 0.5%–20% w/w, a surfactant being a combination of 0.5%–5.0% w/w of poly(oxy-1,2-ethanediyl) and 0.5%–5.0% w/w of stearic acid, sodium lauryl sulfate, sodium laureth sulfate, cetyl trimethyl ammonium chloride or a stearic acid salt, purified water up to 100% w/w	DOW GLOBAL TECHNOLOGIES	EP1758544 A1/ 2005
Ethanol-free essence waterborne dispersion system, particularly to perfumes, body spraying agent, deodorant, waterborne cosmetic composition, domestic detergent, and air freshener	Aroma materials 0.1%–0.6% w/w, phase dispersion agent (dimethyl sulfoxide, *N*-methylpyrrolidone or 1, 4-dioxane) 1%–60% w/w, and water 10%–95% w/w	WANG ZHEMING	CN101283964A/ 2008
Perfume composition with reduced alcohol content (content of monovalent C2–C5 alcohols such as ethanol is in the range of 0–5 wt.%), comprising a transparent or translucent emulsion	Solvent (isohexadecane, C15–C19 alkane, C13–C16 isoparaffin, isoeicosane, or isododecane) 0.1%–20% w/w, emulsifier [glyceryl partial esters, polyglyceryl partial esters, sorbitan partial esters, sorbitol partial esters, carbohydrate esters, (alkylpoly)glycosides, and mixtures thereof.] 0.1%–20% w/w, aromatic oils 1%–35% w/w, water and buffer system	COTY	EP2127632 A1/ 2008
Aqueous fragrancing composition including at least one volatile linear alkane, comprising in a cosmetically acceptable medium	a. At least 5% by weight of water relative to the total weight of the composition b. At least 2% by weight of a fragrancing substance c. At least one volatile linear alkane or a mixture of volatile linear alkanes; the water content ranging from 30% to 80% relative to the total weight of the composition	L'OREAL	EP2324816 A2/ 2009
Ethanol-free perfume oil microemulsion	Water 30%–90% w/w, vicinal diols (1,2-pentanediol, 1,2-hexanediol, and 1,2-octanediol 1,2-heptane), 5%–50% w/w, solvents (2-isobutyric acid 1-hydroxy, 2,4-trimethyl-3-pentyl-2 or isobutyric acid 3-hydroxy, 2,4-trimethyl-1-pentyl), 1%–20% w/w, odorous substances 1%–50% w/w	SYMRISE	EP2316408 A1/ 2010

(Continued)

TABLE 9.1 (Continued)

Patent claims	Perfume composition	Assignee	Patent number/ earliest application date
Alcohol-free flavoring formulation in the form of a nanodispersion	Nanodispersion including a continuous aqueous phase and an oily phase dispersed liquid, wherein said aqueous phase comprises water and at least one nonionic stabilizing agent and the oily phase comprises at least one flavoring agent	CAPSUM	WO201177062A1/ 2010
Ethanol-free perfumes, nonstimulated, color clear, have a refreshment, removing odors	1–3 g of lavender essential oil, 0.6–1.2 g of aromatic oil, 60–80 g of deionized water, 0.5–1 g of rose essential oil, 5–10 g of propylene glycol, and 3–5 g of a traditional Chinese medicinal extract liquid prepared from rosin, camphor, Fructus Gardeniae and mint	AGATE PERFUME	CN104173224A/ 2013
Alcohol-free transparent perfume composition that comprises nanoemulsion with the droplet size ranges 10–100 nm	5–20 parts of essential oil, 8–12 parts of polyoxyethylene surfactants, 5–8 parts of phosphate surfactants, 4–7 parts of polyglycerol ester surfactants, 0.5–1.2 parts of glycerol, and 15–20 parts of water	AGATE PERFUME	CN103637942A/ 2013
Ethanol-free perfumed compositions suitable for leave-on cosmetics, being clear, transparent, and stable during storage. The invention is also directed to ethanol-free perfumed products obtainable by dilution in water of ethanol-free perfume preparation	a. At least one crypto anionic surfactant b. At least one ethoxylated nonionic surfactant c. At least one glycol having from 3 to 8 carbon atoms, an d. Water	CHANEL	US20170298290A1/ 2016
A composition comprising an ethanol-free, aqueous microemulsion	Oil dispersed phase 2.0%–15% w/w, primary surfactant (PEG-40 hydrogenated castor oil, PPG-26-Buteth-26 or a combination) and cosurfactant/hydrotrope (pentylene glycol) 0.1%–25.0% w/w, preservative system (a combination of ocymen5-ol, EDTA, phenoxyethanol and at least one parabens)—1.0%–5% w/w, and water	HOFFMANN WUNDRIARI	EP3372282 A1/ 2018

Based on its unique features, microemulsions have been proposed in the literature as a delivery system for essential oils [26–31]. Yi et al. [29] studied the solubilization of volatile oil from *Houttuynia cordra* in an O/W microemulsion, which was developed by the titration method, but the varieties and amount of surfactant and cosurfactants had effects on the solubilization of the volatile oil. The formulation was composed of medium-chain triglycerides as an oily phase, polyoxyethylene castor oil EL-35 as a surfactant, and propylene glycol as a cosurfactant. This system was able to solubilize the volatile oil from *H. cordra*. Ma and Zhong [30] obtained a soybean oil-based microemulsion containing cinnamon bark essential oils, which was capable of expanding the regime region of microemulsions and reduced the droplet dimensions. The system was stable 90 days. Valoppi et al. [27] obtained transparent microemulsions with lemon oil stabilized with Tween 80 by applying the phase inversion temperature method. To get a high load of lemon oil in microemulsions, lemon oil was mixed with peanut oil because the addition of an oil rich in long-chain fatty acids allowed to include in transparent systems up to 15% lemon oil.

Nanoemulsions also act as a matrix for fragrance substances used in cosmetics [32–34]. They are primarily solutions for emulsification of a single lipophilic component that forms the oily phase of a nanoemulsion, for example, D-limonene. Application of nanoemulsions as a carrier of D-limonene allowed an increase of its bioavailability and chemical stability, including protection against oxidization. Li and Chiang [32] obtained stable nanoemulsions by the ultrasonic emulsification method using mixed surfactants of sorbitane trioleate and polyoxyethylene (20) oleyl ether. The concentration of D-limonene in the formulation was 10% w/w. They found that applied power, ultrasonic time, and D-limonene concentration to mixed surfactant concentration had significant effects on the formation of nanoemulsions droplet and achieve droplet size below 100 nm.

Friberg et al. [8] studied the stability of perfume nanoemulsions containing water, a nonionic surfactant from the group of ethoxylated fatty alcohols (Laureth-4) and phenylethyl alcohol (a component of fragrance compositions with a floral aroma). Studies have shown that the stability of the obtained nanoemulsion is affected by the ratio of phenylethanol to the surfactant. The same team also studied the effects of various aroma compounds (e.g., limonene, a mixture of limonene and phenylethyl alcohol or benzaldehyde) on the stability of the described emulsions. During the emulsification process, liquid crystalline systems of high viscosity were created as a transient stage in the formation of nanoemulsions, which was a technological difficulty. An attempt to eliminate the problem related to the presence of liquid crystals was made by Dumanois and Gueyne [15] using isoprene glycol as a cosolvent in a nonalcoholic fragrance composition. The composition consists of 1%–6% w/w of essential oils, 5%–20% w/w of isoprene glycol, 0.5%–3.5% w/w of PEG-40 Hydrogenated Castor Oil, 0.5%–3.5% w/w of ethoxylated fatty alcohol, 0.1%–0.2% w/w of parabens, and up to 100% water.

Stable perfume oil-in-water (O/W) nanoemulsions, which did not contain ethanol and additional solubilizers and solvents, were obtained by Miastkowska et al. [35]. These systems can be successfully applied as modern carriers of selected fragrance compositions, both with a low-energy method [phase inversion composition (PIC)], at a laboratory scale, and with a high-energy method [ultrasonification (US)], at a production scale. The optimized nanoperfume recipes that were obtained with different methods yielded the same physicochemical properties (stability, medium droplet size of the inner phase, polydispersity, viscosity, surface tension, pH, and density). The simple composition of the formulation is worth mentioning. Stable nanoemulsions were stabilized only by one nonionic surfactant, gentle to the skin, without the need to use additional solubilizers. Stable, transparent nanoperfumes were obtained with a fragrance composition concentration within 6%–15% range. Moreover, the obtained results confirmed the protective role of nanoemulsions. The peroxide number measurement [peroxide value (POV)] showed that the tested fragrance compositions had a high chemical stability. The dermatological test results confirmed the safety of the developed preparations.

9.3 Nanodispersions as a carrier for fragrances

The authors of patent solutions (Table 9.1) and literature reports propose as an alternative to alcohol-based perfumes mainly O/W type microemulsions and nanoemulsions. To explain why those nanodispersions were chosen as an alternative carrier for fragrance composition, authors of this paper briefly summarize their unique properties taking into account similarities and differences.

Microemulsions and nanoemulsions belong to the group of colloidal systems with a high degree of dispersion (radius <100 nm). When it comes to their composition, both oil-in-water type systems contain the oily phase dispersed within the water phase, stabilized by surfactant and possibly cosurfactant. The structure of the particles in both types of nanodispersion is also very similar. The nonpolar tails of the surfactant molecules are inward hydrophobic core formed by the oily phase, while the polar heads are facing toward the aqueous external phase [35–44].

Due to their small droplet size, they are optically transparent or translucent; sometimes, in the case of nanoemulsions, they can be milky. They are also characterized by liquidity and low viscosity. Similar to microemulsions, nanoemulsions increase the expiry date of many products due to their resistance to sedimentation and creaming [35,43,44].

The main difference between discussed colloidal systems is that microemulsions are thermodynamically stable and nanoemulsions are not [36,37,41,42]. This is due to the fact that the free energy of the colloidal dispersion is higher than free energy of the separate phases of nanoemulsion [43]. However, their small droplet size reduces the gravity force, which is overcome by Brownian motion and makes them kinetically stable. Their advantage over microemulsions is that they have a much lower amount of a surfactant (approx. 5%–10%), which allows maintaining adequate stability of the system and makes them safe for human body [35–39].

The most commonly used methods for nanoemulsion preparation are [43,45,46] as follows:

- *High-energy emulsification methods*:
  - High-pressure homogenization
  - Ultrasonification
- *Low-energy emulsification*:
  - Phase inversion temperature
  - Phase inversion composition
  - Spontaneous emulsification

The low-energy emulsification methods rely on chemical energy stored in surfactants, which are able to decrease interfacial tension and thus reduce the energy input required to obtain emulsions with small and uniform droplet size. The high-energy emulsification methods use the mechanical energy, originating mainly from the mechanical devices such as a high-speed mixer, a high-pressure homogenizer, and a ultrasonic generator.

Microemulsions are formed spontaneously by mixing oil, water, and surfactant at an optimal component ratio and temperature without input of external energy. However, sometimes it is necessary to use mechanical or thermal energy to overcome kinetic energy barriers and form microemulsions. The surfactant to oil ratio (SOR) also has great importance. At higher SOR, a microemulsion is formed; at lower SOR, nanoemulsion can be obtained [43]. Moreover, nanoemulsions can be prepared by diluting O/W microemulsions that change the proportions between components [47].

It is difficult to distinguish microemulsions and nanoemulsions taking into account visual observation, particle size distribution, or method of preparation. Detailed examination of the terminology, differences, and similarities between microemulsions and nanoemulsions was provided by McClement [43] and Anton and Vandamme [44].

9.4 The advantage of nanoemulsions over microemulsions

According to the chemical structure of the fragrances compound, they are chemically unstable under exposure to oxygen, light, and heat [48]. Incorporation of fragrances into micelles core of micro- and nanoemulsions turned out to be a great solution for solving problems related to the oxidation and low bioavailability of fragrances [26,35,48–50].

Studies concerning partitioning and localization of fragrances in surfactant mixed micelles were conducted by Fischer et al. [50]. They proved that the localization of the fragrance is strongly related to their $\log(P_{ow})$ value, where P_{ow} is the ratio of concentrations of a compound in octanol and water in equilibrium. In water-based perfumes, micelles built by surfactants serve as carrier systems of hydrophobic fragrance molecules. Investigated fragrance molecules can be classified by three different regions on the log (P_{ow}) scale. Hydrophilic fragrances with $\log(P_{ow}) < 2$ can be localized almost equally between micellar and aqueous phase, depending on the chemical structure. Very hydrophobic fragrances with $\log(P_{ow}) > 4.5$ are fully solubilized in micelles. In the intermediate $\log(P_{ow})$ range, two different sets of fragrances were identified. One of them being bulky structures, predominantly alcohols that are extremely hindered in their isotropic motions and lead to micelle swelling. Another group of fragrances with intermediate $\log(P_{ow})$ are compounds with long alkenyl chains, several of them being aldehydes. They do not cause micelle swelling and are incorporated to a high degree and localized in the micelle core. Results obtained by Fisher et al. [50] are really helpful to understand the exact localization of incorporated fragrance molecules in surfactant aggregates, which determine the product properties and quality.

It should be indicated that from the application point of view, the concentration of high surfactants in the microemulsion formulations (up to 40%) is their main drawback. The presence of large amounts of surfactants is often undesirable for various reasons. Fragrances incorporated in such compositions can react with surfactant and may undergo degradation processes, which affect the stability and organoleptic properties of the product. They increase the cost of the total formulation and may leave a sticky film on a surface to which the perfume is applied. They may have a disagreeable odor and tend to diminish the perceived odor strength of the perfume. Furthermore, especially in perfume formulations that are applied to and intended to be left on skin or hair, the presence of certain surfactants, notably cationic surfactants, should be limited. They are known for their irritating properties and should be avoided in "leave-on products" such as perfumes [11,48].

In the case of nanoemulsions, it is possible to obtain formulations with a much lower amount of a surfactant (approx. 5%–10%), which allows maintaining adequate stability of the system and makes them safe for human body. There is also no need to use cosurfactants, solvents, and solubilizers to obtain stable, clear, and transparent nanoemulsion systems as a vehicle for fragrances [21,35,43,48].

9.5 Conclusions

Nanodispersions open the way for using fragrances that could not be used before because of the insufficient chemical stability. Application of fragrances in the form of oil-in-water type nanosystems protects them against the negative effects of oxidizing factors [35]. It is a good route to control fragrance release, extend its longevity, and make fragrant more durable (during processing, storage, or usage) [33,51]. Thanks to their structure and composition, it is possible to obtain transparent and stable water-based perfumes that are safe, nontoxic, nonflammable, odorless, and with a very good cutaneous tolerance. They can also solve problems related to low bioavailability of fragrances as well as allow to exclude ethanol totally and reduce the amount of surfactants, cosurfactants, and additional solvents. Thanks to this, we could achieve products, with good user properties, dedicated to a wider group of consumers, including children or people with sensitive skin. Aqueous perfumes have benefits such as the environmentally friendly character of water, nonflammable compositions, and societal acceptance. Moreover, water is a much cheaper base than ethanol, which provides cost-saving in production.

References

[1] Marcus J, Klossek ML, Touraud D, Kunz W. Nano-droplet formation in fragrance tinctures. Flavour Fragr J 2013;28:294–9.
[2] Jabłońska-Trypuć A, Farbiszewski R. Sensory and basics of perfumery. Wroclaw: MedPharm; 2008. p. 91–4.
[3] Syariena A, Puziah H. Rapid determination of residual ethanol in perfumery products using headspace gas chromatography-mass spectrometry. Middle East J Sci Res 2014;22:432–7.
[4] Pendlington RU, Whittle E, Robinson JA, Howes D. Fate of ethanol topically applied to skin. Food Chem Toxicol 2001;39:169–74.
[5] Ophaswongse S, Maibach HI. Alcohol dermatitis: allergic contact dermatitis and contact urticaria syndrome, a review. Contact Dermatitis 1994;30:1–6.
[6] Brud SB, Konopacka-Brud I. Podstawy perfumerii. Oficyna Wydawnicza MA: Łódz; 2009, ISBN: 978-83-923517-2-6. p. 231–2.
[7] Herman S. Fragrancing emulsions. Cosmet Toilet 1994;109:71–5.

[8] Friberg SE, Zhang Z, Ganzuo L, Aikens PA. Stability factors and vapor pressures in a model fragrance emulsion system. J Cosmet Sci 1999;50:203–19.
[9] Juszynski M, Azoury R, Raphaeloff R. Fragrance-loaded lyophilized liposomes. SOFW 1992;118:811–15.
[10] Kamada M, Shimizu S, Aramaki K. Manipulation of the viscosity behavior of wormlike micellar gels by changing the molecular structure of added perfumes. Colloids Surf A 2014;458:110–16.
[11] Dartanell N, Breda B, Microemulsion containing a perfuming concentrate and corresponding product. Eur. Patent EP0516508 A1; 31 May 1991.
[12] Behan JM, Ness, JN, Traas PC, Stergios J, Willis, BJ, Aqueous perfume oil microemulsions. Eur. Patent EP0571677 A1; 29 May 1992.
[13] Guenin EP, Trotzinka KA, Smith LC, Warren CB, Munteanu MA, Chung SL, et al., Alcohol free perfume. U. S. Patent US5468725 A; 01 July 1993.
[14] Manzo RP, Kennedy DM, Non-alcoholic perfume or cologne. U.S. Patent US5736505 A; 30 August 1996.
[15] Dumanois M, Gueyne N, Alcohol-free base for aqueous perfume composition, and alcohol-free aqueous perfume composition comprising same. Eur. Patent EP1294350 A1; 30 June 2000.
[16] Piechocki C, Shick RA, Tucker CJ, Gatz LA, Ethanol-free aqueous perfume composition. Eur. Patent EP1758544 A1; 21 April 2005.
[17] Wang Z, No-ethanol essence aquose dispersion system. Chinese Patent CN101283964 A; 27 May 2008.
[18] Bleuez L, Porcu M, Perfume composition with reduced alcohol content. Eur. Patent EP2127632 A1; 29 May 2008.
[19] Grandjon V, Noel-Poquet C, Aqueous flavouring composition including at least one volatile linear alkane; flavouring method. Eur. Patent EP2324816 A2; 19 November 2009.
[20] Wiedemann J, Kaufhold A, Ethanol-free perfume oil microemulsion. Eur. Patent EP2316408 A1; 25 October 2010.
[21] Goutayer M, Courtemanche M, Flavouring formulations in the form of a nanodispersion. WO201177062 A1; 23 December 2010.
[22] Wang C, Ethanol-free perfume. Chinese Patent CN104173224 A; 13 December 2013.
[23] Hu Y, Alcohol-free transparent perfume composition. Chinese Patent CN103637942 A; 2 December 2013.
[24] Diez R, Giblin J, Perfume compositions. U.S. Patent US20170298290 A1; 14 April 2016.
[25] Solinas S, Schmeling M, Hoffmann W, Ethanol-free, aqueous microemulsion perfume compositions. EP3372282 A1; 01 March 2018.
[26] Calligaris S, Manzocco L. Microemulsions as delivery systems of lemon oil and β-carotene into beverages: stability test under different light conditions. J Sci Food Agric 2019. Available from: https://doi.org/10.1002/jsfa.9973.
[27] Valoppi F, Frisina R, Calligaris S. Fabrication of transparent lemon oil loaded microemulsions by phase inversion temperature (PIT) method: effect of oil phase composition and stability after dilution. Food Biophys 2017;12:244–9.
[28] Xavier-Junior FH, Vauthier C, Morais ARV, Alencar EN, Egito EST. Microemulsion systems containing bioactive natural oils: an overview on the state of the art. Drug Dev Ind Pham 2016;43(5):700–14.
[29] Yi H, Gao J, Yang H. Solubilization of O/W microemulsion for volatile oil from *Houttuynia cordra*. Zhongguo Zhong Yao Za Zhi 2010;35:49–52.
[30] Ma Q, Zhong Q. Incorporation of soybean oil improves the dilutability of essential oil microemulsions. Food Res Int 2015;71:118–25.
[31] Ma Q, Zhang Y, Critzer F, et al. Physical, mechanical, and antimicrobial properties of chitosan films with microemulsions of cinnamon bark oil and soybean oil. Food Hydrocoll 2016;52:533–42.
[32] Li P, Chiang B. Process optimization and stability of D-limonene-in-water nanoemulsions prepared by ultrasonic emulsification using response surface methodology. Ultrason Sonochem 2012;19:192–7.
[33] Li PH, Lu WH. Effects of storage conditions on the physical stability of D-limonene nanoemulsions. Food Hydrocoll 2016;53:218–24.
[34] Artiga-Artigas M, Guerra-Rosas MI, Morales-Castro J, Salvia-Trujillo L, Martín-Belloso O. Influence of essential oils and pectin on nanoemulsion formulation: a ternary phase experimental approach. Food Hydrocoll 2018;81:209–19.
[35] Miastkowska M, Lasoń E, Sikora E. Preparation and characterization of water-based nano-perfumes. Nanomaterials 2018;8:981–96.

[36] Usón N, Garcia MJ, Solans C. Formation of water-in-oil (W/O) nano-emulsions in a water/mixed non-ionic surfactant/oil systems prepared by a low-energy emulsification method. Colloids Surf A 2004;250:415–21.
[37] Solans C, Izquierdo P, Nolla J, Azemar N, Garcia-Celma MJ. Nano-emulsions. Curr Opin Colloid Interface Sci 2005;10:102–10.
[38] Mishra B, Patel BB, Tiwari S. Colloidal nanocarriers: a review on formulation technology, types and applications toward targeted drug delivery. Nanomedicine 2010;6:9–24.
[39] Lovelyn C, Attama AA. Current state of nanoemulsions in drug delivery. J Biomater Nanobiotechnol 2011;2:626–39.
[40] Singh V, Bushettii SS, Appala Raju S, Ahmad R, Singh M, Bisht A. Microemulsions as promising delivery systems: a review. Indian J Pharm Educ Res 2011;45(4):392–401.
[41] Olsson U. Characterization of microemulsion structure. Handbook of applied surface and colloidal chemistry. West Sussex: John Wiley & Sons; 2002. p. 338.
[42] Lawrence MJ, Rees GD. Microemulsion-based media as novel drug delivery systems. Adv Drug Deliv Rev 2000;45:89–121.
[43] McClements DJ. Nanoemulsions versus microemulsions: terminology, differences, and similarities. Soft Matter 2012;8:1719–29.
[44] Anton N, Vandamme TF. Nano-emulsions and micro-emulsions: clarifications of the critical differences. Pharm Res 2011;28:978–85.
[45] Miastkowska M, Banach M, Pulit-Prociak J, Sikora E, Głogowska A, Zielina M. Statistical analysis of optimal ultrasound emulsification parameters in thistle-oil nanoemulsions. J Surfactants Deterg 2017;20(1):233–46.
[46] Solè I, Pey CM, Maestro A, González C, Porras M, Solans C, et al. Nano-emulsions prepared by the phase inversion composition method: preparation variables and scale up. J Colloid Interface Sci 2010;344:417–23.
[47] Sole I, Solans C, Maestro A, Gonzalez C, Gutierrez JM. Study of nano-emulsion formation by dilution of microemulsions. J Colloid Interface Sci 2012;376(1):133–9.
[48] Sikora E, Miastkowska M, Wolińska Kennard K. Nanoemulsions as a form of perfumery products. Cosmetics 2018;5:63.
[49] He L, Hu J, Deng W. Preparation and application of flavor and fragrance capsules. Polym Chem 2018;9:4926–46.
[50] Fischer E, Fieber W, Navarro Ch, Sommer H, et al. Partitioning and localization of fragrances in surfactant mixed micelles. J Surfactants Deterg 2009;12:73–84.
[51] van Soest JJG. Encapsulation of fragrances and flavours: a way to control odour and aroma in consumer products. In: Berger RG, editor. Flavours and fragrances. Berlin: Springer; 2007.

CHAPTER

10

Nanocosmetics for broadband light protection sun care products

Paulo Newton Tonolli[1], Thiago Teixeira Tasso[2] and Maurício S. Baptista[1]

[1]Biochemistry Department, Institute of Chemistry, University of São Paulo, São Paulo, Brazil
[2]Chemistry Department, Institute of Exact Sciences, Federal University of Minas Gerais, Minas Gerais, Brazil

10.1 Introduction

The worldwide incidence of skin cancer (melanoma and nonmelanoma) has increased in recent years, creating an epidemiological trend, even with the changes in photoprotection guidelines by regulatory agencies, such as the FDA, and with increasing numbers of people using sunscreens and having "safe" sun exposure habits [1–3]. For example, in the United States, melanoma cases have tripled between the 1970s and 2016: from 7.9 to 23 cases per 100,000 individuals [4]. The number of nonmelanoma cases has also risen in recent decades: in the United States, squamous cell carcinoma and basal cell carcinoma have been increasing at between 3%–10% and 20%–80% per year, respectively [5].

Ultraviolet radiation (UVR) is considered to be the most exogenous factor involved in skin cancer pathogenesis and the target of interest for skin care and sun protection policies [6,7]. The relationship between UV exposure and skin carcinogenesis was established in the early 20th century [8–10]. However, an action spectrum of UVR and skin cancer induction was not elucidated until later, showing a peak in the UVB range (280–315 nm), which was the first described adverse component of sunlight. Curiously, the first sunscreens promised to avoid sunburn (a characteristic of UVB effect), while maintaining the ability of the skin to tan (pigmentation induced by UVA damage), which began to be fashionable in the 1920s.

In 1976, Zigman et al. showed that the longer wavelength of UVR, the UVA range (315–400 nm), was capable of inducing skin cancer in mice [11,12], and also demonstrated that UVA was carcinogenic in haired mice and that it became more carcinogenic when

DOI: https://doi.org/10.1016/B978-0-12-822286-7.00008-5

combined with UVB. In the 1980s, companies started to include chemical blockers for UVA, such as avobenzone, in their formulations. In the next decades, sunscreens have become more specialized, and some ingredients have been included or removed due to suspected harmful effects, reaching a high spectrum of protection from UVR, blocking almost all the solar UVR from penetrating the skin.

However, there is a really serious issue which has been ignored by sun protection and skin care guidelines: the existence of the visible light (400–750 nm) and infrared (>750–2500 nm) range in the sunlight spectrum. The literature has reported that these sunlight ranges are also active photochemically, causing photosensitization reactions, photodamage/aging, triggering melanogenesis, and, probably, promoting mutagenesis and carcinogenesis. Most sunscreens and skin care products marketed today are not effective high-spectrum barriers against sunlight damage. Thus, this chapter presents, discusses, and proposes new possibilities for strategies to protect skin from visible light damage.

10.2 Visible light: should we protect ourselves?

Visible light is the portion of the solar electromagnetic spectrum capable of sensitizing the retinal pigment of the human eye, with a wavelength extending from 400 to 750 nm. In terms of total solar energy that reaches the Earth's surface, visible light corresponds to 43%, which is a similar level of irradiance to that of the infrared (52%), and almost 10 times greater than UVR (5%), which is distributed as 95% UVA and 5% UVB.

10.2.1 Interaction of visible light with the skin

Daily, photons from the sunlight strike and penetrate the skin, where they can be absorbed, reflected, dispersed, and transmitted. Approximately 4%–7% of the light incident on the epithelial tissue is reflected in the stratum corneum due to the difference in the refractive index between the air and the skin surface. The amount of incident light reflected is dependent on the incident angle of the light beam, which normally is <40 degrees, resulting in around 5% reflected light, increasing proportionally with the incident angle.

The 93%–96% remaining radiation entering the skin is scattered or absorbed. Scattering can be elastic, when the light path is changed due to differences in the refraction index, or inelastic, when photons lose energy to vibrational states, generating the emission of lower energy and longer wavelength photons. In the skin, elastic scattering and absorption are the prevalent interactions, which occur due to the intrinsic properties of chromophores and the size of particles in this tissue in a wavelength-dependent way [13].

The position and absorption properties of natural skin chromophores originate the attenuation pattern of the incident light at increasing depth within the skin. Wavelengths shorter than 600 nm are highly absorbed by oxy/deoxyhemoglobin (peaks at 400–420 nm; 540–577 nm) and melanin, or scattered at much smaller wavelengths (<300 nm) by granules (e.g., melanosomes) and subcellular structures. Skin cells have an abundance of

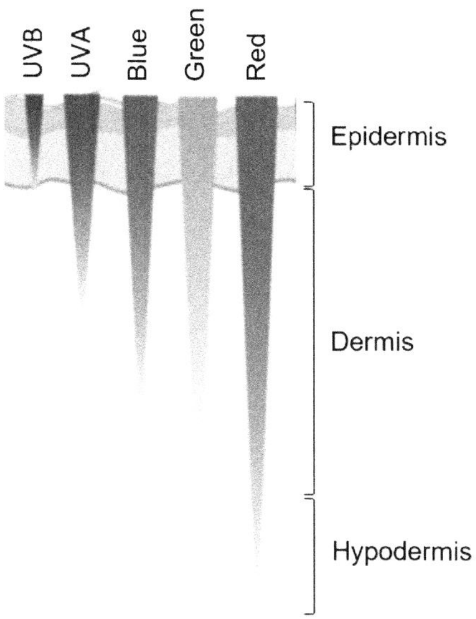

FIGURE 10.1 Penetration of UV and visible light components in the skin layers.

natural chromophores absorbing the visible light range, such as flavin, NADH, NADPH, porphyrins, tryptophan in proteins, lipofuscin, keratin, and carotenoids [14,15].

In contrast to the photons from the UV region, photons from visible light penetrate deep into the dermis. As compared to the ultraviolet range, there are less natural chromophores in the visible range. Photons in the visible range are mainly scattered in dermal layer by collagen and elastin fibers, due to the fibrillar structure with high refractive index. Longer wavelengths than 700 nm show poor scattering and the deepest penetration in the skin. Comparatively, the visible light range has a higher penetration in the skin than UVR. Blue light (400–490 nm) is able to cross through all of the epidermis, while green (490–550 nm) and red light (600–750 nm) reach the dermis layer [16]. Therefore visible light presents a wide network of interactions with the skin, which are poorly understood in regard to its action spectrum wavelength (Fig. 10.1).

10.2.2 Effects of visible light on the skin

The optical properties of the skin are the subject of a dynamic field of investigation, supported by medical areas (e.g., in diagnostic and phototherapies) and industries, such as cosmetics companies.

10.2.2.1 Photosensitizer and photosensitization reactions

For a long time, some misconceptions about the action of visible light on cells have led to visible light being considered as safe, supposedly as it has a low photochemical response compared with UVR [17]. With the advance of photodynamics and laser therapies, many studies into the effects of visible light on the skin have been performed in recent years. Despite this, some of these studies have experienced a problematic

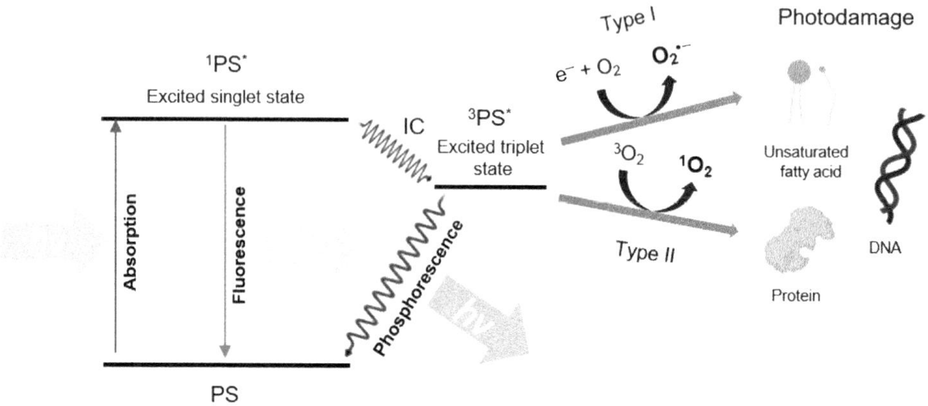

FIGURE 10.2 A simplified Jablonski diagram illustrating the mechanisms of photosensitization. A photosensitizer (PS) in the ground state is excited to a singlet state ($^1PS^*$) after absorbing a photon (blue arrow). Among the possibilities, the $^1PS^*$ may decay to ground state, emitting fluorescence (green arrow) or be converted to the excited triplet state ($^3PS^*$) through the intersystem crossing (IC). In addition, $^3PS^*$ may decay to the ground state emitting phosphorescence (red arrow) or transferring electrons (reaction type I) or energy (reaction type II) to neighboring molecules, such as molecular oxygen ($^3O_2$), forming radicals (e.g., superoxide anion radical, $O_2^{\bullet-}$) or highly reactive species (e.g., singlet oxygen, $^1O_2$), respectively. The reactive species generated can oxidatively damage biomolecules (unsaturated fatty acids, proteins, DNA) and organelles.

interference of UV and infrared components in the light sources used, limiting the specific information on the visible light effects [18]. In reality, photons of visible light are absorbed by a myriad of endogenous molecules, initiating photochemical reactions very similar to those induced by UVA. Indeed, the absorption coefficient of in vivo human skin has a peak at around 400–450 and 540–575 nm [19,20].

Before we progress, it is important to define the photosensitized oxidation reaction. Molecules that absorb light, photon-forming excited states, and that transfer this energy to other molecules are called photosensitizers. Melanin, flavin, vitamins, and several proteins are photosensitizers [21]. Photosensitizers interact with neighboring molecules, which in biological systems can perform two types of oxidation mechanisms: type I, in which the photosensitizer directly transfers electrons to/from substrates, generating radicals or radical-ions [22]; and, type II, in which the photosensitizer reacts with molecular oxygen, transferring its energy and generating singlet oxygen. Through these mechanisms, visible light generates reactive oxygen and nitrogen species, which have several physiological roles, such as immune defense and vasodilatation, but can damage biomolecules and organelles, inducing aging, mutagenesis, or cell death. A simplification of these mechanisms is represented in Fig. 10.2.

10.2.2.2 Reactive species generation in skin cells irradiated with visible light

For decades, the formation of reactive species by UV-induced photochemical process has been known about, being the major agent involved in oxidative damage to the skin [21,23,24]. Either UVB and UVA generate the reactive species and free radicals by photochemical reactions in the skin (melanin, tryptophan, tyrosine, flavin, riboflavin), oxidatively damaging protein, lipids, and membranes [25–27]. However, the other regions that are active photochemically in the sunlight spectrum, such as visible light, remained ignored for a long time considering their ability to produce reactive species.

Haywood [28] demonstrated that visible light has significant implications in the ascorbate radical formation in ex vivo human skin irradiated with solar-simulated light. This work showed that depletion of the ascorbate pool in the antioxidant defense made the skin more vulnerable to UV damage. In addition, the generation of superoxide anion radical ($O_2^{\bullet -}$), hydroxyl radical ($^{\bullet}OH$), and $^{\bullet}$CH-R were detected in human skin biopsies irradiated with visible light [29]. According to this study, visible light is responsible for half of the free radicals produced in the skin due to sun exposure. When irradiating human skin equivalents with visible light from halogen lamps containing UV and infrared filters, Liebel et al. [30] observed a dose-dependent increase in hydrogen peroxide formation. ROS mediates the release of proinflammatory cytokines and matrix-degrading enzymes (metalloproteinases MMP-1 and MMP-9) expression, which implies that visible light is a dangerous agent involved in premature skin aging.

Chiarelli-Neto et al. [31] reported that melanin photosensitized with visible light in hyperpigmented B16-F10 cells generates singlet oxygen in a dose-dependent manner, which does not differ from the effects observed with UVA irradiation. In this work it was observed that pheomelanin generated 30% more $^1O_2$ than eumelanin after visible light irradiation.

Therefore the visible light, similarly to UVA and in a lower scale than UVB radiation, generates reactive species by a photoinduced process involving endogenous chromophores. This reactive species generation has importance in oxidative stress, photodamage, skin aging and, probably, skin cancer.

10.2.2.3 *Photodamage in biomolecules and organelles caused by sunlight*

The reactive species and free radicals generated by photosensitization reactions have several targets inside cells, including biomolecules (proteins, lipids, DNA), organelles (mitochondria, lysosomes), and structural components (cell membrane).

The sunlight range corresponding to UVB (280–320 nm) is highly absorbed by nucleic acids, forming pyrimidine dimers, a direct lesion in DNA, which is responsible for mutagenesis and skin carcinogenesis [32,33]. UVA also was observed inducing cyclobutene pyrimidine dimers in the DNA from murine and human skin melanocytes and keratinocytes, although at a lower rate than with UVB [34–38]. Protein lesions include protein–protein, as well as DNA–protein and lipid–protein cross-linking, which have functional and structural effects, such as inhibition of enzymes, cell signaling, and membrane integrity [39]. Membranes and polyunsaturated fatty acids (PUFAs) are also primary targets of UVR-induced reactive species, such as singlet oxygen insertion in carbon–carbon double bonds, affecting the cell and organelle membrane permeabilization and fluidity [40–42]. The implication of oxidation in lipids is lipid peroxidation, which occurs as a chain reaction containing intermediate radicals, and terminates forming lipid peroxides.

Our group has reported the synergic effect between UVA and visible light, where human skin keratinocytes, after 48 hours of UVA, exposure became more sensitive to visible light due to the accumulation of lipofuscin, a subproduct of oxidative damage, that was photosensitized by visible light, generating singlet oxygen and Fpg-sensitive lesions in nuclear DNA [43].

How about visible light inducing photodamage? If visible light can generate reactive species and free radicals through the photosensitization of chromophores, why has visible light not been considered in sun care products? Studies on the action spectrum of the visible light wavelengths started in the 1980s with Karu et al. [44] investigating the effects on activation of proliferation in HeLa cells, monitoring DNA and RNA synthesis, showing maxima at 610 and 630 nm in continuous red light for DNA synthesis and close to 615 nm for RNA synthesis. The growing interest in visible light wavelength effects on skin cells has grown from the dermatological applications of light sources in photobiomodulation and phototherapies for skin diseases and wounds [45].

Tyrrell et al. [46] were the first to study the action spectra of violet-blue light (380–434 nm) related to cytotoxicity in human epidermal keratinocytes and fibroblasts, showing that the former were more resistant to $\lambda = 405$ nm than the latter.

Kielbassa et al. [47] investigated the action spectrum of visible light wavelengths in the formation of oxidative DNA damage and pyrimidine dimer using AS52 Chinese hamster cells. This paper showed a maximum oxidative DNA base (Fpg-sensitive modifications, e.g., 8-oxo-dG) in the 400–450 nm range, even detecting a low level of T4 endonuclease V-sensitive modifications (e.g., cyclobutene pyrimidine dimers) in the range of 400–430 nm.

Remarkably, violet-blue light (400–450 nm) is the most investigated and harmful fraction of visible light for skin cells. Violet-blue light excites flavins, riboflavins, and cytochrome oxidases, which are prevalent in mitochondria, generating ROS, oxidative stress, and cell death [48–51]. In in vivo conditions, skin exposure to blue-violet light (400–450 nm) is able to promote an erythemal response and reduce the carotenoid concentration, an important component of antioxidant defense in the human skin [52,53].

Nontoxic fluences of blue light reduced the cell proliferation, promoting the differentiation process in human skin keratinocytes [54]. Additionally, human skin immortalized keratinocytes lineage (HaCaT) and in vivo human keratinocytes showed an increase in the expression of genes related to oxidative stress, inflammatory response and photoaging after exposure to LED $\lambda = 405$ nm [55].

Studies into the effects of green light (490–570 nm) on skin cells mainly have focused on phototherapy (hyperbilirubinemia) and photodynamic therapy [56–58]. Green light (LED $\lambda = 520$; 550 nm) was observed to provide an antiinflammatory activity and favored healing in burns on rat skin [59,60]. Despite skin cells having chromophores that absorb at the green light range (i.e., hemoglobin, porphyrins, heme) [61–63], this fraction of the visible spectrum has been poorly investigated for skin photodamage, photoaging, and carcinogenesis.

Red light (620–770 nm) was used for the smallpox and lupus treatment in the late 19th century by Niels Finsen. Besides, the red light effect has been largely tested on skin cells due to its properties, such as high penetration in the tissue and weak absorption by endogenous chromophores, which are attractive for phototherapies and photodynamic therapies. Red light (LED $\lambda = 633$ nm) was reported, in a low-dose exposure for five cumulative days, to retard the photoaging promoted by UVA in human skin fibroblasts [64]. In addition, red light phototherapy has been showed to stimulate cell proliferation in cutaneous wound healing and to reduce inflammation [65–67]. Red light also appears to have an effect on mitochondria, increasing the mitochondrial membrane potential, enhancing oxidative phosphorylation and ATP production, and decreasing ROS formation in RPE cells [68]. Melanogenesis was observed in human skin after visible light exposure from the blue to red light range (470–690 nm, at doses of 40–80 J/cm^2) [53,69,70].

10.3 Functional analysis methods to detect solar damage

Increasingly noninvasive and real-time analyses have been used for measuring in vivo skin properties, which affect dermatologic and skin care. In addition, at the forefront of the advance of clinical relevance concerning photobiomodulation and photodynamic therapies for skin pathologies, specific and effective methods to evaluate biological responses and optimal doses in treatments are important. Through an optical approach, visible light and infrared can be used for the measurement of diffuse reflectance spectra, indicating the skin physiological (e.g., inflammation, erythema, vascular occlusion, hypo/hyperpigmentation, oxygen saturation) and morphological changes, such as the collagen fiber orientation, which is useful for checking sunburn damage and skin resurfacing. However, these techniques do not show where and how the mechanisms of solar damage to skin cell components take place, simplifying the complex scenarios of photochemical reactions and biological consequences in this tissue during and after sun exposure.

In this context, another approach is based on functional analysis investigating solar damage, which is useful in the field of skin photoprotection, allowing the correct interventions against harmful agents to skin health. This an important fact to be considered, since photoaging and carcinogenesis are products of cumulative damage by chronic and repeated overexposure to sunlight. When skin cells suffer oxidative stress and damage after sunlight exposure, functional and structural changes can be detected using methods based on fluorescence and microscopy. Using site-specific fluorophores, fluorescent antibodies, and even autofluorescence of natural and newly generated pigments, it is possible to monitor the loss of membrane integrity, specific sites of oxidation, and damage to biomolecules and organelles. In addition, in the current omics era, alterations to the protein and gene expression levels, as well as the posttranslational modifications and mutational pattern, can be elucidated by exome sequencing, and proteomic and transcriptome analysis [71–74].

An important pigment that was shown to be present in skin cells after interaction with UVA and visible light is lipofuscin [43]. It results from damage to mitochondria and lysosomes, leading to autophagy impairment and exponentially feeding the cycle of ROS generation. Consequently, the oxidized biomolecules and organelles accumulated inside the autophagolysosomes, forming by cross-linking reactions the lipofuscin, an electron-dense, autofluorescent, and lipophilic pigment sensitive to visible light. Thus, the detection of lipofuscin in skin cells is clear evidence of oxidative damage induced by UVR, visible light, or even chemical agents in the environment and in the cosmetic formulations.

10.4 Strategies for protection from visible light

10.4.1 Filters

Considering the carcinogenic and aging effects of prolonged sun exposure to the skin, the search for cosmetics that can effectively protect the user from this type of radiation has remarkably increased in the last few years. In this regard, researchers have focused on developing user-friendly sunscreens with long-lasting effect, minimum toxicity, lightweight, and fast-absorbing formula for everyday use. Although current sunscreens are labeled as "broad spectrum," most cover only the UV region, fully covering the UVB and partially the UVA regions, despite recent studies pointing to the harm of visible light to the skin. Consequently, and unfortunately, the majority of reports on the development of nanocosmetics still focus on UV protection only, and so sunscreen's active compounds are often treated simply as UV filters. From the rare cases where broad-spectrum sunscreens are described, nanoparticle technology is evolved, so this field will most certainly revolutionize sun care products.

UV filters can be divided into two groups: organic and inorganic. Organic filters work by absorbing UV irradiation, followed by a chemical transformation or dissipation of the energy through heat or via emission of longer wavelength light. Inorganic filters can also absorb light but work mainly by scattering and reflecting the radiation back to the environment.

The history of sunscreens starts in ancient times, when natural plant extracts were used as organic filters for sun protection and sunburn healing [75]. The first commercial sunscreen dates from 1928, and was composed of benzyl salicylate and benzyl cinnamate. During World War II, American soldiers used a mixture of hydrocarbons from the crude oil process, called "red vet pet" (short for red veterinary petrolatum), to protected themselves from prolonged sun exposure in tropical and equatorial regions. After that, the development and use of sunscreen reached a wider scale, and inorganic filters such as zinc oxide (ZnO) and titanium oxide (Ti_2O), along with new organic compounds, were also tested [76].

Some problems are associated with organic filters for sunscreen application. Since no single organic molecule provides sufficient UVA and UVB light protection, a mixture of two or more chemicals is often required, and problems of compatibility between these components may occur. The main issue however is the photostability of these molecules, since photodegradation can lead to products with lower efficiency and higher toxicity, possibly leading to allergic reactions in some users [77].

Although inorganic filters do not suffer from photobleaching, they have some drawbacks. The first sunscreens to present inorganic filters in their composition consisted of micrometer-sized zinc oxide particles dispersed in an oil-based medium. This resulted in a white paste with UV and visible light protection that had little appeal due to its thick and opaque nature [78]. Significant improvements in sunscreen formulations came in the 1990s with the advancement of nanoparticle technology. By reducing the size of inorganic oxide particles from the micro- to the nanoscale, transparent sunscreens with better dispersibility were obtained. The combination of nanoparticles and organic filters also resulted in formulations with improved photostability, lower phototoxicity, and a broader spectrum of light protection, including the visible part of the spectrum. The following sections summarize some studies and trends in the field of sunscreen development based on nanotechnology.

10.4.1.1 Nanoparticles

ZnO and TiO_2 are the most frequently used inorganic materials in sunscreen formulations. These materials have high refractive indexes in the UV-vis region, so they can efficiently scatter and reflect light in this wavelength range. However, the interaction with light is highly dependent on the particle size. ZnO and TiO_2 were previously used in particle sizes reaching 10 μm, which resulted in white pigments, but by reducing the particle size to a few dozen nanometers, they become virtually transparent and invisible sunscreen formulations could be prepared [79]. Besides reflection/scattering, these inorganic materials can also absorb UV light, and their absorption spectra are somewhat complementary: ZnO is more UVA-absorbing while TiO_2 absorbs UVB light more effectively [80]. Although ZnO is more transparent and covers a larger UVA spectrum, TiO_2 is considered a better filter due to its higher sun protection factor (SPF).

A main issue related to the use of inorganic NPs as filters lies in their photocatalytic activity. ZnO and TiO_2 are semiconductors, and upon UV absorption, electrons are injected in their conduction band, leading to the formation of radical species, such as hydroxyl ($^{\bullet}OH$) and superoxide ($O_2^{-\bullet}$) radicals, although hydrogen peroxide (H_2O_2) and singlet oxygen ($^1O_2$) production have been also reported [81]. This has proved useful in environmental applications, for example, wastewater treatment, however, production of these species in the skin can lead to cyto- and genotoxicity. Coating of these NPs with silica, aluminum hydroxide, or aluminum oxide is frequently used to suppress NP photocatalytic activity, although some studies have shown that this process may decrease but not completely prevent the formation of oxidative species by NPs [79]. In addition, the thick layer of the coating material needed for photocatalytic suppression of NPs can eventually affect their UV-blocking efficiency.

In this sense, alternative systems have been investigated in order to widen the UV protection range and yet avoid phototoxicity of inorganic filters. Incorporation of ZnO and TiO_2 to an amorphous mesoporous magnesium carbonate (MMC) support is a recent example. The authors reported that no photocatalytic activity was observed for the ZnO-TiO_2-MMC system and discussed the possible action of free carbonate ions in the amorphous structure as radical scavengers [82]. Substitution of zinc and titanium oxides by hydroxyapatite [HA–$Ca_3(PO_4)_2$] is another example. HA is nontoxic and highly biocompatible, and its lack of absorbance in the UV region is fixed by doping the structure with cations such as Zn^{2+}, Fe^{2+}, and Mn^{2+}. Trivalent cations, such as Fe^{3+} and Cr^{3+}, also can be used as dopants, but absorption of visible light by these ions resulted in a formulation with an undesirable visible color effect [83]. Cerium (III) and cerium (IV) ions also absorb UV irradiation due to f→d and charge transfer transitions, respectively, so ceria (CeO_2) and cerium phosphates [$CePO_4$, $Ce_2(PO_4)_2.HPO_4.H_2O$] have also been investigated. Serra and coworkers have shown that although CeO_2 is as phototoxic as ZnO and TiO_2, $CePO_4$ NPs obtained by two different methods showed lower photocatalytic activity [84]. Alternatively, CeO_2 phototoxicity can be reduced by doping the material with M^{2+} cations, such as Ca^{2+} and Zn^{2+} [85], or by coating its surface with inert materials including alumina or silica [86].

Inorganic filters usually offer little visible light protection when in the form of nanoparticles. Iron oxide NPs, however, show a reddish to brown color due to light absorption in the 400–600 nm region, and so they are used in some sunscreen formulations to offer UV-visible protection and also maintain a more visually appealing formulation for the user.

Methylene bis-benzotriazolyl tetramethylbutylphenol (MBBT; marketed under different commercial names) and 1,3,5-triazine, 2,4,6-tris [1,10-biphenyl]-4-yl (TBPT; Tynosorb 2AB) are the only organic filters dispersed in sunscreen formulations in the form of nanoparticles. Their average size is below 200 nm and they attenuate most of the UV radiation by absorption (c. 85%) while the remainder of the rays are scattered or reflected by the particles [87,88].

10.4.1.2 Nanocarriers

Conventional sunscreens are made of emulsions, gels, or oils, which have some washability, stability, and toxicity limitations because of skin permeation of some compounds [89]. In this regard, incorporation of UV-vis filters to nanosized carriers has received increased attention due to a series of benefits including (1) decreased photodegradation of organic filters (and reduced toxicity risks), (2) decreased incompatibility between the formula's ingredients, and (3) increased solar protection compared to UV-vis filters alone. Such carriers can be composed of lipid, gelatin, polymer, or silica nanostructures, which are described in more detail below.

Lipids are excellent materials for the construction of nanocarriers because they are biocompatible/biodegradable, can be produced in large scales, and can form a series of structures with different physicochemical properties, for example, liposomes, solid lipid nanoparticles (SLNs), and nanostructured lipid carriers (NLCs), depending on the methods and lipids used. These last two structures have been extensively investigated in the field of nanocosmetics for dermal applications. Lipid nanoparticles are composed of a lipid matrix which falls in the size range of 100 – 400 nm and is solid at body temperature [89]. SLNs and NCLs are differentiated by their composition: SLNs are formed by solid lipid, while NCLs are composed of a mixture of solid and liquid (oil) lipids with a higher solid content [90]. Since in the SLNs a very ordered and compact lipid crystal structure is formed, their loading capacity is lower, and problems of expulsion of the incorporated compound during storage have been reported. The mixture of liquid lipids in NCLs, on the other hand, increases the number of imperfections or leads to an amorphous structure which results in a material with a higher payload [91]. Organic filters are usually incorporated in such nanostructures, resulting in a material that can absorb and scatter radiation, while delaying photodegradation of the filter. In this sense a lower concentration of the organic filter is needed for the same SPF value [90].

Polymers are also used as nanocarriers. They can be synthetic or natural, and preferably biocompatible/biodegradable in order to minimize skin toxicity and environmental impact. In this sense, different polymers, such as chitosan, polycapralactone (PCL), ethyl cellulose (EC), and poly-D,L-lactide-*co*-glycolide (PGLA) [92–94], as well as synthetic procedures, for example, modification of polyvinyl alcohol (PVA) with fatty acids [95], have been investigated. In these systems, the polymer can encapsulate the UV filter, commonly benzophenone-3 or octyl methoxycinnamate, via intermolecular interactions, or be covalently bonded with the filter, as is the case of the cinnamate-functionalized cellulose nanocrystals reported by Zhang et al. [96]. Encapsulation of organic filters, as exemplified above, with polymers not only increases their photostability and overall light protection but also considerably decreases percutaneous absorption, avoiding toxicity and allergic reactions. In addition, polymeric coating of inorganic filters such as TiO_2 helps stabilize the colloidal dispersion of NPs by changing the physicochemical properties, including particle size and zeta potential, which affects their aggregation tendency [97].

Gelatin is a mixture of water-soluble proteins of high molecular weight and is thus considered a low-cost biopolymer. Some of its amino acid sequences can act as strong radical scavengers, conferring this material some antioxidant properties. De Oliveira and coworkers investigated the encapsulation of organic filters with rutin, a flavonoid that shows UV absorption and antioxidant properties, in gelatin NPs and observed an increase of 74% in the free radical scavenging percentage compared to rutin alone [98]. In vitro permeation studies showed that no rutin-loaded NPs penetrated the skin, and this was attributed to the high molecular weight of both rutin and gelatin and their low lipophilic character [99].

Silica has been used in the cosmetics area for years, as fillers, for a smoother skin, viscosity-enhancing agents, and UV filters in sunscreen formulations [100]. As mentioned previously, silica can be used to coat the surface of TiO_2 and ZnO NPs in order to decrease the photocatalytic activity of these materials, however SiO_2 NPs can also serve as excellent vehicles for organic filters for several reasons: (1) research on the synthesis of silica NPs is highly developed; (2) SiO_2 NPs scatter UV irradiation, so a synergistic effect with other filters can be achieved for higher SPF values; (3) some structures such as mesoporous silica NPs have high loading capacity, incorporating UV filters in its pores; (4) polymerization of organic molecules modified with silane moieties permits the synthesis of organic–inorganic hybrid NPs with different structures and properties; (5) postsynthesis modification of the silica NP surface can be done in a few steps, allowing one to coat them with different materials.

Walenzyk and coworkers reported the incorporation of benzophenone-3 in the matrix and on the surface of mesoporous silica NPs. They showed that the concentration of the UV filter used during the polymerization reaction determined the shape and size distribution of the NPs. Also, the photostability of the filter localized in the matrix was higher than when it localized only on the surface [101]. Considering the still scarce research on visible light protection, Baptista and collaborators investigated the production of silica NPs coated with a thin layer of melanin in order to obtain a material with UV-vis protecting ability. In this work, dihydroxyindolequinone (DHI) or indole-5,6-quinone (IQ), which are intermediates in the melanin synthesis, were attached and polymerized on SiO_2 NP surfaces by the tyrosinase enzyme. Apart from the UV region, the melanin/nanosilica material showed light absorption in the visible (400–750 nm) and infrared regions (750–1400 nm) [102]. Melanin-covered silica NPs show great potential in radiation protection considering that they could also protect bone marrow cells from gamma radiation in radioimmunotherapy sections [103].

10.4.2 Membrane protection

Biological membranes are the main target of the reactive species, which can be generated by a photoinduced process in the skin, affecting structure and function in organelles and skin cells, such as redox imbalance, and changes in membrane permeability and integrity [104,105]. Lipids do not contain chromophores required for photochemical reactions, so such reactions are initiated by photosensitizers in the neighborhood, such as proteins, flavin, and porphyrins, forming peroxy radicals, endoperoxides, and hydroperoxides. In biological membranes, the unsaturated lipids and proteins are the most preferred target of reactive species [106]. Proteins are preferably oxidized in tryptophan, tyrosine, histidine, cysteine, and methionine residues, either side chains or backbone, resulting in fragmentation, protein–protein and lipid–protein cross-linking, and loss of function and structure [107,108].

The large number of oxidized lipids, which are used in lipid peroxidation, affects the hydrophobicity, physical, chemical, and biological activities of membranes, such as the recognition of receptors and target molecules, and membrane area. In addition, oxidation in membrane lipids is an important mediator in many diseases (e.g., atherosclerosis, Alzheimer disease, multiple sclerosis), eliciting inflammatory responses and cell death [109]. In the skin, squalene, sebaleic acid, linoleic acid, and cholesterol are the major unsaturated lipids attacked by singlet oxygen, radicals, and also oxidation mediated by UVR-activated enzymes, such as lipoxygenase and cyclooxygenase, inducing inflammation and several skin diseases [105].

In this context, the membrane protection has been thought of as a promising strategy for minimizing the photodamage by UVR and visible light in skin cells, decreasing the reactive species generation and their access to their targets, ensuring membrane integrity and homeostasis. In addition, the sun-protective power of these photoprotectants is similar to the chemical filters used by the cosmetics and pharmaceuticals industry.

The mechanism underlying the photodefensive effect seems to be centered on membrane integrity in organelles important to the essential process for survival and repair of the photoinduced damage, such as lysosomes. Rodrigues et al. [110] showed that protection of the lysosomal membrane from UVA photodamage and visible light using *Aloe vera* extract, ensured the autophagy activity and decreased the lipofuscinogenesis and cell death after UVA or visible light exposure. Thus, *A. vera* extract can prevent membrane damage, especially in lysosomes, and was shown to be an effective component in sun-protective formulations. Flavonoids, which usually have two maximum peaks in the UVA and UVB region, are also able to protect the lysosomal membrane, as described by Basu-Modak et al. [111], in skin fibroblasts exposed to UVA, preventing iron release, which strongly activates the Fenton reaction, promoting oxidative damage in human skin cells.

Another strategy for protecting the membranes is to use functionalized macromolecules conjugated with photoprotective components. Mertins et al. [112] developed a polymer of gallic acid-labeled chitosan which was able to cover the surface of the mimetic membrane, suppressing the singlet oxygen, and defending the membrane from $^1O_2$.

10.4.3 Antioxidants

As discussed above, many oxidative reactions occur in skin after sunlight exposure, which damage cells, leading to inflammatory responses, erythema, cell death, skin aging, and carcinogenesis. However, skin cells have their own antioxidant defense, when reactive species generation supplants the antioxidant power, oxidative damage starts to accumulate. Haywood et al. [113] reported that current sunscreens reduce ROS generation by nearly 50%, and established a "free-radical protection factor," which could be added to the sunscreen label, indicating the efficacy of sunscreen protecting skin from aging and cancer.

There is a growing interest in including antioxidants and DNA repair enzymes in sunscreen formulations to guarantee high photoprotection against sunlight damage [114]. In addition, there is an inherent challenge in this photoprotection approach: the development of products that hold the chemical and functional stability of these antioxidants, as well as their delivery to the correct target inside the cells [115–117].

Murray et al. [116] showed that a combination of vitamins C and E stabilized by ferulic acid used in a topical product was able to protect skin cells from UVR damage, decreasing the proinflammatory and immunosuppressive responses, thymine dimer formation, and erythema. In fact, synergic action between vitamins C (L-ascorbic acid) and E (α-tocopherol) is remarkably more photoprotective against UVR in topical products than vitamins C and E alone, which is progressively increased when applied for 4 consecutive days [118]. Synergism between vitamins C and E occurs through the regeneration of the oxidized vitamin E by vitamin C in membranes.

In some conditions, vitamin C, water-soluble and accumulated inside cells, is known to neutralize superoxide anions, hydroxyl radicals, singlet oxygen, and peroxynitrite [119,120]. In terms of an antiaging effect, vitamin C topically applied showed a significantly rejuvenescent action on human skin due to the induction of collagen synthesis [121]. Vitamin E is lipid-soluble and shows antioxidant action defending biological membranes and the stratum corneum in the skin layer, preventing protein and lipid oxidation. Vitamin E is a powerful peroxyl scavenger and decreases lipid peroxidation, upregulating the enzymatic and nonenzymatic antioxidants [122]. Antioxidant molecules, such as vitamins C and E, and ubiquinone in sunscreen formulations have additional roles, as a stabilizing filter and by boosting the SPF. Avobenzone used in chemical sunscreens to protect against UVA is effectively photostabilized by ubiquinone [123].

The newer generation of broad-spectrum chemical agents in sunscreen formulations, such as bis-ethylhexyloxyphenol methoxyphenyl triazine (BEMT), is effective in blocking the whole UVR range (290–400 nm), maintaining photostability, and preventing ROS generation, erythema, aging, and cancer [124]. Souza and Campos [115] investigated the stability of antioxidants *Spirulina* and DMC (dimethylmethoxy chromanol), an γ-tocopherol analogue, attached to SLNs in sunscreen formulations in in vivo human skin. These antioxidants showed a significant decrease in the stimulation of pigmentation and collagen degradation, and were nonirritant to skin and stable.

Using an approach based on genic expression regulation, the activation of *Nrf2* transcription factor (nuclear factor erythroid derived 2), a master regulator of antioxidant defense component expression, such as glutathione (GSH), is a photoprotective strategy [125]. In fact, de la Vega et al. [126] showed that apocarotenoid bixin, a natural food colorant extracted from seeds of the tropical American achiote tree (*Bixa orellana*), suppressed acute UV-induced photodamage in mouse skin keratinocytes $Nrf2^{+/+}$ (but not in $Nrf2^{-/-}$) by activation of *Nrf2*, resulting in an increased antioxidant defense and antiinflammatory response. It is interesting that bixin was effective on mouse skin photoprotection against UV, both when used in topical applications and in dietary consumption, making it a promising photoprotectant in sunscreen formulations or food supplementations [127].

10.5 Final remarks

The field of sun protection is set to undergo a revolution in the years ahead. The acceptance that real broad-spectrum protection is the way to go offers a great opportunity for scientists working in this field. With the recent increase in UV radiation levels caused by the ozone layer depletion, banning of several molecules used as sun blockers, and the knowledge that visible light also has a role in ROS production and cytotoxicity, it has became urgent to develop improved sunscreens. This will include broader spectrum coverage, but also antioxidants and membrane photoprotectants, for daily use. In this context, the coupling of chromophores with inorganic or organic nanoparticles has shown promising results, and some technologies have been used in the current sunscreen formulations. Although promising, the field of nanomaterials has some issues, such as production cost, dispersibility and stability of the NPs, loading capacity, toxicity by inhalation and skin absorption, environmental impact, among others, that need to be overcome for the development of better sun-protective agents.

References

[1] Food and Drug Administration. Sunscreen drug products for over-the counter human use. Propos. Rules. 2019;84 (38):6204–75.
[2] Leiter U, Eigentler T, Garbe C. Epidemiology of skin cancer. Adv Exp Med Biol 2014;810:120–40.
[3] Apalla Z, Lallas A, Sotiriou E, Lazaridou E, Ioannides D. Epidemiological trends in skin cancer. Dermatol Pract Concept 2017;7(2):1–6.
[4] National Cancer Institute. Melanoma of the skin—cancer stat facts. Natl. Cancer Inst. 2018.
[5] Wadhera A, Fazio M, Bricca G, Stanton O. Metastatic basal cell carcinoma: a case report and literature review. How accurate is our incidence data? Dermatol Online J 2006;12.
[6] IARC Monographs on the Evaluation of Carcinogenic Risks to Humans. Solar and ultraviolet radiation. IARC monograph on the evaluation of carcinogenic risks on human; 1992.
[7] De Gruijl FR. Skin cancer and solar UV radiation. Eur J Cancer 1999;35:2003–9.
[8] Findlay G. Ultra-violet light and skin cancer. Lancet. 1928;212:1070–3.
[9] Paul C. The influence of sunlight in the production of cancer of the skin. H.K. Lewis & Company, Limited; 1918.
[10] Roffo A. Cancer and the sun: carcinomas and sarcomas caused by the action of the sun in toto. Bull Assoc Fr Etud Cancer 1934;23:590–616.
[11] Zigman S, Fowler E, Kraus AL. Black light induction of skin tumors in mice. J Invest Dermatol 1976;67:723–5.
[12] Strickland PT. Photocarcinogenesis by near-ultraviolet (UVA) radiation in sencar mice. J Invest Dermatol 1986;87:272–5.
[13] Cheong WF, Prahl SA, Welch AJ. A review of the optical properties of biological tissues. IEEE J Quantum Electron 1990;26:2166–85.
[14] Wondrak GT, Jacobson MK, Jacobson EL. Endogenous UVA-photosensitizers: mediators of skin photodamage and novel targets for skin photoprotection. Photochem Photobiol Sci 2006;5(2):215–37.
[15] Besaratinia A, Kim S-i, Bates SE, Pfeifer GP. Riboflavin activated by ultraviolet A1 irradiation induces oxidative DNA damage-mediated mutations inhibited by vitamin C. Proc Natl Acad Sci 2007;104(14):5953–8.
[16] Avci P, Gupta A, Sadasivam M, Vecchio D, Pam Z, Pam N, et al. Low-level laser (light) therapy (LLLT) in skin: stimulating, healing, restoring. Semin Cutan Med Surg 2013;32:41–52.
[17] Smith KC. Common misconceptions about light. In: Pratesi R, Sacchi CA, editors. Lasers in photomedicine photobiology. Berlin, Heidelberg: Springer; 1980. p. 23–5.
[18] Mahmoud BH, Ruvolo E, Hexsel CL, Liu Y, Owen MR, Kollias N, et al. Impact of long-wavelength UVA and visible light on melanocompetent skin. J Invest Dermatol 2010;130(8):2092–7.

[19] Zonios G, Bykowski J, Kollias N. Skin melanin, hemoglobin, and light scattering properties can be quantitatively assessed in vivo using diffuse reflectance spectroscopy. J Invest Dermatol 2001;117:1452–7.
[20] Lister T, Wright PA, Chappell PH. Optical properties of human skin. J Biomed Opt 2012;17:1–15.
[21] Baier J, Maisch T, Maier M, Engel E, Landthaler M, Bäumler W. Singlet oxygen generation by UVA light exposure of endogenous photosensitizers. Biophys J 2006;91(4):1452–9.
[22] Agnez-Lima LF, Melo JTA, Silva AE, Oliveira AHS, Timoteo ARS, Lima-Bessa KM, et al. DNA damage by singlet oxygen and cellular protective mechanisms. Mutat Res—Rev Mutat Res 2012;751:15–28.
[23] Pathak MA, Stratton K. Free radicals in human skin before and after exposure to light. Arch Biochem Biophys 1968;123:468–76.
[24] Ravanat JL, Di Mascio P, Martinez GR, Medeiros MHG, Cadet J. Singlet oxygen induces oxidation of cellular DNA. J Biol Chem 2001;275:40601–4.
[25] Tyrrell RM. Ultraviolet radiation and free radical damage to skin. Biochem Soc Symp 1995;61:47–53.
[26] Wölfle U, Esser PR, Simon-Haarhaus B, Martin SF, Lademann J, Schempp CM. UVB-induced DNA damage, generation of reactive oxygen species, and inflammation are effectively attenuated by the flavonoid luteolin in vitro and in vivo. Free Radic Biol Med 2011;50(9):1081–93.
[27] Jin GH, Liu Y, Jin SZ, Liu XD, Liu SZ. UVB induced oxidative stress in human keratinocytes and protective effect of antioxidant agents. Radiat Env Biophys 2007;46:61–8.
[28] Haywood R. Relevance of sunscreen application method, visible light and sunlight intensity to free-radical protection: a study of ex vivo human skin. Photochem Photobiol 2006;82:1123–31.
[29] Zastrow L, Groth N, Klein F, Kockott D, Lademann J, Ferrero L. Detection and identification of free radicals generated by UV and visible light in ex vivo human skin. Int J Cosmet Sci 2009;11:207–15.
[30] Liebel F, Kaur S, Ruvolo E, Kollias N, Southall MD. Irradiation of skin with visible light induces reactive oxygen species and matrix-degrading enzymes. J Invest Dermatol 2012;132(7):1901–7.
[31] Chiarelli-Neto O, Ferreira AS, Martins WK, Pavani C, Severino D, Faião-Flores F, et al. Melanin photosensitization and the effect of visible light on epithelial cells. PLoS One 2014;9(11):1–9.
[32] Pfeifer GP. Formation and processing of UV photoproducts: effects of DNA sequence and chromatin environment. Photochem Photobiol 1997;65(2):270–83.
[33] Pfeifer GP, You Y-H, Besaratinia A. Mutations induced by ultraviolet light. Mutat Res Mol Mech Mutagen 2005;571(1–2):19–31.
[34] Agar NS, Halliday GM, Barnetson RS, Ananthaswamy HN, Wheeler M, Jones AM. The basal layer in human squamous tumors harbors more UVA than UVB fingerprint mutations: a role for UVA in human skin carcinogenesis. Proc Natl Acad Sci 2004;101:4954–9.
[35] Delinasios GJ, Karbaschi M, Cooke MS, Young AR. Vitamin E inhibits the UVAI induction of "light" and "dark" cyclobutane pyrimidine dimers, and oxidatively generated DNA damage, in keratinocytes. Sci Rep 2018;8(1):1–12.
[36] Douki T, Reynaud-Angelin A, Cadet J, Sage E. Bipyrimidine photoproducts rather than oxidative lesions are the main type of DNA damage involved in the genotoxic effect of solar UVA radiation. Biochemistry. 2003;42(30):9221–6.
[37] Premi S, Wallisch S, Mano CM, Weiner AB, Bacchiocchi A, Wakamatsu K, et al. Chemiexcitation of melanin derivatives induces DNA photoproducts long after UV exposure. Science 2015;347(6224):842–7.
[38] Rochette PJ, Therrien JP, Drouin R, Perdiz D, Bastien N, Drobetsky EA, et al. UVA-induced cyclobutane pyrimidine dimers form predominantly at thymine-thymine dipyrimidines and correlate with the mutation spectrum in rodent cells. Nucleic Acids Res 2003;31(11):2786–94.
[39] Lamore SD, Wondrak GT. Autophagic-lysosomal dysregulation downstream of cathepsin B inactivation in human skin fibroblasts exposed to UVA. Photochem Photobiol Sci 2012;11(1):163–72.
[40] Fruhwirth GO, Loidl A, Hermetter A. Oxidized phospholipids: from molecular properties to disease. Biochim Biophys Acta—Mol Basis Dis 2007;1772:718–36.
[41] Johnson BE, Daniels F. Lysosomes and the reactions of skin to ultraviolet radiation. J Invest Dermatol 1969;53:85–94.
[42] Lamore SD, Qiao S, Horn D, Wondrak GT. Proteomic identification of cathepsin B and nucleophosmin as novel UVA-targets in human skin fibroblasts. Photochem Photobiol 2010;86(6):1307–17.
[43] Tonolli PN, Chiarelli-Neto O, Santacruz-Perez C, Junqueira HC, Watanabe I-S, Ravagnani FG, et al. Lipofuscin generated by UVA turns keratinocytes photosensitive to visible light. J Invest Dermatol 2017;137:2447–50.

[44] Karu TI, Kalendo GS, Letokhov VS, Lobko VV. Biostimulation of HeLa cells by low-intensity visible light. Il Nuovo Cimento D 1982;1:828–40.
[45] Karu TI. Effects of visible radiation on cultured cells. Photochem Photobiol 1990;52(6):1089–98.
[46] Tyrrell RM, Pidoux M. Action spectra for human skin cells: estimates of the relative cytotoxicity of the middle ultraviolet, near ultraviolet, and violet regions of sunlight on epidermal keratinocytes. Cancer Res 1987;47:1825–9.
[47] Kielbassa C, Roza L, Epe B. Wavelength dependence of oxidative DNA damage induced by UV and visible light. Carcinogenesis. 1997;18(4):811–16.
[48] Godley BF, Shamsi FA, Liang F-Q, Jarrett SG, Davies S, Boulton M. Blue light induces mitochondrial DNA damage and free radical production in epithelial cells. J Biol Chem 2005;280(22):21061–6.
[49] King A, Gottlieb E, Brooks DG, Murphy MP, Dunaief JL. Mitochondria-derived reactive oxygen species mediate blue light-induced death of retinal pigment epithelial cells. Photochem Photobiol 2004;79(5):470–5.
[50] del Olmo-Aguado S, Núñez-Álvarez C, Osborne NN. Blue light action on mitochondria leads to cell death by necroptosis. Neurochem Res 2016;41(9):2324–35.
[51] Osborne NN, Núñez-Álvarez C, del Olmo-Aguado S. The effect of visual blue light on mitochondrial function associated with retinal ganglions cells. Exp Eye Res 2014;128:8–14.
[52] Vandersee S, Beyer M, Lademann J, Darvin ME. Blue-violet light irradiation dose dependently decreases carotenoids in human skin, which indicates the generation of free radicals. Oxid Med Cell Longev 2015;2015:1–7.
[53] Porges SB, Kaidbey KH, Grove GL. Quantification of visible light-induced melanogenesis in human skin. Photodermatol. 1988;5:197–200.
[54] Liebmann J, Born M, Kolb-Bachofen V. Blue-light irradiation regulates proliferation and differentiation in human skin cells. J Invest Dermatol 2010;130(1):259–69.
[55] Lawrence KP, Douki T, Sarkany RPE, Acker S, Herzog B, Young AR. The UV/visible radiation boundary region (385–405 nm) damages skin cells and induces "dark" cyclobutane pyrimidine dimers in human skin in vivo. Sci Rep 2018;8:1–12.
[56] Böhm F, Drygalla F, Charlesworth P, Böhm K, Truscott TG, Jokiel K. Bilirubin phototoxicity to human cells by green light phototherapy in vitro. Photochem Photobiol 1995;62:980–3.
[57] Christensen T, Reitan JB, Kinn G. Single-strand breaks in the DNA of human cells exposed to visible light from phototherapy lamps in the presence and absence of bilirubin. J Photochem Photobiol B Biol 1990;7:337–46.
[58] Morton CA, Whitehurst C, Moore JV, Mackie RM. Comparison of red and green light in the treatment of Bowen's disease by photodynamic therapy. Br J Dermatol 2000;143:767–72.
[59] de Vasconcelos Catão MHC, Nonaka CFW, de Albuquerque RLC, Bento PM, de Oliveira Costa R. Effects of red laser, infrared, photodynamic therapy, and green LED on the healing process of third-degree burns: clinical and histological study in rats. Lasers Med Sci 2014;30:421–8.
[60] Catão MHCV, Costa RO, Nonaka CFW, Junior RLCA, Costa IRRS. Green LED light has anti-inflammatory effects on burns in rats. Burns. 2016;42:392–6.
[61] Zhang S, Gerhard GS. Heme mediates cytotoxicity from artemisinin and serves as a general anti-proliferation target. PLoS One 2009;4:1–10.
[62] Horecker BL. The absorption spectra of hemoglobin and its derivatives in the visible and near infrared regions. J Biol Chem 1943;148:173–83.
[63] Robles FE, Chowdhury S, Wax A, Wang HW, Jiang JK, Lin CH. Assessing hemoglobin concentration using spectroscopic optical coherence tomography for feasibility of tissue diagnostics. Biomed Opt Exp 2006;1:310–17.
[64] Niu T, Tian Y, Ren Q, Wei L, Li X, Cai Q. Red light interferes in UVA-induced photoaging of human skin fibroblast cells. Photochem Photobiol 2014;90:1349–58.
[65] Boschi ES, Leite CE, Saciura VC, Caberlon E, Lunardelli A, Bitencourt S, et al. Anti-inflammatory effects of low--level laser therapy (660 nm) in the early phase in carrageenan-induced pleurisy in rat. Lasers Surg Med 2008;40:500–8.
[66] Erdle BJ, Brouxhon S, Kaplan M, Vanbuskirk J, Pentland AP. Effects of continuous-wave (670-nm) red light on wound healing. Dermatologic Surg 2008;34:320–5.

[67] Myakishev-Rempel M, Stadler I, Brondon P, Axe DR, Friedman M, Nardia FB, et al. A preliminary study of the safety of red light phototherapy of tissues harboring cancer. Photomed Laser Surg 2012;30:551–8.
[68] Begum R, Powner MB, Hudson N, Hogg C, Jeffery G. Treatment with 670 nm light up regulates cytochrome C oxidase expression and reduces inflammation in an age-related macular degeneration model. PLoS One 2013;8(2):1–11.
[69] Sklar LR, Almutawa F, Lim HW, Hamzavi I. Effects of ultraviolet radiation, visible light, and infrared radiation on erythema and pigmentation: a review. Photochem Photobiol Sci 2013;12:54–64.
[70] Mahmoud BH, Hexsel CL, Hamzavi IH, Lim HW. Effects of visible light on the skin. Photochem Photobiol 2008;84:450–62.
[71] Yan W, Zhang LL, Yan L, Zhang F, Yin NB, Bin Lin H, et al. Transcriptome analysis of skin photoaging in Chinese females reveals the involvement of skin homeostasis and metabolic changes. PLoS One 2013;8:1–10.
[72] Inman GJ, Wang J, Nagano A, Alexandrov LB, Purdie KJ, Taylor RG, et al. The genomic landscape of cutaneous SCC reveals drivers and a novel azathioprine associated mutational signature. Nat Commun 2018;9:1–14.
[73] Kontogianni G, Piroti G, Maglogiannis I, Chatziioannou A, Papadodima O. Dissecting the mutational landscape of cutaneous melanoma: an omic analysis based on patients from Greece. Cancers (Basel) 2018;10:1–13.
[74] Sengupta D, Tackett AJ. Proteomic findings in melanoma. J Proteom Bioinform 2016;9:1–7.
[75] Shaath NA, editor. Sunscreens: regulations and commercial development. 3rd ed Boca Raton, FL: Taylor & Francis; 2005.
[76] Trivedi M, Murase J. Titanium dioxide in sunscreen. In: Janus M, editor. Application of titanium dioxide; 2017. p. 61–71.
[77] Serpone N, Dondi D, Albini A. Inorganic and organic UV filters: their role and efficacy in sunscreens and suncare products. Inorganica Chim Acta 2007;360(3):794–802.
[78] Faure B, Salazar-Alvarez G, Ahniyaz A, Villaluenga I, Berriozabal G, De Miguel YR, et al. Dispersion and surface functionalization of oxide nanoparticles for transparent photocatalytic and UV-protecting coatings and sunscreens. Sci Technol Adv Mater 2013;14(2).
[79] Smijs T, Pavel S. A case study: nano-sized titanium dioxide in sunscreens. Nanoengineering—global approaches to health and safety issues. Elsevier B.V.; 2015.
[80] Popov AP, Zvyagin AV, Lademann J, Roberts MS, Sanchez W, Priezzhev AV, et al. Designing inorganic light-protective skin nanotechnology products. J Biomed Nanotechnol 2010;6(5):432–51.
[81] Smijs TG, Pavel S. Titanium dioxide and zinc oxide nanoparticles in sunscreens: Focus on their safety and effectiveness. Nanotechnol Sci Appl 2011;4(1):95–112.
[82] Åhlén M, Cheung O, Strømme M. Amorphous mesoporous magnesium carbonate as a functional support for UV-blocking semiconductor nanoparticles for cosmetic applications. ACS Omega 2019;4(2):4429–36.
[83] Pal A, Hadagalli K, Bhat P, Goel V, Mandal S. Hydroxyapatite—a promising sunscreen filter. J Aust Ceram Soc J Australian Ceram Soc 2019;56:345–51.
[84] De Lima JF, Serra OA. Cerium phosphate nanoparticles with low photocatalytic activity for UV light absorption application in photoprotection. Dye Pigm 2013;97(2):291–6.
[85] Yabe S, Yamashita M, Momose S, Tahira K, Yoshida S, Li R, et al. Synthesis and UV-shielding properties of metal oxide doped ceria via soft solution chemical processes. Int J Inorg Mater 2001;3(7):1003–8.
[86] Siddiquey IA, Furusawa T, Hoshi Y ichiro, Ukaji E, Kurayama F, Sato M, et al. Silica coating of CeO_2 nanoparticles by a fast microwave irradiation method. Appl Surf Sci 2008;255:2419–24.
[87] Huber B, Burfeindt J. Nanotechnology in cosmetics. In: Keck Cornier J, Voorde CM, Van de M, editors. Nanocosmetics. Cham: Springer; 2019. p. 17–25.
[88] More BD. Physical sunscreens: on the comeback trail. Indian J Dermatol Venereol Leprol 2007;73(2):80–5.
[89] Puglia C, Damiani E, Offerta A, Rizza L, Giusy G, Stella M, et al. European Journal of Pharmaceutical Sciences Evaluation of nanostructured lipid carriers (NLC) and nanoemulsions as carriers for UV-filters: characterization, in vitro penetration and photostability studies. Eur J Pharm Sci 2014;51:211–17.
[90] Xia Q, Saupe A, Mu RH. Nanostructured lipid carriers as novel carrier for sunscreen formulations. Int J Cosmet Sci 2007;51:473–82.

[91] Müller RH, Radtke M, Wissing SA. Solid lipid nanoparticles (SLN) and nanostructured lipid carriers (NLC) in cosmetic and dermatological preparations. Adv. Drug Deliv. Rev., 1. 2002. p. S131–55.
[92] Jain SK, Jain NK. Multiparticulate carriers for sun-screening agents. Int J Cosmet Sci 2010;32(2):89–98.
[93] Alvarez-Román R, Barré G, Guya RH, Fessi H. Biodegradable polymer nanocapsules containing a sunscreen agent: preparation and photoprotection. Eur J Pharm Biopharm 2001;52(2):191–5.
[94] Muxika A, Etxabide A, Uranga J, Guerrero P, de la Caba K. Chitosan as a bioactive polymer: processing, properties and applications. Int J Biol Macromol Elsevier BV 2017;105:1358–68.
[95] Luppi B, Cerchiara T, Bigucci F, Basile R, Zecchi V. Polymeric nanoparticles composed of fatty acids and polyvinylalcohol for topical application of sunscreens. J Pharm Pharmacol 2004;56(3):407–11.
[96] Zhang Z, Zhang B, Grishkewich N, Berry R, Tam KC. UV-shielding: cinnamate-functionalized cellulose nanocrystals as UV-shielding nanofillers in sunscreen and transparent polymer films (adv. sustainable syst. 4/2019). Adv Sustain Syst 2019;3(4):1970009.
[97] Baek S, Joo SH, Blackwelder P, Toborek M. Effects of coating materials on antibacterial properties of industrial and sunscreen-derived titanium-dioxide nanoparticles on Escherichia coli. Chemosphere 2018;208:196–206.
[98] De Oliveira CA, Peres DDA, Graziola F, Chacra NAB, De Araújo GLB, Flórido AC, et al. Cutaneous biocompatible rutin-loaded gelatin-based nanoparticles increase the SPF of the association of UVA and UVB filters. Eur J Pharm Sci 2016;81:1–9.
[99] de Oliveira CA, Dario MF, Sarruf FD, Mariz IFA, Velasco MVR, Rosado C, et al. Safety and efficacy evaluation of gelatin-based nanoparticles associated with UV filters. Colloids Surfaces B Biointerfaces 2016;140:531–7.
[100] Mihranyan A, Ferraz N, Strømme M. Current status and future prospects of nanotechnology in cosmetics. Prog Mater Sci 2012;57(5):875–910.
[101] Walenzyk T, Carola C, Buchholz H, König B. Synthesis of mono-dispersed spherical silica particles containing covalently bonded chromophores. Int J Cosmet Sci 2005;27(3):177–89.
[102] Miranda A, Baptista MS. Processo de obtenção de nanossílica revestida, nanossílica revestida e seu uso, INPI 2016, BR 10 2016 024262 2.
[103] Schweitzer AD, Revskaya E, Chu P, Pazo V, Friedman M, Nosanchuk JD, et al. Melanin-covered nanoparticles for protection of bone marrow during radiation therapy of cancer. Int J Radiat Oncol Biol Phys 2010;78:1494–502.
[104] Kochevar IE, Lambert CR, Lynch MC, Tedesco AC. Comparison of photosensitized plasma membrane damage caused by singlet oxygen and free radicals. BBA. Biomembranes 1996;1280:223–30.
[105] Niki E. Lipid peroxidation: physiological levels and dual biological effects. Free Radic Biol Med 2009;47:469–84.
[106] Ayala A, Muñoz MF, Argüelles S. Lipid peroxidation: production, metabolism, and signaling mechanisms of malondialdehyde and 4-hydroxy-2-nonenal. Oxid Med Cell Longev 2014;2014:1–31.
[107] Stadtman ER, Levine RL. Free radical-mediated oxidation of free amino acids and amino acid residues in proteins. Amino Acids 2003;25:207–18.
[108] Pattison DI, Rahmanto AS, Davies MJ. Photo-oxidation of proteins. Photochem Photobiol Sci 2012;11:38–53.
[109] Weismann D, Binder CJ. The innate immune response to products of phospholipid peroxidation. BBA. Biomembranes 2012;1818:2465–75.
[110] Rodrigues D, Viotto AC, Checchia RG, Gomide AB, Itri R, Severino D, Baptista MS, Martins WK. Martins, Mechanism of aloe vera extract protection against UVA: shelter of lysosomal membrane avoids photodamage. Photochem Photobiol Sci 2016;15:334–50.
[111] Basu-Modak S, Ali D, Gordon M, Polte T, Yiakouvaki A, Pourzand C, et al. Suppression of UVA-mediated release of labile iron by epicatechin-a link to lysosomal protection. Free Radic Biol Med 2006;41:1197–204.
[112] Mertins O, Mathews PD, Gomide AB, Baptista MS, Itri R. Effective protection of biological membranes against photo-oxidative damage: polymeric antioxidant forming a protecting shield over the membrane. Biochim Biophys Acta - Biomembr 2015;1848:2180–7.
[113] Haywood R, Wardman P, Sanders R, Linge C. Sunscreens inadequately protect against ultraviolet-A-induced free radicals in skin: implications for skin aging and melanoma? J Invest Dermatol 2003;121 (4):862–8.
[114] Matsui MS, Hsia A, Miller JD, Hanneman K, Scull H, Cooper KD, et al. Non-sunscreen photoprotection: antioxidants add value to a sunscreen. J Investig Dermatol Symp Proc 2009;14:56–9.

[115] Souza C, Campos PMBGM. Development and photoprotective effect of a sunscreen containing the antioxidants Spirulina and dimethylmethoxy chromanol on sun-induced skin damage. Eur J Pharm Sci 2017;104:52–64.
[116] Murray JC, Burch JA, Streilein RD, Iannacchione MA, Hall RP, Pinnell SR. A topical antioxidant solution containing vitamins C and E stabilized by ferulic acid provides protection for human skin against damage caused by ultraviolet irradiation. J Am Acad Dermatol 2008;59:418–25.
[117] Pinnell SR, Yang H, Omar M, Riviere NM, DeBuys HV, Walker LC, et al. Topical L-ascorbic acid: percutaneous absorption studies. Dermatologic Surg 2001;27:137–42.
[118] Lin JY, Selim MA, Shea CR, Grichnik JM, Omar MM, Monteiro-Riviere NA, et al. UV photoprotection by combination topical antioxidants vitamin C and vitamin E. J Am Acad Dermatol 2003;48:866–74.
[119] Shindo Y, Witt E, Han D, Epstein W, Packer L. Enzymic and non-enzymic antioxidants in epidermis and dermis of human skin. J Invest Dermatol 1994;102:122–4.
[120] Bodannes RS, Chan PC. Ascorbic acid as a scavenger of singlet oxygen. FEBS Lett 1979;105:195–6.
[121] Crisan D, Roman I, Crisan M, Scharffetter-Kochanek K, Badea R. The role of vitamin C in pushing back the boundaries of skin aging: an ultrasonographic approach. Clin Cosmet Investig Dermatol 2015;8:463–70.
[122] Lopez-Torres M, Thiele JJ, Shindo Y, Han D, Packer L. Topical application of α-tocopherol modulates the antioxidant network and diminishes ultraviolet-induced oxidative damage in murine skin. Br J Dermatol 1998;138:207–15.
[123] Afonso S, Horita K, Sousa E, Silva JP, Almeida IF, Amaral MH, et al. Photodegradation of avobenzone: stabilization effect of antioxidants. J Photochem Photobiol B Biol 2014;140:36–40.
[124] Latha MS, Martis J, Shobha V, Shinde RS, Bangera S, Krishnankutty B, et al. Sunscreening agents: a review. J Clin Aesthet Dermatol 2013;6:16–26.
[125] Kavian N, Mehlal S, Jeljeli M, Saidu NEB, Nicco C, Cerles O, et al. The Nrf2-antioxidant response element signaling pathway controls fibrosis and autoimmunity in scleroderma. Front Immunol 2018;9:1–18.
[126] de la Vega MR, Krajisnik A, Zhang DD, Wondrak GT. Targeting NRF2 for improved skin barrier function and photoprotection: focus on the achiote-derived apocarotenoid bixin. Nutrients 2017;9:1–16.
[127] de la Vega MR, Zhang DD, Wondrak GT. Topical bixin confers NRF2-dependent protection against photodamage and hair graying in mouse skin. Front Pharmacol 2018;9:1–11.

CHAPTER

11

Nanomaterials for hair care applications

Megumi Nishitani Yukuyama, Gabriel Lima Barros De Araújo and Nádia Araci Bou-Chacra

Faculty of Pharmaceutical Sciences, University of Sao Paulo, Sao Paulo, Brazil

11.1 Introduction

Apparently, hair seems to be a basic structure comprised of simple filaments that offer thermal protection and that reflect social status and relations of individuals, including aspects such as healthy appearance, age, and lifestyle trends. But in fact, hair is made up of complex structures that implicated in cell differentiation, proliferation, the lifecycle, and disorders. The daily application of topical components may highly affect the final appearance of hair, considering that different compounds can penetrate and act on different segments of the hair structure.

Hair care products, including shampoo, conditioner, and leave-on products, focus on improving the appearance along hair length, while the dermocosmetic products such as anti-hair loss and antidandruff products focus on treatment based on the absorption of active compounds into the hair scalp.

The efficacy of applied active compounds on hair or scalp depends on their penetration via three different pathways: intercellular, hair follicle, and transcellular. The first pathway is established by diffusion of substances in the stratum corneum of the scalp through the corneocyte surfaces, which are surrounded by lipid layers. The second occurs by the diffusion of compounds through the opening of follicular orifices and can reach the deep region of the scalp. Therein, a dense network of blood capillaries surrounds the hair follicles. This second pathway has gained increasing attention in recent years and is one of the main focuses of this chapter. The third one is the penetration of substances directly through the corneocytes and lipid layers, overcoming the stratum corneum barrier to reach the living cells [1] (Fig. 11.1).

Nanocosmetics
DOI: https://doi.org/10.1016/B978-0-12-822286-7.00010-3

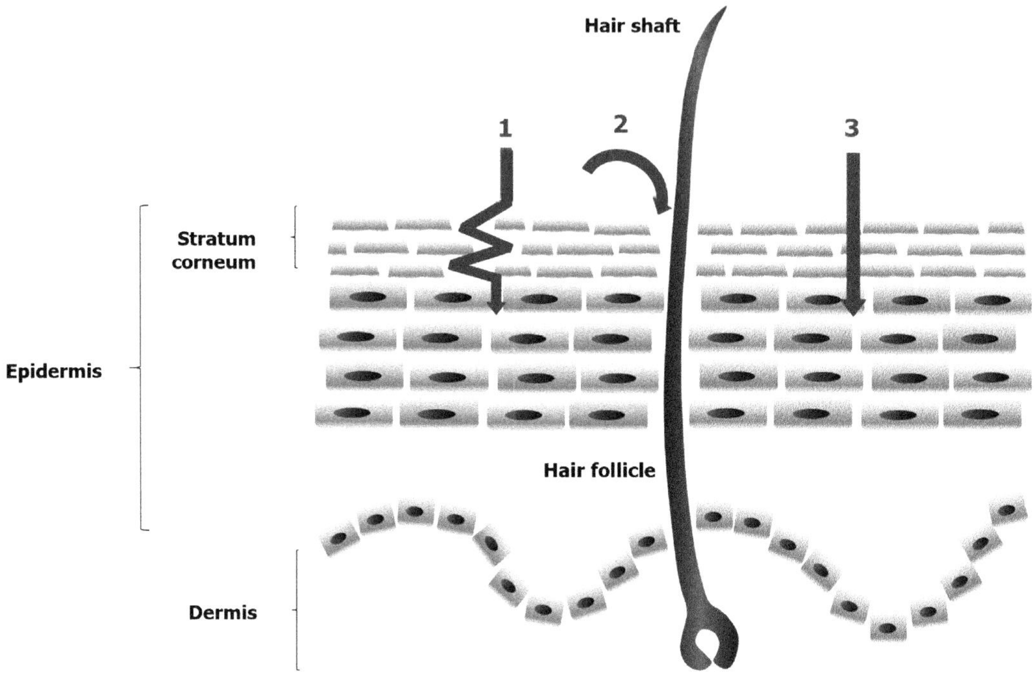

FIGURE 11.1 Penetration of active compounds on hair or scalp.

Hair products are extensively marketed, and several types of nanostructured systems have been developed in recent years. Therefore this chapter aims to present the different disorders and needs related to the hair structure and new tendencies based on the prospective nanosystems.

11.2 Hair structure

To better understand the mechanisms involved in hair application, we focused on the hair structure in two major segments: the hair shaft and hair follicles, which are the main targets of the nanostructures for the current and prospective hair treatments.

The schematic structure of the hair shaft and hair follicles is shown in Fig. 11.2.

11.2.1 Hair shaft

The hair shaft is the upper part of the hair structure, which emerges through the skin surface, and is composed of dead cells. The hair shaft is made up of three concentric regions, named from the external to the internal part: cuticle, cortex, and medulla [2].

The cuticle is the outermost region of the hair, which forms a resistant film that surrounds the cortex. The cuticle is organized as flat overlapping cells in a scale-like structure, and each cuticle cell contains an epicuticle, a thin outer membrane. Under the

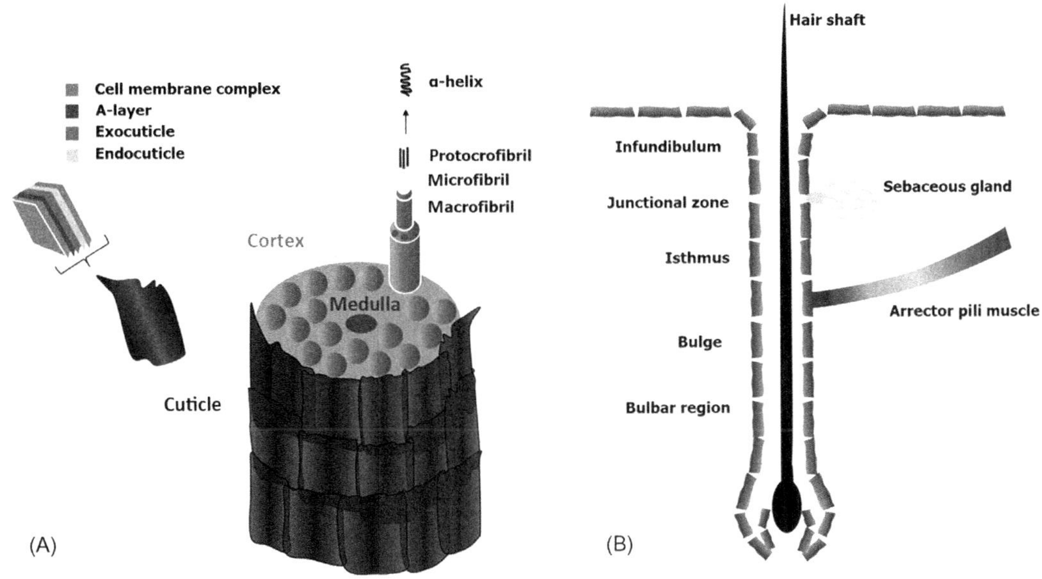

FIGURE 11.2 Schematic structure of the hair shaft (A) and hair follicles (B).

epicuticle, there are three membranes: the A layer, the exocuticle, and the endocuticle. Therein, the content of cystine decreases from the highest ~30% to the lowest ~3% [2]. The cuticle is responsible for protecting the hair from external damage, such as daily brushing, excess heat, or the use of chemical products (e.g., dyeing, relaxing, or straightening processes). The damage of the cuticle leads to loss of natural glow, which reflects the loss of the healthy hair structure, caused by these injuries [2,3].

The cortex comprises the majority of fiber mass of hair, giving mechanical strength to the hair. The cortex, located in the middle of the cuticle and medulla, is composed of cells and the cell membrane complex, also called intercellular binding material, a cement located between cell membranes [2]. Inside the cortex, there are pigment granules (melanin) responsible for the color of hair that undergo influences of the type, distribution, and number of melanin granules [4]. There are also twisted-bundle fibrous cells called macrofibrils or macrofilaments, corresponding to the major portion of the cortex. Each macrofibril contains several microfibrils distributed within a matrix. Microfibrils are highly organized subfilamentous structures, ordered in a spiral formation. Each microfibril is composed of helicoidal subfilamentous units called protofibrils, which are made up of several polypeptide chains of proteins called an alpha-helix [2]. The innermost part of the hair shaft is called the medulla. This porous region, which has a variable diameter, supports the hair structure, but its function remains unclear [3,4].

11.2.2 Hair follicles

Structured as an extended cavity of the epidermis that reaches the depth of the skin, follicles can act as an efficient reservoir of active compounds, compared to the stratum corneum [5]. The density and size of follicles depend on the sites. Impressively, about 10.0% of the total surface area of the skin constitutes follicular orifices of the face and scalp, while only 0.1% is found in other sites of the body [6]. This fact demonstrates the influence of the follicle on the absorption of active compounds when applied topically on those desired sites. The hair follicle is composed of infundibulum, a junctional zone/isthmus, and bulb, which consists of the upper, middle, and lower segments of the hair follicle, respectively [3].

Infundibulum is the most extensive interface between epithelium and the environment. It starts from the follicular orifice of the hair follicle extending to the sebaceous gland. It provides additional absorption of topically applied compounds, promoting their high accumulation [3,5,6]. The infundibulum also influences the hair follicle biology and pathology, harboring a rich residential microflora, and is implicated in the pathogenesis of several skin disorders [7].

One or more sebaceous glands are attached in the hair follicle. The junctional zone is located near the sebaceous gland duct, while the isthmus is located near the bottom of the sebaceous gland duct to the arrector pili muscle. The sebaceous glands are responsible for the production of sebum, a complex lipid mixture, found in the follicular duct [3,5,7].

Below the isthmus, there is a bulging region, where epithelial and melanocyte stem cells are found [3,7]. At the base of the hair follicles, there is a tulip bulb-like structure, which actively produces the hair shaft. This keratinization site determines the morphology of the hair. According to Thibaut et al. [8], the bulb is responsible for programming the hair shape. The asymmetry or symmetry in differentiation programs is responsible for determining the curliness or the straightness of the hair [8]. The bulge and bulb site also have critical roles in melanogenesis, which determines the pigmentation of the hair. The melanocytes are located in the bulb region and the melanocyte stem cells in the bulge–subbulge region. These regions function as a reservoir of melanocytes for hair pigmentation and maintenance, in each hair cycle [9].

A more detailed description of each mechanism of hair pigmentation, hair growth, and the disorder related to the sebum secretion is provided in the following sections.

11.3 Nanostructured systems for hair treatment

In recent years, nanostructured systems have attracted attention in the dermo and cosmetic fields. They may offer treatment or protection of the hair shaft from external damage, in addition to facilitating specific penetration of active compounds through the follicular pathway. It is worth noting that nanoparticles (NPs) of 100 nm are around 1000 times smaller in size than the average diameter of a human hair [10]. In principle, nanostructured systems have several advantages over those of larger particle sizes, as a result of the improved stability of the active compounds, efficient delivery of these compounds to the desired site, safety, and controlled release [11]. A considerable number of nanostructured systems have been developed in recent years, focusing on treating hair shaft and treatment through the hair follicles.

Starting with the hair shaft, the function of the hair does not only refer to the primary function, such as the regulation of body temperature. Healthy, shiny, and silky hair offers daily esthetic features that significantly affect the social and individual quality of life. The flexibility provided by the ability to modify hair, such as hair dyeing, straightening, or curling, may improve the well-being of an individual. However, they lead to increased damage to hair. Also, increased life expectancy and the aging of society may raise the following concerns: thinning hair, hair loss, dryness due to sebum production reduction, and gray hair. Consequently, efficient improvement of the hair shaft using nanostructured systems for an immediate or long-term effect is desirable.

Regarding the second point, targeting hair follicles with active compounds has delivered increasing insights and gained interest of researchers. The use of nanostructured systems offers better guidance in the desired site, increasing their retention and avoiding the side effects and irritations caused by the excessive introduction of these compounds [12]. Evaluation of porcine skin by in vitro testing using NPs with different mean particle sizes indicated that particles between 400 and 700 nm reached deeper sites in the hair follicle. They were compared to those with a mean size higher or lower than this range [13]. Although the correct mean size for the best transfollicular permeation or penetration is still under discussion [6], nanostructured systems are very promising, since they are generally smaller than the size of the follicular openings [14].

The results of in vitro penetration or deposition evaluation of the nanostructured systems might be carefully compared with in vivo performance through the transfollicular pathway. Reduced impact on the penetration and follicular reservoir capacity was observed in in vitro testing due to the contraction of the elastic fibers of the excised skin [15]. Another element to consider is hair movement. The hair follicle and this surface structure perform as a pumping system that conducts particles into the hair follicle, under the movement of hair. This movement, which improves the penetration of the particles, can be simulated by a massage appliance in the in vitro penetration test [16].

The hair follicle is responsible for sebum formation due to its attached sebaceous gland, producing a lipid-enriched environment from the follicular duct to the stratum corneum [3,5,7]. Different types of nanostructured systems, which are mentioned in the next sections, interact differently with sebum for subsequent transport into hair follicles [17]. The hair follicle acts as a reservoir of active compounds, showing the longest accumulation time compared to the stratum corneum. The shorter accumulation of active compounds at stratum corneum is due to the intense turnover of the uppermost cellular layer of the horny layer, where the stratum corneum is located. Moreover, the hair follicle has a deeper extension that reaches the dermal tissue, further enhancing its carrier role [16]. Lademann et al. [18] demonstrated that the reservoir of hair follicles could store particles 10 times longer than in the reservoir of the stratum corneum [18].

11.3.1 Types of nanostructured systems

Different performances are expected in the hair shaft or hair follicles, according to the vehicle, type, surface, or the mean particle size of the applied nanostructure. Several nanostructured types have been developed in the past and recent years, showing specific interactions according to their specificities. Examples of nanostructured systems for hair treatments are liposomes, cyclodextrins, dendrimers, polymeric NPs, metallic NPs, nanocrystals, solid lipid nanoparticles (SLNs) and nanostructured lipid carriers (NLCs), and nanoemulsions.

11.3.1.1 Liposomes

Liposomes are vesicular structures composed of phospholipid bilayers. These nanostructures improve the interaction with sebum and the penetration of the active compounds into the hair follicles. The active compounds can be located in the membrane or the inner part of the liposomes, according to their lipophilicity or hydrophilicity, respectively [19]. The high selectivity and safety of liposomes were successfully demonstrated in 1995 by entrapment of DNA in liposomes for penetration into the hair follicles. This gene therapy was performed by topical application in mice [20].

11.3.1.2 Cyclodextrins

Cyclodextrins are cyclic sugar compounds with a lipidic internal cavity and hydrophilic external surface. Lipophilic drugs are enabled to form nanoscale aggregates by their internalization in the central cavity of cyclodextrins. Using a fluorescence image microscopy, a study with the curcumin–cyclodextrin nano complex showed selective and efficient penetration of curcumin into deep sites of hair follicles [21].

11.3.1.3 Dendrimers

Dendrimers, also called dendritic polymers, are synthetic, tree-like macromolecules with flexibility to modify the architecture according to the desired functionality. They consist of a central core surrounded by repeated branches in the outer part. The unique property of dendrimers is not only their ability to encapsulate a drug in its core but also to conjugate with their surface groups. Different types of dendrimers have been designed since the 1970s, including hyperbranched polymers, perfect dendrimers, dendronized polymers, functional dendrons, dendrimer core shells, and core-multishell particles [22,23].

An example of a dendrimer for hair application is the dendritic core-multishell (CMS) particles, developed to load spin-labeled 5-doxyl stearic acid (5DSA). CMS is a dendritic polar polyglycerol polymer core hyperbranched with C18 diacide building blocks, at the inner shell, and hydrophilic polyethylene glycol (PEG), at the outer shell. 5DSA is a spin-labeled stearic acid, functioning as a hydrophobic spin probe, which enables evaluating the loading and transport capacity of particles into the skin under the use of electron paramagnetic resonance spectroscopy. Results showed the effective penetration of this CMS into hair follicles to a depth of 340 ± 82 μm [11].

Another interesting example is the pH-responsive-dendritic-polyglycerol-nanogel (dPG-NG), developed to evaluate the pH of the hair follicle at different depths, applying an ex vivo porcine-ear model. An inverse nanoprecipitation method was used to prepare this pH-labeled dPG-NG, with a mild and surfactant-free thiol-Michael reaction. Determination of pH within the hair follicle, through analysis of confocal laser scanning microscopy images, showed a pH increase from 6.5 to 7.4, from the skin surface to the deepest part of the hair follicle. It also showed that there is a sudden increase in pH in the first half of the penetration depth. This potential nontoxic dPG-NG may open new possibilities for future pH-triggered nanodrug delivery systems, aiming at the intrafollicular delivery of drugs [24].

11.3.1.4 Polymeric nanoparticles

Another NP that has gained attention especially for follicular delivery is polymeric NPs. These solid-spherical nanosized colloidal systems can be classified as nanospheres, nanocapsules, or nanogels. The first is formed by a polymeric matrix that entraps a drug. The second consists of a polymeric membrane surrounding an encapsulating drug cavity [25]. And the latter are hydrophilic or organic gelling agents, which form three-dimensional networks through physical or chemical cross-links [26].

Two different technologies are mainly used for preparing polymeric NPs: the top-down method and the bottom-up method. The top-down process applies preformed polymers dispersed in nanoscaled structures under several approaches, such as nanoprecipitation, solvent evaporation, emulsion diffusion, or interfacial deposition. The bottom-up method consists of polymerization of monomers, using different approaches, such as emulsion, microemulsion, interfacial polymerization, or molecular inclusion [25,27].

Different polymer materials are used to develop these polymeric NPs: the natural ones, and the synthetic ones, which are preferentially biodegradable or biocompatible. The natural polymers are of plant or animal origin, such as hyaluronic acids, collagen, alginate, chitosan, or dextran. The biodegradable materials are those eliminated or metabolized by the body, after being broken down into nontoxic substances. The biocompatible materials are nondegradable ones, but their final elements, after excretion by renal clearance, are considered safe in in vivo studies. Examples of biodegradable polymers are poly (ε-caprolactone) (PCL), for preparation of polymeric matrixes. Examples of biocompatible materials are the poly(lactic-co-glycolic acid) (PLGA), PEG, or methyl methacrylate, for preparation of hydrogels [26].

Many researchers have extensively studied the penetration and prolonged-release effect of polymer NPs in hair follicles. Lademann et al. [16] compared a 320-nm fluorescent polymeric NP with a nonparticulate form, under in vitro and in vivo topical applications. This polymeric NP, composed of polyvinyl alcohol and PLGA, presented deeper penetration than the nonparticulate form in the in vitro test, after simultaneous application of massage on porcine skin. It also showed a more extended storage period in the in vivo evaluation, by the differential stripping test on human skin (10 days and 4 days, for NP and nonparticulate form, respectively) [16]. Another author tested poly (D,L-lactide) NPs of approximately 150 nm and demonstrated their presence along the entire follicular duct [28].

Size influence studies on the penetration and long-term release of polymer NPs in hair follicles were also performed. Rancan et al. [27] compared fluorescent dye-loaded poly-lactic acid NPs with different particle sizes (228 and 365 nm) in human skin explants. Equal penetration and release profiles were observed for both particle sizes in the hair follicles, indicating that there is no size influence in this range [27]. Another study compared PLGA NPs with the same particle sizes (163–170 nm), however, with different surface modifications: the plain PLGA, chitosan-coated PLGA, and chitosan-coated PLGA with different phospholipids. The in vitro test indicated that negatively charged NPs and lipophilic-surface properties were more favorable for follicular uptake. This study showed the relevance of the surface property over the size in NP follicular distribution [29].

A comparison study between the dexamethasone nanocrystal (221 ± 4 nm) and a polymeric NP, the dexamethasone-loaded ethyl cellulose nanocarriers (147 ± 2 nm), indicated the influence of NP composition on follicle uptake. The former improved the dissolution rate of the drug, showing to be a good candidate for poor solubility drugs, with limited penetration. On the other hand, the latter allowed for prolonged release, and the particles remained for many days in the hair follicle, which is interesting for a drug with a high penetration profile such as dexamethasone [30].

A fluorescein isothiocyanate (FITC)-labeled bovine serum albumin (BSA) hydrogel nanocarrier was used to evaluate the gradient-dependent release of the drug tetramethylrhodamine-dextran (TRITC-dextran), by the in vitro test on porcine ear skin. NPs of approximately 663 nm were evaluated by a confocal laser scanning microscope, and TRITC-dextran was found in the infundibular region of porcine hair follicles [31].

11.3.1.5 Metallic nanoparticles

Metallic NPs are rigid particles that present several benefits, such as increased effectiveness of nanoorganic drugs. They can be made of different metallic materials (e.g., gold NPs and silver NPs) and metal oxide (e.g., iron oxide NPs and zinc oxide NPs). The complexation of the active compounds with the metal particles within this core, or by adsorption on this surface, increases the potential to increase binding to the specific target and reduces toxicity [5,18]. TiO_2NPs and AgNPs are used in shampoo formulations for various purposes, such as preventing hair loss, shine, or antifungal properties [10].

Mahmoud et al. [32] investigated the behavior of gold NPs, with different surface composition and charge, with the follicular compartments of the skin. Hydrophobic gold NPs exhibited higher affinity with hair follicles, providing preferential accumulation within the hair compartments. On the other hand, charged gold NPs presented a physical and chemical skin barrier, in which the penetration percentage was insignificant [32].

AuNP-doped BSA NPs were developed to deliver FITC by a heat-activated release mechanism. These metallic NPs were applied on porcine ear skin, and water-filtered infrared A radiation was used for induction and subsequent release of fluorescent dye. This inducing method for metallic NP enabled deeper penetration of the dye into the follicular duct, with a tolerable radiation level to the skin. The preparation method, without the addition of chemical cross-linkers, provided a more biocompatible nanocarrier with approximately 545 nm in diameter. In vitro testing detected an increased fluorescent signal in the hair follicles and the sebaceous glands for the treatment group, compared to the weak signal in the control group. The strong fluorescent signal is evidence of successful dye release at the targeted site. These findings encourage future development of heat-induced drug-loaded metallic nanocapsules for potential delivery in hair follicles [33].

11.3.1.6 Nanocrystals

Nanocrystals are nanoscale-sized pure drug particles without any matrix material. They are stabilized with adsorption of low quantity of surfactants (e.g., ionic and nonionic surfactants) or polymers (e.g., polyvinyl alcohol, hydroxypropyl methylcellulose, and polyvinyl pyrrolidone/povidone) at their interface, which prevents particle aggregation and allows their suspension in an external liquid phase. Their high drug loading capacity, high dissolution rate, and high saturation solubility make it possible to produce NPs with good stability, delivery, and safety. Nanocrystals can be prepared by two different processes: bottom-up and top-down processes [6,34]. The first is the classic precipitation method, where a solvent-dissolved drug is added to another nonsolvent phase, and the drug precipitation occurs to yield a nanocrystal. The limitation of this process is the need to remove this solvent, which may have stability, safety, or cost issues. The second approach is the reduction of a micrometer drug crystal to a nanometric drug crystal by the use of mechanical processes. These processes include high-pressure homogenization, microfluidization, or media milling, in which high-shear, cavitation, or pressure is applied to convert the drug to a nanometer size [6].

Several experiments have evaluated the affinity of nanocrystals with hair follicles. Corrias et al. [14] prepared a nanocrystal of water-insoluble dye, Nile Red, in combination with Polysorbate 80 or Poloxamer 188 as a stabilizer. The 230-nm nanocrystal was made using the media milling technique. The in vitro skin permeation studies have demonstrated the accumulation of Nile Red nanocrystal in the follicular duct [14]. Another hair follicle location study was performed with curcumin nanocrystal prepared by the smartCrystals process. This process involves a combination of the bead milling technique followed by high-pressure homogenization, yielding a nanocrystal of 200 nm. The alkyl polyglycoside surfactant was used as a stabilizer. An in vitro porcine skin test using fluorescent microscopy was performed by comparing curcumin nanosuspensions of different concentrations, and either the curcumin nanosuspension mixed with high viscosity hydroxypropyl cellulose gel. The results identified curcumin deposition in hair follicles when the concentration of nanosuspension was up to 0.02% by weight. There was no significant influence of the viscosity on the hair follicle deposition, indicating the use of nanocrystals as a promising carrier for water-insoluble compounds for hair follicles [34].

11.3.1.7 Solid lipid nanoparticles and nanostructured lipid carriers

SLNs were introduced in the 1990s as a lipidic NP consisting of solid lipids in its core. Later, Müller et al. introduced NLCs to provide better stability to the encapsulated drug compared to SLNs. This is due to the NLCs core composition, which is comprised of a mixture of solid and liquid lipids with a less organized structure, preventing expulsion of the encapsulated drug from the rigid matrix of SLNs [35]. Both lipid nanostructures are present in a solid state at room temperature. Examples of the solid lipids are fatty acids (e.g., palmitic acid), mono, di, and triglycerides (e.g., glyceryl monostearate, glyceryl behenate, and tristearin), cholesterol, and waxes [36]. Although new components have been emerging for SLN and NLC preparations in recent years, the advantage of these carriers is usually the nature of their lipids, which are mainly biodegradable or biocompatible. A systematic review of SLNs and NLCs was done by Doktorovová et al. [37], with providing data on the safety of these lipid NPs. The authors reported that the in vitro test indicated acceptable SLN/NLC total lipids concentrations up to 1 mg/ml; lower tolerance of mean particle size up to 500 nm in cell culture; greater biocompatibility of lipid NPs stabilized by a single surfactant rather than combinations; and good SLN/NLC skin tolerability by both in vitro and in vivo dermal tests [37].

The efficacy of SLN along the hair follicle was demonstrated by the development of flutamide-loaded SLNs, for treating androgenic alopecia (AGA). Under topical application, this 198 nm SLN showed higher drug deposition in the skin compared to flutamide hydroalcoholic solution and higher growth of new hair follicle in the in vitro evaluation. This study indicated more significant accumulation of SLN in the hair follicles and the slow and controlled release of flutamide along the hair follicles [38].

Podophyllotoxin (POD)-loaded SLNs were prepared by the high-pressure homogenization method. Poloxamer 188 and soybean lecithin were used as stabilizers, yielding SLN with a particle size of 73.4 nm and 48.36 mV zeta potential value. After application of POD-loaded SLN to porcine skin, fluorescence microscopy image indicated POD penetration by two pathways: the stratum corneum and the hair follicle route [39].

Several lipid nanocarriers have been developed for treating alopecia, aiming at safe and effective delivery of drugs to hair follicles. This issue will be discussed in great detail in the next alopecia treatment section.

11.3.1.8 Nanoemulsions

Nanoemulsions are composed of a mixture of the liquid oil phase, surfactants, and water phase. They are classified as oil-in-water (O/W) or water-in-oil (W/O) nanoemulsions, depending on the type and quantity of their components, mainly influenced by the surfactant type. The former consists of the internal oil core surrounded by a hydrophilic surfactant, which is dispersed in the outer water phase. The latter consists of the inner water core surrounded by a hydrophobic surfactant, where this micelle is dispersed in the external oil phase. Most dermal products on the market are O/W emulsions, which provide a lighter touch than W/O emulsions [40].

Nanoemulsions differ from microemulsions since the latter is a thermodynamically stable isotropic phase [41]. Nanoemulsions, being thermodynamically unstable, are not formed spontaneously; therefore some external energy input is required for their formation. There are two different processes for preparing nanoemulsions: high-energy and low-energy ones. The former, the mechanical process, is based on high shear, pressure, or cavitation phenomena, yielding emulsion on the nanosized scale. As a high-energy process, we highlight the high-pressure homogenization, ultrasonication, microfluidizers, and membrane emulsification. The latter, the physicochemical process, is based on lowering the surface interfacial tension during preparation, by the correct balance of the phases. Examples of low-energy processes are spontaneous emulsification, phase inversion temperature (PIT), phase inversion composition (PIC), self-nanoemulsifying method [42], and D-phase emulsification methods [43,44].

Some advantages of nanoemulsion over other nanostructures have been reported. A comparative study of nanosized formulation composed of liquid-lipid state core (such as nanoemulsions) and solid-lipid state core (such as SLNs) was done using testosterone (TP) as a hydrophobic drug. The results showed higher solubility of TP in the liquid-lipid state nanosized formulation than the solid-lipid state, illustrating nanoemulsions as a promising delivery system for poorly soluble components [45]. Nanoemulsions were also cited as more efficient transdermal carriers compared to rigid NPs due to their flexible and fluid microstructure [4].

In another study, improvement of transfollicular delivery of various caffeine-loaded nanoemulsions was investigated by the in vitro permeation test. Skin permeation enhancers such as oleic acid and eucalyptol were part of the composition of these nanoemulsions. Blocked and open follicle models were used: the former model was designed to evaluate transepidermal or transfollicular caffeine permeation; the latter model aimed to evaluate the permeation by transfollicular permeation pathway. The cyanoacrylate skin surface biopsy method was further used to analyze the follicular contents directly. These combined methods demonstrated that nanoemulsions were able to deliver the hydrophilic drug caffeine into and through hair follicles [46].

There is also a study showing the influence of nanoemulsion particle size on different routes and locations in hair follicles. P4-dye-loaded nanoemulsions were used for this proposal and indicated that nanoemulsion particle size of around 80 nm might diffuse along with the viable epidermis and fill the entire hair follicles, while nanoemulsions of around 500-nm particle size have limited penetration into the stratum corneum and easily migrate along the hair follicle [4].

As described, an increasing number of nanostructured systems were developed to address hair treatment. Different advantages or disadvantages were presented for each structure, and the growing number of discoveries shows their potential, targeting hair shafts and hair follicle treatments. It is essential to consider that the regular or daily application of cosmetic products by users can facilitate the retention of the active compounds loaded in the nanostructured systems, in the desired location. Some of the specific uses of nanostructured systems are presented in the next sections.

11.3.2 Hair treatment

11.3.2.1 Hair damage

Hair damage can be caused by different factors: intrinsic, such as genetic origin, and extrinsic, from various external sources. The visible damage of hair shafts may be the first concern of general hair care users, as this is the direct and immediate result of various harmful experiences caused by routine life. Such damage may be induced by physical stress (e.g., daily washing and hair brushing), chemical stress (e.g., hair dyeing and straightening), heat stress (curling iron), or photoaging (ultraviolet light and visible light) [15,47].

The growth of new technologies to easily change hair shape or color has resulted in a number of modern hairstyles that express personality and style. It has also improved understanding of the complexity of distinct hair damage types, as a result of combinations of different technologies to alter hair morphologies. Kim et al. [48] classified the degree of extrinsically damaged hair. The authors used a morphological evaluation based on the use of electron microscopy. They proposed a standard three-stage evaluation grading system to categorize the status of the distinct parts of the hair shaft: the surface, inner layers of the cuticle, and the cortex. The different cuticle damage levels were graded on a scale of 1–4; observation by a scanning electron microscope (SEM) and a transmission electron microscope was graded as "S" and "T," respectively, and the cuticle and cortex by "cu" and "co," respectively. Accordingly, hair damaged by dyeing and permanent wave, for example, indicated irregular cuticle overlap, the presence of bubbles in the cortex, and this damage was classified as Scu1/Tcu2/Tco2. Although this classification grade is not enough to evaluate the entire degree of damage hair, it is a starting point to simplify and schematize assessment and repair degree of extrinsically damaged hair [48].

Developments of nanotechnology-based hair care products have gained attention in recent years, with an increasing number of perspectives and varieties of nanostructured types.

The low solubility of plant extracts can be a challenge to developing rinse-off hair care products. A cationic nanoemulsion of natural quercetin extract of 20 nm was developed by the sub-PIT method. This nanoemulsion indicated, by atomic force microscopy evaluation, high adhesion to the external surface of the hair. The incorporation of this low-solubility extract into a nanoemulsion may offer a treatment product with high antioxidant and conditioning properties for hair application [49].

Another challenge is silicone oil, widely used in cosmetic preparations for lubrication and preservation of hair moisture. This oil has high affinity and deposition on the hair scalp rather than the hair shaft, which can generate build-up and greasy touch after routine application to the hair. The PIC method and nonionic surfactants (Span 80 and Tween 80) were used to prepare silicone oil nanoemulsions with 100–705 nm particle sizes. An X-ray analysis system using an SEM indicated an increased deposition of silicone of nanoemulsion on the surface of the hair shaft compared to the control [50].

Patented products aiming at treating damaged hair have focused on nanostructured systems in recent years. A concentrated nanoemulsion of particle size from 1 to 100 nm using a mechanical dispenser was designed, for easy application as a rinse-off hair care foam [51]. These concentrated nanoemulsions were further prepared using silicone and anionic or cationic surfactant, with the same mechanical dispenser concept. These applications in foam format offered an improved conditioning effect for the hair [52–57].

Epoxy silicone nanoemulsions have been developed to treat damaged hair, mainly caused by chemical processes including straightening, relaxation, dyeing, and lightening. These nanoemulsions, prepared by a microfluidization method, resulted in nanoemulsions with an average particle size of 100–250 nm. The treatment processes, applying these epoxy silicone nanoemulsions, provide greater strength, elasticity, and fatigue resistance upon repeated brushing compared to the control [58]. Another preparation of nanoemulsion for hair conditioning was the use of amino silicone, providing mean particle sizes from 20 to 350 nm and a transparent to the slightly milky visual appearance [59].

Different active compounds, such as sericin, were included in a cationic NP. This NP was prepared by the sericin protein reticulation with quaternized guar gum, followed by cationic surfactant coating. This sericin cationic NP provided gloss, softness, aiming at treating and protecting damaged hair, such as dyed hair [60].

11.3.2.2 Hair graying

Hair graying may be a personal concern that can affect one's social life and self-esteem. Although gray hair has been increasingly accepted in modern society, there has been growing interest in better understanding hair graying factors and alternative solutions for the current dyeing process.

The natural pigmentation process of hair starts in the hair bulb, where melanocyte cells synthesize melanosomes, organelles responsible for synthesis, storage, and transport of melanin. This process of melanin synthesis, induced by several enzymatic or chemical reactions, is called melanogenesis [9].

Managing the hair pigmentation process is quite complex, since melanogenesis takes place only during the active growth phase of the hair cycle, which is called the anagen phase. The duration of hair cycles and the number of cycles throughout life differ from hair to hair in the same individual and from individual to individual. There are also various enzymes involved in melanogenesis, which are activated only after their entry into the melanosome, where they initiate the reaction cascade for melanin formation [61].

Hair color is determined by the amount of melanin type in individuals. Black and brown hair color is due to the high presence of eumelanin, a highly polymerized melanin. Yellow to reddish-brown hair color is the result of the pheomelanin, a less polymerized melanin [12].

Gray hair occurs due to melanogenesis deficiency, from several causes that are not yet completely understood. Possibilities include apoptosis of differentiated follicular melanocytes, discontinuation of melanin production [12], and progressive reduction of melanocyte stem cells in hair follicles [9].

The excessive presence of reactive oxygen species in hair follicles, as a consequence of age-related oxidative stress, is considered to be one of the main triggers of hair graying. The natural defense mechanism of oxidative damage, which is accomplished by the expression of antioxidant proteins (e.g., BCL-2 and TRP-2), is weakened, reducing melanocyte stem cells and repopulation of newly formed anagen follicles. This process results in a gradual depletion of melanocyte in the bulb, and, consequently, the reduced number of melanin content up to the hair shaft [62].

Aiming at protecting melanogenesis, an extract of *Pueraria lobata*, an Asian native plant with antioxidant effects, was used to prevent hair graying. A randomized, double-blind clinical trial in 44 women was performed for 24 weeks. As a result, a slight reduction in new gray hair growth was observed in the test sample compared to placebo. This trial was based on the idea that proper removal of reactive oxygen species may protect the expression of a master transcriptional regulator of melanocytes—the microphthalmia transcription factor—by the use of an effective antioxidant, preventing hair graying [63].

Different antioxidants may be able to influence melanin synthesis, such as vitamin C and E, flavonoids, and phenolic compounds. However, due to several factors (e.g., alternative substrate pathways for tyrosinase inhibition and reducers of melanin synthesis intermediates), they have been limited in achieving satisfactory results [61].

The need for new alternative dye products is due to consumer dissatisfaction (e.g., color fading) with conventional synthetic dyes (e.g., oxidative hair dyes), which are most commonly used, or natural dyes such as dioxyindole and derivatives. Furthermore, synthetic dyes can have a risk of hair damage and the possibility of skin sensitivity [61].

In the last decade, melanin NPs have been highlighted as an attractive material for clinical fields. They have been indicated for radiation therapies aimed at protecting bone marrow for higher dose application [64]. Other research considered them as a pH-responsive nanocarrier, being biocompatible and lacking cytotoxicity in delivery to the colon and intestine [65]. Besides, melanin NPs doped with self-assembled hydrogel networks have been developed as a new potential light-responsive polymeric biomaterial for clinical use [66].

Since active melanocytes for melanin synthesis still remain in gray hair (even in small numbers), unlike white hair [67], boosting this natural coloring of hair through an inside-out approach seems promising. Therefore it has drawn attention to the possibility of a new class of hair dyes with melanin precursors or the use of already identified genes for melanin synthesis [12,61].

In vivo and in vitro human hair follicle assays have identified increased melanin content by silencing the core clock genes, BMAL1 or PER1. The small molecule modulator of these genes, the small interfering RNA, has stimulated human melanogenesis and melanocyte activity, opening new perspectives for future therapeutic strategies for human pigmentary disorders [68].

In addition to the activation of natural hair color, hair color change was also highlighted by the possible use of small interfering RNA capable of interfering with KITLG gene transcription and hair color as a consequence. Also, targeted inhibition of tyrosinase-related protein 1 (Tyrp1) in hair follicles may support cosmetic products capable of inducing pheomelanin production, a non-Tyrp1–dependent pathway, alternatively to the Tyrp1-dependent eumelanin pathway, offering a lighter hair shade [12].

Nevertheless, concerns remain since a deeper understanding of the multiple interferences of biological networks between different genes is required to ensure the safety of future innovative products for cosmetic use.

11.3.2.3 Alopecia

The hair shaft is made up of dead cells, although it originates from a constant supply of materials that continually maintain the hair cycle during their lifetime. This cycle occurs in the hair follicle, which is divided into three main phases: anagen, catagen, and telogen. The anagen phase is the most active stage: matrix cell proliferation and differentiation occur at the follicle base and constant production of hair cells move cells from root to tip. Cell differentiation generates cysteine-rich hair keratins, responsible for the high strength and flexibility of the hair shaft. The duration of the anagen phase determines the hair length. The transition from anagen to catagen phase is promoted by some molecular regulators, such as growth factors (e.g., FGF5 and EGF), neurotrophins (e.g., BDNF), p53, but interactions between them at this transition point are not yet known. During the catagen phase, apoptosis of epithelial cells occurs in the bulb and outermost epithelial layer, discontinuing hair shaft differentiation. The lower part of the hair shaft converts to a rounded structure called the hair club, and the lower follicle retracts, forming a temporary structure called the epithelial strand. Then, in the telogen phase, the hair follicles enter the dormant period. The transition from telogen to anagen phase occurs under the activation of stem cells, located in the bulge, initiating a new cycle of hair differentiation and proliferation [69].

In a normal healthy scalp, anagen is the predominant phase (approximately 80%–85% of hairs). But in alopecia, there is a transition of the dominant period, from the anagen to the telogen phase [2].

Alopecia, or scalp hair loss, is a significant concern of people affected by this disorder. Its origin differs among the different types of alopecia: AGA, alopecia areata (AA), and chemotherapy-induced alopecia (CIA). AGA, the predominant alopecia in Caucasian males, presents balding at the vertex and hairline recession at the templates of the affected population [70]. In AGA, the anagen phase is shortened, leading to hair follicle miniaturization and then to hair loss [5]. The main cause of this anomaly is a decrease in total scalp TP and an increase in androgen hormone dihydrotestosterone (DHT). TP is converted to DHT by the enzymes 5-α reductase (5 α-R). Among them, specifically, isoenzyme 5 α-R1, which is presented at a high level in the scalp of men with AGA, seems to be the leading cause of hair loss [71]. In contrast, AA affects approximately 2% of the general population, which includes children and female adults [72]. AA is considered an autoimmune disease, which may be induced by inflammation caused by certain host proteins acting as autoantigens. CIA is caused by dividing cells apoptosis associated with a side effect of antineoplastic chemotherapy [70].

For treating hair loss by AGA, minoxidil, an antihypertensive vasodilator drug, and finasteride, an antiandrogen with inhibition activity of 5-α reductase, are the only two drugs approved by the US Food and Drug Administration (FDA). The concern about efficiency and safety, due to unpredictable side effects, is the main limitation and barrier for the approval of new drugs for hair loss therapies [73]. However, the use of currently approved antiandrogen can lead to side effects (e.g., sexual dysfunction and feminization), which may result in a deadlock between treating hair loss and avoiding these side effects [70]. Topically applied corticosteroids for AA treatment also have several side effects, such as skin atrophy or irritation, and return to the same stage of hair loss before treatment, when the drug application is discontinued [71,72].

Considering the hair follicle as the primary target in alopecia treatment, the interaction of the active ingredient as well as its vehicle with the hair follicle is the key element for the successful transport of the active ingredient into the follicle. Nanosystems have been favored for topical applications due to their potential to easily penetrate and promote accumulation in hair follicles. Therefore the development of a more effective and safe therapy using them as a carrier may be squarticles, lipid nanocarriers composed of fatty esters and squalene as their matrix. Squalene is a sebum-derived lipid and, due to the high concentration of sebum in the hair follicle, can improve the affinity and accumulation of squarticles in the hair follicle. Squarticles loaded with minoxidil and diphencyprone or conjugated with antiplatelet-derived growth factor-receptor β antibody presented higher follicular accumulation compared to the free control solution. Designed to target dermal papilla cells (DPCs), which are located in the deeper part of the hair follicle and responsible for hair growth and regeneration, squarticles also improved vascular endothelial growth factor expression and proliferation of DPCs [74,75].

Other minoxidil-loaded nanocarriers were developed. NLCs consisting of stearic acid and oleic acid as lipids showed a faster drug release profile compared to minoxidil loaded in the solid-lipid nanostructure, and no erythema was observed in this NLC [76]. Another preparation was the minoxidil encapsulated in 2-hydroxypropyl-β-cyclodextrin which promoted aqueous solubility of the drug to prevent the use of organic solvent, and showed an effect on the hair cycle of mice by in vivo evaluation [77].

Alternative active compounds such as dutasteride were loaded in the NLC coated with chitosan oligomer-stearic acid. This drug is known for treating benign prostate hyperplasia, acting as an inhibitor of type I and type II 5α-reductase. As a result, this nanocarrier showed good stability, slow release over 12 hours and low cytotoxicity, appearing as a potential promoter of hair growth [36]. Natural origin active compounds with 5α-reductase inhibitory effects were either loaded onto nanosystems: hinokitiol and glycyrrhetinic acid were delivered to the hair follicles by incorporation into the PLGA nanosphere carriers [78], PCL nanocapsules were used as a carrier of hinokitiol (H6), and extract of *Platycladus orientalis* leaves was encapsulated in liposomes [79]. These are promising candidates for hair growth therapies [71].

11.3.2.4 *Antidandruff*

Dandruff is a scalp disorder induced by several factors that can lead to social discomfort for the patient. Scaling and redness of the scalp are symptoms, associated with age (usually from puberty), season-related (generally in winter), occur in the same range between men and women, and are caused by different factors. The imbalance between sebum production, lipophilic yeast flora on the scalp, and individual susceptibility are listed as the source of this disorder [80]. Among them, Malassezia yeast (formerly *Pityrosporum ovale*) is highlighted as an essential element, although some studies point to the presence of other microorganisms (e.g., *Propionibacterium* spp. and *Staphylococcus* spp.) as an active contributor to dandruff [81].

The dandruff scalp has a thinner stratum corneum, and the epidermal turnover rate is higher than the healthy scalp [2]. Increased sebum production produces free fatty acids, wax esters, and fatty glycerides, which are the primary lipid substrates for the scalp microorganism. Malassezia secretes lipase, an enzyme that catalyzes lipids, releasing free fatty acids, which increase inflammation. Sebum production in the scalp is increased as microorganisms consume sebum, more free fatty acids are produced, inducing more inflammation in the scalp [81]. Imbalance of sebum production, microorganism growth, and scalp irritation seem to be correlated; therefore dandruff is not correlated with poor hygiene. Commonly used ingredients for antidandruff therapy include antifungal agents (such as zinc pyrithione and coal tar preparations) and antiinflammatory agents (e.g., salicylic acid) [2,81].

The development of nanosystems is proving to be a promising new approach to dandruff treatment. Pant et al. [82] developed silver NPs using different preparation conditions (sunlight, microwave, and room temperature). The 15–20 nm silver NP obtained by sunlight provided high efficacy against various bacterial strains tested (Gram-negative and Gram-negative), fungal strains (*Candida albicans* and *Candida parapsilosis*), and the shampoo containing NPs increased antidandruff effectiveness against *Pityrosporum ovale* and *Pityrosporum folliculitis* [82].

SLNs containing garlic as an antifungal agent were prepared by the hot homogenization method and mixed with shampoo. Tween 80 and soy lecithin were used as surfactants, and the obtained garlic loaded SLN showed up to 90% in vitro drug release profile. The shampoo containing this SLN resulted in excellent appearance, foamability, viscosity, and spreadability [80], showing to be a good alternative to zinc pyrithione, for which it is often difficult to improve the opaque and unattractive appearance when incorporated in shampoo.

Following a homeopathic approach, Kalpana et al. [83] biosynthesized ZnO NPs using *Lagenaria siceraria* pulp extract. ZnO is widely known as an antifungal, antibacterial, and UV filtering agent. *L. siceraria* (Cucurbitaceae) is a well-known plant in India for treating anasarca, ascites, beriberi, and other diseases. This plant extract was identified to participate in the bioreduction process for the conversion of metal ions into metal and metal oxide NPs. The method of biosynthesis of ZnO NPs from *L. siceraria* extract was very simple, and the in vitro test provided potent antidandruff and antimicrobial activities [83].

11.4 Future perspectives

Hair structure involves complex and dynamic mechanisms influenced by several internal and external factors. Interference in the normal dynamic flow of hair growth and maintenance can lead to disorders of the hair shaft or hair follicles. Therefore understanding the specific needs of each disorder is essential to directing consumers to the right treatment.

Studies have shown the efficacy of nanostructured systems in improving active compound penetration and accumulation and their modified release profile. Several factors can influence their final efficacy, such as the particle size, composition, and type of nanostructured systems.

A wide range of nanostructured systems has been developed in recent years, aiming at treating these specific needs. Nevertheless, considering the cosmetics field, careful monitoring of the active compounds to avoid systemic absorption, as well as the right selection of the nanosystems to deliver active compounds to the desired site, remains standard practice. Overcoming these challenges will open possibilities for innovative nanostructured systems, designed for a new level of treatment in the hair field.

Acknowledgments

The authors thank James Joseph Hesson for the English language editing service of this manuscript. The authors thank FAPESP (Fundacao de Amparo a Pesquisa do Estado de Sao Paulo, FAPESP number 2017/08332-3 and 2018/22713-2) for supporting the present work.

References

[1] Mihranyan A, Ferraz N, Strømme M. Current status and future prospects of nanotechnology in cosmetics. Prog Mater Sci 2012;57:875–910. Available from: https://doi.org/10.1016/j.pmatsci.2011.10.001.

[2] Robins CR. Chemical and physical behavior of human hair. 2nd ed. Springer; 1988. <https://doi.org/10.1007/978-1-4757-2009-9>.

[3] Rosen J, Landriscina A, Friedman AJ. Nanotechnology-based cosmetics for hair care. Cosmetics 2015;2:211–24. Available from: https://doi.org/10.3390/cosmetics2030211.

[4] Su R, Fan W, Yu Q, Dong X, Qi J, Zhu Q. Size-dependent penetration of nanoemulsions into epidermis and hair follicles: implications for transdermal delivery and immunization. Oncotarget 2017;8:38214–26.

[5] Fang C, Aljuffali IA, Li Y, Fang J. Delivery and targeting of nanoparticles into hair follicles. Ther Deliv 2014;5:991–1006. Available from: https://doi.org/10.4155/tde.14.61.

[6] Patel V, Sharma OP, Mehta T. Nanocrystal: a novel approach to overcome skin barriers for improved topical drug delivery. Expert Opin Drug Deliv 2018;15:351–68. Available from: https://doi.org/10.1080/17425247.2018.1444025.

[7] Schneider MR, Paus R. Deciphering the functions of the hair follicle infundibulum in skin physiology and disease. Cell Tissue Res 2014;358:697–704. Available from: https://doi.org/10.1007/s00441-014-1999-1.

[8] Thibaut S, Gaillard O, Bouhanna P, Cannell DW, Bernard BA. Human hair shape is programmed from the bulb. Br J Dermatol 2005;152:632–8. Available from: https://doi.org/10.1111/j.1365-2133.2005.06521.x.

[9] Nishimura EK. Pigment cell & melanoma pigmentation. Pigment Cell Melanoma Res. 2011;24:401–10. Available from: https://doi.org/10.1111/j.1755-148X.2011.00855.x.

[10] Niska K, Zielinska E, Witold M, Inkielewicz-stepniak I. Metal nanoparticles in dermatology and cosmetology: interactions with human skin cells. Chem Biol Interact 2018;295:38–51. Available from: https://doi.org/10.1016/j.cbi.2017.06.018.

[11] Lohan SB, Icken N, Teutloff C, Saeidpour S, Bittl R, Lademann J, et al. Investigation of cutaneous penetration properties of stearic acid loaded to dendritic core-multi-shell (CMS) nanocarriers. Int J Pharm 2016;501:271–7. Available from: https://doi.org/10.1016/j.ijpharm.2016.02.004.

[12] Matamá T, Gomes AC, Cavaco-paulo A. Hair coloration by gene regulation: fact or fiction? Trends Biotechnol 2015;33:707–11. Available from: https://doi.org/10.1016/j.tibtech.2015.10.001.

[13] Patzelt A, Richter H, Knorr F, Schäfer U, Lehr C, Dähne L. Selective follicular targeting by modification of the particle sizes. J Control Release 2011;150:45–8. Available from: https://doi.org/10.1016/j.jconrel.2010.11.015.

[14] Corrias F, Schlich M, Sinico C, Pireddu R, Valenti D, Maria A, et al. Nile red nanosuspensions as investigative model to study the follicular targeting of drug nanocrystals. Int J Pharm 2017;524:1–8. Available from: https://doi.org/10.1016/j.ijpharm.2017.03.042.

[15] Poljšak B, Dahmane RG, Godic A. Intrinsic skin aging: the role of oxidative stress. Acta Dermatovenerol APA 2012;21:1–4. Available from: https://doi.org/10.2478/v10162-012-0012-5.

[16] Lademann J, Richter H, Teichmann A, Otberg N, Blume-Peytavi U, Luengo J, et al. Nanoparticles—an efficient carrier for drug delivery into the hair follicles. Eur J Pharm Biopharm 2007;66:159–64. Available from: https://doi.org/10.1016/j.ejpb.2006.10.019.

[17] Rancan F, Afraz Z, Combadiere B, Blume-peytavi U. Hair follicle targeting with nanoparticles. In: Nasir A, Friedman A, Wang S, editors. Nanotechnology in dermatology. Springer; 2013. Available from: https://doi.org/10.1007/978-1-4614-5034-4

[18] Lademann J, Richter H, Schaefer UF, Blume-Peytavi U. Hair follicles—a long-term reservoir. Skin Pharmacol Physiol 2006;19:232–6. Available from: https://doi.org/10.1159/000093119.

[19] Betz G, Imboden R, Imanidis G. Interaction of liposome formulations with human skin in vitro. Int J Pharm 2001;229:117–29. Available from: https://doi.org/10.1016/S0378-5173(01)00824-9.

[20] Li L, Hoffman RM. The feasibility of targeted selective gene therapy of the hair follicle. Nat Med 1995;1:705–6.

[21] Konrádsdóttir F, Ogmundsdóttir H, Sigurdsson V, Loftsson T. Drug targeting to the hair follicles: a cyclodextrin-based drug delivery. AAPS PharmSciTech 2009;10:266–9. Available from: https://doi.org/10.1208/s12249-009-9205-6.

[22] Sun M, Fan A, Wang Z, Zhao Y. Dendrimer-mediated drug delivery to the skin. Soft Matter 2012;8:4301–5. Available from: https://doi.org/10.1039/c2sm07280g.

[23] Zoschke C, Schilrreff P, Romero EL, Brandner JM, Schäfer M. Dendritic nanoparticles for cutaneous drug delivery—testing in human skin and reconstructed human skin. Curr Pharm Des 2015;21:2784–800.

[24] Dimde M, Sahle FF, Wycisk V, Steinhilber D, Camacho LC, Licha K, et al. Synthesis and validation of functional nanogels as pH-sensors in the hair follicle. Macromol Biosci 2017;17:1–8. Available from: https://doi.org/10.1002/mabi.201600505.

[25] Krishnaswamy K, Orsat V. Chapter 2: Sustainable delivery systems through green nanotechnology, In: Nano- and Microscale Drug Delivery Systems Design and Fabrication, Edited by Alexandru Mihai Grumezescu. Elsevier 2017;17–32. Available from: https://doi.org/10.1016/B978-0-323-52727-9.00002-9.

[26] Rancan F, Blume-Peytavi U, Vogt A. Utilization of biodegradable polymeric materials as delivery agents in dermatology. Clin Cosmet Investig Dermatol 2014;7:23–34. Available from: https://doi.org/10.2147/CCID.S39559.

[27] Rancan F, Papakostas D, Hadam S, Hackbarth S, Delair T, Primard C, et al. Investigation of polylactic acid (PLA) nanoparticles as drug delivery systems for local dermatotherapy. Pharm Res 2009;26. Available from: https://doi.org/10.1007/s11095-009-9919-x.

[28] Fernandes B, Silva R, Ribeiro A, Matam T. Improved poly (D, L-lactide) nanoparticles-based formulation for hair follicle targeting. Int J Cosmet Sci 2015;37:282–90. Available from: https://doi.org/10.1111/ics.12197.

[29] Raber AS, Mittal A, Schäfer J, Bakowsky U, Reichrath J, Vogt T, et al. Quantification of nanoparticle uptake into hair follicles in pig ear and human forearm. J Control Release 2014;179:25–32. Available from: https://doi.org/10.1016/j.jconrel.2014.01.018.

[30] Döge N, Hönzke S, Schumacher F, Balzus B, Colombo M, Hadam S, et al. Ethyl cellulose nanocarriers and nanocrystals differentially deliver dexamethasone into intact, tape-stripped or sodium lauryl sulfate-exposed ex vivo human skin—assessment by intradermal microdialysis and extraction from the different skin layers. J Control Release 2016;242:25–34. Available from: https://doi.org/10.1016/j.jconrel.2016.07.009.

[31] Tran NBNN, Knorr F, Mak WC, Cheung KY, Richter H, Meinke M, et al. Gradient-dependent release of the model drug TRITC-dextran from FITC-labeled BSA hydrogel nanocarriers in the hair follicles of porcine ear skin. Eur J Pharm Biopharm 2017;116:12–16. Available from: https://doi.org/10.1016/j.ejpb.2016.09.016.

[32] Mahmoud NN, Alkilany AM, Dietrich D, Karst U, Al-bakri AG, Khalil EA. Preferential accumulation of gold nanorods into human skin hair follicles: effect of nanoparticle surface chemistry. J Colloid Interface Sci 2017;503:95–102. Available from: https://doi.org/10.1016/j.jcis.2017.05.011.

[33] Lademann J, Richter H, Knorr F, Patzelt A, Darvin ME, Rühl E, et al. Triggered release of model drug from AuNP-doped BSA nanocarriers in hair follicles using IRA radiation. Acta Biomater 2016;30:388–96. Available from: https://doi.org/10.1016/j.actbio.2015.11.052.

[34] Vidlárová L, Romero GB, Hanuš J, Štepánek F, Müller RH. Nanocrystals for dermal penetration enhancement—effect of concentration and underlying mechanisms using curcumin as model. Eur J Pharm Biopharm 2016;104:216–25. Available from: https://doi.org/10.1016/j.ejpb.2016.05.004.

[35] Müller RH, Radtke M, Wissing SA. Solid lipid nanoparticles (SLN) and nanostructured lipid carriers (NLC) in cosmetic and dermatological preparations. Adv Drug Deliv Rev 2002;54:S131–55. Available from: https://doi.org/10.1016/S0169-409X(02)00118-7.

[36] Noor NM, Sheikh K, Somavarapu S, Taylor KMG. Preparation and characterization of dutasteride-loaded nanostructured lipid carriers coated with stearic acid-chitosan oligomer for topical delivery. Eur J Pharm Biopharm 2017;117:372–84. Available from: https://doi.org/10.1016/j.ejpb.2017.04.012.

[37] Doktorovová S, Kovacevic AB, Garcia ML, Souto EB. Preclinical safety of solid lipid nanoparticles and nanostructured lipid carriers: current evidence from in vitro and in vivo evaluation. Eur J Pharm Biopharm 2016;108:235–52. Available from: https://doi.org/10.1016/j.ejpb.2016.08.001.

[38] Hamishehkar H, Ghanbarzadeh S, Sepehran S, Adib ZM, Kouhsoltani M, Hamishehkar H, et al. Histological assessment of follicular delivery of flutamide by solid lipid nanoparticles: potential tool for the treatment of androgenic alopecia. Drug Dev Ind Pharm 2016;42:846–53. Available from: https://doi.org/10.3109/03639045.2015.1062896.

[39] Chen H, Chang X, Du D, Liu W, Liu J, Weng T. Podophyllotoxin-loaded solid lipid nanoparticles for epidermal targeting. J Control Release 2006;110:296–306. Available from: https://doi.org/10.1016/j.jconrel.2005.09.052.

[40] Yukuyama MN, Kato ETM, de Araujo GLB, Löbenberg R, Monteiro LM, Lourenço FR, et al. Olive oil nanoemulsion preparation using high-pressure homogenization and D-phase emulsification—a design space approach. J Drug Deliv Sci Technol 2019;49:622–31. Available from: https://doi.org/10.1016/j.jddst.2018.12.029.

[41] Sonneville-Aubrun O, Yukuyama NM, Pizzino A. Application of nanoemulsions in cosmetics. Nanoemulsion. 2018. p. 435–75. Available from: https://doi.org/10.1016/B978-0-12-811838-2.00014-X.

[42] Yukuyama MN, Kato ETM, Löbenberg R, Bou-Chacra NA. Challenges and future prospects of nanoemulsion as a drug delivery system. Curr Pharm Des 2017;23:495–508. Available from: https://doi.org/10.2174/1381612822666161027111957.

[43] Sagitani H, Hirai Y, Nabeta K, Nagai M. Effect of types of polyols on surfactant phase emulsification. J Jpn Oil Chem Soc 1986;35:102–7. Available from: https://doi.org/10.5650/jos1956.35.102.

[44] Yukuyama MN, Oseliero PLF, Kato ETM, Lobënberg R, de Oliveira CLP, de Araujo GLB, et al. High internal vegetable oil nanoemulsion: D-phase emulsification as a unique low energy process. Colloids Surf A Physicochem Eng Asp 2018;554:296–305. Available from: https://doi.org/10.1016/j.colsurfa.2018.06.023.

[45] Wasutrasawat P, Al-obaidi H, Gaisford S, Lawrence MJ, Warisnoicharoen W. Drug solubilisation in lipid nanoparticles containing high melting point triglycerides. Eur J Pharm Biopharm 2013;85:365–71. Available from: https://doi.org/10.1016/j.ejpb.2013.04.020.

[46] Abd E, Benson HAE, Roberts MS, Grice JE. Follicular penetration of caffeine from topically applied nanoemulsion formulations containing penetration enhancers: in vitro human skin studies. Skin Pharmacol Physiol 2018;31:252–60. Available from: https://doi.org/10.1159/000489857.

[47] Bolduc C, Shapiro J. Hair care products: waving, straightening, conditioning, and coloring. Clin Dermatol 2001;19:431–6. Available from: https://doi.org/10.1016/S0738-081X(01)00201-2.

[48] Kim Y, Jeon S, Ji JH, Lee W. Development of a classification system for extrinsic hair damage : standard grading of electron microscopic findings of damaged hairs. Am J Dermatopathol 2010;32:432–8. Available from: https://doi.org/10.1097/DAD.0b013e3181c38549.

[49] Dario MF, Santos MSCS, Viana AS, Arêas EPG, Bou-chacra NA, Conceic M, et al. A high loaded cationic nanoemulsion for quercetin delivery obtained by sub-PIT method. Colloids Surf A Physicochem. Eng Asp 2016;489:256–64. Available from: https://doi.org/10.1016/j.colsurfa.2015.10.031.

[50] Hu Z, Liao M, Chen Y, Cai Y, Meng L, Liu Y, et al. A novel preparation method for silicone oil nanoemulsions and its application for coating hair with silicone. Int J Nanomed 2012;7:5719–24. Available from: https://doi.org/10.2147/IJN.S37277.

[51] Glenn RWJ, Hosseinpour D, Doyle KL, Kaufman KM. Method of treating hair. US 2017/0165162 A1; 2017b.

[52] Glenn RWJ, Hosseinpour D, Doyle KL. Method of treating hair. US 2017/0165157 A1; 2017a.

[53] Glenn RWJ, Kaufman KM, Hosseinpour D. Method of treating hair with a concentrated conditioner. US 2015/0359727 A1; 2015.

[54] Glenn RWJ, Kaufman KM, Iwata T. Method of treating hair with a concentrated conditioner. US 2016/0310372 A1; 2016.

[55] Torres Rivera JV, Kaufman KM, Glenn RWJ. Hair care regimen using liquid concentrated conditioner. US 2016/0310377 A1; 2016a.

[56] Torres Rivera JV, Kaufman KM, Glenn RWJ. Hair care regimen using a mechanical foam concentrated conditioner. US 2016/0310376 A1; 2016b.

[57] Zhao JJ, Glenn RWJ, Mullen SE. Hair care conditioning composition. US 2016/0310371 A1; 2016.

[58] Syed AN, Lenick O', AJ. Methods and compositions for treating damaged hair. US 20190029945A1; 2019.

[59] Oh H, Gartstein V, Bates T. Amino silicone nanoemulsion. US 2015/0030644 A1; 2015.

[60] Pereda MDCV, Polezel MA, de Campos Dieamant G, Nogueira C, Rossan MR, et al. Sericin cationic nanoparticles for application in products for hair and dyed hair. US 8,709,455 B2; 2014.

[61] D'Ischia M, Wakamatsu K, Cicoira F, Mauro ED, Garcia-borron JC, Ghanem G, et al. Melanins and melanogenesis: from pigment cells to human health and technological applications. Pigment Cell Melanoma Res 2015;28:520–44. Available from: https://doi.org/10.1111/pcmr.12393.

[62] Seiberg M. Age-induced hair greying—the multiple effects of oxidative stress. Int J Cosmet Sci 2013;35:532–8. Available from: https://doi.org/10.1111/ics.12090.

[63] Jo SJ, Shin H, Paik SH, Na SJ, Jin Y, Park WS, et al. Efficacy and safety of *Pueraria lobata* extract in gray hair prevention: a randomized, double-blind, placebo-controlled study. Ann Dermatol 2013;25:218–22. Available from: https://doi.org/10.5021/ad.2013.25.2.218.

[64] Schweitzer AD, Revskaya E, Chu P, Pazo V, Friedman M, Nosanchuk JD, et al. Melanin-covered nanoparticles for protection of bone marrow during radiation therapy of cancer. Int J Radiat Oncol Biol Phys 2010;78:1494–502. Available from: https://doi.org/10.1016/j.ijrobp.2010.02.020.

[65] Araújo M, Viveiros R, Correia TR, Correia IJ, Bonifácio VDB, Casimiro T, et al. Natural melanin: a potential pH-responsive drug release device. Int J Pharm 2014;469:140–5. Available from: https://doi.org/10.1016/j.ijpharm.2014.04.051.

[66] Ninh C, Cramer M, Bettinger CJ. Photoresponsive hydrogel networks using melanin nanoparticle photothermal sensitizers. Biomater Sci 2014;2:766–74. Available from: https://doi.org/10.1039/c3bm60321k.

[67] Commo P, Gaillard O, Thibaut S, Bernard BA. Absence of TRP-2 in melanogenic melanocytes of human hair. Pigment Cell Res 2004;17:488–97. Available from: https://doi.org/10.1111/j.1600-0749.2004.00170.x.

[68] Hardman JA, Tobin DJ, Haslam IS, Farjo N, Farjo B, Al-nuaimi Y, et al. The peripheral clock regulates human pigmentation. J Invest Dermatol 2015;135:1053–64. Available from: https://doi.org/10.1038/jid.2014.442.

[69] Alonso L, Fuchs E. The hair cycle. J Cell Sci 2006;2006:391–3. Available from: https://doi.org/10.1242/jcs02793.

[70] Santos Z, Avci P, Hamblin MR. Drug discovery alopecia : gone today, hair tomorrow. Expert Opin Drug Discov 2015;10:269–92. Available from: https://doi.org/10.1517/17460441.2015.1009892.

[71] Herman A, Herman AP. Topically used herbal products for the treatment of hair loss: preclinical and clinical studies. Arch Dermatol Res 2017;309:595–610. Available from: https://doi.org/10.1007/s00403-017-1759-7.

[72] Mathes C, Melero A, Conrad P, Vogt T, Rigo L, Selzer D, et al. Nanocarriers for optimizing the balance between interfollicular permeation and follicular uptake of topically applied clobetasol to minimize adverse effects. J Control Release 2016;223:207–14. Available from: https://doi.org/10.1016/j.jconrel.2015.12.010.

[73] Madaan A, Verma R, Singh AT, Jaggi M. Review of hair follicle dermal papilla cells as in vitro screening model for hair growth. Int J Cosmet Sci 2018;40:429–50. Available from: https://doi.org/10.1111/ics.12489.

[74] Aljuffali IA, Pan TL, Sung CT, Chang SH, Fang JY. Anti-PDGF receptor β antibody-conjugated squarticles loaded with minoxidil for alopecia treatment by targeting hair follicles and dermal papilla cells. Nanomed Nanotechnol Biol Med 2015;11:1321–30. Available from: https://doi.org/10.1016/j.nano.2015.04.009.

[75] Aljuffali IA, Sung CT, Shen F, Huang C, Fang J. Squarticles as a lipid nanocarrier for delivering diphencyprone and minoxidil to hair follicles and human dermal papilla cells. AAPS J 2014;16:140–50. Available from: https://doi.org/10.1208/s12248-013-9550-y.

[76] Wang W, Chen L, Huang X, Shao A. Preparation and characterization of minoxidil loaded nanostructured lipid carriers. AAPS PharmSciTech 2017;18:509. Available from: https://doi.org/10.1208/s12249-016-0519-x.

[77] Chen TC, Yu SC, Hsu CM, Tsai FJ, Tsai Y. Minoxidil—2-hydroxypropyl-β-cyclodextrin inclusion complexes: characterization and in vivo evaluation of an aqueous solution for hair growth in rats. J Incl Phenom Macrocycl Chem 2017;88:27–34. Available from: https://doi.org/10.1007/s10847-017-0705-9.

[78] Tsujimoto H, Hara K, Tsukada Y, Huang CC, Kawashima Y, Arakaki M, et al. Evaluation of the permeability of hair growing ingredient encapsulated PLGA nanospheres to hair follicles and their hair growing effects. Bioorg Med Chem Lett 2007;17:4771–7. Available from: https://doi.org/10.1016/j.bmcl.2007.06.057.

[79] Asili J, Mosallaei N, Shaterzadeh A, Malaekeh-Nikouei B. Preparation and characterization of liposomes containing methanol extract of aerial parts of *Platycladus orientalis* (L.) Franco. Avicenna J Phytomed 2012;2:17–23. Available from: https://doi.org/10.22038/ajp.2011.24.

[80] Rai N, Jain AK, Abraham J. Formulation and evaluation of herbal antidandruff shampoo containing garlic loaded solid lipid nanoparticles. Int J Pharm Res Rev 2013;2:12–24.

[81] Meray Y, Gençalp D, Güran M. Putting it all together to understand the role of Malassezia spp. in dandruff etiology. Mycopathologia 2018;4:893–903. Available from: https://doi.org/10.1007/s11046-018-0283-4.

[82] Pant G, Nayak N, Prasuna RG. Enhancement of antidandruff activity of shampoo by biosynthesized silver nanoparticles from *Solanum trilobatum* plant leaf. Appl Nanosci 2013;3:431–9. Available from: https://doi.org/10.1007/s13204-012-0164-y.

[83] Kalpana VN, Payel C, Rajeswari DV. *Lagenaria siceraria* aided green synthesis of ZnO NPs. Res J Chem Environ 2017;21:14–19.

CHAPTER

12

Nanoparticles in hair dyes

S. Senthil Kumaran[1], *R. Joseph Rathish*[2], *S. Johnmary*[3], *M. Krishnaveni*[4], *Susai Rajendran*[5] *and Gurmeet Singh*[6]

[1]School of Mechanical Engineering, VIT University, Vellore, India [2]PSNA College of Engineering and Technology, Dindigul, India [3]PG and Research Department of Chemistry, Loyola College, Chennai, India [4]PG and Research Department of Chemistry, MVM Government College for Women, Dindigul, India [5]Corrosion Research Centre, St Antony's College of Arts and Sciences for Women, Dindigul, India [6]Pondicherry University, Puducherry, India

12.1 Human hair

Hair is a protein filament which grows from follicles found in the dermis. Hair is one of the defining distinctiveness of mammals. The human body, apart from areas of glabrous skin, is sheltered in follicles which make thick terminal and fine vellus hair. Most general interest in hair is focused on hair growth, hair types, and hair care. However, hair is also a significant biomaterial primarily composed of protein, particularly alpha-keratin.

Attitudes toward diverse forms of hair, such as hairstyles and hair removal, vary widely across different cultures and historical periods. However, hair style is often used to indicate a person's personal beliefs or social position, such as their age, sex, or religion.

12.1.1 Function of human hair

To insulate the human body is the primary function of human hair. Hair does this function in the following two ways: it serves as a physical barrier between external cold air and the skin, and it also traps warm air in between the skin and the hair, keeping the body warmer. It should be mentioned that hairs also act as sense organs.

Nanocosmetics
DOI: https://doi.org/10.1016/B978-0-12-822286-7.00011-5

12.1.2 Chemical composition of hair

The overall chemical composition of hair is 45% carbon, 28% oxygen, 15% nitrogen, 7% hydrogen, and 5% sulfur. The hair shaft is essentially composed of keratin. Hair keratin is hard, compact, and strong. This fibrous protein is gradually formed inside cells from the germinal layer.

12.1.3 Living being found in human hair

There are two main types of hair that the body produces, vellus hair and terminal (or androgenic) hair. Hair fibers or strands grow from an organ in the area under the skin called a follicle, which is found in the dermis skin layer. The only "living" part of a hair is found in the follicle as it grows.

12.1.4 Three major components of the hair shaft

The hair shaft is formed of three layers (Fig. 12.1):

- The medulla—the deepest layer of the hair shaft, only seen in large and thick hairs.
- The cortex—the middle layer of the hair shaft which provides the strength, color, and texture of a hair fiber.
- The cuticle—the outer layer of the hair shaft is thin and colorless.

12.1.5 Anatomy of human hair

Hair is made of a tough protein called keratin. A hair follicle anchors each hair into the skin. The hair bulb forms the base of the hair follicle. In the hair bulb, living cells divide and grow to build the hair shaft.

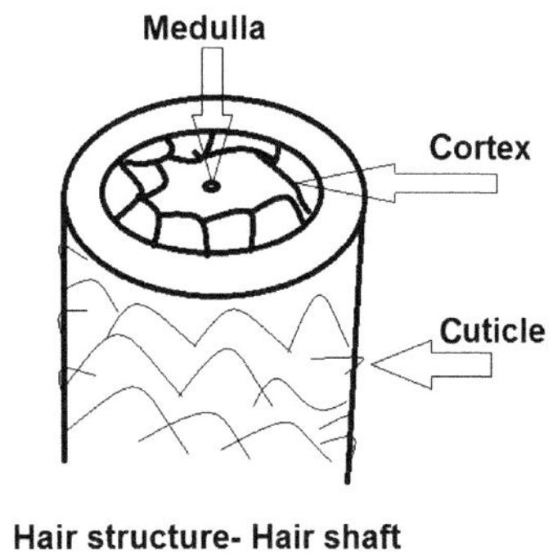

FIGURE 12.1 Layers of hair.

12.2 Hair colors

All natural hair colors are due to two types of hair pigments. They are of melanin types. They are produced inside the hair follicle and packed into granules found in the fibers. Eumelanin is the foremost pigment in brown hair and black hair, whereas pheomelanin is dominant in red hair. Blond hair is the consequence of having little pigmentation in the hair strand. Gray hair occurs when melanin production decreases or stops.

12.3 Melanins

Melanins appear to be heterogeneous, with some small regions of order at the nanometer scale. It is found that the optical properties depend on the ability of monomers and oligomers which make up melanin to absorb light, and the ability of melanin particles to reflect and scatter incident light for different wavelengths. Melanins have some semiconductor properties. However, the proposed band models do not adequately account for this.

12.4 Building blocks of eumelanins and pheomelanins

Eumelanin and pheomelanin play important roles in eye, hair, and skin color. Neuromelanin colors certain characteristic regions of the brain. This coloration is independent of skin and hair type.

12.5 Parkinson's and Alzheimer's diseases

Abnormalities in neuromelanins associate with various neurodegenerative diseases, such as Parkinson's and Alzheimer's.

12.6 Gray hair

Melanin gives color to our skin and hair. When our body stops producing as much melanin, gray hair appears. It is interesting to note that lack of melanin leads to the translucent gray or silvery tone.

12.7 Plucking gray hair is bad

Plucking gray hair will not cause three or more gray hairs to grow back in its place. Nevertheless, plucking is not an advisable activity. Because it can destroy the hair follicle and possibly lead to bald patches.

12.8 Natural turning of gray hair into black hair

Can we convert naturally, gray hair into black hair? The answer is Yes and No. White hair due to old age cannot turn black again naturally. If properly taken care of, white hair appearing due to bleaching, stress, food, pollution, vitamin deficiency, and other physical influence can turn to black again.

12.9 Composition of hair dyes in olden days

In olden days, hair dyes were made from plants, metallic compounds, or a mixture of the two. In Roman times, Rock alum, quicklime, and wood ash were used for bleaching hair. Moreover, herbal preparations included mullein, birch bark, saffron, myrrh, and turmeric. Henna was known in many parts of the world. It produced a reddish dye. The active principle responsible for the red color of the aqueous extract of the henna leaves (*Lawsonia inermis*) is lawsone (Fig. 12.2). King Solomon used to color his beard with the extract of henna leaves.

12.10 Common chemicals used in hair dyes

The following chemicals are commonly used in hair dyes: ammonia, peroxide, p-phenylenediamine (PPD), diaminobenzene, toluene-2,5-diamine, resorcinol, etc.

12.11 Harmfulness of hair dyes

Hydrogen peroxide bleaches the hair and helps PPD, one of the primary coloring agents, to become trapped in the hair. These common dye chemicals are associated with negative health effects.

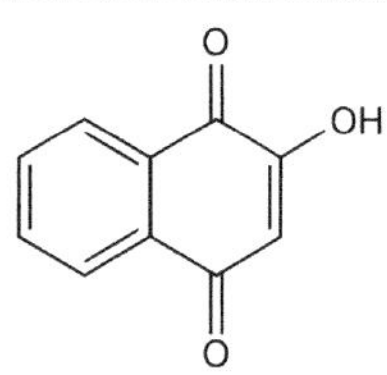

Lawsone, responsible for red color of aqueous extract of henna leaves

FIGURE 12.2 Lawsone.

12.12 Side effects of the chemicals used in common hair dyes

Side effects of the chemicals used in common hair dyes are skin rashes, itching, hair loss, dandruff, irritation, cancer, asthma, allergic reaction, eye sight weak, etc.

12.13 Toxic chemicals in hair dye

The toxic chemicals in hair dyes and the health effects are shown in Fig. 12.3.

12.14 Graphene hair dye

Nanosized graphene finds application in various fields such as electronics, high-efficiency heating systems, water purification technologies, and even golf balls. It was proposed that it could be used as hair dye. "Enough with the toxic hair dyes. We could use graphene instead," was the message given by some researchers. On the other hand, many researches reveal that graphene oxide leads to lung damage.

12.15 Gold nanoparticles as hair dye

Gold nanoparticles (NPs) have been successfully used to dye white hair a deep brown color.

In addition to changing the color, the tiny particles, around 40,000–60,000 times smaller than the width of a human hair, also caused the dyed locks to fluoresce an intense red under blue light. The color change was fashioned by soaking white hairs in a solution prepared from the gold compound chloroauric acid, calcium hydroxide, and distilled water.

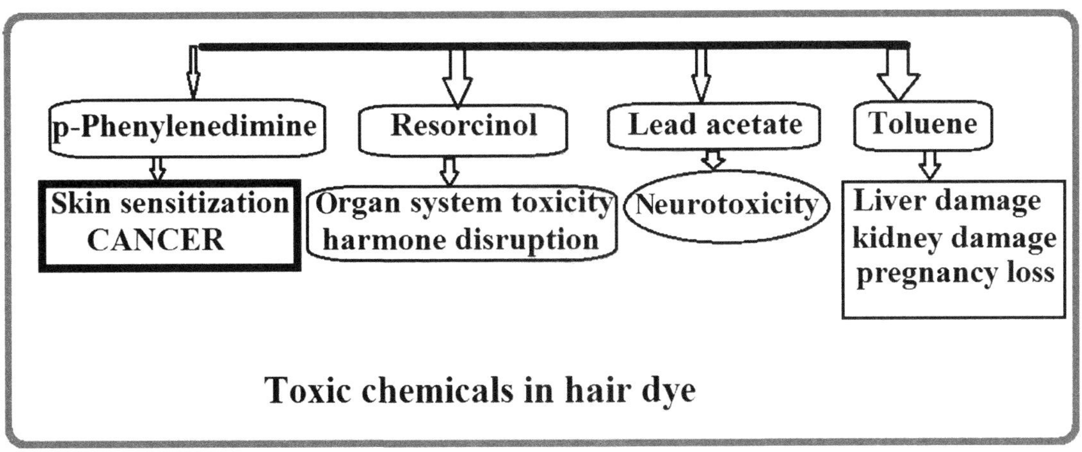

FIGURE 12.3 Toxic chemicals in hair dye.

The ensuing shades appear fade resistant as gold's affinity for the sulfur-containing amino acids found in keratin means the NPs are buried and stabilized inside the hair, remaining unchanged by frequent washings.

12.16 p-Phenylenediamine-incorporated nanoparticles as hair dye

PPD-incorporated NPs were prepared based on ion-complex formation between the cationic groups of PPD and the anionic groups of poly(γ-glutamic acid) (PGA). To reinforce PPD/PGA ion complexes, glycol chitosan (GC) was added. PPD-incorporated NPs can reduce the side toxic effects of PPD as hair dye.

12.17 Tips for faster growth of hair

The following tips will be useful for faster growth of hair (Fig. 12.4).

12.18 Best foods to promote hair growth

The following food items will promote hair growth (Fig. 12.5).

* Have a healthy and balanced diet.
* Give your scalp a massage.
* Don't wash too often.
* Don't brush too often.
* Trim your hair regularly.
* Switch your pillows.
* Use supplementation.
* Lower your stress levels

FIGURE 12.4 Tips for faster growth of hair.

- Eggs: Eggs are a great source of protein and biotin, two nutrients that may promote hair growth.
- Berries; Berries are loaded with beneficial compounds and vitamins that may promote hair growth.
- Spinach.
- Fatty fish.
- Sweet potatoes.
- Avocados.
- Nuts.
- Seeds

Best foods to promote hair growth

FIGURE 12.5 Best foods to promote hair growth.

12.19 Fruits for hair growth

The following fruits will promote hair growth. The best sources are blackcurrants, blueberries, broccoli, guava, kiwi fruits, oranges, papaya, strawberries, and sweet potatoes. Vitamin C helps in the production of collagen, which strengthens the capillaries that supply the hair shafts.

12.20 Recent research on the use of nanoparticles in hair dyes

When nanomaterials are engineered into hair care, they can improve the benefits of active ingredients. Superior knowledge of the composition of the hair fiber and an understanding of follicular targeting pathways are assisting the development of tailored products and new technologies capable of achieving improved hair cosmesis. Recent researches on the use of NPs in hair dyes are discussed in this section.

12.20.1 Estimation of nanoparticles' human skin penetration in vitro by confocal laser scanning microscopy

With quick development of nanotechnology, there is growing interest in NP application and its safety and value on human skin. In their study, Zou et al. [1] have utilized confocal laser scanning microscopy (CLSM) to estimate NP skin dispersion. Three different-sized polystyrene NPs marked with red fluorescence were applied to human skin, and Calcium Green 5N was employed as a counterstain. Dimethyl sulfoxide (DMSO) and ethanol were utilized as substitute vehicles for NPs. Tape stripping was utilized as a barrier-damaged skin model. Skin biopsies dosed with NPs were incubated at 4°C or 37°C for 24 hours and imaged by CLSM. NPs were localized in the stratum corneum (SC) and hair follicles without penetrating the epidermis/dermis. Barrier alteration with tape stripping and change in incubation temperature did not induce deeper penetration. DMSO enhanced NP SC penetration but ethanol did not. It was concluded that except with DMSO vehicle, these hydrolyzed polystyrene NPs did not penetrate intact or barrier-damaged human "viable" epidermis. For additional clinical relevance, in vivo human skin studies and more susceptible analytic chemical methodology are recommended [1].

12.20.2 Phototherapy and multimodal imaging

Targeted phototherapy and multimodal imaging can efficiently develop the therapeutic efficacy and reduce the side effects of theranostics. Herein, Li et al. [2] have constructed novel biocompatible cyanine dye IR808-conjugated hyaluronic acid NPs (HAIR NPs) for photothermal therapy (PTT) with near-infrared fluorescence (FL) and photoacoustic (PA) dual-modal imaging. The NPs produced stable nanostructures under aqueous conditions with even size distribution. The HAIR NPs were quickly taken up by the human lung cancer cells A549 via CD44 [the hyaluronic acid (HA) receptor on the surface of tumor cells] receptor-mediated endocytosis. Upon laser irradiation, the HAIR NPs enabled good near-infrared FA imaging and PA imaging in tumor-bearing mice. Additionally, the tight nanostructure arising from the covalent link between HA and IR808 could extensively enhance the light-thermal conversion efficiency of IR808. Under a low dose of laser power, the HAIR NPs presented more effective PTT for the inhibition of tumor growth than free IR808 in vitro and in vivo. Taken as a whole, these results indicate that the HAIR NPs may be an extremely gifted nanoplatform in cancer theranostics for targeted PTT under FL and PA dual-modal imaging [2].

12.20.3 Assembly of polymer-grafted proteins

Fukui et al. have carried [3] out surface-grafting from proteins and their assembling into objects with exclusive nanostructured materials (nanoobjects). To immobilize polymer-initiating sites, amino groups of bovine serum albumin (BSA) were allowed to react with iniferter groups (BSA-i). Then, graft polymerization of *N*-isopropyl acrylamide (NIPAM) was performed by light-initiated living radical polymerization from immobilized iniferter moieties of BSA-i. The polymer-grafted BSA (BSA-g-PNIPAM) was assembled into nanoobjects through the precipitation of PNIPAM graft chains and their sizes and morphologies were tuned by the chain length, the density, and the chemical structure of graft polymers in accumulation to the ecological conditions such as temperature and pH. It was possible to keep hold of the structures of nanoobjects by thermal denaturation via heat treatment. Fluorescent substances were encapsulated in particulate nanoobjects (NPs) assembled from PNIPAM-g-BSA and their release could be regulated by tuning pH and temperature. Next, additional graft polymerization from PNIPAM-grafted BSA was carried out by living radical polymerization of a cationic monomer, *N,N*-dimethylamino propyl acrylamide methyl chloride quaternary (DMAPAAQ). The grafted polymer was made of a block copolymer of PNIPAM and a cationic polymer (PDMAPAAQ) and the gel-like nanoobject was generated by increasing temperature. On the contrary to PNIPAM-g-BSA, it became insoluble even when the temperature decreased, most likely due to the electrostatic relationship between anionic regions of BSA and cationic regions of graft polymers. Coating of BSA-g-P (NIPAM-b-DMAPAAQ) enabled to form a homogeneous thin layer over a human hair. A free-standing membrane could be obtained by peeling from a water repellent substrate to create a porous membrane [3].

12.20.4 Detection of zinc in human hair by self-assembled NPs

Zinc plays significant roles in regulating physiological and pathological processes. Unfortunately mild to moderate zinc deficiency is common worldwide. Hair Zn^{2+} concentration, which reflects a zinc storage status, is useful for tracking trends in zinc status within populations. In this work, we report BODIPY-based self-assembled NPs as fluorescence turn-on sensor for the selective sensing of Zn^{2+} in human hair. The BODIPY monomers (BAN) self-assemble in aqueous medium to form nonfluorescent NPs. In the presence of Zn^{2+} ions, the NPs selectively show an evident turn-on fluorescence change. This selective response of the NPs allows the determination and quantification of Zn^{2+} in human hair with a finding limit of 61.3 nM. This study demonstrates that the small molecule self-assembled NP is a versatile and useful tool and shows great potential for applications in sensing of important analytes in biological systems [4].

12.20.5 Penetration of nanocarriers into the hair follicles

In humans, topically applied nanocarriers penetrate effectively into the hair follicles where they can be exploited for the localized and targeted treatment of skin disorders. The objective of the research of Knorr et al. [5] was to examine the applicability of particle-based systems for follicular drug delivery in companion animals and livestock, which have a large follicular reservoir. The following items were selected for the study: skin samples from 10 beagle dogs, 14 Wistar rats, and 4 ears from freshly slaughtered cross-bred pigs were used. The following methods were employed: Fluoresceinamine labeled poly (L-lactide-co-glycolide) nanocarriers (256 or 430 nm) were applied on the different skin samples. After penetration, skin biopsies were detached and cryohistological cross-sections prepared and investigated with regard to the follicular penetration depths (in μm ± standard deviation) of the nanocarriers by CLSM. Results: In canine, rat, and porcine hair follicles, the smaller NPs were detected at mean follicular penetration depths of 630.16 ± 135.75 μm, 253.55 ± 47.36 μm and 653.40 ± 94.71 μm, respectively. The larger particles were observed at average follicular depths of 604.79 ± 132.42 μm; 262.87 ± 55.25 μm, and 786.81 ± 121.73 μm, respectively, in canine, rat, and porcine hair follicles. Statistically significant differences ($P < .05$) in the mean follicular penetration depths of the differently sized nanocarriers could be determined for the canine and porcine skin samples. It was concluded that the mean follicular penetration depths of the differently sized nanocarriers were mostly significantly different between the different species, which might be due to diverse species-specific follicular dimensions. This subject needs to be addressed particularly in further studies [5].

12.20.6 Lipid nanoparticles for follicular targeting

Particulate drug carriers, for example, NPs have been shown to penetrate and accumulate preferentially in skin hair follicles creating high local concentration of a drug. In order to develop such a follicle targeting system, we obtained and characterized solid lipid nanoparticles (SLNs) loaded with roxithromycin (ROX). The mean particle size (172 ± 2 nm), polydispersity index (0.237 ± 0.007), zeta potential (−31.68 ± 3.10 mV), and incorporation effectiveness (82.1 ± 3.0%) were measured. The long-term stability of ROX-loaded SLN suspensions was proved up to 26 weeks. In vitro drug release study was performed using apparatus 4 dialysis adapters. Skin impatience test conducted using the EpiDerm tissue model demonstrated no irritation potential for ROX-loaded SLN. Ex vivo human skin penetration studies, employing rhodamine B hexyl ester perchlorate (RBHE) as a fluorescent dye to label the particles, revealed fluorescence deep in the skin, specifically around the hair follicles up to over 1 mm depth. The contrast of fluorescence intensities after application of RBHE solution and RBHE-labeled ROX-loaded SLN was done. Then cyanoacrylate follicular biopsies were obtained in vivo and analyzed for ROX content, proving the likelihood of penetration to human pilosebaceous units and delivering ROX by using SLN with the size below 200 nm [6].

12.20.7 Nanoparticles-based formulation for hair follicle targeting

Hair follicles are widely recognized as the preferential target and site of buildup for NPs after topical application. This feature is of particular significance for hair cosmetics, having the potential to refine the treatment of several hair follicle-related disorders. The intend of this research work of Fernandes et al. [7] was to progress the preparation of poly(D,L-lactide) (PLA) NPs for in vivo follicular target and drug delivery. Envisaging a future industrial scale-up of the process, the nanoprecipitation method was used to prepare PLA NPs: the effect of several processing parameters on their properties was examined and the yield of NPs formation determined. Encapsulation efficiencies and in vitro release profiles of lipophilic and hydrophilic model compounds were also assessed. In vitro cytotoxicity and ex vivo penetration studies were performed on a mentioned skin cell line (NCTC2455, human skin keratinocytes) and porcine skin, respectively. Using acetone:ethanol (50:50, v/v) as the solvent phase, 0.6% (w/w) of Pluronic F68 as a surfactant agent and agitation to mix the solvent and nonsolvent phases, a monodispersed population of noncytotoxic spherical NPs of approximately 150 nm was obtained. The yield of NPs for this formulation was roughly 90%. After encapsulation of model compounds, no significant changes were found in the properties of particles and the sting efficiencies were above 80%. The release kinetics of dyes from PLA NPs indicate an anomalous transport mechanism (diffusion and polymer degradation) for Nile Red (NR) (lipophilic) and a Fickian diffusion of first order for fluorescein 5(6)-isothiocyanate (hydrophilic). Ex vivo skin penetration studies confirmed the presence of NPs along the entire follicular ducts. The study leads to the following conclusions: The optimized method allows the preparation of ideal PLA NP-based formulations for hair follicle targeting. PLA NPs can efficiently transport and release lipophilic and hydrophilic compounds into the hair follicles, and the yields obtained are acceptable for industrial purposes [7].

12.20.8 Nanoparticles as a gene (or drug) carrier to the inner ear

A drug delivery system to the inner ear using NPs consisting of oligoarginine peptide (Arg8) conjugated to poly(amino acid) [poly(2-hydroxyethyl l-aspartamide; PHEA)] was investigated to determine whether the restrictions of low drug transport levels across the round window membrane (RWM) and poor transport into inner ear target cells, including hair cells and spiral ganglion, could be overcome. Three types of carrier materials, PHEA-g-C18, PHEA-g-Arg8, and PHEA-g-C18-Arg8, were synthesized to scrutinize the effects of oligoarginine and morphology of the synthesized carriers. NR was used as a fluorescent indicator as well as to model a hydrophobic (water repellent) drug. Compared with PHEA-g-C18-NR NPs, the oligoarginine-conjugated NPs of PHEA-g-C18-Arg8-NR and PHEA-g-Arg8-NR entered into HEI-OC1 cells at significant levels. Furthermore, the strongest fluorescence intensity was observed in nuclei when PHEA-g-C18-Arg8 NPs were used. The high uptake rates of PHEA-g-C18 and PHEA-g-C18-Arg8 NPs were observed in ex vivo experiments using hair cells. After the delivery of PHEA-g-C18-Arg8 NPs with reporter gene transfer, EGFP (enhanced green fluorescent protein) expression was monitored as an indicator of gene delivery. In the inner ear cells, PHEA-g-C18-Arg8 NPs showed equal or better transfection capabilities than the commercially available Lipofectamine reagent. PHEA-g-C18-Arg8 penetrated in vivo across the RWM of C57/BL6 mice with NR staining and GFP expression in various inner ear tissues. In conclusion, PHEA-g-C18-Arg8 NPs were effectively transported into the inner ear through the intratympanic route and are proposed as hopeful candidates as delivery carriers to address inner ear diseases [8].

12.20.9 Hair-dye assay using graphene/ionic liquid electrochemical sensor

A new analytical approach for detecting diaminopyridine derivatives has been constructed using a molecular imprinting-electrochemical sensor. Opposed to the conservative strategy of employing diaminopyridine as the functional monomer and uracil derivatives as the target analyte, in the current study, the 2,6-diaminopyridine-imprinted core-shell NPs were synthesized with 2,6-diaminopyridine as the template molecule and 6-aminouracil as the functional monomer. Graphene and ionic liquid which can help 2,6-diaminopyridine-imprinted core-shell NPs in electrochemical reaction kinetics by increasing conductivity have been introduced to form one of the electrode modified layers. The planned analytical method has been applied in 2,6-diaminopyridine detection in hair dyes and demonstrated suitable sensitivity and selectivity, with a linear range of 0.0500–35.0 mg/kg and a detection limit as low as 0.0275 mg/kg [9].

12.20.10 A targeted approach for the treatment of psoriasis using polymeric micelle nanocarriers

Tacrolimus (TAC) suffers from poor cutaneous bioavailability when administered topically using conservative vehicles with the result that although it is indicated for the treatment of atopic dermatitis, it has poor efficacy against psoriasis. In their research work, Lapteva et al. [10] have planned to formulate TAC-loaded polymeric micelles using the biodegradable and biocompatible methoxy-poly(ethylene glycol)-dihexyl substituted polylactide (MPEG-dihexPLA) diblock copolymer and to investigate their potential for embattled delivery of TAC into the epidermis and upper dermis. Micelle formulations were characterized with respect to drug content, stability, and size. An optimal 0.1% micelle formulation was residential and shown to be stable over a period of 7 months at 4°C; micelle diameters ranged from 10 to 50 nm. Delivery experiments using human skin and involving quantification by UHPLC-MS/MS demonstrated that this formulation resulted in appreciably greater TAC deposition in skin than that with Protopic (0.1% w/w; TAC ointment) (1.50 ± 0.59 and 0.47 ± 0.20 $\mu g/cm^2$, respectively). The cutaneous biodistribution profile of TAC in the upper 400 μm of tissue (at a resolution of 20 μm) demonstrated that the increase in cutaneous drug levels was due to enhanced TAC deposition in the SC, viable epidermis, and upper dermis. Given that there was no rise in the amount of TAC in deeper skin layers or any transdermal permeation, the results suggested that it would be possible to increase TAC levels selectively in the target tissue without increasing systemic absorption and the risk of side effects in vivo. Micelle distribution and molecular penetration pathways were after visualized with CLSM using a fluorescently labeled copolymer and fluorescent dyes. The CLSM study indicated that the copolymer was unable to cross the SC and that release of the micelle "payload" was dependent on the molecular properties of the "cargo" as evidenced by the different behaviors of DiO and fluorescein. A preferential deposition of micelles into the hair follicle was also confirmed by CLSM. Overall, the results indicate that MPEG-dihexPLA micelles are highly efficient nanocarriers for the selective cutaneous delivery of TAC, superior to the marketed formulation (Protopic). In addition, they may also have considerable potential for targeted delivery to the hair follicle [10].

12.20.11 Quantification of nanoparticle uptake into hair follicles in pig ear and human forearm

Drug delivery via the hair follicle (HF) especially with NPs recently gained awareness due to a depot effect and facilitated absorption circumstances within the lower HF. With the prospect of transdermal drug delivery, it is of attention to optimize the follicular uptake of NP. In this study, Raber et al. [11] have developed a method to quantify NP uptake into HF and applied in vitro in a pig ear model and in vivo in human volunteers. The influence of NP material on HF uptake was investigated using fluorescence-labeled NP based on poly(D,L-lactide-co-glycolide) (PLGA). All NP had similar hydrodynamic sizes (163–170 nm) but different surface modifications: (1) plain PLGA, (2) chitosan-coated PLGA (Chit.-PLGA), and (3) Chit.-PLGA coated with different phospholipids (PL) (DPPC (100), DPPC:Chol (85:15), and DPPC:DOTAP (92:8). Discrepancy stripping was performed, including complete mass balance. The samples were extracted for fluorescence quantification. An effect of the PL coating on follicular uptake was observed as DPPC (100) and DPPC:DOTAP (92:8) penetrated into HF to a higher extent than the other tested NP. The effect was noticed both in the pig ear model and in human volunteers, although it was statistically significant only in the in vitro model. An outstanding in vitro–in vivo correlation (IVIVC, $r2 = 0.987$) between both models was established, further supporting the suitability of the pig ear model as a surrogate for the in vivo situation in humans for quantifying NP uptake into HF. These findings may help to optimize NP for targeting the HF and to improve transdermal delivery [11].

12.20.12 Presentations in conferences

The conference proceedings of "2013 2nd International Conference on Sustainable Energy and Environmental Engineering" contains 363 papers. The special focus in this conference is on Sustainable Energy and Environmental Engineering. The topics include chip-based portable device for recognition of microalgae viability in ballast water; statistical models of predicting SO_2 concentration in shanghai; dew amount in marsh monitoring in the sanjiang plain; mobile vehicle-borne environmental monitoring based on environmental multisensor addition; the performance impact of hair dying sewage on biological treatment and the like [12].

In the conference of "3rd International Conference on Textile Engineering and Materials," its proceedings contains 310 papers. The special focus in this conference is on Fiber Technology, Nonwoven Materials, Structure, Properties and Processes of Textile Materials, Textile Chemistry, Textile Printing, Dyeing, and Finishing Technology, Apparel Design, Manufacturing and Merchandising, Apparel Design, Manufacturing and Merchandising, Metal and Optical Materials, Polymer Materials, Biomaterials, Low Carbon and Environmental Protection, Composites, Micro/Nano Materials, Materials Processing Technology, Testing Technology and Mechanical Dynamics. Photoprotection of Asian human hair exposed to sunlight radiation has been discussed [13].

12.20.13 Polymeric nanoparticles-embedded organogel for roxithromycin delivery to hair follicles

Drug delivery into hair follicles with the use of NPs is gaining more significance as drug-loaded NPs may accumulate in hair follicle openings. Główka et al. [14] developed and evaluated the pluronic lecithin organogel (PLO) with roxithromycin (ROX)-loaded NPs for follicular targeting. Polymeric NPs were evaluated in terms of particle shape, size, zeta potential, deferment stability, encapsulation efficiency, and in vitro drug release. Lyophilized NPs were incorporated into the PLO and rheological dimensions of the NP-embedded organogels were done. The fate of the NPs in the skin was traced by incorporation of a fluorescent dye into the NPs. As a result, ROX was efficiently incorporated into polymeric NPs characterized by the appropriate size (approximately 300 nm) allowing drug delivery to hair follicles. In ex vivo human skin penetration studies, horizontal skin sections exposed fluorescence deep in the hair follicles. Although the organogel has higher affinity to the lipidic follicular area than an aqueous suspension of NPs, it did not seem to improve penetration of the NPs along the hair shaft. The results proved that it was possible to achieve special targeting to the pilosebaceous unit using polymeric NPs formulated into either the aqueous suspension or semisolid topical formulation.

12.20.14 Permanent hair dye-incorporated hyaluronic acid nanoparticles

Lee et al. have prepared PDA-incorporated NPs using HA. PDA-incorporated HA NPs have spherical shapes and sizes were less than 300 nm. The results of Fourier-transform infrared (FT-IR) spectra indicated that PDA-incorporated HA NPs were formed by ion-complex configuration between amine group of PDA and carboxyl group of HA. Furthermore, powder-X-ray diffractogram (XRD) measurement revealed that intrinsic crystalline peak of PDA departed by arrangement of NP with HA at XRD measurement. These results suggested that PDA-incorporated HA NPs were formed by ion-complex formation. In the drug release study, the higher PDA contents induced a faster release rate from NPs. PDA-included NPs showed reduced intrinsic toxicity against HaCaT human keratinocyte cells at MTT assay and apoptosis assay. Lee et al. have recommended that PDA-incorporated HA NPs are promising candidates for novel permanent hair dye [15].

12.20.15 Hair fiber as a nanoreactor in controlled synthesis of fluorescent gold nanoparticles

The synthesis and detailed characterization of gold NPs (AuNPs) inside human hair has been achieved by treatment of hair with $HAuCl_4$ in alkaline medium by Haveli et al. [16]. The AuNPs, which show a strong red fluorescence under blue light, are generated inside the fiber and are given in the cortex in an amazingly regular pattern of whorls based on concentric circles, like a fingerprint. It opens an area of indisputable nanocomposites with novel properties due to AuNPs inside the hair shaft [16].

12.20.16 Hair dye-incorporated poly-γ-glutamic acid/glycol chitosan nanoparticles based on ion-complex formation

PDA or its related chemicals are used more expansively than oxidative hair dyes. Nonetheless, permanent hair dyes such as PDA are known to have potent contact allergy reactions in humans, and ruthless allergic reactions are problematic. PDA-incorporated NPs were produced based on ion-complex formation between the cationic groups of PDA and the anionic groups of PGA. To reinforce PDA/PGA ion complexes, GC was added. PDA-incorporated NPs were characterized by field-emission scanning electron microscopy, FT-IR spectroscopy, dynamic light scattering, and powder X-ray diffractometry (XRD). NPs were formed by ion-complex formation between the amine groups of PDA and the carboxyl groups of PGA. PDA-incorporated NPs are small in size (<100 nm), and morphological observations showed spherical shapes. FT-IR spectra results showed that the carboxylic acid peak of PGA decreased with increasing PDA content, indicating that the ion complexes were formed between the carboxyl groups of PGA and the amine groups of PDA. Additionally, the inherent peak of the carboxyl groups of PGA was also decreased by the addition of GC. Intrinsic crystalline peaks of PDA were observed by XRD. This crystalline peak of PDA was completely nonexistent when NPs were formed by ion complex between PDA, PGA, and GC, representing that PDA was complexed with PGA and no free drug existed in the formulation. During the drug-release experiment, an initial burst release of PDA was observed, and then PDA was continuously released over 1 week. Cytotoxicity testing against HaCaT human skin keratinocyte cells showed that PDA-incorporated NPs had lower toxicity than PDA itself. Furthermore, PDA-incorporated NPs showed reduced apoptosis and necrosis reaction at HaCaT cells. The research suggests that these microparticles are ideal candidates for a vehicle for decreasing side effects of hair dye [17].

12.20.17 Translocation of cell penetrating peptide engrafted nanoparticles across skin layers

Patlolla et al. [18] have evaluated the ability of cell penetrating peptides (CPPs) to translocate the payload into the skin layers. Fluorescent dye (DID-oil) encapsulated nanolipid crystal NPs (FNLCNs) were prepared using Compritol, Miglyol, and DOGS-NTA-Ni lipids by hot melt homogenization technique. The FNLCN surface was coated with TAT peptide (FNLCNT) or control YKA peptide (FNLCNY) and in vitro rat skin permeation studies were performed using Franz diffusion cells. Scrutiny of lateral skin sections obtained using cryotome with a confocal microscope showed that skin permeation of FNLCNT was time dependent and after 24 hours, fluorescence was observed up to a depth of 120 μm which was localized in the hair follicles and epidermis. In the case of FNLCN and FNLCNY formulations, fluorescence was chiefly observed in the hair follicles. This observation was further supported by confocal Raman spectroscopy where higher fluorescence signal intensity was observed at 80 and 120 μm depth with FNLCNT treated skin and intensity of fluorescence peaks was in the ratio of 2:1:1 and 5:3:1 for FNLCNT-, FNLCN-, and FNLCNY-treated skin sections, respectively. In addition, replacement of DID-oil with celecoxib (Cxb), a model lipophilic drug showed similar results and after 24 hours, the CXBNT formulation increased the Cxb concentration in SC by three- and sixfold and in epidermis by two- and threefold as compared to CXBN and CXBNY formulations, respectively. The study reveals that CPP can translocate NPs with their payloads into deeper skin layers [18].

12.20.18 Safety assessment of personal care products/cosmetics and their ingredients

Nohynek et al. [19] reviewed the safety assessment of personal care products (PCPs) and ingredients that are representative and pose complex safety issues. PCPs are generally applied to human skin and mainly produce local exposure, although skin penetration or use in the oral cavity, on the face, lips, eyes, and mucosa may also produce human systemic exposure. In the EU, US, and Japan, the safety of PCP is regulated under cosmetic and/or drug regulations. Oxidative hair dyes contain arylamines, the most chemically reactive ingredients of PCP. Although arylamines have an allergic potential, taking into account the high number of customers exposed, the incidence and prevalence of hair dye allergy appears to be low and stable. A recent (2001) epidemiology study recommended an association of oxidative hair dye use and increased bladder cancer risk in consumers, although this was not confirmed by subsequent or previous epidemiologic investigations. The results of genetic toxicity, carcinogenicity, and reproductive toxicity studies suggest that modern hair dyes and their ingredients pose no genotoxic, carcinogenic, or reproductive risk. Recent reports suggest that arylamines contained in oxidative hair dyes are *N*-acetylated in human or mammalian skin resulting in systemic exposure to traces of detoxified, that is, nongenotoxic, metabolites, whereas human hepatocytes were unable to transform hair dye arylamines to potentially carcinogenic metabolites. An expert panel of the International Agency on Research of Cancer (IARC) concluded that there is no evidence for a causal association of hair dye exposure with an elevated cancer risk in consumers. Ultraviolet filters have important payback by protecting the consumer against adverse effects of UV radiation; these substances undergo a rigorous safety evaluation under current international regulations prior to their marketing. Concerns were also raised about the safety of solid NPs in PCP, mainly TiO_2 and ZnO in sunscreens. Nevertheless, current proof suggests that these particles are nontoxic, do not enter into or through normal or compromised human skin and, therefore, pose no risk to human health. The increasing use of natural plant ingredients in PCPs raised new safety issues that require novel approaches to their protection evaluation similar to those of plant-derived food ingredients. For instance, the Threshold of Toxicological Concern is a promising tool to assess the safety of substances present at trace levels as well as minor ingredients of plant-derived substances. The potential human systemic exposure to PCP ingredients is increasingly predictable on the basis of in vitro skin penetration data. However, new confirmation suggests that the in vitro test may overestimate human systemic exposure to PCP ingredients due to the absence of metabolism in cadaver skin or misclassification of skin residues that, in vivo, remain in the SC or hair follicle openings, that is, outside the living skin. Overall, today's safety assessment of PCP and their ingredients is not only based on science, but also on their respective regulatory status as well as other issues, such as the principles of animal testing. However, the record shows that today's PCPs are safe and offer multiple profits to quality of life and health of the consumer. In the interest of all stakeholders, consumers, regulatory bodies, and producers, there is an urgent need for an international harmonization on the status and safety necessities of these products and their ingredients [19].

12.20.19 Potential novel drug carriers for inner ear treatment: hyperbranched polylysine and lipid nanocapsules

Treatment of sensorineural hearing loss could be superior using novel drug carriers such as hyperbranched polylysine (HBPL) or lipid nanocapsules (LNCs). Scheper et al. [20] have examined HBPL and LNCs for their cellular uptake and possible toxicity in vitro and in vivo as the first step in rising novel nanosized multifunctional carriers. Having incubated HBPL and LNCs with fibroblasts, NP uptake and cell viability were determined by CLSM, fluorescence measurements, and neutral red staining. In vivo, electrophysiology, CLSM, and cytocochleograms were performed for NP detection and also toxicity studies after intracochlear application. The study reveals that both NPs were detectable in the fibroblasts' cytoplasm without causing cytotoxic effects. After in vivo application, they were visualized in cochlear cells, which did not lead to a change in hearing threshold or loss of hair cells. Biocompatibility and traceability were established for HBPL and LNCs. Thus they comply with the basic requirements for drug carriers for potential application in the inner ear [20].

12.20.20 Investigation of polylactic acid nanoparticles as drug delivery systems for local dermatotherapy

The development of particle-based carriers for transepidermal drug delivery has become a field of major interest in dermatology. Rancan et al. [21] have investigated the suitability of biodegradable PLA particles loaded with fluorescent dyes as carriers for transepidermal drug delivery. The penetration profiles of PLA particles (228 and 365 nm) and the release of dye from the particles were investigated in human skin explants by fluorescence microscopy, CLSM, and flow cytometry. The study reveals that PLA particles penetrated into 50% of the vellus hair follicles, reaching a maximal depth corresponding to the entry of the sebaceous gland in 12%–15% of all observed follicles. The addition of particles in the follicular ducts was accompanied by the release of dye to the viable epidermis and its maintenance in the sebaceous glands for up to 24 hours. Kinetic studies in vitro as well as in skin explants revealed that, although stable in aqueous solution, deterioration of the particles and important release of incorporated dye occurred upon contact with organic solvents and the skin surface. These results recommend that particles based on PLA polymers may be ideal carriers for hair follicle and sebaceous gland targeting [21].

12.20.21 Nanoparticles—an efficient carrier for drug delivery into the hair follicles

The infiltration and storage behavior of dye-containing NPs (diameter 320 nm) into the hair follicles was investigated by Lademann et al. [22]. The results were compared with the findings obtained with the same amount of dye in the nonparticle form. In the first part of the experiments, the penetration of the dye into the hair follicles was investigated in vitro on porcine skin, which is an appropriate model for human tissue. It was found that the NPs penetrate much deeper into the hair follicles than the dye in the nonparticle form, if a massage had been applied. Without massage, similar results were obtained for both formulations. Consequently, the storage behavior of both formulations in the hair follicles was analyzed in vivo on human skin by differential stripping. Using the same application protocol, the NPs were stored in the hair follicles up to 10 days, while the nonparticle form could be detected only up to 4 days. Taking into reflection the surface structure of the hair follicles, it was implicit that the movement of the hairs may act as a pumping mechanism pushing the NPs deep into the hair follicles [22].

12.20.22 In vivo drug screening in human skin by femtosecond laser multiphoton tomography

The new femtosecond laser multiphoton imaging system Dermalnspect for in vivo tomography of human skin was used by König et al. [23] to study the diffusion and intradermal accumulation of topically applied cosmetic and pharmaceutical components. Near-infrared 80 MHz picojoule femtosecond laser pulses were employed to excite endogenous fluorophores and fluorescent components of a variety of ointments via a two-photon excitation process. In addition, collagen was imaged by second harmonic generation. A high submicron spatial resolution and 50 ps temporal declaration was achieved using galvoscan mirrors and piezodriven focusing optics together with a time-linked single-photon counting module with a fast microchannel plate detector. Individual intratissue cells, intracellular mitochondria, melanosomes, and the morphology of the nuclei as well as extracellular matrix elements were clearly visualized due to NAD(P)H, melanin, elastin, and collagen imaging and the calculation of fluorescence lifetime images. NPs and intratissue drugs were detected by two-photon-excited fluorescence. In addition, hydration effects and UV effects were investigated by monitoring modifications of cellular morphology and autofluorescence. The system was used to observe the diffusion through the SC and the accretion and release of functionalized NPs along hair shafts and epidermal ridges. The novel noninvasive 4-D multiphoton tomography tool provides high-resolution optical biopsies with subcellular resolution and offers for the first time the option to study in situ the dispersion through the skin barrier, long-term pharmacokinetics, and cellular response to cosmetic and pharmaceutical products [23].

12.21 Concluding remarks

Hair style and color are often used to indicate a person's personal beliefs or social position, such as their age, sex, or religion [24]. All natural hair colors are due to two types of hair pigments. They are of melanin types [24]. Gold NPs have been successfully used to dye white hair a deep brown color [25]. PPD-incorporated NPs were prepared [26]. Best foods are available to promote growth of hair [27]. Nanosized graphene finds application in various fields including hair dye [28].

Although there has been some preliminary success in the development of a multiplicity of nanotechnology platforms for the delivery of hair care products, additional research and a better understanding of toxicity are essential before the technology is readily available to the consumer. Because of the side effects, the tiny particles may create bigger problems when they are used in hair dyes. The use of dyes to color the hair is harmful. The dyes have side effects. We should not give importance to black hair. We have to appreciate other colors of the hair also. Take natural food for better color and growth of the hair.

References

[1] Zou Y, Celli A, Zhu H, Hui X, Maibach H, et al. Confocal laser scanning microscopy to estimate nanoparticles' human skin penetration in vitro. Int J Nanomed 2017;12:8035–41.

[2] Li S, Sun Z, Deng G, Gong P, Cai L, et al. Dual-modal imaging-guided highly efficient photothermal therapy using heptamethine cyanine-conjugated hyaluronic acid micelles. Biomater Sci 2017;5(6):1122–11293.

[3] Fukui Y, Sakai D, Fujimoto K. Preparation of protein nano-objects by assembly of polymer-grafted proteins. Colloids Surf B Biointerfaces 2016;148:503–10.

[4] Jia M-Y, Wang Y, Liu Y, Niu L-Y, Feng L. BODIPY-based self-assembled nanoparticles as fluorescence turn-on sensor for the selective detection of zinc in human hair. Biosens Bioelectron 2016;85:515–21.

[5] Knorr F, Patzelt A, Darvin ME, Ostrowski A, Lademann J, et al. Penetration of topically applied nanocarriers into the hair follicles of dog and rat dorsal skin and porcine ear skin. Vet Dermatol 2016;24:256-e60.

[6] Wosicka-Frąckowiak H, Cal K, Stefanowska J, Sérčič S, Markuszewski MJ, et al. Roxithromycin-loaded lipid nanoparticles for follicular targeting. Int J Pharm 2015;495(2):807–15.

[7] Fernandes B, Silva R, Ribeiro A, Gomes AC, Cavaco-Paulo AM, et al. Improved poly (D,L-lactide) nanoparticles-based formulation for hair follicle targeting. Int J Cosmet Sci 2015;37(3):282–90.

[8] Yoon JY, Yang K-J, Kim DE, Kim D-K, Kim J-D, et al. Intratympanic delivery of oligoarginine-conjugated nanoparticles as a gene (or drug) carrier to the inner ear. Biomaterials 2015;73:243–53.

[9] Zhao P, Hao J. 2,6-Diaminopyridine-imprinted polymer and its potency to hair-dye assay using graphene/ionic liquid electrochemical sensor. Biosens Bioelectron 2014;64:277–84.

[10] Lapteva M, Mondon K, Möller M, Gurny R, Kalia YN. Polymeric micelle nanocarriers for the cutaneous delivery of tacrolimus: a targeted approach for the treatment of psoriasis. Mol Pharm 2014;11(9):2989–3001.

[11] Raber AS, Mittal A, Schäfer J, Hansen S, Lehr C-M, et al. Quantification of nanoparticle uptake into hair follicles in pig ear and human forearm. J Control Release 2014;179(1):25–32.

[12] 2013 2nd International Conference on Sustainable Energy and Environmental Engineering, (ICSEEE 2013), 2014 Applied Mechanics and Materials, p. 522-524.

[13] 3rd International Conference on Textile Engineering and Materials (ICTEM 2013), 2013 Advanced Materials Research, p. 821-822.

[14] Główka E, Wosicka-Frąckowiak H, Hyla K, Jesionowski T, Cal K, et al. Polymeric nanoparticles-embedded organogel for roxithromycin delivery to hair follicles. Eur J Pharm Biopharm 2014;88(1):75–84.

[15] Lee H-Y, Jeong Y-I, Kim D-H, Choi K-C. Permanent hair dye-incorporated hyaluronic acid nanoparticles. J Microencapsul 2013;30(2):189–19716.

[16] Haveli SD, Walter P, Patriarche G, Wang P-A, Kagan HB, et al. Hair fiber as a nanoreactor in controlled synthesis of fluorescent gold nanoparticles. Nano Lett 2012;12(12):6212–17.

[17] Lee HY, Jeong YI, Choi KC. Hair dye-incorporated poly-γ-glutamic acid/glycol chitosan nanoparticles based on ion-complex formation. Int J Nanomed 2011;6:2879–88.

[18] Patlolla RR, Desai PR, Belay K, Singh MS. Translocation of cell penetrating peptide engrafted nanoparticles across skin layers. Biomaterials 2010;31(21):5598–607.

[19] Nohynek GJ, Antignac E, Re T, Toutain H. Safety assessment of personal care products/cosmetics and their ingredients. Toxicol Appl Pharmacol 2010;243(2):239–59.

[20] Scheper V, Wolf M, Scholl M, Lenarz T, Stöver T, et al. Potential novel drug carriers for inner ear treatment: Hyperbranched polylysine and lipid nanocapsules. Nanomedicine 2009;4(6):623–35.

[21] Rancan F, Papakostas D, Hadam S, Blume-Peytavi U, Vogt A, et al. Investigation of polylactic acid (PLA) nanoparticles as drug delivery systems for local dermatotherapy. Pharm Res 2009;26(8):2027–203622.

[22] Lademann J, Richter H, Teichmann A, Wepf R, Sterry W, et al. Nanoparticles - An efficient carrier for drug delivery into the hair follicles. Eur J Pharm Biopharm 2007;66(2):159–64.

[23] König K, Ehlers A, Stracke F, Riemann I. In vivo drug screening in human skin using femtosecond laser multiphoton tomography. Skin Pharmacol Physiol 2006;19(2):78–88.

[24] Available from: <https://en.wikipedia.org/wiki/Hair>.

[25] Available from: <https://www.wired.co.uk/article/gold-nanoparticles-successfully-used-as-hair-dye>.

[26] Available from: <https://www.semanticscholar.org/paper/Hair-dye-incorporated-poly-%CE%B3-glutamic-acid%2Fglycol-Lee-Jeong/d1264e96f990786f7e5db305349f2faea347e47f>.

[27] Available from: <https://www.google.com/search?q = hair + growth + healthy + foods + in + English&sxsrf = ACYBGNQCozvAM0cyXqPjsqZoJ5jIMR1T_A:1569331708896&source = lnms&sa = X&ved = 0ahUKEwj8mqyYyOnkAhUc6nMBHVAhAXgQ_AUIDSgA&biw = 1308&bih = 641&dpr = 1>.

[28] Available from: <https://www.wired.co.uk/article/gold-nanoparticles-successfully-used-as-hair-dye>.

CHAPTER

13

Nanomaterials in fragrance products

N. Vijaya[1], T. Umamathi[2], A. Grace Baby[3], R. Dorothy[4], Susai Rajendran[3], J. Arockiaselvi[5] and Abdulhameed Al-Hashem[6]

[1]Department of Chemistry, Vellalar College for Women, Erode, India [2]Department of Chemistry, Yadava College, Madurai, India [3]Corrosion Research Centre, St Antony's College of Arts and Sciences for Women, Dindigul, India [4]Department of EEE, AMET University, Chennai, India [5]PG and Research Department of Chemistry, SRM University, Chennai, India [6]Petroleum Research Centre, Kuwait Institute for Scientific Research, Al Ahmadi, Kuwait

13.1 Introduction

Fragrance is a sweet or pleasant odor or a scent. Fragrance is felt by the nose, and it can be considered as a sense of smell. We can use the word fragrance to describe the sweet smell of flowers, pine trees, and perfumes. Stink, stench, and reek are some antonyms of fragrance. It is believed that the sense of smell can identify seven types of sensations: camphor, mint, musk, flower, ether, acrid, and putrid. Out of these smells, floral scent can be considered as a fragrance.

The term fragrance is characteristically used by food and cosmetic industries to explain an enjoyable odor. Now and then, the word fragrance can also be used to refer to a perfume. Fragrances are used in the food industry to create different flavors. As mentioned above, the sense of smell also has an effect on the sense of taste. This is why fragrances are used in the food industry. Nanonoses (electronic noses) have been used to study the use of nanomaterials in fragrance products [1–20].

Fragrance and flavor are related terms. They are compared in Table 13.1. Flavor can be influenced by the fragrance, whereas fragrance is not influenced by flavor.

Nanotechnology has entered the production and application of various personal care and cosmetics products. They include sunscreens, antiaging creams, toothpastes, hair care, and perfumes.

Nanocosmetics
DOI: https://doi.org/10.1016/B978-0-12-822286-7.00012-7

TABLE 13.1 Comparison of fragrance and flavor.

Detail	Fragrance	Flavor
Definition	Fragrance is a sweet and pleasant smell.	Flavor is a distinctive taste.
Usage	Fragrance is often used to describe floral scents.	Flavor is usually used to describe food.
Sensory organ	Fragrance is smelt by the nose.	Flavor is felt by the tongue.
Interrelation	Fragrance is not influenced by the flavor.	Flavor can be influenced by the fragrance.

There is no definite definition of applied nanotechnology in terms of production procedures and ingredients. So the current scope and scale of nano-based personal care and cosmetics products are only a wild guess.

Nevertheless, there are about 1000 personal care and cosmetics products on the global market which are nano-based. It has been observed that much less is known about nanotechnology in perfumes, their production, and application.

13.2 Nano-ingredients

Nanomaterials contain particles that are smaller than 100 nm across. Examples range from the DNA molecule (which has a diameter of 2–3 nm) to the flu virus (100 nm). For comparison, a human hair is about 75,000 nm thick.

The nanoparticles found in foods may consist of inorganic (e.g., silver, titanium dioxide, silicon dioxide, iron oxide, and zinc oxide) and/or organic components (e.g., lipids, proteins, and carbohydrates).

13.3 Home fragrance products

There are many home fragrance products. Some of them include scented oil diffusers, incense sticks, oil burners, potpourris, scented oils, incense holders, fragrant room sprays, and home fragrance lamps.

13.3.1 Astonishing ways to make your home smell amazing

There are many astounding ways to make your home smell astonishing. They are

- Sprinkle baking soda on carpets.
- Add essential oils to your air filter.
- Freshen your whole house with ease.
- Simmer potpourri on the stove.
- Make your own room freshening spray.
- Put vanilla in your oven.
- Add an air freshener to your air vent.
- Clean your garbage disposal.

13.3.2 Use of orange peels in making your house smell good

Fill a sauce pan with water, or for a more luscious fruit scent, use apple juice instead. Boil the water or juice and add orange peels from one to three oranges, four cinnamon sticks, and one tablespoon of whole or ground cloves. Pleasant smell will be there in the whole house.

Here are ways to make your dorm room smell good without candles.

- Use a plug-in air freshener.
- Put essential oils in your humidifier.
- Get Febreze scented trash bags.
- Put dryer sheets in your shoe drawer and fan.
- Get a wax melt warmer.
- Put a car freshener on your radiator.

13.3.3 Tips to keep a kitchen smelling so fresh

Here are ways to make your kitchen smelling fresh ever.

- Use dryer sheets all over the place.
- Keep an open box of baking soda in your cabinets and fridge.
- Simmer citrus in water on the stove.
- Hang dried herbs.
- Crush ice in your disposal with lemons or lemonade mix once per week.

13.3.4 Toilet freshener

Toilet freshener can be prepared as follows. Combine vinegar and essential oil in a small spray bottle. Spray vinegar mixture inside the bowl, and also on the toilet seat, lid, and handle. Allow cleaner to sit for several minutes. Sprinkle baking soda inside the toilet bowl and scrub inside of the bowl with a toilet brush.

13.3.5 Fragrance for body

There are many scents that can do wonders for your well-being. They can do wonders for your body and mind. They include lavender can help you sleep, cinnamon can sharpen your mind, pine can alleviate stress, fresh-cut grass can make you more joyful, citrus can help you feel more energized, and vanilla can elevate your mood.

13.4 Natural fragrances

Natural fragrances are complex fragrance compounds made exclusively from natural aromatics as defined by the International Fragrance Association (IFRA). The ingredients used in natural fragrances can be essential oils, oleoresins, distillates, fractions, concretes, absolutes, etc. The ingredients of a natural fragrance can come from any natural source.

There are some natural ingredients having a pleasant fragrance and at the same time cause no irritation and are great for the skin. Many beneficial skincare ingredients (e.g., antioxidants) have a natural fragrance, and some of them even smell great!

"Unscented" is a term that indicates a product is completely scent-free and should contain absolutely zero scent, smell, or odor. Choosing to use products labeled "natural fragrance," "fragrance-free," or "unscented" can help reduce and even eliminate your regular exposure to unnecessary chemicals and toxins.

13.5 Fragrance affects your skin

According to Dermatologists: "Fragrance can be an irritant leading to redness, itchy skin, and sometimes hives" and "Not all fragrances cause irritation to the same degree." Some components of fragrance formulas may have a potential to cause allergic reactions or sensitivities for some people.

13.6 Chemicals present in fragrance

The way most fragrance ingredients impart scent is through a volatile reaction. In fact, research has established that fragrances in skin care products are among the most common cause of sensitizing and other negative skin reactions.

Common ingredients found in perfumes are acetone, ethanol, benzaldehyde, formaldehyde, limonene, methylene chloride, camphor, ethyl acetate, linalool, and benzyl alcohol. Phthalates and synthetic musks are also commonly used, but they are potentially hazardous ingredients.

13.7 Ingredients to be avoided in skin care

The following ingredients have to be avoided in skin care:

- Aluminum.
- DEA (diethanolamine), MEA (monoethanolamine), and TEA (triethanolamine).
- DMDM hydantoin and urea (imidazolidinyl).
- Mineral oil.
- Parabens (methyl, butyl, ethyl, propyl).
- PEG (polyethylene glycol).
- Phthalates.
- Propylene glycol (PG) and butylene glycol.

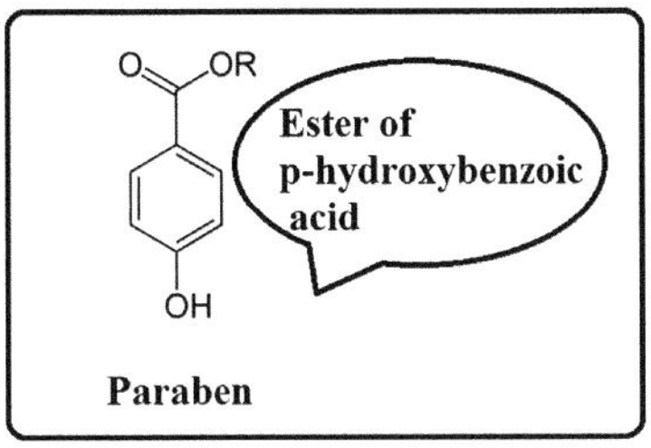

FIGURE 13.1 Paraben.

13.8 Parabens

Parabens (Fig. 13.1) are a class of widely used preservatives in cosmetic and pharmaceutical products. Chemically, they are a series of parahydroxybenzoates or esters of parahydroxybenzoic acid (also known as 4-hydroxybenzoic acid).

Parabens are effective preservatives in many types of formulas. These compounds, and their salts, are used primarily for their bactericidal and fungicidal properties. They are found in shampoos, commercial moisturizers, shaving gels, personal lubricants, topical/parenteral pharmaceuticals, suntan products, makeup, and toothpaste. They are also used as food preservatives.

13.9 Nanotechnology in perfumes

Nanotechnology has entered the production and application of various personal care and cosmetics products. They include sunscreens, antiaging creams, toothpastes, hair care, and perfumes. There are thrills and threats of smelling nano.

Perfume is a mixture of fragrant essential oils or aroma compounds, fixatives, and solvents, used to give the human body, animals, food, objects, and living spaces an agreeable scent. It is usually in liquid form and used to give a pleasant scent to a person's body.

Fragrances are highly toxic. Fragrances commonly contain phthalates, which are chemicals that help the scents last longer. Fragrance chemicals, like other toxic chemicals, can pass from the skin and into the blood. It is surprising to mention that manufacturers are not required to list their fragrance ingredients on product labels.

The most obvious fact when considering the importance of wearing perfume is the main purpose of the perfume is to keep unpleasant body odors at bay and make sure you smell fresh throughout the day. Perfumes also help to boost confidence and enhance the mood.

13.10 Applications of nanotechnology in perfumes

Currently known applications of nanotechnology in perfume production and application are mostly based on nanoencapsulation methods, which involve coating of nanoparticles with different substances [21].

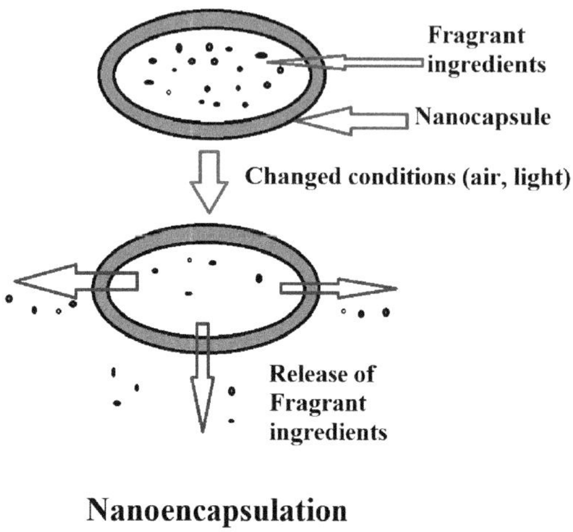

FIGURE 13.2 Nanoencapsulation.

13.10.1 Production of perfume (aroma) compounds

Application of nanotechnology helps in reduction of costs of perfume compounds manufacture, while at the same time making it possible to produce purer and entirely natural perfume compounds.

This can be achieved by using nanoparticles such as gold-palladium that can replace luxurious and potentially toxic reagents that promote oxidation of aromatic primary alcohols to aldehydes, which is one of the crucial processes in the perfume production.

An additional nanoencapsulation procedure proposes the use of nanoparticles coated in natural enzymes in the process of manufacturing expensive perfume compounds. There are no unwanted or harmful residuals.

13.10.2 Time-controlled and prolonged release of scents

Nanoencapsulation (nanodelivery systems) can also help progress the attributes and performance (durability, stability) of substances such as fragrances that can be negatively affected by changed conditions of the environment (light, air) (Fig. 13.2). The use of nanoencapsulation in fragrance products helps more efficient (prolonged) and time-controlled release of the scents.

This can be used in the manufacture of more long-lasting fragrance samples used for marketing purposes, in textile and accessories fashion (e.g., embedding perfume into textiles, shoes, and jewelry) and other materials such as ceramics and baby diapers. Release of scents can be time-controlled by stimuli such as diffusion, pressure, or temperature sensitivity.

13.10.3 Use of nanoencapsulation procedures in development of nanotechnology

As the electronic noses, replication of human olfactory sense promises detection and absorption of a variety of odors, which could be used in detection and absorption of unwanted or hazardous odors (e.g., carbon monoxide). An electronic nose (E-nose) is a device that identifies the specific components of an odor and analyses its chemical makeup to identify it. An electronic nose consists of a mechanism for chemical detection, such as an array of electronic sensors, and a mechanism for pattern recognition, such as a neural network.

Further, this could facilitate electronic sampling and testing of fragrance products, thus reducing the costs of fragrance and fragrance products development, and it could even enable development of artificial noses for people who lost the sense of smell.

Newly, one type of electronic appliances in this direction, nanoperfume ejectors, has been put on market. They are designed to mix nanoparticles with perfume and/or water particles and enable sterilization of air, absorption of unpleasant, and release of pleasant odors.

Bearing in mind the wide range of places where it could be used such as homes, hospitals, public places, this type of nanoappliance certainly has a bright commercial future.

13.10.4 Probable risks of nano-based perfumes

The main concerns of using nanotechnology in perfumes as in all personal care and cosmetics products are associated with potential human health and environmental hazards.

Concerns regarding human health got louder after it has been discovered that it is possible for some nanoparticles to cross the natural blood–brain barrier and that they can lethally damage living cells. Nanoparticles can enter the human bodies in many diverse ways; nanoparticles from nano-based fragrance products, for example, through skin and inhalation.

Due to their small size, nanoparticles are extremely mobile once they enter the body and it is feared to what extent they can penetrate naturally selective barriers in the living cells, which could result in toxic or even lethal consequences. Using nano-based fragrance products could thus allude to a variety of negative health consequences, such as severe damages of DNA, chromosomes, and immune system, toxic accumulation in tissues and organs (e.g., lungs and brain), interference with vital processes and mechanisms.

This is all due to the fact that nanoscaled particles tend to develop properties which cannot be assigned completely to their chemical nature, but to their sizes. Their properties and behavior when interacting with other (living) substances and processes are not yet researched in adequate detail to enable prediction and avoidance of possible negative consequences for human health.

Another concern the application of nano-based fragrance products raises is connected to potential environmental hazards. Scarce information about conducted environmental impact assessments is available regarding the possible impacts of spreading nanoparticles into the environment during the life cycle of nano-based fragrance products.

Parallels and environmental hazard warnings are drawn similar as in the case of introducing genetically modified organisms, nuclear energy, and use of asbestos in construction.

Until recently, a majority of nano-based cosmetic products entered the global market without sufficient safety and risk assessments conducted and transparent product labeling. The producers of nano-based (cosmetics) products advocated this by declaring nanoscaled ingredients to be chemically and thus safety-wise identical to bigger-sized particles of the same substance.

Nevertheless, concerned public, NGOs and even governmental bodies have in the last few years intensified their calls to set up tighter regulatory systems that would more efficiently control the production, risk assessment, handling, and labeling of nano-based (cosmetics) products and that would also apply the so-called precautionary principle already widely applied for the newly introduced medications.

13.10.5 High tech of small serving giant thrills with giant threats

As with any novel cutting edge technology that promises unprecedented benefits and solutions to the existing problems, the same should hold also for the nanotechnology applied in perfumes: "curb your enthusiasm." To avoid unwanted effects on human health and environment, tighter and more efficient regulative rules regarding the manufacture, handling, and labeling of nano-based fragrance products need to be enforced as quickly as possible.

Considering the vast scale and scope at which consumers are directly exposed to fragrance products daily (e.g., perfumes, deodorants, and home fragrances), the unwanted health and environmental consequences of smelling nano could be of unimaginable magnitude in the longer term.

Cosmetics and fragrance industry giants, such as Oreal and Coty, are heavily investing in nanotechnological research, consequently further nanotechnological leaps in the way perfumes and related products are produced and applied can be expected in the very near future. But if the perfume industry wants healthy returning customers, they need to build consumer confidence regarding the safety of using nano-based perfumes.

This could be achieved by conducting meticulous risk assessments and by providing a proficient labeling and consumer information system. Feared or factually proven negative health impacts of using mass products such as perfumes could stigmatize the public image and consequently investments into R&D of nanotechnology as whole.

13.11 Sunscreen

Sunscreen uses zinc oxide nanoparticles to block ultraviolet rays. The proteins are encapsulated in liposome nanoparticles which merge with the membranes of skin cells to allow delivery of the proteins. Sometimes titanium dioxide is also used as sunscreen.

13.11.1 Uncoated zinc oxide

Many sunscreens use zinc oxide particles that have been coated with an inert substance, usually triethoxycaprylylsilane, to make it easier to mix with the other ingredients and less photoreactive (see below). Only uncoated pharmaceutical grade zinc oxide in many products for the following reasons:

- Triethoxycaprylylsilane and the other coatings are not in alignment with our ingredient standards.
- It is not considered as important to the safety of larger particle non-nano zinc oxide that is used.

- An effective way to mix uncoated zinc oxide into sunscreen base has been developed.
- Uncoated zinc oxide has been safely used in topical skin care products such as calamine lotion and diaper rash cream for centuries.

Zinc oxide, like most powders, can be a health risk if inhaled, but this is not a concern with cream and lotion-based sunscreens. Additionally, zinc oxide can be photoreactive, meaning that UV exposure can generate reactive oxygen species, or free radicals, which can damage living cells. This too is not a concern with the zinc oxide in sunscreens because

- The rate of reactivity is still very low compared to that of titanium dioxide, nanoparticle zinc oxide, and many other chemical sunscreen actives.
- Zinc oxide sits on the outer, dead, layer of skin, and any free radicals generated will not affect living cells below.
- The inactive ingredients such as organic sunflower oil, vitamin E, seabuckthorn fruit extract and more provide powerful antioxidants which help scavenge, or absorb, free radicals.

Upon application, zinc oxide particles sit on the outermost layer of your skin, the stratum corneum, where they scatter, absorb, and reflect ultraviolet radiation, protecting your living skin below. Zinc oxide is unique among sunscreen ingredients in that it is truly a broad-spectrum blocker, protecting from UVA, UVB, and even UVC. Titanium dioxide is another mineral active ingredient you may see in other sunscreens. While it protects from UVB rays very well it does not protect from UVA as well as zinc oxide does. Zinc sunscreens are often called "chemical-free sunscreens" [22]. Zinc nanoparticles enhance brain connectivity. Functional magnetic resonance imaging studies lead to the conclusion that zinc nanoparticles enhance brain connectivity in the canine olfactory network [23].

13.11.2 Use of titanium dioxide as sunscreen

Titanium dioxide is safe and effective for protection against UVB (and some UVA) rays; however, titanium dioxide should always be used in combination with zinc oxide to attain true broad-spectrum protection. Truthfully, titanium dioxide received an unfair reputation when some study likened titanium dioxide to asbestos. The study was referring to toxicity when inhaled, which is not a threat with sunscreen creams. The media did not seem to understand this and wrote alarmist stories about the dangers of sunscreens containing titanium dioxide.

For the nanoparticle-free titanium dioxide (small particle size), it has been made extremely hydrophobic through surface treatment with dimethicone. This nanoparticle-free ingredient may be added at up to 3% in finished sunscreen formulations without objectionable whitening of the skin during "rub-in" [24].

Titanium dioxide is also a UV filter and so is an effective active ingredient in sunscreens. It is often used in cosmetic loose and pressed powders, especially "mineral powder" cosmetics, in addition to other cosmetics, lotions, toothpaste, and soap.

Titanium dioxide is used in loose powder cosmetics, sunscreen, and lotions. It leads to cancer when inhaled. Non-nano lotions are safe.

13.12 Recent trends in research on electronic noses, nanoparticles, and fragrant products

Many researches have been carried out on the use of electronic noses to study the presence and use of nanomaterials in fragrant products. Whereas flavor is felt by tongue, fragrance is felt by nose. Nowadays, nanomaterials present in fragrant materials are felt by an electronic nose. Recent trends in research on electronic noses, nanoparticles, and fragrant products are presented in this section.

13.12.1 Titanium dioxide nanoparticle-based indoor antiodor product

Indoor air pollutants and odorants may have psychological and physical impact on showing individuals and the unpleasant room air is measured as one of the factors associated with sick building syndrome including common symptoms such as headache and lethargy. Approaches for improving the quality of indoor air are thus significant as support for human health and well-being. Photooxidation catalyzed by titanium dioxide (TiO_2) is one of the methods used for elimination of volatile organic compounds, which are the cause of odor nuisance in indoor and outdoor air. Mirasoli et al. [1] have explored that the effectiveness of an experimental antiodor air freshener based on TiO_2 nanoparticles was estimated by testing its ability in removing from a small air chamber (200 mL) the odor of triethylamine solutions (50 μL at concentrations between 0.700 and 700 mM), used as a model volatile molecule for simulating fish-like unpleasant indoor environment. The assessment was performed by an electronic nose which provided a holistic and objective data on the efficacy of the product, demonstrating that the effects of triethylamine even at the highest tested concentrations can be completely removed by application of 3.0 g of the product at 25% TiO_2 nanoparticles concentration. The acquired results were confirmed by gas chromatography−mass spectrometry (GC−MS) analysis addressed to the quantitative determination of residual triethylamine in the environment after treatment by the antiodor product [1].

13.12.2 Electronic noses in meat quality assessment

Main factors that are measured by consumers when choosing meat products are color and aroma. It is well known that aroma is a more dependable indicator of quality. Nevertheless, a simple sensory estimate of hedonistic qualities is often not sufficient to govern whether protein is past its shelf-life, and the consumption of spoiled meat can lead to serious health hazards. Some volatile compounds can be used as spoilage indicators, and so a device equipped with a sensor sensitive to particular odorants would prove useful. Regrettably, no such single compound has yet been identified, as the changes taking place in a sample of meat during storage are contingent on several factors. In contrast, a combination of volatile compounds may form a unique "fingerprint," which can be analyzed pattern recognition algorithms with an electronic nose. It can supplement recognized techniques of meat quality assessment by providing results that correlate well with hedonic perception in a short time and at a low cost [2].

13.12.3 Flavor of traditional soup of Chinese Yellow-Feather Chickens

The traditional recipe for Chinese chicken soup generates a popular taste of particular umami and aroma. Qi et al. have explored [3] the effects of stewing time (1, 2, and 3 hours) on the principal taste-active and volatile compounds and the overall flavor profile of traditional Chinese chicken soup by measuring the contents of free amino acids (FAAs), 5′-nucleotides, minerals, and volatile compounds and by evaluating the taste and aroma profiles using an electronic nose, an electronic tongue, and a human panel. The outcome exhibited that the major umami-related compounds in the chicken soup were inosine 5′-monophosphate and chloride, both of which increased suggestively ($P < .05$) during stewing. The taste active values (TAVs) of the equivalent umami concentration increased from 4.08 to 9.93 ($P < .05$) after stewing for 3 hours. Although the FAA and mineral contents increased suggestively ($P < .05$), their TAVs were less than 1. The volatile compounds were mainly hexanal, heptanal, octanal, nonanal, (E)-2-nonanal, (E)-2-decenal, (E, E)-2,4-decadienal, 1-hexanol, and 2-pentyl furan. With the prolonged stewing time, the aldehydes first increased and then decreased suggestively ($P < .05$), while 1-hexanol and 2-pentyl furan increased gradually ($P < .05$). The aroma scores of the chicken soup reached the maximum after stewing for 3 hours. The inconsistency in overall flavor appearances tended to stabilize after 2 hours of stewing. In general, stewing time has a positive effect on refining the flavor profiles of chicken soup, especially within the first 2 hours [3].

13.12.4 Odor intensity assessment

The olfactory assessment function (e.g., odor intensity rating) of E-nose is always one of the most challenging issues in researches about odor pollution monitoring. But odor is normally fashioned by a set of stimuli, and odor interactions among residents knowingly influenced their mixture's odor intensity. Yan et al. [4] have explored the odor interaction principle in odor mixtures of aldehydes and esters, correspondingly. Then, a modified vector model (MVM) was projected and it successfully demonstrated the resemblance of the odor interaction pattern among odorants of the same type. Based on the regular interaction pattern, unlike a determined empirical model only fit for a specific odor mixture in conventional approaches, the MVM distinctly simplified the odor intensity prediction of odor mixtures. In addition, the MVM also provided a way of directly converting constituents' chemical concentrations to their mixture's odor intensity. By combining the MVM with usual data-processing algorithm of E-nose, a new E-nose system was established for an odor intensity rating. Compared with instrumental analysis and human assessor, it showed accuracy well in both quantitative analysis [Pearson correlation coefficient was 0.999 for individual aldehydes ($n = 12$), 0.996 for their binary mixtures ($n = 36$) and 0.990 for their ternary mixtures ($n = 60$)] and odor intensity assessment [Pearson correlation coefficient was 0.980 for individual aldehydes ($n = 15$), 0.973 for their binary mixtures ($n = 24$), and 0.888 for their ternary mixtures ($n = 25$)]. Consequently, the observed regular interaction pattern is measured as an important foundation for accelerating extensive application of olfactory evaluation in odor pollution monitoring [4].

13.12.5 Computer-controlled odor generator

Covington et al. have reported [5] on the on-going development of a simple computer-controlled odor generator. The unit encompasses eight "aroma dispensers" that can be loaded with liquid samples (in their case of fragrances as tea-tree oil). These aroma dispensers use a combination of the capillary effect and thermal heating to release aroma to the user. The instrument also includes a controlled fan and a gas sensor to monitor the release of the aroma. Interaction with the aroma generator is through a custom interface that releases aromas in line with either direct control or a preprogrammed categorization. It is believed that this unit can be used in combination with virtual environments to develop such experiences [5].

13.12.6 Nanoparticle-enzyme sensors for detection of bacteria with olfactory output

Duncan et al. have reported [6] a highly efficient sensor for bacteria that offers an olfactory output, agreeing detection without the use of instrumentation and with a modality that does not need visual identification. The sensor platform uses nanoparticles to reversibly complex and inhibits lipase. These complexes are interrupted in the presence of bacteria, restoring enzyme activity and generating scent from odorless profragrance substrate molecules. This system affords rapid (15 minutes) sensing and very high sensitivity (102 cfu/mL) detection of bacteria using the human sense of smell as an output [6].

13.12.7 Recognition of Chinese Herbal Medicines with electronic nose technology

Zhou et al. have presented [7] a review of the most recent works in machine olfaction as applied to the identification of Chinese Herbal Medicines (CHMs). Owing to the wide variety of CHMs, the complexity of increasing sources, and the diverse specifications of herb components, the quality control of CHMs is a challenging issue. A great deal of researches have established that an electronic nose (E-nose) as an advanced machine olfaction system can overcome this challenge through identification of the complex odors of CHMs. E-nose technology with better usability, high sensitivity, real-time detection, and nondestructive features has revealed better performance in comparison with other analytical techniques such as GC–MS. Although there has been enormous development of E-nose techniques in other applications, there are limited reports on the application of E-noses for the quality control of CHMs. Practical implementation and advantages of E-noses for robust and effective odor identification of CHMs have been discussed. This covers the use of E-nose technology to study the effects of growing regions, identification methods, production procedures, and storage time on CHMs. Furthermore, the challenges and applications of E-nose for CHM identification are explored. Based on the development in E-nose technology, odor may become a new quantitative index for quality control of CHMs and drug discovery. It was also found that more research could be done in the area of odor standardization and odor reproduction for remote sensing [7].

13.12.8 Area identification of Zhongning Goji berries by an electronic nose

High-quality Zhongning Goji (ZNG) berries are illegally adulterated in the market by adding non-ZNG (NZNG). A precise, rapid, and effective approach for the creating area identification of ZNG is needed to protect the geographical indications of Goji berry products and to confirm fair trade. Samples from diverse regions were collected and their odors were noticed by an E-nose. Principal component analysis (PCA), cluster analysis (CA), and linear discriminant analysis (LDA) were employed to build identification models. The E-nose models were further proved by GC–MS. The identification rates of the PCA, CA, and LDA models were 91.0%, 98.9%, and 100%, correspondingly. The PCA and CA models presented superior results, and the LDA model displayed optimum performance. These conditions designate the feasibility of using the E-nose technique for ZNG identification. GC–MS analysis naked differences and similarities in total ion current chromatograms between ZNG and NZNG [8].

13.12.9 Assessment of the indoor odor impact in a naturally ventilated room

Indoor air quality influences people's lives, potentially affecting their health and comfort. Nowadays, ventilation is the only technique generally used for regulating indoor air quality. CO_2 is the reference species measured in order to calculate the air exchange rates of indoor environments. Certainly, regarding air quality, the presence of pleasant or unpleasant odors can strongly affect the environmental comfort. In this paper, Eusebio et al. [9] have reported a case study of indoor air quality monitoring. The indoor field tests were conducted measuring both CO_2 concentration, using a photoacoustic multigas analyzer, and odor trends, using an electronic nose, in order to analyze and compare the information developed. The indoor air monitoring campaign was run for a period of 20 working days into a university room. The work was focused on the determination of both CO_2 and odor emission factors emitted by the human activity and on the estimate of the odor influence in a naturally ventilated room. The consequences highlighted that an air monitoring and recycling system based only on CO_2 concentration and temperature measurements might be insufficient to ensure a good indoor air quality, while its performances could be improved by integrating the prevailing systems with an electronic nose for odor detection [9].

13.12.10 Odor standardization by bioelectronic noses

Odors are perceived differently as a function of individual human experience. Hence communicating about odors between individuals is very difficult. There is a need to classify and standardize odors, but suitable tools have not yet been developed. A bioelectronic nose mimics human olfaction and detects target molecules with high sensitivity and selectivity. This new tool has great potential in several applications and is probable to accelerate odor classification and standardization. In particular, a multiplexed bioelectronic nose can offer complex odor information using pattern recognition techniques and could even reproduce odors via an integrated olfactory display system. Son et al. [10] have proposed that a bioelectronic nose will be a useful tool for odor standardization by providing codes for odors that permit us to communicate odor information [10].

13.12.11 Comparison of volatile fraction of vodkas made from different botanical materials by electronic nose-based technology

Vodka is a spirit-based beverage made from ethyl alcohol of agricultural origin. At present, more and more vodka brands have labels that specify the botanical origin of the product. Until now, the techniques for distinctive between vodkas of different botanical origin have been costly, time-consuming. and inadequate for making a difference between vodka shaped from similar raw materials. Consequently, it is of utmost importance to find a fast and relatively cheap technique for conducting such tests. In their study, Wiśniewska et al. [11] used the two-dimensional gas chromatography (GC × GC) and an electronic nose based on the technology of ultrafast GC with chemometric methods such as partial least square (PLS) discriminant analysis, discriminant function analysis, and soft independent modeling of class analogy. Both techniques allow a difference between the vodkas produced from different raw materials. In the case of GC × GC, the differences between vodkas were more noticeable than in the analysis by an electronic nose; however, the electronic nose allowed the knowingly faster analysis of vodkas [11].

13.12.12 Olfaction as a soldier

Olfaction is one of our five main qualitative sensory abilities. Nagappan et al. [12] have studied the physiology of olfaction from the olfactory receptor to the brain. Through analyzing the physiology of olfaction, they have established that the biochemistry of olfactory nerve stimulation is unique from that of other similar pathways. Upon receiving large amounts of input from the olfactory nerve, the olfactory bulb, followed by several layers of centrifugal and centripetal processing in the brain, has to class the information from the input as well as integrate it with other inputs from the brain to progress a coherent understanding of the input. Nagappan et al. have examined the implications of olfaction in the military, the practical applications of electronic noses and problems associated with injury to olfaction that could affect compensation and combat worthiness of a soldier following injury. In the military, olfaction can allow the army to perform at its best through four main methods, namely, confirming olfaction is consistent with other dimensions of perception (ensuring optimal olfaction ability in all soldiers in combat), understanding the impact of different common combat environments on the sense of smell, utilizing odor as a defense mechanism, and using olfactory aids when necessary. Electronic noses are olfactory aids that have a large potential in the military ranging from saving lives through the finding of explosives to potential methods for improving combustion efficiency. There are numerous problems associated with injury to olfaction that should be considered when conclusive on compensation and combat worthiness of the soldier following an injury [12].

13.12.13 Aromatic wallpaper

Aromatic wallpaper was established in order to provide fragrant atmosphere for inter space with new properties and added value. Microencapsulation technology is an actual method to control the release of fragrance. If wallpaper is treated with microencapsulated fragrance, longer scent holding time is expected. Xiao et al. [13] have reported the detailed preparation methods and analysis of two sorts of wallpapers with GC–MS and electronic nose (E-nose). The morphology of blank wallpaper and the aromatic wallpaper were characterized by scanning electron microscopy. The aroma content of fragrant wallpapers was investigated by head-space solid phase-micro-extraction and GC–MS. The service of an electronic nose to measure the intensity of fragrant release from the wallpaper was also conducted. The release experiments designated that aromatic wallpaper had the eminent and stable sustained release property. What is more, the sustained release time of aroma could reach more than 3 months. In the meantime, the aroma intensity and the comfortable degree were associated with sensory valuation to approve which one is the best fit for human feelings [13].

13.12.14 Bioelectronic noses

A distinguishing feature of human and animal organs of smell is the ability to identify hundreds of thousands of odors. It is accompanied by particular smell sensations, which are a basic source of information about odor mixture. The chief structural elements of biological smell systems are the olfactory receptors. Small differences in a structure of odorous molecules (odorants) can lead to important change of odor, which is due to the fact that each of the olfactory receptors is coded with different genes and usually corresponds to different types of odor. Discovery and characterization of the gene family coding the olfactory receptors contributed to the elaboration and progress of the electronic smell systems, the so-called bioelectronic noses. The olfactory receptors are employed as a biological element in this type of instrument. An electronic system includes a converter part, which allows measurement and processing of generated signals. A suitable data analysis system is also required to visualize the results. Application potentialities of the bioelectronic noses are engrossed on the fields of economy and science where highly selective and sensitive analysis of odorous substances is required. Wasilewski et al. have given a report on the latest achievements and critical estimation of the state of art in the field of bioelectronic noses [14].

13.12.15 A bioelectronic sensor using human olfactory and taste receptors

A multiplexed bioelectronic sensor was established for the use of quick, on-site, and simultaneous detection of several target molecules. Olfactory and taste receptors were produced in *Escherichia coli*, and the reconstituted receptors were immobilized onto a multichannel type carbon nanotube field-effect transistor. This device mimicked the human olfactory/taste system and simultaneously dignified the conductance changes with high sensitivity and selectivity following treatment with various odor and taste molecules commonly known to be indicators of food contamination. Several pattern recognition of odorants and tastants was available with a customized platform for the simultaneous measurement of electrical signals. The simple portable bioelectronic device was suitable for effective monitoring of food freshness and is expected to be used as a rapid on-site sensing platform with several applications [15].

13.12.16 Fragrance profiling of *Jasminum Sambac* Ait. flowers using electronic noses

The extract of Jasmine called concrete is a significant floricultural export commodity of India. The major quality attribute of both jasmine and concrete is fragrance. The fragrance profile of jasmine flower has been considered using a custom-built handheld E-nose (HEN). Three species of jasmine, *J. sambac*, *J. grandiflorum*, and *J. auriculatum* were applied to E-nose one by one. The PCA of the responses exhibited that distinct clusters were formed for different species. Ray et al. [16] made a study on the mostly cultivated *J. sambac* species to find out a relationship between concrete quality with several flowering stages as well as with seasonal variation. Experiments were carried out over 200 samples collected from many gardens of South India. Multivariate sensor data were computed in a single dimensional intensity value termed as fragrance index (FI). Flowering stages were categorized by Industry experts in terms of bud opening indicator (BOI). HEN was trained to formulate a scale of BOI in the range of 0–2 based on the FI. Best quality of concrete is established to be obtained from jasmine flower with a BOI level of 1.5. Jasmine concrete produced at the laboratory from flowers harvested at different time-of-the-day, that is, with different BOI levels was analyzed by GC–MS to check the volatile emission patterns. Results exposed that concrete produced with a BOI level of 1.5 emitted maximum aroma generating volatiles. These observations exhibited that E-nose can be a potential gadget for fragrance profiling as well as selection of best-quality flowers for concrete extraction [16].

13.12.17 Evaluation of an electronic nose for odorant and process monitoring of alkaline-stabilized biosolids production

Electronic noses have been broadly used in the food industry to monitor development performance and quality control, but use in wastewater and biosolids treatment has not been fully discovered. Romero-Flores et al. [17] have studied the feasibility of an electronic nose to discriminate between treatment conditions of alkaline stabilized biosolids and compared its performance with quantitative analysis of key odorants. Seven lime treatments (0%–30% w/w) were prepared and the resultant off-gas was monitored by GC–MS and by an electronic nose equipped with 10 metal oxide sensors. A pattern recognition model was produced using LDA and PCA of the electronic nose data. In general, LDA performed better than PCA. LDA showed clear discrimination when single tests were estimated, but when the full data set was included, discrimination between treatments was reduced. Frequency of accurate recognition was tested by three algorithms with Euclidean and Mahalanobis performing at 81% accuracy and discriminant function analysis at 70%. Concentrations of target compounds by GC–MS were in agreement with those reported in the literature and helped to explain the behavior of the pattern recognition via comparison of individual sensor responses to different biosolid treatment conditions. Results showed that the electronic nose can discriminate between lime percentages, thus providing the opportunity to generate classes of underdosed and overdosed relative to regulatory requirements. Full scale application will need careful evaluation to maintain accuracy under mutable process and environmental conditions [17].

13.12.18 Electronic nose to detect freshness of eggs

Yimenu [18] et al. have explored the potential of a fast GC E-nose for freshness discrimination and prediction of storage time as well as sensory and internal quality changes during storage of hen eggs. All samples were obtained from the same egg production farm and stored at 20°C for 20 days. Egg sampling was conducted every 0, 3, 6, 9, 12, 16, and 20 days. During each sampling time, four egg cartons (each containing 10 eggs) were randomly selected: one carton for Haugh units, one carton for sensory evaluation, and two cartons for the E-nose experiment. The E-nose study comprised two independent test sets; calibration (35 samples) and validation (28 samples). Every sampling time, five replicates were prepared from one egg carton for calibration samples and four replicates were prepared from the remaining egg carton for validation samples. Sensors (peaks) were selected prior to multivariate chemometric analysis; qualitative sensors for PCA and discriminant factor analysis (DFA) and quantitative sensors for PLS modeling. PCA and DFA confirmed the difference in volatile profiles of egg samples from seven altered storage times accounting for a total variance of 95.7% and 93.71%, respectively. Models for predicting storage time, Haugh units, odor score, and overall acceptability score from E-nose data were established using calibration samples by PLS regression. The results exhibited that these quality indices were well predicted from the E-nose signals, with correlation coefficients of $R2=0.9441$, $R2=0.9511$, $R2=0.9725$, and $R2=0.9530$ and with training errors of 0.887, 1.24, 0.626, and 0.629, respectively. As a result of analysis of variance, most of the PLS model results were not significantly ($P>.05$) different from the agreeing reference values. These results verified that the fast GC electronic nose has the potential to evaluate egg freshness and feasibility to predict multiple egg freshness indices during its circulation in the supply chain [18].

13.12.19 Verification of odorants in rose

Rose oil is much too exclusive but very popular. It is well known that the flower oil's aroma profile has not been intensively examined. To verify the aroma profile of rose oil, the synthetic blend of odorants was prepared and then related to the original rose oil using electronic nose analysis (ENA) combined with quantitative descriptive analysis (QDA). The odorants from rose oils were screened out by Gas Chromatography-Olfactometry/aroma extract dilution analysis (GC-O/AEDA) combined with odor activity value (OAV). Both ENA and QDA showed the recombination model derived from OAV and GC-O/AEDA closely look like the original rose oil. The experiment results display that rose oxide, linalool, α-pinene, β-pinene, nonanal, heptanal citronellal, phenyl ethyl alcohol, benzyl alcohol, eugenol, methyl eugenol, β-citronellol, hexyl acetate, β-ionone, nerol, etc., are very vital constituent to rose oil aroma profile [19].

13.12.20 Predicting the growth situation of *Pseudomonas aeruginosa* on agar plates and meat stuffs using gas sensors

A quick method of expecting the growing situation of *Pseudomonas aeruginosa* has been reported by Gu et al. [20]. Gas sensors were used to get volatile compounds generated by

P. aeruginosa on agar plates and meat stuffs. Then, optimal sensors were selected to simulate *P. aeruginosa* growth using modified Logistic and Gompertz equations by odor changes. The results exposed that the responses of S8 or S10 yielded high coefficients of determination (*R*2) of 0.89–0.99 and low root mean square error (RMSE) of 0.06–0.17 for *P. aeruginosa* growth, fitting the models on the agar plate. The responses of S9, S4 and the first principal component of 10 sensors fit well with the growth of *P. aeruginosa* inoculated in meat stored at 4°C and 20°C, with *R*2 of 0.73–0.96 and RMSE of 0.25–1.38. The correlation coefficients between the fitting models as measured by electronic nose responses and the colony counts of *P. aeruginosa* were high, ranging from 0.882 to 0.996 for both plate and meat samples. Also, GC–MS results designated the presence of specific volatiles of *P. aeruginosa* on agar plates. The study confirmed an acceptable feasibility of using gas sensors—a rapid, easy, and nondestructive method for calculating *P. aeruginosa* growth [20].

13.13 Future perspectives

Electronic noses are used to study the use of nanoparticles in fragrant products. Researches in this line will take us to a wonderful fragrant day, while viewing a movie in a theater, or home theater, will smell the fragrance of Jasmine flowers, when a scene of Jasmine garden is shown on the screen. The whole theater will be filled with the fragrance of Jasmine. Similarly, when a scene of eating "Biriyani"—a flavored rice food, for which Dindigul is famous— is played on the screen, the entire theater will be filled with the fragrance of Biriyani.

References

[1] Mirasoli M, Gotti R, Di Fusco M, Fiori J, Roda A, et al. Efficacy of a titanium dioxide nanoparticles-based indoor anti-odor product as assessed by electronic nose and gas chromatography–mass spectrometry. J Pharm Biomed Anal 2017;144:236–41.

[2] Wojnowski W, Majchrzak T, Dymerski T, Gębicki J, Namieśnik J. Electronic noses: powerful tools in meat quality assessment. Meat Sci 2017;131:119–31.

[3] Qi J, Liu D-Y, Zhou G-H, Xu X-L. Characteristic flavor of traditional soup made by stewing Chinese Yellow-Feather Chickens. J Food Sci 2017;82(9):2031–40.

[4] Yan L, Liu J, Jiang S, Wu C, Gao K. The regular interaction pattern among odorants of the same type and its application in odor intensity assessment. Sensors (Basel) 2017;17(7):1624.

[5] Covington JA, Agbroko S, Tiele A 2017. A simple, portable, computer-controlled odour generator. ISOEN 2017—ISOCS/IEEE International Symposium on Olfaction and Electronic Nose, Proceedings 7968848.

[6] Duncan B, Le NDB, Alexander C, Li X, Rotello VM, et al. Sensing by smell: nanoparticle-enzyme sensors for rapid and sensitive detection of bacteria with olfactory output. ACS Nano 2017;11(6):5339–43.

[7] Zhou H, Luo D, Gholamhosseini H, Li Z, He J. Identification of Chinese herbal medicines with electronic nose technology: applications and challenges. Sensors (Basel) 2017;17(5):1073.

[8] Li Q, Yu X, Xu L, Gao J-M. Novel method for the producing area identification of Zhongning Goji berries by electronic nose. Food Chem 2017;221:1113–19.

[9] Eusebio L, Derudi M, Capelli L, Nano G, Sironi S. Assessment of the indoor odour impact in a naturally ventilated room. Sensors (Basel) 2017;17(4):778.

[10] Son M, Lee JY, Ko HJ, Park TH. Bioelectronic nose: an emerging tool for odor standardization. Trends Biotechnol 2017;35(4):301–7.

[11] Wiśniewska P, Śliwińska M, Dymerski T, Wardencki W, Namieśnik J. Qualitative characteristics and comparison of volatile fraction of vodkas made from different botanical materials by comprehensive two-dimensional gas chromatography and the electronic nose based on the technology of ultra-fast gas chromatography. J Sci Food Agric 2017;97(4):1316–25.

[12] Nagappan PG, Subramaniam S, Wang D-Y. Olfaction as a soldier—a review of the physiology and its present and future use in the military. Mil Med Res 2017;4(1):9.

[13] Xiao Z, Zhang Y, Zhu G, Zhou R, Niu Y. Preparation and sustained-releasing characterization of aromatic wallpaper. Prog Org Coat 2017;104:50–7.

[14] Wasilewski T, Gębicki J, Kamysz W. Bioelectronic nose: current status and perspectives. Biosens Bioelectron 2017;87:480–94.

[15] Son M, Kim D, Ko HJ, Hong S, Park TH. A portable and multiplexed bioelectronic sensor using human olfactory and taste receptors. Biosens Bioelectron 2017;87:901–7.

[16] Ray H, Bhattacharyya N, Ghosh A, Biswas SP, Majumdar S, et al. Fragrance profiling of *Jasminum Sambac* Ait. flowers using electronic nose. IEEE Sens J 2017;17(1):160–8 7676374.

[17] Romero-Flores A, McConnell LL, Hapeman CJ, Ramirez M, Torrents A. Evaluation of an electronic nose for odorant and process monitoring of alkaline-stabilized biosolids production. Chemosphere 2017;186:151–9.

[18] Yimenu SM, Kim JY, Kim BS. Prediction of egg freshness during storage using electronic nose. Poult Sci 2017;96(10):3733–46.

[19] Xiao Z, Li J, Niu Y, Liu Q, Liu J. Verification of key odorants in rose oil by gas chromatography – olfactometry/aroma extract dilution analysis, odour activity value and aroma recombination. Nat Prod Res 2017;31 (19):2294–302.

[20] Gu X, Sun Y, Tu K, Dong Q, Pan L. Predicting the growth situation of *Pseudomonas aeruginosa* on agar plates and meat stuffs using gas sensors. Sci Rep 2016;6:38721.

[21] Available from: <http://nanogloss.com/nanotechnology/applications-of-nanotechnology-in-perfumes/#axzz61vWpqJtd>

[22] Available from: <https://www.badgerbalm.com/s-33-zinc-oxide-nanoparticles-clear-zinc-sunscreens.aspx>

[23] Available from: <https://www.frontiersin.org/articles/10.3389/fvets.2018.00127/full>

[24] Available from: <https://www.madesafe.org/science/hazard-list/titaniumdioxide/>

CHAPTER

14

Natural polymers for natural hair: the smart use of an innovative nanocarrier

Pierfrancesco Morganti[1,2] *and G. Morganti*[3]

[1]Academy of History of Health Care Art, Rome, Italy [2]China Medical University, Shenyang, P.R. China [3]ISCD Nanoscience Centre, Rome, Italy

14.1 Introduction: the hair system

The entire surface of the body is supplied with hair, but while on the head they are longer and more pigmented, on two areas, the palm of hand and the sole of feet, they are thinner and devoid of the fibrous, biological material, produced by the follicles [1]. Moreover, both women and men have an equal number of follicles, the structure of which differs in their physiological responses. However, hair represents the "crowning glory of femininity for women and a deep-seated archaic symbol of masculinity for men" [1]. By which material is the hair made?

Hair is a protein-based and hormone-dependent biomaterial which, as an appendage of the epidermis, grows in filament-like structures from the specialized cells of its follicle, beneath the surface of the skin as scales [1] (Fig. 14.1). The only growing or live portion of this structure is the hair root (*papilla*), found at the base of the follicle. As soon as the cell making up hair is produced, it dies becoming harder to form the hair shaft. Successively, the shaft, composed, like the epidermis, of cells filled with the protein keratin, is gradually pushed up the follicle tube toward the surface of the scalp at the rate of about 0.2–0.3 mm a day [1]. Moreover, it is to underline that over 90% of the dry weight of human hair's composition comprises keratins characterized by a high content of disulfide bonds (S-S) which, derived from the amino acid cystine, impart stability to the entire structure. When these disulfide bonds are damaged or broken, the characteristic flexibility of hair strands is lost and hair becomes brittle, snarled, and frayed. However, hair lives according to particular synchronized cycles of three distinct phases [2,3] (Fig. 14.2): *anagen*, or growing

Nanocosmetics
DOI: https://doi.org/10.1016/B978-0-12-822286-7.00013-9

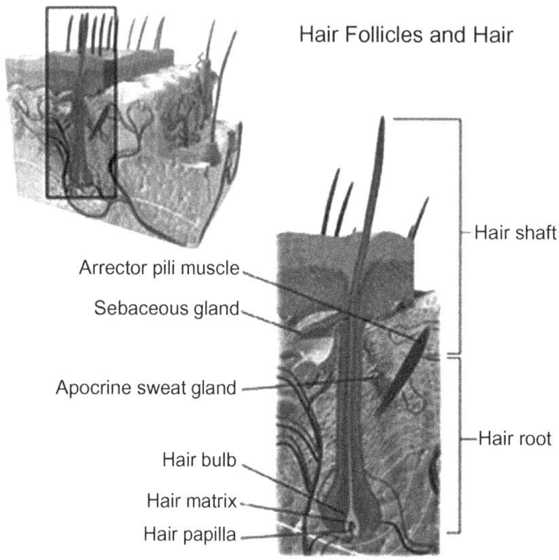

FIGURE 14.1 Hair beneath the skin papilla.

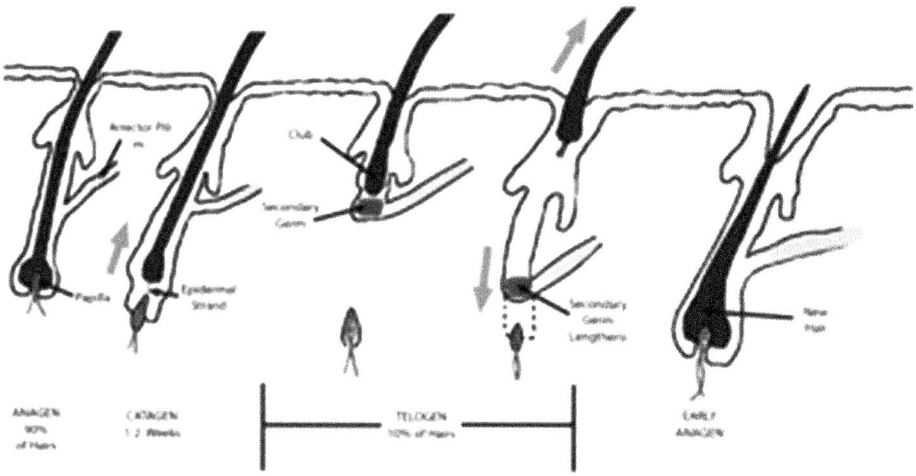

FIGURE 14.2 Hair growth cycle. Schematic representation of normal mature follicle growth cycling from anagen growth, catagen regression, and telogen quiescence [4]. Reproduced with permission from Ref. 4. Copyright 2018 Springer Nature.

phase, that lasts about 3 years and involves 90%–95% of the hair at any one time; *catagen* that, involving daily 13% of the hair, represents the quiescent phase during which the bulb becomes thinner and loses its connection with the papilla; and *telogen* or loss phase, during which the new incoming growth hair pushes out the old ones. It is interesting to remember that 1% of the hair in the telogen phase is lost and, therefore, even in healthy subjects, around 100–200 hair can be lost each day [3].

14.2 Hair structure and damage

Hair is superficially lubricated with lipids (fatty acids) which, predominantly represented by 18-methyl eicosanoic acid (18-MEA), are covalently attached to its epicuticle layer (Fig. 14.3) [5]. These lipids, secreted by the sebaceous glands and opened into the follicle, form a thin superficial coating and lubricant film, providing a hydrophobic surface able to reduce the interfibrillar friction among the hair. Thus lipids and cuticle scales are of basic importance for the protection of the hair's tribological safety barrier against the wear damages, such as the environmental aggressions or the compressive forces of combing [4]. It is, in fact, to remember that the entire outer surface of hair, covered by a cuticle from the root to the tip, is organized as an overlapping scale structure like tiles in a roof, which encloses an inner layer called *medulla* and a middle layer called *cortex* (Fig. 14.4) [6]. Cortex, composed of numerous bundles of fibrous components called macrofibrils, contains large amounts of protein-bound cystine and the pigments responsible for the hair color. Any damage to the hair protective cuticle layer allows the slitting of the cortex and facilitates the further damage caused from usual grooming, such as combing and brushing. The cuticle, therefore, that determines the hair luster also represents the visible portion of the hair, where generally many cosmetic products act as protective compounds toward the environmental aggressions. Thus, while cortex is responsible for the tensile properties of hair, the cuticle scales affect the

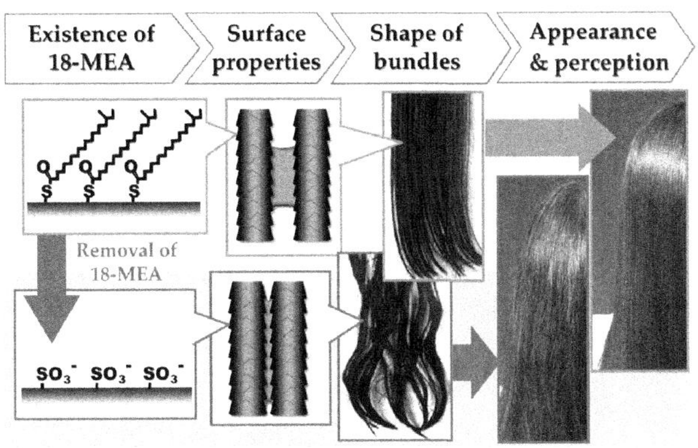

FIGURE 14.3 18-MEA represents the more effective lipid of the hair keratin scales[5].

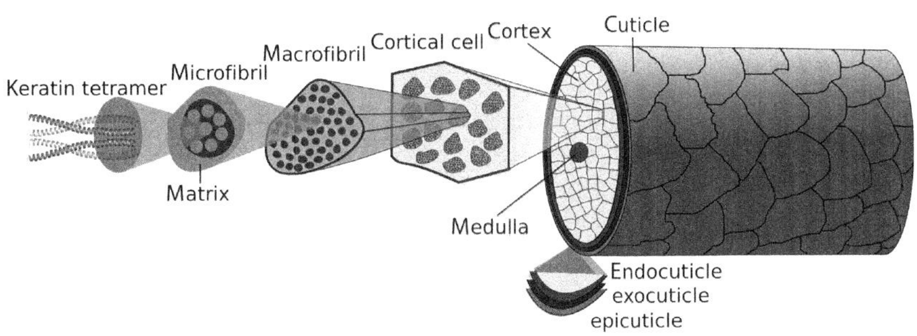

FIGURE 14.4 Hair structure composed fundamentally of cuticle and cortex rich of keratin fibrils[6].

consumer-perceivable properties, such as hair feel, shine, and combing. Macrofibrils of the cortex are made by aggregate keratin polymers, composed of about 18 types of amino acids with a high content of cystine, present as a topographically complex substrate. This sulfur-rich amino acid has an anisotropic directional dependence on the growth mechanism of the hair cuticle and to the complex structure which it forms as a result [5]. Just for its structural organization, therefore, the cystine is primarily responsible for the chemical and structural resilience of hair so that its extensibility seems to be not a function of the outer surface cuticle, but a function of the cortex polypeptide keratin chains [5]. At this purpose, any chemical treatment of the hair, such as perming, bleaching, and permanent dying, based on modifications of the cystine function, can cause significant damage to its fibers, deeply modifying not only their tensile properties but also the entire structure, as previously reported [7]. Many of these treatments, in fact, render hydrophilic the hydrophobic fibers modifying the hair surface, by the introduction of negative charges as a result of oxidation of cystine to cysteic acid. As a consequence, the hair, binding more water molecules, becomes more difficult to comb and more fragile. In addition, other environmental and physical causes can create damages, such as UV light, air particulate, and the hot air of blow dryers. On the one hand, UV light causes photooxidative splitting of cystine linkages and, generating the free radicals formation, is detrimental to the protein matrix, increasing the cuticle porosity and lowering the hair's tensile strength; on the other hand, heat and particulate may denature the hair proteins content. In addition, hair is very sensitive to changes in humidity so that when the air moisture content is too high, its *body* and *hold* characteristics are lost, while when the level is too low, the hair becomes dry and brittle [7]. As a result of the cumulative damages, the hair feels rough and dry, being subject to extensive grooming damage with the change in the number of cuticle scales and the appearance of splits on its surface [7].

14.3 Cosmetic hair care treatments

Specific cosmetic products to treat, protect, and repair the hair fibrils are, therefore, at the basis of many hair care treatments. They provide an interesting modality not only for neutralizing the combing compressive forces between adjacent fibrils, cause of the hair fly-away and the difficult to comb, but also for repairing the wear damages such as the split ends and the many other damages, caused by chemical treatments and environmental aggressions (Fig. 14.5) [8]. However, it has been shown that the relative orientation and number of repeated strokes of contacting hair fibrils as they slide across one another have an influence on the stresses, generating the final hair damage [9]. Thus the necessity to avoid this damage by the use of the right cosmetic care, the effectiveness of which has to be controlled, verifying the tribological properties of hair, and the hair–hair interface activity before, during, and after the cosmetic/medical treatment. By these controls, it is possible to quantify the manageability and the entire hair life cycle, assessing also the performance of the hair care products. In any way, the right cosmetic treatment help to protect the keratinous fiber structure from the environment aggressions of UV rays and air particulate, having also the possibility to stimulate the regular hair regrowth, as reported later. In conclusion, hair is an important beauty component of all the body and together with the skin represents a fundamental barrier which regulates the interchange of water

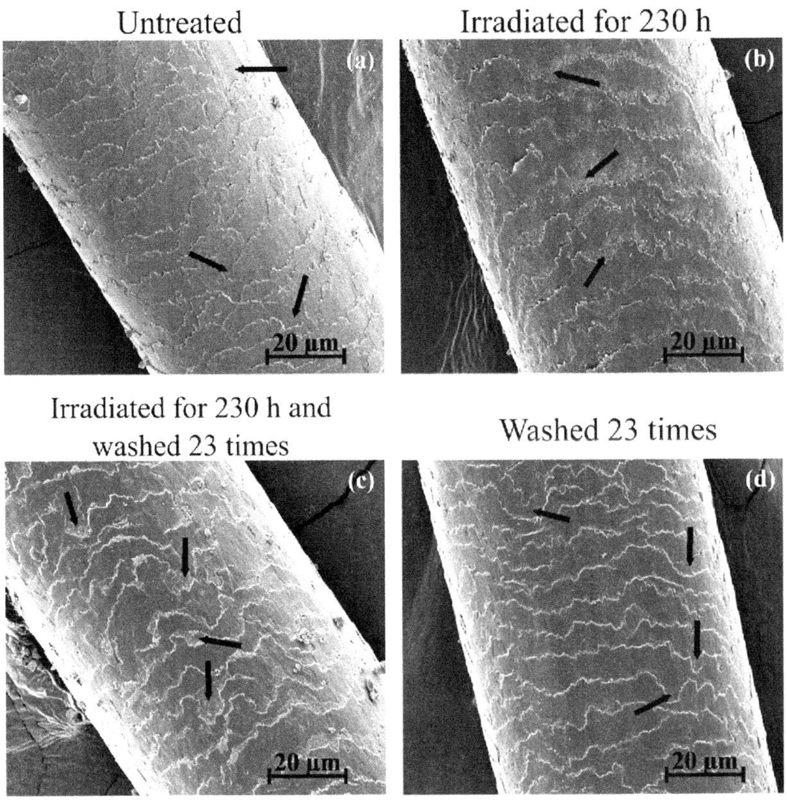

FIGURE 14.5 Images of the dark brown hair surface obtained by FESEM [8]: (a) untreated hair strand; (b) hair UV-irradiated for 230 h; (c) hair UV-irradiated for 230 h and washed after each 10 h UV-irradiation cycle; (d) hair only washed for 23 times. Arrows indicate that some cuticle pieces were broken in (a) and (d) and indicate exposed endocuticles in the hair strand surface in (b) and (c). Reproduced with permission from Ref. 8. Copyright 2016 Elsevier.

loss, being also important to preserve the entrance of pathogenic microorganisms. For all these reasons, everybody hopes to have a smooth, soft, and healthy hair and skin so that they are considered fundamental for the maintenance of wellbeing and a beautiful state. Thus shining, friction, and adhesion properties are considered as the most relevant hair parameters, necessary for protecting its structure from degradation and giving the body appearance of a good health. This is the reason why the realization of more effective and smart nanocosmetics, able to avoid hair damage and improve its quality, became an important must of our society, especially if they are made by natural ingredients, obtainable by waste materials, and made by sustainable technologies [10]. As a consequence, the necessity to standardize the physicochemical characteristics of all the natural materials and polymers used for the hair treatments is born as well as the necessity to verify and considering essential its safeness and effectiveness [11].

14.4 Natural fibers for natural hair

Natural fibers are any hair-like raw material which, produced by plants, animal, and geological processes by sustainable technologies, represents an alternative necessary to substitute the petrol-derived compounds, useful for preserving our health and the environment. These biomaterials, obtained from waste biomass and biodegraded by the enzymatic actions of fungi and bacteria, are also able to easily absorb water [7]. Moreover, their use is considered indispensable to preserve the natural raw materials for the incoming generations, because it is able to reduce the greenhouse gas emissions and maintain the Planet's biodiversity [10–12]. Fortunately, the natural fibers, present in more great quantity in nature and consisting fundamentally of cellulose, hemicellulose, and lignin or chitin and its derived polymers, may be obtained from agricultural biomass or from fishery's by-products, respectively. It is estimated that both agricultural and industrial biomass produce yearly a waste of around 300 billion tons, 20% of which is used to produce energy and unfortunately low-quantity goods [12].

14.4.1 Polysaccharides

Polysaccharides [13], which represent the majority of the natural fibers, are polymeric oligosaccharides or chitooligosaccharides, obtained from agricultural or fishery's by-products, respectively. These sugar-like polymers have recently got attention for their health benefits, possessing particular safeness and effectiveness. They, made of biomaterials in-expensive, non-toxic, and easily degraded, represent, in fact, about 75% of the earth global biomass. On the one hand, the majority of the available waste material is represented from cellulose, lignin, and chitin/chitosan, as previously reported [14–16]. On the other hand, it is estimated that, each year, one-third of all food produced for human consumption in the world (~1.3 billion tons!) is lost, wasted, and underutilized, creating a significant impact on the environment with an estimated footprint of around 3.3 gigatons [17]. Considering the physicochemical properties, it is to underline that solubility, molecular size, stability, as well as the physical morphology and surface charges of these natural polymers, may significantly impact their activity at the skin level. Their effectiveness, in fact, depends on the micro/nanoparticles structure and organization and the relative delivery system that may affect the pharmacokinetic activity. The positively charged particles made, for example, by the chitin nanofibrils may create electrostatic interactions with the negatively charged elements of the skin's interstitial matrix, such as collagen and glycosaminoglycans, as well as may have a specific role in hair cell–cell interactions and adhesion. Moreover, hair scalp at 5.5 pH, hair shaft at pH around 3.67 [18], and the cortex fibers are all characterized by a surface covered by negative charges. Thisis the reason why cosmetic products, characterized by positively charged ingredients, are used to treat scalp and hair, especially when damaged by environmental aggressions, excessive combing, or chemical treatments, which increase the formation of negatively charged surfaces. On the other hand, the anionic alkaline pH, as well as the hair combing, causing friction among the fibers, may lead to cuticle damages and breakages [19]. For these reasons, the majority of hair skin care products are formulated by a pH between 4 and 6. Lignin, accounting for 15%–25% of herbaceous biomass, is present in a significant amount in all plants. Thus raw materials, obtained from the lignocellulosic biomass residue, are estimated to exceed 150 billion tons/year worldwide [20]. However, the global lignin market was estimated at US$ 732.7 million in 2015 and is expected to exceed a revenue of US$ 6.2

billion by 2022 with a value in EU of over 1.6 billion and a production from 50 to 89 billion tons/year [21].

The majority of this production, obtained as a by-product during pulp and paper processing, is used to produce energy with only a small fraction used to make goods [22]. Lignin in fact, also if not well determined in its chemical structure, has an interesting branched polyether, polymeric nature because of the presence in its molecule of antioxidant monomers derived from hydroxy-cinnamic acid and guaiacyl units, connected via aromatic and aliphatic bonds (Fig. 14.6) [23]. For its peculiar characteristic, this ecosustainable polymer offers the potential to produce high-value goods. It has, in fact, the ability to retard and inhibit oxidative reactions, absorb UV rays and, being resistant to decay and biological attacks, possess high stiffness and barrier property, especially when in nanosize dimension [24,25]. At this purpose, it has been shown, for example, that the sun protection factor (SPF) value of sunscreens increases with the decreasing size of lignin [26]. Moreover, lignin has an interesting antimicrobial activity, causing cell membrane damage and lysis of bacteria [27,28]. Thus its use in skin care and hair management has shown a high interest, especially when complexed with chitin nanofibrils (CNs) (Fig. 14.7) to form

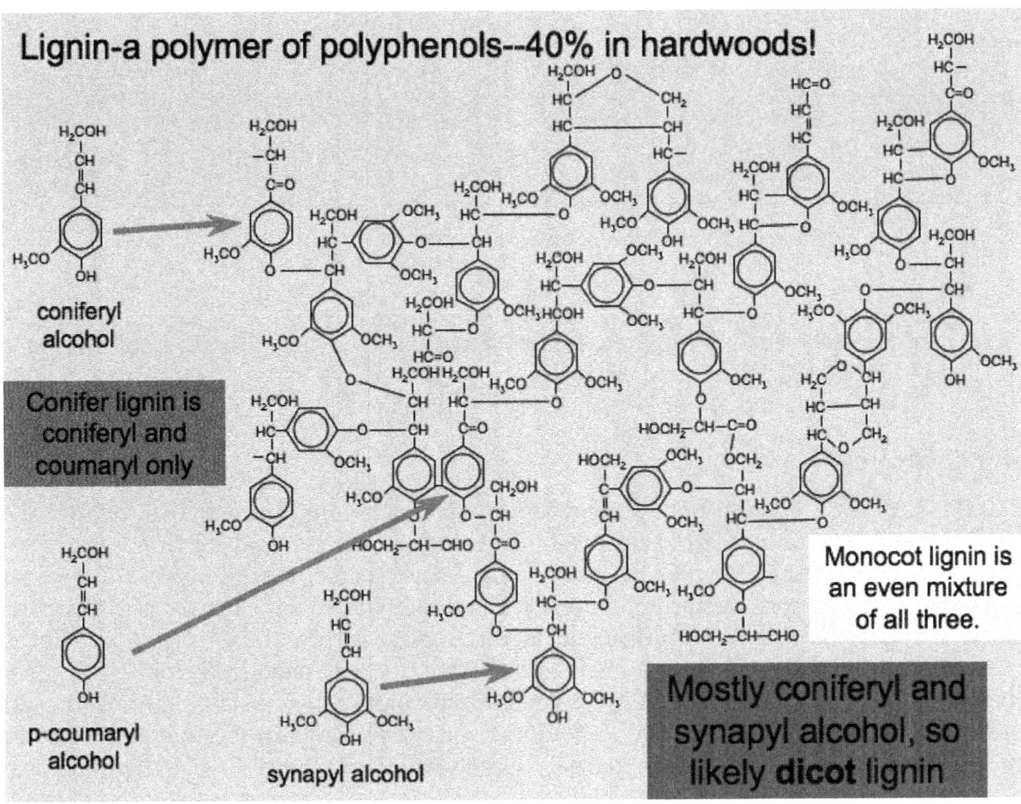

FIGURE 14.6 Supposed lignin chemical structures.

FIGURE 14.7 Chitin nanofibril–nanolignin complex entrapping Vit C.

micro/nanoparticles embedded by specific active ingredients, according to the green chemistry (data in progress not reported).

14.4.2 Chitin

Chitin is the only natural polysaccharide nitrogen-containing, present as structural support of crustaceans and insects exoskeleton, making up the cell wall of fungi and yeasts also [29]. In nature, this polymer predominantly exists as a composite material, consisting of ordered crystalline microfibrils highly hydrated, embedded in a matrix of protein and minerals (Fig. 14.8). The more prominent hierarchical structure of such biological fibrous composite material is the twisted plywood structure, characterized for its excellent biocompatibility and biodegradability [30]. Chitosan composed of beta-(1–4)-linked D-glucosamine and *N*-acetyl-D-glucosamine randomly distributed within the polymer is the linear polysaccharide derived from partial deacetylation of chitin [31]. When the ratio between acetyl and amine groups is higher than 1:1, the polymer is referred to as chitosan, also if the industrial CNs may have also about 60% of amine group [31]. The source of

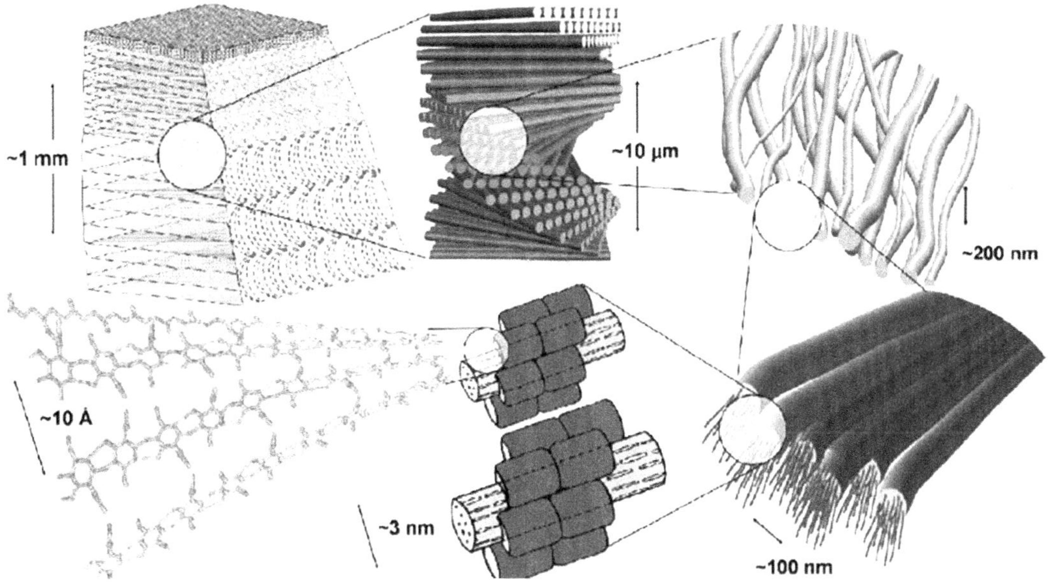

FIGURE 14.8 Hierarchical structure of chitin. [30] Reproduced with permission from Ref. 30. Copyright 2005 Elsevier.

choice of chitin is obtained from crustacean shells, the biomass waste of which represents approximately half of the total weight of shellfish [32,33]. According to FAO [34], in fact, the estimated production of shrimp only in Asian countries is 2.5 million tons, generating 35%–45% of waste [35,36]. If not used, this biomass has a negative impact, representing a potential source of hazards to the coastal and near shore environments [34–36]. However, chitin, chitosan (Fig. 14.9), and its derivatives, for their nontoxicity, low allergenicity, high ecocompatibility and skin-friendly characteristics, have shown biological potential for a wide range of applications also in the cosmetic and medical field [37]. The cationic nature of these natural polymers allows them to form electrostatic complexes or multilayer structures with other negatively charged natural polymers of animal or plant origin, such as hyaluronic acid or lignin, respectively [37,38]. Thus they may be used for skin regenerative purposes in tissue engineering to make, for example, chitin-based polymeric scaffolds having the same structure of extracellular matrix (ECM) (Fig. 14.10) [39], especially when in their nanodimension, which notably increase their utilization and effectiveness [38–40]. Moreover, chitin and its derived compounds have shown to have important antioxidant, antimicrobial, antiinflammatory, immunomodulating, and skin repairing activities [40–43]. These natural polymers, in fact, seem to involve cell lysis and breakdown of the cytoplasmic membrane barrier of microorganisms, while activating the human macrophage via the production of cytokines from intraepithelial lymphocytes [44]. It is interesting to underline, that in their micro/nanosize dimension, CNs (Fig. 14.11), nanolignin (NG) (Fig. 14.12), and their complexes (CN–NG) (Fig. 14.13) embedded by selected active ingredients, seem to easily deliver, load, and carry different active ingredients, thus

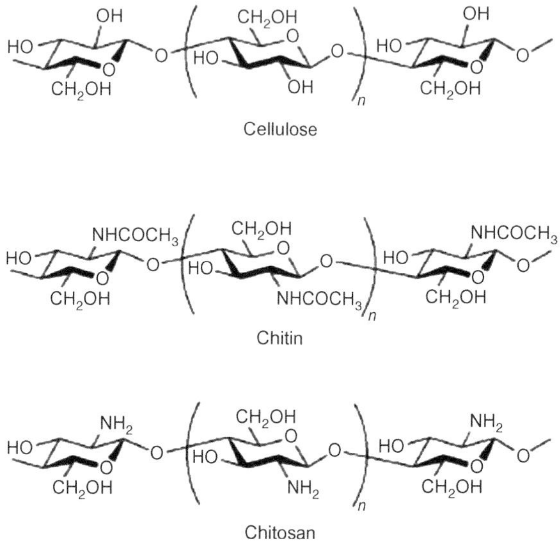

FIGURE 14.9 Chitin and chitosan chemical structure compared to cellulose [50].

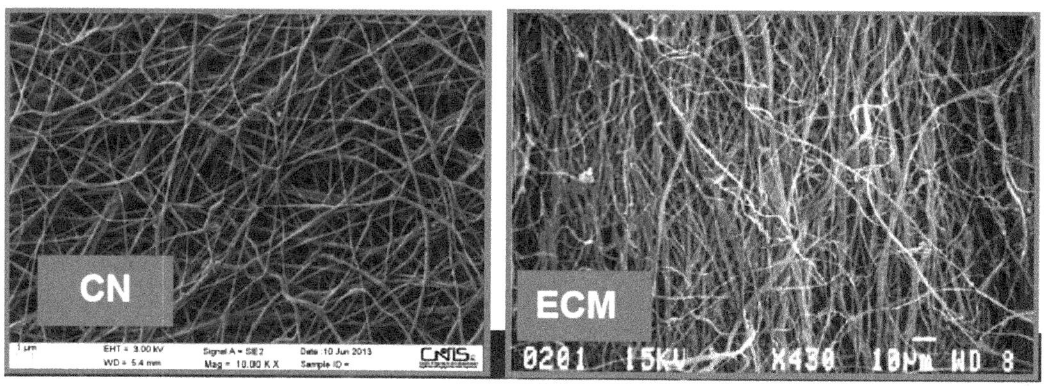

FIGURE 14.10 Chitin nanofibrils structure (left) and natural extracellular matrix (right) at TEM [39].

assuring their carriage to their action sites and maximizing their effectiveness [40–43]. A proper utilization of the CN–NG block copolymeric micro/nanoparticles not only stems from the properties in solving the delivery problems as natural carriers, but also shows an own biological active effectiveness. As previously reported, in fact, CNs and their micro/nanoblock polymeric nanoparticles have shown to possess antimicrobial, antioxidant, anti-inflammatory, and immunoprotective effectiveness [37–43], as well as lignin disclosed antioxidant, antiinflammatory, antimicrobial, and photoprotective activity [22–28]. Thus both CN and NG may be considered interesting active ingredients employable as smart biomaterials in the cosmetic and pharmaceutical field. In fact, they have shown to possess bioimitative characteristics as "active" carriers, able to carry and efficiently release the

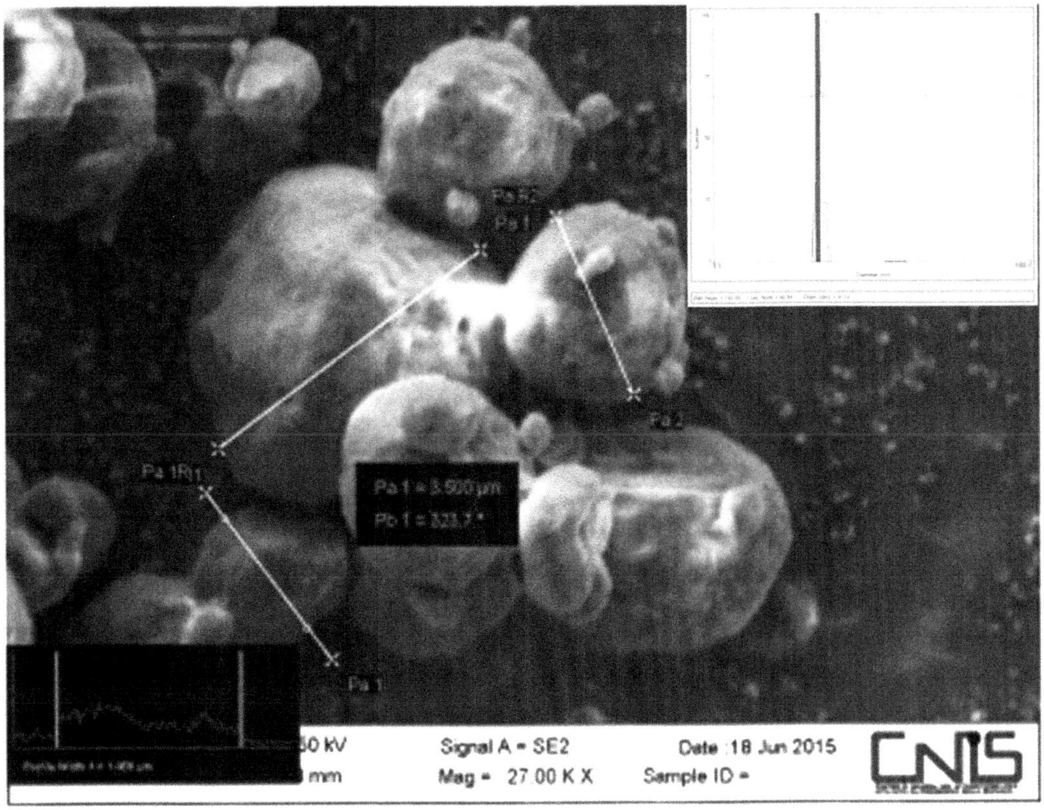

FIGURE 14.11 Chitin nanofibrils at SEM and ZETASIZER Nano.

loaded active ingredients into the designed skin and hair layers, being also easily integrable into engineered tissues [44]. At this purpose, it is also interesting to underline the structural bioimitative characteristic existing between the hair scale and chitosan–CN, shown by SEM (Figs. 14.6 and 14.14) [45]. However, the activity of these micro/nanoparticles depend not only on the form and size of the particle, but also on the dose and quality of the active ingredients bound into their structure or on their surface, as well as on the production cycle selected [46,47]. Naturally, safeness and effectiveness of the final formulation will also depend on the basic vehicle selected into which the nanoparticles are embedded or bound, such as emulsion, gel, film, or nonwoven tissue, which will characterize the cosmeceutical or pharmaceutical product designed [47–50].

14.5 Chitin and hair: final considerations

As reported, the main objective of nanotechnology and nanoparticles is to carry and deliver the active ingredients at the designed level in a controlled dose and time.

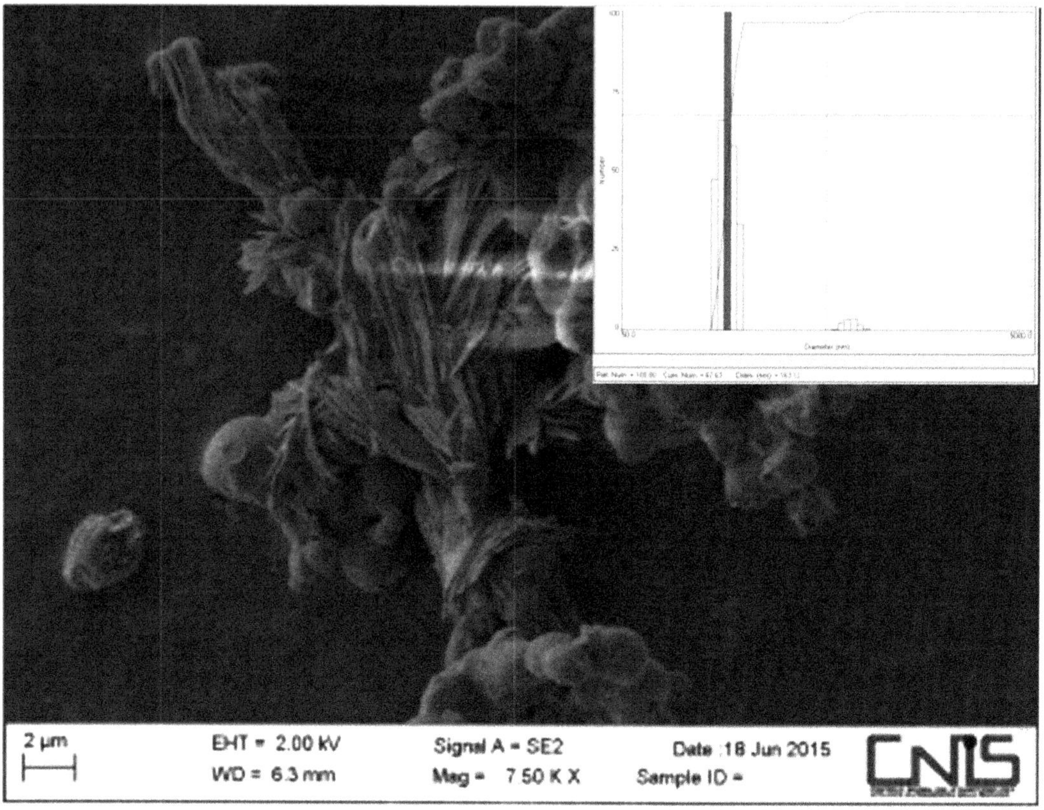

FIGURE 14.12 Nanolignin at SEM and ZETASIZER Nano.

Nanoparticles, in fact, because of their small size and higher surface to weight ratio, show altered physicochemical properties when compared with their larger counterparts, penetrating more easily through the skin and hair structures. The cosmetic product, applied on the hair surface, needs to penetrate until the cortex for being effective; the relative penetration depends on the porosity of its cuticle and the dimensions of the active ingredients. The nanodimension, therefore, can create the opportunity for an increasing uptake and interaction with the biological hair structures, consisting fundamentally of long parallel and folded polypeptide chains, connected by cross-linkages. The addition of the −OH and $-NH_2$ groups of nanochitin and nanolignin into the keratin structure can help to cross-link neighboring keratin molecules and repair broken disulfide and hydrogen bonds, resulting perfectly suited to rebuild stability in damaged hair. Thus the use of hair-like structures, such as CN, NG, and their derived block-polymeric micro/nanoparticles, may be of help to ameliorate the different hair formulations, favoring the penetration of the active ingredients through its structure and rendering the cosmetic treatment more effective. Consumers, in fact, are looking for personal care products which, incorporating

FIGURE 14.13 Chitin nanofibril–nanolignin complex at SEM and ZETASIZER Nano.

natural biomaterial produced by the latest advanced technologies and having the possibility to supply multiple benefits with minimal efforts, are characterized by their high effectiveness and the absence of side effects [51]. This is the reason why bionanotechnology is the fastest developing area of research involved in resolving science-based solutions for smart pharmaceutical and cosmetic industries, looking for the use of innovative biomaterials by innovative techniques [52]. Biomaterials, in fact, indicate natural and synthetic substances that are able to interact with biological systems, without producing toxic side effects, just as the activity shown by the complex of nanochitin–nanolignin at the level of the hair structure (data in progress not reported). As widely recognized, in fact, the unicity of chitin and its derived compounds mainly depends on the many $-NH_2$ and $-OH$ groups present on their backbone, configuring them as natural polycationic polymers in acidic pH. On the other hand, the lignin activity depends prevalently on the many phenolic groups, composing its molecule. Thus, for their specific physicochemical characteristics, CNs, chitosan, lignin, and their block-polymeric nanoparticles could result effective to protect the hair from the environmental aggressions and from the chemicals used for

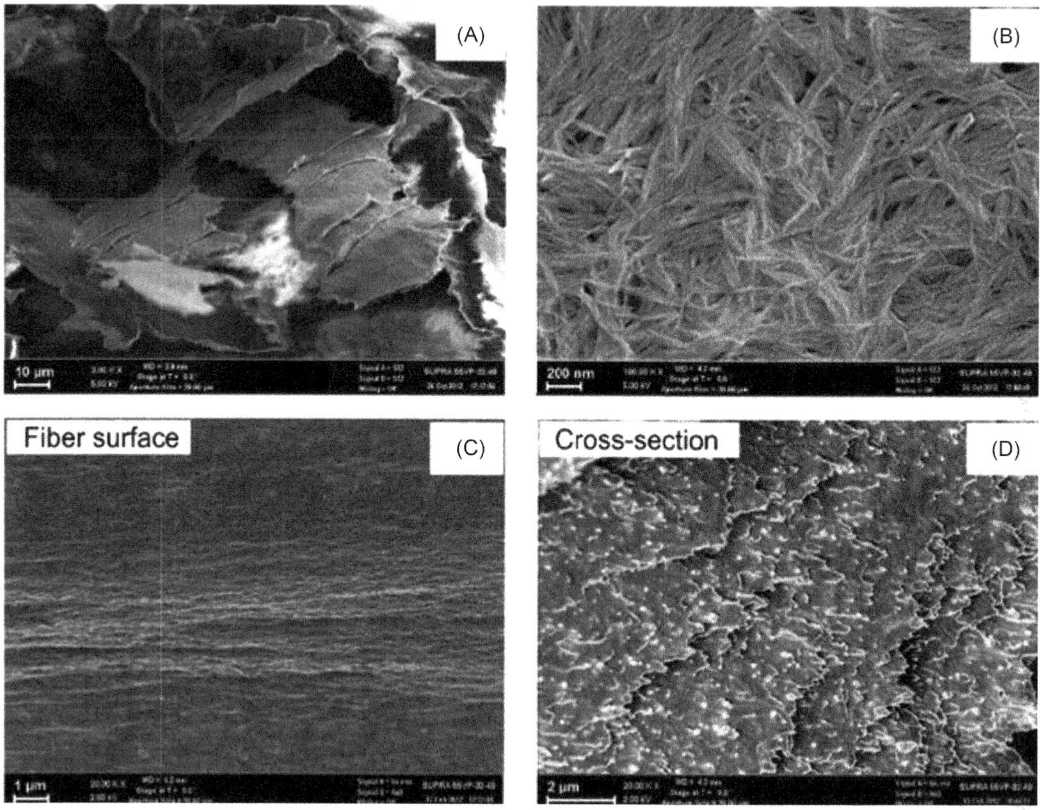

FIGURE 14.14 The organization of chitosan–chitin fiber resembling the hair structure (B–D). Source: *Courtesy Yudin VE, Dobrovolskaya IP, Neelov IM, Elokhovski VY, Kasatkin IA, Okrugin BM, et al. Wet spinning of fibers made of chitosan and chitin nanofibrils. Carbohydr Polym 2014;108:176–82.*

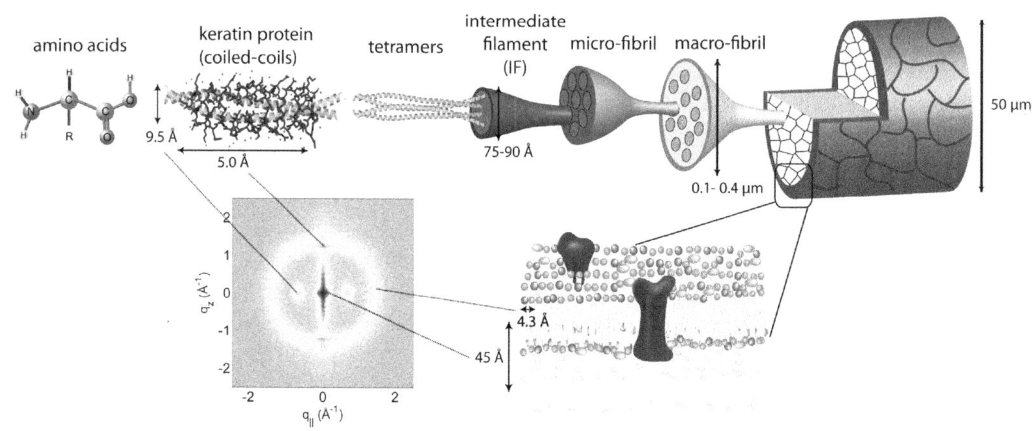

FIGURE 14.15 Structural organization of the hair keratin [54]. Reproduced with permission from Ref. 54. Copyright 2017 Royal Society.

permanent waving, hair straightening, and bleaching, cause of strong depletion of the hair surface lipids with disruption of the protective scales and cortex fibers. These natural polymers are able to enter and interfere with the intricate mechanisms of the hair structure, made by the many hydrogen bonds connecting each other with the different amino acids of the hair, with particular reference to the cystine–cysteine activity. To try to understand their mechanism of activity at the level of the hair, it is better to remember and focus again on the structural organization of the hair. Hair, as previously reported, is a composite structure built from spindle-shaped cortical cells surrounded by a sheath made of several layers of flat cuticle cells overlapping with each other [53]. The cortical cells contain alpha-helical keratin chains [54] which, made from many amino acids bound together by the covalent bonds of cystine, represent 21% of their totality. This complex structure of cells, embedded in an amorphous matrix of keratin-associated protein results, therefore, fundamental for maintaining the mechanical properties of the hair. At this purpose, it is interesting to remember that most of the damages incurred to hair are, directly or indirectly,

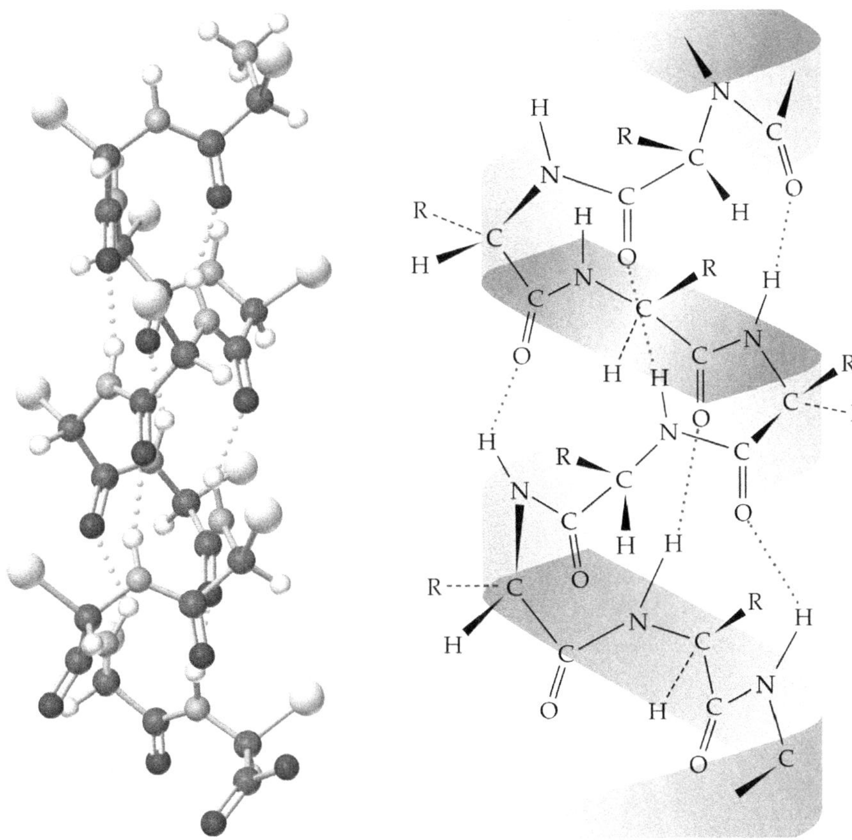

FIGURE 14.16 Hydrogen bonds of alpha-helix keratin.[56]. Reproduced with permission from Ref. 56. Copyright 2019 Elsevier.

related to the chemistry of the cystine [55]. In fact, the ability of this amino acid is to establish disulfide bonds responsible for the high degree of the fibers' cross-linking together with the hydrogen bonds,realizing a bridge between the oxygen atoms of carbonyls and the amide groups from the neighboring chain of the other amino acids. Thus, for example, the temporary chemical rupture of disulfide and the hydrogen bonds leads to temporary hair deformations, namely, the hair set. Cleavage of cystine links and their subsequent reforming in a new position, in fact, is the process that affords the permanent deformation [55]. On the modification of the same linkages is also based on the chemistry of bleaching, necessary to oxidize melanin pigments for obtaining the hair lightening. The bleaching process and the oxidation hair coloring [56], therefore, call a range of chemical reactions involving not only the pigment it is intended to lighten or change, but also the keratin fiber itself. The keratin's polypeptide (side) chains (Fig. 14.15) [56] and the linkages binding them together, in fact, are modified by these cosmetic treatments. In addition, it is to know that during the hair dying operation, also if all the procedures tend to color the hair alone, small quantities of dye, metals, and precursors may reach the scalp, crossing the cutaneous barrier with the risk for the human health [57]. Thus the necessity to use natural ingredients which, able to modulate the many hair treatments, reduce the consequential risks for both hair and the health, complexing and eliminating the residual chemicals remaining on the scalp, for example, after the permanent dying. At this purpose, CN and its derivatives have shown to be useful because of their high capacity to remove and adsorb toxic metals and ingredients at low molecular weight by the activity of the functional

FIGURE 14.17 The planet biodiversity to be preserved (a Chinese garden).

groups of -NH_2 and -OH present on their molecules, being also safe and at low cost [58]. Moreover, due to the positive electrical charges covering their molecular surface, they are able to bind the microfibers to each other, enhancing hair shine and gloss and neutralizing the flyaway phenomenon also. In addition, protecting the hair from the environmental aggressions, CN may normalize its structure, eliminating temporarily the different split ends and safeguarding both its aspect and health. Finally, surface-deacetylated CNs have shown to promote hair growth, increasing the fibroblast growth factor-7 and upregulating the hair follicles [59]. These data confirm the previous results obtained by our group, who evidenced a direct activity of nanochitin on the stem cells of hair papilla [60] (Fig. 14.16).

For all these reasons, a more in-deep knowledge and research studies on these natural polymers are considered of primary importance, for trying to better understand their real mechanism of action and effectiveness. In any way, the major use of these natural polymers to produce nanocosmetics will contribute to reduce the food waste [61], to slow down the CO_2 emission [62], maintaining the natural raw materials for the incoming generations and safeguarding the biodiversity of our planet (Fig. 14.17).

References

[1] Ebling FJ. Biology of hair follicles. In: Fitzpatrick TM, Eisen AZ, Wolf K, Freedberg IM, Austen KF, editors. Dermatology in general medicine. 3rd ed New York, NY.: McGraw Hill; 1987.

[2] Hair follicle, https://en.wikipedia.org/wiki/Hair_follicle.

[3] Kligman AM. The human hair cycle. J Invest Dermatol 1959;33:307.

[4] Sasaki GH. Review of human hair follicle Bbiology: Dynamics of niches and stem cell regulation for possible therapeutic hair stimulation for plastic surgeons. Aesth. Plast. Surg. 2019;43:253.

[5] Tokunaga S, Tanamachi H, Ishikawa K. Degradation of hair surface: Importance of 18-MEA and epicuticle. Cosmetics 2019;6(2):31. Available from: https://doi.org/10.3390/cosmetics6020031.

[6] Cruz CF, Costa C, Gomes AC, Matamá T, Cavaco-Paulo A. Human hair and the impact of cosmetic procedures: A review on cleansing and shape-modulating cosmetics. Cosmetics 2016;3(3):26. Available from: https://doi.org/10.3390/cosmetics3030026.

[7] Robbins C, Kameth Y. Hair breakage during combing. IIII: The effects of bleaching and conditioning on short and long segment breakage by beet and dry combing of stresses. J Soc Cosmet Chem 2007;58:477–84.

[8] Richena M, Rezende CA. Morphological degradation of human hair cuticle due to simulated sunlight irradiation and washing. J. Photoch. Photobio. B 2016;161:430–40. Available from: https://doi.org/10.1016/j.jphotobiol.2016.06.002.

[9] Collina A. Sustainable products: the innovation strategy for the European Chemical Industry. In: Morganti P, editor. Bionanotecnology to save the environment. Plant and Fishery's biomass as alternative to petrol. Basel: MDPI Edition; 2019. p. 13–179.

[10] Morganti P, Hong-Duo C, Li YH. Green bioeconomy for a more sustainable environment. In: Morganti P, editor. Bionanotechnology to save the environment. Plant and Fishery's biomass as alternative to petrol. Basel: MDPI Edition; 2019. p. 39–59.

[11] Cinelli P, Coltelli MB, Mallegni N, Lazzeri A. Biodegradable and biobased polymers: definitions, standards, and future perspectives. In: Morganti P, editor. Bionanotechnology to save the environment. Plant and Fishery's biomass as alternative to petrol. Basel: MDPI Edition; 2019. 195-124.

[12] Morganti P, editor. Bionanotechnology to save the environment. Plant and Fishery's biomass as alternative to petrol. Basel: MDPI Edition; 2019.

[13] Benkeblia N. Polysaccharides: natural fibers in food and nutrition. New York, NY: CRC Press; 2014.

[14] Yan N, Chen X. Don't waste seafood waste. Nature 2015;524:155–7.

[15] Morganti P, Chen H-D, Gao X-H, Morganti G, Febo D. Chitin & lignin: turning food waste into cosmeceuticals. J Clin Cosmet Dermatol 2018;3(1). Available from: https://doi.org/10.16966/2576-2826.135.

[16] Yadav M, Goswami P, Paritosh K, Kumar M, Pareek N, Vivekanad V. Seafood waste: a source for preparation of commercially employable chitin/chitosan materials. Bioresour Bioprocess 2019;6:8. Available from: https://doi.org/10.1186/s40643-019-0243-y.

[17] Gustavsson J, Cederberg C, Sonesson U, Van Otterdijk R, Meybeck A. Global food losses and food waste—extent, causes and prevention. Rome: FAO; 2011.

[18] Gavazzoni Dias MF, de Almeida yay AM, Cecato PM, Adriano AR, Pichler J. The shampoo pH can affect the hair: myth or reality. Int J Trichol 2014;6(3):95–9. Available from: https://doi.org/10.4103/0974-7753.139078.

[19] Robbins CR, Crawford RJ. Cuticle damage and the tensile properties of human hair. J Soc Cosmet Chem 1991;42:59–67.

[20] Zhang ML, Fan YT, Xing Y, Pan CM, Zhang GS, Kay JJ. Enhanced biohydrogen production from cornstalk wastes with acidification pretreatment by mixed aerobic cultures. Biomass Bioenergy 2007;31:250–4.

[21] Morganti P, Stoller M. Chitin and lignin: natural ingredients from waste materials to make innovative and healthy products for human and plant. Chem Eng Trans 2017;60:319–24. Available from: https://doi.org/10.3303/CET1760054.

[22] Vinardell MP, Mitjans M. Lignins and their derivatives with beneficial effects on human health. Int J Mol Sci 2017;(18):1219. Available from: https://doi.org/10.3390/ijms18061219.

[23] Gellerstedt G, Henriksson G. Lignins: major sources, structure and properties. Monomers, polymers and composites from renewable resources. Amsterdam: Elsevier; 2008. p. 201–24.

[24] Laurichesse S, Averous L. Chemical modification of lignins: towards biobased polymers. Prog Polym Sci 2014;39:1266–90.

[25] Cai X, Riedl B, Zhang SY, Wan H. The impact of the nature of nanofillers on the performance of wood polymer nanocomposites. Compos Part A Appl Sci Manuf 2008;39:727–37.

[26] Qian Y, Zhong X, Li Y, Qiu X. Fabrication of uniform lignin colloidal spheres for developing natural broad-spectrum sunscreens with high sun protection factor. Ind Crop Prod 2017;101:54–60.

[27] Telysheva G., Dizhbite T., Lebedeva G., and Nikoleava V. (2005). Lignin products for decontamination of environment objects from pathogenic microorganisms and pollutants. In: Proceedings of the 7th ILI Forum, Barcelona, Spain, 27–28 April, p. 71–74.

[28] Beisl S, Friedl A, Miltner A. Lignin from micro- to nanosize: applications. Int J Mol Sci 2017;2017(18):2367. Available from: https://doi.org/10.3399/ijms18112367.

[29] Named I, Ozogul F, Regenstein JM. Industrial applications of crustacean by-products (chitin, chitosan, and chitooligosaccharides): a review. Trends Food Sci Technol 2016;48:40–50. Available from: https://doi.org/10.1016/j.tifs.2015.11.007.

[30] Raabe D, Romano P, Sachs C, Al-Sawalmih A, Brokmeier H-G, Yi S-B, et al. Discovery of a honeycomb structure in the twisted plywood patterns of fibrous biological nanocomposite tissue. J. Cryst. Growth 2005;283 (1–2):1–7. Available from: https://doi.org/10.1016/j.jcrysgro.2005.05.077.

[31] Ramirez MA, Rodriguez AT, Alfonso L, Peniche C. Chitin and its derivatives as biopolymers with potential agricultural applications. Biotecnol Apl 2010;27:270–6.

[32] Venkatesan J, Kim SK. Chitosan composites for bone tissue engineering. An overview. Mar Drugs 2010;8:2252–66.

[33] Zhang X, Roland M. Engineering strategies for chitin nanofibers. J Mater Chem B 2017;5:2547–59. Available from: https://doi.org/10.1039/c6tb03324e.

[34] FAO. Increased production of farmed shrimp leads to improved international trade. GLOBEFISH-analysis and information on world fish trade, <www.fao.org/in-action/globefish/market-reports/en/c/989543/>; 2017 [accessed 06.04.19].

[35] Mao X, Guo N, Sun J, Zue C. Comprehensive utilization of shrimp waste based on biotechnological methods: a review. J Clean Prod 2017;143:814–23.

[36] Islam MS, Khan S, Tanaka M. Waste loading in shrimp and fish processing effluents: potential source of hazardous to the coastal and near shore environments. Mar Pollut Bull 2004;49:103–10.

[37] Morganti P, Morganti G. Chitin nanofibrils for advanced cosmeceuticals. Clin Dermatol 2008;26:334–40.

[38] Madhihally SV, Matthew HW. Porous chitosan scaffolds for bone tissue engineering. Biomaterials 1999;20 (12):1133–42.

[39] Morganti P, Morganti G, Colairo C. Biofunctional Textiles for Aging Skin. Biomedicines 2019;7(3):51. Available from: https://doi.org/10.3390/Biomedici-nes7030051.

[40] Morganti P. Use of chitin nanofibrils from biomass for an innovative bioeconomy. In: Ebothe' J, Ahmed W, editors. Nanofabrication using nanomaterials. Manchester: One Central Press; 2016. p. 1–22.
[41] Morganti P, Del Ciotto P, Stoller M, Chianese A. Antibacterial and anti-inflammatory green nanocomposites. Chem Eng Trans 2016;47:61–6. Available from: https://doi.org/10.3303/CET1647011.
[42] Morganti P, Fusco A, Paoletti I, Perfetto B, Del Ciotto P, Palombo M, et al. Antiinflammatory, immunomodulatory, and tissue repair activity in human keratinocytes by green nanocomposites. Materials 2017;2017:843. Available from: https://doi.org/10.3390/ma10070843.
[43] Jui YY, Chin TN, Ing TS, Cardillo M, Gagliardini A, Morganti P. Cross-sectional study design and data analysis of the effect of chitin nanofibrils-lignin micro/nanoparticles on Malaysia's subjects with skin disorders. J Clin Cosmet Dermatol 2019;3:1–6 ISSN 2576-2826.
[44] Je J, Kim S. Chitosan derivatives killed bacteria by disrupting the outer and inner membrane. J Agric Food Chem 2006;54:6629–33.
[45] Yudin VE, Dobrovolskaya IP, Neelov IM, Elokhovski VY, Kasatkin IA, Okrugin BM, et al. Wet spinning of fibers made of chitosan and chitin nanofibrils. Carbohydr Polym 2014;108:176–82. Available from: https://doi.org/10.1016/j.carbpol.2014.02.090.
[46] El-Say KM, El-Sawy HS. Polymeric nanoparticles: promising platform for drug delivery. Int J Pharm 2017;528:675–91. Available from: https://doi.org/10.1016/j.ijpharm.2017.06.052.
[47] Rao JP, Geckeler KE. Polymer nanoparticles techniques and size-control parameters. Prog Polym Sci 2011;36:887–913.
[48] Ming LC, Ang WC, Yang Q, Thitilertdecha O, Wong TW, Khan TM. Cosmeceuticals: safety, efficacy and potential benefits. Advances in medical technologies and clinical practice. Australia: IGI Global; 2017. p. 287–9. Available from: https://doi.org/10.4018/978-1-5225-0754-3.ch010.
[49] Shin KO, Park HS. Antiaging cosmeceuticals in Korea and open innovation in the era of the 4th Industrial Revolution: from research to business. Sustainability 2019;11:898. Available from: https://doi.org/10.3399/su11030898.
[50] Morganti P, Coltelli MB. A new carrier for advanced cosmeceuticals. Cosmetics 2019;6:10. Available from: https://doi.org/10.3390/cosmetics6010010.
[51] Mintel. Facial skincare market research reports—consumer behaviour analysis-market trend. Persistent Market Research. London: Mintel Group Ltd; 2018.
[52] Schoenberg FJ, Lemons JE. Biomaterial science: a multidisciplinary endeavor. Biomaterial science: an introduction to materials in medicine. 2nd Ed San Diego, CA: Elsevier, Academic Press; 2004. p. 1–9.
[53] Randall Wickett R. Mechanical properties of hair. In: Evans T, Randall Wickett R, editors. Practical modern hair science. Carol Stream, IL: Allured Books; 2012. p. 223–50. Available from: <www.Alluredbooks.com>.
[54] Asfia S, Alsop Richard J, Atsuko N, Laurent K, Douglas F, Kuczmarski Edward R, et al., Giant axonal neuropathy alters the structure of keratin intermediate filaments in human hair14. J. R. Soc. Interface. https://doi.org/10.1098/rsif.2017.0123.
[55] Popescu C. Hair damage. In: Evans T, Randall Wickett R, editors. Practical modern hair science. Carol Stream, IL: Allured Books; 2012. p. 366–88.
[56] Hassan MM, Carr CM. A review of the sustainable methods in imparting shrink resistance to wool fabrics. J. Adv. Res. 2019;18:39–60. Available from: https://doi.org/10.1016/j.jare.2019.01.014.
[57] Zviak C. Permanent waving and hair straightening. In: Evans T, Randall Wickett R, editors. Practical modern hair science. Carol Stream, IL: Allured Books; 2012. p. 183–212. Available from: <www.Alluredbooks.com>.
[58] Evans T, Randall Wichett R. Practical modern hair science.In: Carol Stream, IL: Allured Books; 2012. Available from: <www.Alluredbooks.com>.
[59] Anastopoulos I, Bhatnagar A, Bikiaris DN, Kyzas GZ. Chitin adsorbents for toxic metals: a review. Int J Mol Sci 2017;2017(18):114. Available from: https://doi.org/10.3390/ijms18010114.
[60] Azuma K, Koizumi R, Izawa H, Morimoto M, Saimoto H, Osaki T, et al. Hair growth-promoting activities of chitosan and surface-deacetylated chitin nanofibers. Science Direct 2018;126:11–17. Available from: https://doi.org/10.1016/j.ijbiomac.2018.12.135.
[61] Biagini G, Zizzi A, Giantomassi F, Orlando F, Lucarini G, Mattioli Belmonte M, et al. Cutaneous absorption of nanostructured chitin associated with natural synergistic molecules (lutein). J Appl Cosmetol 2008;26:69–80.
[62] FAO. Food wastage footprint & climate change. Rome: Food Agricultural Organization; 2011. Available from: <www.fao.org/nr/sustainability/food-loss-and-waste>.

CHAPTER

15

Skin and pollution: the smart nano-based cosmeceutical-tissues to save the planet' ecosystem

P. Morganti[1,2,3], G. Morganti[1] and M.B. Coltelli[4]

[1]Dermatology Unit, Campania University, "L. Vanvitelli", Naples, Italy [2]China Medical University, Shenyang, P.R. China [3]ISCD NanoScience Center, Rome, Italy [4]Department of Civil and Industrial Engineering, University of Pisa, Pisa, Italy

15.1 Introduction

In the world continually and rapidly changing, our actual way of living is not only the main cause of skin irritation, sensitivity, dryness, roughness, and premature aging, but may be dangerous for all the body, provoking at worst cancer [1]. This pathological state can be considered an aging disease that, occurring as a result of cumulative environmental aggressions and altered immune response, takes place throughout the human's life span (Fig. 15.1). The underlying mechanism in both cancer and aging, in fact, is the time-dependent accumulation of cellular damage on the body [2]. Thus, on the one hand, in the last 50 years, the reduction in mortality for the progress in biology and medical advances has led to a great increase of life expectancy. On the other hand, the aging population, by a longer lifetime exposure to various toxic agents (i.e., chemical irritants and environmental pollution factors, such as sun rays, air particulate, and ozone) and their accumulation on skin and mucous membranes, provoked and are still provoking profound effects on the health [3]. Consequently, cancer became worldwide an important health burden for elderly people and it has been forecasted that 43% of men and 38% of women will develop this invasive pathology over a lifetime [1,2].

At this purpose, according to the World Health Organization [4], 91% of the world's population live in places where air quality exceeds the guidelines limits so that 7 million people die prematurely every year for exposure to polluted air [5]: 4.2 million deaths result for

DOI: https://doi.org/10.1016/B978-0-12-822286-7.00014-0

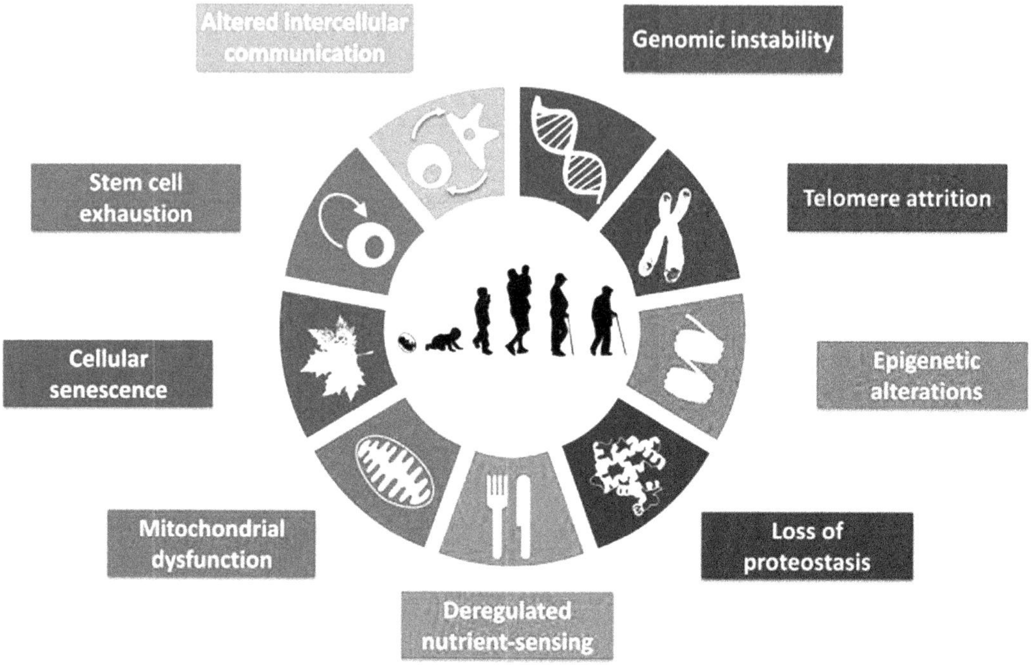

FIGURE 15.1 Different cellular damages depending on time [2]. Reproduced with permission from Lopez-Otin C, Blasco MA, Partridge L, Serrao M, Kroemer G. The hallmarks of aging. Cell 2013;1530:1194–1217. © 2013 Elsevier.

exposure to outdoor pollution, while 3.8 million are due to household nanoparticulates (dirty cookstoves and fuels) (Fig. 15.2).

As a consequence, the correlation between air pollution and the number of people suffering from skin problems such as acne, irritation, eczema topic dermatitis, psoriasis, and premature aging is increasing day-by-day [6,7]. Environmental contaminants, in fact, have a negative impact on the skin which, as the largest organ of human body, represents the most important barrier against pollution.

Therefore, to reduce the cancer occurrence and ameliorate the quality of life, it will be necessary to drastically reduce the air and land pollution, recovering also new means to protect skin and mucous membranes from the environmental aggressions.

15.2 Skin aging

Aging is a progressive and degenerative process that causes functional and structural alterations of the skin, such as marked changes in extracellular matrix (ECM) and integrin expression with appearance of wrinkles, developed as a result of time effects (i.e., *intrinsic aging*) and environmental factors (i.e., *extrinsic aging*) [8]. These alterations are tightly integrated with inflammation, caused from an overproduction of proinflammatory cytokines

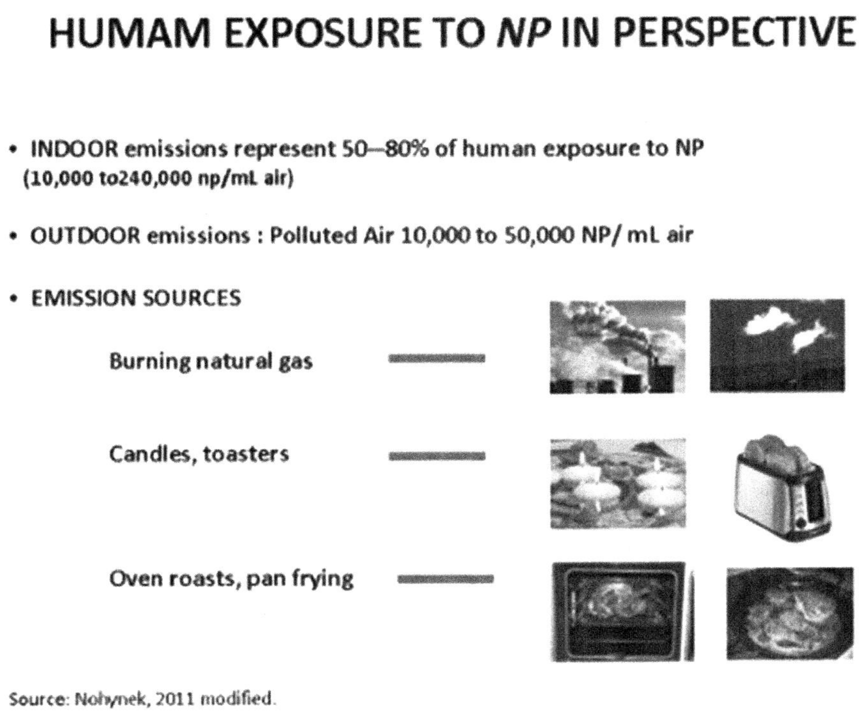

FIGURE 15.2 Outdoor and indoor emissions [65]. Reproduced with permission from Morganti P, Morganti G, Hong-Duo C. Beauty mask, market and environment. J Clin Cosmet Dermatol 2019;3(2). Available from: https://doi.org/10.16966/2576-2826.141. Licensed under a Creative Commons Attribution 4.0 International License.

(Fig. 15.3 [9]) with a dysregulation and alteration of the cellular redox balance due to both destructive reactive oxygen species (ROS) and reactive nitrogen species (RNS) [10].

Consequently, the so-called oxidative stress [11] appears, accompanied by hyperoxia or inappropriate oxygen metabolism. These phenomena, impacting the biological function of the cells, result in skin lesions and premature aging [12,13]. However, it is to remember that, while oxygen is critical for all human tissues and organs, its excessive production is cause of toxicity, also if the human body possesses a variety of direct and indirect systems to prevent and/or minimize the oxidative injury.

At nanoconcentration, in fact, ROS plays a fundamental role in regulating the signaling pathways, necessary for cell proliferation and survival [14,15]. The balance between cellular production of antioxidants, ROS and RNS is, therefore, critical for the maintenance of tissue homeostasis, determining also the degree of oxidative stress. Since these free radicals cause DNA damage, they also play an important role in tumor initiation. They, for example, degrade cartilage and modulate the immunoreactivity of lymphocytes or induce release of chemotactic substance, cause of oxidative stress, disorders, and diseases exacerbation [14].

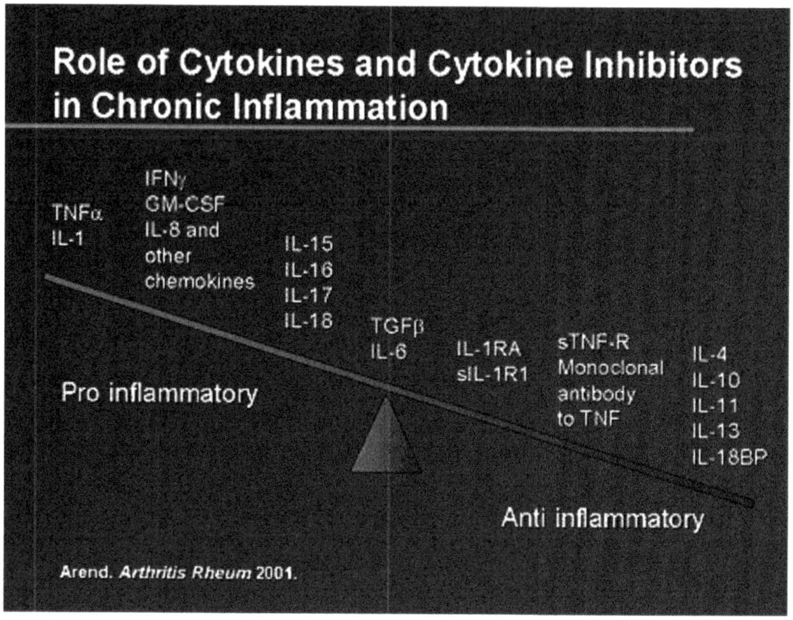

FIGURE 15.3 Proinflammatory and antiinflammatory cytokines. Reproduced with permission from Arend WP. Physiology of cytokine pathways in rheumatoid arthritis. Arthritis & Rheumatism 2001;45:1016. Available from: https://doi.org/10.1002/1529-0131(200102)45 :1,101::AID-ANR90.3.0.CO;2-7.

15.3 Inflammaging and oxidative stress

Oxidative stress occurs in certain disease states or as a result of toxic insult by foreign compounds (xenobiotics), which lead to oxidative changes within the cells. Accordingly, abnormal or chronic inflammation phenomena, termed also *inflammaging* (Fig. 15.4) [15,16], may appear with accumulation over time of detrimental changes of tissues and organs at the molecular and cellular level, resulting in increased risk of morbidity and mortality [12,15]. Inflammation, therefore, provoking modification in the skin barrier functionality, causes lipids peroxidation in membranes or in stratum corneum with degradation of elastic fibers and collagen in dermis [16]. Skin barrier, in fact, consisting of a complex mixture of proteins, carbohydrates, and lipids, known as ECM, provides a specific network able to protect, orient, and define the basal surface of the cells. It minimizes the entrance of noxious compounds and UV radiation into the body, preserving also skin water from the environmental aggressions. It is, therefore, essential to maintain the skin network at the physiological conditions.

It is to remember, in fact, that ECM is an important component for the stem cells anchorage, and during aging and wound healing an extensive interaction between skin proteins and the epidermis/dermis cells is established [17].

As a consequence, aging may be defined as the alteration of molecular mechanisms and accumulation of changes over time. Thus the advanced age is responsible for the chronological and environmental alterations, which result in an increased risk of disease and

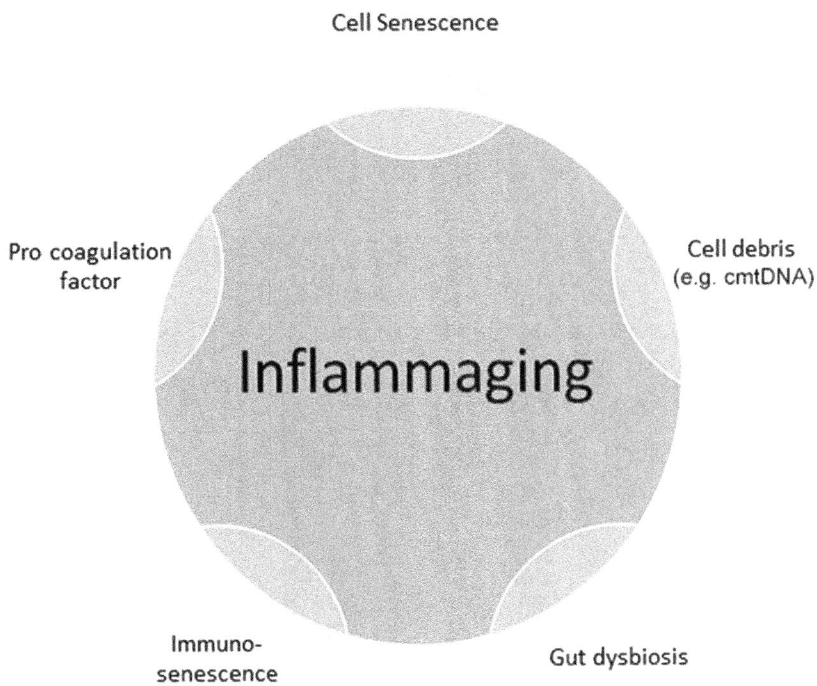

FIGURE 15.4 Causes of inflammaging [16]. Reproduced with permission from [16]. Licensed under Creative Commons Attribution License (CC BY 4.0).

death [12–16,18–20]. Oxidative stress and energy dysregulation seem also to play a key role in immunosenescence, during which the loss of telomeric DNA occurs, impacted by downregulation of immune and endocrine functions [18–20].

At this purpose, telomeres (Fig. 15.5), which are tandem repeats of DNA sequence and form protective caps of chromosomes, are considered as the marker of both biological aging and cancer development [21]. If shortened, in fact, telomeric DNA is not repaired by the enzyme telomerase and the cell becomes senescent [15,18,21]. Thus short telomere length is associated with earlier death, while its overexpression inhibits aging, increasing tumorigenesis [21]. Aging and cancer, therefore, are tightly interconnected so that cancer can be considered as an aging disease, also if aged cells are hypoactive with a cell division inability and cancer cells are hyperactive with a rapid cell division and an upregulated telomerase, compared with normal tissue [20,21]. However until today the

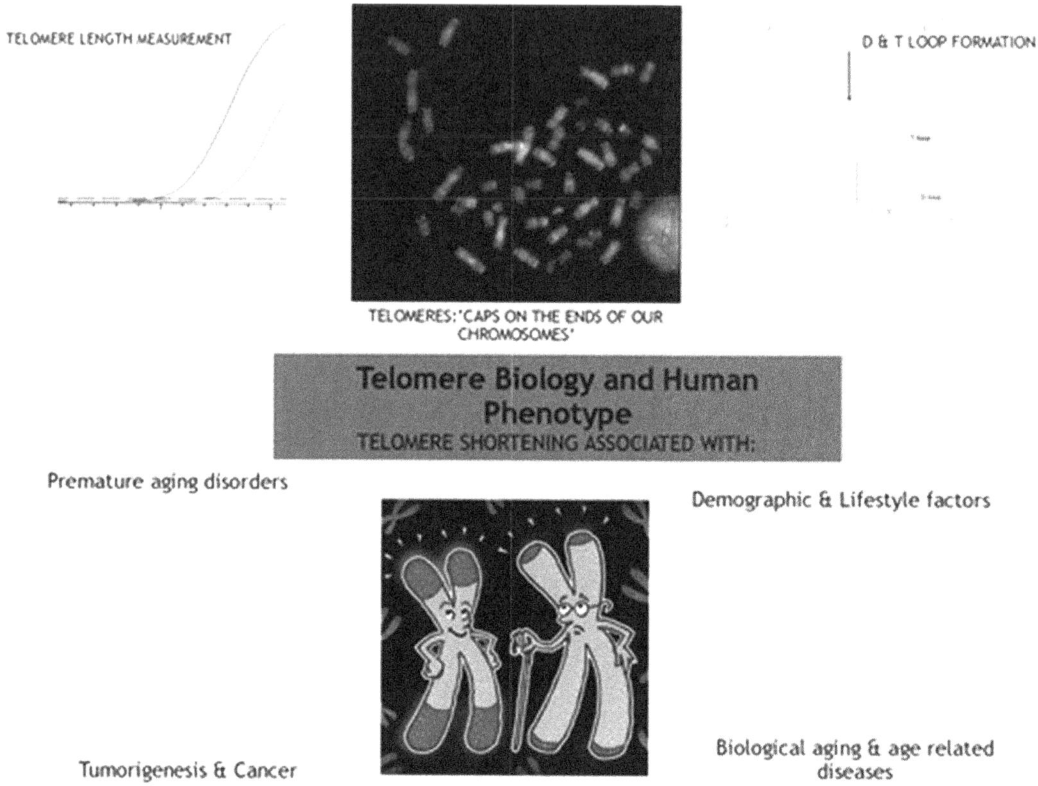

FIGURE 15.5 Telomeres with chromosome cups are the markers of skin aging [19]. Reproduced with permission from Ref. [19]. Licensed under Attribution 4.0 International (CC BY 4.0) License.

physiological mechanisms, underlying the link between telomerase shortening and aging health conditions, have not been completely elucidated.

15.4 Life expectance and cosmetic treatments

In any way, as life expectancy increases, human body is affected by a genetic program with genomic instability, as well as by a cumulative environmental and endogenous insult, characterized by enabling hallmarks, as the reported telomeres shortening [20,21].

However, the changes that occur over time are much more related to the interaction of the skin with the environment than to genetic predisposition and, therefore, they are to be considered a consequence of the personal lifestyle. Consequently, the increased lifetime exposure to sun and pollutants, the decreased immunological surveillance, and the increased free radicals presence with loss of the DNA repair capacity increase the risk of cutaneous malignancy and affect the individuals' quality of life.

Exposure to air pollution [22], for example, has been associated with reduced lung capacity in healthy individuals and with increased risk of hospitalization and mortality in subjects with respiratory chronic diseases, such as asthma and chronic obstructive pulmonary disease, and it has been also recognized as one of the causes of Alzheimer's disease [23]. Particulate matter, in fact, penetrating skin through either hair follicles or transdermally, exerts detrimental effects, generating not only the oxidative stress, which contributes to extrinsic aging, but also being cause of systemic toxicity in other organs [6,22–26]. Moreover, premature skin aging is associated with precocious appearance of deep wrinkling, scaling, roughness, dryness and laxity, telangiectasia, and pigment spots [24,25]. The elderlies, therefore, are looking for dermatological and cosmetic treatments to prevent skin aging and ameliorate their general appearance, by the use of innovative food, diet supplements, and/or cosmetic.

Cosmetic products, in fact, designed and organized with a substrate able to maintain the skin water at the right level and mimic the natural ECM, have to protect the skin from the interior and external aggressions, providing and acting as a new barrier.

15.5 Air pollution, waste, and human health

The confluence of world population growth and the increased air pollution, with the massive presence of atmospheric aerosol and oxidant compounds concentrated at the global scale, raise many challenges in ensuring safety and security for the actual and future human generations.

Outdoor fine particulate (Fig. 15.2) matter <2.5 μm, containing redox-active transition metals (prevalently copper and iron), quinones and volatile organic compounds, has been the fifth ranking global risk factor in 2015, causing annually 4.2 million of premature death (Fig. 15.6) [26–28]. Ambient ozone has further enhanced oxidative stress by depleting the antioxidant' compounds, together with the crucial contribution of indoor particulate (Fig. 15.2), also influenced from fungi, pets, and pest allergens [29].

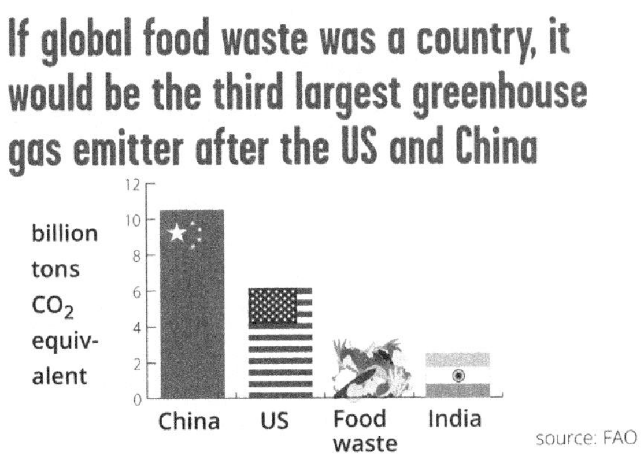

FIGURE 15.6 Global food waste. Source: *Courtesy FAO [65]. Reproduced with permission from [65]. Licensed under a Creative Commons Attribution 4.0 International License.*

Thus the necessity to reduce air pollution by a more sustainable industrial production and a better attitude of consumers to reduce waste, necessary to save the planet' ecosystem.

A third of all food produced for human consumption, for example, is lost or wasted, impacting also on climate and air [30]. Consequently, food loss and waste account for about 4.4 gigatons of greenhouse gas emissions per year. Therefore, according to the World Resource Institute, "If they were its own country, it would be the world's third-largest emitter-surpassed only by China and USA" (Fig. 15.6), impacting not only climate, but personal and national economics [30]. Thus, according to this public institution, by 2050, cutting the rate of this waste material, it will be possible to create environmental and economic benefits sufficient for the food necessities of the previsional 9 billion people living worldwide [30].

Given the increasing level of air pollution and its detrimental effects on the skin and human health, it should be necessary to find the solution for protecting and potentiating the skin barrier activity, contemporary decreasing the air particulate matter' emissions (Fig. 15.7). Moreover, it will be necessary to reduce the food waste during all the industrial production chain and at home, by a better consumers' awareness and increased knowledge on the connection between the environment ecosystem and our health.

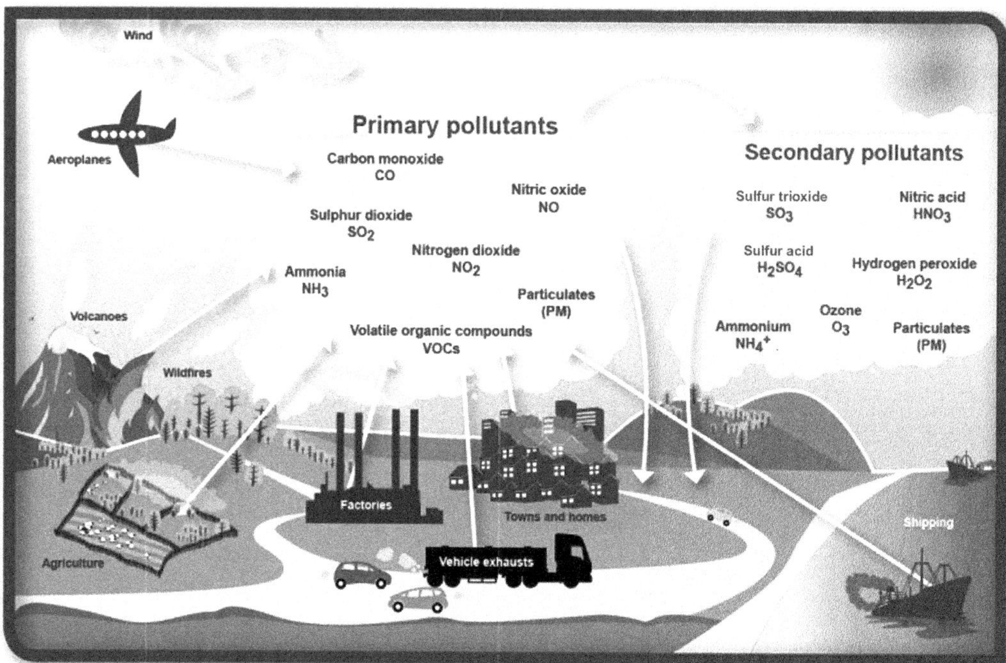

FIGURE 15.7 Primary and secondary pollutants. Reproduced with permission from Photochemical Smog https://www.mrgscience.com/ess-topic-63-photochemical-smog.html. Licensed under Creative Commons Attribution-ShareAlike (CC BY-SA 4.0) International License.

15.6 Innovative smart cosmeceutical-tissues

The introduction in the market and the use of biodegradable and smart *cosmeceutical-tissues* could represent an innovative new barrier to protect and regenerate the skin, contemporary slowing down the premature skin aging. These innovative tissues may be realized by nonwoven tissues, made by natural fibers, obtained by composites of selected polymers, such as nanochitin, chitosan, nanolignin, cellulose, starch, pullulan, PHA, PHB, and PLA, according also to the in progress EU Polybioskin research project [32–35].

Just to remember, cosmeceuticals are considered the hybrid cosmetic-pharmaceutical topical products able to enhance the health and beauty of skin, recording their effectiveness by in vitro and in vivo studies [36,37].

In any way, both synthetic and natural polymers have demonstrated their potentiality to be used for making skin scaffolds [38,39]. Thus, according to the experience of our and other research groups, these *cosmeceutical-tissues* may be produced by the electrospinning technique as porous biodegradable scaffolds, agglomerating into their three-dimensional structures the micro/nanospheres of Nanochitin–Nanolignin (CN–NL) (Fig. 15.8), bound

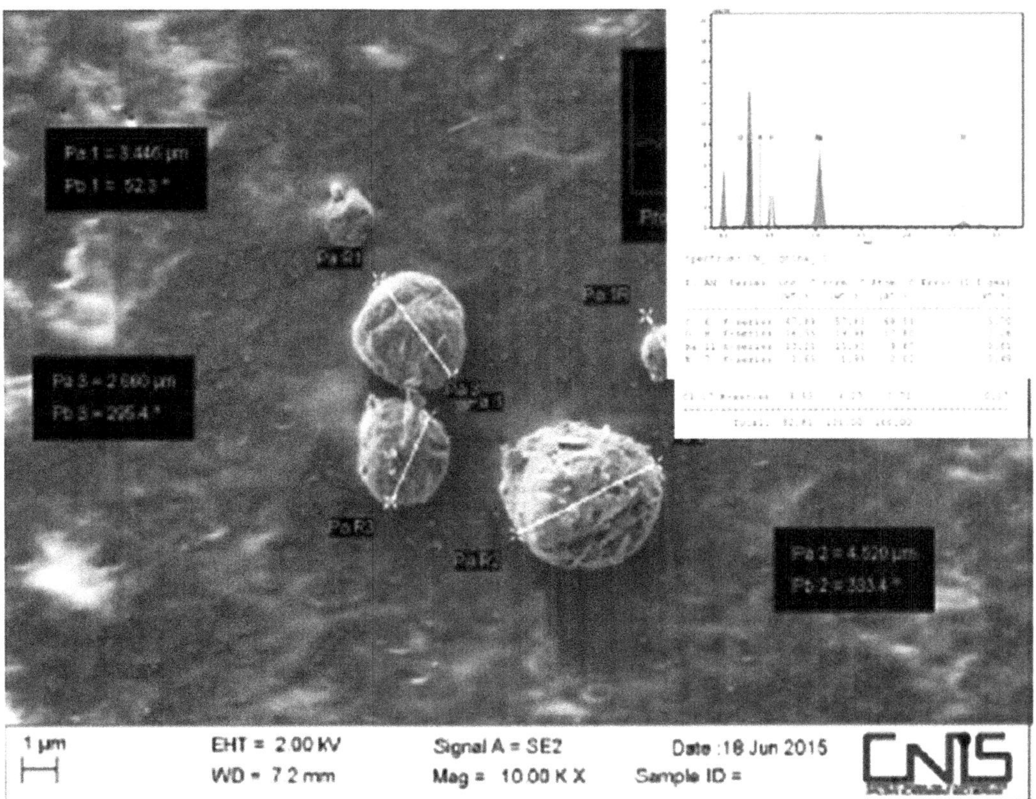

FIGURE 15.8 Nanochitin–nanolignin complex at SEM.

to selected active ingredients, necessary and useful for the temporary support of skin cells [36–41]. The different ingredients, encapsulated and embedded into the tissues and selected, according to their own safeness and effectiveness designed, provide an environment that may orient the basal cell surface, mimicking the skin barrier functions, in dependence of the tissue surface patterning. Moreover, when the active ingredients embedded into the tissue are nanostructured, their effectiveness is further increased because of the interactions they could have with the skin, at the subatomic level [36,37,40,41]. Nanotechnology, in fact [42], is useful to modify the ingredient permeation and penetration by controlling the release of active substances increasing also their period of permanence on the skin. Nanoparticles (NPs) [43,44], in fact, ensure a direct contact with the stratum corneum and skin appendages, not only protecting the active ingredients against chemical and physical instability, but also alleviating the allergic/sensitization reactions they could provoke.

The polymer-based nanoparticles, such as the nanochitin–nanolignin complexes (as nanospheres or nanocapsules) (Fig. 15.8), enriched with selected active ingredients and embedded into the natural fibers of these tissue may be, therefore, of great interest because of the controlled release of the incapsulated active ingredients, which can diffuse through the polymer matrix permeating the skin [39–41]. Thus these innovative tissues, enhancing temporary the incapsulated and loaded active ingredients, have not only the possibility to deliver them across the skin barrier but also to increase their period of permanence on the skin, thus resulting innovative and highly effective carriers for drug and cosmetic products.

The ability they have shown to increase the skin penetration of active ingredients [39–41] could be attributed to the size of the CN–NL NPs used and to the capacity they have to diffuse through the skin surface, increasing the water uptake and enhancing the active ingredients permeation and effectiveness [42–45]. Since water is considered a penetration enhancer, its activity is further increased by the temporary occlusive properties of the tissue, which can enhance the water gradient upper skin layers by avoiding evaporation [45]. Moreover, on the one hand, it has been shown that chitin nanofibrils have an immunomodulatory, antioxidant, bacteriostatic, and skin repairing activity [35,36]. On the other hand, lignin possesses antioxidant and antimicrobial properties thanks to its polyphenolic structure together with an interesting protective activity against the sun rays [46–48]. Consequently, the nanostructured complex CN–NL used as a carrier shows a real effectiveness in this function due to its ability to encapsulate, load, and release the selected different active ingredients. In addition, it may also be considered as an active ingredient, because of the combining activity of chitin and lignin which, easily metabolized from human and the environment enzymes, releases compounds useful for the cell activities [49–51]. Thus the final effectiveness of this innovative tissue has to be attributed to the global activity of all the individual constituents.

15.7 Extracellular matrix and active ingredients

Effectiveness of the active ingredient(s) depends, in fact, not only on the skin absorption capacity, but also on the ingredients' release properties, due to the tissue-fibers' size,

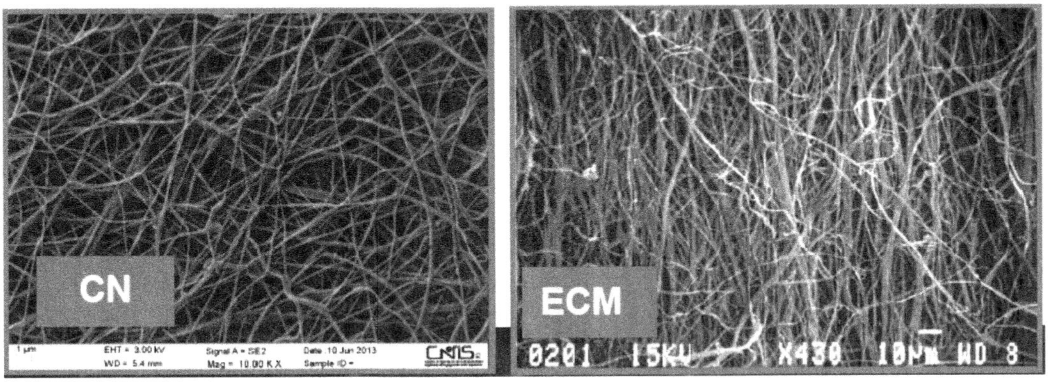

FIGURE 15.9 Chitin nanofibrils structure (left) compared to extracellular matrix (right) at SEM. Reproduced with permission from Ref. [66]. Licensed under Creative Commons Attribution (CC BY 4.0) license.

characterized by their large surface-area-to-volume, high porosity with variable pore-size distribution, and excellent loading capacity [38,39,44]. Thus their relatively large surface area and the smart matrices of these tissues, showing morphological similarities to the extracellular natural matrix (ECM) (Fig. 15.9), are helpful for cell adhesion, growth, and reproduction, and therefore suitable for skin regeneration, as shown by our group [36,40,41]. It has been evidenced, in fact, that the combining activity of chitin nanofibrils and nanolignin, embedded into a biodegradable electrospun tissue, may modulate the expression of proinflammatory cytokines, such as THF-alpha, IL-1alpha, and IL-8, and the metalloproteinases 2 and 9 with beta-defensin-2 expression, at the level of human keratinocytes (Fig. 15.10) [36,37]. This the reason why the polymer composite fibers have been transformed in nonwoven tissues and/or in thin films, by electrospinning or compressing-melting technique, combined with surface modification by plasma treatment, or impregnated by electrospraying powder. Moreover, the selected natural fibers have been functionalized by active ingredients which, bound on their porous surfaces, have been previously encapsulated into nanochitin–nanolignin complexes [34–36,39,40]. These different new technologies have provided a straightforward and versatile approach for the reported Polybioskin project to make innovative nanocomposite scaffolds, skin-friendly and environmentally friendly [32–35]. The aim of this project is to produce at the industrial level biodegradable Baby/Feminine Diapers, Beauty Masks and Advanced Medications, effective and free of toxic side effects.

The technologies, adopted by Polybioskin, have been selected considering the more simple, elegant, and scalable means at disposition today to fabricate nonwoven tissues and/or thin films, made by pure polymers of both natural and synthetic micro/nanofibers with the necessary structural, mechanical, and biological properties [51–53]. Naturally, the physicochemical and biological properties of nanofiber matrices have been controlled and will be modified to meet the requirement of the specific designed applications.

Thus both the tissue carriers and the active ingredients entrapped have been selected to mitigate respectively the baby' skin or the feminine' skin-mucous membranes, protecting them from the irritation phenomena provoked from the aggression of toxic chemicals

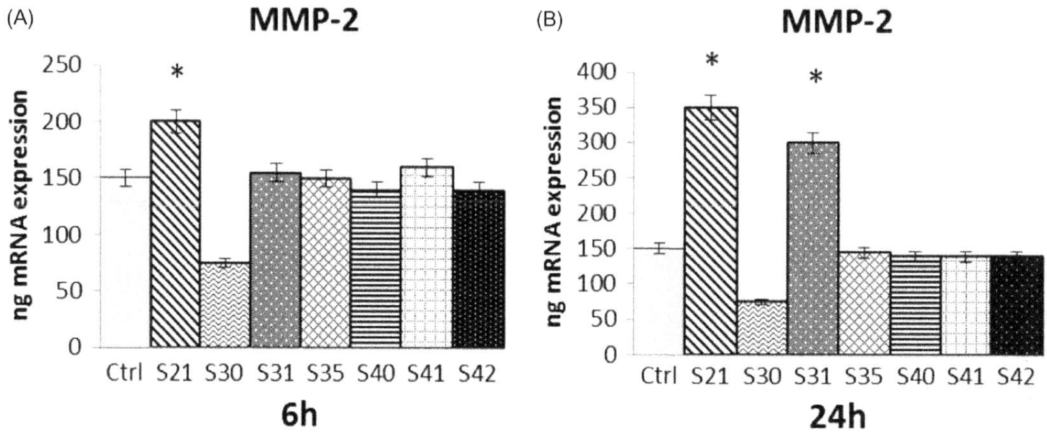

FIGURE 15.10 Beta-defensine-2 expression (hBD2) at the level of human keratinocytes treated by CN–LG complex: mRNA expression of HBD-2 from HaCat cells treated with S21, S30 , S31, S35, S40, S41 and S42 non-woven tissue bio-composites (10 μg/mL) for 6 h (A) and 24 h (B) [37]. Reproduced with permission from Morganti P, Fusco A, Paoletti I, Perfetto B, Del Ciotto P, Palombo M, et al. Antiinflammatory, immunomodulatory and tissue repair activity on human keratinocytes. Materials 2017;2017:843. Available from: https://doi.org/10.3390/ma10070843. Licensed under Creative Commons Attribution (CC BY 4.0) license [37].

and/or pathogen microbiota (Baby/Feminine Diapers) [54,55]; to slowdown fine lines and wrinkles of the prematurely aged people (Beauty Masks) [56,57]; or to regenerate the skin affected by burns and wounds (Advanced Medications) [58,59]. However, these innovative tissues, having all the aim to regenerate and repair the skin of subjects aged or affected by some diseases, have to be realized by fibers characterized for their similarity to the hierarchical structure, orientation, and alignment of the skin collagen fibrils in natural ECM, as previously reported [17,56–58].

These smart tissues for respecting all these parameters would provide the skin cells with the necessary like cues and nourishment, recovered in vivo environment [57,58]. Thus, when necessary, they may also possess a selective antibacterial activity against the pathogenic bacteria, always trying to respect the skin microbiota equilibrium (Fig. 15.11) [56,59–61].

15.8 Conclusive remarks

The proposed innovative tissues, according to the previous studies of our group [36,37] and the first results obtained from the Polybioskin project [32–35,41,55,60,62,63], could represent a new armamentarium to treat both the healthy and aged skin, as well as the skin affected by burns, wounds, or diseases. Their recovered effectiveness and safeness is probably due to the characteristics of the tissue which, used as a carrier, has shown to possess a similar structural organization of the natural ECM [64–66].

These smart dressings probably act on the skin by their own specialized tissues made by the activated block polymeric CN–NG micro-NPs, bound on the surface of the fibers

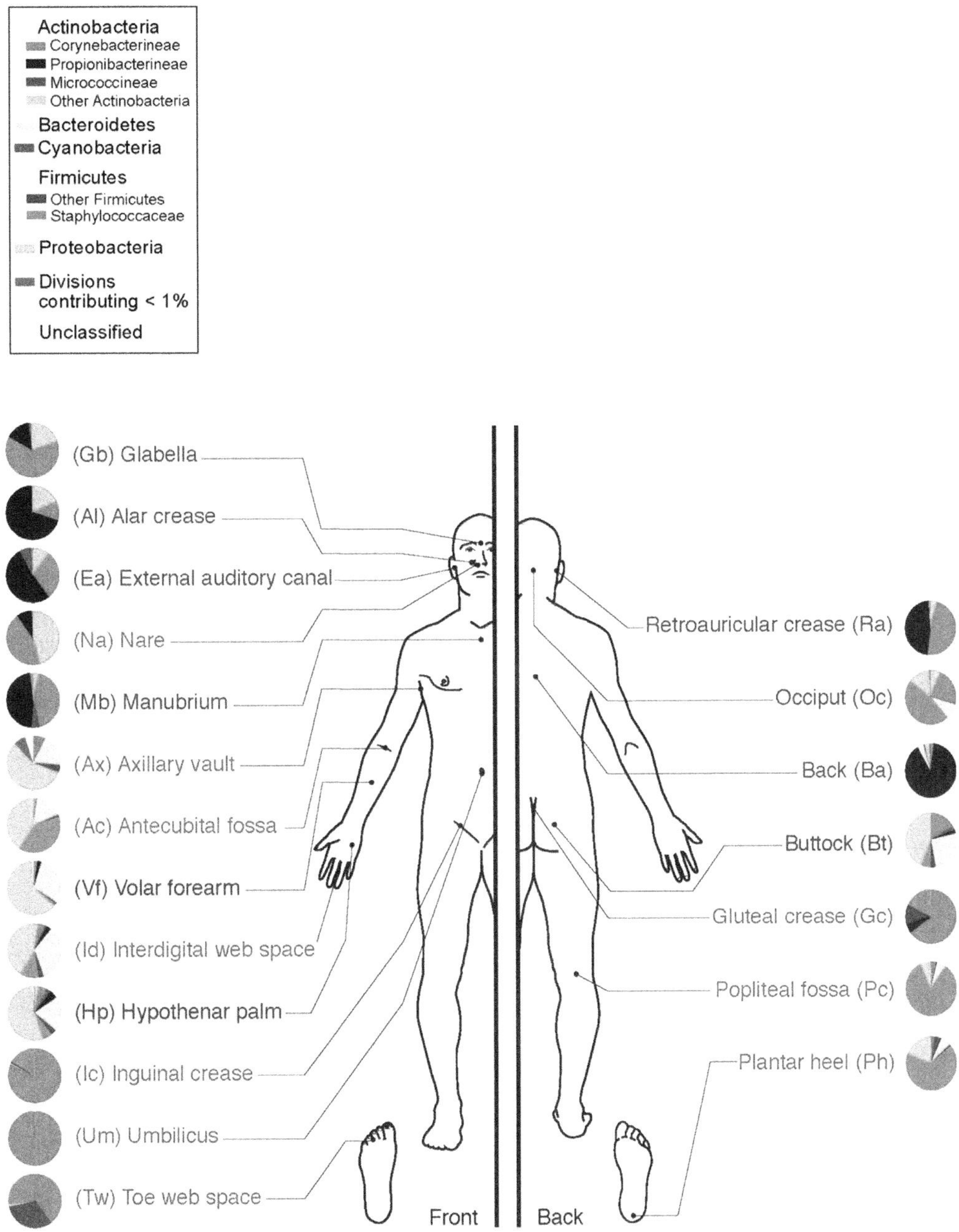

FIGURE 15.11 Human microbiome [61]. Reproduced with permission from Human microbiome, https://en.wikipedia.org/wiki/Human_microbiome. Licensed under Creative Commons Attribution-ShareAlike 3.0 Unported License. Source: *Courtesy Wikipedia.*

or embedded into their structure. Therefore, the used NPs, entrapping specific and selected active ingredients, characterize the mechanism of action of the various tissues, which seem able to substitute the cells microevironment [64–66], as previously reported.

It is again to remember that ECM, main component of a three-dimensional scaffold of collagen interwined with proteoglycans and glycoproteins, has the extrinsic property to self-associate and form offered assembles with other ECM proteins, contributing to the skin repairing activity [67,68].

These innovative tissues, therefore, may represent a new category of cosmeceuticals to be used for protecting the skin against the toxic ingredients present in the air pollution as well as to reduce wrinkling, fine lines, and black spots of a prematurely aged skin. Our group of research and the Polybioskin group are collaborating to control safeness and effectiveness of these innovative cosmeceutical-tissue and produce them at the industrial level. This is the future commune goal.

Aknowledgments

The authors thank Biobased Industries and Horizon 2020 for received fund for the research project Polybioskin necessary to go on with these studies. Moreover, they thank MAVI Sud (Italy) and CIMV (France) for the obtained samples of Chitin nanofibrils and Biolignin, respectively.

Conflicts of interest

The authors declare no conflicts of interest.

References

[1] Aunan JR, Cho UC, Soreide K. The biology of aging and cancer: a brief overview of shared and divergent molecular hallmarks. Aging Dis 2017;8(5):628–42.

[2] Lopez-Otin C, Blasco MA, Partridge L, Serrao M, Kroemer G. The hallmarks of aging. Cell 2013;1530:1194–217.

[3] Kramer U, Schikowski T. Recent demographic changes and consequences for dermatology. In: Gilchrest BA, Krytmann J, editors. Skin aging. Berlin: Springer-Verlag; 2006. p. 1–8.

[4] Pruss-Ustun A, Wolf J, Corvalan C, Bos R, Neira M, editors. Preventing disease through healthy environments. Geneva: WHO Edition; 2016. ISBN: 9789241565196.

[5] WHO. Global conference on air pollution and health. Proceedings. Geneva: World Health Organization Edition; 2018.

[6] Kim KE, Cho D, Park HJ. Air pollution and skin diseases: adverse effects of airborne particulate matter on various skin diseases. Life Sci 2016;152:126–34.

[7] Cui Y, Xie X, Jia F, Li Z, Fu M, Hao H, et al. Ambient fine particulate matter induces apoptosis of endhothelial progenitor cells through reactive oxygen species formation. Cell Physiol Biochem 2015;35:353–63.

[8] Munzel T, Gori T, Al-Kindi S, Deanfield J, Daiber A, Pajagopalan S. Effects of gasous and solid constituents of air pollution on endothelial function. Eur Health J 2018;39:3543–50. Available from: https://doi.org/10.1094/eurhearthj/ehj481.

[9] Arend WP. Physiology of cytokine pathways in rheumatoid arthritis. Arthritis & Rheumatism 2001;45:101–6. Available from: https://doi.org/10.1002/1529-0131(200102)45 :1 < 101::AID-ANR90 > 3.0.CO;2-7.

[10] Di Meo S, Reed TT, Venditti P, Victor V. Role of ROS and RNS sources in physiological and pathological conditions. Oxid Med Cell Longev 2016;2016::1245049. Available from: https://doi.org/10.1155/2016/1245049.

[11] Ludivici M, Bigagli E. Oxidative stress and air pollution exposure. J Toxicol 2011;2011:487074. Available from: https://doi.org/10.1155/2911/487074.
[12] Kammeyer A, Luiten RM. Oxidation event and skin aging. Ageing Res Rev 2015;21(5):16–29.
[13] Xu H., Zheng Y.W., Liu L.P., Luo F.L., Zhou H.C., Isoda H., et al. (2018). Reacti++ve oxygen species in skin repair, regeneration, aging, and inflammation. IntechOpen Ed, Available from: https://doi.org/10.5772interchopen.72747.
[14] Finely T, Holbrook N. Oxidants, oxidative stress and the biology of ageing. Nature 2000;408(6809):239–47.
[15] Jenny NS. Inflammation in aging:cause, effect, or both? Discov Med 2012;13(73):451–60.
[16] Sanada F, Tanisama Y, Muratsu N, Otsu R, Shimuzu H, Takuigi H, et al. Source of chronic inflammation in aging. Front Cardiovasc Med 2018;5(12). Available from: https://doi.org/10.3389/fcvm.2018.00012.
[17] Watt FM, Fujiama H. Extracellular matrix interactions in normal and diseased skin. Cold Spring Harb Perspect Biol 2011;2011(3):a005124.
[18] Makratonaki E, Zouboulis CC. Molecular mechanisms of skin aging: state of the art. Arch NY Acad Sci 2007;1119:40–50.
[19] Kara J Turner, Vimal Vasu, Darren K Griffin. Telomere Biology and Human Phenotype. Cells 2007;8(1):73. Available from: https://doi.org/10.3390/cells8010073.
[20] Aunan JR, Cho WC, Soreide K. The bilogy of aging and cancer: a brief overview of shared and divergent molecular hallmarks. Aging Dis 2017;8(5):628–42.
[21] Blackbum EH, Epel ES, Lin J. Human telomere biology: a contributory and interactive factor in aging, disease risks, and protection. Science 2015;350:1193–8.
[22] Ko FWS, Hui DSC. Effects of pollution on lung health. Clin Pulm Med 2010;17(6):300–4.
[23] Mouton PV, Wei Y. Air pollution, oxidative stress, and Alzheimer's disease. J Environ Publ Health 2012;2012::472751. Available from: https://doi.org/10.1155/2012/472751.
[24] Farange MA, Miller KW, Elsner P, Maibach HI. Intrinsic and extrinsic factors in skin ageing: a review. Int J Cosmet Sci 2008;30:87–95.
[25] Kosmadaki MG, Gilchrest BA. The demographic of aging in the United States: implications for dermatology. Arch Dermatol 2002;138:1427–8.
[26] Puri P, Nandar SK, Kathuria S, Ramesh V. Effects of air pollution on the skin: a review. Indian J Dermatol Venereol Leprol 2017;83:415–23. Available from: https://doi.org/10.4103/0378-6323.199579.
[27] Lakey PSJ, Bekemeier T, Tong H, Arangio AM, Lukas K, Poschi U, et al. Clinical exposure-response relationship between air pollutants and reactive oxygen species in the human respiratory tract. Sci Rep 2016;6:32916. Available from: https://doi.org/10.1038/srep/32916.
[28] Cohen AJ, Brawer M, Burnett R, Anderson HR, Frostad J, et al. 25-year trends of the Global Burden of Diseases Study 2015. Lancet 2017;389:1907–18.
[29] Gang X, Misztal PK, Nazaroff WW, Goldstein AH. Volatile organic compound emissions from humans indoor. Environ Sci Technol 2016;50:12686–94.
[30] Hanson C, Lipinski B, Friedrich J, O'Connor C, James K. What's food loss and waste got to do with climate change? A lot, actually. Washington, DC.: World Resources Institute; 2015.
[31] Photochemical Smog https://www.mrgscience.com/ess-topic-63-photochemical-smog.html.
[32] Morganti P, Coltelli MB, Danti S, Bugnicourt E. The skin: goal of the EU PolyBioSkin Project. Glob Res J Pharm Pharmacol 2017;2(1):7–13.
[33] Morganti P, Coltelli MB, Danti S. Biobased tissues for innovative cosmetic products: PolyBioSkin as an EU Reearch Project. Glob J Nanomed 2018;3(4):555620.
[34] Morganti P, Danti S, Coltelli MB. Chitin and lignin to produce biocompatible tissues. Res Clin Dermatol 2018;1(1):5–11.
[35] Morganti P, Coltelli MB, Morganti G. Natural polymers for a cleaner environment. Int J Nanotechnol Nanomed 2018;3(1):1–4 ISSN: 2476-2334.
[36] Morganti P, Palombo M, Carezzi F, Nunziata ML, Morganti G, Cardillo M, et al. Green Nantechnology: natural beauty masks to save the environment. Cosmetics 2016;2016(3):41. Available from: https://doi.org/10.3390/cosmetic.
[37] Morganti P, Fusco A, Paoletti I, Perfetto B, Del Ciotto P, Palombo M, et al. Antiinflammatory, immunomodulatory and tissue repair activity on human keratinocytes. Materials 2017;2017:843. Available from: https://doi.org/10.3390/ma10070843.
[38] Hubbel JA. Biomaterials in tissue engineering. Nat Biotechnol 1995;13(6):565–76.

[39] Lutolf MP, Hubbell JA. Synthetic biomaterials as instructive extracellular microenvironmentd for morphogenesis in tissue engineering. Nat Biotechnol 2005;23(1):47–55.
[40] Donnarumma G, Fusco A, Morganti P, Palombo M, Anniboletti T, Del Ciotto P, et al. Advanced medications made by green nanocomposites. Int J Res Pharm Nanosci 2016;5(5):261–70.
[41] Morganti P, Hong-Duo C, Xin-Hua G, Morganti G, Febo D. Chitin & lignin: turning food waste into cosmeceuticals. J Clin Cosmet Dermatol 2018;3(1). Available from: https://doi.org/10.16966/2576-2826.135.
[42] Kaul S, Gulati N, Verma D, Mukherjee S, Nagaich V. Role of nanotechnology in cosmeceuticals: a review of recent advance. J Pharm 2018;2018:3420204. Available from: https://doi.org/10.1155/2018/3420204.
[43] Larese Filon F, Mauro M, Adami G, Bovenzi M, Crosera. Nanoparticles skin absorption: new aspects for safety profile evaluation. Regul Toxicol Pharmacol 2015;72(2):310–23. Available from: https://doi.org/10.1016/jyryph.2015.05.005.
[44] Yokota J, Kyotanj S. Influence of nanoparticles size on the skin penetration, SKIN retention and anti-inflammatory activity of non-steroidal anti-inflammatory drugs. J Chin Med Assoc 2018;81:511–19.
[45] Williams AC, Barry BW. Penetration enhancers. Adv Drug Deliv Rev 2004;56:603–18.
[46] Espinoza-Acosta JL, Torres-Chavez PI, Ramirez-Wong B, Lopez-Saiz CM, Montano-Leyva B. Antioxidant, antimicrobial, and antimutagenic properties of technical lignins and their applications. Bioresources 2016;11 (2):1–30. Available from: https://doi.org/10.15376/biores.11.2.Espinoza_Acosta.
[47] Gutierrez-Hernandez JM, Escalante A, Murillo-Vasquez RN, Delegado E, Javier Gonzalez F, Toriz G. Use of agave tequilana-lignin and zinc oxide nanoparticles for skin photoprotection. Photochem Photobiol B Biol 2016;163:156–61. Available from: https://doi.org/10.1016/J.photobiol2016.027.
[48] Vinardell MP, Mitjans M. Lignins and their derivatives with benefical effects on human health. Int J Mol Sci 2017;2017(18):1219. Available from: https://doi.org/10.3390/ijms18061219.
[49] Eide KB, Norberg AL, Heggset EB, Lindbom AR, Varum KM, Eijsink VGH, et al. Humam chitotriosidase-catalyzed hydrolysis of chitosan. Biochemistry 2012;52:487–95. Available from: https://doi.org/10.1021/bi2015585.
[50] Morganti P, Carezzi F, Del Ciotto P, Morganti G, Nunziata ML, Gao XH, et al. Chitin nanofibrils: a natural multifunctional polymer. physicochemical characteristics, effectiveness and safeness. In: Phenix DA, Waqar A, editors. Nanobiotechnology. Manchester: One Central Press; 2014. p. 1–37.
[51] Elmonem MA, van den Heuvel LP, Levtchenko EN. Immunomodulatory effects of chitotriosidase enzyme. Enzyme Res 2016;2016:2682680. Available from: https://doi.org/10.1155/2016/2682680.
[52] Cinelli P, Coltelli MB, Mallegni N, Morganti P, Lazzeri A. Degradability and sustainability of nanocomposites based on polylactic acid and chitin nanofibrils. Chem Eng Trans 2017;60:115–20. Available from: https://doi.org/10.3303/CET1760020.
[53] Coltelli MB, Cinelli P, Gigante V, Aliotta L, Morganti P, Panariello L, et al. Chitin nanofibrils in poly(lactic acid) (PLA) nanocomposites: dispersion and thermo-mechanical properties. Int J Mol Sci 2019;20(3):504. Available from: https://doi.org/10.3390/ijms20030504.
[54] Morganti P, Del Ciotto P, Stoller M, Chianese A. Antibacterial and anti-inflammatory green nanocomposites. Chem Eng Trans 2016;47:61–6. Available from: https://doi.org/10.3303/CET1760001.
[55] Morganti P, Febo P. Problem and solution for biodegradable baby diapers. Eurocosmetics 2017;25:44–6 ISSN: 0944-8942. Available from: <http://www.eurocosmetics-magazine.com>.
[56] Morganti P, Palombo M, Fabrizi G, Guarneri F, Slovacchia F, Cardillo A, et al. New insight on anti-aging activity of chitin nanofibril-hyaluronan blocca copolymers entrapping active ingredients: in vitro and in vivo study. J Appl Cosmetol 2013;41:1–29.
[57] Joergensen P, Tattan SIS. Extracellular matrix modulates morphology, growth, oxidative stress response and functionality of human skin fibroblasts during aging *in vivo*. Aging Sci 2014;2:2. Available from: https://doi.org/10.4172/2329-8847.1000122.
[58] Quan T, Fisher GJ. Role of age-associated alterations of the dermo extracellular matrix microenvironment in human skin aging; a mini-review. Gerontology 2015;2015(61):427–34. Available from: https://doi.org/10.1159/000371708.
[59] Morganti P, Palombo M, Tichenko G, Yudin VE, Guarneri F. Chitin-hyaluronan nanoparticles: a multifunctional carrier to delivery anti-aging active ingredients through the skin. Cosmesis 2014;1:140–58.
[60] Morganti P, Febo P, Cardillo M, Donnarumma G, Barone A. Chitin nanofibril and nanolignin: natural polymers of biomedical interest. J Clin Cosmet Dermatol 2017;1(2). Available from: https://doi.org/10.1696/2576-2826.

[61] Human microbiome, https://en.wikipedia.org/wiki/Human_microbiome.
[62] Morganti P. Chitin nanofibrils: turning fishery's waste into goods. J Clin Rev Case Rep 2018;3(3):1–3 ISSN:2573-9565.
[63] Morganti P, Hong-Duo C, Gao Xin-Hua, Morganti G, Febo D. Chitin & lignin: turning food waste into cosmeceuticals. J Clin Cosmet Dermatol 2018;3(1). Available from: https://doi.org/10.16966/2576-2826.135.
[64] Danti S, Trombi L, Fusco A, Azimi B, Lazzeri A, Morganti P, et al. Chitin nanofibrils and nanolignin as functional agents in skin regeneration. Int J Mol Sci 2019;20(11):2669. Available from: https://doi.org/10.3390/ijms20112669.
[65] Morganti P, Morganti G, Hong-Duo C. Beauty mask, market and environment. J Clin Cosmet Dermatol 2019;3(2). Available from: https://doi.org/10.16966/2576-2826.141.
[66] Morganti P, Morganti G, Colao C. Biofunctional textiles for aging skin. Biomedicines 2019;7:51. Available from: https://doi.org/10.3390/biomedicines7030051.
[67] Mecham RP. The extracellular matrix: an overview. Berlin: Springer; 2011. ISBN: 9783642165542.
[68] Morganti P, editor. Bionanotechnology to save the environment. Plant and fishery's biomass as alternative to petrol. Basel: MPDI; 2019.

CHAPTER

16

Nanocarrier-mediated follicular targeting

B. Betül Gökçe and Sevgi Güngör

Department of Pharmaceutical Technology, Faculty of Pharmacy, Istanbul University, Istanbul, Turkey

16.1 Introduction

Utilizing "nanocarriers" for skin delivery of either active drug molecules or cosmetic substances is one of the most prominent areas in the field of biomedical applications. Novel nanosized delivery systems have been recently pointed out to offer numerous superiorities to the conventional topical formulations in dermatotherapy and cosmetic applications [1–3]. In skin delivery of active substances, either cosmetic or therapeutic purposes, topical administration provides not only direct application of active compounds to the target region on the body, but also avoidance of possible systemic side effects and complications. However, the success of topical treatment depends on the ability of active molecules to effectively delivery into the target layers of the skin, which exhibits an excellent barrier property due to its highly complex morphological structure. In this regard, targeting "actives" via nanocarrier systems to the specific sites in the skin is considered as a strategy to improve their penetration efficacy [2,4].

The potential affinity of nanocarriers into hair follicles, which is being an increasingly highlighted point by many researchers [5], makes such innovative delivery systems promising for their penetration and accumulation features in the skin. In accordance with the subject of this book, the use of nanocosmeceuticals providing such an efficient skin delivery of cosmetic and dermato-cosmetic actives could be an advantageous option to have satisfactory cosmetic outcomes in various topical treatments for cosmetic purposes such as skin nutrition, UV protection, hair growth modulation, wound healing, or antiaging therapies [3,6–10].

Nanocosmetics
DOI: https://doi.org/10.1016/B978-0-12-822286-7.00015-2

This chapter deals with an overview of nanocarrier-mediated follicular targeting of "actives" following topical skin delivery of nanocarriers. The basic barrier characteristics and function of human skin and skin penetration pathways of actives are briefly reviewed to give background information. The follicular pathway and the role of hair follicles in nanocarrier-mediated skin delivery are emphasized. The recent research studies focused on nanosized drug delivery systems for follicular targeting are also reviewed.

16.2 The barrier characteristics of human skin

The skin with an approximate surface area of 1.8 m^2 and 16% of total body weight is known to be the largest organ of the human body [11]. Introducing active molecules to the body via the skin is an alternative route of administration not only for the topical treatment of dermatological conditions and the cosmetic applications but also for the systemic treatment of some chronic diseases [12,13]. Having an excellent barrier property, this biomembrane prevents the entry of exogenous substances including pathogen microorganisms as well as other molecular substances into the organism, and it also keeps from endogenous heat and water loss contributing to thermoregulation and skin moisture [14,15]. The barrier function of the skin is provided by the physical properties with its unique membrane design, and additionally by the chemical and biochemical mechanisms where inflammatory mediators, antioxidants, UV-absorbing molecules, and xenobiotic-metabolizing enzymes are involved [16–18].

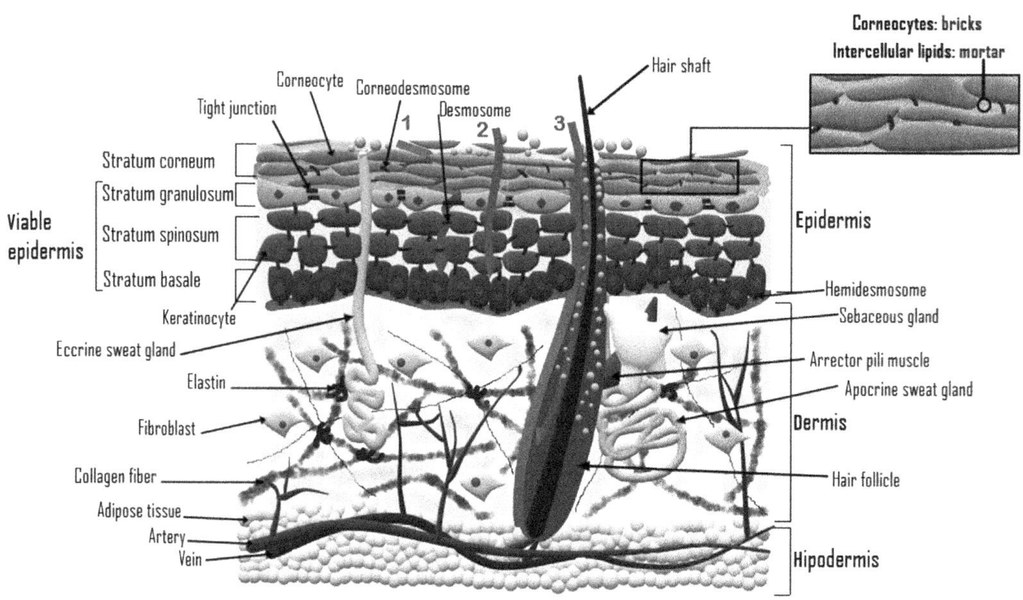

FIGURE 16.1 Skin structure and possible penetration routes_pathways through the skin.

As it can be seen in Fig. 16.1, morphologically the skin is defined with three main layers; epidermis, dermis, and hypodermis, from the outermost to the innermost [19,20]. Epidermis, a keratinized stratified squamous epithelial tissue, is consisting of sublayers with different functionalities: stratum corneum (SC), and *viable* epidermis layers; namely, stratum lucidum (present only in thick skin areas such as on the palms of hands and the soles of feet [21]), stratum granulosum, stratum spinosum, and stratum basale (stratum germinativum). As a self-renewing tissue, the shedding of the SC cells from the skin surface called desquamation and the cell growth in the lower layers of the epidermis demonstrate continuity [22] by means of the process of cell migration and differentiation toward stratum granulosum starting from the stratum basale layer that is the source of the stem cells responsible for epidermal cell production [23].

SC, which is also known as the nonliving epidermis, contributes dominantly to the barrier feature of the skin, as the uppermost layer of the epidermis [24]. Due to its sophisticated structure, SC is considered to be the most important barrier for the penetration of molecules across the skin even though it accounts only for approximately 1% of the total skin thickness [13]. SC is consisted of corneocytes, flattened dead cells, surrounded by the highly impermeable cornified envelope and a lipid matrix [25]. Substantially, the SC barrier is the result of the perfect organization of corneocytes and intercellular lipids, which has been described by the "bricks and mortar" model [26]. According to the model, the structure of the SC is composed of "bricks" that are associated with corneocytes whose cytoplasm consist mainly of fibrous protein (keratin) networks and "mortar" that is the matrix of lipid bilayers where the cell blocks are embedded and held together [27]. Moreover, corneodesmosomes, "mechanical" junctions of the SC [28], support the SC integrity connecting corneocytes to each other. The structure of the SC defined by the bricks and mortar model is shown in Fig. 16.1.

Dermisis, an elastic and compressible connective tissue, is composed of a semigel matrix containing mucopolysaccharides and collagen fibers that provide elasticity and mechanical support to the skin [29]. In addition to blood and lymph vessels, sensory nerve endings and skin appendages such as sweat glands, hair follicles, and associated sebaceous glands are also found in the dermis [14]. Hypodermis as the subcutaneous tissue is attached to the dermis layer with collagen and elastin fibers. It is characterized as a loose connective tissue comprising mainly fat cells called adipocytes. The main functions of hypodermis are connection the skin to the underlying bones and muscles, supply nourishment to the skin through blood vessels and nerves, protection against physical shock, heat insulation, and energy storage to be used when required [29].

Hair follicles and associated sebaceous glands, eccrine and apocrine sweat glands, and nails are defined as "skin appendages." A hair follicle and accompanying sebaceous gland together are called pilosebaceous unit (PSU) [30]. The structure of a PSU with the resident hair shaft can be seen in Fig. 16.2. Sebaceous glands are natural apparatus placed in the skin tissue, working with the specific task assigned to them in connection with hair follicles. They are usually found in hairy regions in the body, producing sebum secretion that prevents drying of the skin and contributes to skin integrity. Eccrine sweat glands are tubular-shaped structures; their secretions are discharged directly to the skin surface. Apocrine sweat glands do not open directly to the skin surface in contrast to eccrine sweat glands; the associated hair follicles are the discharge channels where they open with small ducts [31].

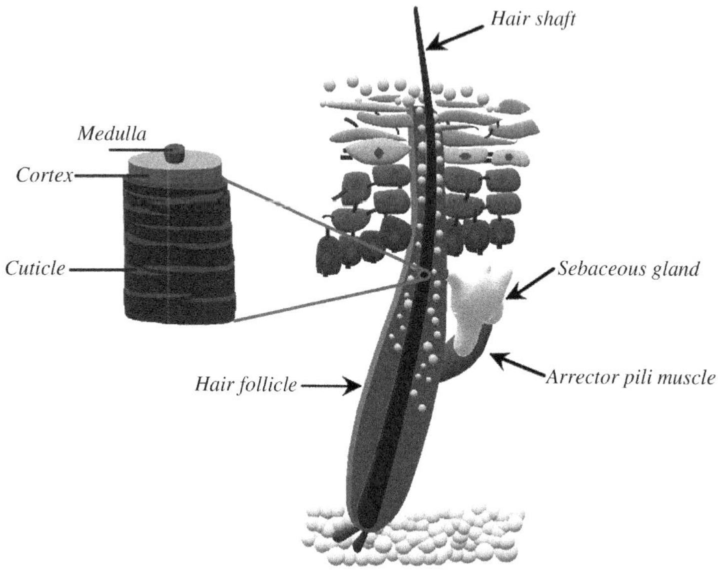

FIGURE 16.2 Representative structure of the pilosebaceous unit and hair shaft.

16.3 Penetration pathways across the skin

As aforementioned, SC, the outermost layer of the skin, stands as a unique barrier to penetration of any substances as well as active drug substances across the skin. The molecules targeted to a specific region in the skin are supposed to pass through the SC barrier. Basically, the penetration phenomenon across SC is subjected to three possible pathways for passively penetrating substances. The interstitial lipid matrix of SC, the intercellular route, provides an uninterrupted pathway for molecules to pass through the skin. The intercellular lipids around corneocytes are the only continuous route starting from the skin surface to the deepest layer of SC. Other dermal passage routes are defined as the intracellular route (transcellular passage), the shunt route or passage by the pores [32,33]. The intracellular route requires penetrants to cross through the environment of both the keratin-rich corneocytes and the surrounding lipids, which is found challenging due to the significant resistance encountered to overcome both the lipophilic and the hydrophilic media [32]. The shunt route also called appendageal route including hair follicles, sebaceous glands, and sweat ducts has recently gained importance as a potential pathway for penetration of therapeutic and/or cosmetic compounds into the skin via nanocarrier-mediated delivery, with the remarkable scientific results demonstrated in recent research studies [32,34–36], despite the fact that they structurally constitute a very small percentage of the skin tissue [37,38].

The penetration properties of nanoparticulates from the skin are *controversial*, which gives prominence to the requirement for comprehensive multidisciplinary approaches to clarify the penetration mechanisms and interactions of the particles with the skin [39]. It is particularly emphasized that nanoparticulates cannot defeat the SC barrier [31–33]. Campbell et al. assessed polystyrene nanoparticles in a range between 20 and 200 nm, applied in aqueous suspension to dermatomed (750 μm-thick) skin in vitro, and emphasized that the nanosized particles could not penetrate through the intact skin, but could only infiltrate superficial parts of SC (corresponding to a depth of 2–3 μm) [40]. However, it has also been considered that nanocarriers find penetration chance to some extent through the transappendageal route including hair follicles and/or furrows, and also the regions in-between corneocyte clusters that are much more permeable than the intercellular pathway between individual corneocytes [41–43].

The potential regions for targeting nanocarriers into the skin have been well illustrated by Prow et al., who point to the skin surface, furrows, and hair follicles [12]. The follicular openings, leading to disruption of the continuity of the SC layer, serve as alternative entry points for nanocarriers to penetrate into the skin [6,44]. Many research studies have shown that nanocarrier systems tend to be deposited in the follicular regions [1,12,45–50]. In fact, the consideration of this shunt route being an potential pathway for the penetration of nanoparticulate systems is accepted realistic [12] even though follicular openings occupy quite a small area all over the total skin surface area (~0.1% in low-density areas such as face and forearm [51,52] and ~10% in high-density areas such as scalp [51]), particularly if the tightly organized structure of SC (intercorneocyte pore ≤10 nm) [43] is taken into account (Fig. 16.2).

16.4 The rationale behind nanocarrier-mediated follicular targeting

The efficacy of a topical treatment is significantly dependent on the penetration capability of the applied drug substances through the skin and the delivery of "actives" to the right place with a sufficient dose. In other words, it is of great importance that the intended molecules should be able to effectively deliver to the target regions of the skin. However, as it has been comprehensively described earlier, the skin barrier creates a critical limitation for topically applied "actives" to overcome. SC plays an important role in limiting the efficacy of most topical treatments in terms of delivery of molecules to the deeper target of the skin layers [32]. Therefore development of more effective strategies aiming to improve the skin penetration of active molecules is crucial in order to provide the required concentrations at the target site and consequently to increase the efficiency in topical treatments.

In this context, nanosized carrier systems for dermal delivery of active compounds offer a great hope [2]. Numerous studies conducted on dermal targeting in recent years have mentioned the superiorities of nanocarrier systems over conventional formulations [2,53–55]. Development of innovative topical dosage forms including nanocarriers is considered promising due to their potential for targeting "actives" to specific sites of the skin [41,46,56], offering a prolonged and controlled release [56], improving topical bioavailability as a consequence of improving cutaneous penetration [57,58], reducing possible side effects [59–62], extending the duration of the residence in the target area of the skin [63,64], increasing chemical stability of the compounds by preventing from degradation [65–68], which are important to achieve an increased treatment efficiency, reduce dosing frequency, and ensure patient compliance.

The point which should be highlighted that hair follicles are found to be a promising route in the case of nanocarrier-based skin delivery strategies. Since nanoparticulate carrier systems are particularly emphasized to accumulate in hair follicles that are considered as target structures on the appendageal route, potentially acting as depots for topically applied actives [1,5,12,45–48,50,69–73].

16.5 The role of hair follicles in nanocarrier-mediated skin delivery

16.5.1 Follicular ducts as reservoirs of nanocarriers

As a matter of fact what makes nanocarriers advantageous is their ability to accumulate in hair follicles which serve as a long-term reservoir [63,74], from which the associated active substances can be released in a controlled/sustained manner. In this regard, the selective accumulation tendency of nanocarriers in the follicular route expectedly makes these appendages valuable assets in terms of passive targeting.

Nanocarriers hold superiority over nonparticulate formulations due to the chance of much longer residence in hair follicles [69]. In fact, it was shown that nanoparticles could be stored in the follicular reservoir up to 10 days [63,69]. In this regard, nanocarrier-mediated follicular targeting is considered as a propitious tool for selective dermal delivery technology with the possibility of ensuring localization in certain compartments and cell populations within the hair follicles via topical application [75,76]. Different types of nanocarriers are currently being investigated within the context of targeting various actives to hair follicles as a topical treatment strategy, including polymeric [77,78] or metal nanoparticles [79], micellar carriers [57,80], lipid nanoparticles [81], microemulsions [81], nanoemulsions [82], and liposomes [83–85].

16.5.2 Follicular deposition ability of nanocarriers

Nanoparticulate systems have been shown to exhibit accumulation with a higher extent into hair follicles than solutes [86]. It has been pointed out in the literature that particles of different sizes and different structures applied to the surface of the skin exhibit tendency to accumulate in the follicular openings where they can penetrate into the follicular duct and importantly the penetration depths are largely dependent on their size [87]. Apart from particle size, it was also reported by Lademann et al. that the vehicle where particulate carriers are suspended may influence the penetration depth of particles in hair follicles [88].

The research studies conducted on nanoparticle-based follicular delivery revealed that the deposition of nanoparticulates into the hair follicles resulted in promising results with greater accumulation depths in contrast to nonparticulate formulations following administration of nanoparticulate systems accompanied with massage [69,89]. The enhancing effect of massage on the deposition efficiency of nanosized particles is attributed to the typical surface structure of hair shaft originated from the cuticle layer, and the movement of the hairs in the course of massaging where the microrelief structure in zigzag geometry is likely to act as a gear pump pushing the nanoparticles into the follicle duct [12,69,90], which can be found reasonable with the assumption of settlement of particles in suitable sizes into the recesses at the surface of cuticular hair to be interlocked. Recently, Radtke et al. demonstrated in the scope of their research that a "ratchet mechanism" as they proposed could be appropriate to explain nanoparticle transport in hair follicles, where the cuticular hair structure was modeled as a periodic asymmetric ratchet-shaped surface and the Brownian dynamics simulation model was used to investigate the motion of a single rigid nanoparticle inside a hair follicle [91].

The findings of these research studies indicated that the deposition of topically applied solid nanosized particulate systems into the follicular duct seems to depend on multiple parameters including the size of the particles, the type and hence the dimensions of the hair follicle, besides the drug loading, and also the vehicle characteristics that may alter the state of aggregation and consequently affect particle size and mechanical transport efficiency [87,88,92,93].

16.5.3 The extent of nanocarrier particle size for follicular delivery

Nanosized carrier systems seem to be favorable compared to microparticles as potential topical carriers for skin delivery, particularly for targeting actives to hair follicles [89]. It was already shown that the penetration of microparticles through the skin by cutaneous appendages (hair follicles, sweat ducts) would be limited in proportion to their size [94]. In this context, it was concluded that particles larger than 10 μm were not able to penetrate from the follicular orifices/openings but remained on the surface of the skin, while particles smaller than 3 μm exhibited higher penetration ability [95].

Toll et al. compared the follicular penetration of particles in size between 750 nm and 6 μm [96] and reported that 750 nm sized particles could penetrate into the hair follicles more efficiently, the particles in sizes of 750 nm vs 1.5 μm could reach greater penetration depths in the hair follicles compared to the microparticles in sizes of 3 and 6 μm. Patzelt et al. investigated the role of particle size for selective follicular targeting [93] and observed that the particles around 300–643 nm exhibited deeper follicular penetration among the particles with varying particle sizes in the range of 122–1000 nm, prepared using different types of materials. In the same study, it was shown that particles in sizes of 230 and 300 nm could penetrate up to the depth where sebaceous glands are found along the hair follicle duct. Rancan et al. investigated biodegradable poly(D,L-lactide) (PLA) nanoparticles of 228 and 365 nm in size [97]; follicular uptake by 50% of the vellus hair follicles was observed maximum penetration depth in the follicular ducts was found to reach the entry of the sebaceous glands in 12%–15% of all observed follicles. They reported no significant difference in follicular penetration between PLA nanoparticles with two different sizes.

An analysis of the surface structure of hairs by electron microscopy showed that the cuticular thickness, which gives a zigzag appearance to the hair structure, was ~530 nm in human hairs (a similar thickness for vellus and terminal hairs) and ~320 nm in porcine ear hairs [90]. It was stated that these findings on cuticular thickness correspond to the suggested optimum size range of about 300–600 nm for follicular penetration [90,93], which were determined in the case of massage application for topically administrated particulate systems. On the other hand, Vogt et al. [50] came up with that 750 nm particles were localized close to the skin surface in the follicular duct of fine hairs, known as vellus hairs, and could only penetrate to a certain depth, whereas 40 nm particles could penetrate deeper regions in the follicles.

16.6 Commonly used methods to examine follicular targeting of nanocarriers

In vitro experimental models have been widely employed to estimate the penetration capability of active substances in skin delivery. These studies are conducted with using different types of excised skin sources, namely, human, pig, rat, and mouse skin. Among other skin models, pig skin, which is known with high resemblance to human skin in terms of morphological structure, thickness, hair follicle content, and permeability, is the commonly used mammalian skin [98]. Nonetheless, even though excised skin models are very useful to understand skin penetration of various molecules for topical application, it should be taken into consideration that in some cases the suitability of the chosen model may be questionable. On this point, it has been reported in a research study comparing the follicular reservoir of human skin in vivo and in vitro on the same donor and same body site that follicular penetration was significantly reduced owing to the contraction of the elastic fibers within the skin due to the cutting during excision and therefore presumably constriction of the follicular openings with the reduction of the follicular reservoir and penetration pathway. Thus it was concluded that excised human skin is likely to be adequate only to a limited extent in follicular penetration studies because of the remarkably diminished follicular reservoir for penetration, gaining importance especially for the substances preferentially penetrate into the hair follicles like nanoparticles [99]. The importance of development noninvasive in vivo methods has been underlined. If an in vitro model is the mandatory choice, then pig ear skin is suggested as a more adequate option to utilize in the follicular delivery studies [99,100].

The most commonly used techniques to investigate the follicular penetration ability of topically applied compounds into the hair follicles are reviewed as briefly defined below.

16.6.1 Imaging techniques

16.6.1.1 Confocal laser scanning microscopy

Confocal laser scanning microscopy (CLSM) has become frequently utilized analysis tool as a noninvasive optical imaging technique to visualize the distribution profile of fluorescent molecules in the skin [101], by which a semiquantitative prediction for skin delivery can be made [101,102]. CLSM analysis provides researchers to obtain information about the localization and penetration pathway of topically applied fluorescent substances within the laser-scanned skin tissue [103]. This technique allows direct visualization of fluorescent compounds in the skin layers at multiple depths without requiring a mechanical sectioning and therefore without requiring any pretreatment process such as cryofixation or placement into an embedding material like paraffin prior an incision [102,103]. Owing to the ability of generating optical sections, this technique eliminates the mechanical sectioning-induced artifact formation. Furthermore, visualization of the compound of interest, which can be either a fluorescent dye or a fluorescently labeled compound within the skin, could be provided by high-resolution images with reduced blurring and a precise quantification of the distribution using a proper image analysis software [104].

Basically, CLSM images can be obtained either in the xy plane (i.e., parallel to the skin surface, SC, which is defined as $z = 0$ plane) on optically sliced sections that are acquired in different depths by scanning the skin at certain intervals (i.e., 1 μm) from the surface to the deeper layers of the skin tissue, or in the xz plane (i.e., *perpendicular* to the skin surface) when vertical cross-sectional slices are optically acquired from the region of interest across the skin sample along a horizontal line on the skin surface ($z = 0$ μm-xy plane) [101,105]. Subsequently, the obtained CSLM images with optical sectioning can be used for qualitative and/or quantitative analyses [106], and three-dimensional (xyz images) images can be generated through acquisition of multiple optical sections in the focal planes (i.e., xy) at different depth levels along the z axis which can also be explained as z-series acquisition of sequential x-y sections as a function of depth [104].

The use of CLSM has been popular in dermal research area investigating the penetration profile and localization of nanoparticles in the skin once topically applied [107]. A diverse range of fluorophores such as fluorescein isothiocyanate (FITC) [105], Nile Red (NR) [108], calcein [109], and rhodamine [110] are available for employment to visualize the distribution within the skin. It can be seen in the literature that fluorophores can be covalently bound to nanocarriers to be tracked within the skin tissue in order to evaluate the penetration and accumulation properties and clarify the penetration depth. FITC is one of the most used fluorescent dyes among many other examples employed for labeling, for instance, the penetration of poly(amidoamine) (PAMAM) dendrimers have often been examined by CLSM technique by binding FITC to their surface groups that are particularly convenient for covalent conjugation [108].

Encapsulating a fluorescent model within the associated nanocarrier is another approach to have information about the penetration properties from nanocarriers to determine the localization in the skin [110]. NR is a well-known compound that is often used as a model for topical active molecules due to its high lipophilicity with the value of log $P_{o/w} > 3$ [111,112]. The fluorescence signal detected at a wavelength of 630 nm, which is far from the wavelength range given by the skin components, is an important reason for preference due to avoidance of autofluorescence coming from the skin itself [97,113].

Besides, it is possible to visualize different fluorescent markers with separate excitation and emission properties in the same sample at the same time. In other words, different compounds can be simultaneously detected based on multiple fluorescence labeling. For example, fluorescently labeled polystyrene nanoparticles with associated NR were investigated by Wu et al. [112], fluorescence from NR and from the fluorescently labeled polymer was imaged independently by CLSM, subsequently confocal images that were overlayed indicated colocalization at the skin surface, at follicular openings and also in skin furrows.

As an important point to note, it is highly important to avoid autofluorescence of skin samples which may interfere with the employed fluorophores that would belie the actual penetration case. A way to separate the fluorescence signal of the fluorophore from that of the skin is to choose fluorescent compounds having distinctly different emission spectra from endogenous fluorescent structures of the skin [104].

16.6.1.2 Fluorescent microscopy

Evaluation of the penetration route of fluorescently marked nanoparticles for dermal delivery can alternatively be carried out using a standard fluorescence microscope. However, a cryosectioning process requirement to perform microscopic imaging of a skin sample due to the lack of capability for visualizing fluorescent probes in the deep layers underlying the skin surface unlike the CLSM technique is a main disadvantage of this tool. Analyzing by a fluorescent microscope, therefore, can be carried out on a specimen that is prepared with mechanical cross-sectioning of the skin. For specimen preparation, the skin sample is often cut into thin cross-sections (e.g., 20 μm) in vertical and/or horizontal direction following a pretreatment with a tissue embedding medium to be frozen. Eventually, fluorescence micrographs are taken basically through the principle of exciting the fluorophore by light at a certain wavelength.

There are many studies available in the literature where fluorescence microscopy was used as a tool for investigating the penetration ability and penetration depths of the fluorescent particles able to accumulate into the hair follicles of either nanocarrier systems themselves or encapsulated model fluorophores [49,114,115].

16.6.2 Quantitative techniques

16.6.2.1 Differential tape stripping

Differential tape stripping is a relatively widespread method to quantify the extent of skin uptake, additionally it gives information about accumulation of drug substances in the SC. The technique was developed with the principle of combining the tape-stripping method and cyanoacrylate skin surface biopsies which are obtained from formerly tape-stripped skin area (i.e., mostly 10–20 times) [116–118]. Tape-stripping technique involves using adhesive tapes precutted into an appropriate size to consecutively strip the SC layer 15–20 times by firmly pressing on the skin surface and pulling off at once [119], followed by which the compound of interest would be extracted from the tapes by a suitable procedure and quantitatively determined by an analytical quantification method.

Differential tape stripping is fulfilled when cyanoacrylate skin surface biopsy is performed as the latter part of the technique subsequent to the removal of the SC, where a sufficient amount (e.g., 1 drop) of cyanoacrylate superglue is applied to the stripped skin, then let total polymerization of the glue (e.g., 5 minutes) complete under a separate tape strip placed onto the glue applied skin tissue using light pressure, and the tape-strip is removed with one quick move. In some of the studies, it can be seen that two cyanoacrylate biopsies were taken one after another [116–118,120]. The rationale behind the latter step, which is also known as follicular biopsy [121], is to obtain the contents of the sebaceous follicles for the evaluation of follicular deposition of topically applied drugs or other substances, by which also follicular casts [122] composed of lipids, bacteria, and a mixture of keratinized material would be extracted. Moreover, cyanoacrylate skin surface stripping technique is found promising for transcutaneous immunization through macromolecules and particle-based vaccines owing to Langerhans cells activation, follicular targeting, and penetration enhancement [123].

16.6.2.2 *Punch biopsy*

Punch biopsy is a relatively recent method proposed by Lapteva et al. to assess the follicular accumulation of drug substances via topically applied nanocarriers [45,80]. The method permits the amount of drug substance of interest in the PSU (the site of follicular deposition) to be determined following in vitro skin penetration studies. In this technique, biopsy punches with a diameter small enough (i.e., 1 mm) to punch out PSUs individually are used and biopsies of nonfollicular regions (bulk) are also taken which act as controls. The biopsy samples, containing either hair follicles or bulk samples, are exposed to an extraction process, subsequently followed by a suitable quantification method to evaluate the follicular accumulation of active substances in PSUs in comparison to the bulk (nonfollicular regions of the skin).

16.7 Nanosized delivery systems for follicular targeting

Different types of nanocarriers for topical administration to the skin have been investigated in order to propose novel delivery systems, offering targeted delivery of active compounds into the hair follicules through their preferential accumulation properties, namely, the follicular deposition ability. When the literature is overviewed, it could be seen that there are many prospective nanosized systems available, having a potential for enhanced follicular delivery that can be potential for topical skin delivery of cosmetic compounds and drug substances. Here, the research studies which deal with the potential skin delivery of nanocarriers into hair follicles and deeper skin layers for mainly aiming antiaging effect, hair growing, photoaging, skin inflammation, and acne treatment have been briefly summarized.

Laredj-Bourezg et al. assessed PLA-block-poly(ethylene glycol) copolymer (PLA-b-PEG) and poly(caprolactone)-block-poly(ethylene glycol) copolymer (PCL-b-PEG)-based block copolymer micelles of 66 and 65 nm in hydrodynamic size [10], aiming to understand the skin delivery of encapsulated all-*trans* retinol (vitamin A), a model with the antiaging effect. They evaluated the distribution characteristics from these biocompatible and biodegradable nanoparticles by confocal microscopy using NR as a model fluorophore, and visualized NR accumulation in the hair follicles of the porcine skin that was examined upon 24 hour exposure following application to the skin in the encapsulated form within the micellar carriers.

Tsujimoto et al. proposed poly(lactic-co-glycolic acid) (PLGA) (75:25 ratio of lactic acid to glycolic acid) nanospheres as beneficial carrier systems for the follicular delivery of hair growing ingredients for hair loss treatment [7]. They encapsulated a combination of hinokitiol with antibacterial effect, glycyrrhetinic acid with antiinflammatory effect, and 6-benzylaminopurine with hair growth promotion effect in PLGA nanospheres. In vitro and in vivo performance of the PLGA nanospheres (about 200 nm) loaded with three different hair growing ingredients was evaluated; the scalp-pore permeability and hair growing effect were tested on human scalp skin in vitro and on mouse skin in vivo, respectively. A total of 2.0- to 2.5-fold greater scalp-pore permeability with the dispersion liquids of PLGA nanospheres encapsulating hair growing ingredient was observed by CLSM analysis of human scalp biopsies, compared to the control solutions of PBS containing either hinokitiol or glycyrrhetinic acid. According to the in vivo results, enhanced hair growing activity and therefore significantly improved hair growth was revealed when the active ingredients were encapsulated in PLGA nanospheres, where the hair growth cycle was transformed from the resting phase to the growing phase.

Arayachukeat et al. prepared poly(ethylene glycol)-4-methoxycinnamoylphthaloylchitosan (PCPLC) nanoparticles (about 227 nm) and investigated the dermal penetration and release of retinyl acetate against skin conditions such as photoaging, skin inflammation, and acne [124]. Retinyl acetate encapsulated in PCPLC nanoparticles was treated to freshly excised baby mouse skin for in-tissue-release assessment of retinyl acetate from PCPLC nanoparticles by CLSM where self-fluorescent retinyl acetate and the PCPLC polymer in the skin samples were able to be followed separately. Their results confirmed the accumulation of nanoparticles and the dermal penetration of retinyl acetate in hair follicles. In-tissue-release of retinyl acetate from nanoparticles was shown by the evidence of an increasingly higher retinyl acetate/PCPLC fluorescence intensity ratio detected deeper into the dermis and distant from the follicles.

Among other recent research studies introducing the potential of nanosized carriers in targeted skin delivery, with the evidence of follicular penetration, nanovesicular systems could certainly be given as a good example as lipid-based nanocarriers. Kumar et al. developed oleic acid-based nanovesicular topical gel formulation of minoxidil for hair loss treatment to obtain a targeted follicular delivery with minimized adverse effects, assuming that fatty acid vesicles would be relevant carriers for enhanced delivery to the hair follicles due to their compositional and physical attributes that could allow fusion with sebum [125]. The researchers incorporated the minoxidil encapsulated oleic acid nanovesicles (about 317 nm) into an emulgel vehicle to increase the contact time to the scalp and also to achieve an enhanced physical stability. The optimized gel system consisted of nanovesicles was evaluated for skin permeation and deposition profile, using a differential tape-stripping technique in comparison with a lotion form of minoxidil as a control. Rhodamine B was employed as the model fluorophore to evaluate the nanovesicles in terms of follicular deposition ability by CLSM. In vitro penetration studies tested on pig ear skin showed an enhanced skin and follicular delivery of minoxidil with nanovesicles. Furthermore, the investigation of the deposition profile of rhodamine in rat skin through CLSM analysis indicated that a deeper penetration with an enhanced rhodamine distribution around the hair follicles was observed with vesicular gel, which was attributed to the flexible nature and sebum compatibility of nanovesicles.

16.8 Iontophoretic technique combined with nanosized delivery systems for follicular targeting

As a sensible approach, loading "actives" to a nanocarrier system can be combined with a physical percutaneous penetration enhancement method like iontophoresis [126–128]. Iontophoresis is a physical method that has been used to increase the number of molecules transported into the skin using a small electrical current ($\leq$0.5 mA/cm^2) [129]. It has already been shown in the literature that follicular pathway could make an important contribution to delivery of some specific molecules across the skin under iontophoretic conditions, depending on physicochemical properties of the penetrant and its affinity for certain environments available in the skin [116,130]. The importance of follicular pathway in electrically enhanced topical/transdermal delivery has been discussed and the lower resistance encountered in this alternative route compared to the other possible transport pathways through the SC (intercellular and intracellular routes) was stated as the most likely reason for the contribution [116,118,131–133].

Importantly, for iontophoretic transport, ionized/charged molecules with good aqueous solubility are known as the particular group of penetrants for which the follicular route may be a relatively significant transport pathway [116,126]. For instance, iontophoresis was shown to provide a greatly enhanced transport through follicular structures over the passive control for calcein, which is a (multiply) charged (−4) fluorescent dye. It was revealed by the presented CLSM images of hairless mouse skin that current application provided low-resistance pathways, particularly via follicular structures for calcein transport [130]. Similarly, in another study minoxidil sulfate, an endogenous metabolite of minoxidil [116], having a hair follicle-stimulation effect against androgenic alopecia [134], was investigated to evaluate whether iontophoresis would improve follicular penetration of this ionizable form with greater aqueous solubility compared to the parent active compound, providing targeted delivery to the follicular route for an improved alopecia treatment. The amount of minoxidil sulfate recovered from the follicles through iontophoresis was found higher compared with its passive delivery [116]. In this way, for example, uncharged and/or large molecules such as protein and peptides, which demonstrate low transport numbers under iontophoretic current due to the fact that their primary electrotransport mechanism is based on convective solvent flow (electroosmosis) [135,136], may be considered to be loaded to a charged nanocarrier for iontophoretic delivery in order to overcome the limited penetration ability by improving the transport efficiency, even to provide targeted follicular delivery with a homogenously controlled release into the skin [49,126,127].

Matos et al. took a similar approach and further investigated their chitosan nanoparticles loaded with minoxidil sulfate in order to test the combined application of these nanoparticles with iontophoresis aiming to see whether an enhanced follicular accumulation of the active compound could be achieved besides a sustained release with reduced dermal exposure [134]. The researchers presumed that iontophoresis treatment with minoxidil sulfate nanoparticles could result in improvement in minoxidil sulfate delivery into the hair follicles due to the positive zeta potential of minoxidil sulfate nanoparticles (originated from amino groups of chitosan). However, unexpectedly the combination of iontophoresis with nanoparticular delivery did not resulted in an increased targeting effect in this study. In fact, follicular accumulation of minoxidil sulfate provided with iontophoresis was found similar statistically with that of provided with passive diffusion and the iontophoretic delivery from free drug solution. Eventually, iontophoretic treatment of minoxidil sulfate nanoparticles caused an increased dermal exposure, represented as minoxidil sulfate amounts found in the viable skin, with significant minoxidil sulfate amounts in the receptor solution [134].

In another study, minoxidil sulfate-loaded microparticles (3.0 ± 1.5 μm) were evaluated for the combination of microencapsulation and iontophoresis technology aiming to investigate the feasibility of follicular targeting of drug substances. Iontophoretic (0.5 mA/cm^2) skin delivery from the minoxidil sulfate-loaded chitosan microparticles was evaluated in comparison with its passive delivery. The follicular accumulation of minoxidil sulfate under passive diffusion conditions exhibits an increase in microparticles compared to control solution containing free minoxidil sulfate. The minoxidil sulfate amount in the follicular cast-containing biopsy samples was increased with the application of iontophoresis for both the control solution and microparticulate formula. Furthermore, it was revealed that 3-hour period of iontophoretic delivery did not provide a significant difference in SC uptake or skin permeation of minoxidil sulfate for control solution and the microparticles, but a significantly better follicular targeting was achieved with the microparticles. It was anticipated that the microparticles, tend to swell when contacted with ethanol/water, are not capable of crossing the skin in their intact form, but likely to overcome the skin barrier by accumulating in follicular openings with relatively small amounts. They stated that the application of the iontophoresis method could provide fast and efficient targeting of minoxidil sulfate-loaded microparticles to follicular openings. The authors emphasized that the appendageal accumulation of microparticles would offer a more significant potential in human scalp skin due to the approximately an order of magnitude higher follicular density than the porcine abdominal skin used in this study. Therefore the need for optimization of current density and application time in vivo was also emphasized for the topical treatment of alopecia [118].

Dendrimers are another group of nanotechnology-based carriers that have been prominently evaluated in skin delivery studies performed with iontophoresis. The potential of charged hydrophilic molecules in iontophoresis to follow the follicular route as a penetration pathway makes this technique important for follicular targeting. From this point of view, since dendrimers have highly functional surface groups, for example, PAMAM dendrimers offered with different ionizable polar surface groups, these potential carriers are reasonable as excellent candidates to take into consideration in terms of combination with iontophoresis as a penetration enhancement method.

In a recent study, PAMAM dendrimers of generation 1, 4, and 7 (G1, G4, and G7, respectively) were evaluated for their effect on electroosmotic flow through mouse skin following the iontophoresis application. Including dendrimers with increasing generations resulted in a decrease in anodal flux, and even change in the direction of electroosmotic flow with increasing sizes, which was assumed to be owing to the decrease in electroosmotic flow with the interaction between dendrimers and net negative charge of the skin. Dendrimer permeation into the skin was visualized by CLSM using G4 PAMAM−FITC conjugate where FITC was covalently attached to amine groups of the dendrimer to make the penetration and distribution of G4 PAMAMs across the skin visible. When anodal iontophoresis was applied, G4 PAMAM−FITC penetrated into the viable epidermis and dermis observed on the vertical cross-sectional images with respect to the skin surface. The authors emphasized that the follicular route is clearly a permeation pathway for G4 PAMAMs into the deeper dermal site of the skin because G4 PAMAM−FITC was detected to be localized on the follicular route reaching the hair bulb region although it did not show a penetration sign in the other regions deeper than 30 μm from the surface within the skin [108].

In another study, based on passive and iontophoretic skin penetration of PAMAM dendrimers as a function of surface charge and molecular weight, it was revealed by CLSM images that the distribution profiles of FITC-labeled PAMAMs with comparable molecular weight and size but different surface-functional groups, G4 PAMAM–FITC; G3.5 PAMAM-COOH–FITC; G4 PAMAM-OH–FITC, were mainly limited to SC following passive in vitro penetration studies on porcine ear skin. However, positively charged G4 PAMAM–FITC displayed a higher penetration into SC compared to the other PAMAM dendrimers having negative and neutral charge at pH 7.4. The fluorescence signal following iontophoretic (0.31 mA/cm^2) treatment of PAMAM dendrimers to the skin also showed that cationic dendrimer penetrated more deeply into the viable skin (up to a depth of 80 μm) than anionic and neutral dendrimers (up to a depth of 40 and 60 μm, respectively). The results indicated that the skin penetration of cationic and neutral dendrimers into viable epidermis was significantly ($P < .05$) enhanced through iontophoresis. As for the determination of the transport pathways for G4 PAMAM–FITC, further confocal analysis performed and skin furrows, intercellular regions, and hair follicles were identified as main distribution points for passive delivery. A similar distribution profile was noticed with iontophoretic delivery, except for the additional observation of bright fluorescence signal from clustered regions repeatedly noticed only with iontophoresis treatment [101].

As an example of the combined use of nanocarriers with iontophoretic technique aiming to achieve an efficient transfollicular delivery by benefiting the penetration enhancement mechanism of iontophoresis, cationic liposomes have been employed for the encapsulation of superoxide dismutase (SOD) [6], which is a highly hydrophilic and negatively charged (under physiological conditions) macromolecule (32.5 kDa) with protective effects against UV-induced skin damage as a potent antioxidant. In this study, iontophoretic delivery of SOD that was encapsulated in 1,2-dioleoyl-3-(trimethylammonium) propane-based cationic liposomes as a noninvasive transfollicular delivery system was investigated. Skin distribution and antioxidant activity of SOD-encapsulated liposomes (SOD-Lipo) were examined in vivo using Sprague Dawley rats. It was revealed that 3-hour iontophoretic treatment lead the liposomes prepared with fluorescent 4-nitrobenzo-2-oxa-1,3-diazoyl-(NBD)-labeled 1,2-dioleoylphosphatidylethanolamine (DOPE) to diffuse widely in the viable skin layer around hair follicles, whereas passive diffusion of the liposomes failed to efficiently penetrate into the skin. The fluorescently labeled SOD-lipo was not able to diffuse or penetrate into deeper skin layers through SC, the lipid marker of SOD-lipo was detected neither inside of skin nor in hair follicles. Iontophoretic delivery of SOD-Lipo was found to induce a significant decrease in the production of oxidative products in the skin induced by UV irradiation. SOD-lipo was proposed for combined use with iontophoresis as a noninvasive technique to routinely apply SOD to the skin, for protection against oxidant factors. In addition, a mean diameter of approximately 250–400 nm for the developed nanocarrier system was determined to be suitable for iontophoretic transfollicular delivery [6].

16.9 Conclusion

The development of effective approaches to improve the delivery of active molecules to the skin for aiming either cosmetic applications or dermatological treatments is crucial in terms of providing the required concentrations at the target sites of the skin.

Nanocarrier-mediated delivery systems targeting "actives" to some specific skin layers offer a great hope among those approaches in this respect. In recent years, nanocarrier-based delivery strategy has drawn great attention with the ability of nanocarriers for selective follicular deposition. Many research studies have shown that nanocarriers tend to be deposited in the follicular regions of the skin, which offer a great potential serving as skin reservoirs. Follicular targeting of actives via nanocarriers could provide an improved skin delivery of various cosmetically active molecules and the drug substances used in the treatment of dermatological diseases. The dose required to obtain expected satisfactory effects could be decreased and possible side effects related applied dose could be minimized via nanocarrier-mediated skin delivery. In addition, the combined application of nanocarrier systems with iontophoresis technique have a potential to provide fast and efficient targeting of active substances to follicular openings. For instance, the selectively appendageal accumulation of nanocarriers could offer a great potential in the treatment of some dermatological conditions such as allopecia, in which the application site of active molecules has high follicular density.

As a conclusion, using the follicular accumulation ability of nanocarriers could be an option for better treatments for many skin conditions including photoaging, UV-induced skin damage, inflammatory reactions of the skin, and other cosmetic problems. However, it should also be stated that the potential of the nanocarrier-based dermal delivery has to be confirmed with in vivo studies in depth and it is also crucial to conduct overall toxicological evaluation and risk assessments and clinical studies in terms of taking into consideration safety issues.

References

[1] Gupta M, Agrawal U, Vyas S. Nanocarrier-based topical drug delivery for the treatment of skin diseases. Expert Opin Drug Deliv 2012;9:783–804.

[2] Schafer-Korting M, Mehnert W, Korting HC. Lipid nanoparticles for improved topical application of drugs for skin diseases. Adv Drug Deliv Rev 2007;59:427–43.

[3] Khezri K, Saeedi M, Maleki Dizaj S. Application of nanoparticles in percutaneous delivery of active ingredients in cosmetic preparations. Biomed Pharmacother 2018;106:1499–505.

[4] Zhang Z, Tsai PC, Ramezanli T, Michniak-Kohn BB. Polymeric nanoparticles-based topical delivery systems for the treatment of dermatological diseases. Wiley Interdiscip Rev Nanomed Nanobiotechnol 2013;5:205–18.

[5] Lademann J, Knorr F, Richter H, Jung S, Meinke MC, Rühl E, et al. Hair follicles as a target structure for nanoparticles. J Innov Opt Health Sci 2015;08:1530004.

[6] Kigasawa K, Miyashita M, Kajimoto K, Kanamura K, Harashima H, Kogure K. Efficient intradermal delivery of superoxide dismutase using a combination of liposomes and iontophoresis for protection against UV-induced skin damage. Biol Pharm Bull 2012;35:781–5.

[7] Tsujimoto H, Hara K, Tsukada Y, Huang CC, Kawashima Y, Arakaki M, et al. Evaluation of the permeability of hair growing ingredient encapsulated PLGA nanospheres to hair follicles and their hair growing effects. Bioorg Med Chem Lett 2007;17:4771–7.

[8] Morgen M, Lu GW, Du D, Stehle R, Lembke F, Cervantes J, et al. Targeted delivery of a poorly water-soluble compound to hair follicles using polymeric nanoparticle suspensions. Int J Pharm 2011;416:314–22.

[9] Wang W, Lu K-J, Yu C-H, Huang Q-L, Du Y-Z. Nano-drug delivery systems in wound treatment and skin regeneration. J Nanobiotechnol 2019;17:82.

[10] Laredj-Bourezg F, Bolzinger M-A, Pelletier J, Valour J-P, Rovère M-R, Smatti B, et al. Skin delivery by block copolymer nanoparticles (block copolymer micelles). Int J Pharm 2015;496:1034–46.

[11] Williams AC. Structure and function of human skin. In: Williams AC, editor. Transdermal and topical drug delivery: from theory to clinical practice. London: Pharmaceutical Press; 2003. p. 1–25.
[12] Prow TW, Grice JE, Lin LL, Faye R, Butler M, Becker W, et al. Nanoparticles and microparticles for skin drug delivery. Adv Drug Deliv Rev 2011;63:470–91.
[13] Prausnitz MR, Mitragotri S, Langer R. Current status and future potential of transdermal drug delivery. Nat Rev Drug Discov 2004;3(2):115–24.
[14] Benson HAE, Watkinson AC. Skin structure, function, and permeation. In: Benson HAE, Watkinson AC, editors. Transdermal and topical drug delivery: principles and practice. Hoboken, NJ: John Wiley & Sons, Inc; 2012. p. 3–22.
[15] Honari G, Maibach H. Skin structure and function. In: Maibach H, Honari G, editors. Applied dermatotoxicology. Boston, MA: Academic Press; 2014. p. 1–10.
[16] Sala M, Diab R, Elaissari A, Fessi H. Lipid nanocarriers as skin drug delivery systems: properties, mechanisms of skin interactions and medical applications. Int J Pharm 2018;535:1–17.
[17] Aksoy B. Derinin doğal bağışıklık sistemi. TURKDERM 2013;47:2–11.
[18] Goldsmith LA, Katz SI, Gilchrest BA, Paller AS, Leffell DJ, Wolff K. Fitzpatrick's dermatology in general medicine. 8th Ed New York, NY: McGrawHill Companies Inc.; 2012.
[19] Forslind B. A domain mosaic model of the skin barrier. Acta Dermato-Venereol 1994;74:1–6.
[20] Brown MB, Martin GP, Jones SA, Akomeah FK. Dermal and transdermal drug delivery systems: current and future prospects. Drug Deliv 2006;13:175–87.
[21] Scott AS, Fong E. Integumentary system. Body structures & functions. 13th Ed Boston, MA: Cengage Learning; 2017. p. 64–84.
[22] Bouwstra JA, Ponec M. The skin barrier in healthy and diseased state. Biochim Biophys Acta 2006;1758:2080–95.
[23] Alberts B, Johnson A, Lewis J, Raff M, Roberts K, Walter P. Specialized tissues, stem cells, and tissue renewal. Molecular biology of the cell. 5th Ed New York, NY: Garland Science; 2008. p. 1417–28.
[24] Maibach H, Patrick E. Dermatotoxicology. In: Hayes WA, editor. Principles and methods of toxicology. 4th Ed Philadelphia, PA: Taylor and Francis; 2001.
[25] Simpson CL, Patel DM, Green KJ. Deconstructing the skin: cytoarchitectural determinants of epidermal morphogenesis. Nat Rev Mol Cell Biol 2011;12:565–80.
[26] Elias PM. Epidermal lipids, barrier function, and desquamation. J Investig Dermatol 1983;80(Suppl.):S44–9.
[27] Madison KC. Barrier function of the skin: "la raison d'etre" of the epidermis. J Investig Dermatol 2003;121:231–41.
[28] Haftek M. Epidermal barrier disorders and corneodesmosome defects. Cell Tissue Res 2015;360:483–90.
[29] Walters KA, Roberts, Michael S. The structure and function of skin. In: Walters KA, editor. Dermatological and transdermal formulations. New York, NY: Informa Healthcare USA, Inc; 2007. p. 1–39.
[30] Rhein LD, Zatz JL, Motwani MR. Targeted delivery of actives from topical treatment products to the pilosebaceous unit. In: Webster GF, Rawlings AV, editors. Acne and its therapy. New York, NY: Informa Healthcare USA, Inc; 2007. p. 223.
[31] Zouboulis CC. Acne and sebaceous gland function. Clin Dermatol 2004;22:360–6.
[32] Trommer H, Neubert RH. Overcoming the stratum corneum: the modulation of skin penetration. A review. Skin Pharmacol Physiol 2006;19(2):106–21.
[33] Barry BW. Lipid-protein-partitioning theory of skin penetration enhancement. J Control Release 1991;15:237–48.
[34] Lauer AC, Lieb LM, Ramachandran C, Flynn GL, Weiner ND. Transfollicular drug delivery. Pharm Res 1995;12:179–86.
[35] Lademann J, Otberg N, Richter H, Jacobi U, Schaefer H, Blume-Peytavi U, et al. Follicular penetration. An important pathway for topically applied substances. Hautarzt Z Dermatol Venerol verwandte Gebiete 2003;54:321–3.
[36] El Maghraby GM, Barry BW, Williams AC. Liposomes and skin: From drug delivery to model membranes. Eur J Pharm Sci 2008;34:203–22.
[37] Benson HA. Transdermal drug delivery: penetration enhancement techniques. Curr Drug Deliv 2005;2:23–33.
[38] Illel B, Schaefer H, Wepierre J, Doucet O. Follicles play an important role in percutaneous absorption. J Pharm Sci 1991;80:424–7.

[39] Baroli B. Penetration of nanoparticles and nanomaterials in the skin: fiction or reality? J Pharm Sci 2010;99:21–50.

[40] Campbell CS, Contreras-Rojas LR, Delgado-Charro MB, Guy RH. Objective assessment of nanoparticle disposition in mammalian skin after topical exposure. J Control Release 2012;162:201–7.

[41] Lapteva M, Santer V, Mondon K, Patmanidis I, Chiriano G, Scapozza L, et al. Targeted cutaneous delivery of ciclosporin A using micellar nanocarriers and the possible role of inter-cluster regions as molecular transport pathways. J Control Release 2014;196:9–18.

[42] Alvarez-Roman R, Naik A, Kalia YN, Guy RH, Fessi H. Skin penetration and distribution of polymeric nanoparticles. J Control Release 2004;99:53–62.

[43] Cevc G, Vierl U. Nanotechnology and the transdermal route: a state of the art review and critical appraisal. J Control Release 2010;141:277–99.

[44] Rotomskis R. Quantum dot migration through natural barriers and distribution in the skin. In: Hamblin MR, Avci P, Prow TW, editors. Nanoscience in dermatology. Boston, MA: Academic Press; 2016. p. 307–21.

[45] Kandekar SG, Del Rio-Sancho S, Lapteva M, Kalia YN. Selective delivery of adapalene to the human hair follicle under finite dose conditions using polymeric micelle nanocarriers. Nanoscale 2018;10:1099–110.

[46] Lapteva M, Mondon K, Moller M, Gurny R, Kalia YN. Polymeric micelle nanocarriers for the cutaneous delivery of tacrolimus: a targeted approach for the treatment of psoriasis. Mol Pharm 2014;11:2989–3001.

[47] Mahmoud NN, Alkilany AM, Dietrich D, Karst U, Al-Bakri AG, Khalil EA. Preferential accumulation of gold nanorods into human skin hair follicles: effect of nanoparticle surface chemistry. J Colloid Interface Sci 2017;503:95–102.

[48] Roberts MS, Mohammed Y, Pastore MN, Namjoshi S, Yousef S, Alinaghi A, et al. Topical and cutaneous delivery using nanosystems. J Control Release 2017;247:86–105.

[49] Takeuchi I, Takeshita T, Suzuki T, Makino K. Iontophoretic transdermal delivery using chitosan-coated PLGA nanoparticles for positively charged drugs. Colloids Surf B Biointerfaces 2017;160:520–6.

[50] Vogt A, Combadiere B, Hadam S, Stieler KM, Lademann J, Schaefer H, et al. 40 nm, but not 750 or 1,500 nm, nanoparticles enter epidermal CD1a+ cells after transcutaneous application on human skin. J Investig Dermatol 2006;126:1316–22.

[51] Roberts MS, Cross SE, Pellett MA. Skin transport. In: Walters KA, editor. Dermatological and transdermal formulations. New York, NY: Informa Healthcare USA, Inc; 2007. p. 89–188.

[52] Otberg N, Richter H, Schaefer H, Blume-Peytavi U, Sterry W, Lademann J. Variations of hair follicle size and distribution in different body sites. J Investig Dermatol 2004;122:14–19.

[53] Masini V, Bonte F, Meybeck A, Wepierre J. Cutaneous bioavailability in hairless rats of tretinoin in liposomes or gel. J Pharm Sci 1993;82:17–21.

[54] Fesq H, Lehmann J, Kontny A, Erdmann I, Theiling K, Rother M, et al. Improved risk-benefit ratio for topical triamcinolone acetonide in transfersome in comparison with equipotent cream and ointment: a randomized controlled trial. Br J Dermatol 2003;149:611–19.

[55] Papakostas D, Rancan F, Sterry W, Blume-Peytavi U, Vogt A. Nanoparticles in dermatology. Arch Dermatol Res 2011;303:533–50.

[56] Matos BN, Reis TA, Gratieri T, Gelfuso GM. Chitosan nanoparticles for targeting and sustaining minoxidil sulphate delivery to hair follicles. Int J Biol Macromol 2015;75:225–9.

[57] Kahraman E, Ozhan G, Ozsoy Y, Gungor S. Polymeric micellar nanocarriers of benzoyl peroxide as potential follicular targeting approach for acne treatment. Colloids Surf B Biointerfaces 2016;146:692–9.

[58] Klang V, Matsko N, Zimmermann AM, Vojnikovic E, Valenta C. Enhancement of stability and skin permeation by sucrose stearate and cyclodextrins in progesterone nanoemulsions. Int J Pharm 2010;393(1-2):152–60.

[59] Schafer-Korting M, Korting HC, Ponce-Poschl E. Liposomal tretinoin for uncomplicated acne vulgaris. Clin Investig 1994;72:1086–91.

[60] Mandawgade SD, Patravale VB. Development of SLNs from natural lipids: application to topical delivery of tretinoin. Int J Pharm 2008;363(1-2):132–8.

[61] Castro GA, Oliveira CA, Mahecha GAB, Ferreira LAM. Comedolytic effect and reduced skin irritation of a new formulation of all-trans retinoic acid-loaded solid lipid nanoparticles for topical treatment of acne. Arch Dermatol Res 2011;303:513–20.

[62] Castro GA, Coelho ALLR, Oliveira CA, Mahecha GAB, Oréfice RL, Ferreira LAM. Formation of ion pairing as an alternative to improve encapsulation and stability and to reduce skin irritation of retinoic acid loaded in solid lipid nanoparticles. Int J Pharm 2009;381:77–83.
[63] Lademann J, Richter H, Schaefer UF, Blume-Peytavi U, Teichmann A, Otberg N, et al. Hair follicles—a long-term reservoir for drug delivery. Skin Pharmacol Physiol 2006;19:232–6.
[64] Lohani A, Verma A, Joshi H, Yadav N, Karki N. Nanotechnology-based cosmeceuticals. ISRN Dermatol 2014;2014:843687.
[65] Hasanovic A, Zehl M, Reznicek G, Valenta C. Chitosan-tripolyphosphate nanoparticles as a possible skin drug delivery system for aciclovir with enhanced stability. J Pharm Pharmacol 2009;61:1609–16.
[66] Brisaert M, Gabriels M, Matthijs V, Plaizier-Vercammen J. Liposomes with tretinoin: a physical and chemical evaluation. J Pharm Biomed Anal 2001;26(5-6):909–17.
[67] Ourique AF, Pohlmann AR, Guterres SS, Beck RCR. Tretinoin-loaded nanocapsules: preparation, physicochemical characterization, and photostability study. Int J Pharm 2008;352:1–4.
[68] Shah KA, Date AA, Joshi MD, Patravale VB. Solid lipid nanoparticles (SLN) of tretinoin: potential in topical delivery. Int J Pharm 2007;345:163–71.
[69] Lademann J, Richter H, Teichmann A, Otberg N, Blume-Peytavi U, Luengo J, et al. Nanoparticles—an efficient carrier for drug delivery into the hair follicles. Eur J Pharm Biopharm 2007;66:159–64.
[70] Takeuchi I, Suzuki T, Makino K. Skin permeability and transdermal delivery route of 50-nm indomethacin-loaded PLGA nanoparticles. Colloids Surf B Biointerfaces 2017;159:312–17.
[71] Rancan F, Blume-Peytavi U, Vogt A. Utilization of biodegradable polymeric materials as delivery agents in dermatology. Clin Cosmet Investig Dermatol 2014;7:23–34.
[72] Hoffman RM. The hair follicle and its stem cells as drug delivery targets. Expert Opin Drug Deliv 2006;3:437–43.
[73] Kumar D, Ali J, Baboota S. Omega 3 fatty acid-enriched nanoemulsion of thiocolchicoside for transdermal delivery: formulation, characterization and absorption studies. Drug Deliv 2016;23:591–600.
[74] Knorr F, Lademann J, Patzelt A, Sterry W, Blume-Peytavi U, Vogt A. Follicular transport route—research progress and future perspectives. Eur J Pharm Biopharm 2009;71:173–80.
[75] Vogt A, Mandt N, Lademann J, Schaefer H, Blume-Peytavi U. Follicular targeting—a promising tool in selective dermatotherapy. J Investig Dermatol Symp Proc 2005;10:252–5.
[76] Lademann J, Otberg N, Jacobi U, Hoffman RM, Blume-Peytavi U. Follicular penetration and targeting. J Investig Dermatol Sympos Proc 2005;10:301–3.
[77] Suwannateep N, Wanichwecharungruang S, Fluhr J, Patzelt A, Lademann J, Meinke MC. Comparison of two encapsulated curcumin particular systems contained in different formulations with regard to in vitro skin penetration. Skin Res Technol 2013;19:1–9.
[78] Pan-In P, Wongsomboon A, Kokpol C, Chaichanawongsaroj N, Wanichwecharungruang S. Depositing α-mangostin nanoparticles to sebaceous gland area for acne treatment. J Pharmacol Sci 2015;129:226–32.
[79] Labouta HI, Schneider M. Interaction of inorganic nanoparticles with the skin barrier: current status and critical review. Nanomed: Nanotechnol Biol Med 2013;9:39–54.
[80] Lapteva M, Moller M, Gurny R, Kalia YN. Self-assembled polymeric nanocarriers for the targeted delivery of retinoic acid to the hair follicle. Nanoscale 2015;7:18651–62.
[81] Aljuffali IA, Sung CT, Shen F-M, Huang C-T, Fang J-Y. Squarticles as a lipid nanocarrier for delivering diphencyprone and minoxidil to hair follicles and human dermal papilla cells. AAPS J 2014;16:140–50.
[82] Wu H, Ramachandran C, Weiner ND, Roessler BJ. Topical transport of hydrophilic compounds using water-in-oil nanoemulsions. Int J Pharm 2001;220:63–75.
[83] Li L, Lishko VK, Hoffman RM. Liposomes can specifically target entrapped melanin to hair follicles in histocultured skin. In Vitro Cell Dev Biol 1993;29:192–4.
[84] Li L, Hoffman RM. Topical liposome delivery of molecules to hair follicles in mice. J Dermatol Sci 1997;14:101–8.
[85] Hoffman RM. Topical liposome targeting of dyes, melanins, genes, and proteins selectively to hair follicles. J Drug Target 1998;5:67–74.
[86] Hansen S, Lehr C-M. Transfollicular delivery takes root: the future for vaccine design? Expert Rev Vaccines 2014;13:5–7.

[87] Lademann J, Knorr F, Richter H, Blume-Peytavi U, Vogt A, Antoniou C, et al. Hair follicles—an efficient storage and penetration pathway for topically applied substances. Summary of recent results obtained at the Center of Experimental and Applied Cutaneous Physiology, Charite-Universitatsmedizin Berlin, Germany. Skin Pharmacol Physiol 2008;21:150–5.
[88] Patzelt A, Richter H, Dähne L, Walden P, Wiesmüller K-H, Wank U, et al. Influence of the vehicle on the penetration of particles into hair follicles. Pharmaceutics 2011;3:307–14.
[89] Neubert RHH. Potentials of new nanocarriers for dermal and transdermal drug delivery. Eur J Pharm Biopharm 2011;77:1–2.
[90] Lademann J, Patzelt A, Richter H, Antoniou C, Sterry W, Knorr F. Determination of the cuticula thickness of human and porcine hairs and their potential influence on the penetration of nanoparticles into the hair follicles. J Biomed Opt 2009;14:021014.
[91] Radtke M, Patzelt A, Knorr F, Lademann J, Netz RR. Ratchet effect for nanoparticle transport in hair follicles. Eur J Pharm Biopharm 2017;116:125–30.
[92] Vogt A, Hadam S, Heiderhoff M, Audring H, Lademann J, Sterry W, et al. Morphometry of human terminal and vellus hair follicles. Exp Dermatol 2007;16:946.
[93] Patzelt A, Richter H, Knorr F, Schäfer U, Lehr C-M, Dähne L, et al. Selective follicular targeting by modification of the particle sizes. J Control Release 2011;150:45–8.
[94] Schaefer H, Watts F, Brod J, Illel B. Follicular penetration. In: Scott RC, Guy RH, Hadgraft J, editors. Prediction of percutaneous penetration: methods, measurements, modelling. London: IBC Technical Services; 1990. p. 163–73.
[95] Rolland A, Wagner N, Chatelus A, Shroot B, Schaefer H. Site-specific drug delivery to pilosebaceous structures using polymeric microspheres. Pharm Res 1993;10:1738–44.
[96] Toll R, Jacobi U, Richter H, Lademann J, Schaefer H, Blume-Peytavi U. Penetration profile of microspheres in follicular targeting of terminal hair follicles. J Investig Dermatol 2004;123:168–76.
[97] Rancan F, Papakostas D, Hadam S, Hackbarth S, Delair T, Primard C, et al. Investigation of polylactic acid (PLA) nanoparticles as drug delivery systems for local dermatotherapy. Pharm Res 2009;26:2027–36.
[98] Summerfield A, Meurens F, Ricklin ME. The immunology of the porcine skin and its value as a model for human skin. Mol Immunol 2015;66:14–21.
[99] Patzelt A, Richter H, Buettemeyer R, Huber HJR, Blume-Peytavi U, Sterry W, et al. Differential stripping demonstrates a significant reduction of the hair follicle reservoir in vitro compared to in vivo. Eur J Pharm Biopharm 2008;70:234–8.
[100] Mak WC, Richter H, Patzelt A, Sterry W, Lai KK, Renneberg R, et al. Drug delivery into the skin by degradable particles. Eur J Pharm Biopharm 2011;79:23–7.
[101] Venuganti VV, Sahdev P, Hildreth M, Guan X, Perumal O. Structure-skin permeability relationship of dendrimers. Pharm Res 2011;28:2246–60.
[102] Alvarez-Roman R, Naik A, Kalia YN, Fessi H, Guy RH. Visualization of skin penetration using confocal laser scanning microscopy. Eur J Pharm Biopharm 2004;58:301–16.
[103] Changez M, Chander J, Dinda AK. Transdermal permeation of tetracaine hydrochloride by lecithin microemulsion: in vivo. Colloids Surf B Biointerfaces 2006;48:58–66.
[104] Rossetti FC, Depieri LV, Bentley VLBM. Confocal laser scanning microscopy as a tool for the investigation of skin drug delivery systems and diagnosis of skin disorders. In: Lagali N, editor. Confocal laser microscopy—principles and applications in medicine, biology, and the food sciences. IntechOpen; 2013. p. 99–140.
[105] Alvarez-Roman R, Merino G, Kalia YN, Naik A, Guy RH. Skin permeability enhancement by low frequency sonophoresis: lipid extraction and transport pathways. J Pharm Sci 2003;92:1138–46.
[106] Chu M, Kollias N, Luedtke M. Confocal scanning laser microscopy: applications for imaging dynamic processes in skin in vivo. In: Wang C-C, editor. Laser scanning, theory and applications. InTech; 2011. p. 31–48.
[107] Zhang LW, Monteiro-Riviere NA. Use of confocal microscopy for nanoparticle drug delivery through skin. J Biomed Opt 2013;18:061214.
[108] Kim HJ, Oh SY. Modulation of electroosmotic flow through skin: effect of poly(amidoamine) dendrimers. Biomol Ther 2018;26:182–90.
[109] Kajimoto K, Yamamoto M, Watanabe M, Kigasawa K, Kanamura K, Harashima H, et al. Noninvasive and persistent transfollicular drug delivery system using a combination of liposomes and iontophoresis. Int J Pharm 2011;403(1-2):57–65.

[110] Gillet A, Lecomte F, Hubert P, Ducat E, Evrard B, Piel G. Skin penetration behaviour of liposomes as a function of their composition. Eur J Pharm Biopharm 2011;79:43–53.

[111] Lombardi Borgia S, Regehly M, Sivaramakrishnan R, Mehnert W, Korting HC, Danker K, et al. Lipid nanoparticles for skin penetration enhancement-correlation to drug localization within the particle matrix as determined by fluorescence and parelectric spectroscopy. J Control Release 2005;110:151–63.

[112] Wu X, Price GJ, Guy RH. Disposition of nanoparticles and an associated lipophilic permeant following topical application to the skin. Mol Pharm 2009;6:1441–8.

[113] Hathout RM, Mansour S, Mortada ND, Geneidi AS, Guy RH. Uptake of microemulsion components into the stratum corneum and their molecular effects on skin barrier function. Mol Pharm 2010;7:1266–73.

[114] Sumian CC, Pitre FB, Gauthier BE, Bouclier M, Mordon SR. A new method to improve penetration depth of dyes into the follicular duct: potential application for laser hair removal. J Am Acad Dermatol 1999;41(2 Pt 1):172–5.

[115] Lohan SB, Icken N, Teutloff C, Saeidpour S, Bittl R, Lademann J, et al. Investigation of cutaneous penetration properties of stearic acid loaded to dendritic core-multi-shell (CMS) nanocarriers. Int J Pharm 2016;501:271–7.

[116] Gelfuso GM, Gratieri T, Delgado-Charro MB, Guy RH, Vianna Lopez RF. Iontophoresis-targeted, follicular delivery of minoxidil sulfate for the treatment of alopecia. J Pharm Sci 2013;102:1488–94.

[117] Ossadnik M, Czaika V, Teichmann A, Sterry W, Tietz HJ, Lademann J, et al. Differential stripping: introduction of a method to show the penetration of topically applied antifungal substances into the hair follicles. Mycoses 2007;50:457–62.

[118] Gelfuso GM, Barros MA, Delgado-Charro MB, Guy RH, Lopez RF. Iontophoresis of minoxidil sulphate loaded microparticles, a strategy for follicular drug targeting? Colloids Surf B Biointerfaces 2015;134:408–12.

[119] Manconi M, Sinico C, Valenti D, Lai F, Fadda AM. Niosomes as carriers for tretinoin. III. A study into the in vitro Cutaneous delivery of vesicle-incorporated tretinoin. Int J Pharm 2006;311(1-2):11–19.

[120] Raber AS, Mittal A, Schäfer J, Bakowsky U, Reichrath J, Vogt T, et al. Quantification of nanoparticle uptake into hair follicles in pig ear and human forearm. J Control Release 2014;179:25–32.

[121] Mills Jr OH, Kligman AM. The follicular biopsy. Dermatologica 1983;167:57–63.

[122] Thielitz A, Helmdach M, Röpke E-M, Gollnick H. Lipid analysis of follicular casts from cyanoacrylate strips as a new method for studying therapeutic effects of antiacne agents. Br J Dermatol 2001;145:19–27.

[123] Vogt A, Hadam S, Deckert I, Schmidt J, Stroux A, Afraz Z, et al. Hair follicle targeting, penetration enhancement and Langerhans cell activation make cyanoacrylate skin surface stripping a promising delivery technique for transcutaneous immunization with large molecules and particle-based vaccines. Exp Dermatol 2015;24:73–5.

[124] Arayachukeat S, Wanichwecharungruang SP, Tree-Udom T. Retinyl acetate-loaded nanoparticles: dermal penetration and release of the retinyl acetate. Int J Pharm 2011;404(1-2):281–8.

[125] Kumar P, Singh SK, Handa V, Kathuria H. Oleic acid nanovesicles of minoxidil for enhanced follicular delivery. Medicines (Basel), 5. 2018. p. 103.

[126] Malinovskaja-Gomez K. Transdermal iontophoresis—delivery control by ion-exchange fibers and nanocarriers [Doctoral thesis]. Helsinki, Finland: University of Helsinki; 2016.

[127] Dragicevic N, Maibach H. Combined use of nanocarriers and physical methods for percutaneous penetration enhancement. Adv Drug Deliv Rev 2018;127:58–84.

[128] Kulkarni SB, Banga AK, Betageri GV. Transdermal lontophoretic delivery of colchicine encapsulated in liposomes. Drug Deliv 1996;3:245–50.

[129] Naik A, Kalia YN, Guy RH. Transdermal drug delivery: overcoming the skin's barrier function. Pharm Sci Technol Today 2000;3:318–26.

[130] Turner NG, Guy RH. Iontophoretic transport pathways: dependence on penetrant physicochemical properties. J Pharm Sci 1997;86:1385–9.

[131] Bath BD, Scott ER, Phipps JB, White HS. Scanning electrochemical microscopy of iontophoretic transport in hairless mouse skin. Analysis of the relative contributions of diffusion, migration, and electroosmosis to transport in hair follicles. J Pharm Sci 2000;89:1537–49.

[132] Cullander C, Guy RH. Visualization of iontophoretic pathways with confocal microscopy and the vibrating probe electrode. Solid State Ion 1992;53-56:197–206.

[133] Uitto OD, White HS. Electroosmotic pore transport in human skin. Pharm Res 2003;20:646–52.
[134] Matos BN, Melo LMA, Pereira MN, Reis TA, Cunha-Filho M, Gratieri T, et al. Iontophoresis on minoxidil sulphate-loaded chitosan nanoparticles accelerates drug release, decreasing their targeting effect to hair follicles. Quím Nova 2017;40:154–7.
[135] Delgado-Charro MB, Guy RH. Characterization of convective solvent flow during iontophoresis. Pharm Res 1994;11:929–35.
[136] Kim A, Green PG, Rao G, Guy RH. Convective solvent flow across the skin during iontophoresis. Pharm Res 1993;10:1315–20.

CHAPTER

17

Nanoemulsion in cosmetic: from laboratory to market

Vikas Pandey[1], *Rajesh Shukla*[1], *Ashish Garg*[2], *Mohan Lal Kori*[3] *and Gopal Rai*[1]

[1]Guru Ramdas Khalsa Institute of Science and Technology (Pharmacy), Jabalpur, India [2]Department of Pharmacy, Rani Durgavati Vishwavidyalaya, Jabalpur, India [3]Department of Pharmacy, Vedica College of B. Pharmacy, Bhopal, India

17.1 Introduction

17.1.1 Cosmetics: the paradigm shift

In today's era, societies attention is substantially directed toward disbursing, waging, and intending on personnel body appearance and amplified looks, as it plays a fundamental role in self-countenance and identity recognition. In the past, looks and beauty were used to be a concern of females, but now males are also not far behind, as looks are now novel deceptive gauge of communal wealth.

Being diligent on luster is part of practice since ancient times, when cleopatra's milk baths used to augment body's expression, Kohl intensifies attraction of eyes, vegetal dyes boosts cheeks and lips, and hair embellishments [1]. However, people regularly laid themselves in critical risky conditions to customarily prescribed approaches of beautification, outcome of which turned into diversified array of serious side effects. To combat with these issues and keeping in view the increasing demand along with progress of cosmetics, regulatory effects came into action. In the later part of the 20th century, significant evolutions in medicine and nutrition blended with growing cognizance toward individual healthcare, which supported regulatory bodies to standardize cosmetic products on grounds of safety and efficacy [2]. As per Federal laws, cosmetics are defined by their intended use, as "articles intended to be rubbed, poured, sprinkled, or sprayed on, introduced into, or otherwise applied to the human body for cleansing, beautifying, promoting attractiveness, or altering the appearance" [FD&C Act, sec. 201(i)] [3].

Nanocosmetics
DOI: https://doi.org/10.1016/B978-0-12-822286-7.00021-8

Over the past one decade, the development of new technologies aimed at betterment for satisfying cosmetic consumer demands, which substituted conventional and perilous merchandise systems. Among progress on various grounds, nanotechnology has gained outstanding popularity, which revolutionized the criteria of formulation and development in the cosmetic industry. Currently, the highest imperative is to fabricate such nanotechnology-based products which could not only amplifies the delivery of active cosmetic ingredients into the skin but also provides enhanced stability with minimal toxicity [4]. Nanotechnology provided new opportunities for novel product development with improvisation in formulations, delivery of active ingredients, interaction with skin, and long-lasting proficient effects. Nanomaterials have boosted the differentiated application of cosmetics, with a hefty array of targeted interaction like skin, hair, and teeth. This interaction may cost safety issues of the product; hence, they must also be kept into deliberation [5]. Accordingly, in order to account side effects and safety concerns associated with nanomaterials, an important field "nanotoxicology" came into existence, which works on the basis of utilizing test models for efficient forecasting of nanocosmetic toxicological effects [6]. Over and all, it is very important to assure the application of regulated methodologies and utilization of cosmetic ingredients to confirm safety in agreement to current legislative laws.

17.1.2 Overview of nanoemulsions

Nanoemulsions are a colloidal particulate composition in the submicron dimension order act as a drug delivery device. Their size ranges differ from 10 to 1000 nm. These vehicles are solid globules and their exterior is amorphous and lipophilic with a negative charge. As a drug delivery tool, they improve the remedial efficacy of the medication and decrease the adverse effect and lethal reactions. The important application involves the treatment of disease of the reticuloendothelial system, catalyst replacement treatment in the liver, treatment of tumor, and vaccination [7]. An emulsion is a biphasic system in which one phase is intimately dispersed in another phase in the construction of tiny droplets varying in diameter from 0.1 to 100 μm. It is a thermodynamically unstable system, which can be maintained by the behavior of an emulsifying agent (emulgent or emulsifier). The dispersed phase is also recognized as the internal phase or the discontinuous phase while the external phase is called dispersion medium, outer phase, or continuous phase. The emulsifying agent is also known as intermediate or interphase. The term "nanoemulsion" also refers to a miniemulsion which is fine oil/water or water/oil dispersion sustained by an interfacial layer of surfactant particle having droplet size range 20–600 nm. Because of minute size, nanoemulsions are transparent. Three varieties of nanoemulsions can be established: (1) oil/water nanoemulsions, in which oil is dispersed in the continuous aqueous phase, (2) water/oil nanoemulsions in which water droplets are dispersed in the continuous oil phase, and (3) bicontinuous nanoemulsions.

This technology later attained its way into popular formulations of cosmeceutical preparation. A cosmeceutical is a cosmetic preparation with the consolidated active component meant for an advantageous physiological activity, following from an improved pharmacological performance when related to an inert cosmetic [8–10]. The significant advantages of utilizing nanotechnology in cosmeceutical applications are the enhanced stability of several cosmetic components, viz., unsaturated fatty acids, vitamins, and antioxidants encapsulated inside the nanoparticles [10], increased penetration rate of specific constituents, such as vitamins and additional antioxidants, enhanced esthetics of the commodity [11], in conjunction with enhanced performance and tolerance of UV filters on the skin covering [12]. For instance, Mayer et al. [13] successfully encapsulated the oil-soluble vitamin (vitamin E) and applied a nanoemulsion as the delivery system. In fact in dermatology, W/O emulsions are a more reliable system, where the oil-soluble active constituents are made more favorable by a lipid film configuration on the skin. Mechanisms describing the enhanced skin penetration by nanoemulsions have so far been linked with their nanosized range and construction [14]. The diffusional boundary of the stratum corneum may be overcome by the surfactant in the formulation, which leads to high penetration of cosmeceuticals. The water content in the system also has an essential functional performance. When the water content is sufficiently more in the preparation, percutaneous penetration of cosmeceuticals is enhanced due to the hydration impact of the stratum corneum. Nanosized droplets dispersed in the continuous phase of the nanoemulsion can now move smoothly into the stratum corneum and distribute the active constituent within the skin barrier [15]. Dirschka et al. [16] revealed that daylight photodynamic treatment with BF-200 ALA gel and MAL cream for the therapy of actinic keratosis and the investigation demonstrates a trend toward greater effectiveness next 3 months and significantly lower recurrence rates after 1-year follow-up. Abd et al. [17] studied transfollicular delivery increase of caffeine from nanoemulsion formulations including oleic acid and eucalyptol as chemical penetration enhancers and this research described efficient transportation of caffeine as a hydrophilic model drug into and by hair follicles and explained that follicles and encircling areas may be targeted by optimized formulations for particular medications. Kong et al. [18] described that hyaluronan (hyaluronic acid, HA)-glycerol α-monostearate–based nanoemulsion results in an attractive colloidal transdermal vehicle fit for purposes in skin care and cosmetic preparation.

It has been currently published that the overall global cosmeceuticals market amounted to a whopping USD 46.93 billion in 2017 and is projected to jump to USD 80.36 billion by 2023 at a composite annual growth rate of 9.38% [19]. Nevertheless, nanotechnology in cosmeceuticals is not specifically a new technology. It was first presented to the cosmeceutical market in the 1980s in the formation of liposomes. Since then, a myriad of other nanotechnology-based preparation that comprises peptides, proteomics, stem cells, and epigenetic factors have been formed and are staying sold to customers [20]. Large cosmetic corporations publish several nanotechnology-related patents every year. Hence, the significance of these patents and investments today.

Many nanotechnology-based cosmeceutical patents and commodities obtainable on the market are cosmetic compositions including retinol stabilized by-porous polymer beads and nanoemulsion (US 20130095157A1, published on 2013-04-18 by Act Co Ltd) was patented as a process for stabilizing retinol (vitamin A) intended for an antiinflammatory and wrinkle-reducing cosmetic developed by nanoemulsification. The retinol polymer nanocapsule is constructed by nanoemulsifying retinol in porous polymer particles (size of 50–200 nm). Mung bean medium-chain triglyceride extract and lecithin are then utilized for the stabilization of retinol [21]. A Japanese company, Ands Corporation, continues a patent which indicates the formation of a nanoemulsion of size 100 nm that comprises nanoparticles of phospholipid and a lysophospholipid, dispersed in an aqueous medium. Iwamoto Shigemi patented a process to formulate a dry collagen face lotion comprising a mix of a collagen or gelatin protein with the establishment of nanoparticle powder of particle sizes between 10 and 40 nm. Nanoemulsion and cosmetic product compounded (JP 2008127327A published on 2008-06-05 by Ands Corporation) [22]. Nanoemulsion containing metabolites of ginseng saponin and a skin-care composition for antiaging including the same (EP 1327434A1 on 2003-07-16 by Pacific Corp) in which glucose and other composites were transformed into liposomes using a dermotropic emulsifier (EP 1327434A1) [23], for enhanced skin penetration to stimulate proliferation of fibroblast and biosynthesis of collagen. Nanotechnology in cosmetics has grown progressively successful, based on the expanding number of patents. It goes to show that customers prefer nano-based cosmetics, as they become more knowledgeable of their benefits as correlated to traditional cosmeceutical preparations.

17.1.3 Nanoemulsions: satisfying cosmetic demands

Ease of formulation, manageable sizing of particles, elevated stability, and comparatively lower surfactant levels make NEs an ideal delivery system for utilization in cosmetics. NEs provide larger platform to formulators to develop a diversified range of modified formulations like spray, fluids, gels, creams, and foams having various efficient characteristics and sensorial advantages [24]. Table 17.1 represents a summary of various NE-based cosmetic formulations.

The capability of carrying active and functional ingredients and grace of appearance decides the proficiency of cosmetic products. Keeping this in view, encompassing hydrophilic, lipophilic, and amphiphilic components becomes a crucial challenge for formulators and NEs furnished ease of incorporation of all three of these. The structure of NEs consists of oil, water, and interfacial regions, which can easily carry various types of media along with active ingredients. Droplet size at the nanolevel renders greater surface-to-volume ration, which amplifies the dispersed phase admission on hair and active ingredient penetration and perviousness into the skin [7]. Through NEs, formulations exclusively designed for topical application can be easily applied, since a thin lipid film forms with ease which allows enhanced performance and increased bioavailability. Several exploratory findings reported about potentials of NEs in cosmetics like Huand team developed cationic NE for hair, results revealed significant improvement on dry hair appearance over an extended period of time [35]. NEs also demonstrated their potential on grounds of sterility, passing through micrometer level filters is a primary and favorable way of sterilization for all companies, and NEs can be easily processed through them because of droplet size at the nanolevel.

17.1.4 Nanoemulsions as a potential carrier in cosmeceuticals

Cosmeceuticals are cosmetic formulations consisting of herbal active agents having therapeutics benefits when applied to skin. Cosmeceuticals bear the biggest advantage of

TABLE 17.1 Summary of various NE-based cosmetic formulations.

Drug	NE-based formulation	Excipient utilized	Method of formulation	Outcome	References
Squalene	Nanocream	Tween 80 and span 80 as emulsifier and stabilized by nonylphenol ethoxylate as surfactant	Shock method via a phase inversion temperature	Developed NE formulation assisted in reducing the droplet from 3000 to 400 nm, which furnished greater solubility and penetration power	[25]
Rapeseed oil	Topical creams	Xanthan gum and carboxyl methyl cellulose		Increase the topical bioavailability of lipophilic active compounds within skin membrane	[26]
Centella asiatica extract	Lotion	Tween 80 and isopropyl myristate	High-pressure homogenization	Stable in physical properties during storage and can penetrate more than the nonnanoemulsion lotion	[27]
Curcumin	Nanoemulsion as a carrier	Clove oil (oil), Tween 80 (surfactant), and PEG-400 (cosurfactant) w	An aqueous microtitration method with high-energy ultrasonication	Enhance the permeation on transdermal drug delivery for wound healing and inflammation	[28]
Carotenoids (beta-carotene)	Nanoemulsion as a carrier	Soy lecithin powder and distilled water	High-speed homogenizer (Box-Behnken Design)	Protect food compounds and system stability along with enhanced bioavailability even after prolonged storage	[29]
Norcanthridin	Nanoemulsion as a carrier	Triacetin, Cremophor EL, and butanol	Spontaneous emulsification method	Improve the insecticidal activity	[30]
Fluvastatin	Hydrogel	Carbopol 940	Aqueous phase titration method	Effective penetration through skin layers	[31]
Thermoresponsive nanoemulsion	Hydrogel	Isopropyl myristate, Tween 80, Span 80, and pluronic copolymers	Gelation mechanism	Potential to serve as an ideal and robust thermogelling vehicles	[32]
Hydrocortisone	Nanoemulsion gel	Palm oil, Tween 80, propylene glycol	Emulsification method	Potent use for topical atopic dermatitis treatment	[33]
Glycyrrhiza glabra extract	gel	Tween and isopropyl alcohol and isopropyl myristate		Faster permeation rate and significant antimicrobial property	[34]

being free from side effects, due to which they are highly popular and preferable in market, along with their claim in enhancing appearance is evidenced over the past few decades. Till date, cosmeceuticals have gained immense popularity in all areas of personnel care ranging from hair treatment, wrinkles, skin care, to hyperpigmentation including medical conditions like psoriasis, bloating, and hair fall [36].

Ease of compatibility with cosmeceuticals, various novel approaches have been coined such solid lipid nanoparticles, liposomes, niosomes, gold nanoparticles, and NEs. NEs have been extensively utilized as a carrier for controlled delivery of cosmeceuticals like shampoos, lotion, nail enamels, conditioners, hair serums, and sunscreens. NEs render fast penetrability and active transport of therapeutically active agents across the skin. Ribeiro and co-workers developed *Opuntia ficus-indica* (L.) Mill extract loaded NE-based moisturizing agent and evaluated it on grounds of thermal stability and moisturizing efficacy. Results depicted enhanced stability up to 60 days and also enhanced water content in *stratum corneum* [37]. Guler and team formulated gold nanoparticles loaded with *Nigella sativa* oil enriched with *Calendula officinalis* extract and lipoic acid using the nanoemulsification system and explored its potential in wound healing and antioxidant activity (cell based and cell free) in in vitro cell culture studies. Results suggested that developed formulation exhibited enhanced antioxidant and wound healing activity in contrast to normal *Nigella sativa* emulsion, this work unbolted further opportunities for other bioactive agents to be turned into more enriched and commanding herbal preparations [38]. Liang et al. formulated peppermint oil loaded NE using medium-chain triglycerol stabilized with food-grade biopolymer further modified to starch. Developed formulation was utilized for determination of antimicrobial properties by using minimum inhibitory concentration and time kill dynamic process against Gram-positive bacterial strains of *Listeria monocytogenes* Scott A and *Staphylococcus aureus*. Results showed that NE formulation presented long-term inhibition of both strains in contrast to bulk peppermint oil [39]. Tatiya and co-workers formulated NE gel containing the extract of *G. glabra* for transdermal application, developed formulation examined for in vitro and ex vivo (antimicrobial) study. Results depicted that NE gel showed substantial increment in permeation of *G. glabra* extract along with significant antimicrobial activity in comparison to aqueous solution of extract [34].

17.2 Challenges associated with cosmetic nanoemulsions

The cosmetic industry always faced some perilous challenges on grounds of safety, efficacy, products appeal, and cost-effectiveness. In today's era, there is a diversified range of excipients available in the market, which assist formulators to fabricate newer and more versatile formulations, but this process consumes lot of time and hard work, and because of which chances of arousal of concerns like products toxicity and safety, environmental hazardous effects, and lackness in products quality and productivity attributes may takes place. Accordingly, it is very much important for formulators to take strict measures at a very early stage especially in the case of new product development.

NE offers high range of easiness in product formulation methodologies and excipient selection like surfactant, lipids, and cosurfactants (Fig. 17.1) [40]. However, the assessment of influence of methods of formulation and effect of constituent's variables on final

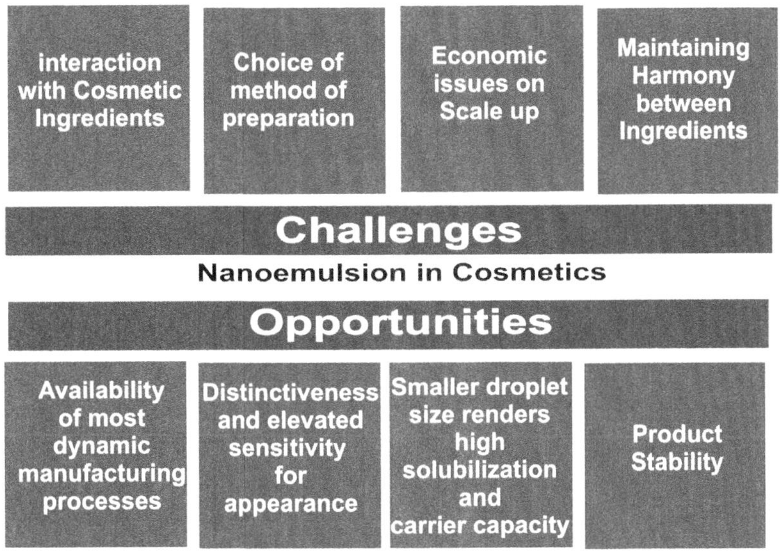

FIGURE 17.1 Challenges and opportunities with nanoemulsions in cosmetics.

products should be done properly [41]. Adoption of a suitable method of formulation basically depends on affluence of scaling up, reproducibility, and optimization of manufacturing time. In the case of NE, it is also very much important for formulators to understand the deportment of surfactant-oil-water phases while processing, storing, and contact with the skin or hair. Understanding the characteristics and attributes of oil/lipid is also significant that is determining the solubilization and effectiveness of hydrophobic active agent. Keeping in view the safety of skin, the developed NE system must guarantee that no other harmful agents will penetrate the skin [42].

Formulation and development of cosmetics requires a diversified range of excipients, because of which, chances of incompatibilities among them rise. Some protruding cosmetics components such as dyes, colorants, pigment, oxidants, and polymers are very much susceptible toward incompatible issues, and this makes NEs a weak candidate for some applications, this area still needs to be explored. For example, antimicrobial activity of preservatives in NEs is still struggling and needs critical attention, since parameters like pH, ethanol concentration, and water behavior affect preservation [43]. Similarly, enriched improvisation in determining deeper insights of effect of particle size on NE systems is also required.

17.3 Methods of production

Formulation of nanoemulsion involves an active drug, additional ingredients, and emulsifying agent. The numerous techniques for the manufacturing of nanoemulsion incorporate two methods: (1) high-energy emulsification method and (2) low-energy emulsification method. The high-energy emulsification technique involves high-energy stirring (homogenization), ultrasonic emulsification technique, and microfluidization [44–46]. The low-energy emulsification technique covers phase inversion temperature (PIT), phase inversion composition (PIC), and spontaneous emulsification.

17.3.1 High-energy emulsification methods

High-energy techniques are widely used to form nanoemulsion [47]. High mechanical energy is utilized that provides strong disruptive strengths, which split up huge droplets to nanosized droplets and create nanoemulsions including high kinetic energy. The disruptive forces are generated by employing mechanical methods such as microfluidizer, ultrasonicators, and high-pressure homogenizers [48]. By applying high-energy processes, achieve greater control of particle size with a selection of formulation design. High-energy designs also produce controls for stability, rheology, and appearance of the emulsion [49,50]. In the case of food constituents, high-energy organizations of nanoemulsion formulation have the benefit of diminishing the risk of decomposition and inactivation of food ingredients without influencing food safety, nutritional, and sensory characteristics [51]. High-energy techniques include subsequent processes.

17.3.1.1 *High-pressure homogenization*

High-pressure homogenizers provide high pressure and give uniform flow to create the minutest particle sizes. Consequently, high-pressure homogenizers are most extensively utilized to formulate nanoemulsions. High-pressure homogenizers are employed to produce intensely disruptive forces that produce nanoemulsions of remarkably low particle size (up to 1 nm) [52]. The coarse emulsion is then transferred through a small orifice with high pressure (500–5000 psi). Various forces, such as intense turbulence, hydraulic shear, and cavitation, are employed collectively during this process, to produce nanoemulsions with extremely small droplet sizes [53]. The particle size of nanoemulsions generated by high-pressure homogenizers depends on sample composition, homogenizer type, and its operating circumstances such as energy intensity, rate, time, and temperature [54]. Enhancing the strength of homogenization reduces the size of droplets of the nanoemulsions. In specific cases, such as when biopolymers are employed as an emulsifier, intense homogenization can lead to an expansion in particle size of resulting nanoemulsion. Consequently, small-molecule surfactants should be employed as emulsifiers in high-pressure homogenizers as they are more efficient and useful than biopolymer for manufacturing nanoemulsions [55–57].

17.3.1.2 *Microfluidization*

Microfluidization is a mixing technique at the micron size level that employs a mechanism called microfluidizer. In microfluidization, solutions (microemulsion) are passing through the microchannels following high pressure (500–20,000 psi). Microchannels are commonly microsize channels that allow processing at the micron size level [58]. The phases of macroemulsion (aqueous and oil phases) are mixed and then transferred into the microfluidizer. The macroemulsion is managed within the microchannels under high

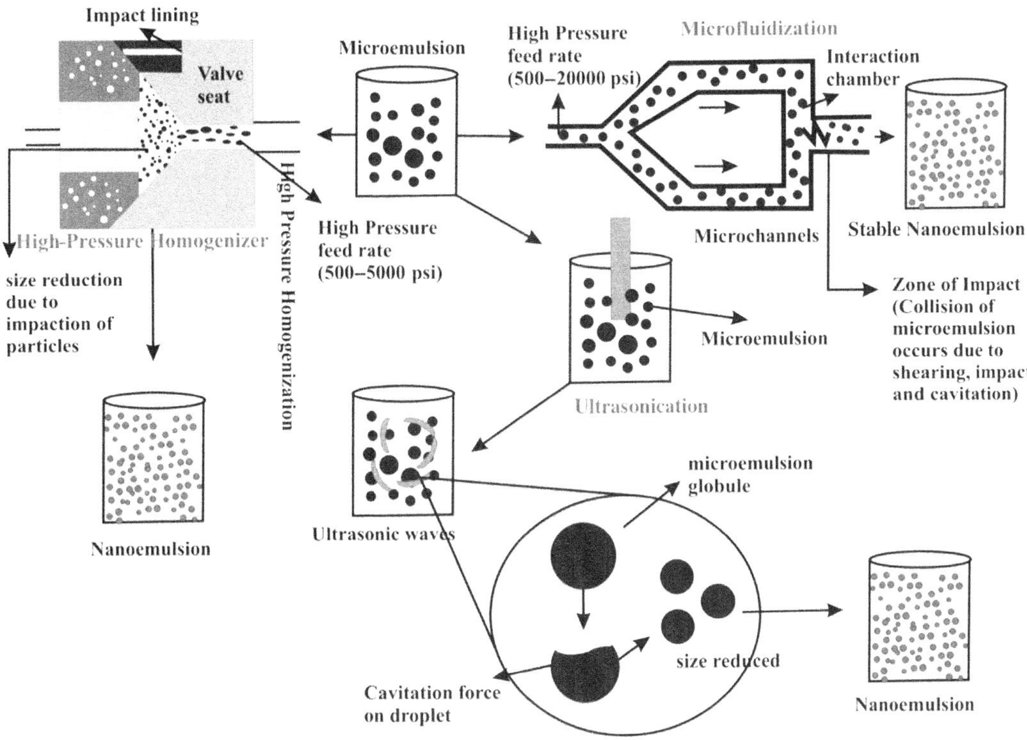

FIGURE 17.2 High-energy emulsification method.

pressure toward the interaction chamber. In the interaction chamber, two streams of macroemulsions strike each other at high velocity. This collision creates forces like shearing, impact, and cavitation, which provide stable nanoemulsions (Fig. 17.2) [47,57]. Microfluidizers originate smaller and narrower nanoemulsion particle size patterns than homogenizers [59,60]. Additionally, microfluidizers fabricate stable nanoemulsions at low surfactant concentrations [61]. Microfluidization techniques have been employed to design food constituent nanoemulsions [62]. Microfluidization systems provide food-grade nanoemulsions with consistent droplet size patterns and greater stabilities [63].

17.3.1.3 Ultrasonication

Ultrasonication is more reliable than other high-energy techniques in terms of process and cleaning [47,64]. In the ultrasonic emulsification method, ultrasonic waves produce cavitation forces that separate the macroemulsion to nanoemulsion. In this technique, ultrasonicators are employed, which consist of a probe that releases ultrasonic waves. By modifying ultrasonic current/energy input and time to accomplish the fancied particle size and stability of the nanoemulsion. In the ultrasonic emulsification process, physical shear is essentially provided by the process of acoustic cavitation [65]. Cavitation is the phenomenon of production of microbubbles and later collapse of microbubbles, which are produced by the stress fluctuations of the acoustic wave. The collapse of microbubbles induces intense turbulence that makes the formation of nanosized droplets [64]. Irradiation of an oil and water method by ultrasound produces cavitation forces and contributes excess strength for new interface configurations, creating nanosized emulsion droplets. Through the ultrasonication method, nanoemulsions can be created in the absence of surfactants [57,66]. In a recent investigation, it was revealed that the efficiency of ultrasonic emulsification is based on ultrasonication intensity and nature of the surfactant. Ultrasonication has been utilized widely for fabricating nanoemulsions of drugs and food constituents. Food-grade ultrasonication nanoemulsion exhibits more excellent stability and less droplet size and requires more limited energy input than other high-energy methods [67]. Fig. 17.2 represents various high energy-based methods of preparations.

17.3.2 Low-energy emulsification method

These techniques require low energy for the production of nanoemulsion. Low-energy emulsification techniques are more energy efficient as certain processes employ the internal chemical potential of the methods and need only moderate stirring for the fabrication of the nanoemulsions [68]. Low-energy emulsification techniques frequently require phase inversion emulsification and self-emulsification method. Commonly, low-energy techniques are not recognized for the formulation of food-grade nanoemulsions as they expect high concentration of surfactant, which adversely affects food formulation taste and protection.

17.3.2.1 Phase inversion emulsification method

In this technique, spontaneous curvature of the surfactant creates phase transition throughout the emulsification method. Differences in spontaneous curvature of the surfactant happen by variations in parameters like composition and temperature [69]. There are two classes of transitional phase inversion (TPI): (1) PIT and (2) PIC. TPI takes place due to the variations in spontaneous curvature or association of the surfactant due to changes in parameters like composition and temperature [68–70].

17.3.2.2 PIT method

In the PIT process, surfactant spontaneous curvature is inversed by varying temperature. Nonionic surfactants (i.e., polyethoxylated surfactants) undergo dehydration of POE groups of polyethoxylated surfactants, which gives it higher lipophilic and leads to alterations in the curvature of the surfactant. Therefore phase inversion happens and nanoemulsion is created [68,71]. In this process, oil, water, and nonionic surfactants are processed at room temperature to create oil-in-water (O/W) emulsions. Then, as the temperature gradually increases, dehydration of surfactant POE organizations happens that executes surfactant more lipophilic and surfactant start bestowing a tremendous affection toward the oily phase. This causes phase inversion from the original O/W emulsion to water-in-oil (W/O) nanoemulsion through intermediate liquid crystalline or bicontinuous arrangements (e.g., lamellar phase).

17.3.2.3 PIC method

The phase inversion composition or PIC process is alike to the PIT method; nevertheless, in PIC, phase inversion is accomplished by modifying the system configuration rather than the system temperature [72]. In PIC, one of the ingredients such as water is added to a compound, and oil–surfactant or oil is added to the water–surfactant mixture of water is added into the oil system. POE type nonionic surfactants are commonly applied in the PIC process to form nanoemulsions, although other classes can also be utilized. When water is added gradually to the oil phase and as the quantity of the water portion extends, surfactant POE chain hydration happens. The surfactant hydrophilic–lipophilic characteristics of the water phase will become stable and spontaneous curvature of surfactant will convert to zero, similar to at the HLB temperature in the PIT method. Throughout this transition, a bicontinuous or lamellar structure is constituted. When supplementary water is added the transition composition is exceeded and the constructions of the surfactant layer with zero curvature change to having high positive curvature. This change in curvature leads to phase inversion and creates nanosize droplet production. Thus modifying the configuration of the system creates phase inversion [68]. Furthermore, other construction parameters, such as the enhancement of salt and pH changes, also create nanosized emulsion droplets by TPI [72,73]. Fig. 17.3 presents the scientific illustration of various low energy-based methods of preparations.

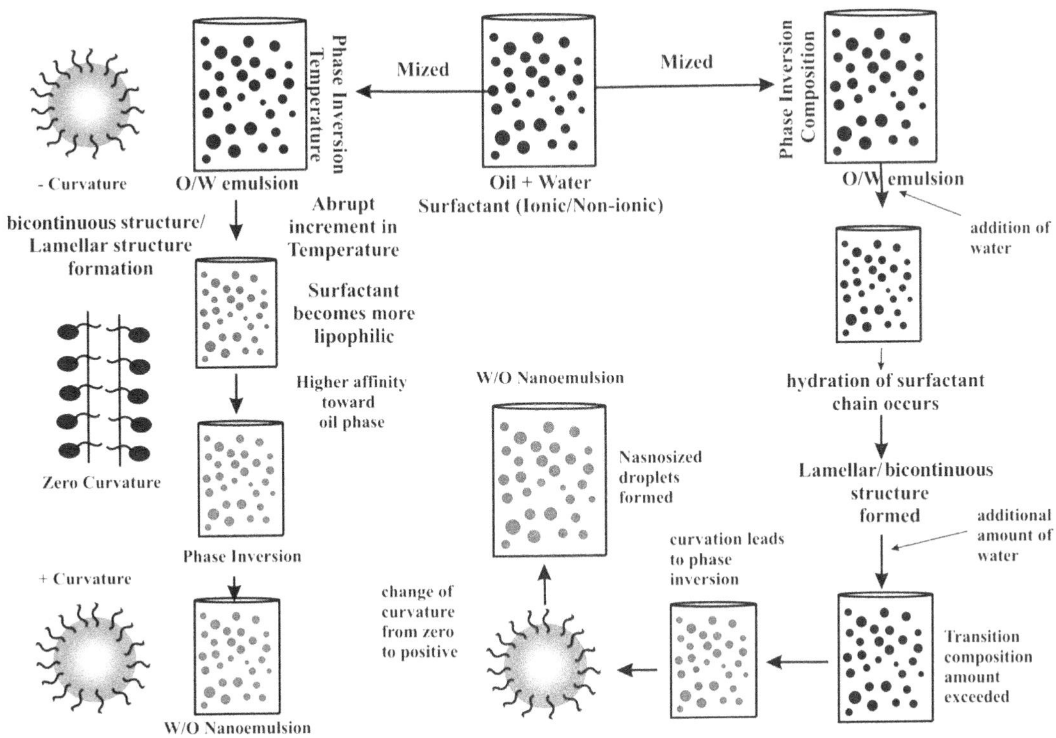

FIGURE 17.3 Phase inversion emulsification method.

17.3.2.4 Self-nanoemulsification method

In the self-emulsification process, nanoemulsion production is accomplished without modifying the spontaneous curvature of the surfactant. Surfactant and/or cosolvent molecules immediately diffuse from the dispersed phase to the continuous phase, which produces turbulence and generates nanosized emulsion droplets. The self-emulsification technique is also introduced as the spontaneous emulsification process [68,69,74]. Self-nanoemulsifying drug delivery systems (SNEDDS) are based on the self-emulsification phenomenon and include more hydrophilic surfactants or cosurfactants (cosolvents) and lower lipid content [75]. SNEDDS can be described as an isotropic composite of oil, cosurfactant, surfactant, and drug molecule. When this mixture is diluted by aqueous solutions in vivo, it forms narrow and optical clear O/W nanoemulsion, supported by moderate agitation produced by the digestive motility of the stomach and intestine (Fig. 17.4) [76,77]. The two commonly described methods of nanoemulsion production from SNEDDS are diffusion of the hydrophilic cosolvent or cosurfactant from the organic phase into the aqueous phase [68], and development of nanoemulsion negative free energy at transient negative or ultra-low interfacial tensions [75]. SNEDDS are additionally the most common and encouraging device for the delivery of hydrophobic drug molecules with low bioavailability. SNEDDS have further been utilized for the delivery of bioactive food ingredients.

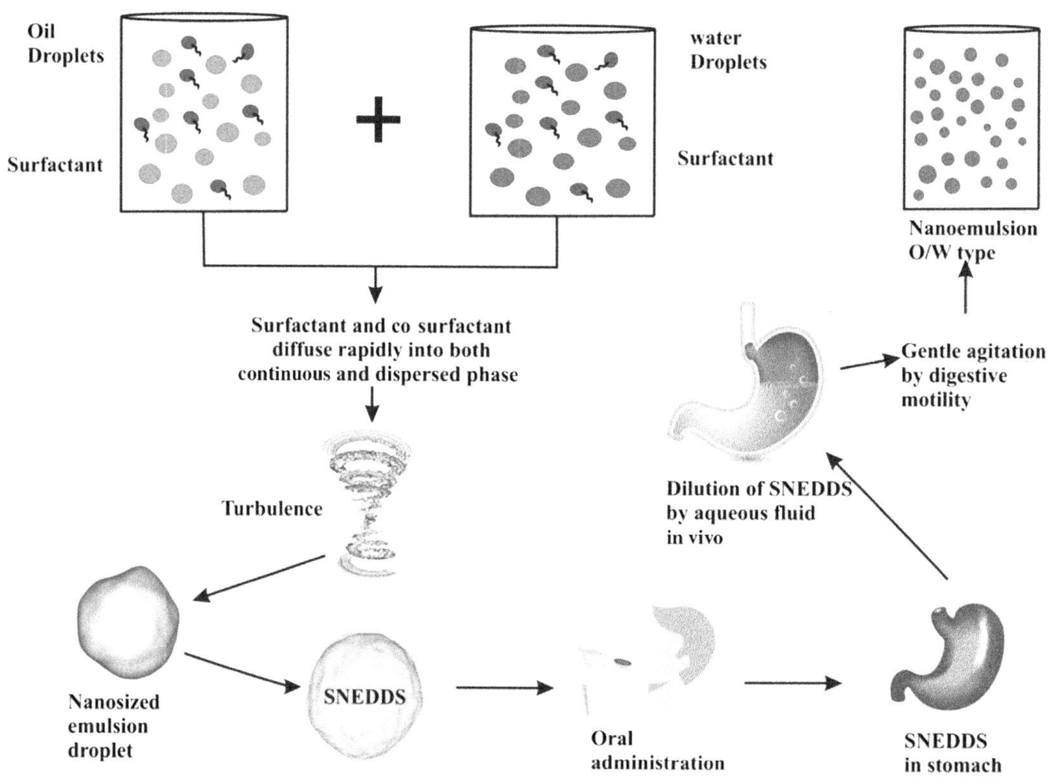

FIGURE 17.4 Self-emulsification method.

17.3.2.5 *Spontaneous emulsification*

It involves three steps: (1) development of uniform organic solution consisting of oil and lipophilic surfactant in water-miscible solvent and hydrophilic surfactant, (2) the organic phase is introduced in the aqueous phase under constant magnetic stirring, o/w emulsion is produced, and (3) the aqueous phase is separated by evaporation under reduced pressure [78,79].

17.4 Nanoemulsion in cosmetic: from laboratory to market

17.4.1 Dermal application

In recent times, nanoemulsions have fascinated consideration as prospective systems for the controlled release of foundation and diffusion of active ingredients in particular dermal layers. The use of nanoemulsion in enhancing topical and transdermal efficacy of medicines has currently well established [52]. The nanoemulsions are systematic encroachment in the delivery of drugs concerning through the dermal and transdermal execution. Topical and transdermal drug delivery systems are noninterference and can be self-administered derogates the adverse reaction, which have obtained improved consideration during the last several years. Nanoemulsions, emulsions sized between 20 and 200 nm with narrow distributions, offer numerous reward for topical and transdermal delivery of pharmaceutical moieties as well as controlled droplet size, the skill to capably liquefy lipophilic medicaments, improved dermal penetration, and extended release of lipophilic and hydrophilic drugs. Furthermore, they comprise excellent sensational and physical possessions such as inclusive dispersion on skin and skin hydration in cosmetic and beauty pharmaceuticals [80]. For nanoemulsions when delivered topically, tiny particles and solubilized hydrophobic moieties offer a pathway to significantly augment the rate of drug dissolution and consequently provide the systemic bioavailability during the transcellular route. Drug releases from nanoemulsion engross separating it from oil into surfactant deposit and then into the aqueous phase, as a result circumvent occlusive effects [7]. This vehicle is an intoxicating substitute to increase drug delivery system on restoration and targeting the explicit of weakly soluble drugs, increase absorption by the skin, enhance retention time of drug in the target region, and an ultimate consequence in fewer side effects. Nanoemulsion is prominent delivery medium in the cosmetics because of the finest droplet size, creaming and flocculation is not observed, most important to a more modish and consequently facilitates uptake by antigen-presenting cells constant drug preparations like moisturizer, lotions, and creams. The intrinsic worth of nanoemulsion with globules in nanoscale size of an emulsion does not rely on the emulsion physical properties itself, however, congregate the bioavailability of remedial drugs in entire. The amalgamations of conventional compound diffusion enhancers to fluidize the stratum corneum lipid bilayers minimize the principal skin barrier and escalate the penetration.

The minute droplet size of nanoemulsions, foundation a huge reduction in gravity force; for that reason, Brownian motion perhaps acceptable for acquiring an elevated permanence beside sedimentation or creaming than an emulsion.

Nanoemulsion in cosmetic preparations offers an accelerated permeation of active constituents through skin caused by the large surface area of droplets. Even occasionally, it is found that nanoemulsion penetrates easily through rough skin. This asset of nanoemulsion diminishes the supplementary employment of unique permeation enhancer which is accountable for inappropriateness of preparation [79].

One speculative benefit of a nanoemulsion is that it necessities a subordinate surfactant concentration for the drug preparations. They are usually valuable since they having low skin irritation, concentrated drug loading competence, and impending for skin hydration and enrichment of medicine permeability. There are quite a lot of category of medicines for nanoemulsion preparations with improved topical and transdermal delivery, for example, steroids, nonsteroidal antiinflammatory, and cytotoxic drugs. Nanoemulsion demonstrates the consequence of lipid content and charge of nanoemulsions on skin hydration, elasticity, and erythema. Nanoemulsions are able to permeate to skin behind the reason is that it has high-quality sensorial assets through fast penetration, integrating consistencies and its biophysical possessions particularly the hydrating capability. Nanoemulsions are highly delicate systems by nature. Because these systems are translucent, the least indication of destabilization emerges visually. The chief resource of unstability was recognized and comprehensively considered are: Ostwald ripening and depletion induced flocculation subsequently the addition of thickening polymer [81]. The management of these two systems is acceptable the industrial fabrication of a great diversity of cosmetic and dermal product products, from water-like fluids to ringing gels acquired through escalating the oil phase contented or by fabricating the polymers.

17.4.2 NEs as a novel vehicle in skin care

Nanoemulsions are projected for abundant appliances in pharmaceutical sciences as drug delivery development because of their competence to solubilize nonpolar dynamic compounds. Nanoemulsions have been extensively used in manufacturing cosmetic medicaments. One of the resourceful relevance of nanoemulsions is in the transdermal drug delivery where nanoemulsion proceeds as a proficient delivery system for bioactive, make ease of administration. Creams, ointments, and lotions are the unsurprising transdermal drug delivery vehicles which indulgences frequently in local skin infection are at present altering at a rapid rate in the directions of a more highly developed level. It is also considered as a prospective system among numerous rewards as well as elevated storage stability, low preparation cost, thermodynamic stability, lack of organic solvents, and excellent manufacturing viability. They encompass the plasma concentration profiles and bioavailability of drugs reproducible. These systems are being used presently to afford dermal and surface property and deeper skin permeation. These are oil-in-water emulsions with an average droplet size ranging from 100 to 500 nm. They have very good stability and they are not participating in phase separation in the period of storage. Nanoemulsions have liquid lipophilic core and be apposite for lipophilic compound carrying. A lot of researches illustrated that minimizing the transepidermal water loss support the obstruction function of the skin [82].

The nanostructured remedies will eventually turn the world of drug delivery upside down. The nanomedicines studied include polymeric nanoemulsions, nanoparticles, liposomes, polymersomes, dendrimers, cubosomes, niosomes, and nanodiamonds, which are used for efficient management of aggressive and largely treatment-resistant ailment.

The most important profit of the use of nanotechnology in cosmeceutical appliances is the improved stability of a variety of cosmetic constituents, viz., unsaturated fatty acids, vitamins, or antioxidants encapsulated within the nanoparticles [83], enhanced penetration rate of all convinced components, such as vitamins and other antioxidants, enhanced esthetics of the preparations, in concurrence by improved concert and acceptance of UV filters on the skin surface. A research illustrated that the fruitfully encapsulated oil-soluble vitamin (vitamin E) used a nanoemulsion as the delivery system [13]. Indeed in dermatology, W/O emulsions have been revealed to be an improved system, where the oil-soluble active moieties are complete more favorable by a lipid film development on the skin. Nanosized droplets dispersed in the continuous phase of the nanoemulsion can travel efficiently by the stratum corneum and disperse the active component through the skin barrier.

17.4.3 NEs in hair system care

Nanotechnology has developed as moderately uncomplicated tools for developing the remedial conclusion of this pathology, encouraging a targeted drug delivery with improved local bioavailability, which might decrease the side effect of the drugs.

Taking into account the impact of hair on worth of existence, it is no revelation that consumers challenge to change their hair for a diversity of artistic rationales. Because of their exclusive and inherent properties, nanomaterials are being more and more exploited in the preparation of superior skin-care medicaments, exclusively targeting the hair follicle and shaft.

Contemporary development in nanotechnology consents to the improvement of carrier through improved capability to penetrate and permeate the skin. The intercellular, transcellular, and follicular permeation are the main routes by which nanoparticles gain penetration. The hair follicles are significantly playing a major role in the penetration and permeation of topically employed nanosized particles. Some confirmation has recommended that follicular allocation is body-region dependent, like forehead, nurturing a considerably increased number of follicular orifices [84]. They put forward that whereas follicular openings could provide as an access for nanoparticulate ingredients to be ecstatic into the hair follicle, there may be considerable differences in percutaneous absorption of appendage-free and copious regions. It is remarkable that these particles are competent in penetrating follicular orifices that are supposed to be no larger than 200 nm in diameter. Whereas outcomes place in compared to previous results, it can elucidate a mechanical effect rather than an effect specific to particle size. Hair progress might be performed to pump the nanoparticles deeper contained by the hair follicles [85]. Consequently, by changing the size of the particles, an assorted variety of cosmetic modalities can be exploiting to selectively target the arrangement surrounded by the follicle.

A good deal with previous researches which were focused in which nanomaterials can be used to improve hair cosmesis—retaining shine, smoothness, and health of hair. Shampoos have incorporated nanomaterials in order to optimize resident contact time with the scalp and hair follicle, permitting active moieties moment to make a defensive film, sealing wetness inside the cuticles (i.e., averting transfollicular water loss). On the other hand, while silicone oil is fabricated into nanomaterials, it is able to diffuse quickly into hair fibers given its minute size. This enhanced hair moisture, shiny finish, and lubrication without damaging the cuticle of the hair fibers, excluding quite its nanosize agrees for diffusion into the hydrolipid emulsion stratum, which confirms that cationic nanoemulsions with a droplet diameter of less than 100 nm is considerably enhanced dry hair, yet following frequent use of shampoo [81]. A further compound included nanoparticles of oxides, hydroxides, carbonates, silicates, and phosphates were competent for managing the lubricant of hair [86].

17.4.4 As potential carrier

In recent times, the use of variable nanocarriers for dermal relevance has been extensively investigated, because as a carrier nanoemulsion recommend quite a lot of rewards compared with conventional topical preparations, including higher skin penetration, controlled and targeted drug delivery, and the accomplishment of advanced therapeutic properties. The carrier system assists to fabricate the active ingredients to the targeted area proficiently [87].

The function of nature/magnitude of zeta potential, categories of emulsifiers, and choice of suitable semisolid bases may well fabricate the nanoemulsion resourceful carriers from the laboratory to industry scale. Nanoemulsion is an appropriate semisolid dosage form for delivery of medicaments by dermal or transdermal route administration and basic concerns for the period of the nanoemulsion absorption into and through the skin.

A colloidal constant cationic nanoemulsion among excellent DNA condensation capability was productively fabricated for resourceful cytosolic delivery and potential for effectively performed in a living organism. Insertion of phospholipids facilitated cationic nanoemulsions to preserve adequate colloidal stability.

A cationic nanoemulsion delivery system was developed to deliver a self-amplifying mRNA vaccine. Induction of immune responses was shown in multiple animal species, including rhesus macaques, at comparable levels to responses elicited by an adjuvanted subunit vaccine [88].

The prospective utilization of nanocarriers for targeting skin inflammation using nanotechnology is a striking approach as it receives high acceptance from patients, resulting in higher compliance and therapeutic results.

The exclusivity of nanoemulsions that oppose physical deterioration by reason of aggregation and gravitational separation can develop the effective action of the nanoemulsion principally in the direction of antimicrobial activity. By the insertion of ester derived from a fungal metabolite O/W nanoemulsion exhibits outstanding consequences as a prospective carrier for hyperpigmentation management [89].

Nanocarriers in dermatology include new direction in medical diagnosis, monitoring, and treatment among the used in non-invasive nanoimaging of high-resolution dermoscopy, microscopy along with the contribution of advanced diagnostic and therapeutic modalities. The nanoemulsion is a carrier system for cosmeceutical uses since it engages submicron sizes ranging from 20 to 200 nm which is able to penetrate into effortlessly deep skin cells.

Using nanoemulsions as a carrier vehicle, effective through the increasing permeation rate of hydrophilic compound along with the shortening the lag time, shows the potential role. A suitable arrangement and percentage of nanoemulsion formulation including the type and level of oil, surfactant, cosurfactant, and water is a chief concern for the transdermal drug delivery.

17.5 Conclusions

Nanotechnology evolved as the most convincing field in the arena of cosmetics over the last one decade. Novel technologies and delivery systems are developing with consistent pace, which successfully assisting formulation scientist in fabricating newer are more versatile cosmetic products. Among various delivery systems, NE came up with immense potential and gained significant popularity in the world of cosmetics. This chapter furnished potential capabilities of NE systems in cosmetics and personal care products utilized for various purposes with evidence. Smaller droplet size in NE satisfies all fundamental necessities like optical stability, rheological, and delivery properties, which is required by an impactful delivery system, as these necessities are not fulfilled by conventional emulsions. Greater penetration aptitudes make NE an ideal choice for both skin and hair care products, whether with usage of synthetic cosmetics or cosmeceuticals. Still, certain areas like impact of components variables, incompatibilities among excipients, and scaling up needs to be explored for NE. Along with this, explorations must be done on developing more cost-effective methods, consumer friendly ingredients, and newer techniques for testing products on in vivo models.

References

[1] Roberta H, Castle DJ. Aging and cosmetic enhancement. Clin Interv Aging 2006;1(2):115–19.
[2] Bloch H. Medicine and science in the 19th and 20th centuries. J Natl Med Assoc 1988;80(2):229–32.
[3] Borchers AT, Hagie F, Keen CL, Gershwin ME. The history and contemporary challenges of the US Food and Drug Administration. Clin Ther 2007;29(1):1–16.
[4] Kaul S, Gulati N, Verma D, Mukherjee S, Nagaich U. Role of nanotechnology in cosmeceuticals: a review of recent advances. J Pharm (Cairo) 2018;2018:3420204.
[5] Panico A, Serio F, Bagordo F, Grassi T, Idolo A, De giorgi M, et al. Skin safety and health prevention: an overview of chemicals in cosmetic products. J Prev Med Hyg 2019;60(1):E50–7.
[6] Duschl A, Windgasse G. A survey on the state of nanosafety research in the European Union and the United States. J Nanopart Res 2018;20(12):335.
[7] Jaiswal M, Dudhe R, Sharmam PK. Nanoemulsion: an advanced mode of drug delivery system. 3 Biotech 2015;5(2):123–7.
[8] Yukuyama MN, Ghisleni DD, Pinto TJ, et al. Nanoemulsion: process selection and application in cosmetics—a review. Int J Cosmet Sci 2016;38:13–24.
[9] Royer M, Prado M. Study of nutraceutical, nutricosmetics and cosmeceutical potentials of polyphenolic bark extracts from Canadian forest species. PharmaNutrition 2013;1:158–67.
[10] Care O, Corner C, Listing J, et al. Nanotechnology in cosmetics. Nanotechnology 2017;2017.
[11] Mu L, Sprando RL. Application of nanotechnology in cosmetics. Pharm Res 2010;27:1746–9.

[12] Weir A, Westerhoff P, Fabricius L, et al. Titanium dioxide nanoparticles in food and personal care products. Environ Sci Technol 2012;46:2242–50.

[13] Mayer S, Weiss J, McClements DJ. Vitamin E-enriched nanoemulsions formed by emulsion phase inversion: factors influencing droplet size and stability. J Colloid Interface Sci 2013;402:122–30.

[14] Boonme P, Junyaprasert VB, Suksawad N, et al. Microemulsions and nanoemulsions: novel vehicles for whitening cosmeceuticals. J Biomed Nanotechnol 2009;5:373–83.

[15] Boonme P. Applications of microemulsions in cosmetics. J Cosmet Dermatol 2007;6:223–8.

[16] Dirschka T, Ekanayake-Bohlig S, Dominicus R, Aschoff R, Herrera-Ceballos E, Botella-Estrada R, et al. A randomized, intraindividual, non-inferiority, Phase III study comparing daylight photodynamic therapy with BF-200 ALA gel and MAL cream for the treatment of actinic keratosis. J Eur Acad Dermatol Venereol 2019;33 (2):288–97.

[17] Abd E, Benson HAE, Roberts MS, Grice JE. Follicular penetration of caffeine from topically applied nanoemulsion formulations containing penetration enhancers: in vitro human skin studies. Skin Pharmacol Physiol 2018;31(5):252–60.

[18] Kong M, Chen XG, Kweon DK, et al. Investigations on skin permeation of hyaluronic acid based nanoemulsion as transdermal carrier. Carbohydr Polym 2011;86:837–43.

[19] Mordor I. Cosmeceuticals market—segmented by product type (skin care, hair care, injectable, oral care), active ingredients (antioxidants, botanicals, exfoliants, peptides, retinoids), and regions—growth, trends, and forecast (2019–2024). Gachibowli, Hyderabad (India): Mordor Intelligence; 2019 [cited 2018 November 10]. Available from: <https://www.mordorintelligence.com/industry-reports/globalcosmeceuticals-market-industry>.

[20] Mihranyan A, Ferraz N, Strømme M. Current status and future prospects of nanotechnology in cosmetics. Prog Mater Sci 2012;57:875–910.

[21] Jeong S-H, Son J-H, Jang S-J, et al. Cosmetic composition containing retinol stabilized by porous polymer beads and nanoemulsion. Google Patents, 2015.

[22] Nanbu T. Skin-revitalizing cosmetic composition. Google Patents, 2009.

[23] Han, S.H., Ju, H.K., Kang, B.Y., Kim, H.K., Sung, D.S., Yeom, M.H., et al. Nanoemulsion comprising metabolites of ginseng saponin and a skin-care composition for anti-aging containing the same. European Patent, EP1327434A1, 2003.

[24] De Souza ML, Oliveira DD, Ribeiro PLL, de Paula Pereira N, Druzian JI. Nanoemulsions for cosmetic applications: what innovation status. Recent Pat Nanotechnol 2018;12(2):101–9.

[25] Charoenkul K, Phromyothin D. Development and characterization of nano-cream preparation containing natural extract using nanoemulsion techniques. Mater Today Proc 2017;4(5):6105–10.

[26] Kaci M, Belhaffef A, Meziane S, Dostert G, Menu P, Velot É, et al. Nanoemulsions and topical creams for the safe and effective delivery of lipophilic antioxidant coenzyme Q10. Colloids Surf B Biointerfaces 2018;1:165–75.

[27] Hanifah M, Jufri M. Formulation and stability testing of nanoemulsion lotion containing *Centella asiatica* extract. J Young Pharm 2018;10(4):404–8.

[28] Ahmad N, Ahmad R, Al-Qudaihi A, Alaseel SE, Fita IZ, Khalid MS, et al. Preparation of a novel curcumin nanoemulsion by ultrasonication and its comparative effects in wound healing and the treatment of inflammation. RSC Adv 2019;9:20192–206.

[29] Moreira JB, Goularte PG, de Morais MG, Costa JAV. Preparation of beta-carotene nanoemulsion and evaluation of stability at a long storage period. Food Sci Technol 2019;30. Available from: https://doi.org/10.1590/fst.31317.

[30] Zeng L, Liu Y, Pan J, Liu X. Formulation and evaluation of norcanthridinnanoemulsions against the *Plutella xylostella* (Lepidotera: Plutellidae). BMC Biotechnol 2019;19(1):16. Available from: https://doi.org/10.1186/s12896-019-0508-8.

[31] Kaur R, Ajitha M. Transdermal delivery of fluvastatin loaded nanoemulsion gel: Preparation, characterization and in vivo anti-osteoporosis activity. Eur J Pharm Sci 2019;136:104956. Available from: https://doi.org/10.1016/j.ejps.2019.104956.

[32] Hashemnejad SM, Badruddoza AZM, Zarket B, Ricardo Castaneda C, Doyle PS. Thermoresponsive nanoemulsion-based gel synthesized through a low-energy process. Nat Commun 2019;10(1):2749. Available from: https://doi.org/10.1038/s41467-019-10749-1.

[33] Sinha P, Srivastava S, Mishra N, Singh DK, Luqman S, Chanda D, et al. Development, optimization, and characterization of a novel tea tree oil nanogel using response surface methodology. Drug Dev Ind Pharm 2016;42(9):1434–45. Available from: https://doi.org/10.3109/03639045.2016.1141931.

[34] Tatiya A, Bhavsar S, Mahajan H, Surana S. Experimental design and characterization of nanoemulsion based topical herbal gel developed for site-specific activity of *Glycyrrhiza glabra* extract: in vitro and ex-vivo studies. Biosci Biotechnol Res Commun 2019;12(2).

[35] Hu Z, Liao M, Chen Y, Cai Y, Meng L, Liu Y, et al. A novel preparation method for silicone oil nanoemulsions and its application for coating hair with silicone. Int J Nanomed 2012;7:5719–24.

[36] Canha M, Staden A, Fibrich F, Lambrechts I, Denga L, Lall N. Book review: cosmeceuticals and active cosmetics, third edition. Front Pharmacol 2019;10:529.

[37] Ribeiro RC, Barreto SM, Ostrosky EA, da Rocha-Filho PA, Veríssimo LM, Ferrari M. Production and characterization of cosmetic nanoemulsions containing *Opuntia ficus-indica* (L.) mill extract as moisturizing agent. Molecules 2015;20(2):2492–509. Available from: https://doi.org/10.3390/molecules20022492.

[38] Guler E, Barlas FB, Yavuz M, Demir B, Gumus ZP, Baspinar Y, et al. Bio-active nanoemulsions enriched with gold nanoparticle, marigold extracts and lipoic acid: in vitro investigations. Colloids Surf B Biointerfaces 2014;121:299–306.

[39] Liang R, Xu S, Shoemaker CF, Li Y, Zhong F, Huang Q. Physical and antimicrobial properties of peppermint oil nanoemulsions. J Agric Food Chem 2012;60(30):7548–55.

[40] Trujillo LS, Belloso OM, McClements DJ. Excipient nanoemulsions for improving oral bioavailability of bioactives. Nanomaterials (Basel) 2016;6(1):17.

[41] Jamali SN, Assadpour E, Jafari SM. Formulation and application of nanoemulsions for the nutraceuticals and phytochemicals. Curr Med Chem 2019;. Available from: https://doi.org/10.2174/0929867326666190620102820.

[42] Shaker DS, Ishak R, Ghoneim A, Elhouni MA. Nanoemulsion: a review on mechanisms for the transdermal delivery of hydrophobic and hydrophilic drugs. Sci Pharm 2019;87(3):17.

[43] Nemes D, Kovács R, Nagy F, Mező M, et al. Interaction between different pharmaceutical excipients in liquid dosage forms—assessment of cytotoxicity and antimicrobial activity. Molecules 2018;23(7):1827.

[44] Tiwari SB, Amiji MM. Nanoemulsion formulations for tumor-targeted delivery. Nanotechnology for cancer therapy. Taylor and Francis Group; 2006. p. 723–39.

[45] Perdiguer AC, Dachs FJG, Carreras N. Nanoemulsion of the oil water type, useful as an ophthalmic vehicle and process for the preparation thereof. Assignee: Laboratorios Cusi, S.A. (Barcelona, ES), 1997.

[46] Banker GS, Lieberman HA, Rieger MM. Pharmaceutical dosage forms. Disperse Syst Marcel Dekker 2002;2 (3):339–40.

[47] Mahdi Jafari S, He Y, Bhandari B. Nano-emulsion production by sonication and microfluidizationa comparison. Int J Food Prop 2006;9:475–85.

[48] Gonçalves A, Nikmaram N, Roohinejad S, Estevinho BN, Rocha F, Greiner R, et al. Production, properties, and applications of solid self-emulsifying delivery systems (S-SEDS) in the food and pharmaceutical industries. Colloids Surf A Physicochem Eng Asp 2018;538:108–26.

[49] Graves S, Meleson K, Wilking J, Lin MY, Mason TG. Structure of concentrated nanoemulsions. J Chem Phys 2005;122:134703.

[50] Gursoy RN, Benita S. Self-emulsifying drug delivery systems (SEDDS) for improved oral delivery of lipophilic drugs. Biomed Pharmacother 2004;58:173–82.

[51] Gharibzahedi SMT, Hernández-Ortega C, Welti-Chanes J, Putnik P, Barba FJ, Mallikarjunan K, et al. High pressure processing of food-grade emulsion systems: antimicrobial activity, and effect on the physicochemical properties. Food Hydrocoll 2019;87:307–20.

[52] Rai VK, Mishra N, Yadav KS, Yadav NP. Nanoemulsion as pharmaceutical carrier for dermal and transdermal drug delivery: formulation development, stability issues, basic considerations and applications. J Control Release 2018;270:203–25.

[53] Floury J, Desrumaux A, Lardières J. Effect of high-pressure homogenization on droplet size distributions and rheological properties of model oil-in-water emulsions. Innov Food Sci Emerg Technol 2000;1:127–34.

[54] Qian C, McClements DJ. Formation of nanoemulsions stabilized by model food-grade emulsifiers using high-pressure homogenization: factors affecting particle size. Food Hydrocoll 2011;25:1000–8.

[55] Azeem A, Rizwan M, Ahmad FJ, Iqbal Z, Khar RK, Aqil M, et al. Nanoemulsion components screening and selection: a technical note. AAPS PharmSciTech 2009;10:69–76.

[56] Azeem A, Rizwan M, Ahmad FJ, Khar RK, Iqbal Z, Talegaonkar S. Components screening and influence of surfactant and cosurfactant on nanoemulsion formation. Curr Nanosci 2009;5:220–6.

[57] Jafari SM, He Y, Bhandari B. Production of sub-micron emulsions by ultrasound and microfluidization techniques. J Food Eng 2007;82:478–88.

[58] Qadir A, Faiyazuddin MD, Talib Hussain MD, Alshammari TM, Shakeel F. Critical steps and energetics involved in a successful development of a stable nanoemulsion. J Mol Liq 2016;214:7–18.

[59] Perrier-Cornet JM, Marie P, Gervais P. Comparison of emulsification efficiency of protein-stabilized oil-in-water emulsions using jet, high pressure and colloid mill homogenization. J Food Eng 2005;66:211–17.

[60] Wooster TJ, Golding M, Sanguansri P. Impact of oil type on nanoemulsion formation and Ostwald ripening stability. Langmuir 2008;24:12758–65.

[61] Pinnamaneni S, Das NG, Das SK. Comparison of oil-in-water emulsions manufactured by microfluidization and homogenization. Pharmazie 2003;58:554–8.

[62] Goh PS, Ng MH, Choo YM, Amru NB, Chuah CH. Production of nanoemulsions from palm-based tocotrienol rich fraction by microfluidization. Molecules 2015;20:19936–46.

[63] Villalobos-Castillejos F, Granillo-Guerrero VG, Leyva-Daniel DE, Alamilla-Beltrán L, Gutiérrez-López GF, Monroy-Villagrana A, et al. Fabrication of nanoemulsions by microfluidization. In: Jafari SM, McClements DJ, editors. Nanoemulsions: formulation, applications, and characterization. San Diego, CA: Academic Press; 2018. p. 207–32.

[64] Leong TSH, Wooster TJ, Kentish SE, Ashokkumar M. Minimising oil droplet size using ultrasonic emulsification. Ultrason Sonochem 2009;16:721–7.

[65] Jayasooriya SD, Bhandari BR, Torley P, D'Arcy BR. Effect of high power ultrasound waves on properties of meat: a review. Int J Food Prop 2004;7:301–19.

[66] Gaikwad SG, Pandit AB. Ultrasound emulsification: effect of ultrasonic and physicochemical properties on dispersed phase volume and droplet size. Ultrason Sonochem 2008;15:554–63.

[67] Ghosh V, Mukherjee A, Chandrasekaran N. Ultrasonic emulsification of food-grade nanoemulsion formulation and evaluation of its bactericidal activity. Ultrason Sonochem 2013;20:338–44.

[68] Solans C, Solé I. Nano-emulsions: formation by low-energy methods. Curr Opin Colloid Interface Sci 2012;17:246–54.

[69] Solè I, Pey CM, Maestro AL, González C, Porras M, Solans C, et al. Nano-emulsions prepared by the phase inversion composition method: preparation variables and scale up. J Colloid Interface Sci 2010;344:417–23.

[70] Ishak KA, Annuar MSM. Phase inversion of medium-chain-length poly-3-hydroxyalkanoates (mcl-PHA)-incorporated nanoemulsion: effects of mcl-PHA molecular weight and amount on its mechanism. Colloid Polym Sci 2016;294:1969–81.

[71] Moreira de Morais J, dos Santos ODH, Delicato T, Gonçalves RA, da Rocha-Filho PA. Physicochemical characterization of canola oil/water nano-emulsions obtained by determination of required HLB number and emulsion phase inversion methods. J Dispers Sci Technol 2006;27:109–15.

[72] Sokolov YV. Nanoemulsion formation by low-energy methods: a review. Vı̄sn Farm 2014;3:16–19.

[73] Maestro A, Solè I, González C, Solans C, Gutiérrez JM. Influence of the phase behavior on the properties of ionic nanoemulsions prepared by the phase inversion composition method. J Colloid Interface Sci 2008;327:433–9.

[74] Solans C, Morales D, Homs M. Spontaneous emulsification. Curr Opin Colloid Interface Sci 2016;22:88–93.

[75] Agrawal S, Giri TK, Tripathi DK, Ajazuddin, Alexander A. A review on novel therapeutic strategies for the enhancement of solubility for hydrophobic drugs through lipid and surfactant based self micro emulsifying drug delivery system: a novel approach. Am J Drug Discov Dev 2012;2:143–83.

[76] Bandyopadhyay S, Katare OP, Singh B. Development of optimized supersaturable self-nanoemulsifying systems of ezetimibe: effect of polymers and efflux transporters. Expert Opin Drug Deliv 2014;11:479–92.

[77] Bandyopadhyay S, Beg S, Katare OP, Sharma G, Singh B. QbD-Oriented development of self-nanoemulsifying drug delivery systems (SNEDDS) of valsartan with improved biopharmaceutical performance. Curr Drug Deliv 2015;2:544–63.

[78] Solans C, Izquierdo P, Nolla J, Azemar N, Garcia-Celma MJ. Nanoemulsions. Curr Opin Colloid Interface Sci 2005;10:102–10. Available from: https://doi.org/10.1016/j.cocis.2005.06.004.

[79] Tadros T, Izquierdo P, Esquena J, Solans C. Formation and stability of nano-emulsions. Adv Colloid Interface Sci 2004;108:303–18. Available from: https://doi.org/10.1016/j.cis.2003.10.023.

[80] Abolmaali SS, Tamaddon AM, Farvadi FS, Daneshamuz S, Moghimi H. Pharmaceutical nanoemulsions and their potential topical and transdermal applications. Iran J Pharm Sci 2011;7(3):139–50.

[81] Sonneville-Aubrun O, Simonnet JT, Alloret FL. Nanoemulsions: a new vehicle for skincare products. Adv Colloid Interface Sci 2004;108–109:145–9.

[82] Müller RH, Benita S, Böhm B. Emulsions and nanosuspensions for the formulation of poorly soluble drugs. Stuttgart: Medpharm; 1998.

[83] Raj S, Jose S, Sumod US, Sabitha M. Nanotechnology in cosmetics: opportunities and challenges. J Pharm Bioallied Sci 2012;4(3):186–93.

[84] Otberg N, Richter H, Schaefer H, Blume-Peytavi U, Sterry W, Lademann J. Variations of hair follicle size and distribution in different body sites. J Invest Dermatol 2004;122:14–19.

[85] Lademann J, Patzelt A, Richter H, Antoniou C, Sterry W, Knorr F. Determination of the cuticula thickness of human and porcine hairs and their potential influence on the penetration of nanoparticles into the hair follicles. J Biomed Opt 2009;14:021014.

[86] Dickhof S, Franklin J, Busch P, Kropf C, Fischer D. Cosmetic composition, for preventing greasy appearance on hair, contains nanoparticles of oxide, oxide-hydrate, hydroxide, carbonate, silicate or phosphate of calcium, magnesium, aluminum, titanium, zirconium or zinc. Patent DE19946784 A12001, 19 April 2001.

[87] Abdel-Mottaleb MM, Try C, Pellequer Y, Lamprecht A. Nanomedicine strategies for targeting skin inflammation. Nanomed (Lond) 2014;9(11):1727–43. Available from: https://doi.org/10.2217/nnm.14.74.

[88] Brito LA, Chan M, Shaw CA, Hekele A, et al. A cationic nanoemulsion for the delivery of next-generation RNA vaccines. MolTher 2014;22(12):2118–29. Available from: https://doi.org/10.1038/mt.2014.133.

[89] Syed Azhar SNA, Ashari SE, Salim N. Development of a kojic monooleate-enriched oil-in-water nanoemulsion as a potential carrier for hyperpigmentation treatment. Int J Nanomed 2018;13:6465–79. Available from: https://doi.org/10.2147/IJN.S171532.

CHAPTER

18

Nanomaterials in sun-care products

Jennifer Gubitosa[1], *Vito Rizzi*[2], *Paola Fini*[1] and *Pinalysa Cosma*[1,2]

[1]Consiglio Nazionale delle Ricerche CNR-IPCF, UOS Bari, Bari, Italy [2]Dipartimento di Chimica, Università degli Studi "Aldo Moro" di Bari, Bari, Italy

18.1 Introduction

The sun−skin relationship is conflicting. Indeed, in addition to important beneficial effects of sunlight, the harmful ones need to be considered, usually linked to the solar spectrum portion, corresponding to type A and B ultraviolet light (UVA and UVB, respectively). In particular, the damage induced by UVA and UVB light can be distinguished in acute, such as solar erythema, sunburn, and photosensitization, and chronic, due to the continuous exposure with poor compensation of the organism. Photoaging, an increased risk of skin cancer, photodermatitis, and eye injuries are worth to mention among the possible severe damages. So, to deal with such damages, the mechanisms of skin defense can be distinguished in endogen and exogen. The tan, the color developed during the UV rays exposure, can be considered the main endogenous defense mechanism: the melanin is formed by UVB stimulation and it is distributed by the melanocytes to all the surrounding keratinocytes with particular localization at the level of the nucleus, the most vulnerable and important part of the cell. Among secondary endogenous defensive mechanisms there are the stratum corneum thickening, the excretion of urocanic acid through sweat, the deposition of β-carotene in the hypodermis, and the activation of DNA repair mechanisms [1]. Despite the endogenous mechanisms of defense actions, the sunlight without appropriate exogenous protection, provided by sunscreens, could severely damage the skin on short as well as long term; [2] so daily sun-care products use should be recommended [3]. Sunscreens are commonly formulated in many cosmetic products with the aim to protect not only the skin but also hair and lips [4]. Indeed, their effectiveness in preventing melanoma or other skin diseases, and in reducing photocarcinogenesis and photoaging deriving from sun exposure, is widely investigated [5]. Further, as well documented, in addition to the primary photoprotection of sunscreen against UVA and UVB, the widening of the protective efficacy toward the oxygen free radicals [reactive oxygen species (ROS)]

Nanocosmetics
DOI: https://doi.org/10.1016/B978-0-12-822286-7.00022-X

is a relatively modern concept. Accordingly, recent scientific evidence point out that the portion of visible (from 400 to 800 nm) and infrared A light (IR-A, from 800 to 1400 nm) induce the formation of free oxygen radicals [1]. If, on one hand, it is expected that a sunscreen has the primary target of minimizing the damage induced by sun exposure allowing the activation of the defensive system of melanogenesis; on the other hand, it is less obvious, and therefore it should be emphasized, that the topical photoprotection is recognized worldwide as one of the major defense strategies, not only against sunburn and edema, but, above all, to prevent carcinogenic manifestations. On this ground, the importance and necessity of diffusion and awareness of a correct sun exposure behavior arise. For example, if the sun exposure during the central hours of the day or under the UV lamps should be avoided, the use of sunscreens with broad spectrum of protection (both UVA and UVB) in suitable quantities, by renewing the application, is strongly recommended [1]. Indeed, these aspects have already been subject of the authorities attention, as reported in the Commission recommendation no. 2006/647/CE of 22 September 2006, in relation to cosmetic sun-care product efficiencies and relative indications. Fortunately, the consumers' demand about the use of sun-care products for protecting themselves from the sun is increasing in the last decade and, interestingly, it goes beyond the simple protection. There is a growing need for ingredients to create multifunctional products, to protect against UV rays, from pollution and against aging, simplifying the integration of protection from sunlight into daily routine [5]. Then, according to the philosophy of enlarged or accessory photoprotection, additional ingredients in sunscreen products are also used to counteract dehydration, photoaging, and immunodepression [1]. From a cosmetic formulation technology point of view, these products contain sunscreens as functional substances and, in addition, ingredients as moisturizing, emollients, antioxidants, and texturizing, able to guarantee their effectiveness and a pleasant skin-feel [6]. Among sunscreens ingredients, nanoengineered materials have been extensively studied. Indeed, the cosmetics technology employs nanomaterials also in sun-care products category to improve photoprotection effectiveness without neglecting the texture or the esthetic acceptability by consumers. For these purposes, nano-TiO_2 and nano-ZnO are commonly used since exhibiting enhanced UV filtering efficiency when formulated in sunscreen products if compared with organic sunscreens [7]. Titanium dioxide and zinc oxide powders consist of primary particles that have crystalline structures held together by atomic or molecular bonding, and characterized by a main particle size of approximately 20 nm. However, it is worth to mention that the typical ranges used in many commercially available products are higher, 60–129 nm for titanium dioxide and 30–200 nm for zinc oxide [8].

The size of the primary particles is determined by the production conditions during the titanium dioxide or zinc oxide crystals formation. Then, due to their high surface energy, the primary particles tend to aggregate forming aggregates (secondary particles) when employed in cosmetic formulations. This behavior is attributable to the occurrence of van der Waals forces, electrostatic forces, and hydrogen bonding between surface hydroxyl groups, besides the involvement of water as a binding agent between the primary particles. These aggregates, then, tend to form agglomerates as a function of formulation processing. Since studies have shown that a smaller particle size can increase the sun protection factor (SPF) of sunscreen products, and since another factor influencing agglomeration is the type of particle carrier, it is possible to reduce agglomeration by modulating the viscosity of the vehicle [8].

About the adapted vehicle for these materials in sun-care formulation, untreated titanium dioxide and zinc oxide have hydrophilic surfaces that can be dispersed in aqueous vehicles, but not so easily in nonaqueous medium. For this reason, various surface treatments are used for improving both particle wetting and dispersion, and their compatibility with various types of formulation, obtaining various further benefits, as, for instance, photostability of the formulation. For example, metal ions in uncoated titanium dioxide can react with the organic sunscreen butyl methoxydibenzoylmethane (Avobenzone), whereas it does not occur if titanium dioxide is appropriately treated. The combination of titanium dioxide and Avobenzone can be marketed in Europe but is currently not permitted in the United States [8].

Another important aspect to be considered in the formulations is the development of suitable carrier systems for preventing the nanoparticles aggregation, enhancing sunscreen photostability and efficacy, but not its cutaneous permeability [9]. With respect to an emulsion formulation, the type of coating will enable the titanium dioxide and the zinc oxide to be incorporated into either the oil phase or the water phase of the emulsion [8]. Indeed, TiO_2 microparticle-coated forms (alumina, stearates, simethicone, dimethicone) are often preferred and possibly predispersed and stabilized in water or in another lipophilic vehicle (C12–15 alkyl benzoate, caprylic/capric triglyceride, isoesadecane, mineral oil, etc.), whose choice is modulated considering the type of formulation [1]. So, the chapter aims to deeply discuss the properties and the physicochemical characteristics of TiO_2 and ZnO, with a general overview on nanomaterials in commercial products, especially in sun-care products, giving important and general concepts about sunscreen and their regularity status and toxicology.

18.2 Solar radiations: characteristics and skin effects

Solar radiations are in part reflected or absorbed by the outer layers of the atmosphere, and in part reaches the earth's soil. The wavelength and energy of the light are linked by a relationship of an inverse proportionality. In detail, the visible radiations occupy a small portion of the electromagnetic spectrum between 400 and 800 nm and are involved in the processes of vision. Infrared (IR) radiations, with wavelengths in the range 700–1 mm, are not perceptible to human eyes and are emitted by any warm body. The IR radiation emitted by the sun reaches the earth's surface in the form of heat and is the main source of energy for biological processes. On the other hand, particular attention is devoted to the UV component of the electromagnetic spectrum, included in a range of wavelengths between 100 and 400 nm, and classified into

- **UVC** (100–280 nm), the most energetic and dangerous portion, but screened by the ozone layer of the stratosphere, thus avoiding it reaches the earth's surface;
- **UVB** (280–320 nm), largely retained by the ozone layer, the troposphere, and the clouds. They are able to stimulate the melanogenesis, promote the erythematogenic manifestations, initiate the direct damage at the level of cellular DNA causing skin tumors;

- **UVA** (320–400 nm) are classified into UVA-I (340–400 nm) and UVA-II (320–340 nm); they induce the maturation of the melanin already present in the skin and are responsible for photoaging and photosensitivity reactions. Due to the high penetration power, UVA rays can cross the epidermal barrier, reaching the dermis and damaging elastic fibers and collagen, essential for supporting the skin. UVA implement the production of free radicals that affect cells' DNA, membranes, and enzyme systems, resulting in the formation of neoplastic processes.

However, besides the negative effects attributed to UV radiations, also positive properties can be listed.

The positive effects of solar radiation include

- Synthesis of vitamin D, which occurs in the skin starting from the 7-dehydrocholesterol precursor; it is essential for bone mineralization and for the regulation of calcium and phosphorus metabolism. Its deficiency leads to rickets, an alteration of the calcification processes of long bones;
- Pigmentation, the so-called tan, a physiological mechanism for protecting the skin from the harmful effects of the sun's rays themselves;
- Antibacterial action;
- Increased sweating and consequent elimination of toxins;
- Release of antioxidant and protective substances, such as urocanic acid, which exhibit filtering activity against UV radiation;
- Improvement of some skin diseases such as psoriasis and eczema;
- Antidepressant effects.

Inattention to the health of skin and the adoption of incorrect behaviors during sun exposure can lead to the occurrence of negative and harmful effects due to UV radiation, reported below:

- **Erythema**: A skin irritation localized mainly at the epidermal level, which manifests itself with some characteristic symptoms (redness, swelling, dryness, itching burning) of variable intensity, depending on the case. UVB radiations, in particular, induce the release of cytokines and many other inflammatory mediators that cause peripheral vasodilation, and an increase in blood capillaries permeability, a typical symptomatology correlated to an inflammatory state.
- **Production of free radicals**: It is well known that the UV radiations promote free radicals production, molecules containing atoms that possess an unpaired electron at the level of the outermost orbital and, therefore, appear to be particularly unstable. Free radicals tend to react quickly with neighboring molecules, subtracting their electrons and transforming them into free radicals. This process entails the initiation of chain reactions that favor the formation of toxic substances, directly responsible for the DNA, proteins, and membrane phospholipid damages. Free radicals are neutralized by specific enzymatic systems (SOD, catalase, glutathione peroxidase) as well as antioxidant substances such as vitamins [6]. The excessive production of radical species or the inefficiency of antioxidant systems determines the beginning of a harmful condition at the cellular level known as oxidative stress that leads also to skin aging [10].

- **Photoaging**: It is a phenomenon that includes some morphological skin alterations, mainly determined by UVA radiations, such as to give it a thickened, rough, and hypohydrated appearance; to this picture, the appearance of deep wrinkles, hyperpigmentations, and signs of atrophy is added. The dermis shows an increase in thickness and a principle of sclerosis, a degradation of the elastic fibers up to a condition, known as elastosis, in which the texture of the skin takes on a noncompact and uniform appearance. These manifestations appear more evident in the skin of the neck, face, and hands.
- **Skin tumors**: The most feared consequence of incorrect behavior during exposure to solar radiation is melanoma, a malignant skin tumor that involves the melanocytes present in the skin, mucous membranes, and nevi. In addition to melanoma, other skin tumors such as basal cell and spinocellular carcinomas are also in part linked to the direct and indirect mutagenic action of UV radiation.
- **Photosensitivity reactions (phototoxicity, photoallergy)**: Some substances applied to the skin can interact with UV radiation, causing photosensitivity reactions. In particular, the phototoxic reactions occur in the form of erythema or hyperpigmentation at the level of the photoexposed areas; photoallergy reactions affect individuals previously sensitized due to the use of certain molecules and occur in the form of eczema, even in unexposed skin areas [6].

18.3 Sunscreen: functional ingredients for sun protection products

With the increased knowledge about UV-induced skin damage, public educational programs are developed to implement the level of primary photoprotection [1], emphasizing the use of sunscreen products [11]. The Regulation (EC) No. 1223/2009 of the European Parliament and of the Council, in Article 2, defines sunscreens as substances intended exclusively or mainly for protecting the skin from UV light through its absorption, reflection, or diffusion [12], avoiding or minimizing the damage that these rays might cause on human health [13], and a list of UV filters authorized in cosmetic products is reported in Annex VI [12]. All the formulated sunscreens must undergo strict efficacy and safety studies [14]. Sun-care products can be commercialized as primary and secondary sunscreens. In detail,

- **Primary sunscreens** protect skin from the effects of the sun. These products are mainly used for outdoor activities.
- **Secondary sunscreens** are products such as daily moisturizing creams, antiwrinkle/antiaging creams, make-up, and skin lightening, for which the sun protection is an additional claim, and not their main objective [11]

Regarding sunscreens, they can be classified into organic, inorganic, and natural filters. Moreover, they are available as W/O or O/W emulsions, lotions, sticks, wipes, and gels [2]. It is important to ensure the effectiveness of the product, as well as the suitability for every skin type. About this purpose, important is also the formulation development. To proceed rationally in the formulation phase, in addition to the considerations relating to the cosmetic pleasantness, it is useful to take into consideration some general requirements of sun protection:

- **Easy spreadability**: the sunscreens must be easily distributed evenly on the skin;
- **Effectiveness in a thin layer**: the amount of product applied on the skin is generally always lower than that actually used in the experimental determination of the SPF (2 mg/cm^2);
- **Poor transdermal absorption**: sunscreens must remain on the surface of the skin, so it is important to avoid ingredients that can act as penetration enhancers;
- **Sustainability on the skin**;
- **Absence of potential phototoxicity–photosensitization**.

Similarly, it is also necessary to underline the requirements for the ideal solar filter; these are the main:

- Maximum absorption in the UVA or UVB solar spectrum region;
- Effective at low dosage;
- Thermal and photostability;
- Nonvolatile;
- Soluble in water or in common oily solvents;
- Compatible with the other ingredients of the formulation [1];
- Easy to formulate;
- Quick drying time after application;
- Less greasy;
- Nonsticky;
- Nonstaining for skin or clothing [14];
- Esthetic appealing;
- Pleasant texture and feeling during and after application;
- Appropriate rheological properties [2];
- Dermatological safety: nontoxic, nonirritant, and nonsensitizing. Even though the product undergoes deterioration, the decomposed products should be nontoxic and nonirritating [14];
- Resistant to secretions (i.e., eccrine sweat);
- Odorless, colorless, tasteless (in particular, for lipstick sunscreens);
- Ecocompatible [1];
- Long-term stability [14];
- Being user-friendly to encourage frequent application and provide reliable protection (compliance) [11];
- Good cost-efficacy ratio

For an effective protection, the photostability of a sunscreen and its homogeneous distribution on the skin and inside the formulation is important. A photostable sunscreen has a low probability of giving rise to secondary reactions (photolysis, isomerizations, dimerizations) during irradiation; however, the photostability of a molecule also depends on what surrounds it. In detail, the photostability and, more generally, the stability of a sunscreen formulation, is a characteristic due to the whole set of ingredients and can influence, even significantly, the final value of SPF. For this purpose, searching for a suitable vehicle ensures that the sun filter remains on the skin surface in order to carry out its function as well as possible, avoiding/limiting transdermal absorption as much as possible [1], without neglecting the esthetic appealing of the product. Therefore substances that can act as penetration enhancers should be avoided and can be introduced film-forming polymers, even when water resistance is not expressly required [1].

It is worth considering that a sunscreen can also be found in lipstick formulations. These products are applied on the mucous membrane of the lips, which differ from the epidermis by their anatomical and physiological characteristics, with greater risks of absorption.

It must not be neglected the protection of hair and scalp from the sun. Erroneously, it is thought that the scalp is covered and protected by hairs; however, it is equally exposed as the total body surface and is easily susceptible to sunburns.

At the same time, the hair may be damaged by the action of wind, sun, salt, or chlorinated water as well as being lightened; for this reason, it is better to protect them with a suitable sun-care routine. Last but not least is the packaging of sunscreen products, essential to guarantee the photoprotection of the filter or of a mixture of them used, and also their protection from the environment. An innovative and smart packaging has been demonstrated by the Blue Lizard sun-care products, commercialized in a patented color change bottle that turns blue when UV levels are increased, providing for consumers a valuable sun damage warning and the need to renew the application of sunscreens [15]. At the endings, to provide a soothing and cooling effect in order to relieve pain resulting from sunburns, and to help smooth and moisturize the skin, even without any signs of redness, after-sun products are usually used. They are marketed frequently as lotions, creams, and gels [2].

18.3.1 Organic and inorganic sunscreens

UV filters can be classified into two main groups, chemical sunscreens (organic filters) and physical sunscreens (inorganic or mineral filters), based on their mechanism of action (Fig. 18.1).

18.3.1.1 Chemical (organic) sunscreens

Chemical (organic) sunscreens are molecules able to absorb the UV light and then disperse the absorbed energy by means of several mechanisms. In detail, once the organic filters absorb incident radiation, they are excited to higher energy levels. Since the excited state is not stable, it emits the excess energy and returns to the ground state (stable state). To achieve this transition, there are two possibilities: the release of energy at a longer wavelength than the incident radiation (as fluorescence) or the release of heat [14]. Unfortunately, upon exposure to UV radiation, the structure of UV filters may be negatively affected and/or destroyed, losing their absorption capacity, instead of returning to the ground state [2]. The different filters are characterized by a specific absorption spectrum (which shows the wavelengths absorbed in the UVA or UVB region) as well as by the extinction coefficient, an important physicochemical parameter, able to quantify the absorbed energy. Among the main UVB filters, there are cinnamates, salicylates, and

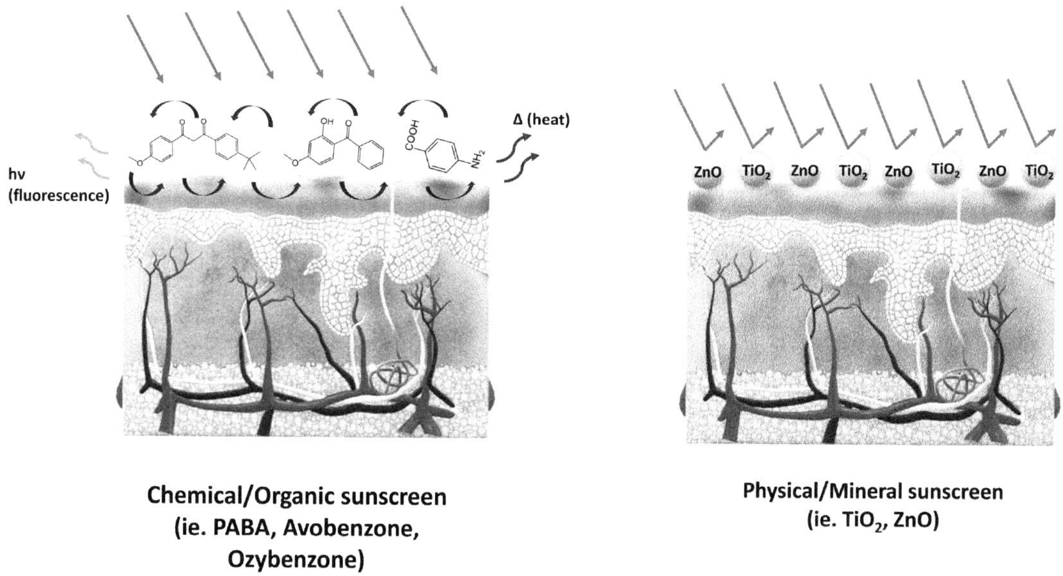

FIGURE 18.1 Classification of sunscreens: organic and inorganic sunscreens, based on their mechanism of action.

camphor derivatives; among the UVA filters instead, it is possible to find dibenzoylmethanes, benzophenones, and anthranilates [6]. However, it is worth to mention that, due to their lipophilic nature, organic sunscreens can penetrate the skin, so may cause safety concerns [2].

18.3.1.2 *Physical (inorganic) sunscreens*

Physical (inorganic) sunscreens provide a broad-spectrum protection since they absorb, reflect, and scatter both UVA and UVB radiations, and thereby prevent the penetration of UV light into the skin [14]. The most common ingredients used as physical sunscreens, often in combination, are TiO_2 that offers UVB and UVA-II protection, and ZnO that provides protection against UVB, UVA-II, and UVA-I radiations [2]. TiO_2 is a semiconductor characterized by a valence band plenty of many densely packed electron states that allow many absorption possibilities, as long as the energy absorption exceeds the bandgap width. TiO_2 crystals absorb more in the UVB, while ZnO absorbs more UVA-I radiation, even though the ZnO bandgap energy exceeds that of TiO_2 [9]. Commonly, ZnO in sunscreens are in the form of nanorods, while TiO_2 are nanoparticles in the rutile crystal structure [13]. The efficiency of the physical sunscreens mainly depends on the film thickness, refractive index, and particle size of the sunscreen formulation [14]. The main drawback of inorganic filters, due to their particle size, is that they reflect and scatter the visible radiation of the solar spectrum (>400 nm), providing a white appearance on the skin after application, hardly appreciated by consumers. Therefore recently, in order to make more esthetically appealing a sunscreen formulation, TiO_2 and ZnO are formulated in micronized (i.e., with a particle size of 1–100 μm) and nanosized (i.e., with a particle size of 1–100 nm) forms; the smaller the particle size, the less reflection occurs in the visible spectrum. Products that contain nanoparticles no longer reflect visible light and appear transparent on the skin. However, particle size reduction leads to a shift in the UV radiation protection profile of inorganic sunscreens toward the UVB and UVA-II range, and the UV protection becomes imbalanced. Using nanosized and micronized particles in combination may improve this situation. Inorganic filters are more often used in combination with organic filters to provide broad-spectrum protection while maintaining good esthetics [2].

18.3.1.3 Natural filtering agents

Natural filtering agents are substances present in some plant extracts, able to protect the skin from UV radiation as already discussed for chemical filters. Carotenoids, flavonoids, polyphenols, walnut oil, rice germ oil, helichrysum oil, and *Punica granatum* extracts are used as antioxidants rather than as primary filtering agents (accessory photoprotection) [6].

It is worth to mention that a trend in sun protection is the increased use of inorganic UV filters, especially in sun-care products for children and people with sensitive skin. Indeed, there are some concerns, expressed by the Food and Drug Administration (FDA), for a potentially systemic absorption of organic sunscreens in infants due not only to their low skin barrier function with respect an adult but also to the surface areas that are, of course, different [16]. This increased use of inorganic UV sunscreens is due to their low potential in producing irritant reactions, and to their cosmetic efficacy. Indeed, cosmetic products with high SPF values are formulated by using only titanium dioxide with different particle sizes. Or, if combined with organic sunscreen agents, TiO_2 improves the absorption in the UVB and UVA regions. Moreover, microfine titanium dioxide is incorporated into sunscreen products for avoiding the decrease of SPF, which can occur due to the photoinstability of organic UV filters [17].

18.3.2 The sun protection factor

The protective role of a sunscreen product is called sun protection factor or protection index. To harmonize the evaluation, since 1994 a European official method has been developed by COLIPA (European Association for Cosmetics and Products for Personal Hygiene, now Cosmetics Europe) which was revised in 2003, in 2006, until 2010, year of publication of the standardized method ISO 24444:2010, "Sun Protection Test Method," and adopted in Europe, Australia, New Zealand, India, Canada, South Africa, Mexico, Chile, Russia, Japan, Mercosur, and Asia [1].

Another important parameter is the sensitivity of the skin to solar radiation, expressed by the erythematous dose (MED), which indicates the lowest dose of UV rays capable of producing visible skin redness after 24–26 hours from the exposure.

The SPF expresses the relationship between the MED of the skin protected by the solar product (MED_p) and the MED of the unprotected one (MED_u); the greater the protective capacity of the product, the higher the MED of protected skin and the higher the SPF [Eq. (18.1)].

$$SPF = \frac{MED_P}{MED_u} \tag{18.1}$$

The SPF is a parameter referring only to the UVB radiation that induces the erythema. The SPF evaluation must take into account that a sun-care product should be used during bathing, with the risk of a drastic reduction in the effective protection following the contact with water (50% of its effectiveness) [1,6]. Indeed, the effectiveness of a sunscreen is dose-dependent, and the commonly applied dose, required to perform the labeled SPF value, is much lower than that used in the tests [1]. To avoid this inconvenience, the SPF is measured before (static SPF) and after (SPF wet) the interested area is immersed in water. The products claimed as "water resistant" on the label must guarantee that the wet SPF is not lower than the static SPF by more than a defined%. In recent years, products with higher SPFs are commercialized; however, there is no direct proportionality between the increase in the protection factor and the increase in the real protective capacity of the product. For example, passing from an SPF 15 to an SPF 30, there is an increase in screened radiation of only 4% [6].

For the evaluation of the protection from UVA radiations that promote the melanin maturation, it is considered the PPD (Persistent Pigment Darkening), a parameter that takes into account the melanin pigmentation after exposure of volunteers with good pigmentation capacity (phototypes III–IV, "mixed" and "Mediterranean," respectively) to UVA radiation. In this case, the ISO 24442:2011 method is followed, and it is adopted in Europe, Canada, Mexico, South Africa, Japan, Mercosur, and Asia [1]. Alternatively, the in vitro evaluation of the UVA protection factor is used [6]. The denominations admitted according to the recent directives of the UNIPRO (Italian Association of Cosmetic Industries, now Cosmetica Italia) are

- "Low protection" (SPF 6–10/PPD 12);
- "Medium protection" (SPF 15/PPD 20–25);
- "High protection" (SPF 50–40–30/PPD 40–20);
- "Very high protection" (50 + /PPD 30) [6].

Regarding the Australian region, since it has one of the highest rates of skin cancer in the world, sunscreens are an important component of a sun protection regime. It is important that sunscreens used in Australia are safe and effective and of good quality. For this reason, the Therapeutic Goods Administration (TGA) regulates some sunscreens as therapeutic goods in Australia.

Published in May 2012 as a procedure for assessing the in vivo and in vitro UVASPF, the methods ISO 24444 and 24443 were indicated. In particular, among the sunscreen products, there are (1) therapeutic sunscreens and (2) cosmetic sunscreens.

Therapeutic sunscreens, regulated by the TGA (*Therapeutic Goods Administration*). They include products that have the primary function of protecting from UV rays with SPF 4 or more, and moisturizers containing sunscreen with SPF $>$ 15.

Cosmetic sunscreens, regulated by NICNAS (*National Industrial Chemicals Notification & Assessment Scheme*). They are protective products that have the secondary function of protecting from UV rays following some requirements:

- Skin-care products with SPF < 15 and size not more than 300 mL;
- Lip-balm and lipstick with sunscreens;
- Make-up products with sun filters.

The **AS/NZ 2604:2012** method indicates a list of 11 SPF levels to be shown on the Australian sun-care product label:

- **Low** protection: 4–6–8–10;
- **Medium** protection: 15–20–25;
- **High** protection: 30–40–50;
- **Very high**: 50 + [1].

18.3.3 Sunscreen regulatory status: safety and effectiveness

Sunscreen products such as UV filters "for which some concern exists with respect to human health" are evaluated for safety at the European Commission level by a Scientific Committee on Consumer Safety (SCCS). The EU process for safety evaluation of cosmetic substances, including sunscreen products, considers all available scientific data: the physical and chemical properties of the compounds under investigation, the results obtained from (Q) SAR [(quantitative/qualitative) structure activity relationship], chemical categories, read-across, physiologically based pharmacokinetics (PBPK)/toxicokinetics (PBTK) modeling, in vitro experiments and data obtained from in vivo studies. Safety of cosmetic products in the EU is based on the safety of the ingredients which are evaluated by toxicological testing [16]. In the United States, any cosmetic product labeled as "sunscreen" is considered Over the Counter (OTC) drug, since it is intended to "prevent, cure, treat or mitigate disease" and/or to "affect a structure or function of the body" (CFR, 1999). Sunscreens also "help to prevent sunburn and may reduce the chance of premature skin aging, skin cancer, and other harmful effects of the sun" by absorbing, reflecting, or scattering the harmful, burning rays of the sun, thereby altering the normalphysiological response to solar radiation, (CFR, 1999) [16]. The two regulatory pathways by which OTC drugs may be approved, the New Drug Application (NDA) process and the OTC Drug Monograph process, differ significantly in terms of defining the needs for safety testing. Under the NDA process, a single final formulation drug product is determined to be safe and effective while under the OTC Monograph process active ingredients are determined to be Generally Recognized as Safe and Effective (GRASE). Once an ingredient is approved as GRASE, there are no requirements to safety test for the subsequent sunscreen product formulations. All currently marketed sunscreen products in the US may contain one or more of the 16 sunscreen active ingredients included in the OTC Monograph with the majority of the ingredients approved prior to 1972. In 2002, FDA promulgated regulations establishing a process for inclusion of active ingredients in an OTC Drug Monograph referred to as Time and Extent Application (TEA). The TEA process applies to sunscreen active ingredients marketed in the US after 1972 under the NDA process (of which there are none) and sunscreen active ingredients marketed in other countries [16].

The UV filters concentration limits allowed in any sunscreen product depend on the different worldwide regulations [13]. The cosmetic regulation underlines the overarching principle that any cosmetic products placed on the EU market must be safe for human health when used under normal or reasonably foreseeable conditions of use and the manufacturers responsibility is to ensure this (EP and CEU 2009, art, 3, 4) [18].

So, in order to guarantee their quality, safety, and effectiveness, any information labeled, and any warning must be simple to understand, helping the consumers in the choice of the suitable product [19].

It is worth to mention that, since there are many differences in regulatory requirements for sunscreen products safety assessment between the US and EU, a scientifically acceptable scheme for the testing methods should be harmonized [16].

Based on several studies, the International Agency for the cancer research of the World Health Organization stressed the importance of correctly applying sunscreen products in order to achieve the effectiveness of the indicated SPF: (1) it is essential to renew frequently the application of these products in appropriate amounts; (2) for reaching the level of protection indicated by the SPF, products must also be applied in quantities similar to those used in the test, 2 mg/cm^2, equivalent to 36 g, a higher quantity than usually applied by consumers. The application of a lower quantity of product significantly reduces protection; therefore consumers should be informed of the risks involved from excessive sun exposure and they also need a guidance on the effectiveness of suitable sunscreen product, which takes into account the frequency of sun exposure and the skin type [19].

18.3.4 Sun-care products labeling

In the way of making communication clearer for consumers, the European and international legislative authorities have been moving for some years with the support of scientific committees such as the COLIPA (European Cosmetic Toiletry and Perfumery Association, now Cosmetics Europe). With the aim of self-regulating the cosmetic industries, Cosmetics Europe periodically drafts recommendations concerning the labeling and the methods to assess the efficacy of the sun-care category products, and to update them about the legislative provisions being implemented. In 2006, the European Commission published the recommendation 2006/647/EC on the efficacy of sunscreen products and related indications. This initiative was intended to standardize and simplify the way in which these products are tested and labeled in Europe. The subsequent recommendation COLIPA no. 23 of 2009 contains key information relating to sun-care products to be reported on the label as already indicated in the European Commission document. It is clear that the fundamental prerequisite that can help consumers to consciously choose a sunscreen product is that the information on their efficacy is universally harmonized, objective, and easy to understand. Among the various fundamental properties of a sunscreen product, there are the following:

- It must protect against both UVA and UVB;
- The minimum SPF value must be equal to 6. Products with SPF < 6 cannot be considered as sun-care products;
- Terms such as "sunblock," "full screen," or "100% protection" should not be used. Since 2008 the EMEA (European Medicines Agency) has banned the indication of

"total screen" or "total protection" on the labels: these terms are considered misleading because there are no products that completely protects from sunlight [20];

- For the determination of UVB protection, the International SPF Test Method CTFA-SA/Colipa/JCIA 2006 (now ISO 24444:2010 method) must be indicated;
- The SPF values to be reported on the label are 8.

  The protection factor obtained with the test is approximated by default to the nearest number among the eight values reported in the SPF classification table.

  The minimum SPF value is 6; the maximum value is 50 + (also according to the FDA method), corresponding to the highest protection category.

  To report SPF 50 + on the label, the value obtained with the test must be equal to or greater than 60;

- The SPF value must also be accompanied by the protection category (low, medium, high, or very high);
- As recommended also by the United States FDA, the protection factor against UVA should be at least one-third of the overall SPF [9] (i.e., an SPF 15 product must have a UVA protection factor of at least 5) [1].

  Moreover, the critical wavelength must be at least equal to 370 nm.

  Compliance with both requirements is indicated on the label [1] with a logo that represents a small circle enclosing the initials UVA, having the same dimensions as the character used to indicate the value of the SPF [6].

Under the new FDA regulations, sun-care products are required to designate UVA protection properties based on a 4-star rating system, which will be derived from in vivo and in vitro tests [1].

The UVA logo and SPF value must be reported on the front of the primary and secondary packaging [1].

Harmonized labeling for UVA protection claim is important in the way to [1] make easier for consumers to choose a sun-care product with a "broad spectrum" that protects against both UVA and UVB rays [20].

Regarding the warnings labeled on a sunscreen product, there are, instead, the following:

- Apply sunscreen product before sun exposure;
- Reapply the product frequently, especially after bathing, sweating, and after drying;
- Apply a correct amount of product;
- Avoid sunbathing for long time even if protected;
- Keep children away from direct sunlight [1].

Another potential point of concern is that sunscreen could be used on skin with disrupted skin barrier function. Labels should report warnings about this possibility [16].

18.3.5 Controversy between sunscreen application and decrease in vitamin D synthesis

Vitamin D synthesis requires exposure to UVB. There is a controversy over the effects of sunscreen use on UV-dependent cutaneous synthesis of vitamin D3 (cholecalciferol).

Since the correct use of SPF 15 sunscreen may reduce vitamin D synthesis by over 98%, it is a common opinion that regular sunscreen applications may cause vitamin D insufficiency; on the other hand, it is also shown that long-term sunscreen use has little to no effect on vitamin D levels and function. Possible explanations for these discrepancies include the fact that a significant portion of vitamin D is obtained through diet and it should consider the fact that most people apply sunscreen insufficiently, and some UV rays still reach the skin [15].

18.4 Nanomaterials in cosmetics

Nanotechnology is a general term describing the design, characterization, production, and application of structures, devices, and systems by controlling size and shape at the nanoscale level. The interesting properties and behaviors of nanomaterials indicate that nanotechnologies could represent a revolutionary opportunity for several fields, from industries to consumers everyday life [21]. So, nanotechnology has rapidly become a subject of global interest.

Nanotechnology in cosmetic products has a long history of use, with publications appearing during the early 1980s. Patent applications exist from 1983 from the Shiseido Company for a skin cosmetic containing hydrophobic particles of titanium dioxide with a maximum size of less than 0.1 μm and an average size of 30–40 nm. Another patent application, published in 1985, focused on an ultrafine zinc oxide powder, able to protect skin from injurious ultraviolet light, with a particle size of less than 0.1 μm and a maximum particle diameter between 10 and 60 nm [8]. Nanomaterials can be found in commercialized make-up, skin care, and sun-care products with the aim to make fragrances last longer, sunscreens more effective or to claim face creams as effective antiaging [21]. Other benefits due to the presence of nanomaterials in cosmetic products are increased efficiency, transparency, unique texture, protection of active ingredient and, overall, higher consumers compliance [22].

Due to the high surface energies, the nanostructured particles tend to agglomerate and aggregate together and/or bind other substances on their surface; this behavior can change in the presence of some stabilizing/capping agents. Nanomaterials are insoluble or poorly soluble nanoparticles dispersed in the medium in the form of nanosuspension rather than a solution [23]. Nanoengineered materials include gold (AuNPs), silver (AgNPs), and silica nanoparticles, titanium dioxide, zinc oxide, and fullerenes [8]. A particular application in cosmetic sun-care products is demonstrated by solid lipid nanoparticles that are used both as sunscreen themselves, due to their ability of scattering and reflecting the UV radiations, and as carrier systems for organic UV filters (i.e., Oxybenzone), in order to enhance SPF, to minimize skin irritation or contact dermatitis, and to improve their phostostability [14].

In 1999, the FDA allowed the use of nanoparticles in sunscreens. Since that time, there are controversies regarding the safety of their use [24]. Due to the frequent use of cosmetics, the safety of nanoscale ingredients has gain importance mainly by means of their dermal exposure [22]. Indeed, debates on their definition and insufficient quantification methods are the major problems still occur in the nanomaterial field [22]. The particles number, the size, their surface area, and their distribution are also required parameters and are considered more relevant [8] than the mass [8].

18.4.1 Regulatory status of nanomaterials in cosmetics

In EU, the cosmetics legislation was adapted in 2009 in order to include nanomaterials [18].

Article 2 of European Regulation 1223/2009 of Cosmetic Products defines nanomaterial as "any insoluble or biopersistant material and intentionally manufactured having one or more external dimensions, or an internal structure, measuring from 1 to 100 nm"; the soluble particles are no longer considered [12].

In Article 16, a different notification procedure for cosmetic products that contain nanomaterials in their formula is described. The so-called "nano-notification" must be carried out 6 months before introducing the product in the market, except when these products are already placed on the market before 11 January 2013.

The information notified to the Commission includes

- The identification of the nanomaterial, including the chemical name (IUPAC) and other descriptors as specified in point 2 of the preamble from Annexes II to VI;
- Description of the nanomaterial including particle size and physical and chemical properties;
- An estimate of the quantity of nanomaterial contained in the cosmetic products that are expected to be placed on the market by year;
- The toxicological profile of the nanomaterial;
- Data on the safety of the nanomaterial relating to the category of cosmetic product in which this nanomaterial is used;
- Reasonably foreseeable exposure conditions.

In the case of concerns about the safety of a nanomaterial, the Commission immediately asks to SCCS, the Scientific Committee on Consumer Safety, to deliver an opinion concerning the safety of these nanomaterials for use in the relevant categories of cosmetic products and reasonably foreseeable exposure conditions. The SCCS delivers its opinion within 6 months of the Commission's request.

The SCCS opinion is publicly available [12].

The information required by the SCCS for the evaluation of nanomaterials as cosmetic ingredients is described in SCCS/1588/17, SCCS/1484/12, and their subsequent revisions.

The following particular aspects are highlighted:

- Methods for the characterization of nanomaterials;
- Where there is evidence of systemic absorption, further studies are needed to confirm whether the absorbed material is in the form of nanoparticles or in the solubilized/ionic/metabolized form;
- Surface modification/surface coating may cause changes in a nanomaterial related to some physicochemical properties;

In addition to the safety data on the basic nanomaterial, the SCCS would require

- Information/data on each material used for surface modification/coating of the nanomaterial to indicate that it is safe for use in the cosmetic product.

Data about the physical-chemical properties of the surface/coated modified nanomaterial, in order to show that they have not changed significantly with respect to the same material when uncoated or with a different surface modification/coating, are already assessed as safe by the SCCS.

- Data on skin penetration, on the stability of the surface modification/coating, and on photocatalytic activity, if relevant [23].

The technical report from International Organization for Standardization (ISO) underlines the importance of physicochemical characterization for identifying materials previous the toxicological testing. Particle size/particle size distribution, aggregation/agglomeration state, shape, surface area, composition, surface chemistry, surface charge, and solubility/dispersibility are listed as the physicochemical parameters.

The US FDA also emphasized the physicochemical properties and aggregation/agglomeration of nanoparticles in final cosmetic products [25].

On 11 January 2014, the Commission published an available catalog containing all of nanomaterials used in cosmetic products placed on the market indicating the categories of cosmetic products and the reasonably foreseeable exposure conditions. Furthermore, the Commission is obliged to prepare an annual status report for the European Parliament and the Council outlining the present and future use of nanomaterials in cosmetic products, the number of notifications, and the progress made nanospecific safety assessment methods (EP and CEU 2009, art. 16) [18].

The Commission periodically review the provisions of this Regulation relating to nanomaterials in the light of scientific progress and, if appropriate, propose appropriate modifications to these provisions. The first revision is carried out on 11 July 2018 [12]. In the United States, FDA regulates nanomaterials in cosmetics to ensure that the products are effective, safe, and properly labeled for the use. For cosmetics, FDA's authority is postmarket only, meaning that the Agency does not evaluate the safety of ingredients, finished products, or the specifics of labeling prior to marketing. FDA exercises its authority within a "risk management" framework, which is applied on a case-by-case basis. In 2007, FDA issued its FDA Nanotechnology Task Force Report, following public meetings and internal discussions aimed at gaining a better understanding of the nature of the science at that time. The conclusion of that report was "The Task Force believes FDA should continue to pursue regulatory approaches that take into account the potential importance of material size and the evolving state of the science." To address the issues of nanotechnology in cosmetics, FDA developed a separate guidance for the nanomaterials safety in cosmetic products, which was published in June 2014: "Guidance for Industry: Safety of Nanomaterials in Cosmetic Products," focused on the safety assessment of nanomaterials in cosmetics [26].

18.4.2 Nanomaterials under REACH and CLP

In 2011, the European Commission issued a specific recommendation on the definition of nanomaterials. The recommendation must be used in various European regulations, including REACH (Registration, Evaluation, Authorization and restriction of Chemicals) and CLP (Classification, Labeling, and Packaging). Although nanomaterials clearly provide unique technical and commercial opportunities, the rapid increase in their use raises questions about their possible effects on human health and the environment. The potential risks posed by these new forms of materials must be assessed and adequately managed; furthermore, although manufacturers, importers, and downstream users are required to ensure the safe use of each substance (regardless of form) under REACH, this involves new challenges for regulators (Commission and ECHA, the European Agency for Chemicals) and for all other interested parties. ECHA works closely with the competent authorities of the Member States, the European Commission, stakeholders, and international organizations such as the Organization for Economic Cooperation and Development (OECD) [1].

18.4.3 ECHA activity

Since the REACH and CLP regulations concern nanomaterials, ECHA must be able to perform its tasks under the various processes covered by REACH and by CLP for nanoforms and all other forms of a substance [1].

18.4.4 Nanoingredients labeling

As established by Article 19 of the regulation 1223/2009, all ingredients in nano form must be clearly indicated in the list of ingredients carrying the word "nano" in brackets, which follows the INCI name of these ingredients.

As an example, referring to a nanosunscreen, nano-TiO_2 is reported as Titanium Dioxide (nano) [12].

18.4.5 Nanomaterials safety concerns

The use of nanomaterials in cosmetics also raises potential safety concerns. Some of these concerns can be addressed by determining the type of nanomaterials used, the stability (also in the final packaging of any cosmetic product), the route of exposure, and how they are formulated in cosmetic products. It has been done considerable effort, at the international level, to harmonize approaches to address definitional issues and safety concerns related to cosmetic field [26]. Indeed, some substances in nano form differ from their conventional materials of bulk in terms of physical-chemical properties, biokinetic behavior, and/or biological effects. The SCCS has published a more detailed specific guide on the risk assessment of nanomaterials (Guidance on the Safety Assessment of Nanomaterials in Cosmetics, SCCS/1484/12), which is currently being revised, a memorandum on the relevance, adequacy and quality of the expected data in nanomaterial safety dossiers (SCCS/1524/13, revision of 27 March 2014) and a checklist for manufacturers that submit dossiers on nanomaterials as cosmetic ingredients (SCCS/1588/17). It is worth noting that, as indicated in the memorandum (SCCS/1524/13, revision of 27 March 2014), the SCCS considers only the data relevant for the nanomaterials, sufficiently complete and of high quality, appropriate to support their safety assessment. The SCCS has also published, in recent years, several scientific opinions on various materials in nano form; these opinions can provide additional information in the light of scientific progress needed in a safety dossier on nanomaterials intended for use as ingredients. Currently, there is inadequate information on the risks associated with nanomaterials. To better assess its safety, the SCCS should provide guidelines, in cooperation with the competent authority, about methodologies that take into account the specific characteristics of nanomaterials [12]. Recently, ECHA established and manages a web portal on nanomaterials called EUON, the European Observatory on Nanomaterials, with the aim of exchanging information and news on uses, industrial sectors interested in nanotechnologies including the cosmetic field, toxicological safety of nanosystems, reference legislation, and updates. These information are addressed not only to employees in the field of regulatory toxicology but also to consumers who have the curiosity or the need to investigate and get information. The committee concluded that new test strategies will be necessary to determine the mechanisms of potential injury related to the cosmetic use of nanomaterials [8]. Another issue needs to be considered is the probability that the physical-chemical characteristics of nanomaterials in the final product and coatings could change and/or the impurities may be released affecting the toxicity of the material. Moreover, the properties of nanomaterials may change also during storage and handling [22]. The objective is to guarantee a high level of consumer and human health protection [12].

18.4.6 Nanotoxicology

Nanotoxicology is proposed as a new branch of toxicology to explore the adverse health effects induced by exposure of nanoparticles [27] and its quickly development extended the knowledge on the phenomena that take place at the interface between nanomaterials and biological systems [5]. The potential skin absorption of nanomaterials if formulated in cosmetic products is an issue that must be addressed in order to understand their safety and usefulness [21].

The physical-chemical parameters, such as size, shape, crystal structure, solubility, surface area, surface charge, and surface coating, may all contribute to toxicity. As established in the definition of the European regulation 1223/2009, cosmetics must be applied on intact skin. However, it can happen that products should be applied on broken skin; so, the purpose of toxicology studies is aimed to check whether the use of a cosmetic product on nonintact skin can cause damage.

To assess the dermal penetration of nanoparticles, future studies will need to focus on:

- complete characterization of the nanomaterials being tested (e.g., size, specific surface area, surface charge, structure phase, monodispersed, or aggregated);
- the analytical method used to detect the nanoparticles (i.e., needs to be able to discriminate between epidermis, dermis, and hair follicles);
- the condition of the skin (e.g., intact, flexed, and broken) before and after testing [27].

Engineered nanomaterials (ENMs), also in sunscreen products, should cause environmental toxicity risks, particularly toward coasts and marine ecosystem.

In the literature, several studies demonstrated that high surface area and reactivity of ENMs make them more dynamic in the environment [28].

However, the role of environmental transformation on the fate of nanomaterials is unclear [29].

For demonstrating nanomaterials environmental safety, further studies must be collected. When applying a new marketing authorization, an environmental risk assessment is required [18]. Material Safety Data Sheets are the key mean to pass information about risk and safety down the supply chain [18].

18.5 Mineral-based nanosunscreens: nano-TiO_2 and nano-ZnO

18.5.1 Titanium dioxide

Titanium dioxide (TiO_2) is reported as a sunscreen agent since 1952. L'Oreal Group was the first company that, in August 1993, patented the use of nanoparticles of metal oxides as ingredients in sunscreens [13].

In Annex VI of the regulation 1223/2009 titanium dioxide is found, among the allowed sunscreens in cosmetic products, as a physical UV filter of inorganic nature. In a cosmetic sun-care product formula, it is preferred in its nanostructured form due to the easy spreadability during the application, its pleasant skin-feel, its light texture, and its transparency, thanks to the different chemical and physical properties if compared to its micronized form; thus avoiding an esthetically unpleasant white color left on the skin after application, without neglecting its effectiveness in sun protection [30].

18.5.1.1 Titanium dioxide characteristics

Titanium dioxide is a white powder with a high refractive index used in cosmetics as a white pigment in make-up products [1]. It exists in different crystalline forms; only the rutile (bandgap energy, 3.0 eV; onset of absorption at 400 nm) and anatase (bandgap energy, 3.2 eV; absorption onset at 387 nm) polymorphs have a significative relevance in sun-care products [17].

Rutile has the most compact atomic structure, the highest refractive index ($n = 2.76$), UV absorption, and better stability (about 8–12 kJ/mol) if compared with anatase ($n = 2.5$), which instead demonstrates a greater photocatalytic potential. This last characteristicis negative in sunscreen products because it involves the formation of free radicals [17].

From the opinion of the SCCS1516/13, 22 July 2013, revision of 22 April 2014, it is also derived that TiO_2 is insoluble in water and organic solvents, has a very low dissociation constant in water and in aqueous systems and therefore can be considered insoluble even in physiological conditions [30].

18.5.1.2 Titanium dioxide: a carefully selected ingredient

The recent attention paid to this ingredient arises from doubts about the suspected toxicity of its nanostructured form, when used as a sunscreen in a cosmetic product intended not only for the adult population but, above all, for children. The main committees and regulatory agencies expert in toxicology have thus begun to deepen their studies and review their opinions reserving a remarkable regard for the application of titanium dioxide as a raw material for the formulation of cosmetic sun-care products. Among these assessments, available in the literature also for consumers, emerged some concerns about the photostability of TiO_2 to solar irradiation in terms of strengthening its photocatalytic activity, its absorption through the skin after the application of a sun-care product in which its nano form is formulated, and its potential exposure by inhalation to TiO_2 nanoparticles if found in sun-care spray products.

18.5.1.3 *Toxicological safety of nano-TiO_2 as UV filter in sun-care products: SCCS Opinion*

In July 2013, the SCCS issued an opinion on titanium dioxide in nano form (SCCS/1516/1311) to assess its safety. The SCCS concluded by arguing that the use of nano-TiO_2 as a UV filter in sunscreen products, with the characteristics indicated in the aforementioned opinion and with a concentration up to 25%, can be considered not to pose any risk of effects on humans after application on healthy, intact skin, or with sunburn. The SCCS admits the presence of the anatase form up to a maximum of 15% [1]. Concerns are limited to spray applications that could lead to exposure of the consumer's lungs to inhaled titanium dioxide nanoparticles; so, the SCCS also considered that, based on the available information, the use of titanium dioxide nanoparticles in cosmetic sun-care spray products cannot be considered safe. The SCCS clarifies, in a further opinion (September 23, 2014), the meaning of the expression "sprayable application/products" for the titanium dioxide and zinc oxide nanoforms. In July 2015, the Commission received new data from the industry to support the safe use of nano-TiO_2 when used as a UV filter in sun-care and personal care spray products at a concentration of up to 5.5%. However, outside the European market, it is reported that nano-TiO_2 is used as a sunscreen in spray formulations (i.e., Brazil) and in powder form [31]. In the absence of specific information on the toxicity by inhalation of TiO_2 nanomaterials in sunscreen spray formulations, the SCCS considerations are based on information available in the literature. As a result, the inhalation exposure to TiO_2 nanoparticle, in general depending on the dose and duration of exposure, can lead to adverse effects in the lungs, being also considered associated with the induction of lung tumors (ECHA, 2016 and references cited therein). Furthermore, the potential toxicological effects are not sufficiently characterized with respect to the route of inhalation [31].

18.5.1.4 *Coated nano-TiO_2*

Several studies agree that the toxicity of nano-TiO_2 is derived by its photochemical properties under solar radiation that depend not only on the crystalline structure but also on the radiation intensity and on the concentration of the nanoparticles [13]. To reduce the photocatalytic potential under solar radiation, avoiding the formation of ROS, nano-TiO_2 is coated with materials suitable to guarantee its inertia and greater safety in sunscreen products. Consequently, it is necessary to evaluate the safety of the materials used for the nano-TiO_2 coating; the opinion of SCCS1516/13 reports this information. Like with other coating materials already in use, not enclosed in this opinion, the SCCS concluded that other cosmetic ingredients can also be used as stable coatings on nanomaterials with TiO_2 cores, with the condition that they are safe and do not influence the properties, behaviors, and/or effects of the particles. The SCCS conclusion clarifies that for the use of a coating substance on TiO_2-based nanomaterials, it must be demonstrated that the properties/behaviors of the particles with the new coating are not significantly different from those already discussed in the SCCS opinion; this requires the availability of data on physicochemical properties and on dermal penetration. Cetyl phosphate, manganese dioxide, and triethoxycaprylylsilane are the substances studied and considered safe and suitable for the intended application, going to support other materials already in commercial use, whose safety is widely demonstrated in terms of chemical inertness, stability (also in terms of storage), and efficiency in performances such as silica and dimethicone [30]. Another report listed the coating materials as follows: alumina/silica, methicone/silica, aluminum hydroxide and dimethicone/methicone copolymer, trimethyloctylsilane, alumina/siliconeand alumina/silica/silicone, simethicone, and dimethoxydiphenylsilane. These coatings are shown to improve the dispersion of TiO_2 nanoparticles, inhibit or control photoactivity, enhance compatibility with other ingredients [25], avoid direct contact with the skin increasing their margin of safety and their effectiveness [7]. Indeed, any new coatings for cosmetic use or sunscreens will need additional tests in case they promote skin penetration of nanoparticles [32].

In the US, the FDA recommends that any safety assessment of cosmetic products containing nanomaterials specifies the aggregation/agglomeration of nanoparticles in the final products [25].

However, despite the SCCS considered safe the materials used for the coating of the nano-TiO_2 and, consequently, the TiO_2 coated nanoparticles, on the basis of studies on their chemical-physical characteristics as well as their toxicological profile, and based on the data available in the literature, it recommends the topical application of sunscreen products in the form of an emulsion; on the other hand, it is not recommended the formulation in spray products, to avoid the possible inhalation of the nebulized particles [31].

18.5.2 Zinc oxide

Zinc oxide has widespread use in cosmetic products with the following functions: bulking, skin protection, and as a UV absorber. It is authorized in all cosmetics as a colorant with the Color Index No. CI 77947 in Annex IV [32]. Zinc oxide is slightly more reactive than titanium dioxide [8]. Nanoscale zinc oxide is only recently approved for use in European sunscreens; [32] however, is not included in Annex VI, but SCCS opinions on its safety assessment as sunscreen ingredient in cosmetic products are available [8].

18.5.2.1 Nano-zinc oxide: SCCS safety assessment opinion

The SCCS is requested to perform a safety assessment of zinc oxide in nano form and delivered an opinion on 18 September 2012, followed by an addendum of 23 July 2013. The SCCS concluded that, on the basis of the available evidence, the use of zinc oxide nanoparticles as UV filter in sun-care products at a concentration up to 25%, cannot pose a risk of adverse effects in humans after dermal application.

In addition, the SCCS indicated that there is no evidence for the absorption of zinc oxide nanoparticles through skin. The SCCS/1489/12 assessment only applies to ZnO nanoparticles, or for similar materials that have the following characteristics:

- ZnO nanoparticles of purity ≥99%, with wurtzite crystalline structure and physical appearance as described in the dossier, that is, clusters that are rod-like, star-like, and/or isometric shapes;
- ZnO nanoparticles with a median diameter (D50: 50% of the number below this diameter) of the particle number size distribution between 30 and 55 nm, and the D1 (1% below this size) above 20 nm;
- ZnO nanoparticles that are either uncoated or coated with triethoxycaprylylsilane, dimethicone, dimethoxydiphenylsilanetriethoxycaprylylsilane cross-polymer, or octyl triethoxy silane;
- ZnO nanoparticles that have a comparable solubility to that reported in the dossier, that is, below 50 mg/L (approximately the maximum solubility of the ZnO nanomaterials for which data are provided in the dossier) [33].

The characteristics indicated by the SCCS are related to the physical-chemical properties of the ZnO (such as purity, structure and physical appearance, particle number size distribution, and water solubility) both uncoated and coated with specific materials. Other cosmetic ingredients can be used as coatings as long as they are demonstrated to SCCS to be stable, safe, and do not affect the particle properties related to behavior and/or toxicological effects, if compared to relevant coated nanomaterials according to the SCCS opinion. Therefore the Commission considers that these physicochemical properties and requirements, regarding the coatings, should be reflected in Regulation (EC) No. 1223/2009. The SCCS also considered that, on the basis of available information, the use of zinc oxide nanoparticles in spray products and powders cannot be considered safe. Considering the SCCS opinions mentioned above, the Commission considers that zinc oxide both in nonnano and in nano form should be authorized as a UV filter in cosmetic products. Both forms of the substance should be authorized at a maximum concentration of 25%, except for applications that may lead to exposure of the end-user's lungs by inhalation [34]. In January 2013, Cosmetics Europe submitted a document in which they proposed their own—broader—interpretation of the characteristics laid out in the scientific opinion on zinc oxide in nano form. In particular, they proposed the purity requirements to be reduced to 96% (as data on one of the material with 96% purity are provided in the submission), the median diameter of the particle number size distribution to be accepted when it is greater than 30 nm, the possible coatings to be extended to all (authorized or not prohibited) cosmetic ingredients, and the omission of the solubility specification [33].

18.5.3 Titanium dioxide and zinc oxide ecotoxicity studies

In the field of the regulatory toxicology, studies on the ecotoxicity of titanium dioxide and zinc oxide are performed with the aim to try to qualify and quantify the damage toward marine ecosystems and terrestrial organisms at different stages of development, taking into consideration the time, the frequency and the route of exposure, the concentration, if any accumulation phenomena occurs, and the ways that favor the uptake into the human organism from the environment. These data are available on the online research portals of the most important scientific organisms involved in the toxicological safety assessment of cosmetic raw materials, which are periodically updated in the light of scientific recent publications.

Among these scientific sources are found:

- CIR, Cosmetic Ingredient Review;
- TOXNET, Toxicology Data Network;
- ECHA, European Agency for Chemicals;
- EFSA, European Food Safety Authority;
- SCCS, Scientific Committee on Consumer Safety;
- NICNAS, National Industrial Chemicals Notification and Assessment Scheme.

Sometimes, information related to ecotoxicity are reported in the SDS.

Even though nanoparticles are usually covered with an inert coating layer to avoid its photoreactivity, this layer is quickly dissolved in aquatic environments after being released from sunscreens [13], enabling the production of ROS [35].

18.6 Conclusions

A daily sun-care products use is recommended to prevent sunlight-induced skin cancer, sunburn, photoaging, and skin wrinkles. To make consumers aware of the need for constant use of sunscreens category products, it is recommended to start educational campaigns with the aim to learn the correct behaviors during sun exposure, to know the light spectrum, in particular, the UV and IR radiations, the consequent damages due to an irrational exposure, to know the protection devices and the sunscreens products available on the market, suitable for every personal skin needs, so targeted for each type of skin, ethnic origin, and age. Mobile apps are also created that allow to recommend the most suitable cosmetic product for every skin type, as well as, based on the input of latitude data, define the amount of sun-care product need to be applied and the frequency of its application. Advances in the field of nanotechnology helped enlightening innovative formulations of UV nanosunscreens like nano-TiO_2 and nano-ZnO in sun-care products, to improve their effectiveness in photoprotection, safety, and consumers compliance. This chapter has focused the attention on the definition of the sunscreens principally from a regulatory point of view, their mechanisms of action in photoprotection, the SPF, and the UVA protection evaluation through the ISO methods adopted with the aim to formulate broad-spectrum sun-care products. Nanomaterials use in cosmetics and as nanosunscreens, in particular, is explored next to their regulatory status and their toxicological safety assessment, in order to guarantee the quality, the safety, and the effectiveness of any sun-care marketed product.

References

[1] Manuale del cosmetologo II edizione. Tecniche nuove.
[2] Baki G, Alexander KS. Introduction to cosmetic formulation and technology. John Wiley & Sons, Inc; 2015.
[3] Mitchnick MA, Fairhurst D, Pinnell SR. Microfine zinc oxide (Z-Cote) as a photostable UVA/UVB sunblock agent. J Am Acad Dermatol 1999;40(1):85–90.

[4] Shi L, Shan J, Ju Y, Aikens P, Prud'homme RK. Nanoparticles as delivery vehicles for sunscreen agents. Colloids Surf A Physicochem Eng Asp 2012;396:122–9.

[5] Monografia solari. Kosmetica. Available from: <https://cdn2.hubspot.net/hubfs/515303/Campagne%20Kosmetica/KOS_2019_005_INT@solari.pdf?__hstc = 221521095.9fb3efca9de6a0fe6a6ed08f45c5413a.1543489977568.1575476452024.1575534625753.59&__hssc = 221521095.2.1575534625753>

[6] Bovero A. Dermocosmetologia. Dall'inestetismo al trattamento cosmetico. Tecniche nuove.

[7] Morganti P. Use and potential of nanotechnology in cosmetic dermatology. Clin Cosmet Invest Dermatol 2010;3:5–13.

[8] Epstein HA, Kielbassa A. Bio-nanotechnology: a revolution in food, biomedical and health sciences. In: Bagchi D, Bagchi M, Moriyama H, Shahidi F, editors. Chapter 23 Nanotechnology in cosmetic products. First Edition John Wiley & Sons, Ltd; 2013.

[9] Smijs TG, Pavel S. Titanium dioxide and zinc oxide nanoparticles in sunscreens: focus on their safety and effectiveness. Nanotechnol Sci Appl 2011;4:95–112.

[10] Wakefield G, Green M, Lipscomb S, Flutter B. Modified titania nanomaterials for sunscreen applications—reducing free radical generation and DNA damage. Mater Sci Technol 2004;20(8):985–8.

[11] Draelos ZD, editor. Cosmetic dermatology: products and procedures. Second Edition John Wiley & Sons, Ltd; 2016.

[12] Regulation (EC) No 1223/2009 of the European Parliament and of the Council of 30 November 2009 on cosmetic products.

[13] Sánchez-Quiles D, Tovar-Sánchez A. Are sunscreens a new environmental risk associated with coastal tourism? Environ Int 2015;83:158–70.

[14] Jose J, Netto G. Role of solid lipid nanoparticles as photoprotective agents in cosmetics. J Cosmet Dermatol 2019;18(1):315–21.

[15] Sambandan DR, Ratner D. Sunscreen: an overview and update. J Am Acad Dermatol 2011;64(4):748–58.

[16] Sargent EV, Travers JB. Examining the differences in current regulatory processes for sunscreens and proposed safety assessment paradigm. Regul Toxicol Pharmacol 2016;79:125–41.

[17] Serpone N, Dondi D, Albini A. Inorganic and organic UV filters: their role and efficacy in sunscreens and suncare products. Inorg Chim Acta 2007;360(3):794–802.

[18] Hansen SF, Baun A. European regulation affecting nanomaterials - review of limitations and future recommendations. Dose Response 2012;10:364–83.

[19] Commission Recommendation no. 265 of 22 September 2006 on the effectiveness of sun protection products and their indications [notified under the number C (2006) 4089] (2006/647/EC).

[20] Tema Farmacia. Available from: <https://pixelbook.tecnichenuove.com/newsstand/temafarmacia/viewer/guest/com.tecnichenuove.temafarmacia.tef.2017.106/>; June 2017; p. 80–83.

[21] Chaudhri N, Soni GC, Prajapati SK. Nanotechnology: an advance tool for nano-cosmetics preparation. Int J Pharm Res Rev 2015;4(4):28–40.

[22] Yapar EA, Inal Ö. Nanomaterials and cosmetics. J Pharm Istanb Univ 2012;42(1):43–70.

[23] The SCCS notes of guidance for the testing of cosmetic ingredients and their safety evaluation 10th revision. SCCS/1602/18 Final version.

[24] Newman MD, Stotland M, Ellis JI. The safety of nanosized particles in titanium dioxide and zinc oxide based sunscreens. J Am Acad Dermatol 2009;61:685–92.

[25] Lu PJ, Fang SW, Cheng WL, Huang SC, Huang MC, Cheng HF. Characterization of titanium dioxide and zinc oxide nanoparticles in sunscreen powder by comparing different measurement methods. J Food Drug Anal 2018;26(3):1192–200.

[26] Katz LM, Dewan K, Bronaugh RL. Nanotechnology in cosmetics. Food Chem Toxicol 2015;85:127–37.

[27] Feng W, Wang B, Zhao Y. Chapter 13. Nanotoxicity of metal oxide nanoparticles in vivo. In: Sahu SC, Casciano DA, editors. Nanotoxicity. John Wiley & Sons, Ltd; 2009.

[28] Zhang P. Ecotoxicity analyses of nanomaterials. In: Zhao Y, Zhang Z, Feng W, editors. Toxicology of nanomaterials. First Edition Wiley; 2016.

[29] Barber DS, Denslow ND, Griffitt RJ, Martyniuk CJ. Sources, fate and effects of engineered nanomaterials in the aquatic environment. In: Sahu SC, Casciano DA, editors. Nanotoxicity: from in vivo and in vitro models to health risks. John Wiley & Sons, Ltd; 2009.

[30] Opinion on titanium dioxide (nano form) coated with cetyl phosphate, manganese dioxide or triethoxycaprylylsilane as UV-filter in dermally applied cosmetic. SCCS/1580/16 Final version of 7 March 2017. Corrigendum of 22 June 2018.
[31] Opinion on titanium dioxide (nano form) as UV-filter in sprays. SCCS/1583/17 final.
[32] Opinion on zinc oxide (nano form). COLIPA S 76. SCCS/1489/12 Revision of 11 December 2012.
[33] Addendum to the opinion on zinc oxide (nano form). SCCS/1518/13 Revision of 22 April 2014.
[34] Commission Regulation (EU) 2016/621 of 21 April 2016 amending Annex VI to Regulation (EC) No 1223/2009 of the European Parliament and of the Council on cosmetic products. 22.4.2016 L 106/4 Official Journal of the European Union EN.
[35] Sánchez-Quiles D, Tovar-Sánchez A. Sunscreens as a source of hydrogen peroxide production in coastal waters. Environ Sci Technol. 2014;48(16):9037–42.

CHAPTER

19

Nanomaterials for lip and nail cares applications

Carla Aparecida Pedriali Moraes and Aline Rocha Vieira

Laboratório de Pesquisa, Tecnologia em Cosméticos, FATEC Diadema—Luigi Papaiz, São Paulo, Brazil

19.1 Introduction

Nanomaterials are materials of metric size ranging from 1 to 100 nm or that have been processed at different scales to reach this size; this is the case of nanoparticles, nanotubes, nanocapsules, and nanofibers. Nanomaterials can also possess either internal or external nanometric structures and can be made from natural or synthetic, organic or inorganic materials. Nanomaterials can have a wide range of applications [1,2].

In a dimensional nanoscale, nanomaterials can be classified as (1) zero—spheres and clusters; (2) one—multilayer covering; (3) two—tubes, fibers, and plaques; (4) three—particles and quantum dots. Nanomaterials are called so due to at least one of their dimensions being in a nanoscale [3–5]. These materials can exist in shapes that are either single of fused, as aggregates or as agglomerates, and can be shaped as spheres, tubes, or be completely irregular. The unique optic, thermal, electric, and magnetic properties of these materials make them usable in a wide range of fields [1,4].

In this chapter, the use of nanomaterials will be approached in lipsticks and nail polishes, which are formulations that are designed to color or protect lips and nails, respectively. To do so, specific characteristics are desirable: ability to form a colored and homogeneous film on applying surface, innocuity, easiness of applying, good adherence (balance between durability and removal), quickness of drying, and protection against external agents; all of these amount to good rheological properties. The most used cosmetic forms or lipsticks are powders, liquids, or molds; for nail polishes, the most used forms are either liquid or gel. External protection agents can also be used in formulations, such as vitamins (E,C, B5, allantoin, hydrating agents, sunscreens, and ceramides) [6–8].

Nanocosmetics
DOI: https://doi.org/10.1016/B978-0-12-822286-7.00029-2

19.2 Nail care: from conventional nail polish to nowadays innovations

Nowadays, there are several nail polish products available in the market way beyond the conventional transparent or colored products, such as topcoats, gel-based products activated by LED or UV lights, and nail strips [7,8]. Nail strips are adhesives made from flexible plastics (vinyl) to which nail polish is applied, being combined or not with a gel-based topcoat; these can also be UV-curable, and thus being processed in a manner that causes monomers such as methyl methacrylate and butyl methacrylate to polymerizate due to the action of free radicals when photoinitiator molecules are exposed to UV light [9,10]. Upon fiber nails, glass fiber nails or plastic films are applied; methacrylate glue is used to keep the fibers firm and allow them to expand in extension [10,11].

Conventional compounds usually found in nail polish formulations include film making compounds (nitrocellulose, methacrylate polymers), thermoplastic resins (p-toluene sulfonamide), plastic agents (di-butylphtalate and camphor), solvents (acetates, alcohols, and toluene), and pigment dispersers (titanium dioxide and iron oxides) [10,12].

Nowadays, there are nail polishes which react to UV light, as their formulations contain curable methacrylate and acrylate oligomers and monomers which, due to the action of a photoinitiator compound, become polymerized forming a thin, colored layer. These products are used either alone or in association with gels used for nail sculpting, such as artificial nails. Conventional nail polish formulations, on the other hand, employ only resins and solvents [12,13].

Artificial nails are used as "extensions" of natural nails. Techniques used for applying these nails involve the use of premade nails and nails sculpted with the aid of liquid formulations containing monomers (ethyl methacrylate 60%–95%) which become curated under UV light. In this specific condition, premade plastic nails and cyanoacrylate glue are used; this glue is made of ethyl cyanoacrylate and polymethylmethacrylate (reticule forming agents used at 3%–5%), hydroquinone (inhibitor used at 100–200 ppm), and stabilizing agents (either any organic sulfonic acid or dimethyl olamine used at 0.75%–1.25%) [7,10,12].

While liquid formulations are the most common, powder formulations which associate polymethyl or ethyl methyl acrylate (50–80 μm) with benzoyl peroxide (1%–2%) also exist [7,10,12].

Polymerization occurs due to reaction of the photoinitiator with monomers and with reticule forming agents, leading to the formation of a sculpted gel; this process depends on exposure to low-intensity UVA light (325–435 nm). Promoters containing mixtures of oligomers, monomer hydroxyls, and carboxyl acid, among others, are used to increase formulation compatibility. Isopropyl acid can also be used on nails to remove oiliness and humidity [10,11].

Base coats do not possess any pigments, are transparent, and are used upon nails to make their surface uniform so they can be covered with regular pigmented nail polish. Topcoats are applied on the final nail polish to confer it more brightness and prevent premature cracking. Some topcoat formulations are treated under low-intensity UV light and their texture differs from that of conventional formulations. Nanotechnology has reduced the use of base coats as nanomaterials have improved adherence and increased the time nail polishes remain on nails [10,11].

The effects topcoat formulations can generate when applied on conventional nail polishes (such as crackling effects) can be considered novel. These effects are due to the fast and uneven drying of the solvent externally to internally which alters surface tension and viscosity of the topcoat layer. Such crackling effect is due to the addition of ethyl alcohol to the topcoat formulation and is more noticeable when the colors of the nail polish and of the top coat are different [10].

There are also nail polishes found in the market which are characterized by the "neon" effect; this happens when quantum dots emit fluorescent light when excited under either visible or UV light. Quantum dots are nanostructured semiconductors such as MgS, ZnS, ZnSe, BaS, and CaS, among others. Fluorescent nanoparticles should be sized between 1.5 and 50 nm. If encapsulated, the nucleus should be sized between 1.5 and 10 nm, and monolayer thickness should be of around 10 layers. Commonly found nail polishes in the average market include Neon Milani Nail lacquer and Neon Color Trend Avon [4,14].

Magnetic nail polishes are formulations which contain liquid-crystal pigments responsible for dispersing light. These pigments should be concentrated in optically active layers (which range from 1 to 100 μm) between two other layers; pigments include usually iron powders [Fe_2O_3, Fe_3O_4, FeOOH, FeO, $Fe(OH)_3$], which can be disposed in a nonsymmetrical layout when in a layer ranging from 50 to 800 nm. This layer must also contain multilamellar electrochromic structures made out of polymers such as polypyrrole or polythiophene [15].

The nail polish must be applied when still humid with the aid of a magnet to "guide" the pigmented particles alongside a magnetic field in a contrasting pattern. This magnet is applied on nail keratin; it has electrosensible batteries and electrodes made of indium tin oxide layers of thickness varying from 50 to 1000 nm [15,16].

Color-changing nail polishes rely on thermochromic or photochromatic properties of chromophore compounds which are inserted in a polymeric matrix, then conferring it to greater thermal stability and aiding it keeping color during photochemical reactions. Most commonly used chromophores with this purpose are spiropyrans, spirooxazines, and diarylethene. These molecules possess photo, thermo, and mecanochromic properties and can be photoactivated by radiation at 300–400 nm (UV light) or 400–700 nm (visible light); even normal system temperature can induce photoactivation. Spiropyrans, when exposed to UV radiation (absorbance between 280 and 400 nm), undergo processes which cleave C–O bindings leading to cis-trans (C=C) isomerization forming the colored merocyanine, which is a monomer of slow thermal regeneration (milliseconds). Merocyanine absorbance occurs at a visible light wavelength of 554 nm [17].

In the work by Su et al. [17], a mixture of spiropyran was prepared by polymerizing an emulsion resulting in particles sized between 30 and 60 nm. This dispersion of nanoparticles resulted in better photoreversibility and photostability in comparison to molecules in aqueous solutions [16,17]. There is a patent by Chromatec Technologies Inc. which describes microencapsulation of thermochromic pigments using solvents of higher molecular weight than what is commonly found in nail polishes, such as methyl isobutyl ketone, 2-butoxy ethanol, and n-butanol; this process also leads to adding stabilizing additives, such as zirconium ethyl hexanoate and melamine resin, maintaining final pH between 6.5 and 7.5 [18,19]. Some commercial thermochromic formulations include In The Mood and Orly topcoat for use upon gel nail polishes and Ruby Wing (Solar Active Technology) and Del Sol Color Change, both change their colors based on UV incidence upon nail polish.

Another important property of nail polish films is their permeability to oxygen and water vapor. Water-based polymers on oil/water emulsions (Syntran 5620 CG) and transparent hydrogel-based hybrid polymers associated with polypeptides are sustainable, nonanimal derived, technologies (Silktran 5621). Both technologies were born from the partnership between AMSilk GmbH and Interpolymer. Sytran (INCI Styrene/Acrylates/Ammonium Methacrylate Copolymer) is an acrylate film-forming emulsion and Silktran (INCI: Styrene/Acrylates/Ammonium Methacrylate Copolymer and Polypeptide-1) is a hybrid polymer registered by The Vegan Society. Both bases are of long duration, water-proof, fast drying, and permeable to oxygen, thus considered part of breathable technologies [20,21].

19.3 Lip care: traditional formulations and innovations

Lipsticks are mixtures of "structure" compounds (such as waxes and polymers), oils, and pigments used at different concentrations; these confer the formulation color, brightness, texture, and lip protection. In the average market, there are several types of lipsticks: bullets and liquid lipsticks, glosses, lip tints, lip pencils, and lip balms. These can be sold in tube packages, cans, and used in different forms, such as with pencils, roll-on, or even with fingers [10,12,22,23].

There are different lipstick products with different functions. They can play roles on hydration, volume, opaqueness (matte effect), brightness, long duration, and transfer resistance [23].

Nanotechnology not only is present in pigments or fillers, but also plays a role in cosmeceutical lipsticks, as it promotes controlled release of cosmetic active compounds while protecting them from external agents while inside nanocarriers, such as liposomes. Liposomes are vesicle structures of aqueous core revested by a lipid bilayer of variable size (ranging from 20 nm up to micrometric scales) and are ideal for transport of either hydrophobic or hydrophilic compounds. Niosomes are also types of nanocarriers; these are bilayered vesicles made of anionic surfactants, containing or not cholesterol in their composition. These nanocosmeceuticals can be fused into lipsticks or glosses increasing lip volume, delineating and filling lip wrinkles, and generally improving lip hydration. One such example of this type of product is Fillderma Lip Volumizer (Sesderma), which contains low molecular weight hyaluronic acid (nanosomes), mimetic peptides, niacinamide, and collagen [9,24]. Another example is Primordiale Optimum Lip (Lancôme), which contains vitamin E-filled nanocapsules alongside alpha-hydroxyacids which diminish the fine lines of the lips increasing lipstick adherence [25].

Nanotechnology also improves lipstick properties such as duration, transfer resistance, breathability, film brightness, and softness; it is very difficult for all of these characteristics to be equally optimal in a formulation. Avon Products Inv registered a patent in November 2012 which describes the synthesis of a nanocompound (particle size between 0.1 and 100 nm) of high elasticity by reacting graftable polyhedral oligo silsequioaxanes (POSS) or extended polyhedral oligo silsequioaxane (EPOSS) with a functional polymer alongside a cross-linking agent [26]. There are patents published in 2017 and 2019 for nail polishes with this POSS nanocomposite by Criative Nail Design and Revlon to improve performance in terms of durability and film brightness [27,28].

Nanoscale particles used in lipsticks are also vital for reducing the visual aspect of wrinkles, as these lipsticks have skin-like tones (nude effect). This improvement in optical performance is due to the revesting ability of iron oxides and titanium dioxides which form microplaques [23].

To obtain proper nanotechnological lipsticks, a mixture of structure agents or crystalline/amorphous waxes should be used; these allow for the formation of smaller internal crystals after the final product is cooled, leading to a better internal arrangement, which improves lip adherence. Waxes can be mixed at a concentration of around 20%, being of vegetal origin, such as candelilla and carnauba which confer brightness and toughness, of mineral origin, such as ozokerite, which confers adhesion, or of synthetic origin, such as polyethylene, compatible with silicones and prevents oil exudation due to being an agglutinizing agent, increasing formulation toughness and ease of applying the lipstick [10,12,22–24].

Some copolymers such as diisostearyl malate (and) ethylenediamine/hydrogenated dimer dilinoleate copolymer bis-DI-C14-18 alkyl amide and dibutyl lauroyl glutamine mixtures function as oil phase thickeners. These compounds form nanofibers in oils and aids solid particles to remain in suspension, such as pigments, producing solid oil-gel lipsticks at a concentration from 0.2% to 5.0%. This is a patent from Ajinomoto [23,29].

Humidifying or pigment disperser agents are the major compounds of lipsticks which influence applying, color, spreadability, and brightness. Humidifying agents of high molecular weight, viscous, and of a refraction index higher than 1.49 are the most recommended, such as mineral oil, ricin oil, vegetal oils, pure lanoline, or synthetic lanoline (bis-diglyceryl polyacyladipate-2), karite butter, hydrogenated polyisobutene, phenylthrimeticone, and phenyl isopropyl dimethicone. These compounds prevent bleeding and feathering or wicking of the final product into skin creases around lips. Also, these compounds also form homogeneous thin films upon lips evidencing a brighter color [10,12,22].

Specific lipstick colors are achieved by using organic pigments, such as Red 7 and Red 28, or inorganic pigments, such as red.

Traditionally shaped lipsticks are composed of 30%–70% w/w of emollients; 15%–25% w/w of structure agents (up to five different compounds are usually mixed); 2%–20% w/w of pigments; 0%–15% w/w of pearl agents; up to 5% of matte agents; up to 0.5% w/w of fragrances/essences and from 0.2% to 0.5% w/w of antioxidants. Filler materials can be used up to 2%, and can be nylon 12, silica, BiOCl, lauroyl lysine, and copolymer acrylate [10,12,22].

Long duration and transfer-resistant lipsticks usually carry in their formulations similar amounts of compounds, but their classes do tend to differ. For example, resin from Shorea robusta can be used as an alternative emollient to poly-isobutene or to poly-vinylpyrrolidone; isododecane, cyclopentasiloxane, and dimethicone are also alternatives. Structure agents can include silicone polymers, such as dilauroyl trimethylolpropane siloxyl silicate, diisostearoyl trimethylolpropane siloxyl silicate, trimethylsiloxysilicate, polymethylsilsesquioxane (PMSQ), and bis-hydroxypropyl dimethicone/SMDI copolymer. This interaction between emollients and structure agents increases formulation adherence upon skin or mucosae, forming a film which maintains color for a prolonged time period. These lipsticks are usually liquids of average viscosity [23,24,30].

19.4 Other nanotechnological innovations used in nail and lip care formulations

Nanocolorants are a class of nanocompounds which combine the best properties of conventional pigments with multilayer coatings, leading to alterations in optical properties; formulations can acquire satin, pearly, metallic, or crystalline appearances. These compounds can modify lighting effects (which leads to color visualization based on modified angles) and brightness properties. Sensorial properties can also be modified and adherence to the applied surface usually increases. These nanocolorants usually confer bright colors without the need for heavy metals conventionally used, such as chromium oxide and ferric ferrocyanide [31,32]. Inorganic and organic nanopigments are insoluble, inert, range from 100 to 200 nm in size, and can be associate with polymers [33] and ceramics [34].

The development of pigments using multilayered structures upon a substrate has seen plenty of advances in the last 50 years. In the last 30 years, technology on nanostructure materials of varying size, morphology, distribution, and dimensions has received significant funding [4,6,31,32]. These properties of these materials allow for the formulation to be better suited to skin, lip, and nails, modifying

- Sensorial properties, making the formulation more pleasing;
- Affinity with the applying and thus improving product adherence;
- Duration of the product on the applying area;
- Spreadability of the product on the applying area, making it more uniform;
- Optical properties of the formulation (refraction index, absorption, transmittance) and its visual effects;
- Filling of skin and lip wrinkles;
- Stability under direct light;
- Ability for other compounds to be better incorporated into the formulation.

According to Anselmann [3], the most suited materials for coating with inorganic nanoparticles in industry are majorly synthetic platelets and spherical silica particles. Natural and synthetic mica, silica (SiO_2), calcium and sodium borosilicate, and alumina (Al_2O_3) can also be used. These raw materials function as "substrates" for coating using inorganic nanoparticles or high refraction metallic oxides; the coating is made in multilayers (TiO_2, Fe_2O_3, Ba_2SO_4), conferring pearly or metallic brightness to eye shadows, lipsticks, and nail polish.

As for products of metallic effects in the average market, Metasheen vacuum-metallized aluminum pigments (VMPs), from Basf, are a good example. These VMPs are aluminum agglomerates of smooth and plain surface (lamellar structure, nonfoliated, and disperse), which confers reflexiveness to the final product. Due to the structures not being foliated, colored metallic effects can be achieved [35].

Transparent or pearly pigments can generate surface optical effects, such as pearlescence, iridescence, color travel, interfering metallic color, and metallic brightness. These effects rely on light reflection, refraction, diffraction, absorbance, or transmittance from the translucent layers that coat the substrate [36–38].

Opaque pigments (bronze, copper, and Russet) which generate metallic brightness are used upon mica and borosilicate substrates coated with a thick layer of iron oxides, thus generating higher refraction indexes. The final visual effect is a high light reflection of lustrous appearance [36–38].

On the other hand, thinner (mostly ranging from 5 to 150 μm) and more numerous and complex coating layers (layers alternating mica, silica, borosilicate, titanium dioxide, and iron oxide) lead to more pronounced color of the final product (iron oxide better absorbs color); this effect relies on combining light interference with brightness (sparkling) and shining (high brightness and transparence) effects. As such, colors per particle size include copper (20 μm), pink (80 μm), and blue (120 μm) [36,37].

All of the aforementioned characteristics are the major properties for substrate-based effect pigments used in cosmetics; natural mica, however, can be cleaved in thin flakes of average thickness ranging from 200 to 500 nm [38,39].

Combination of light interference and color absorption by iron oxides, when greater in the final layers, leads to pearlescence. When the same effect occurs in all layers, the combination of different refraction indexes leads to complementing colors of varying brightness, thus leading to iridescence. The selective reflection of light at different wavelengths caused by a high ratio among refraction indexes of the coating layers leads to color travel, or "chameleon" effect; this effect relies on the interaction between several thin coating layers which causes specular transmission, where light is transmitted at a certain wavelength which generates the same color when light is reflected (double light interference). In the common market, these effects are seen in formulations such as Reflecks Multidimensions (Basf), where sodium and calcium borosilicate coated with silica, titanium dioxide, and iron oxide is used (color travel of metallic effects, particle size ranging from 20 to 130 μm), and Reflecks Multireflections Soft Sparkle Pigments (Basf), where mica coated with silica and titanium dioxide is used (soft brightness, particle size ranging from 15 to 80 μm) [31,35,38,39].

In the cosmetic market, pigments which promote light interference are often used. For such, substrates are used as flakes or plaques made of sodium calcium borosilicate; this material is known as "flake glass" and can have from a single up to several coating layers of semitransparent metallic material where the refraction index is higher than 1.8. This coating can be composed of 65%–75% silica, or 2%–9% aluminum oxide, or 5%–12% calcium oxide, or 0.1%– 5% titanium dioxide [31,32,40].

Glass flakes or plaques have a very smooth surface and can be made in a manner their overall thickness is much more uniform than flakes or plaques made of mica substrates, which are traditionally used as substrates for obtaining pearly pigments. Pearly pigments made of glass flakes are brighter and of an intense color than those made of mica. These substrates can be prepared with an initial surface no thicker than 10 nm, and the core of the structure is called "nucleation"; the final substrate contains tin oxide or aluminum hydroxide oxide. Upon the nucleation core, there is adhesion of coating layers of metallic oxides. Titanium dioxide used for coatings can be shaped as rutile or anatase at thickness ranging from 500 to 1200 nm. Besides pearly effects, these coatings can also generate metallic or light interference effects, which rely on the number of layers, their material, and their thickness, which ranges from 1000 to 2200 nm [31,40].

Such "effect pigments" available in the market include products of the Moonshine line, from Glassflake Ltd., part of the company group Croda International Plc. In this product line, Astral effect Series made of sodium and calcium borosilicate (particle size ranging from 10 to 100 μm) coated with titanium dioxide at a thickness ranging from 500 to 1200 nm generates light interference of a single color. The product Colour Travel Series, on the other hand, is made of the same substrate (particle size ranging from 30 to 200 μm), but the coating is of amorphous silica and titanium dioxide of thickness ranging from 1210 to 2210 nm which also causes light interference, but three colors are generated instead of a single one; these pigments are named "color pigments of multicolor effect" [40].

Other examples include Colorona (Merck KGaA), where the formulation employs pigments of absorption effects which confers "depth" and "shiny brightness" (pearlescence). Other pigments which confer gilded appearance are also used in the formulations Colorona Bright Gold (10–60 μm) and Colorona Fine Gold MP-20 (5–25 μm), such as mica flakes coated with titanium dioxide and iron oxides [31,32,37,39].

Timiron formulations contain silvery white pigments, composed of pearly microparticles made of natural mica and coated with titanium dioxide or alumina; the particles are of 50 nm of size and can provide satin brightness (5–25 μm), pearly or metallic effects (10–60 μm) such as seen in Timiron Supersilk MP-1005, or can provide crystalline effects, such as seen in Timiron Diamon Cluster MP-149 (10–150 μm). Timiron Synwhite 40 contains in its base formulation plaques made of synthetic mica (fluoroflogopita). Timiron Glam Silver contains in its base formulation thin alumina flakes (Al_2O_3) 5–40 μm thick; the surface upon applying becomes highly reflective due to the presence of shiny silvery white pigments [31,32,37,39] (Fig. 19.1).

Products from the Xirona line use pigments of color travel or chameleon effect. The thickness of the titanium dioxide coating on both sides of the mica plaque ranges from 80 to 250 nm; this leads to light interference, as seen in Xirona Caribbean Blue and Xirona Volcanic Fire. The anatase form of titanium dioxide can be directly placed upon mica while the rutile form requires an intermediary phase of tin (SnO_2). Xirona Le Rouge is a pigment made of silica flakes of thickness ranging from 300 to 400 nm, coated with iron oxide, a pearl-red pigment which confers colors which interchange between red-yellow and blue-red; the formulation is also of increased stability when under the effects of light, water, or pH variations [31,32,37,39].

Products from the Ronastar line use highly colored and luminous pigments, of matte effect alongside crystal-bright effects, composed by aluminum and sodium borosilicate

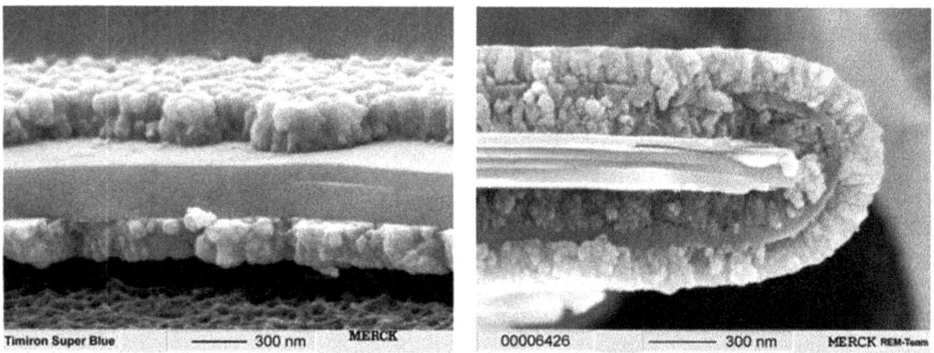

FIGURE 19.1 Pearlescent pigments (left: TiO_2 on synthetic mica, right: $TiO_2/SiO_2/TiO_2$ on mica). Source: *Reproduced with permission from Anselmann R. Nanoparticles and nanolayers in commercial applications. J Nanopart Res 2001;3:329–36 [3]. Copyright 2001 Springer Nature.*

flakes; this material is squamous and silvery-white, combining multicolor effects (particle size ranging from 10 to 100 μm or from 20 to 200 μm) with transparency. Such an example is the product Ronastar Golden Jewel (particle size ranging from 20 to 200 μm), which confers intense brightness in association with saturated gilded color [37,39,41].

Other companies, such as Kobo, possess lines of pigments made of mica flakes associated with iron oxides and titanium dioxides, which are absorption and interference pigments, conferring 3D effects to the formulations; examples of formulations containing these pigments include K-Ray (natural mica, ranging from 10 to 60 μm), K-Ray SM (synthetic mica, ranging from 5 to 1000 μm), and K-Ray G (calcium and aluminum borosilicate, ranging from 20 to 200 μm). This line of products not only utilizes special pigments; they also undergo hybrid surface treatment, which confers them high resistance to water, high spreadability (either in esthers, oils, silicones or in hydrocarbons), greater skin adherence and lip fixation up to 12 hours. The hybrid treatment confers hydrophobic and lipophilic properties to the pigments; it involves the use of titanate derivatives and dimethicone [INCI: Isopropyl Titanium Triisostearate (and) Triethoxysilylethyl Polydimethylsiloxyethyl Dimethicone] [42].

Overall, the resulting effect of using nanolayers on mica and bismuth oxychloride (BiOCl) plaques is obtaining pigments of pearly effects, improving the sensorial effects when applied on skin. Silica spheres have increased color intensity when coated with metallic oxides [43]. These pigments can also be composed of synthetic nanoparticles, such as BiOCl and SiO_2, and can be used as substrates for pearly pigments. BiOCl particles exist as monocrystalline plaques, either spherical or lamellar, dense, white, and of size ranging from 50 to 150 nm. These plaques are made from crystals of bismuth precursors, which, depending on reaction conditions, can be tetragonal bipyramidal, flake-shaped, or rose-shaped structures. These plaque structures confer to the formulation pearl-like metallic brightness due to light interference; these properties make formulations containing these structures very useful for nail polishes. The effect is due to the high refraction index (close to 2.15); the material can also be coated with titanium dioxide. Structure of microplaques also influences the visual effects of the formulation; for example, BiOCl microplaques of rose-like shape promote soft focus and matte effects (high light diffusion and lesser effect of specular light reflection). This is a very desirable characteristic for products such as eye shadows, nail polishes, lighting agents, glosses, and lipsticks [43].

As for synthetic silica, flakes are made under a screen-revesting process and can be of thickness ranging from 150 to 500 nm, forming a multilayer coating system which possesses higher optical activity in comparison to other pearly pigments [3,44].

Examples of inorganic particles used for pearly pigments include titanium dioxide, chromium oxide, and iron oxides. Certain organic particles can also be used as pigments. These pigments are made from the stacking of inorganic films of inorganic particles on plaque substrates [37,38,45,46]. Thickness of the layer usually varies from 60 to 240 nm, depending on the desired optical effect. The more metallic oxide layers of varying refraction indexes are stacked on mica plaques, the easier light interference effects take place, as this effect relies on dispersion of light in several angles [38,45,46].

Microspheres coated with nanoparticles acquire shapes which vary from granules to spheres (oblong). These structures are used in cosmetics to confer pleasurable touching sensations to the formulation, absorb oiliness (with the aid of pores), and diffuse light in several angles (soft focus or optical blurring effects) and thus hide skin or lip imperfections; for this last function, spherical microspheres of lower diameters are usually better. These structures are classified as functional fillers. Particles sized lesser than 15 μm confer soft and satin optical aspects to the formulation [45–47].

According to Delrieu and House, soft focus and optical blurring effects act as a sort of "fog," as the dispersion of light generated by different angles "hides" skin and lip imperfections to the naked eye. Ideal conditions for these effects to take place derive from reduced particle size, distribution of particle size, porosity, microsphere composition, and refraction index; these effects, when ideal, confer to the skin natural appearance masking its opaqueness [46–48].

If the refraction index of a homogeneous particle is the same as the medium it is in, there will be no light dispersion and the suspension will be transparent. However, when the difference between refraction indexes is considerable, the formulation becomes opaque. The refraction index changes when the formulation is applied to the skin due to interactions with surface lipids or due to transepidermal evaporation of water [48].

To eliminate transparency or opacity of a suspension, it is ideal to mix two or three types of microspheres composed of different materials, and thus with very distinct refraction indexes. Microsphere mixtures are of very advantageous use, as they optimize formulation development time. While microspheres composed of a single material are commonly used, microspheres composed of more than one material can improve sensorial touching properties and adherence to the skin of a formulation, as well as providing soft focus effects. Microspheres composed of different materials are made of particles coated with substances different from that of the particle core, which will undergo physical-chemical transformations. For example, PMSQ microspheres coated with silica confer to the core hydrophilic properties, allowing the pigment to better disperse in aqueous media. Depending on the formulation, microparticles composed of PMSQ can be coated with even smaller metallic oxide particles generating a silk-like texture to the final formulation [49].

Nanospheres, besides being fillers in cosmetic formulations, can also influence rheological properties of emulsions. Their size ranges from 50 to 1000 nm, but a material patented by LDP Merck KGaA named Ronasphere (light diffusor pigment, refraction index ranging from 1.45 to 1.65) has a particle size of less than 50 μm. This material is a silica microsphere (80.5%–84.5% w/t) coated with nanoparticles of sizes ranging from 10 to 50 nm, which are composed of titanium dioxide (14.4%–18.5% w/t) and iron oxides (0.1%–1% w/t); the ending result is the formation of highly colored microspheres and pigments of high light diffusion. These pigments can be used in lipsticks and nail polishes as pigmentary dispersions of 0.1%–70%, usually used in emulsions aimed at masking wrinkles. The association of titanium dioxide and iron oxide particles with silica coating causes light to be reflected by wrinkles, increasing their reflection in comparison to the normal skin surface, being very useful for lipsticks [3,38] (Fig. 19.2).

Flake-shaped mica plaques can be used as makeup fillers at sizes varying from 5 to 150 μm and thickness of 500 nm. They can be processed when humid and form agglomerates, causing their surfaces to become very smooth. Emulsion cosmetic formulations can have their sensorial and optical properties improved [44,50].

Mica plaques can also be modified by coating with inorganic nanoparticles. The process of depositing nanoparticles upon a substrate varies according to the absorption or pearly pigments used and according to the type of coating desired as well [39,51]. Examples of

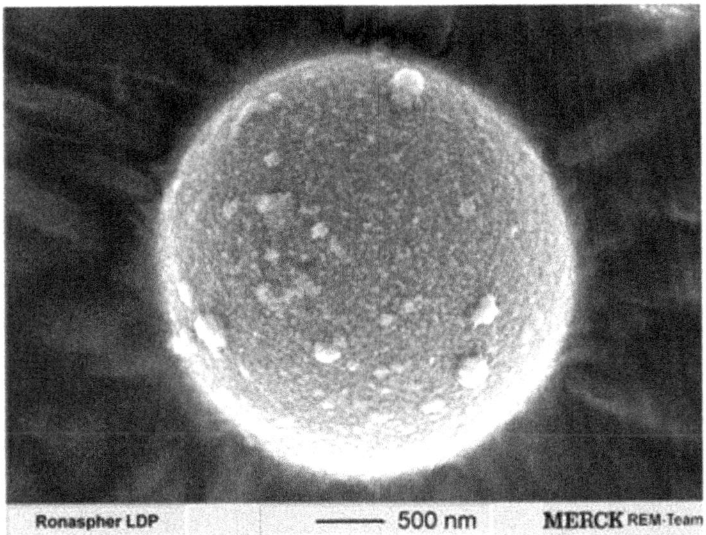

FIGURE 19.2 Ronasphere LDP. Source: *Reproduced with permission from Anselmann R. Nanoparticles and nanolayers in commercial applications. J Nanopart Res 2001;3:329–36 [3]. Copyright 2001 Springer Nature.*

inorganic nanoparticles used as absorption pigments include titanium dioxide, barium sulfate, and iron oxides. The size of primary nanoparticles varies from 50 to 150 nm and can be used in formulations such as eye shadows and iron oxides. Inorganic nanoparticles dispersed on flake surface can confer to the cosmetic formulation transparency and the ability to mask wrinkles and lip/skin imperfections [38] (Fig. 19.3).

There are studies which explore the disposition and the characteristics of coating layers on mica or silica substrates. Flake-shaped substrates of a circular-factor from 1.2 to 2.0 can have up to eight coating layers, and these can also be mixed with different nanoparticles of high (RI higher than 1.8 and thickness ranging from 40 to 350 nm) and low (RI lower than 1.8 and thickness ranging from 10 to 80 nm) refraction index, the core of the structure usually containing the particles of higher refraction index. The total particle size ranges from 5 to 60 μm and its total thickness ranges from 0.2 to 0.6 μm [37,39].

Titanium dioxide is polycrystalline and has a squamous form, shaped in very thin flakes, useful for providing transparency effects. Nanoparticles of titanium dioxide are sized from 200 to 400 nm and their surface can be modified by organic molecules and metallic oxides [52], used to confer pigments hardness. Also, titanium dioxide particles interact with light at wavelengths not seen by the naked eye, making them transparent [53,54]. The German brand Misslyn employs nanoparticles of titanium dioxide in some of its topcoat and liquid and gel nail polish formulations [55].

Carbon black is a carbon-based substance, similar in size to fullerene (around 1 nm). Among plenty of applications, carbon black is used as a catalyst, absorbing agent, pigment, conduit, and can modify thermal, optical, and mechanical properties of the formulation it is used in [56]. In the cosmetic market, there are nail polishes containing carbon black used as a nanothin spherical powder (0.4–12 μm), conferring to the formulation intense color and high resistance to UV light degradation. One example of such product is Models Own Bluebelle [57].

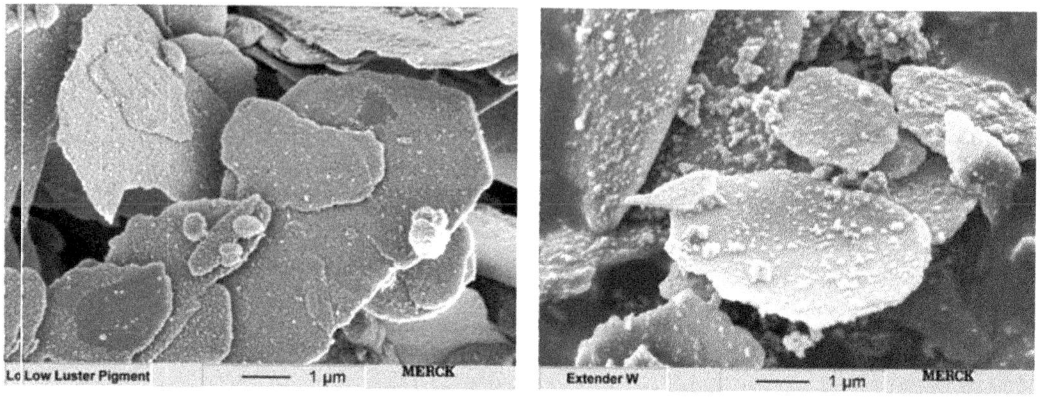

FIGURE 19.3 Mica platelets coated with TiO_2 (absorption pigments). Source: *Reproduced with permission from Anselmann R. Nanoparticles and nanolayers in commercial applications. J Nanopart Res 2001;3:329–36 [3]. Copyright 2001 Springer Nature.*

The interaction of carbon black with grafted surface polymers makes for an efficient UV filter [58,59]. Carbon black nanoparticles are sized of 150 nm and are stable in saline solutions and can be easily dispersed [59]. When in formulations, it is named Cl77266 Black 2 (nano). Misslyn is one of the companies which most often use this pigment [55]. Titanium dioxide and carbon black are described in a report by the European Union in 2018 which lists and characterizes nanopigments [60].

Ceramic pigment nanoparticles sized less than 5.0 nm can better reflect light, cover surfaces, and be dispersed, allowing for homogeneous mixtures to be more easily obtained; not only so, but duration of adherence to nail also increases [50]. In the market, nail polishes containing ceramic particles associated with polymeric nanoparticles are patented by Artdeco under commercial name Ceramic Nail lacquer [61].

Clay materials can be divided into two major types: kaolinite (1:1), composed of silica and alumina, and phyllosilicate (2:1), composed of a layer of alumina placed between two layers of silica called smectite. Both types of clay materials are subdivided into groups: kaolinite, smectite, chlorite, illite, and vermiculite. Smectites are the most commonly used clay materials in cosmetic formulations (hectorite, bentonite, and montmorillonite, among others) [62–65].

Clay particles can be modified by physical methods which do not alter their overall structure, leading only to adsorption of other compounds on their surface, but improving viscosity and the stability of the formulation systems where they are used. Chemical methods lead to interactions of the clay surface with polymers and anionic/cationic functional groups, leading to the formation of organoclays. These nanometric compounds improve clay materials as thermal barriers and their mechanical and optical properties in a polymer/clay system used in nail polishes [63,64].

Smectites are more commonly used for the preparation of nanoclays, usually modified by chemical methods which form conventional nanomaterials, nanocomposites intercalating polymer and clay and nanocomposites of exfoliated polymer and clay; this structure of nanocomposites has better dispersion and homogeneity properties in comparison to clay [63,64].

Montmorillonite of the smectite group is 0.96 nm thick and easily reacts under modifying processes, which allows for exfoliation and stacking of polymers. When modified by quaternary ammonium surfactants, it is called cloisite. Cloisites are of different structure than montmorillonites and generally improve overall properties of the systems they are used in [64–67].

Nanoclays modify rheological properties, as described by the patent by Coty Inc., which describes the use of modified bentonite (Cloisite15A) [68]. The patent describes a surface treatment with a cationic emulsifier, making the nanoclay more hydrophilic. A formulation containing clay nanoparticles disposed in two layers forming a 70–150 nm wide, 1 Å thick plaque is also described [65,68].

Nanoclays can also influence viscosity, elasticity, and stability of suspensions containing pigment particles, as demonstrated by Grigale-Sorocina and Birks [67], which assessed hectorite and bentonite in a nail polish formulation forming a film by UV polymerization.

References

[1] Behera A, Mohapatra SS, Verma DK. Nanomaterials: fundamental principle and applications. In: Verma DK, Goya MR, Suleria HAR, editors. Nanotechnology and nanomaterial applications in food, health, and biomedical sciences. 1st ed. Palm Bay, FL: Apple Academic Press Inc; 2020. p. 168–94.

[2] SCCS (Scientific Committee on Consumer Safety). Guidance on the Safety Assessment of Nanomaterials in Cosmetics, 30–31 October 2019, SCCS/1611/19.

[3] Anselmann R. Nanoparticles and nanolayers in commercial applications. J Nanopart Res 2001;3:329–36.

[4] Mordorski B, Landriscina A, Friedman A. An overview of nanomaterials in dermatology. In: Hamblin M, Avci P, Prow T, editors. Nanoscience in dermatology. Cambridge: Academic Press; 2016. p. 31–46.

[5] Sharma P, Bhargava M. Applications and characteristics of nanomaterials in industrial environment. Int J Civ Struct Environ Infrastruct Eng Res Dev 2013;3(4):63–72.

[6] Fukui H. Development of new cosmetics based on nanoparticles. In: Naito M, Yokoyama T, Hosokawa, Nogi K, editors. Nanoparticle technology handbook. 3rd ed. 2018. p. 399–405.

[7] Koichi R, Tranchant JF. Decorative products. In: Barel AO, Paye M, Maibach H, editors. Handbook of cosmetic science and technology. 4th ed. CRC Press Taylor & Francis Group; 2014. p. 439–51.

[8] Chaudhri N, Soni GC, Prajapati SK. Nanotechnology: an advance tool for nano-cosmetics preparation. Int J Pharm Sci Rev Res 2015;4(4):28–40.

[9] Kaul S, Gulati N, Verma D, Mukherjee S, Nagaich U. Role of nanotechnology in cosmeceuticals: a review of recent advances. J Pharm 2018;2018:1–19.

[10] André J, Baran R. Nail cosmetics: handle of skin care. In: Barel AO, Paye M, Maibach HI, editors. Handbook of cosmetic science and technology. 3rd ed. Informa Healthcare; 2009. p. 745–67.

[11] Iorizzo M, Piraccini B, Tosti M. Nail cosmetics in nail disorders. J Cosmet Dermatol 2007;6(1):53–8.

[12] Giancola G, Schlossman ML. Decorative cosmetics. In: Sivamani RK, Jagdeo JR, Elsner P, Maibach HI, editors. Cosmeceuticals and active cosmetics. 3rd ed. CRC Press Taylor & Francis Group; 2016. p. 191–220.

[13] Rich P, Kwak HSR. Nail physiology and grooming. In: Draelos ZD, editor. Cosmetic dermatology: products and procedures. 1st ed. Wiley-Blackwell; 2010. p. 197–221.

[14] Dubertret L, Dubertret B, inventors. Cosmetic composition comprising fluorescent nanoparticles as pigments. US patent 20060165621A1, 2006 July 27.

[15] Hollman AM, Schottland P, inventors. Magnetic pigments and process of enhancing magnetic properties. US patent 20120261606A1, 2012 October 18.

[16] Kergosien G, Thevenet L, inventors; L'OREAL, assignee. Cosmetic article including an electrochromic multilayer structure. US patent 2012/0152273A1, 2012 June 21.

[17] Su J, Chen J, Eng F, Chen Q, Wu S, Tong Z. Synthesis and photochromic property of nanoparticles with spiropyran moieties via one-step miniemulsion polymerization. Polym Bull 2008;61:425–34.

[18] Small LD, Highberger G, inventors; Chromatec Technologies Inc, assignee. Thermochromic ink formulation, nail lacquer and methods of use. US patent 5591255A, 1997 January 07.

[19] Owen TJ, Clayton TS, Siske BM, Roberts CW, Tomovich V, inventors; Chromatic Technologies Inc, assignee. Stabilizing additives for thermochromic pigments.US patent 9216397B2, 2014 July 03.
[20] Piotrowska AB. Comparison of the properties of nail polish bases depending on the film-forming ingredient used [Thesis]; 2019.
[21] Schlay S, Schacht K, Lohmann A, Storzer U. Breathable nail polish on the basis of a new blend: a complex of water-based polymer and functional vegan silk. SÖFW J 2017;143:10–17.
[22] Endlein E, Peleikis KH. Natural waxes—Properties, compositions and application. SÖFW J 2011;137(4):1–8.
[23] Barba C, Abergel A, inventors; LOreal AS, assignee. Method for making up the lips. US patent 20100291015A1, 2010 November 18.
[24] de Gennes PG. Ultradivided matter. Nature. 2001;412(6845):385.
[25] PEN. The Project on Emerging Nanotechnologies. Primordiale optimum lip [Internet]. 2007 [accessed 05.03.20]. Available from: <http://www.nanotechproject.org/cpi/products/primordiale-optimum-lip/>.
[26] Maitra P, Zheng T, inventors; Avon Products Inc, assignee. Cosmetic nanocomposites bases on in-situ cross-linked POSS materials. US patent 8133478B2, 2012 March 13.
[27] Valia D, Ellis J, inventors; Creative Nail Design, assignee. Nail coatings having enhanced adhesion. US patent 9713585B2, 2017 July 25.
[28] Welton J, Dixon K, Lichtenhan J, Hait S, Schwab J, inventors; Revlon Consumer Products Corporation, assignee. Silsesquioxanes and use in nail coatings thereof. US patent 2019/0192411A1, 2019 June 2.
[29] Yamato N c/o Ajinomoto Co Inc, inventor; Ajinomoto Co Inc, assignee. Gel-state composition. EP patent 1609836A1, 2005 December 28.
[30] Lochhead R, Lochhead M. Two decades of transfer-resistant lipstick. Cosm & Toil [Internet]. 2015 [accessed 05.03.20]. Available from: <https://www.cosmeticsandtoiletries.com/research/chemistry/Two-Decades-of-Tr ansfer-resistant-Lipstick-290207561.html>.
[31] Zhang Q. Nanocolarants. In: Sattler KD, editor. Handbook of nanophysics: functional nanomaterials. 1st ed. CRC Press Taylor & Francis Group; 2010.
[32] Gürses A, Açikyildiz M, Günes K, Gürses MS. Dyes and pigments. Switzerland: Springer; 2016. p. 83.
[33] Hu Z, Xue M, Zhang Q, Sheng Q, Liu Y. Nanocolorants: a novel class of colorants, the preparation and performance characterization. Dye Pigm 2008;76(1):173–8.
[34] Nejad FM, Baghshahi S, Bakhtiari L. Advantages of nano pigments over micro pigments in obtaining larger spectra of colours in CMYK system. Trans Indian Ceram Soc 2011;70(2):93–9.
[35] Bagala F, inventor; BASF Catalysts LLC, assignee. Effect pigment. US patent 20040123778A1, 2004 July 01.
[36] Carrion D, Coleman-Nally D, Graves G, inventors; LOreal SA, assignee. Cosmetic compositions comprising nanoparticles and processes for using the same. US patent 20030064086A1, 2003 April 03.
[37] Pfaff G, Andes S, Ambrosius K, Petry R, Roesler M, Schoen S, inventors; Merck Patent GmbH, assignee. Effect pigments. US20130164356A1, 2013 June 27.
[38] Yadav T, inventor; Nano Products Corp, assignee. Color pigments nanotechnology. US patent 20040156986A1, 2004 August 12.
[39] Pfaff G, Becker M. Special effect pigments in cosmetic applications. An amazing development for bright future. H&PC Today (Monographic Suppl series: Color Cosmetics) 2012;1:12–15.
[40] Kaupp G, Schmidt U, Schumacher D, Steinbach K, inventors; Eckart GmbH and Co KG and Glassflake Ltd, assignee. Effect pigments comprising a glass flake substrate. US patent 20100116169A1, 2010 May 13.
[41] Anselmann R, Hillgärtner U, Knapp M, inventors; Merck Patent GmbH, assignee. Pigment mixture having a high light diffusion. EP patent 1115796B1, 1999 September 13.
[42] Schlossman D, Shao Y, inventors; Kobo Products Inc, assignee. Hybrid coated cosmetic powders and methods of making and using same. US patent 20080014233A1, 2008 January 17.
[43] Anselmann R, Hillgärtner U, Schoen S, inventors; Merck Patent GmbH, assignee; Pigment mixture comprising BiOCl pigments. US patent 6743285B1, 2004 January 06.
[44] Guerrero-Martínez A, Pérez-Juste J, Liz-Marzán LM. Recent progress on silica coating of nanoparticles and related nanomaterials. Adv Mater 2010;22(11):1182–95.
[45] Patel A, Prajapati P, Boghra R. Overview on application of nanoparticles in cosmetics. Asian J Pharm Sci Clin Res 2011;1(2):40–55.
[46] Ealias AM, Saravanakumar MP. A review on the classification, characterisation, synthesis of nanoparticles and their application. IOP Conf Ser Mater Sci Eng 2017;263(3):032019.

[47] Delrieu YS, House S. Combinational use of microspheres for soft focus effect. In: IFSCC Congress, 29th, 2016, Orlando, FL, Kobo Products, 2016.

[48] Kobo Products. KoboBlur™ 100 natural plastic-free powder complex for multimedia soft focus effect, <http://www.koboproductsinc.com/Downloads/Kobo-KoboBlur100Natural.pdf> and Solid microsphere complexes <http://www.koboproductsinc.com/Downloads/Kobo-SolidMicrospheresComplexes.pdf>.

[49] Kobo Products. Composite microspheres, <http://www.koboproductsinc.com/Products.aspx?p = Microsphere%20Optical%20Enhancers>.

[50] Noebel RD, Elliott L. inventors; Globe Technology Corp, assignee. Hydrophobic silica coating. US patent 5204088A, 1993 April 20.

[51] Schmid R, Mronga N, inventors; BASF SE, assignee. Luster pigments based on multiply coated plateletlike metalic substrates. US patent 5364467A, 1994 November 15.

[52] Dréno B, Alexis A, Chuberre B, Marinovich M. Safety of titanium dioxide nanoparticles in cosmetics. J Eur Acad Dermatol 2019;33(S7):34–46.

[53] Cavalcante PMT, Dondi M, Guarani G, Raimondo M, Baldi G. Colour performance of ceramic nano-pigments. Dye Pigm 2009;80(2):226–32.

[54] Piri N, Shams-Nateri A, Mokhtari J. The effect of TiO_2 nanopigment on the optical properties of polyester fabric in UV–VIS–NIR regions. Color Res Appl 2019;44:257–63.

[55] Prospect INCI Misslyn. Gel effect color (INCI 40, 61, 79 and 114) [Internet]. 2019 [accessed 07.12.19]. Available from: <https://www.misslyn.com/files/inci_misslyn.pdf>.

[56] Soares MC, Viana MM, Schaefer ZL, Gangoli VS, Cheng Y, Caliman V, et al. Surface modification of carbon black nanoparticles by dodecylamine: thermal stability and phase transfer in brine medium. Carbon. 2014;72:287–95.

[57] Nanotechnology Products Database. BlueBelle: Nailpolish–carbon (nanoparticle/nanopowder) [Internet]. 2015 [accessed 05.03.20]. Available from: <https://product.statnano.com/product/2981/bluebelle>.

[58] Tsubokawa N. Functionalization of carbon black by surface grafting of polymers. Prog Polym Sci 1992;17(3):417–70.

[59] Roy N, Sengupta R, Bhowmick AK. Modifications of carbon for polymer composites and nanocomposites. Prog Polym Sci 2012;37(6):781–819.

[60] ECHA (European Chemicals Agency). Literature study on the uses and risks of nanomaterials as pigments in the European Union Study [Internet]. 2018 September [accessed 24.12.19]; p. 56–84. Available from: <https://op.europa.eu/en/publication-detail/-/publication/dd38af13-bca7-11e8-99ee-1aa75ed71a1/language-en/format-PDF>.

[61] Artdeco. Nanoproject ceramic nail lacquer. Artdeco [Internet]. 2011 [accessed 05.03.20]. Available from: <https://www.nanotechproject.org/cpi/products/ceramic-nail-lacquer/>.

[62] Kar B, Rout D. Nanoclays: synthesis, properties and applications. In: Mohanty S, Nayak SK, Kaith BS, Kalia S, editors. Polymer nanocomposites based on inorganic and organic nanomaterials. 1st ed. Bervely: Scrivener Publishing LLC; 2015. p. 195–214.

[63] Usmani MA, Khan I, Ahmad N, Bhat AH, Sharma DK, Rather JA, et al. Modification of nanoclay systems: an approach to explore various applications. In: Jawaid M, Qaiss A, el K, Bouhfid R, editors. Nanoclay reinforced polymer composites. Singapore: Springer; 2016. p. 57–83.

[64] Valapa RB, Loganathan S, Pugazhenthi G, Thomas S, Varghese TO. An overview of polymer–clay nanocomposites. In: Jlassi K, Chehimi MM, Thomas S, editors. Clay-polymer nanocomposites. Cambridge: Academic Press; 2017. p. 29–81.

[65] Guo F, Aryana S, Han Y, Jiao Y. A review of the synthesis and applications of polymer–nanoclay composites. Appl Sci 2018;8(9):1–29.

[66] Seyidoglu T, Ulku Y. Modification and characterization of bentonite with quaternary ammonium and phosphonium salts and its use in polypropylene nanocomposites. J Thermoplast Compos Mater 2015;28(1):86–110.

[67] Sorocina ZG, Birks I. Hectorite and bentonite effect on water-based polymer coating rheology. C R Chim 2019;22(2-3):169–74.

[68] Amato SW, Farer A, Hoyte WM, Pavlovsky M, Smith R, Valdiviezo G, inventors; Coty Inc, assignee. Coatings for mammalian nails that include nanosized particles. US patent 20100196294A1, 2010 August 05.

PART 3

Toxic risks of nanocosmetics

CHAPTER

20

Current legal frameworks and consumer protection in nanocosmetics

Sonia Selletti

Studio Legale Astolfi e Associati, Milan, Italy

20.1 Legal definition of nanomaterial

Nanomaterials are substances found in different categories of products (from food and cosmetics to medical devices), whose use must be carefully considered in terms of the safety of products in which they are used.

To guarantee a high level of protection of consumer health, diverse European regulations exist that govern the use of nanomaterials, with specific reference to the various product categories. Thus these represent special sectoral regulations.

The need to ensure a standard definition of nanomaterial at a European level led the European Commission to adopt a unique definition of nanomaterial with Recommendation 2011/696 of October 18, 2011,[1] which must be used by Member States, EU agencies, and economic operators as a reference in the adoption and application of legislation and strategic and research programs related to products derived from nanotechnologies.

As it is a Recommendation, it is not a legally binding act for its recipients, but rather a proposal to adopt a certain behavior. In the case under exam, it is a useful point of reference to ensure the standardization of the definition of the term "nanomaterial" used by the legislation and regulatory provisions at a European level.

As specified by the Commission in the recitals to the Recommendation, the definition of "nanomaterial" used by European legislation is based exclusively on the size of the constituent particles of a material and covers natural, incidental, or manufactured materials, without

[1] Published in the EU Official Journal No. L.275/38 of October 20, 2011.

Nanocosmetics
DOI: https://doi.org/10.1016/B978-0-12-822286-7.00016-4

regard to hazard or risk that have to be carefully evaluated at a second level in assessing the safety of the product containing "nanomaterial."

According to that established by the Recommendation, "nanomaterial" therefore means "a natural, incidental or manufactured material containing particles, in an unbound state or as an aggregate or as an agglomerate and where, for 50% or more of the particles in the number size distribution, one or more external dimensions is in the size range 1 nm–100 nm."

The European Commission also states that "In specific cases and where warranted by concerns for the environment, health, safety or competitiveness the number size distribution threshold of 50% may be replaced by a threshold between 1% and 50%," and provides for some exceptions to this definition, admitting that fullerenes, graphene flakes, and single-wall carbon nanotubes with one or more external dimensions below 1 nm should be considered as nanomaterials.

The Recommendation also clarifies the meaning of terms used in the definition, specifying that "particle" means "a minute piece of matter with defined physical boundaries," while the word "agglomerate" identifies "a collection of weakly bound particles or aggregates where the resulting external surface area is similar to the sum of the surface areas of the individual components," and the term "aggregate" indicates "a particle comprising of strongly bound or fused particles."

Where technically feasible and requested in specific legislation, compliance with the definition formulated by the Commission may be determined on the basis of the specific surface area by volume: a material should be considered as falling under the definition of nanomaterial where its specific surface area by volume is greater than 60 m^2/cm^3. However, a material which, based on its number size distribution, is a nanomaterial should be considered as complying with the definition even if it has a specific surface area lower than 60 m^2/cm^3.

This definition is based on available scientific knowledge; considering that technological development and scientific progress are rapidly advancing, according to the provisions of the Recommendation of 18 October 2011, a first review of the definition of nanomaterial should have been made by December 2014, in order to evaluate if the number size distribution threshold of 50% should be increased or decreased.

To date, the procedure for reviewing the definition of nanomaterial is not yet complete.

The European Commission plans to open a public consultation before proceeding with the revision of the Recommendation of October 18, 2011 and has published reports on the revision prepared in 2014 and 2015 by the Joint Research Centre (JRC) on its website, which describe the purposes of the revision and the proposed changes:

- "Compilation of information concerning experience with the definition";
- "Assessment of collected information concerning the experience with the definition";
- "Scientific-technical consideration to clarify the definition and to facilitate its implementation".[2]

[2] The reports are available on the website ec.europa.eu, respectively, at the following links: <https://ec.europa.eu/jrc/en/publication/eur-scientific-and-technical-research-reports/towards-review-ec-recommendation-definition-term-nanomaterial-part-1-compilation-information>; 2020 [accessed 30.03.2020]; <https://ec.europa.eu/jrc/en/publication/eur-scientific-and-technical-research-reports/towards-review-ec-recommendation-definition-term-nanomaterial-part-2-assessment-collected>; 2020 [accessed 30.03.2020]; <https://ec.europa.eu/jrc/en/publication/eur-scientific-and-technical-research-reports/towards-review-ec-recommendation-definition-term-nanomaterial-part-3-scientific-technical>; 2020 [accessed 30.03.2020].

The revision of the Recommendation is not intended to modify the main elements of the definition of nanomaterial (such as the reference to size), but rather to provide clarifications in order to ensure increasingly greater uniformity in the interpretation and implementation of the definition.

From this point of view, on February 25, 2019, the JRC of the European Commission published the report "An overview of concepts and terms used in the European Commission's definition of nanomaterial" on the website ec.europa.eu in order to clarify the meaning of key concepts used in the definition of nanomaterial referred to in the Recommendation of October 18, 2011.[3]

The report, in clarifying the meaning of the terms used by the European Commission in the definition of nanomaterial as formulated with the Recommendation of October 18, 2011, defines the scope of the latter,[4] first clarifying that the definition of nanomaterial is cross-sectorial, "horizontal," that is, that its scope of application is not limited to a particular sector, but is extended to any area.

It is also clarified that the definition applies to all materials, regardless of their origin (natural, manufactured, or incidental) and that it is formulated on the basis of the only feature common to all nanomaterials: the nanoscale dimension of their external particles, with the consequence that "nanostructured materials"—materials that have only internal nanoscale particles—are excluded from the definition. Again, even in the recent revision report, the criteria for the definition of "nanomaterial" remain for the horizontal perspective the nanoscale dimension of the external particles while the cosmetic product "nano" definition considers the dimension of the internal particles as well.

With particular reference to the term "particles," the report states that such indicates a minute piece of matter up to 100 mm. The definition of nanomaterial referred to in the recommendation only refers to solid particles at normal temperature and pressure (i.e., 298.15 K and 101.325 Pa).

20.2 Nanomaterials in EC Regulation No. 1223/2009 on cosmetic products

20.2.1 Cosmetic definition of "nanomaterials"

The definition provided by the European Commission in the Recommendation of October 18, 2011 is aimed at giving a standard meaning to the term "nanomaterial," as used and defined by different regulatory and legislative provisions at a European level (i.e., cross-sectorial recommendation).

[3] The report is available at the link <https://ec.europa.eu/jrc/en/publication/overview-concepts-and-terms-used-european-commissions-definition-nanomaterial>; 2020 [accessed 30.03.2020]. A second report dedicated to measurements was in progress at the time of publication of this document, which is aimed at providing specific indications on the identification of materials according to the criteria identified by the European Commission. This report is currently not yet available.

[4] The content of the report is very technical in nature. For what concerns the topic discussed in this chapter, only a few passages are examined that can help to clarify the concepts contained in the Recommendation of October 18, 2011, and the scope of application of the definition contained therein.

With reference to the cosmetic sector, Article 2(k) of EC Regulation No. 1223/2009 defines nanomaterials as "an insoluble or biopersistent and intentionally manufactured material with one or more external dimensions, or an internal structure, on the scale from 1 to 100 nm."[5]

The definition provided by the Regulation on cosmetic products thus only includes nanomaterials that are intentionally manufactured and insoluble (or poorly soluble) or biopersistent (such as metals, metal oxides, and carbon materials) and excludes those that are completely soluble and nonbiopersistent (such as liposomes and oil/water emulsions).[6]

This definition precedes that adopted by the Commission with the Recommendation of October 18, 2011 and has some points in common and differences with the former.

First, it should be noted that the Regulation on cosmetic products, though it uses the size of nanomaterials (which must be on the scale from 1 to 100 nm) as parameter to define nanomaterials in the same manner as the Recommendation, takes into consideration not only their external dimension, but also the internal structure, unlike the Recommendation, which takes only the external dimension into consideration for qualification as a nanomaterial.

In the definition adopted by the EC Regulation No. 1223/2009, no reference is made to particles, number size distribution, and origin of nanomaterials. Unlike the Recommendation, which includes any natural, incidental, or manufactured material, the

[5] Several definitions are contained in other European regulations on other product categories.
For example, with reference to novel foods, EU Regulation 2015/2283 contains a definition of "engineered nanomaterial," understood as "any intentionally produced material that has one or more dimensions of the order of 100 nm or less or that is composed of discrete functional parts, either internally or at the surface, many of which have one or more dimensions of the order of 100 nm or less, including structures, agglomerates, or aggregates, which may have a size above the order of 100 nm but retain properties that are characteristic of the nanoscale." [Article 3(f) of EU Reg. No. 2015/2283].
EU Regulation No. 528/2012 concerning the making available on the market and use of biocidal products contains a definition of "nanomaterial," which is identified as "a natural or manufactured active substance or non-active substance containing particles, in an unbound state or as an aggregate or as an agglomerate and where, for 50% or more of the particles in the number size distribution, one or more external dimensions is in the size range 1–100 nm." [Article 3(z) of EU Reg. No. 528/2012].
EU Regulation No. 2017/745 on medical devices, on the other hand, uses a definition of nanomaterial that is in line with that established by the Recommendation of the European Commission of October 18, 2011, defining nanomaterial as "a natural, incidental or manufactured material containing particles in an unbound state or as an aggregate or as an agglomerate and where, for 50% or more of the particles in the number size distribution, one or more external dimensions is in the size range 1–100 nm. Fullerenes, graphene flakes and single-wall carbon nanotubes with one or more external dimensions below 1 nm shall also be deemed to be nanomaterials."
The Annexes to EC Regulation No. 1907/2006 concerning the Registration, Evaluation, Authorisation, and Restriction of Chemicals (REACH), as amended by EU Regulation No. 2018/1881, use the term "nanoform," to which it is attributed the same meaning as in the definition of nanomaterial introduced by the European Commission with the Recommendation often cited above.

[6] See "The SCCS notes of guidance for the testing of cosmetic ingredients and their safety evaluation—10th revision" adopted by the Scientific Committee on Consumer Safety in the plenary meeting of October 24–25, 2018, available at the link <https://op.europa.eu/en/publication-detail/-/publication/1d5f1653-38ce-11e9-8d04-01aa75ed71a1/language-en/format-PDF/source-87840423>; 2020 [accessed 30.03.2020].

Regulation limits the scope of application of the definition in Article 2(k), to insoluble, biopersistent, and intentionally manufactured materials.

As regards the meaning of "intentionally manufactured" it means that a material is considered to be a nanomaterial if it is manufactured, through a physical or chemical process, with a view to intentionally producing nanometric-scale objects, structures, or associated properties.

20.2.2 The use of nanomaterial in cosmetic products

In addition to the definition of "nanomaterial," EC Regulation No. 1223/2009 includes provisions on their presence in cosmetic products in order to ensure a high level of consumer protection, as well as the free movement of goods and legal certainty for manufacturers (see recital no. 29 of the Regulation).

It must be considered that the rules of the Regulation on cosmetic products first and foremost serve to guarantee their safety based on the "precautionary principle." Article 3 of the Regulation expressly provides that "a cosmetic product made available on the market shall be safe for human health when used under normal or reasonably foreseeable conditions of use," taking account, in particular, of presentation, labeling, and instructions for use and disposal, as well as any other indication or information provided by the responsible person.

Considering that the main objective of EC Regulation No. 1223/2009 is to guarantee the safety of cosmetic products placed on the market, it is not surprising that this legislation lists specific provisions on products containing nanomaterials in consideration of the relevance of the potential impact on human health.

That highlighted above has already been covered by the recitals of the Regulation, which call for an intervention by the European Commission to "regularly review the provisions on nanomaterials in the light of scientific progress" (see recital no. 31), as well as the Scientific Committee for the Consumer Safety (SCCS), which should "provide guidance in cooperation with relevant bodies on test methodologies which take into account specific characteristics of nanomaterials" in order to better assess their safety (see recital no. 30), and opinions on the safety of use of nanomaterials in cosmetic products, based on full information made available by the responsible person (see recital no. 35).[7]

[7] In compliance with the provisions of EC Regulation No. 1223/2009, the SCCS has adopted the following documents:
"Guidance on the safety assessment of nanomaterials in cosmetics" (SCCS/1611/19, which is an up-to-date revision of the previous guidance SCCS/1484/12, currently under review) of October 2019, 30–31, available at the link <https://ec.europa.eu/health/sites/health/files/scientific_committees/consumer_safety/docs/sccs_o_233.pdf>; 2020 [accessed 30.03.2020];
MEMORANDUM on "Relevance Adequacy and Quality of Data in Safety Dossiers on Nanomaterials" (SCCS/1524/13), revision of March 27, 2014, available at the link <https://ec.europa.eu/health/scientific_committees/consumer_safety/docs/sccs_o_142.pdf>; 2020 [accessed 30.03.2020];
"Checklists for Applicants submitting dossiers on Cosmetic Ingredients to be evaluated by the SCCS," revision of May 16, 2018, available at the link <https://ec.europa.eu/health/sites/health/files/scientific_committees/consumer_safety/docs/sccs_o_210.pdf>; 2020 [accessed 30.03.2020].

20.2.3 Declaration of nanomaterial use and nanonotification

As concerns the provisions of EC Regulation No. 1223/2009, which examine the presence of nanomaterials in cosmetic products, it is first and foremost necessary to reference Article 13 as regards notification to the competent authorities of the presence in the products of nanomaterials.

This rule states that before placing a cosmetic product on the market, the responsible person must submit, by electronic means, a series of information to the European Commission (through the web-based Portal—Cosmetic Products Notification Portal, CPNP)[8]. This information includes:

1. The category of cosmetic product and its name or names, enabling its specific identification;
2. The name and address of the responsible person where the product information file (PIF) is made readily accessible;
3. The country of origin in the case of import;
4. The Member State in which the cosmetic product is to be placed on the market;
5. The contact details of a physical person to contact in the case of necessity;
6. The presence of substances in the form of nanomaterials;
7. The name and the Chemicals Abstract Service (CAS) or EC number of substances classified as carcinogenic, mutagenic, or toxic for reproduction (CMR), of category 1A or 1B, under Part 3 of Annex VI to Regulation (EC) No. 1272/2008; and
8. The frame formulation allowing for prompt and appropriate medical treatment in the event of difficulties.

With reference to nanomaterials, the Commission must be notified as to the identification (including chemical name and other descriptors, as specified in point 2 of the recitals to Annexes II–VI of Regulation),[9] as well as the reasonably foreseeable exposure conditions to nanomaterials.

[8] It should be recalled that, pursuant to Article 4 of EC Regulation No. 1223/2009, the responsible person is the natural or legal person who guarantees the compliance of a cosmetic product with the relevant obligations established by the Regulation. The responsible person is usually the "manufacturer," although delegation to another person is permitted provided that the proxy is given and accepted in writing. The responsible person is not required to possess specific subjective requisites, but is responsible for fulfilling the many obligations prescribed by the legislation on cosmetics.

[9] For the sake of completeness, we report the text of point 2 of the recital to Annexes II–VI of EC Regulation No. 1223/2009:
"In order to facilitate substance identification, the following descriptors are used:
The Non-proprietary Names (INN) for pharmaceutical products, WHO, Geneva, August 1975;
The Chemical Abstracts Service numbers (CAS);
The EC number which correspond to either the European Inventory of Existing Commercial chemical Substances (EINECS) numbers or the European List of Notified Chemical Substances (ELINCS) numbers or the registration number given under Regulation (EC) No. 1907/2006;
The XAN which is the name approved by specific country (X), e.g., USAN which correspond to the United States approved name;
The name in the glossary of common ingredient names referred to in Article 33 of this Regulation."

In addition to stating that a high level of protection of human health must be ensured for every cosmetic product that contains nanomaterials in the first paragraph, Article 16—specifically entitled "Nanomaterials"—also states that, in addition to Article 13, cosmetic products containing nanomaterials shall be notified to the Commission by the responsible person by electronic means 6 months prior to being placed on the market (except where they have already been placed on the market by the same responsible person before January 11, 2013)[10].

So, Article 16 of the Regulation provides a separate and additional notification as a requirement for the responsible person to notify to the Commission cosmetic products containing nanomaterials.

Exemptions are listed in the second paragraph of Article 16 and concern nanomaterials used as colorants, UV filters, or preservatives pursuant to Article 14 (relating to substances that cosmetic products cannot contain) from the scope of application of Article 16, unless expressly specified.

The information that must be communicated to the Commission pursuant to Article 16 must include the following, at minimum:

1. The identification of the nanomaterial, including the chemical name (IUPAC) and other descriptors, as specified in point 2 of the recital to Annexes II–VI;
2. The specification of the nanomaterial including size of particles, physical, and chemical properties;
3. An estimate of the quantity of nanomaterial contained in cosmetic products intended to be placed on the market per year;
4. The toxicological profile of the nanomaterial;[11]
5. The safety data of the nanomaterial relating to the category of cosmetic product, as used in such products;
6. The reasonably foreseeable exposure conditions.

The European Commission can amend the list of information to be notified by introducing additional requirements, while taking scientific progress into account.

Pursuant to the provisions of Article 16(4) of EC Regulation No. 1223/2009, if the Commission has concerns about the safety of a nanomaterial, it immediately asks the SCCS to express an opinion on the safety of use of this nanomaterial in the relevant category of cosmetic products, as well as in reasonably foreseeable exposure conditions.

The SCCS issues its opinion within 6 months of the Commission's request.

If the Committee deems the data to be insufficient, the Commission asks the person responsible to provide such data within an explicitly stated reasonable time, which cannot be extended. The SCCS then issues its final opinion within 6 months of the submission of additional data.

The opinions of the Committee are made available to the public.

[10] In this case, cosmetic products containing nanomaterials had to be notified, by electronic means, to the Commission by the responsible person between 11 January 2013 and 11 July 2013, in addition to the notification pursuant to Article 13 [see Article 16(3), Reg. EC No. 1223/2009].

[11] The Commission provides a reference number for presentation of the toxicological profile, which can replace the information to be notified pursuant to letter (d).

The procedure described above under Article 16(4) can be activated by the Commission whenever it has any concerns about safety, such as subsequent to new information provided by third parties.

Taking the opinions of the SCCS into account, if there are risks to human health or insufficient data, the Commission may amend Annexes II and III of EC Regulation No. 1223/2009, which respectively list substances prohibited in cosmetic products and substances which such products must not contain, except subject to the restrictions laid down. In other words, the use of a certain "nanomaterial" could be restricted or limited to specific conditions.

20.2.4 Labeling

Article 19 of Reg. EC No. 1223/2009 introduces specific provisions relating to labels of cosmetic products containing nanomaterials.

This regulation establishes that cosmetic products can be made available on the market only if the container and the packaging of the same bear certain indications in indelible, easily legible, and visible lettering, including the name of the responsible person, the nominal content at the time of packaging (expressed in weight or volume), the minimum duration of the product, special precautions to be observed in use, the batch number, the product function, as well as the list of ingredients.

The new provision concerns the information to consumer when nanomaterials are used in the cosmetic products. In such cases, all of the ingredients present in the form of nanomaterials must be clearly indicated in the list of ingredients by placing the suffix "nano" in brackets after the INCI name of the (nano) ingredients.

This rule applies to all nanomaterial without exemptions.

20.2.5 Safety assessment (report) of nanomaterials

Article 10 of EC Regulation No. 1233/2009 provides that, in order to demonstrate the safety of cosmetic products and in accordance with the provisions of Article 3 (saying "a cosmetic product made available on the market shall be safe for human health when used under normal or reasonably foreseeable conditions of use..."), the responsible person must, prior to placing a cosmetic product on the market, ensure that the cosmetic product has undergone a safety assessment on the basis of the relevant information and that a cosmetic product safety report is set up in accordance with Annex I of the Regulation.

The cosmetic product safety report shall also contain a reference to nanomaterials.

The contents of the safety report are governed by Annex I of Reg. EC No. 1223/2009 on which the European Commission has issued specific Guidelines on November 25, 2013,[12] which states that this report is comprised of two parts, the first of which (Part A) contains the cosmetic product safety information and the second (Part B) contains the cosmetic product safety assessment.

[12] Published in EU Official Journal No. L.315/82 on 26 November 2013.

Part A is intended to gather the data necessary to prove that the cosmetic product is safe. The information should enable the safety assessor to clearly identify and quantify, based on the identified hazards, the risks a cosmetic product may present to human health. A hazard may arise, for example, from the raw materials, the manufacturing process, the packaging, the conditions of use of the product, the microbiological specifications, the quantities used, and the toxicological profile of the substances.

The "particle size" of the substances used in the product can have an influence on the toxicity of a substance. As a consequence, when it cannot be excluded that they have an impact on the safety of the finished product, they should be included among its physicochemical characteristics and be taken into account during the safety assessment according to the most recent scientific opinions on the subject.[13]

Then, the safety report must contain a section (n. 8) relating to the toxicological profile of substances whose assessment must also include any effects on the toxicological profile due to particle sizes, including nanomaterials.

The expert who drafts the safety assessment ("safety assessor") must possess specific subjective requisites (degree or diplomas in the pharmaceutical, toxicological, medical, or similar disciplines, or courses recognized as equivalent by a Member State).

The cosmetic product safety report forms integral and substantial part of the PIF that the responsible person is requested to keep when a cosmetic product is placed on the market. The PIF shall be kept for a period of 10 years following the date on which the last batch of the cosmetic product was placed on the market.

The responsible person is also responsible for ensuring that the cosmetic product safety report is kept up to date, taking additional relevant information generated subsequent to placing the product on the market into account.

As a matter of facts since nanomaterials are involved by ongoing scientific progress, the update of the safety report is a sensitive issue.

20.3 Catalog of nanomaterials

According to the provisions of Article 16(10) of Reg. EC No. 1223/2009, "by January 11, 2014, the Commission shall make available a catalogue of all nanomaterials used in cosmetic products placed on the market, including those used as colorants, UV filters and preservatives in a separate section, indicating the categories of cosmetic products and the reasonably foreseeable exposure conditions. This catalogue shall be regularly updated thereafter and be made publicly available."

The European Commission has long postponed the compilation of the catalog due to the difficulties acknowledged in the data notified by responsible persons through the CPNP. So, the first edition of the catalog (dated December 31, 2016) was published on its website for the first time on June 15, 2017 and it has been updated on November 21, 2019.[14]

[13] SCCS, Guidance on safety assessment of nanomaterials in cosmetics, SCCS/1484/12; SCENIHR (Scientific Committee on Emerging and Newly Identified Health Risks), Opinion on the scientific basis for the definition of the term "nanomaterial," December 8, 2010.

[14] The updated version of the catalog is available on the website of the European Commission at the link <https://ec.europa.eu/docsroom/documents/38166>; 2020 [accessed 30.03.2020].

The updated version of the catalog includes 29 (while the previous version included 43 nanomaterials), of which 3 are colorants, 4 are UV filters, and 22 have different functions.

The catalog is compiled and updated based on information communicated by responsible persons through the CPNP and contains the list of nanomaterials present in cosmetics placed on the market, identified by their name, category of cosmetic products in which they are present, methods of exposure, and it is specified whether the product in which the nanomaterial is present is rinsed off or left on for each.

As expressly specified by the European Commission, the catalog has a merely informative value and does not constitute in any way a list of "authorized nanomaterials." By the way, the catalog represents a reference for the manufacturers even if it is important to underline that the issue has to be considered a "work in progress" thanks to the fact that the scientific and technical progress will define new boundaries.

20.4 Conclusions

As reported by the European Commission in the catalog of nanomaterials, at the end of 2016, cosmetic products containing these substances represented less than 1% of all the cosmetics communicated through the Cosmetic Products Nanomaterial Portal.

Based on the aforementioned legislation and the provisions contained in EC Regulation No. 1223/2009 on cosmetic products, attention given by European institutions is evident as regards the presence of these substances in everyday products, such as cosmetics.

In this context, the responsible person who has the task of guaranteeing the safety of cosmetic products placed on the market plays a fundamental role, also and in particular with reference to the nanomaterials that such products contain.

The rapid evolution of technical-scientific progress entails the need for continuous and constant monitoring by the European Commission, whose intervention is necessary in order to ensure that the definition of "nanomaterial" used at a European level with the Recommendation of 18 October 2011 is updated and applied and interpreted as a standard in all product sectors.

Even the criteria for assessing the safety of products containing nanomaterials must be updated in a real time on the basis of available technical-scientific knowledge, in particular considering that risks to human health attributable to nanomaterials are still uncertain.

CHAPTER

21

Nanoparticle toxicological risks on intact-skin dermal exposures

Tasleem Arif[1] *and Konchok Dorjay*[2]

[1]Ellahi Medicare Clinic, Kashmir, India [2]Department of Dermatology, Dr. Ram Manohar Lohia Hospital and Post Graduate Institute of Medical Education and Research, New Delhi, India

21.1 Introduction

Nanoscience is the study of particles on an atomic or molecular scale, whose size is measured in nanometers, with at least one dimension in the range of 1–100 nm. Nanotechnology is the application of this scientific knowledge for various diagnostic and therapeutic purposes [1]. Nanoparticles may show physical and chemical properties quite distinct from a similar material with particles of a larger scale, opening the exciting gateway for novel uses of already existing substances [2]. Nanodermatology is the use of nanotechnology in the prevention, diagnosis, and treatment of various skin disorders and in the field of cosmetology.

The concept of nanotechnology is fairly new. The word "Nano" is derived from a Greek word "Nanos," which means "dwarf" [3]. Richard Feynman is considered as the father of nanotechnology [4]. However, the term nanotechnology was coined by Professor Norio Taniguchi and according to him, "Nanotechnology is a process of separating, consolidating, and deforming materials atom by atom or molecule by molecule" [3].

Nanotechnology has revolutionized the treatment of several skin diseases by its armamentarium of nanoparticles each with different physicochemical properties. In the present scenario, an increasing demand is felt by scientists and pharmaceutical companies throughout the world for nanodermatology.

Nanoparticles possess novel physical and chemical properties in contrast to the same material.

Nanodermatology has since then emerged as a fast-growing field with great interest shown by pharmaceutical industries, medical professionals, and patients alike. The various uses of nanotechnology in dermatology and cosmetology are as follows: antiseptics,

Nanocosmetics
DOI: https://doi.org/10.1016/B978-0-12-822286-7.00023-1

sunscreens, cosmetics and antiaging products, moisturizers, hair diseases, inflammatory skin diseases, wound healing, sebaceous gland diseases, light-based therapies and laser, vaccines, diagnostic uses, and transdermal drug therapy [5,6].

21.2 Factors mediating toxicity

The 2004 report of the Royal Society and Royal Academy of Engineering reviewing nanoparticles is one of the earliest accounts of nanotoxicity to human and environmental health [7]. This triggered a spurt in the research on risk assessment of nanotechnology and causes thereof. Several factors influence the hazards of nanoparticles. These are dependent on the physicochemical properties of nanoparticles and on host factors.

Particle factors

Size and chemical composition are the most important causes of toxicity. There is an inverse correlation between particle size and toxicity [8]. This is explained on the basis of a greater surface area of these particles in relation to volume, exposing a large number of atoms for biological interactions in an exponential fashion [9]. A decrease in particle size can lead to the formation of new allergens, haptens, and irritants [10]. Small size can enable these particles to cross biological membranes disrupting functions of cells. Only particles of size less than 500 Da can penetrate the intact skin [11]. The addition of solvents enhances permeability. These particles may further disintegrate into smaller toxic fragments [12]. Chemical composition is the obvious determinant of toxicity. An example is fullerenes that generate free radicals because they attract electrons. Charge, concentration, shape, solubility, and surface coating are the other major determinants of toxicity. It has been demonstrated that the addition of hydroxyl and carbonyl groups decreased the cytotoxic potential of carbon nanoparticles. The cause of oxidative stress in cells is explained by the negative charge on the surface of the carbon particles [13]. Impurities in nanoparticles can occur during manufacturing due to solvents or chemical reagents or due to nanoparticle themselves, both can produce potential toxins. The absence of a natural excretion pathway, as is the case with carbon nanotubes, can lead to accumulation in the body and even small doses over long term might cause toxic levels in the body [14].

Host factors

Impaired capability to excrete various wastes of the body may predispose an individual to toxicity by accumulation [9]. This is likely to occur in patients with renal and hepatic diseases. Failure to excrete gadolinium was responsible for several cases of nephrogenic systemic fibrosis in renal failure patients [15]. Infants are likely to absorb a greater amount of nanoparticles as they do not have a fully developed skin. Further, the penetration of nanoparticles has been shown to be higher in the skin with an impaired barrier, as in eczemas, in comparison with the healthy intact skin. Some nanomedicines may trigger defense mechanisms of the body like complement activation and inflammation that can cause damage to various tissues [16].

21.3 Routes of entry and kinetics

Nanoparticles might gain entry in the body by various routes, important are percutaneous absorption, inhalation, ingestion, and intravenous administration.

The abundance of dermatologic and cosmetic products intended for topical use makes the percutaneous route one of the most important portals of entry of nanoparticles in the body. The hair follicles act as small channels through which particles as large as 10 nm can permeate [17]. The observation that polystyrene nanoparticles could pass through follicles pretreated with the cyanoacrylate skin surface stripping method, and were ingested by the Langerhans cells, provided one of the first hints of potential toxicity through the dermal route. Concerns about the safety of sunscreens were laid to rest by the Royal Society and Royal Academy of Engineers, UK, with their report that titanium oxides do not penetrate deeper than the epidermis [18]. Disturbed skin barrier seems to be the key to nanoparticle penetration.

Inhalation is an important route of entry of aerosolized particles. Exposure can occur with the use of aerosolized perfumes, deodorants, etc., at the consumer end or in persons working in industries producing nanoparticles. It is believed that the number of particles has a more deleterious effect on health than the mass on inhalation. Spherical particles of less than 10 μm are inhaled while those less than 2.5 μm reach the alveoli [19]. Inhaled nanoparticles have been detected in blood in humans. They have the potential to consequently reach the central nervous system, liver, spleen, and the fetus as well, but there is not enough evidence for the same [20].

Inhaled carbon nanoparticles can also travel along the nasal mucosa through the first cranial nerve to reach the olfactory bulb in the brain. Some inhaled nanoparticles have also been shown to cause adverse effects on the cardiovascular system in animal and human studies, but the exact component has been unidentifiable [12].

Exposure through ingestion occurs with cosmetic products applied to lips or from elsewhere by hand to mouth contact. This route is also relevant in clearing nanoparticles from the lungs through mucociliary elimination. While most of the ingested nanoparticles clear rapidly, a small fraction might be retained. Carboxylated nanoparticles of latex can easily penetrate the intestinal mucous layer and reach enterocytes, with smaller particles crossing more readily [21].

Other possible modes of entry of nanoparticles are intravenous, ocular, genital mucosal, and peritoneal routes. The mechanism of toxicity has been proposed to be mostly dose-dependent cytotoxicity and oxidative stress leading to inflammatory responses [22].

21.4 Nanoparticles and their hazards

All the health effects of the nanoparticles used in consumer products are not yet known, though nanotoxicology has revealed adverse health effects of materials previously considered safe. The toxicity of various nanoparticles has been summarized in Table 21.1.

Nanoparticles of the oxides of titanium and zinc have been extensively used in sunscreens to make them more esthetically pleasing. In vivo studies have now demonstrated that these particles do not pose a risk as they are unable to penetrate the intact skin [23].

TABLE 21.1 Toxicity of nanoparticles.

Nanoparticles	Possible toxic effects
Nanoparticles of titanium oxide	• Damage to the nucleic acids, cell organelles, and mutations in chromosomes • Pulmonary toxicity • Liver cell necrosis • Raised markers of cardiac damage
Nanoparticles of zinc oxides	• Brisk inflammatory response
Silica nanoparticles	• Acute inflammatory response and consequent granuloma formation in the lungs
Nanoparticles of cobalt and chromium	• Destruction of human fibroblasts
Nanoparticles of silver	• Toxic to keratinocytes and skin fibroblasts as well as of the lungs • Inflammation in the alveoli and granuloma formation
Carbon nanotubes	• Granulomatous tissue response
Nanoparticles of aluminum	• Alzheimer's disease

However, concerns regarding penetration when sunscreens are applied to burnt, diseases, and damaged skin still remain. Titanium oxide produces free radicals on exposure to ultraviolet radiation and this can lead to damage to the nucleic acids, cell organelles, and mutations in chromosomes. Pulmonary exposure to titanium oxides leads to pulmonary toxicity by producing inflammatory cytokine profile and goblet cell hyperplasia, having a potential to aggravate pulmonary disease [24]. These particles have been detected in the liver after oral ingestion and produce liver cell necrosis and also raised markers of cardiac damage [25]. Teratogenic effects such as brain damage and decreased sperm count have been seen in experimental studies on mice with titanium oxide. Studies on rats have shown that chronic exposure to inhaled titanium oxide can lead to tumors in the lungs [23]. Zinc oxides have also shown brisk inflammatory response that resolves after the exposure ceases. Silica nanoparticles have been reported to produce an acute inflammatory response and consequent granuloma formation in the lungs, but other studies do not support this finding [26].

Nanoparticles of cobalt and chromium can penetrate the intact skin barrier and cause destruction of human fibroblasts [27]. Silver, widely used as an antibacterial agent, proves to be toxic to humans or animal cells when it used in a nanoparticle form. Inhalation of silver nanoparticles leads to their migration to the olfactory bulb, where they locate in mitochondria, as well as translocation to the circulatory system, liver, kidneys, and heart [28]. Silver nanoparticles have been found in the blood of patients with blood diseases and in the colon of patients with colon cancer [29]. Silver particles have been shown to be toxic to keratinocytes and skin fibroblasts as well as of the lungs. These particles after ingestion are taken up by intestines and gain entry in the peripheral circulation. It is the subacute and chronic exposure to silver particles, and not an acute exposure, that produces inflammation in the alveoli and granuloma formation [30]. Silver nanoparticles deposit in the laminae of the skin and produce argyria, a blue black skin discoloration. Liver is proposed to be the site most affected by nanosilver and nanogold toxicity [31]. Silver nanoparticles have also been shown to cause a disruption of the blood brain barrier, neural degeneration, and astrocyte swelling in rats. It has also been shown to increase neurotransmitters such as dopamine and serotonin, resulting in behavioral abnormalities.

Carbon nanotubes behave like asbestos fibers owing to their long fine nature and can induce inflammation and mesothelioma. The presence of iron in the carbon nanotube was found to be an important predictor of toxicity in an in vitro study. Granulomatous tissue response is also seen after percutaneous exposure. Fullerenes at a lower concentration have antiinflammatory effects but they elicit inflammation in the lungs at a higher concentration. In vitro studies have shown that fullerenes might be toxic to the cardiovascular system by being prothrombogenic. Contradictory reports exist in the literature regarding the genotoxic potential of fullerenes, for which the photoactivity of fullerenes is believed to be responsible [31].

The superparamagnetic nanoparticles have shown neurotoxic effects by the generation of reactive oxygen species, impairment of mitochondrial processes, and leakage of lactate dehydrogenase [32]. Quantum dot nanoparticles have been detected in the brain after intraperitoneal exposure in animal studies, though they do not produce any inflammation [33].

A controversial subject is the association between the uptake of aluminum and Alzheimer's disease. Epidemiological studies researching the connection between aluminum in antiperspirants, antacids, or drinking water and Alzheimer's disease are conflicting, some finding positive associations and others none [34]. Due to their latent evolving nature and multipart etiology, these neurological diseases are difficult to associate with specific factors. For example, the exposure takes place much earlier than the disease occurrence; hence, the subjects may not recall possible exposure; their memory is already affected by the disease. Moreover, subjects that suffer from advanced neurodegenerative diseases are not likely to participate in epidemiological studies due to their reduced ability to communicate and remember. In addition, multiple factors are known to contribute to Alzheimer's disease, such as genetics, increasing age, endocrine conditions, oxidative stress, inflammation, smoking, infections, pesticides, and electromagnetic fields [34].

Conclusion

Nanomedicine has opened hitherto unknown arenas in diagnosis and therapy of diseases as well as in cosmetology. This has led to a boom in the nanomedicine industry with a large number of products being produced each day. However, there is also a growing alarm that nanoparticles are associated with considerable toxic effects on the human body and the environment. Thus appropriate studies regarding their safety must be undertaken before granting permission to introduce these particles on a commercial scale. The safety of nanoparticles already in use should also be appropriately assessed. These steps might help in the judicious use of nanoparticles in the field of medicine, without compromising the safety of the consumer or patient and the environment.

References

[1] Arif T, Nisa N, Amin SS, Shoib S, Mushtaq R, et al. Therapeutic and diagnostic applications of nanotechnology in dermatology and cosmetics. J Nanomed Biotherapeutic Discov 2015;5:134.
[2] Saraceno R, Chiricozzi A, Gabellini M, Chimenti S. Emerging applications of nanomedicine in dermatology. Skin Res Technol 2013;19:13–19.
[3] Bangale MS, Mitkare SS, Gattani SG, Sakarkar DM. Recent nanotechnological aspects in cosmetics and dermatological preparations. Int J Pharm Pharm Sci 2012;4(2):88–97.
[4] Drexler KE. Machine-phase nanotechnology. Sci Am 2001;285(3):74–5.
[5] Arif T, Adil M. Nanotechnology dermatological perspective. Int J Nanomed Nanosurg 2016;2(2).
[6] Arif T, Sami M. Nanodermatology: a blessing or bane? Clin Dermatol Open Access J 2016;1(1):000105.
[7] Nanoscience and nanotechnologies: opportunities and uncertainties. The Royal Society and the Royal Academy of Engineering, London, UK. 2004.
[8] Vishwakarma V, Samal SS, Manoharan N. Safety and risk associated with nanoparticles: a review. J Miner Mater Charact Eng 2010;9:455–9.
[9] Nasir A. Nanodermatology: a glimpse of caution just beyond the horizon—part II. Skin Ther Lett 2010;15:4–7.
[10] Oberdörster G, Oberdörster E, Oberdörster J. Nanotoxicology: an emerging discipline evolving from studies of ultrafine particles. Environ Health Perspect 2005;113:823–39.
[11] Bos JD, Meinardi MM. The 500 Dalton rule for the skin penetration of chemical compounds and drugs. Exp Dermatol 2000;9:165–9.
[12] Papakostas D, Rancan F, Sterry W, Blume-Peytavi U, Vogt A. Nanoparticles in dermatology. Arch Dermatol Res 2011;303:533–50.
[13] Sayes CM, Fortner JD, Guo W, Lyon D, Boyd AM, Ausman KD, et al. The differential cytotoxicity of water-soluble fullerenes. Nano Lett 2004;4:1881–7.
[14] MonteiroRiviere NA, Nemanich RJ, Inman AO, Wang YY, Riviere JE. Multiwalled carbon nanotube interactions with human epidermal keratinocytes. Toxicol Lett 2005;155:377–84.
[15] High WA, Ayers RA, Chandler J, Zito G, Cowper SE. Gadolinium is detectable within the tissue of patients with nephrogenic systemic fibrosis. J Am Acad Dermatol 2007;56:21–6.
[16] Boraschi D, Costantino L, Italiana P. Interaction of nanoparticles with immunocompetent cells: nanosafety considerations. Nanomedicine 2012;7:121–31.
[17] Toll R, Jacobi U, Richter H, Lademann J, Schaefer H, Blume-Peytavi U. Penetration profile of microspheres in follicular targeting of terminal hair follicles. J Invest Dermatol 2004;123:168–76.
[18] Yah CS, Iyuke SE, Simate GS. A review of nanoparticles toxicity and their routes of exposures. Iran J Pharm Sci 2011;8:299–314.
[19] Hoet PHM, Bruske-Hohlfield I, Salata OV. Nanoparticles—known and unknown health risks. J Nanobiotechnol 2004;2:12.
[20] MacNee W, Li XY, Gilmour P, Donaldson K. Systemic effect of particulate air pollution. Inhal Toxicol 2000;12 (Suppl. 3):233–44.
[21] Szenkuti L. Light microscopic observations on luminally administred dyes, dextranes, nanosheres and microspheres in the pre-epithelial mucus gel layer of the rat distal colon. J Control Release 1997;46:233–42.
[22] Xia T, Kovochich M, Brant J, Hotze M, Sempf J, Oberley T, et al. Comparison of the abilities of ambient and manufactured nanoparticles to induce cellular toxicity according to an oxidative stress paradigm. Nano Lett 2006;6:1794–807.
[23] Heinrich U, Fuhst R, Rittinghausen S, Creutzenberg O, Bellmann B, Koch W, et al. Chronic inhalation exposure of Wistar rats and two different strains of mice to diesel engine exhaust, carbon black, and titanium dioxide. Inhal Toxicol 1995;7:533–6.
[24] Chen HW, Su SF, Chien CT, Lin WH, Yu SL, Chou CC. Titanium dioxide nanoparticles induce emphysema-like lung injury in mice. FASEB J 2006;20:2393–5.
[25] Wang J, Zhou G, Chen C, Yu H, Wang T, Ma Y, et al. Acute toxicity and biodistribution of different sized titanium dioxide particles in mice after oral administration. Toxicol Lett 2007;168:176–85.
[26] Barbarin V, Xing Z, Delos M, Lison D, Huaux F. Pulmonary overexpression of IL-10 augments lung fibrosis and Th2 responses induced by silica particles. Am J Physiol Lung Cell Mol Physiol 2005;288:L841–8.

[27] Lohani A, Verma A, Joshi H, Yadav N, Karki N. Nanotechnology based cosmeceuticals. ISRN Dermatol 2014;2014:843687.

[28] Takenaka S, Karg E, Roth C, Schulz H, Ziesenis A, Heinzmann U, et al. Pulmonary and systemic distribution of inhaled ultrafine silver particles in rats. Environ Health Perspect 2001;109(Suppl. 4):547–51.

[29] Gatti AM, Montanari S, Monari E, Gambarelli A, Capitani F, Parisini B. Detection of micro- and nano-sized biocompatible particles in the blood. J Mater Sci Mater Med 2004;15:469–72.

[30] Sung JH, Ji JH, Yoon JU, Kim DS, Song MY, Jeong J, Han BS, et al. Lung function changes in Sprague-Dawley rats after prolonged inhalation exposure to silver nanoparticles. Inhal Toxicol 2008;20:567–74.

[31] Johnston HJ, Hutchison G, Christensen FM, Peters S, Hankin S, Stone V. A review of the in vivo and in vitro toxicity of silver and gold particulates: particle attributes and biological mechanisms responsible for the observed toxicity. Crit Rev Toxicol 2010;40:328–46.

[32] Yildirimer L, Thanh NTK, Loizidoua M, Seifalian AM. Toxicological considerations of clinically applicable nanoparticles. Nano Today 2011;6:585–607.

[33] Arora M, Katoch K, Natrajan M, Kamal R, Yadav VS. Changing profile of disease of leprosy patients diagnosed in a tertiary care centre during years 1995-2000. Indian J Lepr 2008;80:257–65.

[34] Antonini JM, Santamaria AB, Jenkins NT, Albini E, Lucchini R. Fate of manganese associated with the inhalation of welding fumes: potential neurological effects. Neurotoxicology 2006;27:304–10.

CHAPTER

22

In vitro standard methods for cellular toxicity of nanocosmetic

Swati Gajbhiye

Department of Cosmetic Technology, LAD & SRP College for Women, Nagpur, India

22.1 Introduction

In recent year, the interest in nanocosmetic has increased as it offers several advantages over the convectional cosmetic, and thus the nanotechnology in the field of cosmetic has emerged as a part of modern era cosmetic [1]. The advance in the field of nanotechnology has also led to the advancement in the field of nanocosmetic as it also offers the revolutionize treatment of several skin diseases including antiaging treatments. With the advancement in the nanotechnology, the cosmetic has got the new and improved properties such as better skin penetration, better entrapment, enhanced performance, and long-lasting effects. This has proved to stand with the expectations of both consumers and manufacturers in terms of better quality, effectiveness, and improved stability of the nanocosmetic [2]. Nanotechnology has led to several innovating products with nanoparticles or nanocolloidal carriers that can penetrate deeper into the skin for the better effectiveness of the cosmetic with improved results [3]. As this technology has been used in several skin care cosmetic including sunscreens and antiaging creams on large scale which includes natural and physical agents successfully in nanocosmetic [4]. This is due to the smaller dimension and large surface area which has greater potential ability for quick and deeper penetration. At the same time, safety aspect of the nanocosmetic should also be considered. In general, smaller particles have got large surface area to volume ratio; this has also led to increase the chemical reactivity which likely to make it more toxic in nature [5]. Due to the smaller size of the nanoparticles, it accesses easy penetration but it has also got the drawback of easily taking up by the cell mitochondria and the cell nucleus which results in increased oxidative stress, inflammatory cytokine production, and cell death [6]. Nanoparticles can cause cytotoxicity by adherence of particles to cell membrane, degradation, and subsequent release of cytotoxic degradation product. It has also been reported that nanoparticles can cause DNA mutation which results in the structural damage to

Nanocosmetics
DOI: https://doi.org/10.1016/B978-0-12-822286-7.00018-8

mitochondria and leads to cell death. Toxicity is mainly due to the production of reactive oxygen species which includes free radicals results in oxidative stress, inflammation, and damage to the DNA [6]. Due to the smaller size nanoparticles, it can easily access to the blood stream through skin and inhalation and can transport to the various organs [7]. It is also reported that high dose and long residence time of nanocosmetic can result in dysfunction of vital organs. This is because the nanoparticles that are smaller in size have the ability to alter the distribution and the bioavailability of the same material in comparison to the same material that are larger in size with the same chemical composition. The increase in biological interaction may be due to the smaller size of the material, which has got the large surface area to the mass of the particle [8]. Certain studies have also shown that the nanoparticles can penetrate through the pig skin within 24 hours, thus concerns have been raised that the nanoparticles in sunscreen can penetrate through the damaged skin [9]. However, there are several studies done on the safety of the nanoparticles which ruled out the penetration of nanoparticles into systemic circulation. This might be due to the safe elimination of the nanoparticles from systemic circulation [10].

So far, nanoscale materials are used in cosmetic ingredients very small scale but their use is growing [11]. Nowadays, nanomaterials are mostly used in sunscreen, antiaging, lipsticks, toothpaste, color cosmetics, etc. Some of the active ingredients such as ascorbic acid, vitamins like A, E, and K including essential oils which tends to oxidized and loses its activity are now being encapsulated by incorporating it into the nanosized vesicular systems such as niosomes, liposomes, transferosomes, and microspheres [12]. Nanopigments which may be metallic and nonmetallic are extensively used in cosmetics. Solid lipid nanoparticles which are nonmetallic nanoparticles are now being used. Gold and silver nanoparticles which are technically inert and nontoxic are widely used in lipsticks as nontoxic colorants. Titanium dioxide and zinc oxide are used as sunscreen agent and using it in nanosize helps the cosmetic formulation to give it a lighter texture, better dispersibility, and transparency [13]. In general, most of the nanocosmetic including sunscreens, antiaging cream, and toothpaste now contain one or the other type of nanoparticles [14]. No doubt that the nanotechnology has so much to offer in cosmetics but the technology and its safety are need to considered for the safety of the consumers [15]. When we talk about the toxicity of the nanomaterial where traditional methods for the testing of nanomaterial are used, the manufacturer must also consider the test methods that require the modification in nanomaterial testing such as solvent and dosing formulations, aggregation, and agglomeration of nanoparticles and most important the stability condition that are associated with the nanocosmetic [16]. This is because there are some nanomaterials that are soluble, insoluble, and partially soluble in the particular solvent which are likely to affect the traditional methods that are used for toxicity tests [17]. In case if the traditional toxicity methods cannot be modified then FDA do not restrict the development in the new methods for the safety assessment of the nanocosmetics. As the nanoparticles have the tendency to aggregate and agglomerate that affects the solubility in the particular solvent, this factor is likely to affect the traditional methods for toxicity testing. Therefore toxicity testing should be conducted on both individually on free nanoparticles and their aggregates and agglomerates. Thus the manufacturer should incorporate the appropriate modification in the testing methods so that it can reflect the range of free particles and any other aggregates and agglomerates that may be found in the nanocosmetic formulation.

These modifications of the testing methods assess the toxicity of the nanocosmetics and ensure the safety of nanocosmetics [18].

Cosmetic those are applied by spray such as perfumes, deodorants, hair spray, etc., has the risk of inhalation and the toxic nanomaterial that might present in such cosmetic has the possibility that might get into systemic circulation from inhalation along with ocular and oral exposures. Therefore the toxicity testing of exposure organs should be considered while evaluating the toxicity of such cosmetic formulations [19]. Most commonly aerosolized cosmetic containing nanoparticles may result in the deposition of nanoparticles in the respiratory system via the respiratory tract. This might be due to the aerosol nanocosmetic ingredients interaction with the respiratory epithelium [20]. Such nanomaterials especially which are soluble nanoparticles might get metabolized and transfer to the internal organs and blood whereas the insoluble nanoparticles may get retained by the respiratory tract and result in pulmonary effects. Thus there are some potential adverse effect that might not only affect the respiratory system but also the heart and blood vessels along with the CNS and immune system of the body [21].

In general, the exposure through oral route is said to be limited as such products which includes mouthwash, lips cosmetics are used limitedly but the studies have shown that the uptake of nanoparticles has the limited uptake into the gastrointestinal tract but the translocation of such materials in the intestinal barrier can be increased substantially [22].

As discussed earlier, nanoparticles have got the unique physicochemical properties and different from larger particles, and thus the potential toxicity of a compound might also increase. For example, reduction in particle size increases the intake or the absorption properties of the nanoparticles into the deeper layer which further can get into cells and across the blood–brain barrier; this might also alter the bioavailability or biological half-life [23]. Due to the smaller size of nanoparticles and large surface area, small quantity of dose can show great result and thus dose delivery to the sensitive tissues also increases [24]. The toxicity level of nanocosmetic also increases due to the higher dose level, that is, greater than the desired level and also due to the long duration exposure of the nanoparticles. Thus the toxicity evaluation of nanocosmetic should be based on the absorption, distribution, metabolism, and excretion parameters for the safety of the nanocosmetic [25].

The following part of the chapter discusses about the toxicity assessments of nanocosmetics. The Cosmetic, Toiletry and Fragrance Association (CTFA) and Organization for Economic Cooperation and Development (OECD) has given the guideline for the toxicity testing, which is based on toxicological profile of the ingredients and their routes of exposure. Some of the toxicological testings that are based on the type of cosmetic products are skin irritation and skin sensitization test, phototoxicity test, ocular irritation test, and test for genotoxicity and mutagenicity [26]. These testing methods can change with the time as the new technologies are developed; and thus it is mandatory to use the latest test methods that are available in the guideline. The safety assessment of the cosmetic in 7th Amendment has imposed the ban on the animal testing; this prohibits the use of in vivo testing of the finished cosmetics and thus only the in vitro validated methods can be used with the modification in the traditional methodology for testing nanomaterials [27]. Though there are several methods that are available for studying biological activity of compounds, only validated methods are permitted for cosmetic products. These validated methods must be used for the safety assessment of cosmetic ingredients.

In recent years, there has been a great development in the in vitro toxicology studies which work on the principles based on the 3Rs strategy, that is, refinement, reduction, and replacement. Although there are various alternative methods that are available for studying the biological compounds but only a few validated methods are permitted for regulatory purposes. The validated 3R methods are mostly used in cosmetic for the safety assessment of the cosmetic ingredients [28]. Following are the validated in vitro methods that can be used in the safety assessment of cosmetic ingredients.

22.2 In vitro skin irritation test

Skin irritation is defined as the reversible damage produce to the skin after the exposure to the chemical or nanomaterial. This test guideline provides in vitro procedure that is used to detect the irritant chemical or nanomaterial, by replacement of in vitro testing of skin irritation using animals by alternative methods or as a partial replacement testing strategy. This test is based on the in vitro test that uses the reconstructed human epidermis (RhE), which resembles the biochemical and physiological properties of upper part of human skin known as epidermis. This is reconstructed from human epidermis obtained from human-derived nontransformed epidermal keratinocytes, which closely mimic the histological, morphological, biochemical, and physiological properties of the epidermal layer of human skin conforms to OECD regulatory guidelines [29].

22.2.1 Initial considerations and limitations

This test guideline does not provide the information on the skin corrosion test; thus OECD TG 431 should be used which is based on the same RhE test system. This guideline represents the in vitro target organ of species of interest, which is based on the RhE model that uses human keratinocytes [30]. Various three-dimensional (3D) human skin models are used for the alternative test methods such as EpiSkin, EpiDerm, and SkinEthic for regulatory purpose. This guideline is used for the variety of substance ranges from liquid to solid, in the case of solid substance it must be grounded or other pretreatment is required before application [31].

22.2.2 Principle

Test substance is tropically applied on the three-dimensional RhE model, which is comprised of nontransformed human-derived epidermal keratinocytes, which cultured into multilayered similar to human epidermis. It consists of organized basal, spinous, and granular layers along with stratum corneum multilayered with intercellular lamellar lipid layers [32].

Skin irritation is normally induced by the chemical or the substance which result in the erythema and edema. This is due to the cascade events that arise due to the chemical penetration through stratum corneum which damages the lower layer of keratinocytes and other skin cells [33]. Inflammatory mediators are released by the damage cells or they may also induce inflammatory cascade which may act on the cells in dermis and on the stromal and endothelial cells of blood vessels [34]. Thus erythema and edema are produced by the dilation and increased permeability of these endothelial cells. But in the absence of any vascularization in the in vitro test, RhE based test measures the events of cascade, that is, cell/tissue damage by cell viability assay. The RhE based test measures the cell viability by enzymatic conversion of vital dye MTT into blue formazan which is measured quantitatively after extraction from the tissues. The test substance that produces the cell viability above the threshold value must be consider nonirritant, that is, >50% as the irritant substance tends to decrease the cell viability below threshold value, that is, <50% [35].

22.2.3 Procedure

22.2.3.1 General condition

The epithelium should be reconstructed by nontransformed human keratinocytes. The EpiDerm or any epithelial cell should consist of multiple viable cell layers under the functional stratum corneum containing essential lipid profile so that it can withstand the functional barrier to resist rapid penetration of cytotoxic chemicals or substances. The RhE model should be free of contamination by bacteria, mycoplasma, etc., as these may result in poor passage of material through stratum corneum and result in false result [36].

22.2.3.2 Functional condition

22.2.3.2.1 Viability

The MTT assay quantifies the viability of the cells of RhE tissue by reducing vital dye MTT into blue MTT formazan precipitate which is then extracted by using isopropanol from the tissue. Then, the optical density of the extracted MTT formazan should be quantified by standard absorbance (OD) measurement or by HPLC/UPLC spectrophotometry procedure. An acceptable range (upper and lower limit) for negative control OD value for HPLC/UPLC spectrophotometry should be in between 0.6 and 1.5.

22.2.3.2.2 Quality control

The data should be provided by the RhE model developer to the user for the acceptable range (upper and lower limit) for IC_{50} value. The normal range for the EpiSkin for the 18 hour treatment with SDS is in between 1 and 3 mg/mL [37].

22.2.3.2.3 Application of test chemical and control substance

For each test chemical or nanomaterial three replicates should be used along with the control for each. The test substance should be used in the sufficient quantity so that it can be applied to the epidermis surface uniformly. The recommended range is from 26 to 83 μL/cm^2 or mg/cm^2. The epidermis surface should be moistening for the solid substance application with the water and should be used as fine powder. The test substance or nanomaterial should be washed from the surface of the epidermis with buffer or 0.9% NaCl, for the exposure period of 15–60 minutes and incubated at 20°C–37°C. The negative and positive control should be used in each run to demonstrate the viability.

The recommended positive control should be 5% aqueous SDS and for negative control water of phosphate buffer saline should be used [38,39].

22.2.3.2.4 Cell viability measurements

It is recommended to measure the viability after the posttreatment incubation period when the tissue is rinsed in fresh medium. This allows the recovery period from weak cytotoxic to clear cytotoxic effect. This period is usually recommended as 42 hour post-treatment incubation period.

The MTT assay is compatible with three-dimensional tissues to measure the cell viability. When the tissue is placed in the MTT solution in the concentration of 0.36–1 mg/mL for about 3 hours, it converts MTT into blue formazan. This precipitated formazan is extracted by solvent such as isopropanol and is measured by OD at 570 nm or by HPLC/UPLC spectrometry procedure.

False result may occur in the test if chemical is not removed from the tissue by rinsing or when it penetrates the epidermis. Thus the freshly prepared MTT solution is introduced into the test chemical. In the case of colored test chemical or the substance is used, it may not be compatible with OD measurement as it may interfere with MTT assay; then HPLC/UPLC spectrophotometry can be used as this system allows the separation of MTT formazan from the test chemical or substance before quantification [40].

22.2.3.2.5 Interpretation of in vitro test results

The percentage of viability is calculated from the OD values obtained from each chemical and for negative control it is set as 100%. The percentage viability for HPLC/UPLC spectophotometry is calculated as percentage MTT formazan peak obtained with living tissue to the MTT formazan peak obtain with concurrent negative control. If the tissue viability after exposure and posttreatment incubation is less than or equal to 50%, then the chemical or substance is considered as irritant to skin in accordance with OECD TG439 [41].

22.3 In vitro skin corrosion test

22.3.1 General description

Skin corrosion is define as the irreversible skin damage with a visible necrosis through epidermis and then into the dermis by the application of the cosmetic or any other test substance. This validated in vitro test avoids the pain and suffering of the animals. This test is based on the method called as rat skin transcutaneous electrical resistance (TER), which identifies the corrosive by using skin discs by the ability to produce loss of normal stratum corneum integrity and barrier function [42].

22.3.2 Initial considerations

The TER test method can discriminate between skin corrosive and noncorrosive with 94% of sensitivity and 71% of specificity but not the subcategorization of corrosive substance [43].

22.3.3 Principle of the test

The test substance is applied to the epidermal surfaces of skin discs in two compartment test systems for 24 hours in which skin disc functions as separation between the compartments. The corrosive substance is identified by the ability of the skin disc to produce a loss of normal stratum corneum integrity and barrier function; this is detected by the reduction in TER below the threshold level. Rat skin TER cutoff value is 5 KΩ or in between 10 and 3 KΩ. And the value is confirmed as a positive result by the dye binding step, which may be due to the increase in permeability due to the physical destruction of stratum corneum [43,44].

22.3.4 Procedure

Rat skin is used as it is sensitive and has been validated. The age and the strain of the rats are taken into consideration at the time of collection as it ensures the hair follicles at the dormant phase before the hair growth begins.

22.3.4.1 Preparation of skin disc

The skin disc is placed over the (PTFE) polytetrafluoroethylene tube in such a way that the epidermal surface is in contact with the tube. A rubber "O" ring is fitted over the end of the tube so that it can hold the skin in place and can be trimmed off the excess tissue. This ring is carefully sealed with the petroleum jelly. The spring clip inside the tube supports the tube with the receptor containing $MgSO_4$ solution. With this technique from the single rat skin 10–15 skin disc can be obtained. Before the evaluation of the test, the disc is checked for the electrical resistance value which should be greater than 10 KΩ that is recommended.

22.3.4.2 Application of the test and control substance

Positive and negative controls are used in each experiment with the use of single animal skin disc obtained in each experiment. For positive control, 10 M hydrochloric acid is used and for negative control water is used. In the case of liquid test substance, it should be around 150 μL applied to epidermal surface within the tube. Whereas for solid it should be evenly applied or powdered so as to cover the entire epidermal surface if needed deionized water around 150 μL should be used and agitated gently. If needed, the test substance can be warmed at 30°C. For this experiment, three skin discs for each test are used. Test substances are applied for 24 hours at 20°C–23°C and then the test substance is removed by the jet of tap water at room temperature [45].

22.3.4.3 Transcutaneous electrical resistance measurements

The TER is measured by the low voltage, alternating current wheatstone bridge with 1–3 V operating voltage, alternating current of 50–1000 Hz, and measuring range up to 0.1–30 KΩ. For TER corrosivity measurement resistance is recorded at the frequency of 100 Hz. Before measuring electric resistance, the skin surface resistance should be reduced by adding 70% ethanol and then the tissue is hydrated by adding 3 mL of $MgSO_4$. If the resistance value is greater than 20 KΩ, then the $MgSO_4$ solution is discarded and fresh $MgSO_4$ solution is used [46].

22.3.4.4 *Dye binding method*

The electrical resistance is reduced by the noncorrosive substance as allow the passage of ion through stratum corneum and can result in low resistance below the cutoff value 5 KΩ. Substance such as surface active agent can remove the skin lipids, which makes the barrier more permeable to ions. Thus both control and treated tissue should be determined for the TER value by the dye penetration method. In the case of skin corrosion where stratum corneum is disrupted then dye sulforhodamine B can be used, which is suitable for wide range of substance.

22.3.4.5 *Sulforhodamine B dye application and removal*

After the TER evaluation, the magnesium sulfate is discarded from the tube and the tissue is examined for any kind of damage. Then, 150 μL of 10% dye sulforhodamine B diluted in distilled water (w/v) is exposed to the skin disc for 2 hours and then it is washed with tap water and finally skin is removed and placed in vial containing 8 mL deionized water. This vial is then agitated for 5 minutes to remove the additional dye. This vial is washed repeatedly with 5 mL of 30% (w/v) sodium dodecyl sulfate in distilled water and then incubated overnight at 60°C after incubation vials are discarded with the remaining solution and centrifuged for 8 minutes at 21°C. Then, 1 mL sample of supernatant is diluted with 1 in 4 parts of 30% SDS in distilled water and measure for the optical density at 565 nm [47].

22.3.4.6 *Calculation of dye content*

The dye contain is determined by the calibration curve and mean dye content.

22.3.4.7 *Acceptability criteria*

The recommended acceptable resistance range of mean TER result is as follows [45]:

Control	Substance	Resistance range (KΩ)
Positive	10 M hydrochloric acid	0.5–1
Negative	Distilled water	10–25

The recommended acceptable resistance range of mean dye binding result is as follows:

Control	Substance	Dye content range (μg/disc)
Positive	10 M hydrochloric acid	40–100
Negative	Distilled water	15–35

22.4 In vitro phototoxicity test

Photoxicity is defined as the toxic effect from the dermal application of the cosmetic or any other dermal application which increases after the exposure to the sunlight; it can also experience skin irradiation after the systemic administration of the substance. The in vitro 3T3 NRU phototoxicity test is generally used to identify the potential of the phototoxic substance, which is increased after the cosmetic or the chemical exposed to the light. The photocytotoxicity is evaluated by the reduction in viability of cells that are exposed to the cosmetic or the chemical in the presence versus absence of light [48].

22.4.1 Initial consideration

There are various types of substances that induce phototoxic effects including nanoparticles used in cosmetic. This is due to the ability to absorb the light energy within the sunlight range as it requires the absorption of the light quanta for photoreaction according to the first law of photochemistry. Therefore it is recommended to determine UV/Vis absorption spectrum of test chemical according to OECD 101 before biological testing is considered. If the molar absorption coefficient is less than $10\ L \times mol^{-1} \times cm^{-1}$, the substance is unlikely to be photoreactive than such substance may not be used for testing in vitro 3T3 NRU phototoxicity or any other biological test for adverse photochemical effects [49].

22.4.2 Principle of the test method

This test is based on the comparison of cytotoxicity of substance or nanomaterial that is needed to be tested in the presence and absence of noncytotoxic dose exposure to simulate solar light. Cytotoxicity in this test is based on the reduction of uptake in Neutral Red dye which is concentration dependent when measured after 24 hours treatment. This weak cationic Neutral Red dye penetrates into the cell membrane by nondiffusion and accumulating intracellularly on lysosomes. This alteration in the cell leads to the lysosomal fragility and irreversible changes. The action of xenobiotics results in such changes, which gradually decrease the uptake of Neutral Red binding. Thus this is the base of the test which is why one can distinguish between viable cells, damage cells, or dead cell.

Balb/c cells are used in the test and maintained in culture for 24 hours which forms the monolayers. In two 96-well plates, the test substance is preincubated for 1 hour with eight different concentrations. One plate is exposed to the noncytotoxic irradiation high dose and other is kept in dark. The treatment medium is replaced by the culture medium in both the plates and after 24 hours it is determined for the cell viability by uptake of Neutral Red dye and the viability is evaluated as percentage of control compared to the percentage of each test concentration. This is in both the presence and absence of irradiation at the IC_{50} level [50,51].

22.4.3 Description of the test method

22.4.3.1 Preparations

22.4.3.1.1 Cells

Mouse fibroblast cell line Balb/c 3T3, clone 31, either can be used along with it; other cells can also be used if proper culture condition can be maintained. Cell should be checked for the mycoplasma contamination. UV sensitivity of the cell can be checked regularly because UVA sensitivity increases with the number of passages.

22.4.3.1.2 Media and culture

For routine cell passage and during test procedure, appropriate culture media and incubation condition should be used and humidified along with incubation at 37°C, 5%–7.5% CO_2 to assure cell cycle time and normal historical range of cells.

22.4.3.1.3 Preparation of culture

Cell must be seeded and subcultured at least once before the test. And the cell density should be 1×10^4 cells per well in 96-well plates where one plate is irradiated and other is kept in dark.

22.4.3.1.4 Preparation of test substance

The nanomaterial test substance used for the testing should be prepared fresh and the treatment must be performed under light condition so as to avoid photoactivation or degradation of the test nanomaterial prior to irradiation. Salt solution must be used to dissolve test substance and free from protein and light-absorbing component such as vitamins.

Test substance must be dissolved in appropriate solvent and should be noncytotoxic and concentration should be such so as to avoid precipitation. Solvents such as DMSO and EtOH of low cytotoxicity may be used appropriately such as to avoid the test substance reaction, quenching of phototoxic effect, radical scavenging properties, and chemical instability. The test substance must be vortex, sonicated, or warmed up for the solubilization and stability of test substance.

22.4.3.1.5 Irradiation conditions

Light source: UVA and visible regions are associated with phototoxic reaction in vivo but UVB is cytotoxic in nature and increases 1000-fold as wavelength goes from 313 to 280 nm. Light source must be such that test substance should absorb the emitted wavelength and should be sufficient to detect phototoxic substance.

Xenon arcs and (doped) mercury-metal halide arcs are used as solar simulators as it emits sufficient UVB with suitable filter to attenuate cytotoxic UVB wavelengths. For Balb/c cells, 5 J/cm^2 for 50 minutes is used and irradiance was adjusted to 1.7 mW/cm^2. For another cell line, irradiation dose must be calculated in the following way.

$$t(\text{min}) = \frac{\text{irradiation dose}\left(\frac{\text{J}}{\text{cm}}\right)^2 \times 1000}{\text{irradiance}\left(\frac{\text{mW}}{\text{cm}^2}\right) \times 60} (1\text{J} = \text{Ws})$$

22.4.3.2 *Test conditions*

22.4.3.2.1 Test substance concentrations

The test substance concentration in the presence (+lrr) and absence (−lrr) should be in the range 6.5–7.8 to avoid toxicity induced by culture condition or pH of cell culture. The maximum test concentration should not exceed 1000 g/mL and osmolality not more than 10 mM.

22.4.3.2.2 Test procedure

First, dispense 100 μL culture medium into a 96-well plate and labeled it as blank. In another 96-well plate, dispense 100 μL of cell suspension of 1×10^4 cells/well. Two plates should be prepared one for solvent and another for positive control. Then incubate the cells for 24 hours so as it forms the monolayer. This incubation allows the cell to grow in the exponential form.

Second day, culture medium is decanted and wash with 150 μL of buffer solution and then add test concentration containing 100 μL of buffer. Apply eight different concentrations of test substance and incubate it in dark for 60 minutes from two plates; one is used for detecting cytotoxicity (−lrr) for control and other for treatment for photocytotoxicity (+lrr). (+lrr) is exposed to irradiate cell at room temperature for 50 minutes and (−lrr) for 50 minutes in dark box at room temperature. After 50 minutes, decant the test solution and washed twice with 150 μL of buffer solution and replace it with culture medium and incubate overnight.

On third day, microscopic evaluation is done to check the cells growth and change in cell morphology.

22.4.3.2.3 Neutral Red uptake test

Wash the cell with 150 μL of prewarmed buffer and remove the washing if any by taping and add 100 μL of 50 μg/mL of Neutral Red dye and incubate for 3 hours. After incubation, remove the Neutral Red medium and wash the cells with 150 μL of buffer. Decant and remove excess by blotting or centrifugation. After that add 150 μL of Neutral Red desorb solution which is freshly prepared as 49 part of water + 50 part ethanol and 1 part acetic acid. Now shake the microliter plate for 10 minutes so as to homogenize the solution, that is, Neutral Red extracted from cells and then measure optical density of Neutral Red extracted at 540 nm in spectrophotometer [49,50,52,53].

22.4.3.2.4 Result

Test data should be analysis for concentration response obtained from the presence and absence of irradiation and calculate photoirritation factor.

A Photo-Irritation-Factor (PIF) is calculated using the following formula:

$$\mathrm{PIF} = \frac{\mathrm{IC}_{50}(-\mathrm{Irr})}{\mathrm{IC}_{50}(+\mathrm{lrr})}$$

22.5 In vitro dermal absorption measurements of cosmetic ingredients

The combination of OECD guideline 427 on skin absorption in vivo method and the OECD guideline 428 in vitro methods came up with the in vitro method for dermal absorption study in toxicological material of cosmetic.

22.5.1 General principle

The general principle of the dermal absorption includes three steps penetration, permeation, and resorption. The purpose of dermal absorption study is to obtain qualitative and quantitative information on amount of substance that may be entering into systemic compartment of human body. In vitro the microcirculation is compromised, thus potential of compound cannot be adequately studied in such setting.

22.5.2 Principle of the test

At present, skin preparation of natural origin may be used. OECD guideline 428 should be followed. The test substance should be applied in formulated skin sample, which is placed in diffusion cell which is positioned between upper and lower chambers of diffusion cell. The test sample should be in contact with skin donor side for a defined period of time. The receptor fluid should be sampled after 30 minutes. The skin and the fluid sample should be analyzed by analytical methods such as HPLC GC or other suitable methods. Information on sensitivity and time different should be provided [54].

22.5.3 Factors affecting dermal absorption and methodology

There are several factors that may affect the dermal absorption such as physical and chemical properties of substance, concentration of the substance, exposure pattern, and technical aspect in vitro test. Following are the factors that may affect in vitro dermal absorption studies.

22.5.3.1 *Diffusion cell design*

Diffusion cell consists of upper donor and lower receptor chamber, which is separated by the skin. Stratum corneum faces the donor chamber. Diffusion chamber should be made up of inert nonadsorbing material. Temperature of the receptor fluid should be maintained throughout the experiment. The temperature in diffusion cell at in vitro skin should be at $32^\circ \pm 1^\circ C$. The receptor fluid in static cells should be stirred throughout the study.

22.5.3.2 *Receptor fluid*

The composition of the receptor fluid should be such that it does not diffuse the test substance, that is, the solubility and stability of the chemical in receptor fluid should be maintained. For hydrophilic substance, buffer saline solution must be used, and for lipophilic molecules, serum albumin or suitable emulsifier should be added so that it does not interfere with the integrity of the membrane. Care must be taken during analytical procedure regarding the interference of the fluid. The receptor should be degassed to avoid the formation of air bubble at the time of experiment, thus it should be stirred (static cells) or should be continuously replaced (flow through cells) during experiment. The condition of the receptor cell should depend on the absorption properties and objective of the study. The amount of penetrated substance should not be more than 10% of saturation level in the receptor fluid so that the interference with free diffusion process should be minimized for underestimation of dermal absorption. It is important that the substance should be stable in the receptor fluid during in vitro test analysis [55].

22.5.3.3 *Skin preparation*

Human skin is normally used but it is not always available so the alternative pig skin can be used as it has the similar permeation characteristics with human skin. Rat skin is not recommended as it has 2–10 times more permeable than human skin and may result in false result. Also, guinea pig or rabbit skin is not recommended as it does not have the adequate barrier function similar to human skin

22.5.3.4 *Skin integrity*

Integrity of the barrier is crucial point for experiment. This can be measured by the penetration marker such as titrated water, sucrose, or caffeine or by physical method such as TER or TEWL (transepidermal water loss).

22.5.3.5 *Skin temperature*

The rate of skin absorption depends on temperature, and thus it should be maintained constant at 32° ± 1°C, which is similar to the normal human skin surface temperature.

22.5.3.6 *Test substance*

The physical and chemical data should be given such as solubility, stability, and pKa along with the nanoparticles absorption.

22.5.3.7 *Preparation of dose and vehicle*

The concentration should be such that the range of liner curve of concentration versus dermal absorption is demonstrated.

22.5.3.8 *Dose and volume of test substance*

Usually doses between 2 and 5 mg/cm^2 are used for the solid and 10 $\mu L/cm^2$ for liquids is applied at the same time the volume of the substance should be such that it should be spread evenly over the skin surface.

22.5.3.9 *Study period and sampling*

The exposure time of the substance should be 24 hours with regular sampling intervals. Longer duration may result in membrane deterioration, and thus it requires careful control of membrane integrity. The frequency of sampling should depend on the rate of absorption profile.

22.5.3.10 Analytical method

Analytical techniques such as HPLC or GC should be used along with the qualitative or semiquantitative methods such as microautoradiography may be used for skin distribution assessments.

22.5.4 Result

Dermal absorption should be expressed as $\mu g/cm^2$ of skin surface. The amount of penetrated substance present in receptor fluid is considered to be systemically available. The epidermis and dermis are considered as a sink; therefore the amounts found in these tissues are considered as absorbed and are added to the receptor fluid. The amounts that are retained by the stratum corneum at the time of sampling are not considered to be dermally absorbed, and thus they are not considered to be the part of the systemic dose. For each diffusion cell, the calculation must be done separately for the absorption rate and mass balance. Only then, the mean ± SD and median with 10% and 90% percentiles should be calculated [56].

22.6 Genotoxicity/mutagenicity testing

The potential genotoxicity assessment of nanomaterials is evaluated by standard genotoxicity assays. Test methods that are mostly used are given by OECD guideline these are as: Bacterial Reverse Mutation Test OECD 471, in vitro Mammalian Chromosome Aberration test OECD 473, in vitro Mammalian Cell Micronucleus Test OECD 487, and in vitro Mammalian Cell Gene Mutation Test OECD 476 [57]. Most commonly used test for genotoxicity of nanoparticles is micronuclei and comet assay [58]. The Ames test is most commonly used in detection of genotoxicity. Although many nanoparticles are found to be negative in the Ames test but found to be positive in comet assay and micronucleus assay [59]. Thus these two assays are frequently used for genotoxicity of nanoparticles. The positive result might be due to the presence of cytochalasin B (cytoB) in micronucleus assay which prevent cytogenesis as a result divided cell remains binucleated. CytoB inhibits phagocytosis which results in the intracellular dose, and thus geotoxicity. It is also reported that (S9 mix) rat liver homogenate forms nanoparticles micelles or coat the nanoparticles. The negative result in the Ames test might be due to the specific nature of the end point detection, difference in metabolic activation, or it might be due to the difference in bioavailability. Thus there are concerns regarding using cytoB and S9 mix in micronucleus assay [60]. Some of the in vitro tests are discussed below.

22.6.1 Ames test (bacterial reverse mutation test)

The bacterial reverse mutation test is the biological assay also known as Ames test which is used to assess mutagenic potential of substance. It utilizes bacteria to test if the given substance causes the mutation in the DNA of the test organism. This test was given by Bruce N. Ames in the 1970s to determine mutagenic material [61].

The amino acid requiring strain of *Salmonella typhimurium* and *Escherichia coli* are used in the bacterial reverse mutation test to detect point mutations. The principle of the test is that it detects mutation which revert mutations that are present in the test strain of bacteria which is capable of synthesizing essential amino acid. The revertant bacteria are detected by the parent test strain as they have the ability to grow in the absence of amino acid. This mutation results in his- or trp-organisms that cannot grow in the absence of histidine or tryptophan. Culturing His- salmonella in a media containing certain chemicals regains the ability to synthesis of histidine (His +) and causes mutation in histidine encoding gene. This is where mutagenic event occurs which is based on base substitutes or frameshifts in the gene which result in the reversion to amino acid prototrophy which is called reverse mutation. These reverted bacteria then grow in histidine or tryptophan deficient media. The potential of mutagenic material is assessed by exposing amino acid requiring organisms to the varying concentration of chemical and selecting for the reversion event. Media that lack certain amino acid are used in the selection which results in the cell growth which have undergone the reversion to histidine/tryptophan prototrophy. Thus if the material is mutagenic then it will cause this reversion.

22.6.1.1 *Principle of the method*

In this test, the bacterial cells suspension is exposed to the test substance in the exogenous metabolic activation system. There are two standard techniques of plating: one is the plate incorporation method and second is the preincubation method. In the first method, the suspension is mixed with the overlay agar and plated immediately onto minimal medium. In the second method, before plating onto the minimal medium the mixture is incubated and then overlay agar is mixed. In both the methods, revertant colonies are counted and it is compared with the number of revertant colonies of the control solvent plate. This is done after the 2 or 3 days of incubation.

22.6.1.2 *Description of method*

- First, the fresh bacteria culture should be grown (approximately 10^9 cells per mL) for the determination of viable cell numbers by a plating experiment.
- Temperature of the culture should be 37°C.
- Following strain of bacteria can be used single or in combination depending upon the mutagens results: (1) *S. typhimurium* TA1535; (2) *S. typhimurium* TA1537 or TA97 or TA97a; (3) *S.* typhimurium TA98; (4) *S.* typhimurium TA100, and (5) *E. coli* WP2 uvrA, or *E. coli* WP2 uvrA (pKM101), or *S. typhimurium* TA102 [62].
- Second, an auxotrophic strain of *S. typhimurium* is isolated for the histidine and tryptophan for *E. coli* strains which required amino acid to grow.
- Then the minimal agar containing Vogel-onner minimal medium E and glucose along with overlay agar which may contain histidine and biotin or typhimurium is used for the cell division.
- Metabolic activation: Bacteria should be exposed to the material that is been tested both in the presence and absence of metabolic activation systems. Most commonly used post mitochondrial fraction is S9 mix in the range of 5%–30% v/v.

22.6.1.3 *Preparation of test substance*

The solid substance that has been tested needs to be dissolved in an appropriate vehicle. The vehicle that is used should be compatible with the bacterial survival and S9 activity.

22.6.1.3.1 Exposure condition

The recommended test concentration is 5 mg/plate or 5 μL/plate. And, it should be soluble in mixture and the precipitate should not interfere with the scoring.

22.6.1.3.2 Control

- A positive and negative control with and without metabolic activation should be used in assay. The positive control reference substance should be selected on the basis of bacterial strain. For example, 2-aminoanthracene and 9,10-dimethylanthracene.
- Negative controls, consisting of vehicle alone, which are without the test substance.
- A control should also be used which contain the vehicle that is used in the preparation of test substance.

22.6.1.3.3 Procedure

- For the first method, that is, plate incorporation method usually (0.05 or 0.1 mL of test solution is taken along with 0.1 mL fresh culture of bacteria with the viable cells up to 10^8 and mix 0.5 sterile buffer with 2 mL of overlay agar) is used for without metabolic activation assay. For assay with metabolic activation, (0.5 mL metabolic activation mixture containing postmitochondrial fraction mixed with overlay agar 2 mL, along with bacteria and test substance) is used. This mixture is then poured over the minimal agar plate and allowed to solidify before incubation.
- For the second method, the test substance is preincubated with test strain with the viable cell count up to 10^8 and sterile buffer or metabolic activation system 0.5 mL for 20 minutes at 37°C. This is preincubated before mixing with the overlay agar and then poured over the minimal agar plate (In this normally 0.05 or 0.1 mL of test substance, 0.1 mL bacteria, and 0.5 mL of S9 mix or sterile buffer is mixed with 2 mL overlay agar).
- Note: Triplicate plating should be used at each dose level.

22.6.1.3.4 Incubation

All the plates should be incubated at 37°C for 48–72 hours; after incubation, revertant colonies per plate are counted.

22.6.1.3.5 Result and interpretation

The mutagenicity of test material is proportional to the number of colonies observed.

If the number of colonies in the test plate is greater than the control plate, then such material is said to be mutagens.

If the control plate contains low number of colonies, then this might be the case of spontaneous point mutation on histidine encoding gene.

Note: if the bacterial reverse mutation test do not work or if it is not appropriate then in such case mammalian mutation tests may be more appropriate [63–65].

22.6.2 In vitro mammalian cell gene mutation test

The gene mutation induced by the nanoparticle can be detected by the in vitro mammalian cell gene mutation test. Cell lines that are suitable for the testing are mouse lymphoma cells L5178Y, Chinese hamster cells AS52 and V79, and the CHO and human lymphoblastoid cells TK6. To detect the different spectra of genetic event, the above cells are used to measure the genetic endpoint mutation at thymidine kinase (TK) and hypoxanthine guanine phosphoribosyl transferase (HPRT) along with transgene of xanthine guanine phosphoribosyl transferase (XPRT). Culture from cell lines or the cell strain can be used for in vitro test. Chinese hamster cell V79 has a single function copy of HPRT gene, which induces phosphoribosylation of hypoxanthine and guanine base. These cells can be used on the basis of spontaneous mutation frequency and on the basis of growth ability in culture. While performing the test one should take care of the possibility of intrinsic mutagenicity that may not give positive result, this may be due to the pH change, osmolality, or change in the cytotoxicity level. Conflicting result can be seen in the literature for the mutagenicity of nanoparticles, which may be due to the use of different cell lines [66].

22.6.2.1 *Principle of the test*

TK proficient cells are sensitive to trifluorothymidine (TFT), as the cell deficiency in TK is caused by the mutation $TK^{+/-}$ to $TK^{-/-}$, which are resistant to cytotoxic effects of pyrimidine analog TFT. Thus, in the presence of TFT, the mutant cells can proliferate but normal cells cannot which contain tymidine kinase. The deficiencies of cell in HPRT or XPRT are selected in the same way by resistance to 6-thioguanine (TG) or 8_ azaguanine (AG). Thus on the basis of analog or the selective agent for the compound that are tested in the test should be considered carefully along with the properties of test substance that are structurally related to selective agent for the toxicity test in mutant and nonmutant test.

The test substance is exposed to the cell suspension or the monolayer culture in both the presence and absence of metabolic activation. These are exposed for a suitable period of time along with the subculture for detection of cytotoxicity and allow phenotypic expression prior to mutant selection. Cytotoxicity is detected by measuring the relative total growth of culture or the cloning efficiency after the treatment period. To detect the mutant frequency, cells are seeded with the known number in the medium containing selective agent to detect mutant cells, along with the cloning efficiency (viability) determination containing medium without selective agent. Colonies are counted after the suitable incubation time. Frequency of mutant is derived from the number of mutant colonies in selective medium and in nonselective medium.

22.6.2.2 *Description of method*

22.6.2.2.1 Preparation

Cells Cells and subclones of cells used for test should be sensitive to chemical mutagens, high cloning efficiency, and should have a stable spontaneous mutant frequency. Cells should also be checked for contamination such as mycoplasma contamination and should not be used.

The number of cells, culture, and concentration of test substance should have defined parameters. The viable cells should be based on the spontaneous mutation frequency it is recommended to use cell number at least 10 times the inverse of spontaneous mutation frequency or up to 10^6 cells.

Media and culture conditions Media should be appropriate to the selective system and the type if cell used in the test. Similarly, the culture condition should be such that it allows the growth of cells and ability of colony forming for both mutant and nonmutant cells for which the culture media and incubation condition such as culture vessels, temperature, CO_2, concentration, and humidity should be appropriate.

Preparation of cultures Prior to use in this test, cells are propagated from stock cultures and seeded in culture medium and incubated at 37°C. Preexisting mutant cells may need to cleanse the culture.

Metabolic activation The test substance should be introduced to the cell in both the absence and presence of metabolic activation system. The cofactor supplemented postmitochondrial fraction (S9) is the common system used in the test. This fraction is used in the range 1%–10% v/v in final test medium. But the use of the system should be appropriate to the test substance being used for testing. The cell lines that are used in the test should be scientifically justified (e.g., by relevance of cytochrome P450 isoenzyme to metabolism of test substance).

Test substance/preparations Solid test substance should be dissolved in appropriate diluents, that is, case of nanomaterial it should be sonicated if required before the treatment of the cells. Liquid test substance can be used directly or diluted if needed.

22.6.2.2.2 Test conditions

Solvent/vehicle The diluents which are used to dissolve the test substance should be compactable to the test substance for the cell survival and S9 activity. If possible, aqueous diluents like water should be used. Organic solvent/diluents can be used if needed and should be free of water, if needed water can be removed by molecular sieve.

Exposure concentrations Cytotoxicity and solubility of test system and change in pH or osmolarity should be considered while determining the highest concentration. Cytotoxicity should be used in both metabolic and nonmetabolic systems using suitable indicator of cell integrity and growth such as cloning efficiency or total growth of the cells.

For analyzing the test, four concentrations should be used and should cover the wide range of toxicity. It should result in the 10%–20% but not less than 10% of cloning efficiency for maximum cytotoxicity concentration. Range of concentration that can be used is 5 mg/mL, 5 μL/mL, or 0.01 M.

Controls In each experiment, positive and negative controls both along with and without metabolic activation should be included.

Positive control such as ethylmethanesulfonate for HPRT, methylmethanesulfonate for TK, ethylmethanesulfonate for XPRT can be used for the absence of metabolic system where as in the presence of metabolic system 3-Methylcholanthrene for HPRT, Cyclophosphamide (monohydrate) for TK, *N*-Nitrosodimethylamine (for high levels of S-9) for XPRT can be used. Negative control is generally the diluents which are used for the test substance.

22.6.2.2.3 Procedure

Treatment with test substance Test substance should be introduced to the proliferating cells in both with and without metabolic activation for specific periods of time generally up to 4–6 hours. It can be extended for more cell cycles if required. In this test, duplicate or single treated culture can be used for each test concentration.

Measurement of survival, viability, and mutant frequency At the end of exposure period, cells are washed and cultured for survival determination. Cytotoxicity is measured by determining the cloning efficiency or relative total growth of culture after the treatment period. It usually takes 6–8 days to induce the mutant for HPRT and XPRT and 2 days for TK. The viability of cells is measured at the end of expression time. If the test substance is positive and negative, then the colony sizing may be performed [67].

22.6.2.2.4 Result

The number of mutant cells per number of surviving cells is expressed as the mutant frequency. Verification of a clear positive response is not normally required. Statistical methods should be used for the evaluation of the test results. Positive result indicates the substance induce gene mutation whereas negative do not induce gene mutation. Classification of the colonies is based on the size, large colonies indicate the gene mutation whereas small colonies indicate chromosomal breakage [68,69].

The cloning efficiency (CE, %) for each test group:

$$\text{CEx} = \frac{\ln\,[\text{total number of empty wells}/\text{total number of seeded wells}]}{\text{Number of seeded cells per well} \times 100}$$

$$\text{RCEx} = [\text{CEx of the test group}/\text{CEx of the vehehicle control}] \times 100$$

Mutant frequency calculated as:

$$\text{MFuncorrected} = \frac{-\ln\,[\text{total number of empty wells}/\text{total number of seeded wells}]}{\text{Number of seeded cells per well}} \times 10^6$$

$$\text{MFcorrected} = \frac{\text{MFuncorrected}}{\text{CE2}} \times 100$$

22.6.3 In vitro mammalian cell micronucleus test

The in vitro micronucleus test is used to detect the micronuclei present in cytoplasm of interphase cells. Micronuclei may be derived from the acentric chromosome fragments or from whole chromosome, which are not able to migrate toward the pole during cell division in anaphase stage. Thus this test provides the potential evaluation of chromosome damage, that is, aueugens and clastogens both can be detected in cell which has gone under cell division during and after introduction of test substances.

The micronucleus test is effective against several types of cells, in the presence and absence of cytoB.

22.6.3.1 Description of the method

22.6.3.1.1 Cells

Cell culture from primary human or mammalian peripheral blood lymphocytes may be used like CHO, V79, CHL/IU, and L5178Y cells or human cell lines such as TK6 can be used. Also, L5178Y cell line has reported to have an impact on CytoB, and thus this cell is not recommended to use. The cell culture is maintained in the exponential growth phase or stimulated with the mitogenic agent so that it does not synchronized during the exposure of test substance, as the synchronized cell is not recommended [70].

22.6.3.1.2 Media and culture condition

The cell culture is incubated at 5% CO_2 atmosphere at 37°C to maintain culture. Cell line should be checked regularly for the chromosome number and absence of mycoplasma contamination as such contaminated cell is not recommended for the use. In case if cytokinesis block method is used, then cytokinesis inhibitors can be used for the yield of binucleated cells score.

22.6.3.1.3 Preparation of culture

Cells are seeded in the culture medium so that the cell grows in the exponential manner and confluence should be avoided for the cell growing in monolayer. In the case of whole blood lymphocytes, anticoagulant must be used and in the case of separated lymphocyte culture mitogen can be used to induce cell division before the exposure to test substance and cytoB.

22.6.3.1.4 Test chemical preparation

Solid substance may be dissolved in appropriate solvent or diluents. Solid nanomaterials mostly need to be sonicated. Liquid can be used directly or if needed used as diluents.

22.6.3.2 Test condition

22.6.3.2.1 Solvents

Solvents that are used for the test substance should not affect the assay, that is, change in cell growth, affecting the integrity of the test chemical, impairing the metabolic activation system. It is always best to use aqueous solvent or culture medium solvent such as water and dimethyl sulfoxide (DMSO) organic solvent if used it should not exceed 1% v/v. If DMSO is used to dissolve cytoB, then the total amount of test substance and cytoB should not exceed 1% v/v or else it might have an adverse effect. In the case of aqueous solvent such as saline and water, it should not exceed 10% v/v in final treatment medium. If solvent like ethanol or acetone is used, it should be compatible with test substance.

22.6.3.2.2 Use of cytoB as a cytokinesis blocker

The important criteria in the micronucleus test are to ensure the cells being scored have mitosis completed during or after the treatment incubation period. cytoB is mostly used to block cytokinesis as it is an inhibitor which results in binucleate cell by preventing separation of daughter cell after mitosis.

The concentration of cytoB should be between 3 and 6 μg/mL.

Measuring cell proliferation and cytotoxicity and choosing treatment concentration:

- The concentration of the substance should be such that it should not interfere with the excessive cytotoxicity, precipitation in culture medium, or change in pH. In case if the substance that is to be tested changes the pH of the medium, then it might be adjusted by buffer.
- The treatment of culture with cytoB and the measurement of frequency of mononucleate, binucleate, and multinucleate cell in the culture give the accurate method of quantifying the cell proliferation and cytotoxic or cytostatic activity. The recommended estimate of cytotoxic and cytostatic activity by the cytokinesis block proliferation index (PI) and replication index should be from at least 500 cells per culture in comparison with control culture.
- Study done without cytoB should give the result that the cells in culture have divided so that the cell scored have undergone division during or following treatment along with the test substance, or the false negative response may be developed. The measurement for evaluation of relative population doubling (RPD) or relative increase in cell count (RICC) is recommended for the evaluation of the cytotoxicity and cytostatic activity. At the time of late sampling option B must be used for treatment 1.5–2 normal cell cycle length and harvest after additional 1.5–2 normal cell cycle lengths which lead the sampling time longer than 3–4 normal cell cycle length in total.
- At least three concentrations should be evaluated.
- If the cytotoxicity is based on the maximum concentration, the highest concentration should aim for 55% cytotoxicity.
- The cytotoxicity and genotoxicity evaluation of nanomaterials can affect due to some factors, which includes type of cell lines with different p53 status, fluorescence of nanomaterial, nanomaterial coating, and fetal bovine serum used in cell treatment medium [71].

22.6.3.3 Procedure

- It is important to treat the test substance with the sufficient number of cells during all stages of their cell cycles.
- The aneugens and clastogen can be detected in short period of treatment at 3–6 hours in the presence and absence of S9, after that test substance is removed and the growth period of 1.5–2 cell cycle length is followed.
- Cell should be exposed to the metabolic activation for about 1.5–2 normal cell cycle length. Recovering time can be extended up to 3–4 cell cycle length.

22.6.3.4 Cell harvest and slide preparation

- Each culture is harvested and processed separately. Cell preparation may involve hypotonic treatment.

- The slide can be stained with the appropriate fluorescent stain for non-DNA–specific stain along with the DNA counter stain for identifying the contents.

22.6.3.5 Analysis

- All slide including solvent and the control should be coded first and then proceed for the microscopic analysis for the micronucleus frequencies.
- Automated scoring system such as flow cytometry, laser scanning cytometry, or image analysis can be used for the CBPI, RI, RPD, or RICC assessment.
- Micronucleus frequencies should be analyzed as 2000 binucleate cells per concentration and control, equally divided among the replicates [72].

22.6.3.6 Result

- Measures of cytotoxicity and cytostasis for all treated, positive, and negative control culture should be determined.
- In the presence of cytoB, CBI or RI should be calculated for treated and control culture.
- In the absence of cytoB, RPD or RICC should be used.

22.6.3.6.1 Formulas for cytotoxicity assessment

When cytoB is used:

$$\%\,\text{Cytostasis} = 100 - 100\{(\text{CBPI}_{\text{T}} - 1) \div (\text{CBPI}_{\text{C}} - 1)\}$$

where T = test chemical treatment culture and C = control culture.

$$CBPI = \frac{[(\text{No. of mononucleate cells}) + (2 \times \text{No. of binucleate cells})] + (3 \times \text{No. of multinucleate cells})}{(\text{Total number of cells})}$$

Thus a CBPI of 1 (all cells are mononucleate) is equivalent to 100% cytostasis.
Cytostasis = 100 − RI

$$RI = \frac{(\text{No. of binucleate cells}) + (2 \times \text{No. of multinucleate cells}) \div (\text{Total number of cells})_{\text{T}}}{(\text{No. of binucleate cells}) + (2 \times \text{No. of multinucleate cells}) \div (\text{Total number of cells})_{\text{C}}} \times 100$$

where T = treated cultures and C = control cultures.

Thus an RI of 53% means that, compared to the numbers of cells that have divided to form binucleate and multinucleate cells in the control culture, only 53% of this number divided in the treated culture, that is, 47% cytostasis.

When cytoB is not used:

$$RICC = \frac{\text{Increase in number of cells in treated cultures}(\text{final} - \text{starting})}{\text{Increase in number of cells in control cultures}(\text{final} - \text{starting})} \times 100$$

$$RPD = \frac{(\text{No. of Population doublings in treated cultures})}{(\text{No. of Population doublings in control cultures})} \times 100$$

where

$$\textit{Population Doubling} = [\log(\text{Post treatment cell number} \div \text{Initial cell number})] \div \log 2$$

Thus an RICC or an RPD of 53% indicates 47% cytotoxicity/cytostasis.

By using a *PI*, cytotoxicity may be accessed via counting the number of clones consisting of 1 cell (cl1), 2 cells (cl2), 3–4 cells (cl4), and 5–8 cells (cl8) [72].

$$\mathrm{PI} = \frac{(1 \times \mathrm{cl1}) + (2 \times \mathrm{cl2}) + (3 \times \mathrm{cl4}) + (4 \times \mathrm{cl8})}{(\mathrm{cl1} + \mathrm{cl2} + \mathrm{cl4} + \mathrm{cl8})}$$

22.7 Embryotoxicity testing via three test: embryonic stem cell line, micromass, whole embryo culture

An alternative method to test the toxicity of the chemical an in vitro test has been developed which ranges from cell, tissue, and organ culture to whole embryo culture. For the validation of the test for toxicity, the European Centre for the Validation of Alternative Method (ECVAM) along with European Teratology Society (ETS) recommended several testing for identifying embryotoxic chemical. Thus the validation of three embryotoxic tests, that is, test employing embryonic stem cell line (EST), the micromass (MM) test, the postimplantation rat whole embryo culture assay (WEC) test, was determined whether these three tests are capable of predicting embryotoxic potencies in vivo. The three embryotoxicity tests which were validated for discriminating strong embryotoxic from weakly embryotoxic to nonembryotoxic compounds demonstrated a predictivity of 100% with strong embryotoxic chemicals in each in vitro test [73]. Thus these three tests can be used for detecting strong embryotoxic chemicals for screening new substance. Following are the method discussed in detail for evaluating nanomaterials for embryotoxicity [74].

22.7.1 Embryonic stem cell test

This test was developed by the Scholz et al. for in vitro enbryotoxicity testing in drugs. For the EST, two permanent murine cell lines are used such as mES-D3 cells, which are used to detect the effect of drug on developing embryo and BALB/c 3T3 cells to detect the effect of drug on adult tissues and organs. mES cell is cultured by the hanging drop suspension method so as to form embryoid bodies (EBs) which is in the absence of antidifferentiation agent, for example, leukemia inhibitory factor and embryonic fibroblasts. These EBs are then seeded in dish which can spontaneously differentiate into contracting cardiomyocytes. Thus the mES cells can differentiate into three germ layers within EBs which can also express specific tissue proteins. This resembles closely to the developing embryos in the in vivo processes. Thus the negative effect of embryonic development can be eliminated by treating mES cells in vitro. Thus the embryotoxic potential of test compounds can be classified by three different endpoints that can detect the cytotoxicity of (1) mES cells, (2) 3T3 cells, and (3) inhibition of differentiation of mES cells [74,75].

22.7.1.1 Material and methods

22.7.1.1.1 Cell culture

mES-D3 cells and BALB/c 3T3 cells are cultured at 37°C at 5% atmosphere. The mES cell needs to be cultured routinely on mouse embryonic fibroblast feeder in the presence of leukemia inhibitory factor to maintain undifferentiated status and passaged every second.

22.7.1.1.2 Test compound

Test substance can be dissolved in ethanol and DMEM.

22.7.1.1.3 Assessment of cytotoxicity

The cytotoxicity of the test substance on 3T3 cells and mES-D3 cells can be determined by MTT cytotoxic assay repeated three times for each test substance. In 96-well plates, a total of 500 cells in 50 μL of culture medium without Leukemia inhibitory factor (LIF) is seeded with the density of 1×10^4 cells/mL and incubated for 2 hours at 37°C at 5% CO_2 atmosphere. Culture medium 150 μL along with the test substance with various concentrations added in each well except for solvent control and positive control. After 3 and 5 days, the culture medium is replaced with new culture containing appropriate test substance concentration. After the 10 days of culture, new medium is replaced with 0.5 mg/mL of MTT and incubated at 37°C at 5% atmosphere for 2 hours and then the MTT is removed from the cell by flipping the 96-well plate and adding 100 μL DMSO per well, which helps in extracting formazan from the cells followed by agitating plates on an incubator shaker at 200 rpm for 15 minutes. Finally, at the wavelength of 570 nm using 630 nm reference wavelength optical density is measured. The inhibition viability of ES cells and 3T3 fibroblast is compared with the solvent-treated cells.

22.7.1.1.4 Differentiation of ES cells

Undifferentiated mES cells are incubated at 37°C at 5% CO_2 atmosphere in vitro by the hanging drop suspension method without LIF, 20 μL droplets of cells suspension placed into the inner side of lid of 10 cm petri dish to form EBs. The EBs are transferred into sterile petridish after 3 days of culturing and again culture in suspension of 5 mL differentiation medium for 2 days. Subsequently, EBs are then seeded in 6-well plates at the density of 100 EBs per well for 5 days which can differentiate into beating cardiomyocytes. Then, the EBm cells are exposed to the test substance concentrations. The setting of the concentration is based on the results of cytotoxicity assay and preliminary experiments along with the control and solvent control in each experiment.

22.7.1.1.5 RNA isolation and (RT-PCR) analysis

On day 10, approximately EBs 100 per sample were harvested and total RNA is isolated with TRIzol reagent. cDNA is synthesized from 1000 ng RNA per reaction. The α/β MHC gene is expressed specially in cardiomyocyte differentiation, which is chosen as a marker gene and GAPDH as a housekeeping gene. The polymerase chain reaction is performed with the specific primer by initial denaturation at 94°C for 3 minutes, followed by 32 cycles of PCR amplification: Denaturation at 94°C for 30 seconds, annealing at 62°C (α/β

MHC) or 56°C (GAPDH) for 30 seconds, and completed by a final extension of 72°C for 5 minutes. PCR fragments were run on a 3% agarose gel, which contain 0.2 μg/mL ethidium bromide and visualized under UV light with molecular imager and analyzed. The ID_{50} of the test substance is calculated from concentration response curve as the concentration suppressed the expression of MHC by 50% in comparison with control. This assay is repeated for three times. The mean ID_{50} value is taken as the result.

22.7.1.2 *Classification of embryotoxicity*

The potential of each test substance is classified into three grades (strong, weak and nonembryotoxic), which are based on IC_{50}3T3, IC_{50}ES, and ID_{50} mean values of test substance [76–78].

22.7.2 Micromass test

Micromass is based on the technique devised by Umansky (1966) to study the development and differentiation of culture chick embryo limb cells. A foci of differentiating chondrocytes in MM culture is formed by the undifferentiated mesenchyme cells of limb buds in the test. This test is based on the ability of a chemical or substance to inhibit the formation of foci. Thus in the test the number of foci is reduced or number of cells within the foci by the positive chemical or substance. The endpoint is detected by the teratogenic potential of chemical substance, which is based on the interference with basic cell development functions such as cell proliferation, differentiation, cell to cell communication, and cell to extracellular matrix interaction.

22.7.2.1 *Procedure*

The limb bud is isolated from the embryos obtained from the 14 days of gestation of Wistar rats. Single cell suspension is prepared by the trypsin action where the limb bud is transferred into dish containing HBSS and count the number of limb bud. Then, add 1% trypsin in CMF and incubate for 20 minutes at 37°C, then the trypsin action is stopped by the addition of Ham's F12 with 5% FCS. After that, the cells are spotted in the center of the 96-well plates and placed in the incubator for 2–3 hours to allow the cells to attach the plates. After 2–3 hours, add 300 μL of medium with or without test substance and leave them for 5 days. At the end, total number of viable cells, IC_{50} (50% inhibition of cell viability and growth along with differentiated cells), or ID_{50} (50% inhibition of cells differentiation) is determined, as well as the determination of the number of foci.

22.7.2.2 *Prediction model*

The MM test determines two endpoints, that is, ID50 for cells differentiation and IC50 for cell viability. However, both the tests provide the same information, and thus in ECVAM validation study only ID50 is used.

This study classifies the test compound into three classes, that is, nonembryotoxic, weak embryotoxic, and strong embryotoxic.

In the in vitro testing of teratogenicity, the system has two approaches: (1) modeling of event that occurs during development such as intercellular communication, reaggregation, and regeneration; and (2) monitoring of development for a limited period or within a defined organ system.

The advantage of MM test is that it provides ample cells to set up multiple replicates culture from two or three animals. Thus it is more economical. The drawback of this test is that the system lacks the drug metabolizing enzyme which can be overcome by the use of metabolic activation system such as S9 fraction; however, this might be toxic to some cells.

22.7.2.3 *Range finding*

There must be at least three concentrations within the range of 90%–10% control differentiation. For negative control, frozen stock of penicillin G may be used, and for positive control, freshly prepared 5FU must be used in the experiment.

In each experiment, one plate should include a full range of positive control concentration and a column of negative control.

Measurement of number of differentiated cells and foci:

Remove the acid alcohol/eluted neutral red rinse three times with saline. Add 200 μL alcian blue in 0.1 N HCl, living the stain overnight. After that, remove the alcian blue and rinse it three times with saline. Number of foci can be counted at this point. After that, add 200 μL 6 M guanidine hydrochloride which is freshly prepared and leave for at least for 2 hours to elute stain. Finally, read absorbance in a multiscan spectrophotometer at 620 nm.

22.7.2.4 *Prediction model*

The log ID50 uses the three linear functions, which classified into three classes:

Function 1	$6.65 \times \log(\text{ID50}) - 9.49$
Function 2	$6.16 \times \log(\text{ID50}) - 8.29$
Function 3	$-1.31 \times \log(\text{ID50}) - 1.42$

Classification:

The test substance or chemical can be classified as follows:

- Nonembryotoxic: result of function 1 exceeds function 2 and function 3.
- Weak embryotoxic: result of function 2 exceeds function 1 and function 3.
- Strong embryotoxic: result of function 3 exceeds function 1 and function 2 [79–81].

22.7.3 Whole embryo culture

This test is validated by ECVAM and can be used for detecting embryo toxicants.

Any rat strains which are used in the positive and negative control compound must have a large number of embryos. Rats must be kept at the specific condition 20°C–24°C at 50%–70% humidity. In this method, rat embryos with 1–5 somites are used as these are sensitive for xenobiotics as compared to older embryos. A major aspect of organogenesis occurs during 48 hours of culture which includes development of heart, ear, and eye, closure of neural tube, brachial bars, and limbs buds. During this period, interference of any substance may lead to general retardation of growth and development or specific malformations can be observed in organ anlagen.

Embryos are cultured in a rotating container so that culture vessels are subsequently gassed for the exchange of culture medium with gaseous phase above. Oxygenation of the culture is essential as it increases the oxygen demand for the culture. The morphology of the embryos is assessed carefully after the culture. The conclusion of the embryotoxicity test in based on the comparison of the control embryo and the exposed embryo.

22.7.3.1 *Procedure*

On day 9.5 of gestation, rat embryos are cultured then 2 mL of rat serum is added to each embryo culture vessel. In the cultured vessel containing culture serum, a standard volume of compound solution which is diluted with proper solvent such as DMSO hanks' balanced salt solution is added before the addition of embryo. The same volume is added to the control culture. The maximum solvent culture used are 1% aqueous medium, 0.125% DMSO. Embryos are distributed evenly in the control and the test concentration so that the average of somite number is the same for each concentration. After 48 hours culture, each embryo is transferred to the Petridish containing 37°C and embryos are scored in the same manner in which they were put into culture.

22.7.3.2 *Prediction model*

Morphological scoring system is used in the WEC assay after the embryos exposed for 48 hours for the signs of malformation or retardation as well as embryonic death. This scoring system applied for two prediction models. The second model for prediction was developed because the first model can only assess the parameters of differentiation and development. The second model incorporates the cytotoxic data for the different mouse fibroblast cell line 3T3, which is derived from the Embryotoxic Stem Cell test. This prediction model classifies the potential of test substance of validation study into three classes of embryotoxic, that is, non, weak, and strong. Two endpoint values are accepted as functional variables:

For first prediction model:

- $IC_{50\ Mal}$ for malformations: where the concentration at which 50% of all tested embryos of the group are malformed.
- IC_{NOEC} for total morphological score: where the lowest concentration has no effect on TMS.

Function 1	$18.08 \times \log(IC_{50\ Mal}) - 11.56 \times \log(IC_{NOEC\ TMS}) - 10.19$
Function 2	$21.55 \times \log(IC_{50\ Mal}) - 15.31 \times \log(IC_{NOEC\ TMS}) - 10.65$
Function 3	$8.70 \times \log(IC_{50\ Mal}) - 8.53 \times \log(IC_{NOEC\ TMS}) - 2.53$

For second prediction model:

- IC_{max} for malformation: the lowest concentration at which maximum malformation rate is obtained.

- Relative distance between $IC_{50\ 3T3}$ (inhibition of growth of 3T3 cells by 50% of control in Neutral Red Uptake test) and IC_{NOEC} for TMS.

Function 1	$0.21 \times (IC_{50\ 3T3-IC\ NOEC\ TMS} / IC_{50\ 3T3}) \times 100 + 15.37 \times \log(IC_{max}) - 23.58$
Function 2	$0.27 \times (IC_{50\ 3T3-IC\ NOEC\ TMS} / IC_{50\ 3T3}) \times 100 + 17.71 \times \log(IC_{max}) - 32.37$
Function 3	$0.093 \times (IC_{50\ 3T3-IC\ NOEC\ TMS} / IC_{50\ 3T3}) \times 100 + 4.21 \times \log(IC_{max}) - 4.23$

Classification:

The test substance or chemical can be classified as follows.

- Nonembryotoxic: result of function 1 exceeds function 2 and function 3.
- Weak embryotoxic: result of function 2 exceeds function 1 and function 3.
- Strong embryotoxic: result of function 3 exceeds function 1 and function 2.

The advantage of this test is that many relevant development endpoints in the critical phase of organogenesis can be measured and the limitation of this test is the lack of biotransformation capacity. The ECVAM validated study has shown the accuracy of 80% including second prediction model of cytotoxicity data with the 100% accuracy for strong embryotoxic substance, 70% for nonembryotoxic substance, and 65% for weakly embryotoxic substance.

22.7.3.3 *Dose–response assessment*

Toxicity range finding is carried three embryos per compound concentration. The highest ineffective concentration results in at least 50% reduction in control. Concentration test may differ by two factors at the minimum. The highest concentration of test substance should be 1000 mg/L and if no effect occurs in seven embryos then no additional concentration is needed to be tested.

22.7.3.4 *Endpoint assay*

After the 48 hours culture, each embryo is transferred to Petridish which contains HBSS at 37°C and on the basis of development of embryos these are scoured in the manner they were put into the culture. With the help of microscope, the diameter of the yolk sac is measured along with the ocular micrometer yolk sac vascularization, circulation, and heartbeat are assessed. After that with the help of microsurgical forceps, an yolk sac is removed and all the morphological development of all the organ anlagen are determined along with the embryo crown rump length, head length, number of pairs of somites, and embryo flexion. Somites are counted in caudal direction beginning at the forelimb bud, assuming the somite located next to middle of forelimb bud is number 9. Morphology development of the organ anlagen is scored as follows.

It can be divided into four groups: (1) growth parameter which includes yolk sac diameter, (2) functional parameter which includes heartbeat, yolk sac circulation, and allantoic circulation, (3) general morphology parameter which includes final minus initial somite number, and (4) morphological parameter which includes yolk sac blood vessels, flexion, heat, brain, branchial bars, and limb.

Before testing substance or compound, the assay is checked using 5-fluorouracil as positive control at the concentration of 0.03, 0.1, and 0.3, and 1 mg/L and Penicillin G sodium as negative control at the concentration of 1000 mg/L [82–84].

References

[1] Singh P, Nanda A. Nanotechnology in cosmetics: a boon or bane? Toxicol Environ Chem 2012;94. Available from: https://doi.org/10.1080/02772248.2012.723482.

[2] Nanda S, Nanda A, Lohan S, Kaur R, Singh B. Chapter 3: Nanocosmetics: performance enhancement and safety assurance. In: Grumezescu AM, editor. Nanobiomaterials in galenic formulations and cosmetics—Applications of nanobiomaterials, 10. Elsevier; 2016. p. 47–67. Available from: http://dx.doi.org/10.1016/B978-0-323-42868-2.00003-6.

[3] Nanda S, Madan K. Nanotechnology driven cosmetic products: commercial and regulatory milestones. Appl Clin Res 2018;5. Available from: https://doi.org/10.2174/2213476X05666180530093111.

[4] Ajazzuddin M, Jeswani G, Jha A. Nanocosmetics: past, present and future trends. Recent Pat Nanomed 2015;5. Available from: https://doi.org/10.2174/1877912305666150417232826.

[5] Poland CA, Larsen PB, Read SAK, Varet J, Hankin SM, Lam HR. Assessment if nano-enabled technologies in cosmetics. Environmental Project 2016;1825.

[6] Ajdary M, Moosavi MA, Rahmati M, Falahati M, Mahboubi M, Mandegary A, et al. Health concerns of various nanoparticles: a review of their in vitro and in vivo toxicity. Nanomaterials (Basel, Switzerland) 2018;8(9):634. Available from: https://doi.org/10.3390/nano8090634.

[7] Sukhanova A, Bozrova S, Sokolov P, Berestovoy M, Karaulov A, Nabiev I. Dependence of nanoparticle toxicity on their physical and chemical properties. Nanoscale Res Lett 2018;13(1):44. Available from: https://doi.org/10.1186/s11671-018-2457-x.

[8] Raj S, Jose S, Sumod US, Sabitha M. Nanotechnology in cosmetics: opportunities and challenges. J Pharm Bioallied Sci 2012;4(3):186–93. Available from: https://doi.org/10.4103/0975-7406.99016.

[9] Kaul S, Gulati N, Verma D, Mukherjee S, Nagaich U. Role of nanotechnology in cosmeceuticals: a review of recent advances. J Pharm 2018;2018:3420204. Available from: https://doi.org/10.1155/2018/3420204.

[10] Larese F, Mauro M, Adami G, Bovenzi M, Crosera M. Nanoparticles skin absorption: new aspects for a safety profile evaluation. Regul Toxicol Pharmacol 2015;72. Available from: https://doi.org/10.1016/j.yrtph.2015.05.005.

[11] <https://www.fda.gov/cosmetics/cosmetics-science-research/cosmetics-nanotechnology>.

[12] Nanda S, Nanda A, Lohan S, Kaur R, Singh B. Chapter 3 - Nanocosmetics: performance enhancement and safety assurance. Appl Nanobiomater 2016;10:47–67. Available from: https://doi.org/10.1016/B978-0-323-42868-2.00003-6.

[13] Lohani A, Verma A, Joshi H, Yadav N, Karki N. Nanotechnology-based cosmeceuticals. ISRN Dermatol 2014;2014:843687. Available from: https://doi.org/10.1155/2014/843687.

[14] Melo A, Amadeu MS, Lancellotti M, Hollanda L M de, Machado D. The Role of nanomaterials in cosmetics: national and international legislative aspects. Quím Nova 2015;38(4):599–603. Available from: https://doi.org/10.5935/0100-4042.20150042.

[15] <https://www.cosmeticsbusiness.com/technical/article_page/Nanotechnology__keeping_cosmetics_out_of_the_courtroom/60306>.

[16] International Conference health and safety issues related to nanomaterials for a socially responsible approach, Minatec, Grenoble, France, 7–10 November, 2016. Available from: <http://www.cea.fr/ceatech/pns/nanosafe/en/Documents/Book_Of_Abstracts_Nanosafe_2016.pdf>.

[17] Scientific Committee on Consumer Safety (SCCS) guidance on the safety assessment of nanomaterials in cosmetics. Available from: <https://ec.europa.eu/health/scientific_committees/consumer_safety/docs/sccs_s_005.pdf>.

[18] John Jr. CM. FDA's evolving approach to nanotechnology. Food Drug Law J 2012;67(4). Available from: <https://www.porterwright.com/content/uploads/2017/11/jmonica2013fda-fdliarticle.pdf>.

[19] The SCCS's notes of guidance for the testing of cosmetic substances and their safety evaluation 8th revision. The SCCS adopted this opinion at its 17th plenary meeting of 11 December 2012. Available from: <https://ec.europa.eu/health/scientific_committees/consumer_safety/docs/sccs_s_006.pdf>.

[20] Grumezescu, A. Chapter 9: Route of exposure and translocation of nanoparticles. Nanobiomaterials in galenic formulations and cosmetics: applications of nanobiomaterials. Elsevier, vol. 10. p. 164

[21] De Jong WH, Borm PJ. Drug delivery and nanoparticles: applications and hazards. Int J Nanomed 2008;3 (2):133–49.

[22] Guidance for Industry Safety of Nanomaterialsin Cosmetic Products. U.S. Department of Health and Human Services Food and Drug Administration Center for Food Safety and Applied Nutrition, June 2014. Available from: <https://www.fda.gov/media/83957/download>.

[23] Hoshyar N, Gray S, Han H, Bao G. The effect of nanoparticle size on in vivo pharmacokinetics and cellular interaction. Nanomedicine (London, England) 2016;11(6):673–92. Available from: https://doi.org/10.2217/nnm.16.5.

[24] Singh R, Lillard Jr JW. Nanoparticle-based targeted drug delivery. Exp Mol Pathol 2009;86(3):215–23. Available from: https://doi.org/10.1016/j.yexmp.2008.12.004.

[25] Nanotechnology and human health: scientific evidence and risk governance. Report of the WHO expert meeting 10–11 December 2012, Bonn, Germany. Available from: <https://apps.who.int/iris/bitstream/handle/10665/108626/e96927.pdf;jsessionid = F231EB59729D48215F5C4289DA5501E0?sequence = 1>.

[26] Grumezescu, A. Chapter 3: Nanocosmetic: performance enhancement and safety assurance. 3.14 toxicity testing. In Nanobiomaterials in galenic formulations and cosmetics: application of nanobiomaterials. vol. 10. p. 64.

[27] Eskes C, Zuang V. Alternative (non-animal) methods for cosmetics testing: current status and future prospects. A report prepared in the context of the 7th Amendment to the Cosmetics Directive for establishing the timetable for phasing out animal testing. Altern Lab Anim 2005;33:X–I + 1.

[28] Thomas Hartung. Food for thought ... on alternative methods for cosmetics safety testing. Available from: <http://www.altex.ch/resources/altex_2008_3_147_162_FFT_HartungE.pdf>.

[29] Roger M, Fullard N, Costello L, Bradbury S, Markiewicz E, O'Reilly S, et al. Bioengineering the microanatomy of human skin. J Anat 2019;234(4):438–55. Available from: https://doi.org/10.1111/joa.12942.

[30] Kandarova H, Liebsch M, Gerner I, Schmidt E, Genschow E, Traue D, et al. The EpiDerm test protocol for the upcoming ECVAM validation study on in vitro skin irritation tests—an assessment of the performance of the optimised test. Altern Lab Anim 2005;33:351–67. Available from: https://doi.org/10.1177/026119290503300408.

[31] Kolle SN, Ravenzwaay B, Landsiedel R. Regulatory accepted but out of domain: in vitro skin irritation tests for agrochemical formulations. Regul Toxicol Pharmacol 2017;89:125–30.

[32] OECD guidelines for the testing of chemicals. In vitro skin irritation: reconstructed human epidermis test method. 439 Adopted: 26 July 2013.

[33] Pedrosa TDN, Catarino CM, Pennacchi PC, Assis SR, Gimenes F, Consolaro MEL, et al. A new reconstructed human epidermis for in vitro skin irritation testing. Toxicol In Vitro 2017;42:31–7. Available from: https://doi.org/10.1016/j.tiv.2017.03.010.

[34] Chen L, Deng H, Cui H, Fang J, Zuo Z, Deng J, et al. Inflammatory responses and inflammation-associated diseases in organs. Oncotarget 2017;9(6):7204–18. Available from: https://doi.org/10.18632/oncotarget.23208.

[35] Terry LR, RA Moravec, AL Niles, S Duellman, HA Benink, TJ Worzella, et al. Cell viability assays. Assay guidance manual. Published May 1, 2013; Last Update: July 1, 2016. Available from: <https://www.ncbi.nlm.nih.gov/books/NBK144065/>.

[36] Drasler B, Sayre P, Steinhäuser KG, Petri-Fink A, Rothen-Rutishauser B. In vitro approaches to assess the hazard of nanomaterials. NanoImpact 2017;8:99–116.

[37] Lin YC, Hsu HC, Lin CH, Wu CY, Chen W, Lai HM. Testing method development and validation for in vitro skin irritation testing (SIT) by using reconstructed human epidermis (RhE) skin equivalent - EPiTRI®. In: Kojima H, Seidle T, Spielmann H, editors. Alternatives to animal testing. Singapore: Springer; 2019. Available from: <https://doi.org/10.1007/978-981-13-2447-5_2>.

[38] Explanatory background document to the OECD Draft Test 5 guideline on in vitro skin irritation testing. Available from: <http://www.oecd.org/chemicalsafety/testing/43670220.pdf>.

[39] Vinardell OrcID MP, Llanas H, Marics L, Mitjans M. In vitro comparative skin irritation induced by nano and non-nano zinc oxide. Nanomaterials 2017;7(3):56. Available from: https://doi.org/10.3390/nano7030056.

[40] Kim H, Choi J, Lee H, Park J, Yoon B-I, Jin S, et al. Skin corrosion and irritation test of nanoparticles using reconstructed three-dimensional human skin model, EpiDerm TM. Toxicol Res 2016;32:311–16. Available from: https://doi.org/10.5487/TR.2016.32.4.311.

[41] Choi J, Kim H, Choi J, Oh SM, Park J, Park K. Skin corrosion and irritation test of sunscreen nanoparticles using reconstructed 3D human skin model. Environ Health Toxicol 2014;29:e2014004. Available from: https://doi.org/10.5620/eht.2014.29.e2014004.

[42] Guest R. The rat skin transcutaneous electrical resistance (TER) test. Alternatives for dermal toxicity testing. Springer; 2017. Available from: <https://doi.org/10.1007/978-3-319-50353-0_13>.

[43] C. Eskes, E van Vliet, H.I. Maibach. 13: The rat skin transcutaneous electrical ressistance (TER) test. Alternatives for dermal toxicity testing. Springer. vol. 178.

[44] A. Wallace Hayes. 21: validation and regulatory acceptance of new, revised and alternative Toxicological methods. Principles and methods of toxicology, 5th ed. CRC Press.

[45] OECD. Test no. 431: in vitro skin corrosion: reconstructed human epidermis (RHE) test method, OECD guidelines for the testing of chemicals, Section 4. Paris: OECD Publishing; 2016. Available from: <https://doi.org/10.1787/9789264264618-en>.

[46] Botham PA, Hall TJ, Dennett R, McCall JC, Basketter DA, Whittle E, et al. The skin corrosivity test in vitro. Results of an inter-laboratory trial. Toxicol in Vitro 1992;6(3):191–4.

[47] Grindon C, Combes R, Cronin M, Roberts D, Garrod J. Integrated decision-tree testing strategies for skin corrosion and irritation with respect to the requirements of the EU REACH legislation. Altern Lab Anim 2008;35:673–82. Available from: https://doi.org/10.1177/026119290703500612.

[48] Kim K, Park H, Lim KM. Phototoxicity: its mechanism and animal alternative test methods. Toxicol Res 2015;31(2):97–104. Available from: https://doi.org/10.5487/TR.2015.31.2.097.

[49] OECD guideline for testing of chemicals. In vitro 3T3 NR U phototoxicity test. Available from: <https://www.oecd-ilibrary.org/docserver/9789264071162-en.pdf?expires = 1573206779&id = id&accname = guest&checksum = 7044F21FFB16C30A94FFECF453E0C940>.

[50] Choi JE, Park Y-H, Lee E, Jeong S, Kim S, Kim M-K, et al. A safety assessment of phototoxicity and sensitization of SiO_2 nanoparticles. Mol Cell Toxicol 2011;7:171–6. Available from: https://doi.org/10.1007/s13273-011-0023-z.

[51] Spielmann H, Balls M, Brand M, Doering BE, Holzhütter H, Kalweit S, et al. EEC/COLIPA project on in vitro phototoxicity testing: first results obtained with a Balb/c 3T3 cell phototoxicity assay. Toxicol In Vitro 1994;8(4):793–6.

[52] Towle KM, Fung ES, Monnot AD. Phototoxicity evaluation of hair cleansing conditioners. Cosmetics 2019;6(3):53. Available from: https://doi.org/10.3390/cosmetics6030053.

[53] Spielmann H. DB-ALM Protocol no. 78: 3 T 3 Neutral Red Uptake (NRU) Phototoxicity Assay; 2018.

[54] Diembeck WH, Beck H, Benech F, Courtellemont P, Dupuis J, Lovell W, et al. Test guidelines for in vitro assessment of dermal absorption and percutaneous penetration of cosmetic ingredients. Food Chem Toxicol 1999;37:191–205. Available from: https://doi.org/10.1016/S0278-6915(98)00114-8.

[55] Guidance Notes on Dermal Absorption Series on Testing and Assessment No. 156. Available from: <https://www.oecd.org/chemicalsafety/testing/48532204.pdf>.

[56] OECD. Test no. 428: skin absorption: in vitro method, OECD guidelines for the testing of chemicals, Section 4. Paris: OECD Publishing; 2004. Available from: <https://doi.org/10.1787/9789264071087-en>.

[57] Doak SH, Manshian B, Jenkins GJ, Singh N. In vitro genotoxicity testing strategy for nanomaterials and the adaptation of current OECD guidelines. Mutat Res 2012;745(1-2):104–11. Available from: https://doi.org/10.1016/j.mrgentox.2011.09.013.

[58] Di Bucchianico S, Cappellini F, Le Bihanic F, Zhang Y, Dreij K, Karlsson HL. Genotoxicity of TiO_2 nanoparticles assessed by mini-gel comet assay and micronucleus scoring with flow cytometry. Mutagenesis 2017;32(1):127–37. Available from: https://doi.org/10.1093/mutage/gew030.

[59] Azqueta A, Dusinska M. The use of the comet assay for the evaluation of the genotoxicity of nanomaterials. Front Genet 2015;6:239. Available from: https://doi.org/10.3389/fgene.2015.00239.

[60] Kim HR, Park YJ, Shin DY, Oh SM, Chung KH. Appropriate in vitro methods for genotoxicity testing of silver nanoparticles. Environ Health Toxicol 2013;28:e2013003. Available from: https://doi.org/10.5620/eht.2013.28.e2013003.

[61] Wessner D, Maiorano P, Kenyon J, Pillsbury R, Campbell A. Spot-overlay Ames test of potential mutagens. Assoc Biol Lab Educ.; 2000. 22.

[62] Sugiyama K, Yamada M, Awogi T, Hakura A. The strains recommended for use in the bacterial reverse mutation test (OECD guideline 471) can be certified as non-genetically modified organisms. Genes Environ 2016;38:2. Available from: https://doi.org/10.1186/s41021-016-0030-3.

[63] Clift M, Raemy D, Voss C, Ali Z, Lehmann A, Brandenberger C, et al. Can the Ames test provide an insight into nano-object mutagenicity? Investigating the interaction between nano-objects and bacteria. Nanotoxicology 2012;7. Available from: https://doi.org/10.3109/17435390.2012.741725.

[64] Heshmati M, ArbabiBidgoli S, Khoei S, Rezayat SM, Parivar K. Mutagenic effects of nanosilver consumer products: a new approach to physicochemical properties. Iran J Pharm Res 2015;14(4):1171–80.

[65] OECD. Test no. 471: bacterial reverse mutation test, OECD guidelines for the testing of chemicals, Section 4. Paris.: OECD Publishing; 1997. Available from: <https://doi.org/10.1787/9789264071247-en>.

[66] Senapati VA, Kansara K, Shanker R, Dhawan A, Kumar A. Monitoring characteristics and genotoxic effects of engineered nanoparticle–protein corona. Mutagenesis 2017;32(5):479–90. Available from: https://doi.org/10.1093/mutage/gex028.

[67] Rees BJ, Tate M, Lynch AM, Thornton CA, Jenkins GJ, Walmsley RM, et al. Development of an in vitro PIG-A gene mutation assay in human cells. Mutagenesis 2017;32(2):283–97. Available from: https://doi.org/10.1093/mutage/gew059.

[68] OECD. Test no. 490: in vitro mammalian cell gene mutation tests using the thymidine kinase gene, OECD guidelines for the testing of chemicals, section 4. Paris: OECD Publishing; 2016. Available from: <https://doi.org/10.1787/9789264264908-en>.

[69] Murphy S, Ellis-Hutchings R, Finch L, Welz S, Wiench K. In vitro genotoxicity studies: n-Butyl acrylate L5178Y mouse lymphoma (TK + / − locus assay), 2-Ethylhexyl acrylate gene mutation assay in Chinese hamster V79 cells, and 2-Ethylhexyl acrylate micronucleus test in human lymphocytes. Data Brief 2018;20:316–25. Available from: https://doi.org/10.1016/j.dib.2018.06.008.

[70] Kirsch-Volders M, Decordier I, Elhajouji A, Plas G, Aardema MJ, Fenech M. In vitro genotoxicity testing using the micronucleus assay in cell lines, human lymphocytes and 3D human skin models. Mutagenesis 2011;26(1):177–84. Available from: https://doi.org/10.1093/mutage/geq068.

[71] li Y, Doak S, Yan J, Chen D, Zhou M, Mittelstaedt R, et al. Factors affecting the in vitro micronucleus assay for evaluation of nanomaterials. Mutagenesis. 2016;32:gew040. Available from: https://doi.org/10.1093/mutage/gew040.

[72] OECD. Test no. 487: in vitro mammalian cell micronucleus test, OECD guidelines for the testing of chemicals, section 4. Paris: OECD Publishing; 2016. Available from: <https://doi.org/10.1787/9789264264861-en>.

[73] Genschow E, Spielmann H, Scholz G, Seiler A, Brown N, Piersma A, et al. The ECVAM international validation study on in vitro embryotoxicity tests: Results of the definitive phase and evaluation of prediction models. European Centre for the Validation of Alternative Methods. Altern Lab Anim 2001;30:151–76. Available from: https://doi.org/10.1177/026119290203000204.

[74] Ward, S. L. AltTox Contributing Editor. Available from: <http://alttox.org/mapp/toxicity-endpoints-tests/reproductive-developmental-toxicity/>.

[75] Liu H, Ren C, Liu W, Jiang X, Wang L, Zhu B, et al. Embryotoxicity estimation of commonly used compounds with embryonic stem cell test. Mol Med Rep 2017;16(1):263–71. Available from: https://doi.org/10.3892/mmr.2017.6552.

[76] <https://ecvam-dbalm.jrc.ec.europa.eu/public/datasheet/?id = DC5ABDF7AC30F1B7DF7EF27E87D68AAC7180BB0BC12CB10496CDA74B54630A05A3291B895581F634>.

[77] Hofmann MC. Stem cells and nanomaterials. Adv Exp Med Biol 2014;811:255–75. Available from: https://doi.org/10.1007/978-94-017-8739-0_13.

[78] Liu H, Ren C, Liu W, Jiang X, Wang L, Zhu B, et al. Embryotoxicity estimation of commonly used compounds with embryonic stem cell test. Mol Med Rep 2017;16(1):263–71. Available from: https://doi.org/10.3892/mmr.2017.6552.

[79] <https://ecvam-dbalm.jrc.ec.europa.eu/public/datasheet/?id = 7DDBC8A6A03A1DB626729D0A17A0D2FF7180BB0BC12CB10496CDA74B54630A05A3291B895581F634>.

[80] Flint OP. A micromass culture method for rat embryonic neural cells. J Cell Sci. 1983;61:247–62.

[81] Singh P, Sinha N, Singh RK. Micromass culture: a recent in vitro system for testing embryotoxic potential of chemicals. Indian J Sci Technol 2009;2(8).

[82] <https://ecvam-dbalm.jrc.ec.europa.eu/public/datasheet/?id = 15BFE9AC6ECC8ED3D7F6DF03B5D5BBE97180BB0BC12CB10496CDA74B54630A05A3291B895581F634>.

[83] Lee HY, Inselman AL, Kanungo J, Hansen DK. Alternative models in developmental toxicology. Syst Biol Reprod Med 2012;58(1):10–22. Available from: https://doi.org/10.3109/19396368.2011.648302.

[84] Piersma A, Genschow E, Verhoef A, Spanjersberg M, Brown N, Brady M, et al. Validation of the postimplantation rat whole-embryo culture test in the international ECVAM validation study on three in vitro embryotoxicity tests. Altern Lab Anim 2004;32:275–307. Available from: https://doi.org/10.1177/026119290403200307.

CHAPTER

23

Current commercial nanocosmetic products

Tuan Anh Nguyen[1] *and Susai Rajendran*[2]

[1]Institute for Tropical Technology, Vietnam Academy of Science and Technology, Hanoi, Vietnam [2]Department of Chemistry, St Antony's College of Arts and Sciences for Women, Dindigul, India

23.1 Introduction

As beauty products, cosmetics can temporally protect us from the attacks of time and environment. Market consumption of cosmetic products is accelerating. OrbisResearch.com reported that the global cosmetic products market was valued at $532.43 billion in 2017, and it is expected to grow to $805.61 billion by 2023. Nowadays, there is a significant demand for innovative products in the worldwide cosmetics market. Customers seek and prefer the new products that can resolve multiple beauty-related concerns within a limited period of time.

Harmonization of nanoscience and beauty produces the "nanocosmetics." Nanotechnology is key to next-generation cosmetics. It was reported that nanotechnology could enhance the activities and properties of cosmetics, such as color, transparency, solubility, texture, and durability. It was reported that the sales of cosmetics containing nanomaterials were about $2.6 billion in 2015, and it is expected to grow to $55.3 billion by 2022.

In the case of sunscreen products, such as UV nanofilters, nano-TiO_2 and nano-ZnO particles can offer the advantage over their traditional counterparts due to their broad UV-protection and noncutaneous side effects. In perfumery products, it was reported that the presence of nanodroplets could be expected in "Eau de Toilette" and "Eau de Parfum."

In the case of cosmeceuticals, nanomaterials can be used widely in cosmetics as transdermal drug delivery systems. By using the smart nanocontainers, the active compounds, such as "vitamins/antioxidants/nutrients/antiaging" or "antiinflammatory/antiinfective agents," can be delivered effectively. These smart nanocontainers are typically related to the smart releasing property for their embedded active substances. These smart releases could be obtained by using the smart coatings as their outer nanoshells. These nanoshells could prevent the direct contact between these active agents and the adjacent local environments.

Nanocosmetics
DOI: https://doi.org/10.1016/B978-0-12-822286-7.00019-X

According to the EU cosmetic regulation 1223/2009, from July 11, 2013, all ingredients present as nanomaterials have to be indicated on the package of cosmetic products with the term 'nano' in brackets [1]. This chapter emphasizes the currently available nanocosmetics in the global market.

23.2 Current commercial nanocosmetic products

In this book, Faria-Silva et al. [2] introduce nanoemulsions as vehicles to be used in dermocosmetics and cosmetics for skin and hair applications. Svarc and Hermida [3] also reported the use of transdermal and bioactive nanocarriers for skin care. Hari et al. [4] focus on polymeric nanocarriers for topical drug delivery in skin cream. Table 23.1 summarizes several current commercial nanocosmetics as a moisturizer for skin care. As seen in this table, nanoemulsion is a breakthrough moisturizer formulated to improve the appearance of tone and texture for skin [6].

In the case of antiwrinkle and antiaging commercial nanocosmetics, Table 23.2 presents various nanoantiwrinkle and nanoantiaging products in the market. As seen in this table, some products have been manufactured by major cosmetic companies, including Lancôme, L'Oreal, and Kara Vita.

For hair care applications, different types of nanomaterials can be used in cosmetics [10,11]. Table 23.3 shows the main types of nanocosmetics for hair care in the market.

TABLE 23.1 Various nanomoisturizer products in the market.

Commercial product	Manufacturer	Marketing claims	References
Hydra Flash Bronzer Daily Face moisturizer	Lancôme	Nanocapsules of pure vitamin E provide powerful antioxidant protection. A light touch of self-tanner ensures a natural, healthy glowing skin.	[5]
Hydra Zen Cream	Lancôme	Containing nanoencapsulated triceramides, Hydra Zen helps restore perfect comfort and softness and renew skin's healthy look. Protected from signs of daily stress and fully hydrated, your skin is beautifully soft and smooth all day long.	[5]
Nano-In Hand and Nail Moisturizing Serum and Foot Moisturizing Serum	Nano-Infinity Nanotech	Fine crystals of ZnO nanoparticles will go straight into skin tissue to prevent hand and nails from being hurt and restore skin health.	[5]
Nano Emulsion Multi-Peptide Moisturizer	Hanacure	Packed with a remarkably high concentration of scientifically proven peptides and exclusive "Ruby Of The Forest" mushroom extract, the nanoemulsion improves the appearance of skin's tone and texture. By using the nanotechnology process, the skin is instantly rejuvenated with hydration that lasts.	[6]

Reproduced with permission from Lohani A, Verma A, Joshi H, Yadav N, Karki N. Nanotechnology-based cosmeceuticals. ISRN Dermatol 2014;2014:843687. Licensed under Creative Commons Attribution License (CC BY 3.0).

TABLE 23.2 Various nanoantiwrinkle and nanoantiaging products in the market.

Commercial product	Manufacturer	Marketing claims	References
Lancôme RenergieMicrolift	Lancôme	Formulated with colloidal silica and soy protein nanoparticles to provide the closest possible face-lift effect.	[5]
RevitaLift Anti-Wrinkle and Firming Face and Neck Contour Cream	L'Oreal	The RevitaLift formula is enriched with Pro-Retinol A, a powerful antiwrinkle agent, which is encapsulated in nanosomes. Nanosomes penetrate deep into the epidermis to work at the heart of wrinkles.	[5]
RevitaLift Double Lifting	L'Oreal	It contains nanosomes of Pro-Retinol A. RevitaLift Double Lifting is a unique dual action treatment that instantly retightens skin and effectively fights wrinkles.	[5]
Eye Tender	Kara Vita	It contains nanospheres, delivers 13 bioactives including proven, wrinkle-reducing peptides to stimulate fibroblasts, build collagen, brighten skin, and reduce inflammation for a younger, healthier appearance	[5]
Eye Contour Nanolift	Euoko	It is based on nanocapsules technology. Lifting nanocapsules join seven other immediate and long-term fighters of fine lines, wrinkles, and puffiness. It provides instant and long-term smoothness, gives the eye area more radiance, and diminishes the appearance of dark circles and puffiness.	[5]
Soleil Soft-Touch Anti-Wrinkle Sun Cream SPF 15	Lancôme	It contains vitamin nanocapsules that help to preserve skin's youth effectively. SPF 15 offers optimal protection against the sun. It contains exclusive ingredients to guarantee a long-lasting effect.	[5]
Nano Gold Firming Treatment	Chantecaille	Infinitely small nanoparticles of pure gold are bound to silk microfibers to firm and tone skin, while delivering incredible antiinflammatory, healing, and age-defying power.	[5]
Nanosphere Plus	DermaSwiss	A stem cell revolutionary antiaging therapy Nanosphere Plus serum has been specially formulated to allow natural stem cells to preserve and protect skin cells. Using the cells from a rare Swiss apple (UttwilerSpatlauber), Nanosphere Plus protects longevity and combats chronological aging.	[5]
Zelens Fullerene C-60 Night Cream	Zelens	Fullerene C-60 is a naturally occurring microscopic form of carbon, which was found to have remarkable antioxidant properties.	[5]
Chantecaille Nano Gold Energizing Cream	Chantecaille	Nanoparticles of 24-karat gold are bound to silk microfiber, a natural protein that is moisturizing, antioxidant, and antiinflammatory. These elements reach the cellular level where they act as the ultimate healing and preserving force.	[7]

(Continued)

TABLE 23.2 (Continued)

Commercial product	Manufacturer	Marketing claims	References
Nano antiaging cream	PHYTOMO	Nano3X has been developed by using nanotechnology and age-retarding technology for skin therapy. It is an antiaging hydra complex with the highly potent active ingredients in nanoformat. It contains a highly active substance of hyaluronic acid, which has both high and low molecular contents. Regular use result is smooth, firm pulmped-up skin. Fine lines and wrinkles caused by drying out of the skin are visibly reduced.	[8]
Nano Anti-Aging Serum	Nano by Saurina	Cream of semilight texture based on nanotechnology that treats aging and improves skin elasticity.	[9]

TABLE 23.3 Nanocosmetic products for hair care in the market.

Commercial product	Manufacturer	Marketing claims	References
Shampoo Hidratante com Batana Oil	FacinatusCosméticos	Batana oil in a nanoencapsulated form contributes to the permeation of the product, potentializing its action. Unique, ultraemollient formulas that promote differentiated hydration and repair	[12]
Nanover	Nanogist	Containing silver nanoparticles	[13]
Nano Collagen	Braziliss	Nano Collagen enriched with thermoactivated nanotechnology. Due to its easy absorption of active nanocapsules, it protects the hair, eliminates the porosity, and promotes the balance of the hydration level of each fiber resulting in healthier hair	[14]
Nutraliss Nano Repair Shampoo + Nano Repair Conditioner Liter	Nutraliss	Shampoo: Low pH without salt and enriched with nanoparticles of sericin, organic extracts, and an exclusive blend of 12 nourishing-functional oils prolongs the duration of smoothing and maintains the color for more time indicated for all types of hairConditioner: Nutraliss Nano Repair conditioner regenerates, hydrates, and deeply nourishes yarn enriched with Nano sericin particles	[15]

For sun care applications, also in this book, Baptista [16] introduces the broadband light protecting sun care products. Table 23.4 presents various products for lip and sun cares in the market.

Tables 23.5 and 23.6 show some typical nanocosmetic products for face and body cares, respectively. As shown in Table 23.5, noble metal nanoparticles, such as silver, gold, and platinum, can be used in the skin care formulations [5,17,18,23,24].

TABLE 23.4 Various nanocosmetics for lip and sun cares in the market.

Commercial product	Manufacturer	Marketing claims	References
DiorSnow Pure UV Base SPF 50	Dior	Contains nano-UV filters for ultraprotection against the damaging effects of UVA and UVB rays	[5]
Soleil Instant Cooling Sun Spritz SPF 15	Lancôme	Contains vitamin nanocapsules. Instant cooling sun spray SPF 15 immediately offers a sensation of freshness. SPF 15 provides optimal protection against the sun	[5]
Primordiale Optimum Lip	Lancôme	Delivers 100% botanically pure vitamin E via nanocapsule technology to reduce lip bleeding and feathering due to fine lines and wrinkles	[5]
Lip Tender	Kara Vita	Ten bioactive ingredients are precisely calculated to work within lyphazomes, delivering a 4-in-1 formula and bringing long-lasting hydration for fast and dramatic lip repair	[5]

Reproduced with permission from Lohani A, Verma A, Joshi H, Yadav N, Karki N. Nanotechnology-based cosmeceuticals. ISRN Dermatol 2014;2014:843687. Licensed under Creative Commons Attribution License (CC BY 3.0).

TABLE 23.5 Face care products in the market.

Commercial product	Manufacturer	Marketing claims	References
Cosil Whitening Mask	Natural Korea	Made with nanocolloidal silver used for the effect of getting rid of germs from your face, compressing pores, soothing the skin condition, and keeping your skin radiant and soft	[5]
Nanorama—Nano Gold Mask Pack	LexonNanoTech	It contains pure nanosized gold that is highly effective in penetrating small pores and disinfecting skin, helps to reduce pore size, and prevents and treats acne. It is well known that nanogold is very effective disinfectants	[5]
LifePak Nano	Pharmanex	LifePak Nano is a nutritional antiaging program formulated to nourish and protect cells, tissues, and organs in the body with the specific purpose of guarding against the ravages of aging. Nanoencapsulation increases bioavailability coenzyme Q10 by 5–10 times	[5]
Nano Gold Repair Collagen Face Mask	Knesko	It promotes age-defying healing by helping to minimize dryness, redness, puffiness, fine lines, and enhance overall skin maintenance. The doctor-formulated GEMCLINICAL serum is infused with colloidal gold that quickly absorbs deep into skin and bolsters self-rejuvenation for superb skin care results	[17]
Majestic Pure, 24 K Gold Facial Mask	Majestic Pure	To remedy skin problems with its powerful healing and antiaging properties. As one of the softest metals, it is easily absorbed into the skin; giving it influence as a cellular level. Also, it helps reverse oxidation damage to the collagen fibers, aiding the regeneration of healthy new cells, bringing natural luster and youthful radiance to all skin types	[18]

(Continued)

TABLE 23.5 (Continued)

Commercial product	Manufacturer	Marketing claims	References
Nano Corrector lifting effect	tianDe	It eliminates the necessity of surgical face lifting and collagen injections. This complex saturates lower layers of dermis with collagen, activates the synthesis of native collagen fibers, and restores elasticity of skin and clear facial contoursThe complex includes two products: regenerative meso-cocktail (collagen peptide, β-glucan, glycosamine, hyaluronic acid) and microcapsules with collagen	[19]
Nano Pearl Powder	Ancient Purity	Pearl Powder (Nano)-A Grade 150,000 Mesh. Enrich the skin, preserve a youthful appearance, and secure longevity with this ancient skin and health tonic. The makeup of Pearl participates in DNA and RNA metabolic activities and thus can support cell regeneration.	[20]
Nano Pearl Powder	Zhejiang Phoenix	Pearl nanometer powder facial mask. Daily Nursing (apply on face, make up, eliminate eye pouch, lighten dark eye socket, block sun). External application: whitening, controlling oils, removing blackheads, eliminating freckles	[21]
Nanowhite and nanowhite men	Nanowhite	Advanced nanotechnology produces 1000 times more effective in inhibiting melanin synthesis for lasting fairnessNanoactive technology brightens and hydrates the skin while deeply cleansing to remove excess oil and impurities	[22]
Nano platinum serum	Chantal Le Beau	Regenerate skin texture, skin is firmer and lifted skin becomes brighter and glowing after treatmentA formidable advance formula using nanotechnology and active ingredients that can be used during or with massage, masques, ultrasonic treatment, or galvanic treatment to stimulate and improve cellular circulation. It can also be applied directly on skin as treatment usage to correct problem areas	[23]
Stem Cellue Nano Platinum Skin Rejuvenating Cream	Cada Suissesse	With the scientific revolutionary ingredients and formulation (nanoconjugation with platinum particles), it actively generates skin cells renewal system, boosts the production of human stem cells to rejuvenate aging skin, thus decreases in number and depth of lines and wrinkles, strengthens skin to resist the environmental assaults, and also lessens the appearance of unsightly wrinkles. In a clinical trial for 30 days with 20 subjects, using nanoconjugation with platinum particles and the antiaging ingredients skin care products, improvement of elasticity and hydration was found, whereas elasticity 55%, hydration 45%, reduction of wrinkles by 90%	[24]

Reproduced with permission from Lohani A, Verma A, Joshi H, Yadav N, Karki N. Nanotechnology-based cosmeceuticals. ISRN Dermatol 2014;2014:843687. Licensed under Creative Commons Attribution License (CC BY 3.0).

TABLE 23.6 Nanocosmetics of body lotion and cleanser in the market.

Commercial product	Manufacturer	Marketing claims	Ref.
Fresh As A Daisy Body Lotion	Kara Vita	This lotion uses nanospheres to quickly penetrate, moisturize, and nourish all types of skin	[5]
Nano Cyclic Cleanser Silver	Nano Cyclic	Cyclic cleanser is a scientifically balanced blend of nanosilver and natural ingredients. It kills harmful bacteria and fungi, treats acne, exfoliates dead skin on all parts of the body, diminishes age spots, deodorizes the body, and fights wrinkles	[5]
Cosil Nano Beauty Soap	Natural Korea	Silver nanoparticles are highly effective as a disinfectant and guarantee protection of skin	[5]

Reproduced with permission from Lohani A, Verma A, Joshi H, Yadav N, Karki N. Nanotechnology-based cosmeceuticals. ISRN Dermatol 2014;2014:843687. Licensed under Creative Commons Attribution License (CC BY 3.0).

TABLE 23.7 Oral care nanocosmetics in the market.

Commercial product	Manufacturer	Marketing claims	References
Nano Silver toothpaste	Hanil	By the action of the calcium carbonate and the silver nanoparticles, it is good for the prevention of almost all dental disease sand for the bad breath and whitening teeth	[25]
Coral White Nano Silver	Coral White	Traditional calcium toothpaste can be ineffective because it cannot penetrate the bacterial biofilm. Our Nano Silver toothpaste penetrates through to deliver calcium straight to the tooth surface for stronger enamel and remineralization. The Nano Silver helps prevent gingivitis and bleeding gums and reduce inflammation	[26]
Nano Gold Particle Whitening Toothpaste	Whitewash	It contains gold particles along with green tea and vitamin E. Also featuring special micropolishers to help brighten teeth and restore them to their natural whiteness	[27]
Nano whitening Toothpaste	Whitewash	It utilizes patented hydroxyapatite which is an extremely small type of hydroxyapatite crystal which has the ability to reharden softened enamel	[27]
APAGARD M plus	Sangi	Contains nano Medical Hydroxyapatite (nano) for protection against caries	[28]
Nano Silver Xylitol Mouthwash	Elementa Silver	It is the first biofilm penetrating nanotechnology. Traditional calcium mouthwash and calcium toothpaste can be ineffective because they cannot penetrate the bacterial biofilm. This mouthwash penetrates through to deliver calcium straight to the tooth surface for stronger enamel and remineralization	[29]

Finally, in Table 23.7, different oral care nanocosmetics in the market are summarized. As shown in Table 23.7, most products used nanosilver in their cosmetic formulations [25,26,29]. In addition, nanohydroxyapatite is very promising ingredient for oral care applications [27,28].

In the case of perfumery, nanoperfumes can be achieved via the nanoencapsulated fragrance systems. In the previous chapter, Miastkowska and Lasoń presented the water-based nanoperfumes [30]. In the market, the Seo-Kwang company developed the nanoperfume ejector. In the model SKC-SE 500II, it mixed the silver nanoparticles with both water particles and lavender/lemon fragrances. This product could sterilize and eliminate the air and bad odors [31,32].

In conclusion, nanomaterials have featured in cosmetic formulations for more than 10 years, with enhancing characters and offering advantages over the traditional formulations. However, only a few types could be used commercially, such as liposomes and metal oxide nanoparticles. The reason why their development in cosmetics is rather limited is the regular bans and halts on nanocosmetic products by many establishments [33]. In the previous chapter of this book, Selletti and Ardemagni [34] discussed on the current legal frameworks and consumer protection in nanocosmetics.

References

[1] Regulation (EC) No. 1223/2009 on cosmetic products. <https://ec.europa.eu/growth/sectors/cosmetics/legislation_en>.

[2] Faria-Silva AC, et al. Chapter 4. Nano-emulsions for cosmetic products December In: Nanda A, Nanda S, Nguyen TA, Rajendran S, Slimani Y, editors. Nanocosmetics: fundamentals, applications and toxicity. Elsevier; 2020. ISBN: 9780128222867.

[3] Svarc F, Hermida L. Chapter 3. Transdermal and bioactive nano-carriers for skin care December In: Nanda A, Nanda S, Nguyen TA, Rajendran S, Slimani Y, editors. Nanocosmetics: fundamentals, applications and toxicity. Elsevier; 2020. ISBN: 9780128222867.

[4] Hari BN, et al. Chapter 6. Polymeric nano-carriers for topical drug delivery in skin cream December In: Nanda A, Nanda S, Nguyen TA, Rajendran S, Slimani Y, editors. Nanocosmetics: fundamentals, applications and toxicity. Elsevier; 2020. ISBN: 9780128222867.

[5] Lohani A, Verma A, Joshi H, Yadav N, Karki N. Nanotechnology-based cosmeceuticals. ISRN Dermatol 2014;2014:843687. Available from: https://doi.org/10.1155/2014/843687.

[6] Nano Emulsion Multi-Peptide Moisturizer, <https://www.hanacure.com/products/multi-peptide-nano-emulsion>.

[7] Chantecaille Nano Gold Energizing Cream, <https://www.saksfifthavenue.com/chantecaille-nano-gold-energizing-cream/product/0435297663735?sid = 120A0B312FA8&R = 656509702909&FOLDER%3C%3Efolder_id = 2825744927119868&ASSORTMENT%3C%3East_id = 1408474399545537&bmUID = 1239646676269&P_name = Chantecaille>.

[8] Nano 3x Anti-Aging Cream, <https://www.indiamart.com/proddetail/nano-3x-anti-aging-cream-13953309530.html>.

[9] Nano-anti-aging-serum, <https://www.nanobysaurina.eu/product/nano-anti-aging-serum/>.

[10] Yukuyama MN, et al. Chapter 10. Nanomaterials for haircare applications December In: Nanda A, Nanda S, Nguyen TA, Rajendran S, Slimani Y, editors. Nanocosmetics: fundamentals, applications and toxicity. Elsevier; 2020. ISBN: 9780128222867.

[11] Rajendran S, et al. Chapter 11. Nanoparticles in hair dyes December In: Nanda A, Nanda S, Nguyen TA, Rajendran S, Slimani Y, editors. Nanocosmetics: fundamentals, applications and toxicity. Elsevier; 2020. ISBN: 9780128222867.

[12] Shampoo Hidratante com Batana Oil, <https://product.statnano.com/product/10448>.

[13] Nanover, <https://product.statnano.com/product/6980>.

[14] Nano Collagen, <http://ngct.ae/product/nano-collagen-1000ml/>.

[15] Nutraliss Nano Repair Shampoo + Nano Repair Conditioner Liter, <https://www.amazon.com/Nutraliss-Repair-Shampoo-Conditioner-Liter/dp/B076TRQ28S>.

[16] Baptista MS. Chapter 9. Nanocosmetic for broad band light protecting sun care products December In: Nanda A, Nanda S, Nguyen TA, Rajendran S, Slimani Y, editors. Nanocosmetics: fundamentals, applications and toxicity. Elsevier; 2020. ISBN: 9780128222867.
[17] <https://knesko.com/product/gold-face-mask/>.
[18] <https://www.iherb.com/pr/Majestic-Pure-24K-Gold-Facial-Mask-8-8-oz-250-g/89179?gclid = CjwKCAiAh5_uBRA5EiwASW3Iahr7xmj9eCSPymqdfg9x2_qH4uzrtoKdyy32E8peS3ip4mXgehjTCRoCarQQAvD_Bw-E&gclsrc = aw.ds>.
[19] Nano Corrector lifting effect, <http://www.tiande-global.com/en/product/10345/>.
[20] Pearl Powder (Nano) A Grade (Beauty) 50g, <https://www.ancientpurity.com/pearl-powder-nano-a-grade-beauty-50g>.
[21] 100% Nano Pearl Powder, <https://www.globalsources.com/si/AS/Zhejiang-Phoenix/6008819503684/pdtl/100-Nano-Pearl-Powder/1020437926.htm>.
[22] Nanowhite and nanowhite men products, <https://www.nanowhitesg.com/nanowhite/>.
[23] Nano platinum serum, <https://goldenqueencosmetics.com/products/nano-platinum-serum>.
[24] Stem Cellue Nano Platinum Skin Rejuvenating Cream, <http://www.cadasuissesse.com/en/products/%E5%B9%B9%E7%B4%B0%E8%83%9E%E7%B4%8D%E7%B1%B3%E7%99%BD%E9%87%91%E5%A4%9A%E8%82%BD%E8%82%8C%E9%B9%B7%E9%80%86%E8%BD%89%E9%9D%A2%E9%9C%9C-%252d-10-ml.html>.
[25] Nano Silver toothpaste (Clean world Ace toothpaste), <https://www.tradekorea.com/product/detail/P369483/Clean-world-Ace-toothpaste.html>.
[26] Coral White Nano Silver, <https://www.amazon.com/Coral-White-Toothpaste-Fluoride-Whitening/dp/B07M7LDGXB>.
[27] Nano Gold Particle Whitening Toothpaste, <https://whitewashnano.co.uk/products/nano-gold-particle-whitening-toothpaste>.
[28] <https://www.sangi-co.com/en/dental_products/index.html>.
[29] Nano Silver Xylitol Mouthwash, <https://www.amazon.com/Formulated-Silver-Mouthwash-Balanced-Fluoride/dp/B07F79DFCT?th = 1>.
[30] Miastkowska M, Lasoń E. Chapter 7. Water based nano-perfumes December In: Nanda A, Nanda S, Nguyen TA, Rajendran S, Slimani Y, editors. Nanocosmetics: fundamentals, applications and toxicity. Elsevier; 2020. ISBN: 9780128222867.
[31] <https://www.nanotech-now.com/news.cgi?story_id = 31151>.
[32] DY Putro, R Saputra, MF Albab, S Nano, Perfume Ejector to Destroy Bacteria for Clothes, International Conference on Natural Science, Engineering, and Technology Proceedings, Bangkok, 4–5 November 2013, p. 72–5. ISBN: 978-616-907-493-9 <http://aasic.org/proc/aasic/article/view/50/49>
[33] Abbasi BH, et al. Chapter 5. Nanomaterials for cosmeceuticals December In: Nanda A, Nanda S, Nguyen TA, Rajendran S, Slimani Y, editors. Nanocosmetics: fundamentals, applications and toxicity. Elsevier; 2020. ISBN: 9780128222867.
[34] Selletti S, Ardemagni E. Chapter 16. Current legal frameworks and consumer protection in nanocosmetics December In: Nanda A, Nanda S, Nguyen TA, Rajendran S, Slimani Y, editors. Nanocosmetics: fundamentals, applications and toxicity. Elsevier; 2020. ISBN: 9780128222867.

CHAPTER

24

Nanocosmetics: future perspective

P. Morganti[1,2], Hong-Duo Chen[3] and G. Morganti[4]

[1]Accademy of History of Health Care Art, Rome, Italy [2]China Medical University, Shenyang, P.R. China [3]Dermatology Department, Key Laboratory of Immunodermatology, Hospital No. 1, China Medical University, Shenyang, P.R. China [4]ISCD Nanoscience Center, Rome, Italy

24.1 Introduction

Degradation and depletion of natural raw materials for food and personal care systems [1] due to population growth rate [2], the increased production of CO_2 and greenhouse gas (GHG) emissions [3], and the use of indestructible plastic packaging waste with the consequent climate change [3] are providing the necessary inputs for a more sustainable development and a different way of living. By a sustainable development, therefore, it is possible to ensure the right food and skincare products for health, social protection, and equality of opportunity (Fig. 24.1) [4]. This smart and "green" development is fundamentally based on the circular economy turned to an innovative use of bionanotechnology, which deals with the self-assembling of biomolecules. The final purpose is to build functional nanostructure, able to incorporate nanomolecules into biological systems [5].

Nanoscience and nanotechnology, therefore, are considered to have huge potential for bringing benefits to many areas of research and applications, attracting increasing investments from Governments and businesses worldwide, especially in the medical and cosmetic field [6]. However, production, use, and application of nanocompounds require new challenges for versifying their effectiveness and safeness and organizing innovative regulatory rules [7]. Moreover, these smart technologies permit to utilize waste materials for producing every kind of goods, food, cosmetic, and personal care products based on natural ingredients and recyclable packagings. It is possible, for example, to produce not only material and devices such as scaffolds for cell and tissue engineering, and sensors able to monitoring aspects of human beauty and health, but also packaging materials biodegradable and compostable to save the environment [8]. Thus the so-called circular green economy, based on the three R (reuse, reduce, recycle), represents a radical, transition for more efficient, environmentally friendly products and resources, and more safe technologies, able to reduce GHG emissions and mitigate the effects of climate changing. By this new

Nanocosmetics
DOI: https://doi.org/10.1016/B978-0-12-822286-7.00020-6

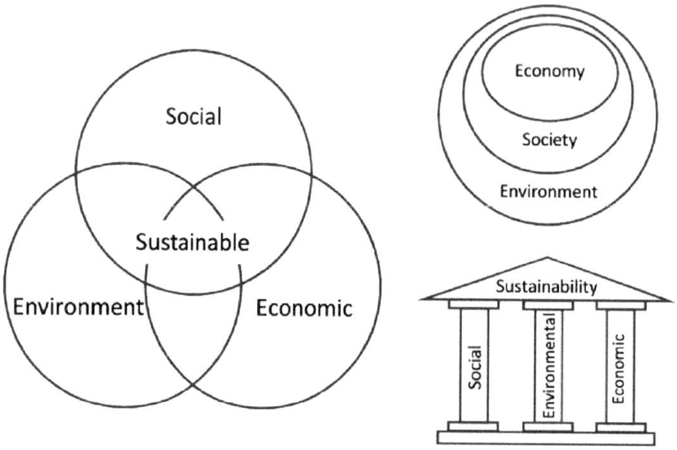

FIGURE 24.1 Social, environmental, and economic equality for a sustainable development. Source: *Reproduced with permission from Purvis B, Mao Y, Robinson D. Three pillars of sustainability: in search of conceptual origins. Sustain Sci 2019;14:681. Licensed under Creative Commons Attribution 4.0 International License.*

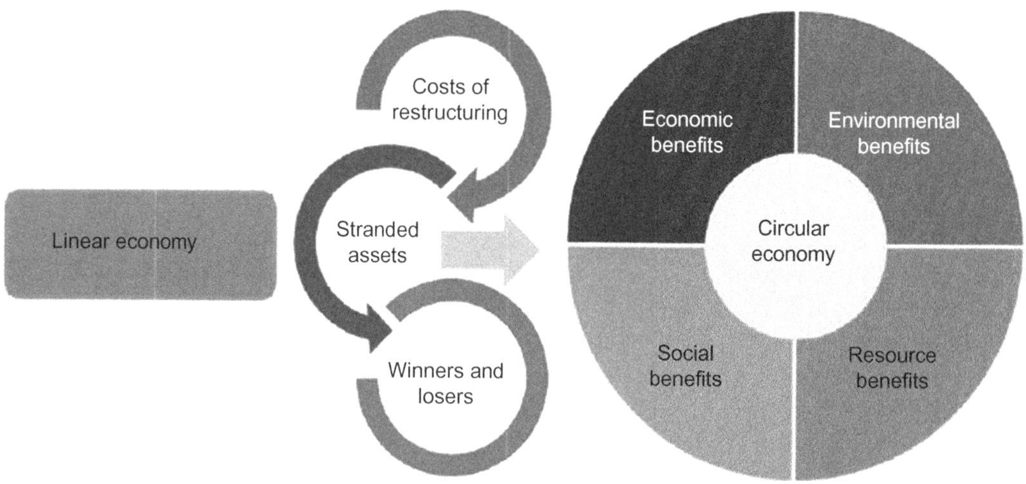

FIGURE 24.2 Linear economy versus circular economy. Source: *Reproduced with permission from EEA Report No. 2/2016, Circular economy in Europe, <https://www.eea.europa.eu/publications/circular-economy-in-europe>. Licensed under Creative Commons Attribution 2.5 Denmark (CC BY 2.5 DK).*

way of thinking and working, it will be possible to realize six main results: renewable energy, green buildings, sustainable transportation, and a more efficient water, waste, and land management [8–10]. Converting the linear economy to the recycling economy (Fig. 24.2) has to be considered essential to reduce the actual waste and save natural raw materials and biodiversity of our planet [11]. The most interesting waste material, recovered by the fishery's and agroforestry biomass results of around 300 billion tons/year ($\sim 150 + 150$, respectively) [12], represented from chitin/chitosan and lignin/cellulose, in their nanodimension also. These biopolymers in nature possess a unique 3D structure that, loosed during their extraction method, is fixed as hemicellulose in the wood architecture (Fig. 24.3) and as peptides in crab (Fig. 24.4) [13,14]. However, their original tridimensional

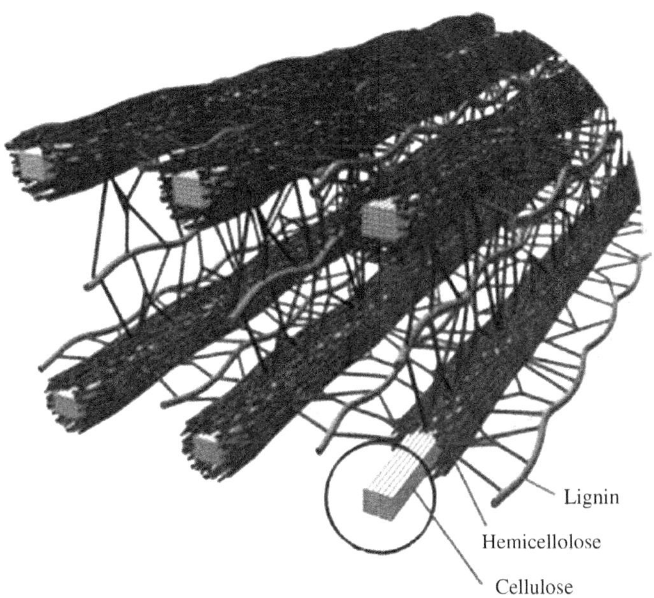

FIGURE 24.3 3D structure of hemicellulose and lignin in wood. Source: *Reproduced with permission from Doherty WOS, Mousavioun PA, Fellow CM. Value-adding to cellulosic ethanol: lignin polymers. Ind Crop Prod 2011;33:259–76. Copyright 2011 Elsevier.*

organization may be obtained again by the application of additional templates during, for example, the electrospinning methodology by which it is possible to obtain the same structure of the native extracellular matrix (ECM) (Fig. 24.5) [16]. Thus chitin nanofibrils (CNs), nanolignin (NG), and nanocellulose (NCE) and their complexes can be easily converted to specialized structures which, characterized for their highly ordered porous architecture stabilized by noncovalent interactions, also show an interesting biocompatibility and high mechanical strength. These natural polymers, therefore, for the porosity that permits reproduction and growth of human cells, may be used to make active carriers for innovative smart cosmeceutical-tissues, affecting positively the active aging processes as reported also in Chapter 1, Nanocosmetics: an introduction (Fig. 24.6) [15,17,18]. For all these reasons, polysaccharides and chito-oligosaccharides seem to represent a new and *green* raw material for the future innovative cosmetic products, also because obtainable as nanostructured polymers.

24.2 Polymeric natural compounds and biopolyesters

Polymeric natural compounds, such as CN, nanochitosan, and NG, microstarch, and NCE, nanocellulose, derived from waste feedstocks, and biopolyesters such as polylactic acid (PLA) and polyhydroxyalkanoates (PHAs) synthesized by renewable resources, can be considered the future natural raw materials of great interest as suggested in Chapter 1, Nanocosmetics: an introduction. These naturally derived biopolymers, in fact, obtained by the use of bionanotechnologies might represent interesting basic *bricks* to formulate future innovative nanocosmetics. They can be used indifferently as cosmetic carriers and active

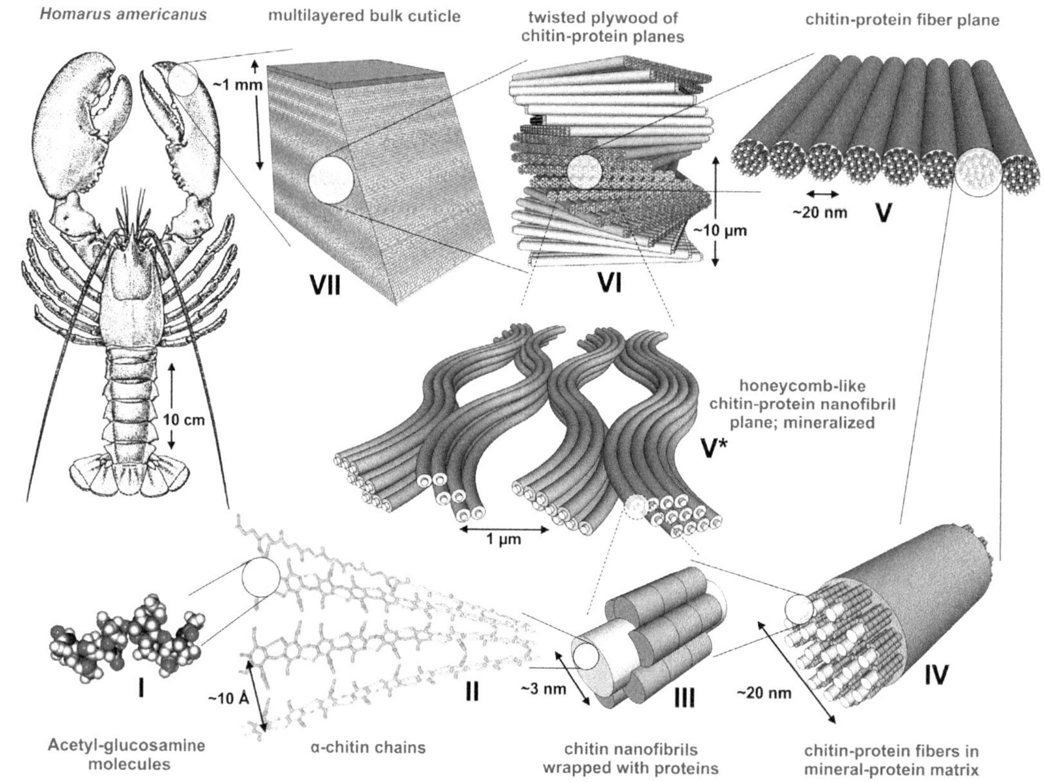

FIGURE 24.4 3D structure of alpha-chitin in crab. Source: *Reproduced with permission from Nikolov S, Fabritius H, Petrov M, Frisk M, Lymperakis L, Sachs C, et al. Robustness and optimal use of design principles of arthropod exoskeletons studied by ab initio-based multiscale simulations. J Mech Behav Biomed Mater 2011;4:129–45. Copyright 2011 Elsevier.*

ingredients [15–20] as well as packaging material [21,22], being not only useful for the skin cell nourishing and mitotic division [16,23–25], but also enough shock- and heat-resistant to be used as biodegradable material for cosmetic and food active edible packagings [19–22].

24.2.1 Chito-oligosaccharides and polysaccharides

24.2.1.1 Chitin and chitosan

Alpha-chitin is a polymer of cationic nature which, composed of beta-(1-4)-D-glucosamine units and *N*-acetyl glucosamine, results in long antiparallel chains forming a polysaccharide represented by a crystalline amorphous structure, just resembling the silk and cellulose organization (Fig. 24.7) [24]. This biopolymer in nature is reinforced by cross-linking with proteins, mineral, and waxes, giving rise to different shapes and functions as diverse as the insect cuticle, wings, and exoskeleton, the fungal cell wall, and helminth

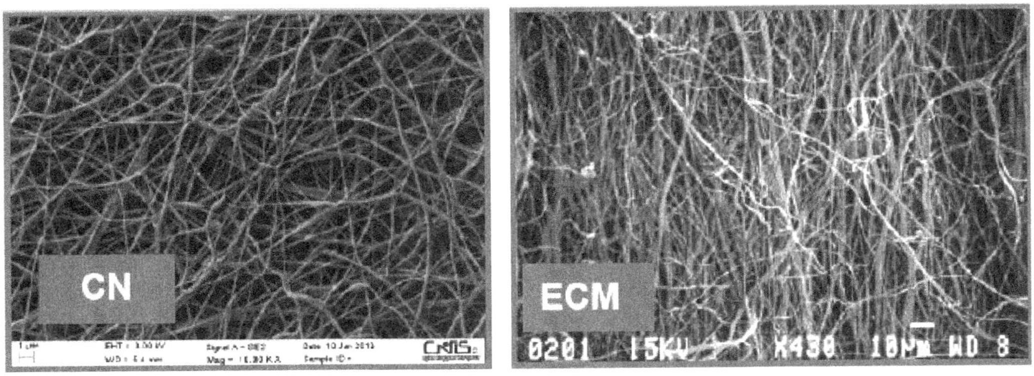

FIGURE 24.5 SEM 3D structure of electrospun CN tissue (left) compared to natural extracellular matrix (right). Source: *Reproduced with permission from Morganti P, Morganti G, Colao C. Biofunctional textiles for aged skin. Biomedicines 2019;7(3):51. Licensed under Creative Commons Attribution (CC BY 4.0) license.*

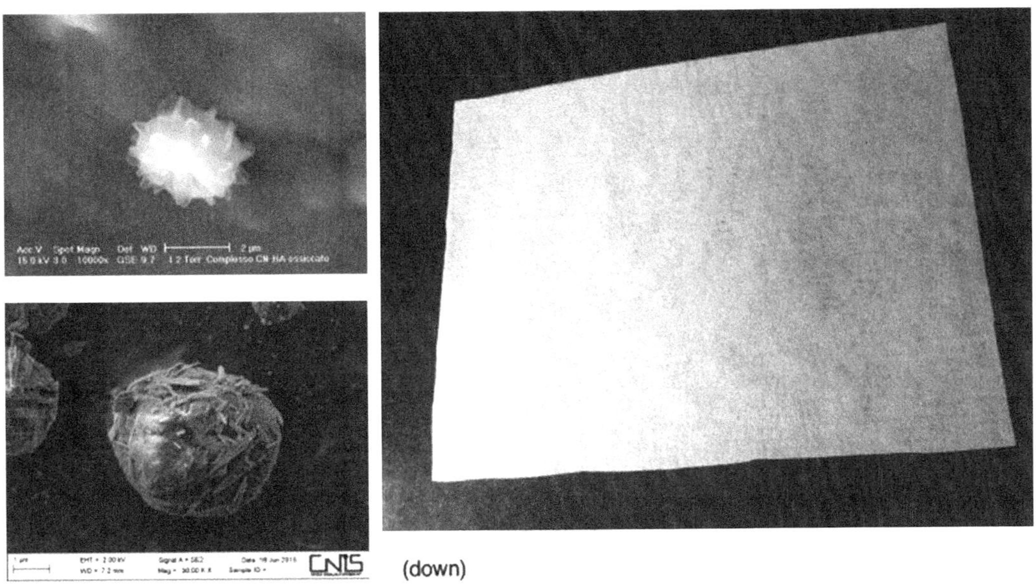

FIGURE 24.6 Cosmetic biodegradable tissue (right) embedded by active aging complexes (left) CN hyaluronic acid (up) and CN lignin.

eggs (Fig. 24.8) [26]. Chitin mainly recovered from the crustaceans processing industry is the second most available polysaccharide after cellulose, being produced by living organisms on the order of 100–1000 billion tons/year [26–28]. Chitosan is a biopolymer which obtained by alkaline deacetylation of chitin possesses nanochitin, acid-base properties, and chelation abilities for the presence of free –OH and NH_2– groups, which lead to the capability of forming hydrogen and covalent bonding. However, being the deacetylation degree of chitosan between 60% and 90% and CNs between 50% and 60%, both the

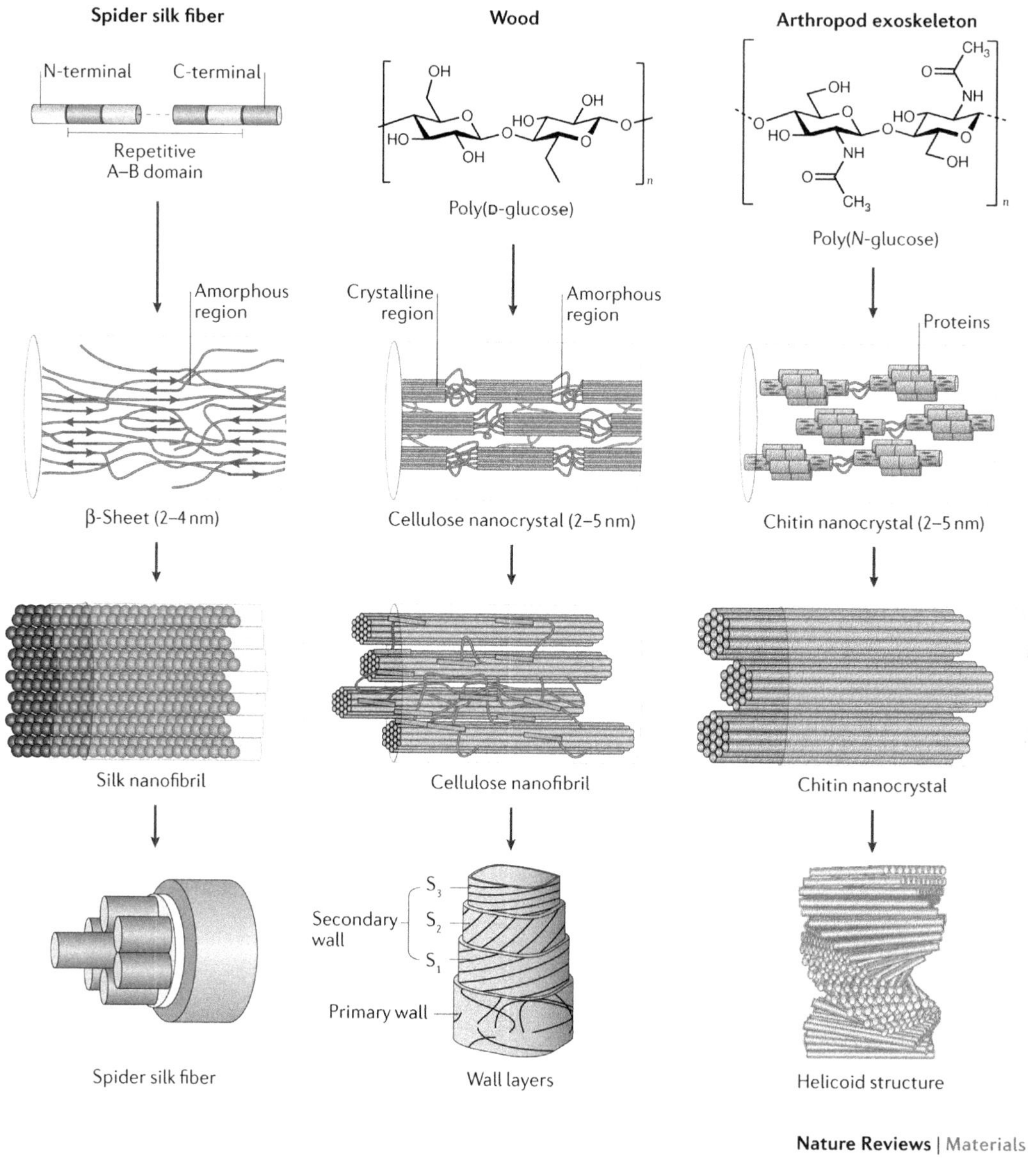

FIGURE 24.7 Crystalline/amorphous structure of chitin, cellulose, and silk. Source: *Reproduced with permission from Ling S, Kaplan DL, Buehel MJ. Nanofibers in nature and materials engineering. Nat Rev Mater 2018;3:18016. Copyright 2018 Springer Nature.*

polymers possess an interesting moisturizing activity, resembling the hyaluronic acid structure (Fig. 24.9). Moreover, they exhibit also an enormous surface improvement which allows them to impart strength to several materials such as nanocomposites and nonwoven tissues. Also if more than 28,000 tons of chitin have been produced in 2015, they are

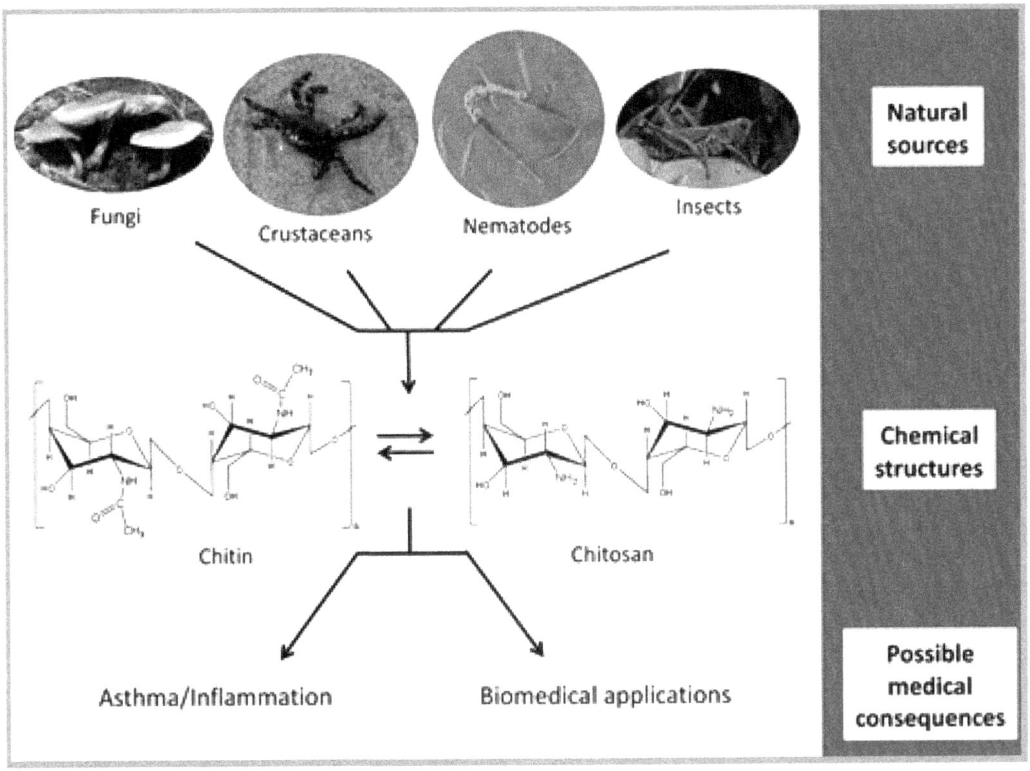

FIGURE 24.8 Chitin in insect and fungi and its activity in human diseases. Source: *Reproduced with permission from Morganti P, Febo P, Cardillo M, Barone A. Chitin nanofibril and nanolignin: natural polymers of biomedical interest. Clin Cosmet Dermatol 2017;1(2). Licensed under Creative Commons Attribution 4.0 International License.*

not sufficient to cover the world demand of around 60,000 tons (Fig. 24.10). It has been estimated, in fact, that its global market of US$ 893 million in 2017 will increase to US$ 2.9 billion at a Compound Annual Growth Rate (CAGR) from 4.0% to 12.7% throughout the forecast period 2017–27, with the healthcare segment estimated to range a value of US$ 1 billion [28–30]. Therefore it is to underline the necessity to process in chitin–chitosan all the fishery's byproducts, transforming them in goods, also to eliminate the actual environment hazard created by this dangerous waste. All around the world, in fact, it is thrown away at sea, burned, landfilled, or simply left out to spoil, creating a great problem of pollution [26,27]. For the many possibilities of its use, the industries are producing not only industrial grade chitin but also food grade chitin characterized by its high purity, obtained through improved nanotechnologies. By a green methodology, in fact, it has been possible to produce chitin nanocrystals/nanofibrils [31–33] which, for their structure, purity, and electrical positive charges covering their surface, are able to easily form block polymeric complexes with electronegative polymers [34–37] (Fig. 24.11). CNs are, therefore, widely used for medical and cosmetic applications, due not only to their noncytotoxicity and high biocompatibility, but also to the flexibility of their fibrils which play a key role in determining the mechanical performance of the relative network. Thus these natural fibrils are

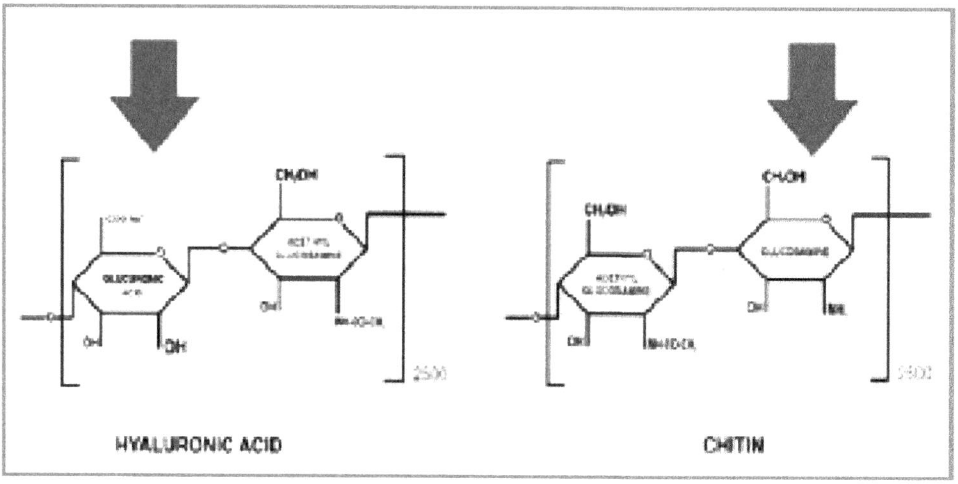

FIGURE 24.9 The same backbone of chitin and hyaluronic acid. Source: *Reproduced with permission from Morganti P, Febo P, Cardillo M, Barone A. Chitin nanofibril and nanolignin: natural polymers of biomedical interest. Clin Cosmet Dermatol 2017;1(2). Licensed under Creative Commons Attribution 4.0 International License.*

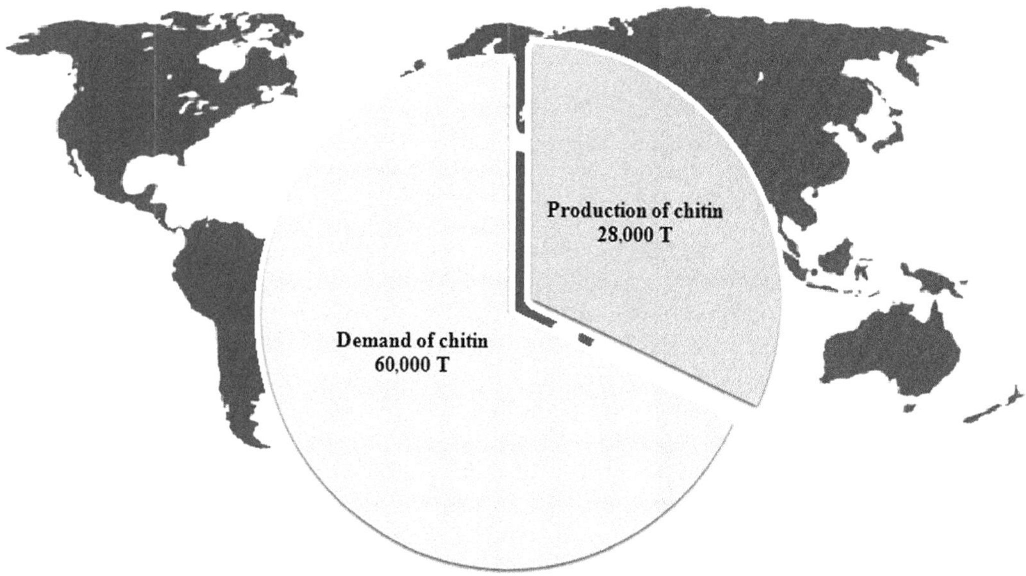

FIGURE 24.10 Production and worldwide demand of chitin. Source: *Reproduced with permission from Tyliszczak B, Drabczyk A, Kudłacik-Kramarczyk S, Sobczak-Kupiec A. Sustainable production of chitosan. In: Królczyk G, Wzorek M, Król A, Kochan O, Su J, Kacprzyk J, editors. Sustainable production: novel trends in energy, environment and material systems. Studies in systems, decision and control. Cham: Springer; 2020. p. 198. Copyright 2018 Springer Nature.*

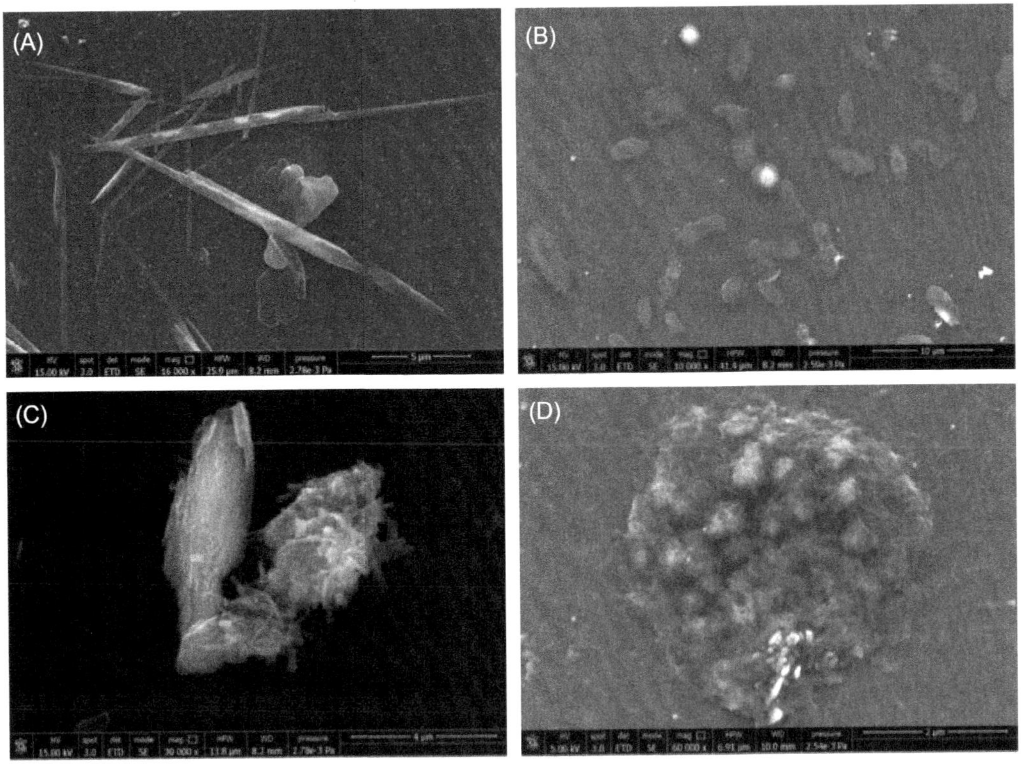

FIGURE 24.11 Chitin nanofibril at SEM. (A) Chitin nanofibrils (CN) and (B–D) its complex with nanolignin (NG). Source: *Reproduced with permission from Danti S, Trombi L, Fusco A, Azimi B, Lazzeri A, Morganti P, et al. Chitin nanofibrils and nanolignin as functional agents in skin regeneration. Int J Mol Sci 2019; 20(11):2669. Licensed under Creative Commons Attribution (CC BY 4.0) license.*

widely utilized as tissue engineered scaffolds, drug and cosmetic delivery carriers, wound and active aging dressings, film separation membranes, and antibacterial coatings [31–37]. These crystalline fibers, in fact, interact with enzymes, platelets, growth factors, and other cell compounds existing in living tissues. They, induce formation of granulation tissues and slow down the aging processes, being thus active in skin regeneration of wounded and burned skin as well as in slow down of fine lines and wrinkles [15–21]. Other growing factors in the use of CNs are coming from agriculture and agro-chemical activities, while demand for the biomedical and pharmaceutical sector has been evaluated to increase its market by 17% by 2024 [28]. Moreover, as reported recently, the use of antiaging formulations, based on nanochitin/chitosan, is switching hyaluronic acid as part of moisturizing creams, lipsticks, buccal care, and conditioning hair, increasing the market of nanocosmetics [28]. In conclusion, chitin and chitosan, especially in their nanosize dimension, seem to represent the more effective and safe polymers for formulating advanced nanocosmetics. They, in fact, may be used as *active* and safe carriers free of toxic side effects, because metabolized from the human chitotriosidases to produce glucosamine, acetyl glucosamine, and glucose useful for survival and growth of the cells [32–36].

24.2.1.2 Lignin

Lignin, naturally occurring in woody and no woody plants, represents from ~15% to 30% by weight of their total dry mass. This complex molecule, originated from the polymerization of three basic precursors pcoumaryl, coniferyl, and sinapyl alcohols (Fig. 24.12), plays the dual role in giving stability to the plant cells and protect them from the attacks of microbes or saprophytic organisms [38]. Unlike the well-defined cellulose and hemicellulose, the lignin structure not displays a regular repetitive motif, as one might expect for a natural polymer. Its structural formula not only changes between different plant types but also may result different within the same plant [38]. However, the abundance of lignin recovery is directly connected with the pull and paper processing, representing the residual part (black liquor of the Kraft process) of the wood not used as cellulose. Recovered from the plant biomass, the industrial Kraft liquor production with an interesting higher phenolic content is represented by a quantity of 130 million tons/year and a value of US$ 793.7 in 2018, projected to expand at a CAGR between 1.9% and 4% from 2029 to 2025 [39,40]. Unfortunately, the global production is predominately burnt, while less than 2% is actually sold to make value-added goods, also if these coproducts

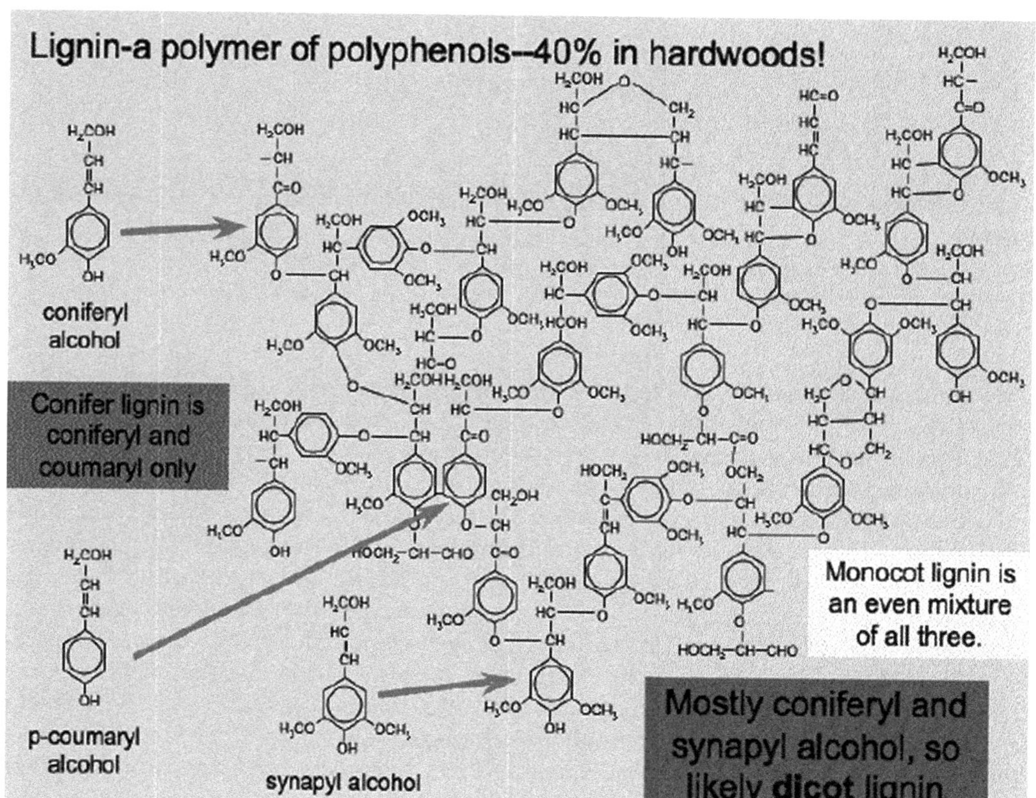

FIGURE 24.12 Lignin structure. Source: *Reproduced with permission from Morganti P, Coltelli MB. A new carrier for advanced cosmeceuticals. Cosmetics 2019;6:10. Licensed under Creative Commons Attribution (CC BY 4.0) license.*

could result necessary to make the processes more cost-effective. However, the increasing use of lignin for the production of naturally derived raw materials, such as aromatics, polymers, binders and adhesives, carbon fibers, and sunscreen agents, is expected to drive the lignin market over the next few years. Extensive R&D, to replace the nonrenewable energy sources in developing countries including China and India, therefore is expected to offer novel growth opportunity avenues for the market of lignin, produced from the Kraft, organosolv and lignosulfonate processes. In any way, lignocellulose sulfonate was the largest lignin segment in 2018 and is estimated to maintain its dominance throughout the forecast period [41]. Just as for chitin, micro/nanolignin [40] has been proposed and used for the generation of bioplastics, control and slow release of fertilizers [42], and production of new, high-value materials such as the nanochitin–NG complexes, previously reported as useful as an ingredient to produce innovative cosmetics [36,37,43,44]. The field of polymer nanoparticles, in fact, is quickly expanding playing a pivotal role, because of the capacity to meet a wide range of applications and market needs in many areas, including the pharmaceutical and cosmetic ones, thus contributing to fostering the future bio-based industry [45–47]. As previously reported in Fig. 24.11, both nanochitin and nanolignin (used by the right conditions) tend to form spherical particles and block polymeric complexes able to entrap active ingredients either hydrophobic or hydrophilic, appositely selected, for example, to make moisturizing and antiaging nanocosmetics [12,15,17,18,33–37].

24.2.1.3 Starch

Starch, as one of the most abundant biomass material in nature, is a basic and important raw material for natural polymers, extensively used in both food and nonfood applications [48]. Starch-derived compounds, obtainable from different natural sources such as potatoes, rice, and other plants, are prepared treating the polymer by different processes, such as hydrolysis, extraction, purification, and chemical treatments. In the form of microcrystalline granules, it is a polysaccharide that serves in plants as supply of energy. This compound is composed of two polymers: D-glucose, accounting for around 20%–30% of linear amylose and branched amylopectin for ~70%–80% [49–55]. Just as chitin, it has a structure organized by two regions of crystalline and amorphous lamella, but forming rings (Fig. 24.13) [50] and because of its great safety it is a polymer skin-friendly and eco-friendly, being free of any toxicity. Thus, for their specific physicochemical characteristics, both starch and its derivatives may be used as a drug and cosmetic carrier, being able to increase the solubility of nonwater-soluble active ingredients and control their release and targeting [49–53]. Starch nanocrystals, in fact, are useful to stabilize oil-in-water cosmetic nanoemulsions rich in oils and, together with nanocellulose and nanochitin, are used as a reinforcement filler in polymeric matrices for the bioplastic packagings [49–53]. For this purpose and to increase the low mechanical and barrier properties of this polymer, it is necessary to modify its structure, blending the starch with biodegradable or synthetic polymers and composites for trying to overcome its water sensibility and hard workability [51]. This natural and low cost polymer, therefore, may be used to produce bioplastic nanocomposites together with PLA and PHA. The global production of bioplastics, in fact, represents an interesting business because it has expected to increase roughly 25% over the next 5 years, increasing from around 2.1 million tons in 2018 to 2.6 million in 2023 [54].

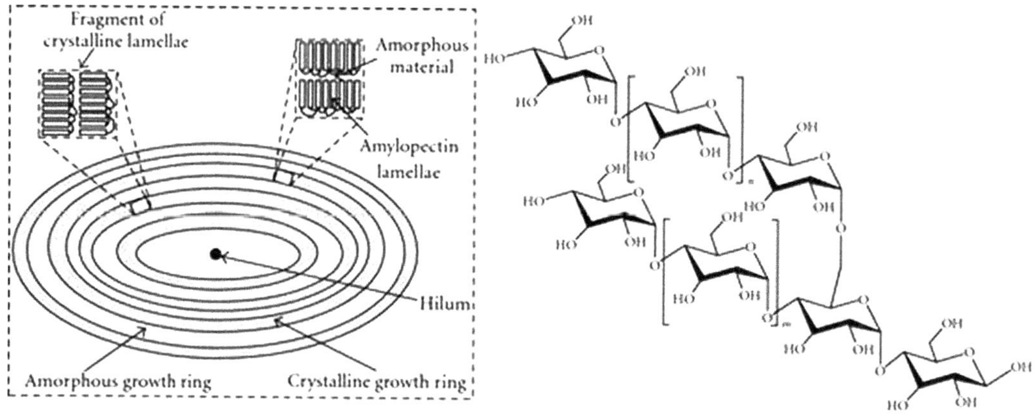

FIGURE 24.13 Starch structure organization by crystalline and amorphous lamellae forming rings. Source: *Reproduced with permission from Lin N, Chang PR, Anderson DP, Yu J. Preparation, modification, and application of starch nanocrystals in nanomaterials: a review. J Nanomater 2011;2011:573687. Licensed under Creative Commons Attribution 3.0 Unported (CC BY 3.0).*

Regarding the global starch derivatives market size, it was estimated at 47.17 billion tons in 2014, and 16 billion tons in 2015, with a value estimated to be worth around US $ 55 billion in 2017, and being likely to grow with a CAGR between 3.5 and 7.2% until 2024 [55,56]. Unfortunately, raw starch for industrial production has limitations in its applications due to its easily decaying and retrograding properties. Thus it is normally modified physically, chemically, or enzymatically for obtaining better properties.

Europe's starch derivatives market has the second major share, while the fastest growth in demand is likely to be registered by South America and rest of the world markets. North America has the lowest growth rate. However, starch nanocrystals may be considered as biodegradable and biomimetic material to be used to produce innovative future nanocosmetics.

24.2.1.4 *Cellulose*

Cellulose (Fig. 24.14) is comprised of beta-(1,4)-ᴅglucose polymeric chains organized as a linear renewable biopolymer [57,58]. This biodegradable polymer, present naturally in flowering plants, possesses excellent biocompatibility, lower density, interesting strength, and mechanical characteristics with a relative low cost and a structure similar to chitin and chitosan. It, in fact, has organized by the same fibrillar structure, which comprises both crystalline and amorphous components. Each microfibril of the cell plant, existing as a mesh-like assembly, is connected to one another via binding agent, namely, hemicellulose while lignin, acting as the matrix phase, occupies the void space between cellulose, hemicellulose, and pectin compounds (Fig. 24.3) [13,58]. With the increasing environmental concerns on industrial sustainability materials derived from renewable resources, such as cellulose nanocrystals (NCEs), have been advocated as potential alternative to petroleum-based compounds, for making "green" carriers and biodegradable packagings for drugs and cosmetics [60,61]. Ligno-cellulosic feedstock, in fact, is composed of a

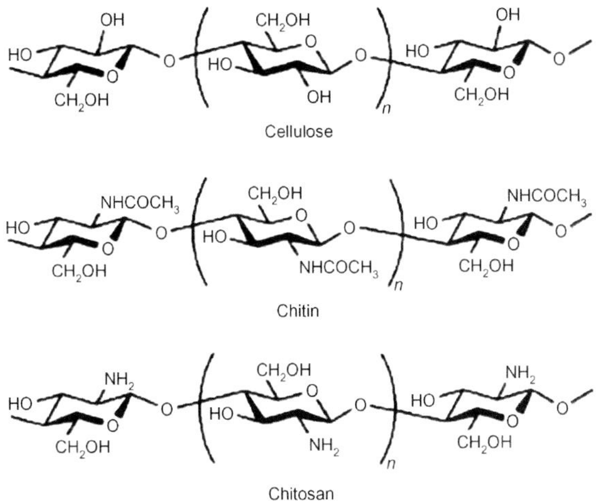

FIGURE 24.14 Cellulose backbone compared to chitin and chitosan. Source: *Reproduced with permission from Morganti P, Coltelli MB. A new carrier for advanced cosmeceuticals. Cosmetics 2019;6:10. Licensed under Creative Commons Attribution (CC BY 4.0) license.*

natural biopolymer complex built up by three main constituents—cellulose, hemicellulose, and lignin, which vary by plant species, cultivation regions, and growth conditions [60]. Chemically cellulose and hemicellulose are polysaccharides, while lignin is a phenolic polymer.

NCEs, derived from agro-biomass and with a dimension of around 5 nm in diameter and 150–200 nm in length, just as CNs, have attracted increasing attention in biomedical and cosmetic applications as reinforcement in bionanocomposite materials [58,61,62]. By the use of the senano-crystals, it is possible to obtain natural, nontoxic biodegradable, and ecocompatible films and 3D porous scaffolds with different architectures, pore sizes, and properties, necessary for cell adhesion, proliferation, migration, and differentiation [62].

According to the global GHG emissions study, in fact, 18.5% of the environmental pollution is due to industrial petrol-based processes, whereas 1.5% is due to the fossil fuel retrieval processing and distribution [54,61]. Thus the necessity to use natural materials as cellulose and derived compounds to produce biodegradable and skin-friendly biopolymers under the form of nanoparticles and nanofibers. This cellulosic material may be transformed, by mechanical or chemical technique, into nanofibers or nanocrystals which possess major biocompatibility in comparison with the native cellulose. Due to the green nature, they may be used as environmentally friendly and sustainable material for different applications, such as paper processing, biodegradable composites, pharmaceutical and personal care products, oil and gas, paints and coatings materials, food and beverages and other products, such as wound dressings and sanitary napkins. In terms of revenue, this polymer, segmented into bacterial NCE, nanofibrillated cellulose, and nanocrystalline cellulose, has shown the nanocrystalline product market surpassing US$ 35 million in 2016 with a CAGR of over 35% up to 2024, while nanofibrillated accounted for the major market share in 2017 and is anticipated to continue its dominance until 2024 with a share of over US$ 540 million. Bacterial CN is estimated to grow at the highest CAGR of over 40% for capturing a market share close to US$ 27 million by 2024 [63]. However, the global cellulose market was evaluated at approximately US$ 232 million in 2017 and is expected to surpass US$ 1 billion by 2024, growing at a CAGR of around 18% between 2018 and 2024 [63,64]. On the basis of geography, North America contributed for the major share in terms of revenue in 2017, followed by Europe, while Asia Pacific showed a considerable growth in the use of CN applications in the personal care sector [63]. Due to its biocompatibility and the right mechanical properties, CN-based biomaterials can provide a cell-friendly environment to encourage cell attachment and proliferation when used to make special tissues for the pharmaceutical and cosmetic sector [59,62]. For this purpose, it is interesting to remember that, according to recent studies [59], it has been evidenced how CN, anchoring itself on the cell structure, seems to stimulate mesenchymal stem cell expansion and formation, tuning the collagen-1 production. Thus there is a possibility of its use in aging cosmetics also. However, because the mechanisms between cell and CN remain until not clear, further studies are necessary to better understand both mechanism of action and interactions, verifying its effectiveness and safeness. Moreover, if it is considered broadly biocompatible, CN is not readily degraded by the human body that lacks cellulosic enzymes [59,65,66]. In conclusion, a more use of cellulose nanofibrils and/or nanocrystals, enhanced by new studies, may be of help to realize innovative future nanocosmetics in combination with other chito-oligosaccharides or polysaccharides, such as CNs.

24.2.1.5 Pullulan

Pullulan (PUL) (Fig. 24.15) is a natural occurring polysaccharide, water-soluble, showing low viscosity, stability over a broad range of pH, good adhesiveness and film-forming properties, and heat resistant [67,68]. This polymer, produced from starch by microbial fermentation of the fungus *Aureobasidium pullulans* and by advanced technology is a white amorphous powder, tasteless, and amorphous. It is a linear uncharged sugar-like polymer consisting of three glucose units that connected by alpha-1,4 glycosidic bonded maltotriose are linked via alpha-1,6 glycosidic linkage associated with randomly ordered chains [68]. This polysaccharide for its interesting characteristics of biodegradability and nontoxicity can be used as an effective and stable carrier. PUL, in fact, having the capacity to load and release active ingredients at different skin layers when used as nonwoven tissue with other polysaccharides, may be useful to formulate and make smart skin care products and innovative diet supplements for slowing down fine lines, wrinkles, and black spots (data not reported). Moreover, it has also shown the possibility to produce transparent and strong films, which unfortunately became easily sticky at high humidity conditions [69]. On the other hand, it seems possible to eliminate this problem, opportunely modifying the

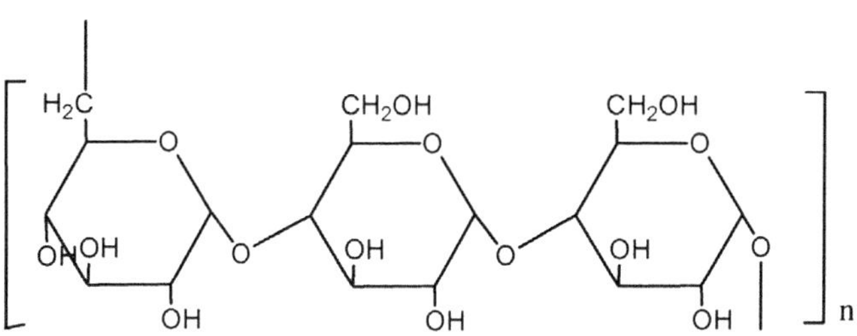

FIGURE 24.15 Pullulan' chemical formula. Source: *Courtesy Wikipedia.*

structure of these films by the use of nanochitin–nanolignin complexes (data in progress not reported). In addition, being also inert to mammalian amylases, PUL possesses prebiotic properties so that its films or tissues can be used as antioxidant and low calorie dietary products as health and/or functional food, binding into their fibers effective active ingredients and using them in combination with the right cosmeceuticals [70]. In our laboratory, in fact, it has been possible to make nonwoven tissues and films by both the electrospinning and casting technology by the use of a gel composed of pullulan and the nanochitin–nanolignin complexes, encapsulating different active ingredients. These smart cosmeceutical-tissues will be used to make innovative and smart Facial Beauty Mask and diet supplements (data in progress not reported). The global potential demand of pullulan is far larger than the actual global production and is expected to grow at a CAGR of roughly 2.2% over the next 5 years, reaching US$ 130 million in 2024, from US$ 120 million in 2017 [70,71]. Japan is the biggest consumption region with 667 million tons consumed in 2015, while China is the second ones with a share of 20.65%, and USA the third with a share of 20.16 in the same year. Regarding the market, the pharmaceutical industry accounted for the largest ones with about 40.74% of the global consumption, while food was the second market to use pullulan with over 36.24% share [71,72]. Certainly, the cosmetic sector could be the future more important market for this sugar-like polymer, if the industries will be able to realize innovative formulations by new ideas and technologies.

24.2.2 Biopolyesters

24.2.2.1 *Polylactic acid*

Development and production of polymers metabolized from the human body have shown a great increase in the last 20 years for their biomedical and cosmetic applications. Among these polymers, PLA (Fig. 24.16), produced by polymerization of lactic acid, has received a significant attention because easily metabolize from the human enzymes [73,74]. It, in fact, is produced from lactic acid naturally occurring in human body. However, the more important benefits of this polymer are its renewability, biocompatibility, process capability, and energy saving.

Lactic acid can be polymerized to produce PLA polymers by different polycondensation methods under controlled temperature and pressure without catalyst, solvent, or initiation [73]. Moreover, because of the favorable characteristics, it has been utilized as general ecological material, as well as active ingredient in medical and cosmetic applications, that is, for surgical implant, carrier for delivery systems, and micro/nanoparticles and scaffolds for the growth of new tissues. In the cosmetic personal and skin care sector, this biodegradable polymer seems to improve the skin lightening effect, accelerate cell renewal and exfoliation, and enhance collagen and elastin synthesis. Additionally, rapid invention of innovative products, such as the reported cosmeceutical-tissues made by our group

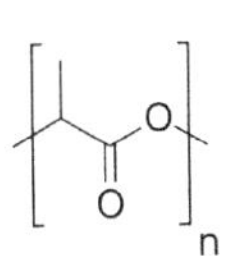

FIGURE 24.16 Polylactic acid chemical formula. Source: *Courtesy Wikipedia.*

[15–18,35–37], could improve the market of this interesting polymer, when used to formulate and produce nanocosmetics. The utilization of PLA, in fact, is due not only to its biodegradability, but also to its effective and safe properties and low cost. It, compared to other traditional biodegradable polymers, possesses particular and specific characteristics of biodegradability to make degradable sutures, releasing micronanoparticles and porous scaffolds, as previously reported [74,75]. On this topic, recently a clinical paper has been published, reporting some data and interesting results on an injected suspension of PLA, used as rejuvenation treatment for the hands [76]. Additionally, PLA is finding an interesting use in food and cosmetic packagings because of its excellent physicochemical characteristics, such as mechanical, thermal, and barrier properties, when processed by the traditional technologies, such as extrusion, injection, and compression and blow molding [74,75]. Unfortunately, the processed polymer results quite rigid and brittle so that its properties have to be modulated by proper additives useful for increasing its flexibility. For this purpose, chitin nanofibers as filler and/or other different materials as renewable plasticizers are normally used [77,78] to make innovative natural fibers, nanocomposites and nonwoven tissues, useful for realizing, for example, innovative cosmetic products [79,80].

The global PLA market [80] is expected to register a CAGR between about 20.5% and 29.49% during the forecast period 2018–26 [77–79]. It was valued at US$ 2.23 billion in 2017 due to its use in creams, shampoos, and body care products, as well as in manufacturing green bottles, films, and medical implants made by the ongoing advancements in science and technology [81–83]. The use of this green eco-friendly polymer for packagings resulted, to be the largest segment, accounting for a volume-based share of 35.7% in 2017, while its consume in textile and medical market will boost especially in China for the subsidies offered from the state by its 12th five-year Plan. However, packaging, agriculture, transport, electronics, and textiles represent today some of the key application areas for this polymer. In terms of region, North America dominated the global market, because of the rinsing consumption of biodegradable polymers, while Asia Pacific and specifically China, India, and Japan are likely to exhibit the fastest growth over the coming years.

24.2.2.2 *Polyhydroxyalkanoates*

PHAs are polyesters of different hydroxyalkanoate units, representing constituents of cell membrane of yeasts, plants, and animals. They are synthesized by many Gram-positive and Gram-negative bacteria, assisting their survivability under environmentally challenging conditions [84]. The polyesters are produced from microorganisms's culture placed in a medium suitable for their synthesis. Thus they are synthesized and deposited in the cells as granules with a yield as high as 80% of the organism's dry weight. The obtained granules are successively disrupted together with the bacteria cells and purified to obtain a safer product. PHA, in fact, is a natural polymer accumulated as water insoluble granule in the cytoplasm of numerous bacteria and several extremophiles archaea [84,85]. Due to the existence of more than 100 different monomer units, it is possible to create by biosynthesis different types of biodegradable polymers with an extensive range of properties. Technologically, PHAs attract attention because of their thermoplastic activity, their production starting from abundantly available and renewable resources, and their biodegradability and biocompatibility [84]. By their use, in fact, it is possible to produce bioplastics which, in combination or not with other biodegradable biopolymers, have characteristics similar to conventional plastics such as polypropylene. However, the reason of the rapid increase in the production of bioplastics is due to the increasing public concern over the harmful effects in the environment of the petrol-derived plastics, which are invading lands and oceans so that 5250 billion plastic particles are floating on the world's seas and oceans, equivalent to 2,68,940 metric tons of waste [86]. Unfortunately, PHA still displays drawback characteristics such as low mechanical stability, unfavorable (bio) degradation rate, and too high or too low degree of crystallinity and purity. Thus the necessity to develop this polymer with better biologic approaches for its use in medicine, eliminating the microbial component from their crude compounds, such as the lipopolysaccharide endotoxins which, located in the outer cell membrane of Gram-positive microbes, can be the cause of inflammatory processes [84,85]. In any way, PHA are used in different fields such as pharmaceutical, biomedical, cosmetic, and agricultural under the form of biodegradable plastics and packaging materials [83]. In the cosmetic field, our group together with other research groups from industries and universities, such as the components of the EU PolyBioSkin research project (www.polybioskin.eu) studies are in progress for realizing advanced medications, baby/elderly/feminine diapers, and beauty masks, all based on biodegradable tissues made by polysaccharides such as starch cellulose, lignin, and chito-oligosaccharides such as chitin/chitosan, or the polyesters PLA and PHA. While in the cosmetic sector, PHAs are used for biodegradable microbeads and innovative eco-friendly packaging materials, in the biomedical field they are further categorized into drug release means, sutures, and other biomedical applications, while in agriculture are divided among plant pots, mulch films, and others uses, and in packaging are categorized into flexible and/or rigid packaging, and other packaging applications [86,87]. It is also interesting to remember the possibility to make micro/nanoparticles of PHA to be used as smart nanocarriers able to better load, transport, and release bioactive compounds into the target tissue. These innovative carriers may be of help for both the pharmaceutical and cosmetic sector [15–18,34–37]. PHA market, according to MRC [84], is accounted for US$ 78.20 million in 2017 and is expected to reach US$ 135.78 million by 2026 with a CAGR of 6.3%. In conclusion, apart the use of PHAs as active ingredients in the pharmaceutical and cosmetic fields, the packaging sector appears to be the future biggest market for cosmetics and food so that it is increasing in Europe in the order of millions of tons per year [87,88]. This is the reason why our and other research groups have realized interesting hard and soft containers made by PHA, PHB, PLA, chitin, and other biodegradable polymers to be used, for example, as biodegradable and compostable packaging for cosmetics, coffee, and eggs [89].

24.3 Progress and future perspective of the cosmetic products

24.3.1 Natural green cosmetics and biodiversity

As reported in Chapter 1, Nanocosmetics: an introduction, consumers are looking for natural, skin-friendly cosmetics made by advanced technologies and able to preserve the environment. These products, therefore, have preferably to contain natural ingredients derived from plant and animals, such as essential oils, surfactants, pigments, and other substances of biological origin [90]. Thus became a *must* the necessity to pave the way for discovering new molecules and building block polymers which are not only innovative and effective, but also naturally derived and friendly for both humans and the environment, in comparison with the existing synthetic petrol-derived products. To buy *green* represents a progressive and sustainable consumer movement that worldwide is finding out nature, economic sustainability, and biodiversity as part of a new sense of pride [8–11,89]. As a consequence, for example, the new Brazilian biodiversity regulations "intends to create a secure regulatory environment to stimulate the use of Brazilian biodiversity" (Law No. 13, 123/15 and Decrete 8772/16), generating more research and cultivation of indigenous plant species [91]. On the one hand, Chinese consumers show to prefer organic and natural cosmetics, maintaining the holistic approach of their own culture which makes the link between health and beauty in a natural environment from 6000 years [92]. This the reason why many scientific papers have been published in China on natural products, between the years 2015–16 (Fig. 24.17) [92].

On the other hand, the Indian organic skincare market, emerged as an alternative to Ayurveda products, is valued at US$ 125 million, growing at 25% yearly, while the herbal category is expect to grow at 8%–10% [93]. However, the rising consumer demand for natural ingredients and cosmetic products made by sustainable and innovative technologies is increasing in the spam way in Asia Pacific and North and South America, as well as in Russia, East Asia, and Europe. It means that natural ingredients, biodiversity, and nanotechnology are becoming worldwide an increasing strategic issue for the cosmetic industry and consumer. While recent research clearly demonstrates the future potential of the nature' biodiversity, the relative market opportunities for industry and consumer will depend on the trends of the nature conservation status in the different countries. Moreover, our knowledge of biodiversity is still very limited, also if many different flowering plants will certainly provide a wide variety of building materials for innovative cosmetics evidenced by new research studies and smart bionanotechnologies [5–7]. For this purpose, it is to underline that, for example, over 50% of the modern prescription

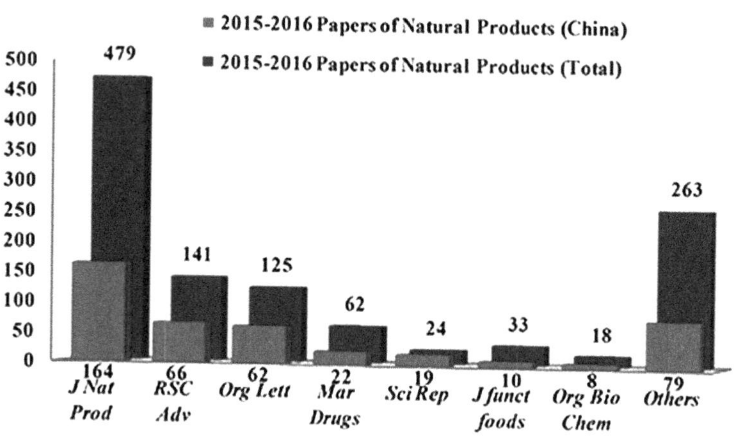

FIGURE 24.17 Scientific papers published in China in the years 2025–26 (with an impact factor of greater than 3.0) Source: *Reproduced with permission from Liu H, Zhu G, Fan Y, Du Y, Lan M, Xu Y, et al. Natural products research from 2015 to 2016. Front Chem 2018;6:45. Licensed under Creative Commons Attribution 4.0 International (CC BY 4.0).*

medicine and the majority of active ingredients used in cosmetic dermatology were originally discovered and continue to be recovered in plants [90,92,94]. Moreover, it is also interesting to remember that the high quantity of raw materials, generated from agroforestry processing as waste, contains active ingredients proven more valuable than the main products [94]. This means that agro-industry should not only guarantee the final food at high quality and safety, but may also represent a remunerable solution to recycle its byproducts, realizing a *green* alternative to the regular derived extract, through a zero waste strategy [95].

24.3.2 Advanced nanomaterials and nanotechnologies

Nanomaterials and nanotechnology have the potential to produce the more interesting changes in material research, forming the new generation of fibers and composites which represent an emerging class of nanomaterials with new structural and functional applications, obtainable, for example, by the electrospinning technology [96]. Regarding the modern technologies applied to cosmetics, they could have four distinct kinds of advancements [6]: (1) at the level of material by the property of which it may be possible to enhance the incorporation of "passive nanostructures" (i.e., coating) on fibers or nonwoven tissues (obtainable by electrospinning) and/or on the skin, by the use of a spray solution/emulsions; (2) making use of specialized structures able to provide active ingredients at a specific skin level (i.e., nanocapsules or smart tissues), releasing them at dose and time designed; (3 and 4) starting to produce advanced nanosystems (i.e., nanorobotics or 3D printing techniques for scaffold fabrication such as microcontact printing, Inkject fabrication, and robotic deposition) able to control the skin type variability and growth. Naturally, for a better knowledge of the different technologies, it seems important to remember scope and definition of *Nanotechnology, Nanobiotechnology, and Bionanotechnology* [97,98].

Nanotechnology was defined in 1974 in Japan and successively popularized in USA as "a process of separation, consolidation and deformation of materials by one atom or by one molecule," showing novel properties and applications [99]. It is based on an engineering approach working at the nanometer scale, that is, at dimensions less than 100 nm. This approach, for example, may go toward the regenerative medical therapy that, by the use of bionanomaterials, has the aim to rejuvenate an aged skin and repair defective tissues [100]. Naturally, these nanomaterials have to be selected for creating a right environment able to mimic the surface properties of natural tissues, promoting proliferation, adhesion, migration, and differentiation of the cells. Among all the different nanostructures selected, chitin nanofibers have shown to have the capacity to replace the natural ECM, creating more advantageous scaffolds, compared to other nanoforms [12,15–18,32–37].

Nanobiotechnology, integrating biological science with engineering, is a biological method which incorporates nanomolecules into biological systems to achieve a greater efficacy. It is, however, a branch of nanotechnology that studies existing biological events in nature in order to realize new devices for biochemical applications or uses.

Bionanotechnology, used interchangeably with Nanobiotechnology, refers to the use of nanotechnology for creating new nanostructures by self-assembling biomolecules, utilized for different nanotechnological applications, such as the manufacture, for example, of the reported nanochitin–nanolignin complexes [25,36,37,43]. Last but not least it is also to be known the innovative 3D printing manufacturing process by which 3D structures can be created, by adding different kinds of biomaterial, such as biomaterial polymers and bioplastics, on a layer-by-layer basis [101–103]. By this emerging technique seems possible to restore or replace defective or injured living organs and tissues. Thus the possibility to regenerate, for example, an injured or aged skin, finding the right and advanced biomaterial and polymerization techniques for engineering structures, able to mimic cell components and ECM. Biopolymers, in fact, in their different form (tissue, fiber, film, gel) may provide not only support but also the correct structural architecture essential for tissue regeneration [100–104]. Additionally, it will be possible to create multifunctional delivery systems as well as cell-based materials for regenerative medicine. These innovative systems, characterized by an accelerated release of active ingredients, could be realized making anatomically specialized and personalized wound dressings and/or antiaging beauty dressings more affordable than conventional flat ones [15–17]. However, the 3D printing methodology seems also useful to understand the locomotive mechanism of microorganisms, modulating and driving the ecology of the skin surface (i.e., the skin microbiome) [102,103,105]. For all these reasons, this method is gaining importance in the field of pharmaceutical and medical applications and certainly will be used in future in the cosmetic sector also. By this innovative technique, in fact, it will be possible the realization and rapid preparation of tailor-made scaffolds, objects, and products to be used in personalized medicine or beauty treatments customized to each individual subject [105]. Thus, in a world of mass consumerism where everyone has access to the same product, the need for individuality has never been greater, while the necessity of health and beauty is under the influence of research, technology, and the environment. As a consequence, consumers make use of apps personalization services to create a product uniquely for them. Moreover, as the general public becomes more educated to the causes of aging by a better knowledge of the ingredients that can combat its visible signs, customers are turning to meet natural and technologically advanced tools and cosmetics able to give answers in their quest for a younger looking skin [105–107]. Thus, the 3D, smart, simply, and low cost technology, combined with innovative formulations and overlapped by the use of nanocosmetics and specialized nutraceuticals, could allow to individualize the products, verifying the consumer body weight and lifestyle, as requested today from 61% of women worldwide [100].

24.3.3 Smart nanocosmetics and future perspectives

In any way, the future innovative cosmetics, continually requested from the consumers' majority, could be formulated by the recovery and use of innovative, effective, and safe ingredients of natural origin. These ingredients have to be possibly nanostructured for enhancing their effectiveness, and incorporated into carriers made by micro/nanoemulsions at a low level of emulsifiers and preservatives, as previously reported. However, on the one hand, the nanoemulsions provide great advantage as carriers for formulating innovative, safe, and effective cosmetics, due to their own controlled delivery system and bioactive effects; on the other hand, they require an expensive process method, a selective use of surfactants and a better understanding of the interfacial chemistry involved in the production [108]. Alternatively, the carrier could be composed of biodegradable tissues or films which, made by fibers binding active ingredients, are free of emulsifiers, preservatives, and other chemicals, as reported elsewhere from our group [15–18,34–37]. The natural micro/nanofibers utilized to make nonwoven tissues and films, produced respectively by the electrospinning or casting technology, have been used, in fact, as effective skin scaffolds, showing to enhance keratinocytes adhesion and proliferation especially in the presence of collagen type I coating [100–104]. For this purpose, it has been shown from our group [109] that chitin–lignin nanoparticles have the same biological activity of hyaluronan in eliciting epithelial beta-defensins [110], probably for the capacity to form a structure ECM-like [15,17,18].

However, nanoparticles and nanocarriers used to produce the innovative nanocosmetics have to work by specific and well-understood mechanisms, being characterizing by their better performances, free of potential side effects also. The basic point to be focused in the use of these smart cosmetics, therefore, is to maximize their bioavailability by an efficient delivery system, able to transport and release the loaded active ingredients to the different skin target layers, at the time and dose designed. According to the customer' request, nanocosmetics could be at the base of a new category of skin care products able to regenerate and rejuvenate the skin damages provoked from environment-induced injuries and the aging phenomena [15–18,111]. By the use of these innovative cosmetics, it seems possible to quickly restore function and morphology of the aged skin layers [111], regenerating the complex orchestrated sequence of events mediated by cells and grow factors [112], but contemporary respecting the skin microbiome equilibrium [113]. In conclusion, the global skin care industry has an enormous opportunity to improve the cosmetic efficacy, also if the cost investment and risks associated with product development are high for the changing of the customer needs, who are too frequently looking to change and bettering the face and body appearance. In a world where consumer appreciation and trend versus the cosmetic products change rapidly, skin care manufacturers need to embrace all the technological necessities and the environmental and social aspects connected with the modern concept of health and beauty (i.e., bionanotechnology, biodiversity, robotics, artificial intelligence, big data, intelligent supply chain management, and the Internet of things). Thus the necessity of a strategic management in the use of natural raw materials to produce innovative nanocosmetic lines, based on a successful biodiversity of the ingredients used and the sustainable and environmentally friendly processes [114]. As first step, therefore, it will be necessary to analyze type, quantity, and origin of all the natural raw materials used, evaluating their biodiversity value, directly connected with the geographic origin and the relative cultivation methods of the plants. For this purpose, a deep knowledge on the regions, where the plants are cultivated and the methodologies adopted for the ingredients' extraction, is considered an important prerequisite for the high quality of the biological material and their trace ability. The right selection and dose of the natural ingredient(s) adopted for the cosmetic product will determine, therefore, its (their) effectiveness and safeness. The second step regards the carrier selected (i.e., nanoemulsion or nonwoven tissue) for loading and transporting the active agents. It is considered of vital importance, in fact, to release these fundamental ingredients at the level of the designed skin layers and cells, in a predetermined dose and time. Thus the carrier structure has to be engineered so that the combination and topography of the employed materials with the typology of the active ingredients selected may favor reproduction and survival of the different skin cell type. Scaffold materials and the effectiveness of the nanoingredients, in fact, influence significantly the nature of cell interactions. As previously reported, they play a critical role in tissue engineering by serving a synthetic ECM to provide a temporary mechanical support for the cells, mimicking their native environment that lead to tissue formation [115]. Naturally also because of their increasing use, both the active nanoingredients and nanocarriers together with the final nanocosmetics have to be controlled in vitro and in vivo by scientific and validated technologies to verify their effectiveness and safeness, thus safe-guarding the human health. Additionally, new and effective laws have to be promulgated worldwide for governing nanocompounds and nanocosmetics in the same way [7]. Third and fundamental step is the marketing communication that has to clearly shown to the consumer, the environmental sustainability of these smart nanocosmetics nature-oriented. Finally, it will be necessary to underline by the different mass media their more high effectiveness, safeness, and healthy-oriented activity, in comparison with the normal cosmetic formulations distributed in the worldwide market, especially when used in combination with nanostructured specific nutraceuticals. If all these rules will be respected, these innovative and smart nanocosmetics could represent, in my opinion, successful skin care products for the future generations with particular regard to the category of the cosmeceutical-tissues, which are skin-friendly, completely degraded to safe secondary ingredients, and made from biodegradable and biocompatible polymers able to create the same scaffold of the natural ECM [15–18,32–37].

References

[1] Westhoek H, Ingram J, Van Berkum S, Ozay L, and Hajer M Food systems and natural resources. Report of the Working Group on Food Systems. UNEP Environmental, Netherland, 2016. ISBN: 978-92-807-3560-4.

[2] UNDESA. World population prospects 2019: highlights, <https://population.un.org/wpp>; 2019 [accessed 15.08.19).

[3] Richie H, Roser M. CO2 and greenhouse gas emissions. Published online at Our-WorldInData.org, Available from: <https://ourworldindata.org/co2-other-greenhouse-gas-emissions>; 2019.

[4] Purvis B, Mao Y, Robinson D. Three pillars of sustainability: in search of conceptual origins. Sustain Sci 2019;14:681. Available from: https://doi.org/10.1007/s11625-018-0627-5.

[5] Liji T Nanotechnology and bionanotechnology—similarities and differences. AZOnano, <www.nanowerk.com>; 2018 [accessed 15.08.19].

[6] The Royal Society. Nanosize and nanotechnologies: opportunities and uncertainties. Science policy section. London: The Royal Society; 2004.

[7] Snir R. Trends in global nanotechnology regulation: the public-private interplay. Vand J Ent & Technol Law 2014;17:107–69.

[8] Janicke M. Green growth: from a growing eco-industry to economic sustainability. Energy Policy 2012;48:13–21.

[9] EEA Report No. 2/2016, Circular economy in Europe, <https://www.eea.europa.eu/publications/circular-economy-in-europe>.

[10] Ali A. Green Industry towards Green Economy in the KMG Sector in Bangladesh. BH-MEA, Dhaka Office, 12/07/2015; 2015 [accessed 15.08.19].

[11] EPA. EnviroAtlas Benefits Category: Biodiversity Conservation. Biodiversity underpins allecosystems and their services. NewYork, NY: United States Environmental Protection Agency; 2018. Available from: <https://www.epa.gov/enviroatlas/enviroatlas-benefit-category-biodiversity-conservation> [accessed 15.08.19].

[12] Morganti P. Use of chitin nanofibrils from biomass for an innovative bioeconomy. In: Ebothe J, Ahmed W, editors. Nanofabrication using nanomaterials. Manchester: One Central Press; 2016. p. 1–22.
[13] Doherty WOS, Mousavioun PA, Fellow CM. Value-adding to cellulosic ethanol: lignin polymers. Ind Crop Prod 2011;33:259–76. Available from: https://doi.org/10.1016/j.indocrop.2010.10.022.
[14] Nikolov S, Fabritius H, Petrov M, Frisk M, Lymperakis L, Sachs C, et al. Robustness and optimal use of design principles of arthropod exoskeletons studied by ab initio-based multiscale simulations. J Mech Behav Biomed Mater 2011;4:129–45. Available from: https://doi.org/10.1016/j.jmbbm2010.09.015.
[15] Morganti P, Morganti G, Colao C. Biofunctional textiles for aged skin. Biomedicines 2019;7(3):51. Available from: https://doi.org/10.3390/Biomedici-nes7030051.
[16] Morganti P, Anniboletti T, Pollastrini C, Morganti G. Natural polymers for body care to save the environment. Biomed J Sci Tech Res 2019;17:12570–4. Available from: https://doi.org/10.26717/BJSTR.2019.17.002955.
[17] Morganti P, Coltelli MB. A new carrier for advanced cosmeceuticals. Cosmetics 2019;6:10. Available from: https://doi.org/10.3390/cosmetics6010010.
[18] Morganti P, Morganti G, Coltelli MB. Chitin nanomaterials and nanocomposites for tissue repair. In: Choi AH, Ben-Nissan, editors. Marine-derived biomaterials for tissue engineering applications. Singapore: Springer; 2019. p. 523–44. ISBN: 978-981-13-8855-2 (eBook).
[19] Morganti P, Fabrizi G, Palombo P, Palombo M, Ruocco E, Cardillo A, et al. Chitin nanofibrils: a new active carrier. J Appl Cosmetol 2008;26:113–28.
[20] Morganti P, Morganti G, Morganti A. Transforming nanostructured chitin from crustacean waste into beneficial health products: a must for our society. Nanotechnol Sci Appl 2011;4:123–9.
[21] Muzzarelli RAA, Muzzarelli C. Chitosan chemistry: relevance to the biomedical sciences. Adv Polym Sci 2005;186:151–209.
[22] Cinelli P, Coltelli MB, Signori F, Morganti P, Lazzeri A. Cosmetic packaging to save the environment: future perspectives. Cosmetics 2019;6:26. Available from: https://doi.org/10.3390/cosmetics6020026.
[23] Howling GI, Dettmar PW, Goddard PA, Hampson FC, Dornish M, Wood EJ. The effectof chitin and chitosan on the proliferation of human skin fibroblasts and keratinocytes in vitro. Biomaterials 2001;22:2959–66.
[24] Ling S, Kaplan DL, Buehel MJ. Nanofibers in nature and materials engineering. Nat Rev Mater 2018;3:18016. Available from: https://doi.org/10.1038/naturemats.2018.16.
[25] Morganti P, Febo P, Cardillo M, Barone A. Chitin nanofibril and nanolignin: natural polymers of biomedical interest. Clin Cosmet Dermatol 2017;1(2). Available from: https://doi.org/10.1696/2576-2826.
[26] Revathi M, Saravanan R, Shanmugam A. Production and characterization of chiti-nase from Vitro species, a head waste of shrimp *Metapenaeous dobsonii* and chitin of *Sepiella inermis*. Adv Biosci Biotechnol 2012;3:392–7.
[27] Tyliszczak B, Drabczyk A, Kudłacik-Kramarczyk S, Sobczak-Kupiec A. Sustainable production of chitosan. In: Królczyk G, Wzorek M, Król A, Kochan O, Su J, Kacprzyk J, editors. Sustainable production: novel trends in energy, environment and material systems. Studies in systems, decision and control, 198. Cham: Springer; 2020. Available from: <https://doi-org.elib.vast.vn:2443/10.1007/978-3-030-11274-5_4>.
[28] SFly. A woldwide market with a strong demand. Comgraf R&D; 2017 [accessed 17.08.19].
[29] FMI. Chitin market: agrochemical end use industry segment inclined towards high growth-moderate value during the forecast period: global industry analysis (2012-2016) and opportunity assessment (2017-2027). Future Market Insight, London, UK; 2017 [accessed 17.08.19].
[30] KTVN. Chitin Market 2018 global analysis, growth, trends and opportunities. Research report forecasting to 2023. KTVN EEO Report, <www.ktvn.com>; 2018 [accessed 17.08.19].
[31] Jakunar R, Prabaharan M, Nair SV, Tamura H. Novel chitin and chitosan nanofibers in biomedical applications. Biotechnol Adv 2010;28(1):142–50. Available from: https://doi.org/10.1016/j.biotecadv.
[32] Morganti P, Morganti G, Nunziata ML. Chitin nanofibrils, a natural polymer from fishery waste: nanoparticle and nanocomposite characteristics. In: Morganti P, editor. Bionanotechnology to save the environment. Plant and fishery's biomass as alternative to petrol. Basel: MDPI; 2019. p. 69–81.
[33] Chen H-D, Li YH, Morganti P. Chitin-hyaluronan block copolymeric nanoparticles for innovative cosmeceuticals. In: Morganti P, editor. Bionanotechnology to save the environment. Plant and fishery's biomass as alternative to petrol. Basel: MDPI; 2019. p. 260–71.
[34] Morganti P, Hong-Duo C, Li YH. Green-bio-economy and bio-nanotechnology for a more sustainable environment. In: Morganti P, editor. Bionanotechnology to save the environment. Plant and fishery's biomass as alternative to petrol. Basel: MDPI; 2019. p. 39–81.

[35] Tishchenko G, Morganti P, Stoller M, Kelnar I, Mikesova J, Koravova J, et al. Chitin nanofibrils-chitosan composite films: characterization and properties. In: Morganti P, editor. Bionanotechnology to save the environment. Plant and fishery's biomass as al-ternative to petrol. Basel: MDPI; 2019. p. 191–226.

[36] Morganti P, Coltelli MB, Danti S. Chitin and lignin to produce biocompatible tissues. Res Clin Dermatol 2018;1(1):5–11. Available from: https://doi.org/10.35841/clin-dermatology.1.1.5-11.

[37] Danti S, Trombi L, Fusco A, Azimi B, Lazzeri A, Morganti P, et al. Chitin nanofibrils and nanolignin as functional agents in skin regeneration. Int J Mol Sci 2019;20(11):2669. Available from: https://doi.org/10.3390/ijms20112669.

[38] Lange H, Crestini C. Lignin: isolation, structure and valorisation. In: Morganti P, editor. Bionanotechnology to save the environment. Plant and fishery's biomass as alternative to petrol. Basel: MDPI; 2019. p. 125–71.

[39] Business Wire. Global lignin products market-marker segmentation and analysis 2017-2022-Research and Markets, BusinessWire, Berkshire Hathaway, Dublin Co, September 11, <press@researchandmarkets.com>; 2017 [accessed 19.08.19].

[40] Abbati de Assis C, Greca LG, Ago M, Balakshin MY, Jameel H, Gonzales R, et al. Techno-economic assessment, scalabikity, and applications of aerosol lignin nanoparticles. ACS Sustain Chem Eng 2018;6:11853–68.

[41] Grand View Research. Lignin market, share & trends analysis report by product, by application, by region, and segment forecasts, 2019-2025. Grand View Research, San Francisco, CA, USA, <www.grandviewresearch.com> [accessed 19.08.19].

[42] Capanema EA, Balakshin MY, Chen CL, Gratzl JS, Kirkman AG. Oxidative ammonolysis of technical lignins. Part 1. Kinetics of the reaction under isothermal condition at 130°C. Holzforschung 2001;55:397–404.

[43] Morganti P, Di Massimo G, Cimini C, Del CP. Chitin nanofibril-hyaluronan block polymer characterized. Personal Care Europe 2013;6(4):61–6 and Personal Care Asia 14(6):49–54.

[44] Morganti P. The meaning of nanodimension involving the cosnetic chemist from lab to the industrial process. J Sci Res Rep 2014;4(2):79–100.

[45] Yearla SR, Padmasree K. Preparation and characterization of lignin nanoparticles: evaluation of their potential as antioxidants and UV protectants. J Exp Nanosci 2016;11(4):289–302.

[46] Ago M, Tardy BL, Wang L, Guo J, Khakalo A, Rojas OJ. Supramolecular assemblies oflignin into nano- and micro-particles. MRS Bull 2017;42:371–8.

[47] Morganti P, Carezzi F, Del Ciotto P, Morganti G, Nunziata ML, Gao X, et al. Chitin nanofibrils: a natural multifunctional polymer. In: Phonix DA, Ahmed W, editors. Nanobiotechnology. Manchester: One Central Press; 2014. p. 1–31.

[48] Przetaczek-Roznowska I. Physicochemical properties of starches isolated from pumpkin compared with potato and corn starches. Int J Biol Macromol 2017;101:536–42.

[49] Le Correct D, Bras J, Duffesne A. Starch nanoparticles: a review. Biomacromolecules 2010;11(5):1139–53.

[50] Lin N, Chang PR, Anderson DP, Yu J. Preparation, modification, and application of starch nanocrystals in nanomaterials: a review. J Nanomater 2011;2011:573687. Available from: https://doi.org/10.1155/2011/573687.

[51] Rub AL, Garcia NL, D'Accorso N, Goyanes S. Disadvantages of starch-based materials, feasible alternatives in order to overcome these limitations. doi: 10.1016/B978-0-12-809439-6-00003-0, <www.researchgate.net/pubblication318031272>; 2017.

[52] Chena Aldao D, Sarka E, Ulbrich P, Mesikova E. Starch nanoparticles—two ways of their preparation. Czech Food Sci 2018;36:1–6. Available from: https://doi.org/10.17221/371/2017-CJFS.

[53] Chases da Silva NM, Freitas de Lima F, Lopes Lima Fialho R, de Magalhaes Cabral Albuquerque EC, Velasco JI, Fakhouri M. Production and characterization of starch nanoparticles 2018;. Available from: https://doi.org/10.5772/interchopen.74362.

[54] EU-Bioplastics. New market data: the positive trend for the bioplastcs industry remain-stable. European Biopastics, Berlin, Germany, December, 5, <www.european-bioplastic.org/news/multimedia-pictures-videos/>; 2018 [accessed 25.08.19].

[55] Global Market Insights. Starch derivatives market size potential, industry outlook, regional analysis, application development, competitive landscape & forecast. Selbyville, DE: Global Market Insights, Inc; 2019. p. 2019–25. Available from: <www.giminsights.com/industry-analysjs/starch-derivatives-market> [accessed 18.08.19].

[56] Mordor Ingelligence. Starch derivatives market-growth, trends, and forecast (2019-2024). Gachibowli, Hyderabad, <info@mordoringelligence.com>; 2019 [accessed 18.08.19].

[57] Nair SS, Zhu JY, Deng Y, Ragauskas AJ. High performance green barriers based onnanocellulose. Sustain Chem Proc 2014;23:1–17.
[58] Kumar Mishra R, Sabu A, Tiwari SK. Material chemistry and the futurist eco-friendly applications of nanocellulose: status and prospect. J Saudi Chem Soc 2018;22:949–78. Available from: https://doi.org/10.1016/j.jscs.2018.02.005.
[59] Lin N, Dufresne A. Nanocellulose in biomedicine: current status and future. Eur Polym J 2014;59:302–25. Available from: https://doi.org/10.1016/j.europolymj.2014.07.025.
[60] Palmisano P, Piglione MC. Cellulose and lignin: the abundant renewable polymers from plant biomass. In: Morganti P, editor. Bionanotechnology to save the environment. Plant and fishery's biomass as alternative to petrol. Basel: MDPI; 2018. p. 82–104.
[61] Kangas H. Guide to cellulose nanomaterials. Research Report VTT-R-05013-05014, Tecnical Research Centre, Finland; 2014.
[62] Fortunati E, Kenny JM. Cellulose nanocrystals: multifunctional systems for biomedical applications. In: Morganti P, editor. Bionanotechnology to save the environment. Plant and fishery's biomass as alternative to petrol. Basel: MDPI; 2019. p. 172–90.
[63] openPR. Nanocellulose market is thriving world wide with leading players LikeSappi, Stora Ensi, American Process Inc, FPInnkvations, Kruger Inc. Global MarketInsight, Inc, Selbylle, DA, USA, <sales@gminsights.com>; 2019 [accessed 21.08.19].
[64] Zion Market Research. Global nanocellulose market expected to reach US$ 784 millionby 2024. New York, NY: Zion Market Research; 2018. Available from: <www.zionmarketresearch.com> [accessed 21.08.19].
[65] Vielreicher M, Kralisch D, Volkl S, Sternal F, Arkudas A, Friedrich O. Bacterial nanocellulose stimulates mesenchymal stem cell expansion and formation of stable collagen-I networks as a novel biomaterial in tissue engineering. Sci Rep 2018;8:9401. Available from: https://doi.org/10.1038/s41598-018-27760-z.
[66] Rhim JW. Characteristics of pullulan-based edible films. Food Sci Biotechnol 2003;12(2):161–5.
[67] Cheng MC, Demircj A, Catchmark JM. Effect of plastic composite support and pH profiles on pullulan production in biofilm reactor. Appl Microbiol Biotechnol 2010;86:853–61.
[68] Leathers TD. Biotechnological production and applications of pullulan. Appl Microbiol Biotechnol 2003;62:468–73.
[69] Li R, Tomasuka P, Moreira de Sousa AM, Liu SC, Tunick M, Liu K, et al. Electrospinning pullulan fibers from salt solutions. Polymers 2017;9:32. Available from: https://doi.org/10.3390/polym9010032.
[70] MarketWatch. Pullulan Market 2019 Industry research, share, trend, global industry size, price, future analysis, fegional outlook to 2024 research report. MarketWatch Co, NewYork, USA, <www.360marketupdates.com/purchase/12485408>; 2019 [accessed 21.08.19].
[71] Journal Dairy Global Pullulan Market Report-top players, market size, share, sales, growth and forecast to 2024, <www.resarchformarkets.com/sample/global-pullulan-market-365377>; 2019 [accessed 21.08.19].
[72] Lopes MS, Jardini AL, Filho M. Synthesis and characterization of poly(lactic acid) by ring-opening polymerization for biomedical applications. Chem Eng Trans 2014;38:331–6. Available from: https://doi.org/10.3303/CET1438056.
[73] Hamad K, Kaseem M, Yang HW, Deri F, Ko YG. Properties and medical applications of polylactic acid: a review. Express Polym Lett 2015;9(5):435–55. Available from: https://doi.org/10.3144/expresspolymlett.2015.42.
[74] Al Saaeb RAA, Aladjb H, Othman NZ, Malek RD, Leng OM, Aziz R, et al. Recenr applications of polylactic acid in pharmaceutical and medical industries. J Chem Pharm Res 2015;7(12):51–63.
[75] Redaelli A. Cosmetic use of polylactic acid for hand rejuvenation on 27 patients. J Cosmet Dermatol 2018;5:233–8.
[76] Coltelli MB, Della Maggiore I, Bertoldo M, Signori F, Bruno S, et al. Poly(lactic acid) properties as a consequence of poly(butylene adioste co-terephthalate) blending and acetyl tributylcitrate plasticization. J Appl Polym Sci 2008;119(2):1259–62.
[77] Fehri MK, Mugoni C, Cinelli P, Anguillesi I, Coltelli MB, et al. Composition dependence of the synergistic effect of nucleating agent and plasticizer in poly(lactic acid): a mixture design study. Express Polym Lett 2017;10(4):274–88.
[78] Cinelli P, Coltelli MB, Mallegni N, Morganti P, Lazzeri A. Degradability and sustainability of nanocomposites based on polylactic acid and nanochitin. Chem Eng Trans 2017;60:115–20. Available from: https://doi.org/10.3303/CET1760020.

[79] Morganti P, Coltelli MB, Danti S. Biobased tissues for innovative cosmetic products: Polybioskin as an EU Research Project. Glob J Nanomed 2018;3(4):555620. Available from: https://doi.org/10.19080/GJN.2018.03.55562.

[80] Grand View Research. Polylactic acid (PLA) market size, share & trends analysis report by application (packaging, agriculture, transport), by region (North America, APAC, Europe, CSA, MEA), and segment forecasts, 2019–2025, <https://www.grandviewresearch.com/industry-analysis/polylactic-acid-pla-market>; 2019 [accessed 25.12.19.].

[81] Mordor Intelligence. Bio-polylactic acid (PLA) market-growth, trends, and forecast (2019-2024). Mordor Intelligence, Gachibowli, Hyderabad, <info@mordorintelligence.com>; 2016 [Accessed 23.08.19].

[82] GLOBE NEWSWIRE. Polylactic acid (PLA) market worth US$ 4.32 Bn by 2026. CAGR20.5% -Acumen Research and Consulting, GlobeNewswire, Inc, Pune, <www.acumenre-searchandconsulting.com/Polylactic-acid-market>; 2019 [accessed 23.08.19].

[83] Koller M. Biodegradable and biocompatible polyhydroxyalkanoates (PHA): auspicious microbial macromolecukes for pharmaceutical and therapeutic applications. Molecules 2018;23:362. Available from: https://doi.org/10.3390/molecules23020362.

[84] Reddy CSM, Ghai R, Kalia RVC. Polyhydroxyalkanoates: an overview. Bioresour Technol 2003;87:137–46.

[85] Abreu A, Pedrotti ML. Microplastics in the oceans: the solution lie on land. Field Action Sci Rep 2019;19:62–7. Available from: <http://Journals.openedition.org/factsreports/5290>.

[86] MRC. Polyhydroxyalkanoate market outlook. Global trends, forecast and opportunity assessment (2017-2026), <info@strstegymrc.com>; 2019 [accessed 23.08.19].

[87] Research and Markets Global polyhydroxyalkanoate (PHA) market analysis &trends industry forecast to 2025. Research an Market, USA, <www.researchandmarkets.com>; 2017 [accessed 23.08.19].

[88] Bugnicourt E, Cinelli P, Lazzeri A, Alvarez V. Polyhydroxyalcanoate (PHA): review of synthesis, characteristics, processing and potential applications in packagings. Express Polym Lett 2014;8(11):791–808. Available from: https://doi.org/10.3144/expresspolymlett.2014.82.

[89] Naman CB, Leber CA, Gerwick WH. Modern natural products drug discovey and its relevance to biodiversity conservation. In: Hurtboke I, editor. Microbial resources. Elsevier Inc; 2017. p. 103–20. ISBN 978-0-12-804765-1.

[90] da Silva M. The new Brazilian legislation on access to the biodiversity (Law 13, 123/15and Decree 8772/16). Braz J Microbiol 2017;49(1):1–4. Available from: https://doi.org/10.1016/j.bjm.2017.12.001.

[91] Newman N, Crag GM, Snader KM. Natural products as sources of new drugs over the period 1981-2002. J Nat Prod 2003;66:1022–37.

[92] Liu H, Zhu G, Fan Y, Du Y, Lan M, Xu Y, et al. Natural products research from 2015to 2016. Front Chem 2018;6:45. Available from: https://doi.org/10.3389/Chem.2018.00045.

[93] Barbulova A, Colucci G, Apone F. New trends in cosmetics: by-products of plant origin and their potential use in cosmetic active ingredients. Cosmetics 2015;2:82–92. Available from: https://doi.org/10.3390/cosmetics2020082.

[94] Song Q, Li J, Zeng X. Minimizing the increasing solid waste through zero waste strategy. J Clean Prod 2015;104:199–219. Available from: https://doi.org/10.1016/j.jclepro2014.08.027.

[95] Papkov D, Delpouve N, Delbreilh L, Araujo S, Stockdale T, Mamedov S, et al. Quantifying polymer chain orientation in strong and tough nanofibers with low crystallinity: toward next generation nanostructured superfibers. ACS Nano 2019;13:4893–927. Available from: https://doi.org/10.1021/acsnano.8b08725.

[96] Alan Khan F, editor. Bionanotechnology fundamentals. Boca Raton, FL: CRC Press; 2012.

[97] Yubing X. The nanobiotechnology handbook. CRC Press; 2017. ISBN 9781138075856.

[98] Phoenix DA, Ahmed W, editors. Nanobiotechnology. Manchester: One Central Press; 2014.

[99] Koltson D. Industrial nanotechnology in the public eye. In: Proceedings of Nanotechnology and Biotechnology meeting, Milano, Italy, March 30; 2011.

[100] Tripathy N, Ahmad R, Khang G. Nanomaterial-assisted tissue engineering and regenerative medical therapy. In: Khang G, editor. Intelligent scaffolds for tissue engjneering and regenerative medicine. 2nd Ed. Singapore: Pan Stanford Publishing; 2017. p. 6/1–56.

[101] Kumar R, Chakraborty A, Mukherjee A, Roy A. 3D printing in biomedical applications. In: Ranjan M, editor. Chemical engineering world. Jasubhai Media Pvt. Ltd; 2018. p. 40–4. Available from: <www.researchgate.net/publication/323627975>.

[102] Ahagar P, Cooke ME, Weber MH, Rosenzweig DH. Current biomedical applications of 3D printing and additive manufacturing. Appl Sci 2019;9:1713. Available from: https://doi.org/10.3390/app9081713.

[103] Jamroz W, Szafraniec J, Kurek M, Jachowicz R. 3D Printing in pharmaceutical and medical applications—recent achievements and challanges. Pharm Res 2018;35:176. Available from: https://doi.org/10.1007/sl1095-018-2454-x.

[104] Nair LS, Laurencin C. Biodegradable polymers as biomaterials. Prog Polym Sci 2007;32(8-9):762–98.

[105] Siegel E. The future of personalized beauty lies in the Tech industry. Allure, Conde' Nast., CA, USA, April 17; 2019 [accessed 29.08.19].

[106] Westbrook G. Top 10 global consumer trends 2019. London: Euromonitor International; 2019.

[107] Cossma. Marketing trends at in-cosmetics Global 2019, in-cosmetics global' meeting presentations, Barcelona, 31 March-2 April 2020; 2019 [accessed 29.08.19].

[108] Chime SA, Kendchokwo FC, Attama AA. Nanoemulsions—advances in formulation, characterization, and applications in drug delivery. InTech 2014;. Available from: https://doi.org/10.5772/15371.

[109] Morganti P, Fusco A, Paoletti I, Perfetto B, Del Ciotto P, Palombo, et al. Antiinflammatory, immunomodulatory and tissue repair activity on human keratinocytes. Materials 2017;10:843. Available from: https://doi.org/10.3390/ma10070843.

[110] Bellar A, Kessler SP, Obery DR, Sangwan N, Welch N, Nagy LE, et al. Safety of Hyaluronan 35 in healthy human subjects: a pilot study. Nutrients 2019;11:1135. Available from: https://doi.org/10.3390/nu11051135.

[111] Krutmann J, Gilchrest BA, editors. Skin aging. Berlin: Springer; 2006. ISBN-103-540-24443.

[112] Yamaguchi Y, Hearing VJ, Itami S, Yoshikawa K, Katayama I. Mesenchymal-epithelial interactions in the skin: aiming for site-specific tissue regeneration. J Dermatol Sci 2005;40:1–9.

[113] Grice EA, Segre JA. The skin microbiome. Nat Rev Microbiol; 2011;9(4):244–53. Available from: https://doi.org/10.1038/nrmicro2537.

[114] Bartholott W, Elisabetsky E, Farrel R, Kheng CT, Prance I, Laud S. New products and industries from biodiversity. In: Beattie AJ, editor. Ecosystems and human well-being: current state and trends. Island Press; 2014. p. 271–95. 9781559632270.

[115] Lanza R, Langer R, Vacanti JP, editors. Principles of tissues engineering. New York, NY: Academic Press; 2007.

Index

Note: Page numbers followed by "*f*" and "*t*" refer to figures and tables, respectively.

A

AA. *See* Alopecia areata (AA)
Abraxane, 82–83
Acceptability criteria, 418
"Aclaim", 160–161
Acrylonitrile, 22
Active ingredients, 296–298
Active molecules, 41, 319–320
Adhesive, 23
Adipocytes, 307
Advanced Medications, 297–298
Advanced nanomaterials, 473–474
AFM. *See* Atomic force microscopy (AFM)
AG. *See* 8_Azaguanine (AG)
AGA. *See* Androgenic alopecia (AGA)
Aging, 288–289
  and cancer, 291–292
  population, 7, 287
AgNPs. *See* Silver nanoparticles (AgNPs)
Air pollution, 293–294
Alginate nanoemulsion, 53
Alkyl polyglycoside surfactant, 213
Aloe vera, 196
Alopecia, 218–220
Alopecia areata (AA), 219
5-α reductase (5α-R), 219
Alpha-chitin, 458–463
  in crab, 455–457, 458*f*
α/β MHC gene, 434–435
Alumina (Al_2O_3), 380
Alzheimer's diseases, 229, 407
Ambient ozone, 293
Ames test, 424–426
  description, 425
  preparation of test substance, 426
    control, 426
    exposure condition, 426
    incubation, 426
    procedure, 426
    result and interpretation, 426
  principle, 425
Amino acid, 425
Anacystic nidulans, 87–88, 148
Analytical method, 424
Anatase, 367
Androgenic alopecia (AGA), 214
Ands Corporation, 329–330
Antiaging
  creams, 154–159
  molecules, 35–36
Antiaging, 445
Anticellulite compounds, 35–36
Antidandruff, 220–221
Antiinflammatory cytokines, 288–289, 290*f*
Antiinflammatory/antiinfective agents, 445
Antioxidants, 35–36, 88, 133, 196–197
Antioxidants, 445
APAGARD M plus, 451*t*
Apocrine sweat glands, 307
Appendageal route. *See* Shunt route
Aromatic wallpaper, 260–261
Artificial nails, 376
AS/NZ 2604:2012 method, 359
AS52 (Chinese hamster cells), 427
Ascorbic acid, 412–413
Associative/nonassociative polymers, 69
Asthma, 293
Asymmetric oxygen carrier system liposomes, 149
Atomic force microscopy (AFM), 44
Au fraiche, 173
Aueugens, 429–430
AuNPs. *See* Gold nanoparticles (AuNPs)
Aureobasidium pullulans, 468–469
Authorized nanomaterials, 402
Avobenzone, 351
Avon Products Inv, 378
8_Azaguanine (AG), 427

B

Baby/Feminine Diapers, 297–298
Backing laminate, 23
Bacterial reverse mutation test. *See* Ames test
Bacterial Reverse Mutation Test OECD 471, 424
BALB/c 3T3 cells, 434
Balb/c cells, 419
Base coats, 376

Base note, 173
Beauty Masks, 297–298
BEMT. *See* Bis-ethylhexyloxyphenol methoxyphenyl triazine (BEMT)
Bentonite, 387
Beta-defensine-2 expression (hBD2), 298*f*
Bicontinuous nanoemulsions, 150, 328
Bioactive nanocarriers, 25–28
  cosmetic agents loaded in nanocarriers, 26*t*
  for cosmetic products, 26*f*
  liposomes, 25–27
  nanoemulsions, 27–28
  niosomes, 27
  SLNPs, 27
Bioactive nanocarriers, 446
Biodegradable PLA nanoparticles, 311
Biodiversity, 471–473
Bioelectronic noses, 261
Biological membranes, 195–196
Biologically active substances, 22
Biomagnification, 130
Bionanotechnology, 7, 473–474
Biopolyesters, 457–471
  PHA, 470–471
  PLA, 469–470, 469*f*
Biopolymers, 8–10, 455–457, 473–474
Bis-DI-C14–18 alkyl amide, 379
Bis-ethylhexyloxyphenol methoxyphenyl triazine (BEMT), 197
Bis-hydroxypropyl dimethicone/SMDI copolymer, 379
Bismuth oxychloride (BiOCl), 383
Black tea (*Camellia sinensis* Linnaeus), 132
Blond hair, 229
Blue Lizard sun-care products, 355
BOI. *See* Bud opening indicator (BOI)
Bovine serum albumin (BSA), 212, 234
Brazilian biodiversity regulations, 471–472
"Bricks and mortar" model, 307
Broadband light protection sun care products
  final remarks, 198
  functional analysis methods, 191
  visible light, 186–190
BSA. *See* Bovine serum albumin (BSA)
Bud opening indicator (BOI), 262
2-Butoxy ethanol, 377
Butyl methoxydibenzoylmethane, 351

C

CA. *See* Cluster analysis (CA)
Caffeine (CAF), 47–48
CAGR. *See* Compound Annual Growth Rate (CAGR)
Calcein, 313, 317
Calcium, 380
Calendula officinalis extract, 332
Camellia sinensis Linnaeus. *See* Black tea (*Camellia sinensis* Linnaeus)
Cancer, aging and, 291–292
Candelilla, 379
Candida albicans, 220
Candida parapsilosis, 220
Carbon black, 385
Carbon fullerenes, 95–96
Carbon nanotubes (CNTs), 80–82, 96, 153, 407
Carboxylated nanoparticles of latex, 405
Cardiomyocyte differentiation, 434–435
Carnauba, 379
Carotenoids, 331*t*, 357
Carrier systems, development of, 351
CAS. *See* Chemicals Abstract Service (CAS)
Catalog of nanomaterials, 401–402
Cationic nanoemulsion delivery system, 342
Cavitation, 335–336
CE. *See* Cloning efficiency (CE)
Celecoxib (Cxb), 240–241
Cell penetrating peptides (CPPs), 240–241
Cell viability measurements, 416
Cellular damages depending on time, 287, 288*f*
Cellulose, 466–468
  crystalline/amorphous structure, 460*f*
Cellulose nanocrystals (NCEs), 466–467
Centella asiatica, 27
  extract, 331*t*
Ceramic pigment nanoparticles, 386
Cetyl phosphate, 368–369
Chantecaille Nano Gold Energizing Cream, 447*t*
Chemical composition, 404
Chemical sunscreens, 355–356
Chemicals Abstract Service (CAS), 398
Chemotherapy-induced alopecia (CIA), 219
CHI. *See* Chitosan (CHI)
Chinese hamster cells (CHO), 427
Chinese Herbal Medicines (CHMs), 258
  recognition of, 258
Chinese Yellow-Feather Chickens, flavor of traditional soup of, 257
Chitin, 6, 8–10, 274–277, 458–463
  backbone of, 462*f*
  chitosan, 276*f*
  crystalline/amorphous structure, 460*f*
  and hair, 277–283
    hydrogen bonds of alpha-helix keratin, 281*f*
    planet biodiversity, 282*f*
    structural organization of hair keratin, 280*f*
  hierarchical structure, 275*f*
  in insect and fungi, 461*f*
  nanofibril-hyaluronic acid block polymer, 10*f*

nanofibril–nanolignin complex, 279*f*
nanofibrils, 463*f*
at SEM and ZETASIZER Nano, 277*f*, 278*f*
structure, 276*f*
tissue, 9*f*
organization of chitosan–chitin fiber resembling hair structure, 280*f*
production and worldwide demand, 462*f*
Chitin nanofibrils (CNs), 8–10, 9*f*, 273–274, 277*f*, 297*f*, 455–458
electrospun CN tissue, 459*f*
Chito-oligosaccharides, 458–469
Chitosan (CHI), 53–54, 458–463
nanoparticles, 317
Chlorite, 386
CHMs. *See* Chinese Herbal Medicines (CHMs)
CHO. *See* Chinese hamster cells (CHO)
Cholesterol, 142
Chromium nanoparticles, 406–407
Chromium oxide, 380
Chronic obstructive pulmonary disease, 293
CIA. *See* Chemotherapy-induced alopecia (CIA)
Cinnamates, 129
CIR. *See* Cosmetic Ingredient Review (CIR)
Circular economy, 455–457, 456*f*
Circular green economy, 455–457
Cl77266 Black 2, 386
Classification, Labeling, and Packaging (CLP), 364–365
nanomaterials under, 364–365
Clastogens, 429–430
Clay materials, 386
Cloisite15A, 387
Cloisites, 387
Cloning efficiency (CE), 429
CLP. *See* Classification, Labeling, and Packaging (CLP)
CLs. *See* Conventional liposomes (CLs)
CLSM. *See* Confocal laser scanning microscopy (CLSM)
Cluster analysis (CA), 259
CMS. *See* Core-multishell (CMS)
CN–NL complexes. *See* Nanochitin–nanolignin complexes (CN–NL complexes)
CNs. *See* Chitin nanofibrils (CNs)
CNTs. *See* Carbon nanotubes (CNTs)
CO_2 emissions, 455
Coated nano-TiO_2, 368–369
Cobalt, nanoparticles of, 406–407
Coenzyme Q10, 43, 135
COLIPA, 360–361
Colloidal constant cationic nanoemulsion, 342
Color-changing nail polishes, 377
Colorants, 333
Colorona, 382
Colour Travel Series products, 381–382
Comet assay, 424
Commercial nanocosmetic products, 445–452
face care products in market, 449*t*
nanoantiwrinkle and nanoantiaging products in market, 447*t*
nanocosmetics
of body lotion and cleanser in market, 451*t*
for lip and sun cares in market, 449*t*
products for hair care in market, 448*t*
nanomoisturizer products in market, 446*t*
oral care nanocosmetics in market, 451*t*
Compound Annual Growth Rate (CAGR), 3–6, 458–463, 465–468
Confocal laser scanning microscopy (CLSM), 233
detection of zinc, 234
estimation of NPs human skin penetration, 233
phototherapy and multimodal imaging, 233–234
polymer-grafted proteins, 234
Confocal laser scanning microscopy (CLSM), 312–313
Control substance application, 415–416
Conventional liposomes (CLs), 41
Copper nanoparticles, 100
Coral White Nano Silver, 451*t*
Core-multishell (CMS), 210
Corneocytes, 36, 307
Corneodesmosomes, 307
Cosil Nano Beauty Soap, 451*t*
Cosil Whitening Mask, 449*t*
Cosmeceutical-tissues, 295–296
Cosmeceuticals, 17–18, 83, 141–142, 328–329
nanoemulsions as potential carrier in, 330–332
Cosmetic, Toiletry and Fragrance Association (CTFA), 413
Cosmetic Ingredient Review (CIR), 371
Cosmetic nanoemulsions, challenges associated with, 332–333, 333*f*
Cosmetic products, 293
advanced nanomaterials and nanotechnologies, 473–474
nanomaterials in EC Regulation No. 1223/2009 on, 395–401
natural green cosmetics and biodiversity, 471–473
smart nanocosmetics and future perspectives, 474–476
use of nanomaterial in, 397
Cosmetic Products Notification Portal (CPNP), 398
Cosmetics, 141–142, 413, 445
consumer concerns and regulatory guidances
European Union Guidelines, 165–166
guidance document issued by FDA for industries, 166–167
routes of nanocosmetics exposure, 163–165
definition of "nanomaterials", 395–397

Cosmetics (*Continued*)
- liposomes, 142
- nanocarriers/nanomaterials in, 142–154
- nanoemulsion in, 327–328
- nanomaterials in, 362–367
  - ECHA activity, 365
  - nanoingredients labeling, 365
  - nanotoxicology, 366–367
  - under REACH and CLP, 364–365
  - regulatory status, 363–364
  - safety concerns, 365–366
- nanotechnology in, 83–86
  - nanocarriers used in cosmeceuticals, 85*f*
  - nanomaterials used in cosmetics, 86*t*
- nanotechnology-based cosmetics, 154–163
- sciences, 71–72
- sunscreens, 359

Cosurfactants, 62
Coty Inc., 387
CPNP. *See* Cosmetic Products Notification Portal (CPNP)
CPPs. *See* Cell penetrating peptides (CPPs)
Crackling effect, 377
Creams, 109, 124*t*, 340
Criative Nail Design, 378
Crocus sativus L., 109, 114–115, 118*t*
Cryosectioning process, 314
CTFA. *See* Cosmetic, Toiletry and Fragrance Association (CTFA)
Cubosomes, 94–95, 151, 340–341
Curcumin, 331*t*
Cuticle, 206–207
Cxb. *See* Celecoxib (Cxb)
Cyanoacrylate
- skin surface
  - biopsy, 37–38
  - stripping technique, 314
- superglue, 314

Cyclodextrins, 132–133, 210
Cytochalasin B (cytoB), 424, 431–432
Cytokines, proinflammatory and antiinflammatory, 288–289, 290*f*
Cytotoxicity, 419
- formulas for cytotoxicity assessment, 432–433

D

Dandruff, 220
Daphania magna, 101
DEA. *See* Diethanolamine (DEA)
Del Sol Color Change, 377
Dendrimers, 50–52, 51*f*, 93–94, 152, 210–211, 318, 340–341
Dendritic structures, 50–52, 51*f*
Dendrons, 50
Dermal absorption, 422
Dermal papilla cells (DPCs), 219
Dermal passage routes, 308
Dermal routes of nanomaterials, 100
Dermatotherapy, 305
Dermis, 18–19, 36, 307
Dermis–epidermis junction, 19
Dermisis, 307
Design Expert Software (Version 8. 0), 110
Desquamation, 307
Destabilization mechanisms, 66–68
DFA. *See* Discriminant factor analysis (DFA)
DHI. *See* Dihydroxyindolequinone (DHI)
DHT. *See* Dihydrotestosterone (DHT)
Diarylethene, 377
Dibutyl lauroyl glutamine, 379
Dicetyl phosphate, 41
Diclofenac sodium, 112
Diethanolamine (DEA), 250
Differential skin stripping, 37–38
Differential tape stripping, 314
Diffusion cell design, 422
Dihydrotestosterone (DHT), 219
Dihydroxyindolequinone (DHI), 195
Diisostearoyl trimethylolpropane siloxyl silicate, 379
Diisostearyl malate, 379
Dilauroyl trimethylolpropane siloxyl silicate, 379
Dimethyl sulfoxide (DMSO), 174, 233, 430
1,2-Dioleoyl-3-(trimethylammonium) propane-based cationic liposomes, 319
1,2-Dioleoylphosphatidylethanolamine (DOPE), 319
DiorSnow Pure UV Base SPF 50, 449*t*
Diphencyprone (DPCP), 71–72
Discriminant factor analysis (DFA), 263
DMAPAAQ. *See* *N,N*-dimethylamino propyl acrylamide methyl chloride quaternary (DMAPAAQ)
DMSO. *See* Dimethyl sulfoxide (DMSO)
DOPE. *See* 1,2-Dioleoylphosphatidylethanolamine (DOPE)
Dose–response assessment, 438
Double-walled nanotubes, 153
5-Doxyl stearic acid (5DSA), 210
DPCP. *See* Diphencyprone (DPCP)
DPCs. *See* Dermal papilla cells (DPCs)
dPGNG. *See* pH-responsive-dendritic-polyglycerol-nanogel (dPGNG)
Drug entrapment, 51
5DSA. *See* 5-Doxyl stearic acid (5DSA)
Dyes, 333
- binding method, 418
- content calculation, 418

E

E-nose. *See* Electronic nose (E-nose)
Eau de parfum, 173
Eau de toilette, 173
EBs. *See* Embryoid bodies (EBs)
EC. *See* Ethyl cellulose (EC); European Commission (EC)
Eccrine sweat glands, 307
ECHA. *See* European Agency for Chemicals (ECHA)
ECM. *See* Extracellular matrix (ECM)
Ecotoxicity studies of TiO_2 and ZnO, 370–371
ECVAM. *See* European Centre for Validation of Alternative Method (ECVAM)
Edema, 414–415
EE. *See* Entrapment efficiency (EE)
EFSA. *See* European Food Safety Authority (EFSA)
EGFP. *See* Enhanced green fluorescent protein (EGFP)
EINECS. *See* European Inventory of Existing Commercial chemical Substances (EINECS)
Elastosis, 353
Electronic nose (E-nose), 258
  area identification of Zhongning Goji berries by, 259
  detect freshness of eggs, 263
  evaluation of, 262
Electronic nose analysis (ENA), 263
Electroosmosis, 317
ELINCS. *See* European List of Notified Chemical Substances (ELINCS)
Embryoid bodies (EBs), 433
Embryonic stem cell line test (EST cell line test), 433–435
  embryotoxicity, 435
  material and methods, 434–435
    cell culture, 434
    cytotoxicity assessment, 434
    differentiation of ES cells, 434
    RNA isolation and (RT-PCR) analysis, 434–435
    test compound, 434
Embryotoxicity testing, 433–439
  EST cell line test, 433–435
  MM test, 435–436
  whole embryo culture, 436–439
EMEA. *See* European Medicines Agency (EMEA)
Emissions, outdoor and indoor, 288–289
Emollient, 70
Emulsion, 328
  delivery systems in cosmetics
    emulsion generalities, 60–61
    microemulsions and nanoemulsions, 61–62
ENA. *See* Electronic nose analysis (ENA)
Endpoint assay, 438–439
Engineered nanomaterials (ENMs), 366
Enhanced green fluorescent protein (EGFP), 236
ENMs. *See* Engineered nanomaterials (ENMs)
Entrapment efficiency (EE), 66, 112, 115
Environmental contaminants, 288
Epidemiological studies, 407
EpiDerm, 414–415
Epidermis, 18, 36, 307, 414
EpiSkin, 414–415
Epithelium, 415
EPOSS. *See* Extended polyhedral oligo silsequioaxane (EPOSS)
Epoxy silicone nanoemulsions, 216
Erythema, 352, 414–415
Erythematous dose (MED), 357
Escherichia coli, 261, 425
  E. coli WP2 uvrA, 425
EST cell line test. *See* Embryonic stem cell line test (EST cell line test)
Ethanol, 173
Ethosomes, 39, 41, 148–149
Ethyl cellulose (EC), 194
Ethylenediamine/hydrogenated dimer dilinoleate copolymer, 379
ETS. *See* European Teratology Society (ETS)
EU. *See* European Union (EU)
EU PolyBioSkin research project, 470–471
Eumelanin, 229
  building blocks of, 229
EUON. *See* European Observatory on Nanomaterials (EUON)
European Agency for Chemicals (ECHA), 364–365, 371
  activity, 365
European Centre for Validation of Alternative Method (ECVAM), 433
European Commission (EC), 165–166, 399, 401
European Food Safety Authority (EFSA), 371
European Inventory of Existing Commercial chemical Substances (EINECS), 398
European List of Notified Chemical Substances (ELINCS), 398
European Medicines Agency (EMEA), 360–361
European Observatory on Nanomaterials (EUON), 365–366
European Teratology Society (ETS), 433
European Union (EU), 165–166
  Guidelines, 165–166
Exclusivity of nanoemulsions, 342
Exposure to air pollution, 293
Extended polyhedral oligo silsequioaxane (EPOSS), 378
Extracellular matrix (ECM), 8–10, 288–290, 296–298, 455–457
Extrait de parfum, 173
Extrinsic aging, 288–289
Eye care

Eye care (*Continued*)
benefits of nanocosmetics, 159*f*
skin care and under, 155*t*
Eye Contour Nanolift, 447*t*
Eye Tender, 447*t*

F

FAAs. *See* Free amino acids (FAAs)
Face care products in market, 449*t*
Factorial design for polymeric nanocarriers formulation, 113–114, 115*t*
FDA. *See* Food and Drug Administration (FDA)
Federal Food, Drug, and Cosmetic Act (FD&C Act), 35–36, 87, 166
Femtosecond laser multiphoton tomography, 243
Ferric ferrocyanide, 380
Ferric Reducing Antioxidant Power method (FRAP method), 112
FI. *See* Fragrance index (FI)
Fillderma Lip Volumizer, 378
FITC. *See* Fluorescein isothiocyanate (FITC)
FL. *See* Fluorescence (FL)
Flake glass, 381
Flake-shaped mica plaques, 384
Flavonoids, 357
Florescent nanoparticles, 80–82
Fluorescein isothiocyanate (FITC), 212, 313
Fluorescence (FL), 233–234
Fluorescent dye (DID-oil) encapsulated nanolipid crystal NPs (FNLCNs), 240–241
Fluorescent microscopy, 314
Fluvastatin, 331*t*
FNLCNs. *See* Fluorescent dye (DID-oil) encapsulated nanolipid crystal NPs (FNLCNs)
Folic acid, 51–52
Follicular biopsy, 314
Follicular deposition ability of nanocarriers, 310–311
Follicular ducts as reservoirs of nanocarriers, 310
Food and Drug Administration (FDA), 219, 357
guidance document, 166–167
Nanotechnology Task Force Report, 364
Food-grade ultrasonication nanoemulsion, 335–336
Foot Moisturizing Serum, 446*t*
4-star rating system, 361
Fourier-transform infrared (FT-IR), 239
Fourier-transform infrared spectroscopy, 68
4R Approach, 3–6
Fragrance, 173, 247
affects skin, 250
chemicals present in, 250
home fragrance products, 248–249
astonishing ways to make home smell amazing, 248
fragrance for body, 249
orange peels in making house smell good, 249
tips to keep kitchen smelling so fresh, 249
toilet freshener, 249
natural, 249–250
Fragrance index (FI), 262
FRAP method. *See* Ferric Reducing Antioxidant Power method (FRAP method)
Free amino acids (FAAs), 257
Free radical production, 352
Fresh As A Daisy Body Lotion, 451*t*
FT-IR. *See* Fourier-transform infrared (FT-IR)
Fullerenes, 100, 155*t*, 362, 407
Fullerenes/buckyballs, 95–96

G

G3. *See* Generation 3 (G3)
G3.5 PAMAM-COOH–FITC, 318–319
G4 PAMAM-OH–FITC, 318–319
G4 PAMAM–FITC conjugate, 318
GAO. *See* US Government Accountability Office (GAO)
Gas Chromatography-Olfactometry/aroma extract dilution analysis (GC-O/AEDA), 263
Gas chromatography–mass spectrometry (GC–MS), 256
Gelatin, 194–195
Generally Recognized as Safe and Effective (GRASE), 359–360
Generally recognized safe (GRAS), 88
Generation 3 (G3), 51–52
Generation 6 (G6), 51–52
Genotoxicity/mutagenicity testing, 424–433
Ames test, 424–426
in vitro mammalian cell gene mutation test, 427–429
in vitro mammalian cell micronucleus test, 429–433
GHG emissions. *See* Greenhouse gas emissions (GHG emissions)
Glass ionomers cement (GIC), 80–82
Global food waste, 293, 293*f*
Glutathione (GSH), 197
Glycereth-7-triacetate, 174
Glycerosomes, 148–149
Glycyrrhiza glabra extract, 331*t*, 332
Gold nanoparticles (AuNPs), 92–93, 93*f*, 94*t*, 153, 155*t*, 231–232, 239, 362, 412–413
nanoreactor in controlled synthesis of fluorescent, 239
Gooris and Bouwstra protocol, 47–48
Granulomatous tissue response, 407
GRAS. *See* Generally recognized safe (GRAS)
GRASE. *See* Generally Recognized as Safe and Effective (GRASE)
Gray hair, 229

natural turning of, 230
plucking, 229
Green energy, 80–82
Greenhouse gas emissions (GHG emissions), 455
GSH. *See* Glutathione (GSH)

H

HA. *See* Hyaluronic acid (HA); Hydroxyapatite (HA)
HaCaT. *See* Human keratinocyte cell line (HaCaT)
Hair, 227, 267–268
color, 217, 229
damage, 215–216
fibers, 228
graying, 216–218
growth cycle, 268*f*
hair system care, NEs in, 341–342
shaft, 206–207, 207*f*, 307, 308*f*
structure and damage, 269–270, 269*f*
tips for faster growth of, 232, 232*f*
Hair care, 162–163
applications
hair structure, 206–208
nanostructured systems, 208–221
penetration of active compounds, 206*f*
perspectives, 221
benefits of nanocosmetics, 164*f*
nanocosmetic products in market, 448*t*
of nanoemulsions, 71–72
nanomaterials for, 98
nanoproducts, 164*t*
products, 205
Hair dyes
best foods to promote hair growth, 232, 232*f*
common chemicals, 230
composition of, 230
fruits for hair growth, 233
gold NPs, 231–232
graphene, 231
graphene/ionic liquid electrochemical sensor, 237
harmfulness of, 230
poly-γ-glutamic acid/glycol chitosan nanoparticles, 240
PPD-incorporated NPs, 232
side effects, 231
targeted approach for psoriasis treatment, 237
toxic chemicals, 231, 231*f*
Hair follicles, 207–208, 235–236, 238, 305, 307, 310, 341
hair dye-incorporated hyaluronic acid nanoparticles, 239
lipid nanoparticles for, 235
in nanocarrier-mediated skin delivery, 310–312
extent of nanocarrier particle size for follicular delivery, 311–312
follicular deposition ability of nanocarriers, 310–311
follicular ducts as reservoirs of nanocarriers, 310
nanoparticles, 242–243
as gene carrier to inner ear, 236
nanoparticles-based formulation, 235–236
nanoreactor in controlled synthesis of fluorescent AuNPs, 239
penetration of nanocarriers into, 235
polymeric nanoparticles-embedded organogel, 239
potential novel drug carriers for inner ear treatment, 242
presentations in conferences, 238
quantification of nanoparticle, 238
translocation of CPPs engrafted nanoparticles, 240–241
Hair-like structures, 277–283
Handheld E-nose (HEN), 262
hBD2. *See* Beta-defensine-2 expression (hBD2)
HBPL. *See* Hyperbranched polylysine (HBPL)
Heart note, 173
Hectorite, 387
Helianthus annuus, 132
Helichrysum oil, 357
Hemicellulose, 455–457, 457*f*
HEN. *See* Handheld E-nose (HEN)
High-energy emulsification methods, 179, 333–336, 335*f*. *See also* Low-energy emulsification methods
high-pressure homogenization, 334
microfluidization, 334–335
ultrasonication, 335–336
High-pressure
homogenization, 64, 334
homogenizers, 334
Higuchi and Korsemeyer–Peppas model, 121
Histidine (His +), 425
HPLC/UPLC spectrophotometry, 415–416
HPRT. *See* Hypoxanthine guanine phosphoribosyl transferase (HPRT)
Human hair, 227–228
anatomy of, 228
chemical composition, 227–228
function of, 227
layers of hair, 228*f*
living being found, 228
major components, 228
Human health
air pollution, waste, and, 293–294
Human keratinocyte cell line (HaCaT), 42
Human microbiome, 298, 299*f*
Human-derived nontransformed epidermal keratinocytes, 414

Humidifying agents, 379
Hyaluronic acid (HA), 35–36, 328–329
backbone of, 462*f*
Hydra Flash Bronzer Daily Face moisturizer, 446*t*
Hydra Zen Cream, 446*t*
Hydrated silica, 47
Hydrating compounds, 35–36
Hydrocortisone, 331*t*
Hydrogels, 95
Hydrogen peroxide (H_2O_2), 193
Hydrophilic fragrances, 180
Hydrophilic lipolytic molecules, 48
Hydroxyapatite (HA), 193
Hydroxyl (•OH), 193
Hyperbranched polylysine (HBPL), 242
Hyperbranched polymers, 52
Hypodermis, 19, 307
Hypoxanthine guanine phosphoribosyl transferase (HPRT), 427

I

IARC. *See* International Agency on Research of Cancer (IARC)
IC50 for cell viability, 435
ID50 for cells differentiation, 435
IFRA. *See* International Fragrance Association (IFRA)
Illite, 386
Imaging techniques, 312–314. *See also* Quantitative techniques
CLSM, 312–313
fluorescent microscopy, 314
In vitro 3T3 NRU phototoxicity test, 419
In vitro dermal absorption measurements of cosmetic ingredients, 422–424
factors affecting dermal absorption and methodology, 422–424
analytical method, 424
diffusion cell design, 422
dose and volume of test substance, 423
preparation of dose and vehicle, 423
receptor fluid, 422–423
skin integrity, 423
skin preparation, 423
skin temperature, 423
study period and sampling, 423
test substance, 423
principle, 422
of test, 422
result, 424
In vitro dissolution studies, 119–120
In vitro experimental models, 312
In vitro mammalian cell gene mutation test, 427–429
description of method, 427–429
preparation, 427–428
procedure, 429
result, 429
test conditions, 428–429
principle of test, 427
In vitro Mammalian Cell Gene Mutation Test OECD 476, 424
In vitro mammalian cell micronucleus test, 429–433
description of method, 430
cells, 430
media and culture condition, 430
preparation of culture, 430
test chemical preparation, 430
test condition, 430–431
analysis, 432
cell harvest and slide preparation, 431–432
cytoB, 431
procedure, 431
result, 432–433
solvents, 430
In vitro Mammalian Cell Micronucleus Test OECD 487, 424
In vitro Mammalian Chromosome Aberration test OECD 473, 424
In vitro phototoxicity test, 419–421
initial consideration, 419
preparations, 420
cells, 420
irradiation conditions, 420
media and culture, 420
preparation of culture, 420
preparation of test substance, 420
principle, 419
test conditions, 421
Neutral Red uptake test, 421
result, 421
test procedure, 421
test substance concentrations, 421
In vitro skin corrosion test, 416–418
general description, 416
initial considerations, 416
principle, 417
procedure, 417–418
acceptability criteria, 418
application of test and control substance, 417
dye binding method, 418
dye content calculation, 418
skin disc preparation, 417
sulforhodamine B dye application and removal, 418
TER measurements, 417–418
In vitro skin irritation test, 414–416
initial considerations and limitations, 414

principle, 414–415
procedure, 415–416
functional condition, 415–416
general condition, 415
In vitro test results, 416
In vitro toxicology studies, 414
In vitro UVASPF, 358
In vivo studies, 405–406
In vivo UVASPF, 358
Indole-5,6-quinone (IQ), 195
Inflammaging, 290–292
causes of, 291*f*
Infrared radiations (IR radiations), 351–352
Infundibulum, 208
Ingestion of nanomaterials, 100
Inhalation, 405
of nanomaterials, 99–100
Inhaled carbon nanoparticles, 405
Innovative smart cosmeceutical-tissues, 295–296
Inorganic
filters, 193
nanocarriers, 47–50
nanoparticles, 132–133
particles, 383
sunscreens, 355–357, 356*f*
UV blockers, 35–36
Intact-skin dermal exposures, nanoparticle toxicological risks on
factors mediating toxicity, 404
host factors, 404
particle factors, 404
nanoparticles and hazards, 405–407
routes of entry and kinetics, 405
toxicity of nanoparticles, 406*t*
Intercellular lipids, 36, 307–308
route, 38
International Agency on Research of Cancer (IARC), 241–242
International Fragrance Association (IFRA), 249
International Organization for Standardization (ISO), 364
Intrinsic aging, 288–289
Invasomes, 149
Iontophoresis, 23–24, 316–317, 319
Iontophoretic technique combined with nanosized delivery systems, 316–319
IQ. *See* Indole-5,6-quinone (IQ)
IR radiations. *See* Infrared radiations (IR radiations)
IRA. *See* Near infrared A (IRA)
Iridaceae, 109
Iron oxides, 383
Irradiation dose, 420–421
ISO. *See* International Organization for Standardization (ISO)
Isopropyl acid, 376

J
Joint Research Centre (JRC), 394–395

K
K-Ray, 383
K-Ray G, 383
K-Ray SM, 383
Kaolinite, 386
Keratinocytes, 18
Kobo, 383
Kraft process, 464–465

L
L5178Y (mouse lymphoma cells), 427
Labeling, 400
Lactic acid, 469–470
Lagenaria siceraria, 221
Lancôme RenergieMicrolift, 447*t*
Langerhans cells, 18
Laser radiation, 25
Lawsone, 230, 230*f*
Lawsonia inermis, 230
LDA. *See* Linear discriminant analysis (LDA)
LE. *See* Loading efficiency (LE)
Leave-on products, 180–181
Life expectance and cosmetic treatments, 292–293
LifePak Nano, 449*t*
Light source, 420
Lignin, 6, 455–457, 457*f*, 464–465, 464*f*
Ligno-cellulosic feedstock, 466–467
Linear discriminant analysis (LDA), 259
Linear economy, 455–457, 456*f*
Lip care, 160, 378–379
benefits of nanocosmetics, 161*f*
nanocosmetic products in market, 449*t*
nanocosmetics, 161*t*
nanomaterials for, 97
nanotechnological innovations in nail and lip care formulations, 380–387
Lip Tender, 449*t*
Lip volumizer, 160
Lipid nanocapsules (LNCs), 242
Lipophilic drugs, 53
Liposomal carrier systems, 133
Liposomes, 25–27, 87–88, 88*f*, 142–150, 155*t*, 161*t*, 210, 340–341, 378, 412–413
marketed formulations, 89*t*
and related particles, 39–42
Lipsticks, 378

Lipsticks (*Continued*)
formulations, 355
Listeria monocytogenes, 332
LNCs. *See* Lipid nanocapsules (LNCs)
Loading efficiency (LE), 66
Lotions, 340
Low-energy
emulsification, 179–180
techniques, 65
Low-energy emulsification methods, 333–334, 336–339. *See also* High-energy emulsification methods
phase inversion emulsification method, 336
PIC method, 337
PIT method, 336
self-nanoemulsification method, 338
spontaneous emulsification, 339
Lyphazome, 164*t*
nanospheres, 155*t*

M

Macroemulsions, 61
Magnetic nail polishes, 377
Majestic Pure, 449*t*
Manganese dioxide, 368–369
Marinosomes, 148
MBBT. *See* Methylene bis-benzotriazolyl tetramethylbutylphenol (MBBT)
MEA. *See* Monoethanolamine (MEA)
18-MEA. *See* 18-Methyl eicosanoic acid (18-MEA)
MED. *See* Erythematous dose (MED)
MED of skin protected by solar product (MEDp), 358
MED of unprotected one (MEDu), 358
Melanins, 229
maturation, 358
Melanocytes, 18, 208
Melanogenesis, 190
Membrane protection, 195–196
Mercury-metal halide arcs, 420–421
Merkel cells, 18
Merocyanine, 377
mES cells, 433
mES-D3 cells, 434
Mesoporous magnesium carbonate (MMC), 193
Mesoporous silica nanoparticles, 134
Metallic nanoparticles, 212
Methoxy-poly (ethylene glycol)-dihexyl substituted polylactide (MPEG-dihexPLA), 237
Methoxycinnamate, 130
18-Methyl eicosanoic acid (18-MEA), 269–270, 269*f*
Methyl isobutyl ketone, 377
Methylene bis-benzotriazolyl tetramethylbutylphenol (MBBT), 193–194
MF. *See* Mutant frequency (MF)
$MgSO_4$ solution, 417–418
Mica, 383
flakes, 383
plaques, 384–385
Micro/nanoemulsions, 11–12
Microbubbles, 335–336
Micrococcus luteus, 87–88, 148
Microdermabrasion, 25
Microemulsions. *See* Nanoemulsions (NEs)
Microfibrils, 207
Microfluidization, 64, 334–335
Microfluidizer, 334–335
Microgel, 69
Micromass test (MM test), 433, 435–436
prediction model, 435–436
procedure, 435
range finding, 436
Microneedles, 24–25
Micronized polycarbonated resin, 164*t*
Micronized titanium dioxide, 164*t*
Micronized topaz powder, 161*t*
Micronuclei, 424, 429–430
Microspheres, 383–384, 412–413
Microstarch, 457–458
Milani Nail lacquer, 377
Mineral-based nanosunscreens, 367–371
TiO_2, 367–369
ZnO, 369–370
Minerals, 154
Miniemulsion, 328
Minoxidil, 71–72
Minoxidil sulfate, 317
Misslyn, 385
MM test. *See* Micromass test (MM test)
MMC. *See* Mesoporous magnesium carbonate (MMC)
Models Own Bluebelle, 385
Modified bentonite, 387
Modified vector model (MVM), 257
Moisturizers, 159–160
Monoethanolamine (MEA), 250
Montmorillonite, 387
Moonshine line products, 381–382
Mouth wash, 162
MPEG-dihexPLA. *See* Methoxy-poly (ethylene glycol)-dihexyl substituted polylactide (MPEG-dihexPLA)
MTT assay, 415–416
Multimodal imaging, 233–234
Multiple walled-carbon nanotubes (MWCNTs), 80–82
Multiwalled nanotubes, 153
Mutant frequency (MF), 429
MVM. *See* Modified vector model (MVM)

MWCNTs. *See* Multiple walled-carbon nanotubes (MWCNTs)

N

N,N-dimethylamino propyl acrylamide methyl chloride quaternary (DMAPAAQ), 234
n-butanol, 377
N-isopropyl acrylamide (NIPAM), 234
Nail care, 163, 376–378
 benefits of nanocosmetics, 165*f*
 nanomaterials for, 97
 nanoproducts, 164*t*
 nanotechnological innovations, 380–387
Nail Moisturizing Serum, 446*t*
Nail polish, 376–378, 386
Nail strips, 376
Nano Anti-Aging Serum, 447*t*
Nano antiaging cream, 447*t*
Nano Collagen, 448*t*
Nano Corrector lifting effect, 449*t*
Nano Cyclic Cleanser Silver, 451*t*
Nano Cyclic Inc., 160
Nano Emulsion Multi-Peptide Moisturizer, 446*t*
Nano Gold Firming Treatment, 447*t*
Nano Gold Mask Pack, 449*t*
Nano Gold Particle Whitening Toothpaste, 451*t*
Nano Gold Repair Collagen Face Mask, 449*t*
Nano Pearl Powder, 449*t*
Nano platinum serum, 449*t*
Nano Silver toothpaste, 451*t*
Nano Silver Xylitol Mouthwash, 451*t*
Nano whitening Toothpaste, 451*t*
Nano-In Hand, 446*t*
Nano-notification, 363
Nano-TiO_2, 349–350, 365, 367–371
 coated, 368–369
 particles, 445
 toxicological safety as UV filter in sun-care products, 368
Nano-zinc oxide, 161*t*, 369–370
Nano-ZnO, 349–350, 367–371
 particles, 445
Nanoantiaging products, 446, 447*t*
Nanoantiwrinkle products, 446, 447*t*
Nanobiotechnology, 3–6, 82–83, 473
Nanocapsules (NCs), 92, 134, 155*t*, 161*t*, 457–458
Nanocarrier-mediated follicular targeting, 305–306
 barrier characteristics of human skin, 306–307
 hair follicles in nanocarrier-mediated skin delivery, 310–312
 iontophoretic technique combined with nanosized delivery systems for, 316–319
 methods, 312–315
 nanosized delivery systems for follicular targeting, 315–316
 penetration pathways across skin, 308–309
 rationale behind, 309–310
Nanocarrier-mediated skin delivery, hair follicles in, 310–312
Nanocarriers (NCs), 3–6, 29, 37–39, 133, 162–163
 in cosmetics, 142–154
 in dermatology, 342–343
 for sun protection, 132–136
 system for skin care, 39–54
Nanocellulose (NCE), 455–457, 475–476
Nanochitin–nanolignin complexes (CN–NL complexes), 8–10, 295–296, 295*f*
Nanochitosan, 457–458
Nanoclays, 387
Nanocolorants, 380
Nanocomposites, 8–10
Nanocosmeceuticals, 25, 83, 305
Nanocosmetic(s), 411–412, 445
 biopolyesters, 457–471
 of body lotion and cleanser in market, 451*t*
 chito-oligosaccharides, 458–469
 consumer requests and cosmetics, 6–7
 controls for cosmetic products based on use of nanoingredients, 6*f*
 embryotoxicity testing via three tests, 433–439
 exposure routes, 163–165
 formulations, 141–142, 163
 genotoxicity/mutagenicity testing, 424–433
 innovative nanocarriers, 5*f*
 for lip and sun cares in market, 449*t*
 micro/nanoemulsions, 11–12
 nanodimension and use in nature and technology, 4*f*
 packaging material, 12–13
 polymeric natural compounds, 457–471
 polymers, nanocomposites, and nonwoven tissues, 8–10
 polysaccharides, 458–469
 products for hair care in market, 448*t*
 progress and future perspective of the cosmetic products, 471–476
 sectors involving nanotechnology, 4*f*
 in vitro dermal absorption measurements of cosmetic ingredients, 422–424
 in vitro phototoxicity test, 419–421
 in vitro skin corrosion test, 416–418
 in vitro skin irritation test, 414–416
Nanocream, 71
Nanocrystals, 92, 212–213
Nanodermatology, 403
Nanodiamonds, 340–341
Nanodispersions as carrier for fragrances, 179–180

Nanoemulsions (NEs), 27–28, 44–47, 59, 61–62, 89–90, 132–133, 135, 150, 155*t*, 164*t*, 178–179, 214–215, 328–330, 446
advantage of, 180–181
challenges and future perspectives, 72
challenges associated with cosmetic, 332–333, 333*f*
characterization, 65–69
quality control assays of dispersed systems, 67*t*
chemical categories for cosmetics, 63*t*
cosmetics, 327–328, 339–343
dermal application, 339–340
NEs in hair system care, 341–342
as novel vehicle in skin care, 340–341
as potential carrier, 342–343
emulsion delivery systems in cosmetics, 60–62
formulation and production of, 62–65
methods of production, 333–339
NE-based cosmetic formulations, 331*t*
needs in cosmetics, 69–72
hair care, 71–72
skin care, 71
as potential carrier in cosmeceuticals, 330–332
satisfying cosmetic demands, 330
Nanoencapsulation, 155*t*, 252, 252*f*
procedures in development of nanotechnology, 252–253
of sunscreen, 132–133
Nanoengineered materials, 362
Nanogold, 92–93, 153
Nanohydroxyapatite, 451
Nano-ingredients, 248
labeling, 365
Nanolignin (NL), 8–10, 274–277, 455–458
Nanomaterials, 83, 328, 375, 412–413, 445, 452
catalog of, 401–402
classes of nanocosmeceuticals, 97–98
hair care, 98
lip care, 97
nail care, 97
skin care, 97–98
cosmetic definition, 395–397
in cosmetics, 142–154, 362–367
declaration, 398–400
in EC Regulation No. 1223/2009 on cosmetic products, 395–401
ecological hazardous issues, 100–101
in fragrance products
chemicals present in, 250
comparison of fragrance and flavour, 248*t*
fragrance affects skin, 250
home fragrance products, 248–249
ingredients to be avoided in skin care, 250
nano-ingredients, 248
nanotechnology in perfumes, 251
natural fragrances, 249–250
parabens, 251
sunscreen, 254–255
future prospects and opportunities, 101–102
human health insecurities, 98–100
routes of exposure, 99–100
legal definition of, 393–395
nanotechnology, 79–80
in cosmetics, 83–86
in multidisciplinary field, 80–83
nanotechnology-based cosmetics, 87–97
perspectives, 264
safety assessment, 400–401
in cosmetic products, 166–167
safety of cosmetic products, 167
trends in research on electronic noses, nanoparticles, and, 256–264
aromatic wallpaper, 260–261
assessment of indoor odor impact, 259
bioelectronic noses, 261
bioelectronic sensor, 261
comparison of volatile fraction of vodkas made, 260
computer-controlled odor generator, 258
electronic noses in meat quality assessment, 256
flavor of traditional soup, 257
fragrance profiling, 262
nanoparticle-enzyme sensors, 258
odor intensity assessment, 257
olfaction as soldier, 260
verification of odorants in rose, 263
Nanomedicine, 95, 340–341, 404
Nanomicelles, 155*t*
Nanomoisturizer products, 446, 446*t*
Nanonotification, 398–400
Nanoparticles (NPs), 3–6, 37, 83, 110, 154, 155*t*, 164*t*, 208, 231, 295–296, 340–341, 403, 405, 411–413, 475–476
of cobalt and chromium, 406–407
in hair dyes
building blocks of eumelanins and pheomelanins, 229
gray hair, 229
hair colors, 229
human hair, 227–228
melanins, 229
Parkinson's and Alzheimer's diseases, 229
investigation of polylactic acid, 242
research on use of, 231–232
toxicity, 406*t*
Nanoparticulate systems, 310
Nanoperfumes, 452

Nanopigments, 154, 412–413
Nanorama, 449*t*
NanoRepair Q10 cream, 151
Nanoscale materials, 80–82, 412–413
Nanoscale particles, 379
Nanoscience, 403, 455–457
Nanoshells, 445
Nanosilver, 92–93, 152, 155*t*
Nanosized carrier systems, 311
Nanosized delivery systems for follicular targeting, 315–316
Nanosomes, 149, 155*t*, 164*t*
Nanosphere Plus, 447*t*
Nanospheres, 97, 153, 155*t*, 161*t*, 384
Nanosponges, 152
Nanostructured lipid carriers (NLCs), 39, 87, 96–97, 132–133, 151, 155*t*, 194, 209–210, 213–214
Nanostructured materials, 395
Nanostructured polymeric lipid carriers (NPLCs), 134
Nanotechnological innovations in nail and lip care formulations, 380–387
Nanotechnology, 3–7, 79–80, 109, 141, 247, 328, 341, 362, 376, 403, 411–412, 445, 455–457, 473–474
  in cosmetics, 83–86
  in multidisciplinary field, 80–83
  nanoencapsulation procedures in development, 252–253
  nanotechnology-based cosmetics, 87–97
    carbon nanotubes, 96
    cubosomes, 94–95
    dendrimers, 93–94
    fullerenes/buckyballs, 95–96
    gold and silver nanoparticles, 92–93
    hydrogels, 95
    liposomes, 87–88, 88*f*
    nanocapsules, 92
    nanocrystals, 92
    nanoemulsions, 89–90
    nanomedicine, 95
    nanospheres, 97
    niosomes, 90–91
    NLCs, 96–97
    polymersomes, 96
    SLNs, 91–92
  in perfumes, 251
    applications of, 251–254
    high tech of small serving giant thrills, 254
    time-controlled and prolonged release of scents, 252
Nanotechnology-based cosmetics, 154–163
  hair care, 162–163
  lip care, 160
  nail care, 163
  oral care, 160–163
    mouth wash, 162
    toothpastes, 160–161
  skin care, 154–160, 155*t*
    antiaging creams, 154–159
    moisturizers, 159–160
    skin cleansers, 160
    sunscreens, 154
Nanotoxicity, 404
Nanotoxicology, 328, 366–367
Nanover, 448*t*
Nanovitamins, 155*t*
Nanowhite and nanowhite men, 449*t*
National Industrial Chemicals Notification and Assessment Scheme (NICNAS), 359, 371
Natural antioxidants, 134
Natural enhancers, 22–23
Natural fibers for natural hair, 271–277
  chitin nanofibril–nanolignin complex, 274*f*
  polysaccharides, 272–274
  supposed lignin chemical structures, 273*f*
Natural filtering agents, 357
Natural fragrances, 249–250
Natural green cosmetics, 471–473
Natural ingredients, 3–6, 472–473
Natural mica, 380
Natural moisturizing factor (NMF), 35–36
Natural polymers, 22
  for natural hair
    chitin, 274–277
    chitin and hair, 277–283
    cosmetic hair care treatments, 270–271
    hair beneath the skin papilla, 268*f*
    hair system, 267–268
    natural fibers for, 271–277
NCE. *See* Nanocellulose (NCE)
NCs. *See* Nanocapsules (NCs); Nanocarriers (NCs)
NDA. *See* New Drug Application (NDA)
Near infrared A (IRA), 132
Neon Color Trend Avon, 377
Neon effect, 377
NEs. *See* Nanoemulsions (NEs)
Neurodegenerative disorder, 80–82
Neuromelanin colors, 229
Neutral Red dye, 419
Neutral Red uptake test, 421
New Drug Application (NDA), 359–360
NG. *See* Nanolignin (NL)
NICNAS. *See* National Industrial Chemicals Notification and Assessment Scheme (NICNAS)
Nigella sativa, 332
Nile Red (NR), 235–236, 313

Niosomes, 27, 41, 90–91, 91*t*, 147, 155*t*, 164*t*, 340–341, 378, 412–413
NIPAM. *See* N-isopropyl acrylamide (NIPAM)
NL. *See* Nanolignin (NL)
NLCs. *See* Nanostructured lipid carriers (NLCs)
NMF. *See* Natural moisturizing factor (NMF)
Non-ZNG (NZNG), 259
Nonionic surfactants, 336
  vesicles. *See* Niosomes
Nonliving epidermis. *See* Stratum corneum (SC)
Nontoxic fluencies, 190
Nontransformed human-derived epidermal keratinocytes, 414
Nonwoven tissues, 8–10
Norcanthridin, 331*t*
Novasomes, 148
Novel nanosized delivery systems, 305
NPLCs. *See* Nanostructured polymeric lipid carriers (NPLCs)
NPs. *See* Nanoparticles (NPs)
NR. *See* Nile Red (NR)
Nucleation, 381
Nutraliss Nano Repair Shampoo + Nano Repair Conditioner Liter, 448*t*
Nutrients, 445
NZNG. *See* Non-ZNG (NZNG)

O

O/W. *See* Oil-in-water (O/W)
O/W nanoemulsions. *See* Perfume oil-in-water nanoemulsions (O/W nanoemulsions)
Odor activity value (OAV), 263
Odor standardization by bioelectronic noses, 259
OECD. *See* Organization for Economic Cooperation and Development (OECD)
Oil, 173–174
Oil-in-water (O/W), 214
  emulsion, 60, 336
  emulsion-based cream, 111
  nanoemulsions, 150
Oil/water nanoemulsions, 328
Ointments, 340
Oleosomes, 149
Olfaction as soldier, 260
Olfactory assessment function, 257
Olive oil, 70
Opaque pigments, 380
Opuntia ficus-indica (L.), 332
Oral care, 160–163
  mouth wash, 162
  nanocosmetics in market, 451*t*
  nanoproducts, 162*t*
  toothpastes, 160–161
OrbisResearch. com, 445
Organic sunscreens, 355–357, 356*f*
Organic UV blockers, 35–36
Organic UV filter loaded nanocarriers, 128–129. *See also* Polymeric nanocarriers
  conventional products for sun protection, 130–132, 131*t*
  protection filters, 128–130
  for sun protection, 132–136
Organization for Economic Cooperation and Development (OECD), 364–365, 413
OTC drug. *See* Over the Counter drug (OTC drug)
Outdoor fine particulate, 293
Over the Counter drug (OTC drug), 359–360
Oxidants, 333
Oxidative stress, 289–292, 352
Oxybenzone, 130, 362
Ozokerite, 379

P

p-aminobenzoic acid (PABA), 129–130
P-phenylenediamine (PPD), 230
PA dual-modal imaging. *See* Photoacoustic dual-modal imaging (PA dual-modal imaging)
PABA. *See* *p*-aminobenzoic acid (PABA)
Packaging material, 12–13
PAMAM. *See* Polyamidoamines (PAMAM)
Parabens, 251, 251*f*
Parkinson's diseases, 229
Partial least square (PLS), 260
Particles, 395
Patent applications, 362
Patents based on nanocosmetics, 143*t*
PBPK. *See* Physiologically based pharmacokinetics (PBPK)
PBTK. *See* Physiologically based pharmacokinetics/toxicokinetics (PBTK)
PC. *See* Polycarbonate (PC)
PCA. *See* Principal component analysis (PCA)
PCL. *See* Polycaprolactone (PCL)
PCL-b-PEG. *See* Poly(caprolactone)-block-poly(ethylene glycol) copolymer (PCL-b-PEG)
PCPLC. *See* Poly(ethylene glycol)-4-methoxycinnamoylphthaloylchitosan (PCPLC)
PCPs. *See* Personal care products (PCPs)
PdI. *See* Polydispersity index (PdI)
Pearlescent pigments, 382, 382*f*
PEG. *See* Polyethylene glycol (PEG)
Peptides, 317
Perfluorodecaline (PFD), 44
Perfume, 251
  nanotechnology in, 251
  probable risks of nano-based, 253–254

production of, 252
Perfume oil-in-water nanoemulsions (O/W nanoemulsions), 178–179
Permeation enhancers, 22–23
Peroxide number measurement (POV), 178–179
Persistent Pigment Darkening (PPD), 358
Personal care products (PCPs), 241–242
safety assessment of, 241–242
PFD. *See* Perfluorodecaline (PFD); Probe to quantify F (PFD)
PG. *See* Propylene glycol (PG)
PGLA. *See* Poly-D,L-lactide-*co*-glycolide (PGLA)
pH-responsive-dendritic-polyglycerol-nanogel (dPGNG), 210–211
PHAs. *See* Polyhydroxyalkanoates (PHAs)
Phase inversion composition (PIC), 65, 178–179, 214, 333–334, 337
Phase inversion emulsification method, 336, 337*f*
Phase inversion temperature (PIT), 46, 65, 214, 333–334, 336
PHB. *See* Poly-hydroxy-alkanoates (PHB)
PHEA. *See* Poly(2-hydroxyethyl l-aspartamide) (PHEA)
Pheomelanins, building blocks of, 229
Phosphatidylcholine, 25–27
Phospholipids (PL), 40, 238
Photo-irritation-factor (PIF), 421–422
Photoacoustic dual-modal imaging (PA dual-modal imaging), 233–234
Photoaging, 349–350, 353
Photoallergy, 353
Photodynamic therapy, 190
Photosensitivity reactions, 353
Photosensitization reactions, 187–188
Photosensitizer, 187–188
Photosomes, 87–88, 148
Photostable sunscreen, 354–355
Phototherapy, 190, 233–234
Photothermal therapy (PTT), 233–234
Phototoxicity, 353
Photoxicity, 419
Phyllosilicate, 386
Physical sunscreens, 356–357
Physiologically based pharmacokinetics (PBPK), 359–360
Physiologically based pharmacokinetics/toxicokinetics (PBTK), 359–360
Phytosomes, 149
PI. *See* Proliferation index (PI)
PIB. *See* Polyisobutylene (PIB)
PIC. *See* Phase inversion composition (PIC)
PIF. *See* Photo-irritation-factor (PIF); Product information file (PIF)
Pig skin, 312
Pigment, 333, 380
disperser agents, 379
Pilosebaceous unit (PSU), 307, 308*f*
PIT. *See* Phase inversion temperature (PIT)
Pityrosporum folliculitis, 220
Pityrosporum ovale, 220
PL. *See* Phospholipids (PL)
PLA. *See* Polylactic acid (PLA)
PLA-block-poly(ethylene glycol) copolymer (PLA-b-PEG), 315
Plasticizers, 23
Plate incorporation method, 425
PLGA. *See* Poly(lactic-co-glycolic acid) (PLGA)
PLO. *See* Pluronic lecithin organogel (PLO)
PLS. *See* Partial least square (PLS)
Plucking gray hair, 229
Pluronic lecithin organogel (PLO), 239
PMSQ. *See* Polymethylsilsesquioxane (PMSQ)
Podophyllotoxin (POD), 214
Pollutants, primary and secondary, 294*f*
Poly (2-hydroxyethyl l-aspartamide) (PHEA), 236
Poly-D,L-lactide-*co*-glycolide (PGLA), 194
Poly-hydroxy-alkanoates (PHB), 8–10
Poly(caprolactone)-block-poly(ethylene glycol) copolymer (PCL-b-PEG), 315
Poly(ethylene glycol)-4-methoxycinnamoylphthaloylchitosan (PCPLC), 316
Poly(lactic-co-glycolic acid) (PLGA), 211, 238, 315–316
nanospheres, 52
Polyamidoamines (PAMAM), 51, 313
dendrimers, 318
Polycaprolactone (PCL), 52, 110, 194, 211
polymeric nanocarriers formulation, 110
Polycarbonate (PC), 47–48
Polydispersity index (PdI), 64, 114–115
Polyethoxylated surfactants, 336
Polyethylene glycol (PEG), 22, 210, 250
Polyhedral oligo silsequioaxanes (POSS), 378
Polyhydroxyalkanoates (PHAs), 457–458, 470–471
Polyisobutylene (PIB), 22
Polylactic acid (PLA), 8–10, 52, 235–236, 311, 457–458, 469–470, 469*f*
Polymer matrix, 22
Polymer-based nanoparticles, 296
Polymeric micelle nanocarriers, 237
Polymeric nanocarriers, 446. *See also* Organic UV filter loaded nanocarriers
calibration curve of saffron in distilled water, 116*f*
characterizations, 111–112
cream evaluation containing saffron-loaded polymeric nanoparticles, 113
evaluation of cream formulation, 119–121

Polymeric nanocarriers (*Continued*)
release kinetics, 121, 124*t*
in vitro dissolution studies, 119–120
formulation design, 115*t*
3^2 factorial design for preparation of nanoformulation, 115*t*
cream formulation, 111
factorial design experiment, 110
PCL polymeric nanocarriers formulation, 110
release kinetics of *C. sativus L.* nanoparticle formulation, 122*t*
materials, 110
nanotechnology, 109
saffron-loaded polymeric nanoparticles evaluation, 113–119
SEM for optimized nanoformulation, 117*f*
Polymeric nanoemulsions, 340–341
Polymeric nanoparticles, 52–53, 80–82, 132–133, 211–212
polymeric nanoparticles-embedded organogel, 239
Polymeric natural compounds, 457–471
Polymerization, 376
Polymerosomes, 150
Polymers, 8–10, 194, 333
Polymersomes, 96, 340–341
Polymethylsilsesquioxane (PMSQ), 379
microspheres, 384
Polyphenols, 357
Polypropylene, 22
Polypyrrole, 377
Polysaccharides, 272–274, 458–469
nanocarriers, 53–54
Polystyrene nanoparticles, 308–309
Polytetrafluoroethylene (PTFE), 417
Polythiophene, 377
Polytrap microparticles, 52, 53*f*
Polyunsaturated fatty acids (PUFAs), 189
Polyvinyl alcohol (PVA), 110, 194
Polyvinylchloride (PVC), 22
Porous inorganic nanocompounds, 39
POSS. *See* Polyhedral oligo silsequioaxanes (POSS)
POV. *See* Peroxide number measurement (POV)
Powder-X-ray diffractogram (XRD), 239–240
PPD. *See* P-phenylenediamine (PPD); Persistent Pigment Darkening (PPD)
Prediction model, 435–438
Preincubation method, 425
Primary sunscreens, 353
Primordiale Optimum Lip, 378, 449*t*
Principal component analysis (PCA), 259
Probe to quantify F (PFD), 44
Product information file (PIF), 398
Proinflammatory cytokines, 288–289, 290*f*, 296–297
Proliferation index (PI), 431
Propionibacterium spp., 220
Propylene glycol (PG), 250
Protection filters, 128–130
Protection index. *See* Sun protection factor (SPF)
Protein, 317
lesions, 189
Pseudomonas aeruginosa, 263 264
predicting growth situation of, 263–264
Psoriasis treatment, targeted approach for, 237
PSU. *See* Pilosebaceous unit (PSU)
PTFE. *See* Polytetrafluoroethylene (PTFE)
PTT. *See* Photothermal therapy (PTT)
PUFAs. *See* Polyunsaturated fatty acids (PUFAs)
Pullulan (PUL), 468–469
chemical formula, 468*f*
Punch biopsy, 315
Punica granatum extracts, 357
PVA. *See* Polyvinyl alcohol (PVA)
PVC. *See* Polyvinylchloride (PVC)

Q

Quality control, 415
tests, 66
Quantitative descriptive analysis (QDA), 263
Quantitative techniques, 314–315. *See also* Imaging techniques
differential tape stripping, 314
punch biopsy, 315
Quantum dots, 377
nanoparticles, 407

R

Rapeseed oil, 331*t*
Rat skin TER, 416
Ratchet mechanism, 310–311
RBHE. *See* Rhodamine B hexyl ester perchlorate (RBHE)
REACH. *See* Registration, Evaluation, Authorisation, and Restriction of Chemicals (REACH)
Reactive nitrogen species (RNS), 288–289
Reactive oxygen species (ROS), 288–289, 349–350
Receptor fluid, 422–423
Reconstructed human epidermis (RhE), 414–415
Refinement, reduction, and replacement strategy (3Rs strategy), 414
Reflecks Multireflections Soft Sparkle Pigments, 381
Registration, Evaluation, Authorisation, and Restriction of Chemicals (REACH), 364–365, 396–397
nanomaterials under, 364–365
Regression equation, 110
Relative increase in cell count (RICC), 431
Relative population doubling (RPD), 431

Response Surface Methodology (RSM), 110
Retinol, 329–330
Reuse, reduce, recycle (three R), 455–457
Reverse mutation, 425
RevitaLift Anti-Wrinkle and Firming Face and Neck Contour Cream, 447*t*
RevitaLift Double Lifting, 447*t*
Revlon, 378
RhE. *See* Reconstructed human epidermis (RhE)
Rhodamine, 313
Rhodamine B, 316
Rhodamine B hexyl ester perchlorate (RBHE), 235
RICC. *See* Relative increase in cell count (RICC)
Rice germ oil, 357
"Risk management" framework, 364
RNS. *See* Reactive nitrogen species (RNS)
Ronasphere LDP, 384, 385*f*
Ronastar Golden Jewel product, 382–383
Ronastar line products, 382–383
ROS. *See* Reactive oxygen species (ROS)
Round window membrane (RWM), 236
Roxithromycin (ROX), 235, 239
RPD. *See* Relative population doubling (RPD)
RSM. *See* Response Surface Methodology (RSM)
Ruby Wing, 377
Ruscus aculeatus, 47–48
Rutile, 367
RWM. *See* Round window membrane (RWM)

S

Safety assessment of nanomaterials, 400–401
Saffron-loaded polymeric nanoparticles evaluation, 113–119
  antiinflammatory activity, 117–119
  anti-oxidant activity, 117
  entrapment efficiency, 115
  factorial design for polymeric nanocarriers formulation, 113–114
  measurements of particle size and zeta potential, 114–115
  morphological studies, 116–117
Salmonella typhimurium, 425
  S. typhimurium TA98, 425
  S. typhimurium TA100, 425
  S. typhimurium TA102, 425
  S. typhimurium TA1535, 425
  S. typhimurium TA1537 or TA97 or TA97a, 425
SC. *See* Stratum corneum (SC)
Scaffold materials, 475–476
Scanning electron microscopy (SEM), 44, 215–216
Scientific Committee for the Consumer Safety, 397
Scientific Committee on Consumer Safety (SCCS), 165–166, 359–360, 363, 365–366, 368–369, 371
  safety assessment opinion, 369–370
Sclerosis, 353
Sebaceous glands, 208, 307
Second-generation lipid nanoparticles. *See* Nanostructured lipid carriers (NLCs)
Secondary sunscreens, 353
Self-emulsification method, 338, 338*f*
Self-nanoemulsification method, 338
Self-nanoemulsifying drug delivery systems (SNEDDS), 338
SEM. *See* Scanning electron microscopy (SEM)
Shampoo Hidratante com Batana Oil, 448*t*
Shampoos, 341–342
Shunt route, 308
Silica (SiO_2), 47, 380
  nanoparticles, 362, 405–406
Silicon rubber, 22
Silicone polymers, 379
Silk, crystalline/amorphous structure of, 460*f*
Silver nanoparticles (AgNPs), 92–93, 152, 362, 406–407, 412–413
Single-walled carbon nanotubes (SWCNTs), 80–82
Singlet oxygen ($^1O_2$), 130, 193, 208
Skin, 18–19, 36–37, 59–60, 165, 306
  aging, 288–289
  appendages, 307
  barrier characteristics, 306–307
  cleansers, 160
  corrosion, 416
  dermis, 18–19
  dermis–epidermis junction, 19
  disc preparation, 417
  epidermis, 18
  hydration, 70
  hypodermis, 19
  integrity, 423
  irritation, 414–415
  layers, 60*f*
  lipids, 418
  microbiome, 473–474
  nanocarrier system, 39–54
    aggregation of phospholipid to generate bilamellar phases, 40*f*
    dendrimers and other dendritic structures, 50–52
    inorganic nanocarriers, 47–50
    liposomes and related particles, 39–42
    nano-and microemulsions, 44–47
    polymeric nanoparticles, 52–53
    polysaccharide nanocarriers, 53–54
    SLNs and structured lipid nanocarriers, 42–44
  nanocarriers and skin penetration, 37–39
    nanoparticle skin penetration pathways, 38*f*
  penetration, 37–39, 423

Skin (*Continued*)
pathways, 308–309
scents, 173–174
structure and penetration routes–pathways, 306*f*
temperature, 423
tumors, 353
Skin care, 154–160, 155*t*
antiaging creams, 154–159
benefits of nanocosmetics, 159*f*
moisturizers, 159–160
of nanoemulsions, 71
nanomaterials for, 97–98
NEs as novel vehicle in, 340–341
skin cleansers, 160
sunscreens, 154
SkinEthic, 414
SLNPs. *See* Solid lipid nanoparticles (SLNs)
SLNs. *See* Solid lipid nanoparticles (SLNs)
Smart nanocontainers, 445
Smart nanocosmetics, 474–476
Smectites, 386
SNEDDS. *See* Self-nanoemulsifying drug delivery systems (SNEDDS)
Social media channels, 7
SOD. *See* Superoxide dismutase (SOD)
SOD-Lipo. *See* Superoxide dismutase-encapsulated liposomes (SOD-Lipo)
Sodium borosilicate, 380
Solar radiations, 351–353
Soleil Instant Cooling Sun Spritz SPF 15, 449*t*
Soleil Soft-Touch Anti-Wrinkle, 447*t*
Solid lipid nanocarriers. *See* Solid lipid nanoparticles (SLNs)
Solid lipid nanoparticles (SLNs), 17–18, 27, 39, 42–44, 43*f*, 85–86, 91–92, 132–134, 151, 155*t*, 194, 209–210, 213–214, 235, 412–413
Solvents, 430
SOR. *See* Surfactant to oil ratio (SOR)
SPF. *See* Sun protection factor (SPF)
Sphingosomes, 149–150
Spironolactone, 23–24
Spirooxazines, 377
Spiropyrans, 377
Spirulina, 197
Spontaneous emulsification, 333–334, 339
Sprayable application/products, 368
Squalene, 219, 331*t*
Staphylococcus spp., 220
S. aureus, 332
Starch, 465–466
structure organization, 466*f*
Stem Cellue Nano Platinum Skin Rejuvenating Cream, 449*t*
Stimuli-responsive patches, 28
Stober sol-gel process, 49
Stratum basale, 307
Stratum corneum (SC), 18, 35–36, 59, 233, 307, 313, 332
barrier, 8–10
Stratum granulosum, 307
Stratum lucidum, 307
Stratum spinosum, 307
Stripping techniques, 37–38
Structured lipid nanocarriers, 42–44
Subcutaneous tissue, 36
Submicron emulsion. *See* Nanoemulsions (NEs)
Sulforhodamine B dye, 418
application and removal, 418
Sun Cream SPF 15, 447*t*
Sun protection
conventional products for, 130–132, 131*t*
nanocarriers for, 132–136
Sun protection factor (SPF), 52, 92, 127, 192–193, 273–274, 350, 357–359
for body creams, 130–131
Sun-care
nanocosmetic products in market, 449*t*
products, 349–350
mineral-based nanosunscreens, 367–371
nanomaterials in cosmetics, 362–367
solar radiations, 351–353
sunscreens, 353–362
Sunscreens, 46–47, 127, 133, 154, 254–255, 349–350, 353–362, 356*f*
sunscreen application *vs.* decrease in vitamin D synthesis, 361–362
cosmetics, 130
nanoencapsulation, 132–133
organic and inorganic sunscreens, 355–357, 356*f*
chemical sunscreens, 355–356
natural filtering agents, 357
physical sunscreens, 356–357
photoprotective formulation with chemical, 136
photoprotective potential, 128
regulatory status, 359–360
sun protection factor, 357–359
sun-care products labeling, 360–361
titanium dioxide use as, 255
uncoated zinc oxide, 254–255
Superoxide ($O_2^{-}\bullet$), 193
Superoxide dismutase (SOD), 319
Superoxide dismutase-encapsulated liposomes (SOD-Lipo), 319
Superparamagnetic nanoparticles, 407
Surface active agent, 418
Surfactant to oil ratio (SOR), 180
Surfactants, 62, 180–181

Survismeter, 68
Sustainable development, 455
  social, environmental, and economic equality, 456*f*
SWCNTs. *See* Single-walled carbon nanotubes (SWCNTs)
Synthetic elastomers, 22
Synthetic enhancers, 22–23
Synthetic mica, 380
Synthetic polymers, 22
Sytran, 378

T

3T3 cells, 433
Tacrolimus (TAC), 237
Tan, 352
Tape Stripping (TS), 37–38, 314
Targeted approach for psoriasis treatment, 237
Taste active values (TAVs), 257
TEA. *See* Time and Extent Application (TEA); Triethanolamine (TEA)
TEAH. *See* Triethanolamine hydroiodide (TEAH)
Telomeres, 291–292
  with chromosome cups, 292*f*
TEM. *See* Transmission electron microscope (TEM)
TEOS. *See* Tetraethyl orthosilicate (TEOS)
TER. *See* Transcutaneous electrical resistance (TER)
Test chemical application, 415–416
Testosterone (TP), 214
Tetraethyl orthosilicate (TEOS), 49
Tetramethyl orthosilicate (TMOS), 49
Tetramethylrhodamine-dextran (TRITC-dextran), 212
TFT. *See* Trifluorothymidine (TFT)
TG. *See* 6-Thioguanine (TG)
TGA. *See* Therapeutic Goods Administration (TGA)
Theranostic nanoparticles, 80–82
Therapeutic Goods Administration (TGA), 358
Therapeutic sunscreens, 358
Thermoresponsive nanoemulsion, 331*t*
6-Thioguanine (TG), 427
Three-dimension (3D)
  human skin models, 414
  RhE model, 414
Thymidine kinase (TK), 427
  proficient cells, 427
Time and Extent Application (TEA), 359–360
Timiron Diamon Cluster MP-149, 382
Timiron formulations, 382
Timiron Glam Silver, 382
Timiron Supersilk MP-1005, 382
Timiron Synwhite 40, 382
Tin (SnO_2), 382
Titanium dioxide (TiO_2), 154, 192, 254–255, 356–357, 362, 367–369, 383, 385, 405–406, 412–413
  carefully selected ingredient, 367–368
  characteristics, 367
  coated nano-TiO_2, 368–369
  ecotoxicity studies, 370–371
  mica platelets coated with, 386*f*
  nanoparticle-based indoor antiodor product, 256
  powders, 349–350
  sunscreen, 255
  toxicological safety of nano-TiO_2 as UV filter, 368
Titanium oxide. *See* Titanium dioxide (TiO_2)
TK. *See* Thymidine kinase (TK)
TK6 (Human lymphoblastoid cells), 427
TMOS. *See* Tetramethyl orthosilicate (TMOS)
Toilet freshener, 249
Toothpastes, 160–161
Top note, 173
Top-down approach, 79–80
Topcoats, 376–377
Topical skin delivery, 305–306, 315
Toxicity, 411–412
Toxicological testings, 413
Toxicology Data Network (TOXNET), 371
TP. *See* Testosterone (TP)
TPI. *See* Transitional phase inversion (TPI)
Traditionally shaped lipsticks, 379
Transcutaneous electrical resistance (TER), 416
  measurements, 417–418
Transdermal bioactive nanocarriers, 35–36
Transdermal delivery, 19–20
  advantages, 21
  components of transdermal patch, 21–23, 22*f*
    adhesive, 23
    backing laminate, 23
    biologically active substances, 22
    excipients, 23
    permeation enhancers, 22–23
    polymer matrix, 22
    release liner, 23
  development in cosmetics, 21
  evolution, 20–21
  novel technologies toward development of transdermal system, 23–25
    electroporation, 24
    iontophoresis, 23–24
    laser radiation, 25
    microdermabrasion, 25
    microneedles, 24–25
Transdermal drug transport, 39
Transdermal nanocarriers, 446
Transferosomes. *See* Ultradeformable liposomes (UDLs)
Transitional phase inversion (TPI), 336
Transmission electron microscope (TEM), 49–50

Triethanolamine (TEA), 250
Triethanolamine hydroiodide (TEAH), 47–48
Triethoxycaprylylsilane, 368–369
Trifluorothymidine (TFT), 427
Trimethylsiloxysilicate, 379
TRITC-dextran. *See* Tetramethylrhodamine-dextran (TRITC-dextran)
TS. *See* Tape Stripping (TS)
24 K Gold Facial Mask, 449*t*
Tyrosinase-related protein 1 (Tyrp1), 218

U

Ultradeformable archaeosomes (UDAs), 42
Ultradeformable liposomes (UDLs), 39, 41, 147–148, 412–413
Ultrasomes, 87–88, 148
Ultrasonication, 335–336
Ultrasonification (US), 178–179
Ultraviolet (UV)
  filters, 192, 359–360
  nanofilters, 445
Ultraviolet light A (UVA), 127, 349–350, 352
  UVA-I, 352
  UVA-II, 352
Ultraviolet light B (UVB), 127, 349–351
Ultraviolet light C (UVC), 351
Ultraviolet radiation (UVR), 127, 185
  penetration of, 187*f*
Urocanic acid, 352
US. *See* Ultrasonification (US)
US Government Accountability Office (GAO), 100
UVA. *See* Ultraviolet light A (UVA)
UVB. *See* Ultraviolet light B (UVB)
UVC. *See* Ultraviolet light C (UVC)
Uvinul A Plus Granular, 135
Uvinul T150, 135
UVR. *See* Ultraviolet radiation (UVR)

V

V79 (Chinese hamster cells), 427
Vacuum-metallized aluminum pigments (VMPs), 380
Vegetable oils, 70
Vellus hairs, 311–312
Vermiculite, 386
Vesicular gel, 316
Vesicular nanosystems, 132–133
Viability, 415
Visible light (VIS), 132, 186–190
  effects, 187–190
    photodamage in biomolecules and organelles, 189–190
    reactive species generation in skin cells, 188–189
    simplified Jablonski diagram, 188*f*
  interaction of, 186–187
  strategies for protection from, 191–197
    antioxidants, 196–197
    filters, 191–195
    membrane protection, 195–196
    nanocarriers, 194–195
    nanoparticles, 192–194
Vitamin B9. *See* Folic acid
Vitamin D synthesis, 352
  controversy between sunscreen application and decrease in, 361–362
Vitamin K1, 52
Vitamins, 445
Vitamins A, E, and K, 412–413
VMPs. *See* Vacuum-metallized aluminum pigments (VMPs)
Vodka, 260

W

W/O. *See* Water-in-oil (W/O)
Walnut oil, 357
Waste, 293–294
Water-based nanoperfumes
  advantage of nanoemulsions over microemulsions, 180–181
  nanodispersions as carrier for fragrances, 179–180
  water-based or/and alcohol-free perfumes, 174–179, 175*t*
Water-based polymers, 378
Water-in-oil (W/O), 214
  emulsion, 60
  nanoemulsions, 150, 328, 336, 341
Water-miscible compounds, 65
Waxes, 379
Whiteners, 35–36
Whole embryo culture assay test (WEC test), 433–439
  dose–response assessment, 438
  endpoint assay, 438–439
  prediction model, 437–438
  procedure, 437
Wholistic approach, 6–7

X

X-ray diffraction patterns, 41
Xanthine guanine phosphoribosyl transferase (XPRT), 427
Xenobiotics, 290
Xenon arcs, 420–421
Xirona Caribbean Blue, 382

Xirona Le Rouge, 382
Xirona line use pigments, 382
Xirona Volcanic Fire, 382
XPRT. *See* Xanthine guanine phosphoribosyl transferase (XPRT)
XRD. *See* Powder-X-ray diffractogram (XRD)
XRF (SPECTRA plus) analysis, 111–112

Y

Yeast-based liposomes, 149

Z

Zelens Fullerene C-60 Night Cream, 447*t*
Zhongning Goji (ZNG), 259
Zinc oxide (ZnO), 154, 192, 356–357, 362, 369–370, 405–406, 412–413
  ecotoxicity studies, 370–371
  nano-zinc oxide, 369–370
  powders, 349–350

9780128222867